T0300922

THE ROUTLEDGE HANDBOOK OF SMART TECHNOLOGIES

This *Handbook* provides a thorough discussion of the most recent wave of technological (and organisational) innovations, frequently called "smart" and based on the digitisation of information. The acronym stands for "Self-Monitoring, Analysis and Reporting Technology". This new wave is one in a row of waves that have shaken up and transformed the economy, society and culture since the first Industrial Revolution and have left a huge impact on how we live, think, communicate and work: they have deeply affected the socioeconomic metabolism from within and humankind's footprint on our planet. The *Handbook* analyses the origins of the current wave, its roots in earlier ones and its path-dependent nature; its current forms and actual manifestations; its multifarious impact on economy and society; and it puts forward some guesstimates regarding the probable directions of its further development. In short, the *Handbook* studies the past, the present and the future of smart technologies and digitalisation.

This cutting-edge reference will appeal to a broad audience, including but not limited to: researchers from various disciplines with a focus on technological innovation and their impact on the socioeconomic system; students across different fields but especially from economics, social sciences and law studying questions related to radical technological change and its consequences; as well as professionals around the globe interested in the debate of smart technologies and socioeconomic transformation, from a multi- and interdisciplinary perspective.

Heinz D. Kurz is Emeritus Professor of Economics at the University of Graz, Austria, and a Fellow of the Graz Schumpeter Centre.

Marlies Schütz works as a postdoctoral researcher at the Graz Schumpeter Centre, University of Graz, Austria.

Rita Strohmaier is an economic researcher at the German Development Institute / Deutsches Institut für Entwicklungspolitik (DIE) in Bonn, Germany.

Stella S. Zilian, M.A. (Econ.), works as a researcher at the Graz Schumpeter Centre, University of Graz, Austria, and at the Institute for Heterodox Economics, Vienna University of Economics and Business.

THE ROUTLEDGE HANDBOOK OF SMART TECHNOLOGIES

An Economic and Social Perspective

Edited by Heinz D. Kurz, Marlies Schütz, Rita Strohmaier and Stella S. Zilian

LONDON AND NEW YORK

Cover image: © Getty Images

First published 2022
by Routledge
4 Park Square, Milton Park, Abingdon, Oxon OX14 4RN

and by Routledge
605 Third Avenue, New York, NY 10158

Routledge is an imprint of the Taylor & Francis Group, an informa business

British Library Cataloguing-in-Publication Data
A catalogue record for this book is available from the British Library

Library of Congress Cataloging-in-Publication Data
Names: Kurz, Heinz D., editor. | Schütz, Marlies, editor. | Strohmaier, Rita, editor. | Zilian, Stella S., editor.
Title: The Routledge handbook of smart technologies : an economic and social perspective / edited by Heinz D. Kurz, Marlies Schütz, Rita Strohmaier and Stella S. Zilian.
Description: New York, NY : Routledge, 2022. |
Series: Routledge international handbooks | Includes bibliographical references and index.
Identifiers: LCCN 2021037529 (print) | LCCN 2021037530 (ebook) | ISBN 9780367369231 (hardback) | ISBN 9781032130811 (paperback) | ISBN 9780429351921 (ebook)
Subjects: LCSH: Technological innovations—Economic aspects. | Information technology—Management. | Economic development—Technological innovations. | Social change.
Classification: LCC HC79.T4 R688 2022 (print) | LCC HC79.T4 (ebook) | DDC 338/.064—dc23
LC record available at https://lccn.loc.gov/2021037529
LC ebook record available at https://lccn.loc.gov/2021037530

ISBN: 978-0-367-36923-1 (hbk)
ISBN: 978-1-032-13081-1 (pbk)
ISBN: 978-0-429-35192-1 (ebk)

DOI: 10.4324/9780429351921

Typeset in Bembo
by Apex CoVantage, LLC

CONTENTS

FIGURES

TABLES

CONTRIBUTORS

Jeremias Adams-Prassl is a professor of law at Magdalen College, Oxford. He is particularly interested in the future of work and innovation, and has won numerous prizes for his research, teaching and public impact. Recent books include *Humans as a Service: The Promise and Perils of Work in the Gig Economy* (OUP 2018) and *Great Debates in EU Law* (Macmillan 2021). From April 2021, Jeremias leads a five-year research project on Algorithms at Work, funded by the European Research Council and a 2020 Leverhulme Prize. He tweets at @JeremiasPrassl.

Karishma Banga is a research fellow in digital development at the Institute of Development Studies (IDS), University of Sussex, UK. Her research examines new models of digital-led development, changing nature of global value chains and digital trade negotiations, with a focus on development implications for low- and middle-income countries. Previously, she was a research fellow at the International Economic Development Group, ODI working on international trade in the digital age with a focus on Africa and Asia, and a visiting research fellow at the Centre for Trade and Economic Integration, Graduate Institute of Geneva. She has led and worked on a range of projects with international organisations and stakeholders, including UNECA, Commonwealth Secretariat, UNCTAD, Afreximbank, African Union, WTO, and Pathways for Prosperity, University of Oxford. Karishma holds a PhD in global value chains from the Global Development Institute, University of Manchester and a MPhil in economics from the University of Cambridge.

Nadja Berseck has many years of experience in the fields of innovation, urban development and digitalisation. She is part of the Hybrid City Lab at zero360 and applies innovation and design methods to public organisations. Previously, she led the foresight team at the corporate strategy department of Deutsche Bahn and worked as a research manager at izt. She completed her PhD in 2017 in the field of economic urban development at Technical University Berlin.

Martina Bisello is a research manager at the European Foundation for the Improvement of Living and Working Conditions (Eurofound). Her research interests include gender gaps in the labour market, occupational change and the impact of technology on work. Prior to joining Eurofound in 2014, she was visiting researcher at the Centre for Market and Public

Organisation at the University of Bristol and at the Institute for Social and Economic Research at the University of Essex. She holds an MSc in International Economics from the University of Padova and a PhD in Economics (Doctor Europaeus mention) from the University of Pisa.

Ernst Brudna, MSc, is an economist who studied political and empirical economics at the University of Graz. His research focuses on the digitalisation of labour markets, blockchain and crypto assets. In his master thesis 'Tacit Collusion and the Blockchain', he showed the increased possibility of collusion in theoretical blockchain markets.

Bernardo Buarque is a PhD candidate at the Spatial Dynamics Lab, University College Dublin (UCD). His doctoral research project investigates the determinants of local knowledge creation. Using data from scientific publications and patents, he seeks to estimate the competence of regions to produce certain technologies. Bernardo utilises a selection of dynamics models to study the patterns of regional specialisation, and to appraise which places have the core competence to produce new emergent technologies.

Uwe Cantner is a professor of economics and holds the chair of economics/microeconomics at the Department of Economics and Business Administration of the Friedrich Schiller University Jena in Germany. He is also professor at the I2M Group of the Department of Marketing and Management at the University of Southern Denmark in Odense (DK). Furthermore, he is chairmen of the Commission of Experts for Research and Innovation (EFI) to the German Federal Government. His research majors are economics of innovation, evolutionary economics, productivity and efficiency measurement, as well as research and innovation policy. He has been publishing papers and books on a broad range of topics from entrepreneurship to innovation-driven macro growth, from innovation networks to industrial dynamics and from productivity studies to policy analyses on various levels.

René Carraz is an associate professor at Tokyo University, Department of Global Innovation Studies, Japan. He holds a PhD in economics and teaches and researches on science, technology and innovation using an economic perspective. His recent work includes studies of science and technology policy, university-industry linkages in Japan and urban creativity.

Thomas Czypionka, MD, Mag.rer.soc.oec., is the head of the Health Economics and Health Policy Unit at the Institute for Advanced Studies and Visiting Senior Research Fellow at the London School of Economics and Political Science. His main areas of interest are policy relevant health services research, comparative health systems research and finding solutions for current challenges to health systems. He has published in medical as well as social science journals and has been on international missions to improve healthcare systems and is currently vice-president of the Austrian Health Economics Association. He earned his MD under the auspices of the president of the republic from the Medical University of Vienna and his Magisterium in economics from the Vienna University of Economics and Business.

Bernhard Dachs is a senior scientist at the Center for Innovation Systems and Policy of Austrian Institute of Technology, Vienna. His areas of expertise are the economics of innovation and technological change, in particular with regard to the effects of information and communication technologies, innovation in services and the analysis of national and international technology policy.

Ulrich Dolata is a professor of Organizational Sociology and Innovation Studies at Stuttgart University (Germany). His work and research focus on the sociology of technology and organisations, on innovation research and technology policy as well as on the sociology and political economy of the internet. He is the author of *The Transformative Capacity of New Technologies: A Theory of Sociotechnical Change* (Routledge, 2013) and, most recently, *Collectivity and Power on the Internet: A Sociological Perspective* (Springer, 2018 – together with Jan-Felix Schrape).

J. Carlos Domínguez is a full-time professor at Instituto Mora in Mexico City. He holds a BA in economics from the Monterrey Institute of Technology (ITESM); a BA in philosophy from Birkbeck College, UK; an MPhil and a DPhil in development studies from Oxford University, UK. He is currently co-coordinator of PRODIGEES (Promoting Research on Digitalisation in Emerging Powers and Europe towards Sustainable Development), an EU-funded project, aimed at consolidating a network of knowledge creation and knowledge sharing between North and South partners, on issues related to digitalisation, technological change and sustainable development.

Susanne Drexler, BA, BSc, is a graduate student in health economics and management at the Erasmus School of Health Policy and Management, Rotterdam and a research assistant in the field of health economics and health policy at the Institute for Advanced Studies, Austria.

Tobias Eibinger, MSc, is a PhD student and lecturer at the Department of Economics at the University of Graz. His research interests focus on the blockchain technology, crypto assets and environmental economics. In his master's thesis, he discussed the blockchain technology from an institutional perspective and re-evaluated the boundaries of the firm.

Xiaolan Fu is the founding director of the Technology and Management Centre for Development (TMCD), a professor of technology and international development, University of Oxford and a Fellow of the Academy of Social Sciences. Her research interests include innovation and technology policy and management, trade, foreign direct investment and economic development. She is appointed by the Secretary-General of the United Nations to the Council of the Technology Bank of the UN and to the Ten-Member High Level Advisory Group of the UN Technology Facilitation Mechanism.

Thomas Grebel is a professor of economics and head of the economic policy group at the Faculty of Economic Sciences and Media at the Ilmenau University of Technology. Economics of innovation, energy and resource economics, productivity measurement, as well as open access are his major fields of research. His publications cover, among others, topics in entrepreneurship research, knowledge flows and networks, industrial organisation, R&D policy and environmental policy.

Martin Gruber-Risak is an associate professor at the Department of Labour Law and Law of Social Security at the University of Vienna (Austria). He is the co-editor of the 'HSI Report on European Labour and Social Security Law' and the national expert for Austria with the European Centre of Expertise (ECE) in the field of labour law, employment and labour market policies that advises the European Commission. He has published and edited 25 books and more than 200 articles and book chapters on Austrian and European labour law and social security, industrial relations and alternative dispute resolution.

Yuko Harayama is an executive director charged of international affairs at Riken. Prior to joining Riken, she spent five years at the Cabinet Office of Japan as an executive member of the Council for Science, Technology and Innovation; two years at the OECD as the deputy director of the Directorate for Science, Technology and Innovation and ten years at the Graduate School of Engineering of Tohoku University as a professor of management science and technology. She is a Legion D'Honneur recipient (Chevalier) and was awarded an honorary doctorate from the University of Neuchâtel. She holds a PhD in education sciences and a PhD in economics, both from the University of Geneva.

Fredrik Heintz is a professor of computer science at Linköping University, Sweden. His research focus is artificial intelligence, especially the intersection between knowledge representation and machine learning. He is the director of the Graduate School for the Wallenberg AI, Autonomous Systems and Software Program (WASP), the coordinator of the TAILOR H2020 ICT-48 network of AI research excellence centres, the president of the Swedish AI Society, a member of the European Commission High-Level Expert Group on AI and a fellow of the Royal Swedish Academy of Engineering Sciences (IVA).

Dan van der Horst is an early climate change migrant and holds a personal chair in Energy, Environment & Society at the University of Edinburgh. He has extensive experience in leading and supervising interdisciplinary, citizen-oriented and policy-relevant research on the adoption of low carbon technologies and lifestyles. Dan has (co-)authored more than 100 research papers and reports in the last 16 years and has started an interdisciplinary MSc programme on just energy transitions, called Energy, Society & Sustainability.

Sampsa Hyysalo is a professor of CoDesign at the Aalto University School of Art, Design and Architecture in Helsinki, Finland. His research focuses on designer-user relations in sociotechnical change and takes place in the intersection of science and technology studies, innovation studies and collaborative design. He has published over 70 full-length articles and authored several books.

Angela Jäger is a researcher at Fraunhofer Institute for Systems and Innovation Research ISI in Karlsruhe, Germany. In 2006, she joined Fraunhofer ISI as an expert in empirical methods in social and economic sciences, focusing mainly on research design, surveying manufacturing firms and quantitative data analysis in the field of innovation research and technological development. Moreover, she is coordinating the participation of Fraunhofer ISI in the European Manufacturing Survey (EMS) network.

Muztoba Ahmad Khan is an assistant professor at Carroll University. His primary research interests are in topics that arise from the convergence of manufacturing, digital transformation and business model innovation. He takes an interdisciplinary perspective to study the two emerging trends in manufacturing: (1) servitisation, i.e. the shift from selling products to selling integrated Product-Service Systems (PSS) and (2) smart manufacturing and industry 4.0 adoption.

Mark Knell is a research professor specialising in the economics of innovation at NIFU. His research focuses on the link between effective demand and economic growth, knowledge creation (R&D activity and innovation) and future technology. Mark has published on various aspects of technology and innovation, history of economic thought and the economic transformation of Eastern Europe in international peer-reviewed journals and edited books. He taught these subjects, as well as economic theory, at several universities in the United States and Europe.

Dieter F. Kogler is the academic director of the Spatial Dynamics Lab at University College Dublin. His research focus is on the geography of innovation and evolutionary economic geography, with particular emphasis on knowledge production and diffusion, and processes related to technological change, innovation and economic growth. He is currently an ERC Starter Grant holder with the following project title: Technology Evolution in Regional Economies (TechEvo).

Heinz D. Kurz is an emeritus professor of economics at the University of Graz and a Fellow of the Graz Schumpeter Centre. He is the author (together with Neri Salvadori) of *Theory of Production* (Cambridge University Press) and of *Economic Thought: A Brief History* (Columbia University Press). In *Chips und Jobs* (Metropolis) (co-authored by Peter Kalmbach), he studied the impact of automation on the West German economy using a dynamic input-output model.

Vivian Anette Lagesen is professor in Science and Technology Studies (STS) at the Department of Interdisciplinary Studies of Culture, the Norwegian University of Science and Technology (NTNU). She has published widely on the topic of gender, science and technology. Her present research is focused on gender in academia.

Han Li, PhD is an associate professor of the Institute of Quality Development Strategy at the Wuhan University, China. His specialties are government regulation, international trade and development economics, and he published widely in these areas. Dr Li holds a PhD in economics, and he has taught different courses in the bachelor, master and doctoral levels at Wuhan University.

Chiehyeon Lim is an associate professor in the Department of Industrial Engineering and the Graduate School of Artificial Intelligence at UNIST. He obtained his BS and PhD from the Department of Industrial and Management Engineering at POSTECH. As part of his postdoctoral experience, he served as an assistant project scientist and lecturer in the School of Engineering at UC Merced. His research interests include smart service systems, service engineering and knowledge discovery with data mining.

Simone Lucatello is a full-time researcher at Instituto Mora, a public research centre based in Mexico City. He holds a joint BA in history from the University of Venice Cá Foscari, Italy, and University College London (UCL), an MA in international relations from the London School of Economics and Political Science (LSE) and a PhD in governance for Sustainable Development from Venice International University (VIU), Italy. His research interests deal with disaster relief, climate change, humanitarian action and sustainability. He is an IPCC (Intergovernmental Panel for Climate Change) Leading Author for the Ar6 and coordinating Leading Author for the North American Chapter (Chapter 14).

Paul P. Maglio is a professor of management and cognitive science at the University of California, Merced. He holds a bachelor's degree in computer science and engineering from MIT and a PhD in cognitive science from the University of California, San Diego. One of the founders of the field of service science, Dr Maglio was the editor-in-chief of INFORMS Service Science (2013–2018) and is lead editor of the *Handbook of Service Science*, Volumes I and II. He has published more than 125 papers in computer science, cognitive science and service science.

Irene Mandl is head of Unit Information and EURES at the European Labour Authority (ELA). Before joining ELA in September 2021, she was head of Unit Employment at the European Foundation for the Improvement of Living and Working Conditions (Eurofound). In that role, she was mainly involved in applied policy research on labour market developments (including new forms of work and employment, digitalisation, restructuring and related public policy approaches) and topics at the intersection of employment and entrepreneurship (such as job creation, workplace practices, small and medium-sized enterprises, business start-ups and scale-ups and internationalisation). She holds master's degrees in international business administration and in business and law.

Luigi Marengo is a professor of economics in the Department of Business and Management of LUISS University in Rome. He got his PhD from the Science Policy Research Unit (SPRU), Sussex University, UK. His main research areas include economics of innovation and technological change, organisational economics, industrial economics, evolutionary economics.

Brendan Markey-Towler is an associate at Economists Without Borders Australia. Prior to this he was a senior advisor at Evidn, and before that a researcher, educator and communicator at the University of Queensland, RMIT University and University College London. He researches the behavioural, institutional and evolutionary economics of new technologies.

Sameer Mittal is currently working as postdoctoral research fellow at the Unit of Information and Knowledge Management, Faculty of Management and Business, Tampere University, Finland. His research area includes Industry 4.0, lean management, product-service systems and pay-per-X business models. Broadly, his research is focused towards investigating the change in operations management practices with the integration of technologies.

Syed Mohib Ali Ahmed is a doctoral student at the University of Siena. He has graduated with masters' degree from the University of Hyderabad, Kingston University and Université de Paris Sorbonne and works in the areas of macroeconomics and economic development.

Claudia Ortiz Chao is a full-time professor at Universidad Nacional Autónoma de Mexico (UNAM). She coordinates the Multidisciplinary Working Group on Smart Cities within the Iberoamerican Universities Union (UIU) and is co-responsible for the Architecture + Design and Experimental Technology Lab (LATE). Her academic activity is centred on spatial problems, analyses and solutions, particularly those related to complexity, emergent urban processes, urban morphology and smart urban planning. She holds an undergraduate degree in architecture (UNAM) and a master's in science (Bartlett School of Graduate Studies, UCL). She is also a PhD candidate (Bartlett, UCL).

Elke Pahl-Weber has been holding the chair for sustainable urban development and urban renewal at TU Berlin from 2004–2020. Between 2009 and 2011 she was head of the Federal Office for Building and Regional Planning. She is a member of a broad range of commissions and institutions and has been working on the development of smart cities in Berlin for several communities in Germany and for the Federal Ministry. She has been a member of the international expert group on urban and regional planning at the UN, she was head of several international research projects and developed the method of Urban Design Thinking. She has been awarded for urban research and development projects in Germany and on the international level.

Eleonora Peruffo is a research officer at the European Foundation for the Improvement of Living and Working Conditions (Eurofound). She works with the European Working Conditions Survey (EWCS) team providing data analysis and data visualisations. She holds an MA in Diplomatic Sciences and International Relations awarded by the University of Trieste and an MSc in Computer Science – Data Analytics from the National College of Ireland. She joined Eurofound in 2014, and she has also worked on topics related to restructuring, digitalisation and upward convergence.

Werner Plumpe studied history and economics in Bochum, has been a professor of economic history at Goethe University in Frankfurt am Main since 1999. From 2008 to 2012 he was chairman of the Association of Historians of Germany. Latest publications: 'Carl Duisberg 1861–1935. Anatomie eines Industriellen' (Munich 2016), 'Unternehmensgeschichte in Deutschland im 19. Und 20. Jahrhundert' (Boston/Munich 2018), 'Das Kalte Herz. Kapitalismus: Geschichte einer andauernden Revolution' (Berlin 2019) and *Deutsche Bank 1870–2020: The Global House Bank* (Munich 2020, co-authored with Alexander Nützenadel and Catherine R. Schenk).

Charles Raab is a professor of politics and international relations, University of Edinburgh, a Fellow of the Alan Turing Institute and a co-director of the Centre for Research into Information, Surveillance and Privacy (CRISP). His research and publications include the subjects of privacy, surveillance, data protection, identity, information systems and technology, data ethics, policy and regulation.

Viktoria H.S.E. Robertson is a professor of commercial law, competition law and digitalisation at the Vienna University of Economics and Business, as well as a professor of international antitrust law at Graz University. Her current research focuses on the application of competition law in the digital economy, as well as on the intersection of competition law and sustainability. She holds a law diploma and a doctorate from Graz University and an MJur from Oxford University. She tweets at @VRobCompLaw.

Ricardo Rodriguez Contreras is a research manager at the European Foundation for the Improvement of Living and Working Conditions (Eurofound). He focuses on comparative industrial relations, social dialogue and collective bargaining. He is also involved in research on restructuring and the implications of technology and digital change in employment relations. Prior to joining Eurofound in 2014, he worked as a consultant for the European Commission and the European Parliament conducting studies at EU level assessing the implementation of social legislation throughout the EU, the EEA and the candidate countries. For years, he performed training activities for multinational companies and European Works Councils in the area of comparative industrial relations and labour legislation in the EU. He has also worked for the Inter-American Development Bank and the Central American Bank for Economic Integration in some countries of Latin America. He holds a BSc in Law from the Complutense University in Madrid.

David Romero is a full professor of advanced manufacturing at the Tecnológico de Monterrey, Mexico. His research interests include circular manufacturing, service engineering, cyber-physical systems and human systems, advanced production management systems, and digital lean manufacturing. He has published over 100 articles and serves on different academic, industry and government boards and committees in the disciplines of business and industrial engineering supporting digital transformation. He is recognised as the father and driving force of the Operator 4.0 vision focused on promoting a socially sustainable Industry 4.0.

Kjetil Rommetveit is an associate professor at the Centre for the Study of the Sciences and Humanities, University of Bergen. His main research interests include the public dimensions and governance of technoscience. He focuses on issues relating to privacy, autonomy and democracy and the roles of design, engineering and interdisciplinarity in governance.

Marlies Schütz works as a postdoctoral researcher at the Graz Schumpeter Centre, University of Graz, Austria. She holds a PhD in economics from the Department of Economics at the School of Business, Economics and Social Sciences of the University of Graz. Her main areas of research are the economics of innovation and technological change, socioeconomic transformation processes as well as technology and innovation policy.

Antti Silvast is an associate professor at the Technical University of Denmark (DTU), DTU Management, Department of Technology, Management and Economics. His research examines energy, infrastructures and technology from a science and technology studies perspective. He is an editor of Science & Technology Studies, the official journal of the European Association for the Study of Science and Technology (EASST). He held postdoctoral appointments at Princeton University (Princeton Institute for International and Regional Studies), University of Edinburgh (Science, Technology and Innovation Studies), Durham University (Department of Anthropology), and Norwegian University of Science and Technology (Interdisciplinary Studies of Culture).

Robert Skidelsky is an emeritus professor of political economy at Warwick University and sits as a crossbencher in the House of Lords. His latest book, *Money and Government: A Challenge to Mainstream Economics*, was published by Allen Lane in 2018.

Knut H. Sørensen is professor emeritus, engaged in Science and Technology Studies (STS) at the Department of Interdisciplinary Studies of Culture, the Norwegian University of Science and Technology (NTNU). He has published extensively in STS and gender studies, including studies of information and communication technology. His current research focuses on universities, interdisciplinarity and quantification.

Andreas Stamm is an economic geographer by training. Since 1998 he is working in research, policy advice and postgraduate training at German Development Institute / Deutsches Institut für Entwicklungspolitik (DIE). In his research, he focuses on value chain development, sustainability standards and science and innovation for sustainable development, including technology assessment.

Rita Strohmaier is an economic researcher at the German Development Institute / Deutsches Institut für Entwicklungspolitik (DIE) in Bonn, Germany. Prior to this, she worked as a postdoctoral researcher at the Graz Schumpeter Centre, University of Graz, and the Institute for Comprehensive Analysis of the Economy, University of Linz, Austria. Her research mainly revolves around radical technological change, innovation policy, climate change and sustainability.

Richard Sturn is a Joseph A. Schumpeter professor for Innovation, Development and Growth and the director of the Institute of Public Economics and the Graz Schumpeter Centre at the University of Graz. He holds a PhD in economics from Vienna University. His main research interests include history of economics, foundations of public good theory and nonmarket institutions, tax-and-transfer systems, transformation processes and normative theory.

Simone Vannuccini is a lecturer in the economics of innovation at the Science Policy Research Unit (SPRU) of the University of Sussex Business School, University of Sussex (UK), and an associated fellow at the Graduate College 'The Economics of Innovative Change', Friedrich Schiller University Jena (Germany).

Beat Weber is an economist at Oesterreichische Nationalbank, Austria's central bank. His current research focuses on applied monetary theory. In his book *Democratizing Money?* (Cambridge, 2018) and a number of papers, he discussed the claims of concepts like Bitcoin, regional currencies, sovereign money and modern monetary theory (MMT) and compared them to the current architecture behind official currencies.

Karsten Weber co-heads the Institute for Social Research and Technology Assessment (IST) and is one of three directors of the Regensburg Center of Health Sciences and Technology (RCHST) at the OTH Regensburg/Germany. He teaches at the BTU Cottbus-Senftenberg/Germany as honorary professor for culture and technology. In his scholarly work, Karsten Weber primarily is concerned with ethical and social implications and impacts of information and communication technology on individuals and societies.

Adam Whittle is a postdoctoral research fellow at the Spatial Dynamics Lab, University College Dublin (UCD). He received his PhD in economic geography from the School of Geography (UCD) and since then has been a visiting researcher at both the School of Human Geography and Spatial Planning at Utrecht University (January–May 2017) and the Agglomeration and Social Networks Research Lab in Budapest (January 2019–present). His research interests are primarily as an evolutionary economic geographer. In particular, his research focuses on evolutionary connotations of technological change, knowledge complexity, network analysis and regional diversification.

Robin Williams is a professor of social research on technology and the director of the Institute for the Study of Science, Technology and Innovation at the University of Edinburgh. His research, over three decades, has centred around the development, evolution and implications of information infrastructures.

Ariel L. Wirkierman is a lecturer in economics at Goldsmiths, University of London (UK). Previously a research fellow at the University of Sussex (UK) and a postdoc researcher at the Catholic University of the Sacred Heart (Italy), his research focuses on the historical roots, theory advancement and applied analysis of structural interdependence.

Thorsten Wuest is an assistant professor and J. Wayne and Kathy Richards Faculty Fellow in Engineering at West Virginia University and globally recognised as one of SME's 20 most influential professors in smart manufacturing. His research focuses on smart manufacturing, machine learning and AI, hybrid analytics, Industry 4.0, digital supply networks, and product service systems. Wuest received several awards including an outstanding teacher of the year in 2019, SME Journal Award, and a number of best papers from journals and conferences.

Wei Zhang is a research assistant in the Institute of Quality Development Strategy at the Wuhan University, China. His specialties are international trade and industrial economics.

Laura S. Zilian, MSc, works as a researcher at the Institute of Systems Sciences, Innovation and Sustainability Research, University of Graz and at the Institute for Heterodox Economics, Vienna University of Economics and Business. Her main research interests are innovations and technological change, digital inequality, gender inequality and labour markets. Laura S. Zilian holds a master's degree in environmental system sciences from the University of Graz and is currently enrolled in an interdisciplinary PhD programme at the University of Graz.

Stella S. Zilian, M.A. (Econ.), works as a researcher at the Graz Schumpeter Centre, University of Graz and at the Institute for Heterodox Economics, Vienna University of Economics and Business. Her research focuses on different forms and dimensions of social inequality in the digital age, in particular digital inequality, gender inequality and labour market inequality. Stella S. Zilian holds a master's degree in economics from the University of Graz and is currently enrolled in a PhD programme at the Vienna University of Economics and Business.

INTRODUCTION AND CONTENTS OF THE VOLUME[1]

The Handbook provides a thorough discussion of the most recent wave of technological (and organisational) innovations, frequently called "smart" and based on the digitisation of information. The acronym stands for "Self-Monitoring, Analysis and Reporting Technology". Smart technologies allow inanimate objects such as tools and machines to communicate with humans, and vice versa: they integrate digital and non-digital functionalities. The technological transformation thereby not only relates to the digital (such as blockchain or the platform economy) and mechanical spheres (e.g., advanced robotics, additive manufacturing), but also pervades other areas, such as biology (e.g., gene editing). This new wave is one in a row of waves that have shaken up and transformed economy, society and culture since the first Industrial Revolution and even before and have left a huge impact on how we live, think, communicate and work: they have deeply affected the socio-economic metabolism from within and mankind's footprint on our planet. The Handbook analyses the origins of the current wave, its roots in earlier ones and its path-dependent nature; its current forms and actual manifestations; its multifarious impact on economy and society; and it puts forward some guesstimates regarding the probable directions of its further development. In short, the Handbook studies the past, the present and the future of smart technologies and digitalisation.

It hardly needs to be stressed that the future is uncertain and that we cannot pretend to be able to foretell it. However, current developments shape the potential for future ones, so that the uncertainty we face is contingent. As time goes by, the contingency will change, but at any instant of time it may perform the function of a guardrail that narrows the corridor within which future change can reasonably be assumed to take place. This does not rule out further "disruptive" technological breakthroughs that are currently in the making in R&D departments of companies, other private and public research organisations, such as universities, military research laboratories and so on, and will be followed by new thrusts of "creative destruction" (Schumpeter). Several of the organisations under consideration take pains to protect the emerging knowledge from transpiring fully, and so we can say little about its probable shape and impact. This limitation must not, however, prevent us from investigating as best as we can the

1 A background to this Handbook is the paper of the editors on "Riding a new wave of innovations. A long-term view at the current process of creative destruction", *Wirtschaft und Gesellschaft* 44(4): 545–583, 2018. In the following we repeatedly draw freely on it.

DOI: 10.4324/9780429351921-1

kind of information and knowledge available to us and to draw out what are presumably its most important implications. It goes without saying that we have every reason to be humble about our ability to anticipate the future.

The Handbook is subdivided in six parts.

Part 1 is on "Disruptive technological change: historical record, economic analysis, methods and tools" and has six chapters. It sets the stage by spanning a large space of dimensions – epistemological, methodological, historical, technical, theoretical and instrumental, to name but the most important ones – along which the authors in the Handbook approach the grand themes of the problem under consideration.

Chapter 1 is by Lord Robert Skidelsky, expert on John Maynard Keynes's writings and far beyond, and asks the fundamental question: "Is technological progress inevitable?" For thousands of years there was hardly any persistent and discernible technological innovation at all. How come that since only very recently things have changed dramatically and technological change is now regarded as something quite normal, even natural? What causes and drives technological change? Skidelsky argues that technological change is not caused by institutions, but reflects an attitude of mind, a "culture of innovation"; institutions matter however in regard to how technological change is received and processed, the speed at which new technological knowledge is absorbed and the impact it has on economy and society and culture at large. The attitude under consideration was absent during long periods of time in human history. It was the result of a complex interaction of different factors and may vanish again in the distant future. There is no presumption that humans are under the spell of some sort of technological determinism. The view that technology determines the future misses an important fact, namely, that technology may deliberately be shaped by a reflection about "the place of technology in the great scheme of things the future opens up. Humans choose their own destinies; they are not imprisoned by them".

Chapter 2 is by the economic historian Werner Plumpe and deals with "Disruptive technological change in recent economic history". It provides a succinct account of different patterns of technological change since the emergence of modern capitalism in the 17th century. This is the time when a growth gap began to emerge between Western Europe and its oversea offsprings, on the one hand, and the rest of the world, on the other. Plumpe stresses that this was not just a one-off event but inaugurated a new era of persistent technological change and economic dynamism. Disruptive change typically gives rise to pronounced economic cycles, therefore its association with long waves of development or "Kondratieffs", as Joseph A. Schumpeter called them. Plumpe stresses that the success of technical inventions in revolutionising the economy depends crucially on the environment into which they are born. Several inventions have been known for a long time but lay idle until socio-economic conditions became favourable to their utilisation. In some cases, complementary innovations were a precondition for their eventual success. That is to say, whether an invention becomes a path-breaking innovation, we do not generally know ante factum but only post factum after it has effectively played out its potential.

In Chapter 3 Heinz D. Kurz provides a brief account of how some early and a few modern political economists responded to the "Machine ages" that began to unfold before their eyes, that is, their analyses of the "Causes, forms and effects of technological change". The first machine age refers to the (first) Industrial Revolution and how Adam Smith, David Ricardo, Charles Babbage and Karl Marx attempted to come to grips with the progressive mechanisation of production, replacing labour and animal power by machine power, and the way this transformed economy and society. The chapter concerns, second, the current machine age characterised by machine learning and artificial intelligence and new forms of dynamically increasing returns to scale. It is shown that the seeds to our world were sown some two centuries ago, when authors like Ricardo contemplated already a fully automated system of production and Babbage, the "father of the computer", spoke of manufacturing as founded on principles deriving from an investigation of the human mind. The impact of different forms of technological progress is studied inter alia in terms of movements of the constraint-binding changes in the distributional variables (or "factor price frontier") over time. The current forms of technological change involve a double segmentation, amongst "superstar firms" and the competitive part of the economy, on the one hand, and skilled and unskilled workers, on the other.

Chapter 4 by Mark Knell and Simone Vannuccini is on "Tools and concepts for understanding disruptive technological change after Schumpeter". It continues with the inspection of some of the methods economists elaborated in an attempt to come to grips with the properties of the prime mover of socio-economic change, technological progress. (See, however, also several other chapters, such as, for example, the one by Ariel L. Wirkierman in Part 3.) It deals with radical innovations that affect the socio-economic system in a fundamental way. New technologies are said to be "radical" when the functions they perform are novel and when they give rise to entirely new configurations with known functions. In the literature several concepts have been put forward in an attempt to specify the genuine significance of the involved novelties and their major innovations. These include Schumpeter's concept of "long waves" by means of which he attempted to characterise various phases of development since the Industrial Revolution up until the world economic crisis starting in 1929. He saw five such waves revolving around cotton textiles, railroads, steel, automobiles and electric power. Schumpeter's approach triggered a rich literature that critically scrutinised his argument and improved upon it in several respects. New analytical concepts were forged, including inter alia that of "techno-economic paradigms", "great surges of development" and "general purpose technologies". Knell and Vannuccini bring clarity into a bewildering variety of concepts, including the distinction between five technological revolutions or techno-economic paradigms, with a sixth one on the way.

Based on Schumpeter's early perception that entrepreneurial behaviour is a key element of innovation, Chapter 5 by Uwe Cantner and Thomas Grebel is on "Entrepreneurship and industrial organisation". Cantner and Grebel aim to shed light on the role of entrepreneurship against the background of industrial organisation and industry dynamics from the perspective of innovation economics. The authors review main contributions on entrepreneurship and industrial organisation, which they subsequently confront with empirical evidence, mainly from the German automotive industry. These include the "structure-conduct-performance approach", "tournament and non-tournament models", the "industry life cycle theory" and the concept of "entrepreneurial ecosystems". The authors find that apart from profit opportunities, also technology- and industry-specific factors are key to entrepreneurship and innovation activities. As long as profits can be obtained from a market and entry barriers are not too high, innovation activities won't come to a rest. Cantner and Grebel conclude by pointing to a gap that needs to be closed in future research, as there is lack of an understanding of the way how entrepreneurs create for themselves new (profit) opportunities, giving eventually rise to new industries. This is found particularly important with regard to the development of smart technologies as some of them are in their very infancy and most likely bring about radical new industry patterns and dynamics.

Finally, in Chapter 6 Luigi Marengo focuses on the employment effects of new technologies by asking: "Is this time different? A note on automation and labour in the fourth Industrial Revolution". According to one view, the economic system is currently in a transition phase in which there will be a net loss of jobs, which, however, will be made good and may even get overcompensated in a later phase, as historical experience has shown time and again. According to another view, this time is different, with displacement exceeding compensation not only in the short, but also in the long run. Hence the spectre of technologically induced unemployment can be expected not to go away in the foreseeable future. The net losses of jobs will not be limited to low-skill manual and routine jobs but will more and more concern also cognitive medium- and higher-skill jobs. That is to say, the age of technological unemployment that John Maynard Keynes envisaged in his essay "Economic possibilities for our grandchildren" in 1930, might finally come true. In Marengo's view there are already clear indications of this, viz. the decrease in the share of wages in national income and the increase in wage polarisation, both of which reflect a growing increase of income inequality in Western advanced economies.

Marengo's chapter provides a bridge to Part 2 of the Handbook dedicated to "Smart technologies and work". Its five chapters address some of the crucial issues when studying the impact of technological change on work and socio-economic inequality from a multidisciplinary perspective.

There is no doubt that the task and skill content of occupations has already been significantly affected by the diffusion of the new Information and Communication Technologies (ICTs) since the 1970s. It is often argued that this trend is reflected in the empirically observed polarisation of employment and wages in developed countries. However, as the economist Karishma Banga points out in Chapter 7, "Smart technologies and the changing skills landscape in developing countries", the evidence on labour market polarisation in developing countries is weak up until now. Banga argues that developing countries, which are still at early stages of using smart technologies, may actually have "a window of opportunity" in less digitalised sectors, as they still require more traditional, industrial skills. The potential of manufacturing-led development regarding the creation of sustainable new jobs is, however, limited by shifting patterns of globalisation. Banga therefore maintains that countries will need to focus on attracting and developing the right set of skills to increase the competitiveness of their workforce, namely, job-neutral and job-specific digital skills but also job-neutral soft skills.

In Chapter 8, "The impact of disruptive technologies on work and employment", researchers from the European Foundation for the Improvement of Living and Working Conditions (Eurofound) Irene Mandl, Ricardo Rodriguez Contreras, Eleonora Peruffo and Martina Bisello analyse the expected impact of automation, digitalisation and the platform economy on work with a focus on the European Union. They do not only cover the quantitative aspects of employment (i.e., job creation and destruction), but also discuss important dimensions related to job quality, such as work organisation and working conditions, occupational health, and safety. Finally, they investigate potential implications for labour market institutions, concentrating on industrial relations and social dialogue. The chapter concludes with policy suggestions aimed at the EU level, but also at the national level, highlighting how conditions can be created that enable the use of modern technologies to the benefit of both employers and workers.

In Chapter 9 on "The fourth industrial revolution and the distribution of income", the economist Stella S. Zilian and the system scientist Laura S. Zilian provide an overview of the key relationships that are commonly discussed in economics regarding technological change and income inequality. They summarise the extensive economics literature on skill-biased and routine-biased technological change before assuming a sociological perspective, according to which the distribution of digital skills ("the skills of the future") is itself characterised by social stratification and may thus exacerbate economic inequality as the fourth Industrial Revolution – amounting to the digital transformation of the economy – evolves. Finally, they describe how digitalisation leads to the emergence of highly concentrated markets and affects the labour share as well as top income inequality.

It is evident that the global decline of the labour share indicates that workers' and their trade unions' bargaining power has significantly deteriorated in recent years. This is partly owed to the effect of technological change on the organisation of work, viz. the organisation of platform work in the gig economy. The legal aspects of platform work are the topic of Chapter 10, "The legal protection of platform workers", written by legal scholars Jeremias Adams-Prassl and Martin Gruber-Risak. The authors argue that operators of these platforms make the most of the obvious trade-off between flexibility and insecurity characteristic of gig economy workers. This is especially the case as the latter deftly circumvent employment law by insisting on acting merely as intermediaries between independent contractors and their customers. Adams-Prassl and Gruber-Risak explore various options to bring platform work (back) under the umbrella of labour law. They conclude with a very topical discussion of some of the challenges posed by the Covid-19 pandemic, highlighting the societal importance of ensuring the protection of platform workers.

Part 2 closes with Chapter 11, "Smart technologies and gender: a never-ending story", by Knut H. Sørensen and Vivian Anette Lagesen in which they address the issue of gender and technology – a topic that has received broader public attention in recent years but has been studied intensively in the social sciences already since the 1980s. Sørensen and Lagesen, both experts in science and technology studies, use analytical tools from the field of feminist technoscience to discuss the causes of the gender imbalance in ICT, the gendering of smart technologies and their harmful societal consequences. The authors draw on the notion of co-production – a form of dynamic knowledge production that describes how knowledge and associated technologies shape the world we live in and vice versa – and argue that due to the exclusion of women from designing smart technologies and the lack of interest in women as users of smart technologies, the gendering of smart technologies remains highly resilient to change. Sørensen and Lagesen particularly highlight the roots of and problems associated with the "invisibility of women" in this context. They conclude that "the most important problem is not the exclusion of women but the feeble efforts to include them". Overall, the chapter shows that (more) serious action is needed on the government and industry level to counteract the gendering of smart technologies.

It is not only the gendering of smart technologies but also their impact on employment, the labour force and the workplace that highlights their transformative capacity. Yet, this goes far beyond this impact and it is reasonable to assume that smart technologies will shape almost *every* realm of society, not to forget the speed at which inventions occur.

Against this background, Part 3 of this Handbook is on "Smart technologies and social and economic transformation". It contains eight chapters and in Chapter 12 starts with an introduction to "Artificial intelligence" from a non-technical perspective, written by the computer scientist Fredrik Heintz. The chapter provides a summary account of artificial intelligence (AI) and its diverse research sub-fields, and discusses central features of it, such as reasoning and learning (including the different models of machine learning: supervised, unsupervised and reinforcement learning). Beyond these technical aspects, Heintz also approaches important institutional topics that shape the co-evolution of technology and society: trustworthy AI; Europe's mission to secure a human-centred development of artificial intelligence that does not focus merely on technical features but conceives AI as a sociotechnical construct; and human-AI interaction, that reflects upon the co-existence of, and collaboration between, man and machinery. In this way, the chapter sets the stage for the remaining chapters in this part, which focus on the economic and social dimensions of AI and other smart technologies.

In Chapter 13 on "The science space of artificial intelligence knowledge production: global and regional patterns, 1990–2016", Dieter F. Kogler, Adam Whittle and Bernardo Buarque approach the AI technology domain from an evolutionary economic geographic perspective. To enhance the understanding of where and how scientific knowledge of AI is created, the chapter seeks to measure its creation and diffusion across space and time. For the empirical and scientometric part of their chapter, a relational database composed of scientific publications on AI (and related terms) from Web of Science is constructed, to which they subsequently apply network analysis as well as co-occurrence analysis. For different spatial levels, including the global, the country and the regional level, the authors trace how the science space, its structure and the collaborative network of knowledge creation in AI have evolved since 1990. Comparing finally two European Union capital regions (i.e., Dublin and Vienna) with regard to the embeddedness of AI knowledge, they shed light on the region-specific capacity to generate new knowledge in this disruptive technology domain.

In the following three chapters, the focus is on the diffusion and deployment of smart technologies in the economy, and, especially the production system.

In Chapter 14 the economist Ariel L. Wirkierman provides an in-depth analysis of the "Structural dynamics in the era of smart technologies". He focuses especially on three aspects of how advanced digital production technologies affect the system of production and the labour process: first, they alter profoundly the relationship between fixed capital (tools and machines) and human labour and re-define the human-robot complementarity; second, they necessitate the mutual adaptation of labour tasks and skills and the characteristics of the new technologies; and, third, they call for a distinction between physical and digital output and assets and the replacement of the received system of socio-economic accounting that is ill-adapted to a world of bits and bytes. The author asks, among other things: what will be the role of human activity in value generation and as regards income distribution in the world of smart machines? With consumption data becoming an input in production processes, how will the received distinction between consumption and investment have to be adapted to the new situation? How will the advanced digital production technology affect the international division of labour? How will it affect the economic performance of less developed countries that are possessed of large amounts of unskilled labour that will be made redundant? Which policy measures are suited to cope with the numerous challenges the new technologies pose?

In Chapter 15 the economists Bernhard Dachs and Xiaolan Fu together with the social scientist Angela Jäger study "The diffusion of industrial robots" and their economic effects. The authors present a rich empirical material on aspects of the geographical distribution of robot shipments and its development since the 1990s, the inter-country distribution of robot intensity and, among other things, the use of robots across sectors and changes therein over time, with a special focus on the manufacturing industries. The authors also discuss some firm-level evidence on how the deployment of robots relates to firm size, production characteristics, etc. They find industrial robots to be very unevenly distributed across these dimensions. One of the main conclusions they draw is that industrial robots have not turned into a general purpose technology yet, though their future potential is large. Moreover, since findings from the literature on the productivity and employment effects of robots are mixed, the authors consider this an important topic that needs to be further researched.

Then, in Chapter 16 by Thorsten Wuest, David Romero, Muztoba Ahmad Khan and Sameer Mittal on "The triple bottom line of smart manufacturing technologies: an economic, environmental and social perspective", an engineering and technology-management perspective is coupled with the triple bottom line model of sustainable development to answer the question: "What is the impact of new smart manufacturing technologies on the 'Triple Bottom Line' of the sustainability of businesses and supply chains?" Based on a systematic literature review of an earlier work of theirs, they identify and discuss the economic, social and environmental impact of ten key smart technology clusters, including automation and robotics, additive manufacturing as well as artificial intelligence, machine learning and advanced simulation, etc.

They also survey the typical obstacles to the adoption of smart manufacturing technologies and find that smart manufacturing systems have the potential to support and reach the triple-bottom line. The authors further argue that due to the economic, social and environmental dimension being contingent upon each other, reaching this goal requires that organisations adopt a holistic approach.

While smart technologies initially revolutionised in particular the organisation of production in manufacturing, by now they have become an integral part of services as well: cases in point are mobile devices, smart energy management or smart transport that includes the optimisation of freight routes (by means of advanced data analytics) and self-driving vehicles. In short: smart services are nowadays to be found everywhere.

In Chapter 17, "From smart technologies to value creation: understanding smart service systems through text mining", Chiehyeon Lim and Paul P. Maglio undertake an in-depth literature review of the topic of smart service systems. The authors apply machine learning algorithms to almost 5,500 scientific articles and around 1,200 news articles published until 2016 to detect the most important research topics (such as wireless networks/communications, cloud computing/environment or Internet/Web of Things), technology factors (e.g., context-aware computing, connected network) and application areas (e.g., smart home, smart city, smart energy or smart education). This work thus provides detailed insights into the formation phase of smart service systems that have become a pivotal research field in recent years. From their analysis, they derive five common characteristics for a unified understanding of smart service systems: (1) connection of things and people, (2) collection of data for context awareness, (3) computation in the cloud, (4) communications to automate or facilitate and (5) value co-creation between customers and providers. These five "Cs" help to demarcate the different types of smart service systems and may also give guidance to private sector actors and the government regarding their research and development in the future. Last but not least, through the use of smart technologies (machine learning), the chapter comes to a new conceptualisation of "smartness" and thus also shows the value of these technologies as methodological tools and for research design.

Smart technologies will also facilitate daily life: The "connected home" is no longer science fiction, as smartphones supported by sensors already switch on and off lights, regulate the room temperature and monitor home appliances. Sensor technologies allow the management of traffic, logistics, resource use and waste, amongst others, for whole urban areas, paving the way for "smart cities".

Elke Pahl-Weber and Nadja Berseck, both experts in the field of urban development, in Chapter 18 on "Smart cities, a spatial perspective: on the 'how' of smart urban transformation", provide insights into urban planning by focusing on the process of how to develop smart cities. The authors analyse cities in the context of complex systems and use the urban design thinking approach, developed by Pahl-Weber, to deal with the challenges associated with smart city transitions, such as reconciling competing interests. Urban design thinking uses human-centred design principles known from, for example, product design or industrial innovation, and applies them to the shaping of urban space. Pahl-Weber and Berseck argue that smart cities are complex systems with wicked and tangled problems (i.e., housing bubbles, ethnic segregation, automation of human learning, traffic congestion). Tackling these requires a holistic approach and an integrated intervention logic, an emphasis on citizens, taking into account place specifics and new levels of collaboration among urban actors. Finally, in terms of multiple case studies, Pahl-Weber and Berseck show how urban design thinking can help to identify challenges, which really matter for urban citizens and highlight the potential of their approach for creating, testing and iterating solutions through multi-stakeholder collaboration.

Another example are smart electricity grids that allow the decentralisation of the energy market and let the single household become a "prosumer" of electricity, a combined producer and consumer. The altering role of users in the development of smart energy technologies and the potential of these technologies for energy sustainability are the focus of Chapter 19 by Antti Silvast, Robin Williams, Sampsa Hyysalo, Kjetil Rommetveit and Charles Raab, titled "Producing the 'user' in smart technologies: a framework for examining user representations in smart grids and smart metering infrastructure". Anchoring their work in the sociology of user representations, the authors stress the importance of understanding the variety of user types in the development of a new generation of energy services: is the anticipated user the end-user of smart meter technologies or rather the intermediate user of an entire new spectrum of energy services and applications (such as time-of-use tariffs or more complex demand-side flexibility services that control the energy use of household appliances)? Addressing this question, the authors investigate the perceptions of European smart-grid developers and technology use experts and study the lessons learnt from the smart meter roll out in the UK and Finland. Reaping the benefits of smart grids and smart metering and accessing all layers of novel services associated with them requires not only a stronger differentiation of user profiles in the design process, but also a continuous demand-response system where the consumer is actively involved in the development of these new energy technologies.

Apart from all these dimensions of social and economic transformation tied to smart technologies, another key characteristic of their gradual absorption is that they give rise to data-driven, digital markets, in which data assume two important functions: first, it serves as a kind of new "money" – private information – in terms of which the customer pays the platform firm. (If the collection of data is costly, the platform firm can be expected to recover the extra costs by charging its customers higher prices for the products it sells.) Second, and correspondingly, data is a productive resource that fuels the learning of machines. Data gains in importance relative to labour, land, capital and conventional money and finance. These developments do not leave unaltered the role of the state, raising inter alia fundamental questions about the design of the legal, political and institutional framework.

Against this backdrop, Part 4, which contains six chapters, is devoted to "Smart technologies, governance and institutions".

Numerous political philosophers from David Hume, Adam Ferguson and Adam Smith to Friedrich August Hayek argued that limiting the power of members of society or groups is the most important problem of social order. Government has to protect all members of society against coercion and violence from others. Yet when government assumes effectively the monopoly on the use of force, it unavoidably becomes itself a potential threat to individual freedom. This threat still exists and with new surveillance technologies and methods of monitoring peoples' behaviour has assumed nightmarish dimension. They ask whether government is still the "main threat" to individual freedom. Given the amount of personal data available to them, rogue platform companies are able to secretly manipulate their customers in ways and to an extent never seen before in history. The sovereignty of the state on the one hand and the legal framework of data-rich markets under the new economic order of data capitalism are addressed in Chapters 20 and Chapter 21, respectively.

In Chapter 20 Richard Sturn turns to the problem of "Digital transformation and the sovereignty of nation states". Digital technologies affect fundamentally state sovereignty, and while some commentators are of the opinion that a post-sovereign state of affairs will result, in which markets take over and the role of the state gets marginalised, he is convinced that this will not be the case. A sovereign public agency will be indispensable in order to cope successfully with the numerous challenges the new technologies and their consequences pose, including "Medici vicious circles of economic and political power" and socially undesirable lock-ins due to incomplete contracts and coordination failures. He discusses the most important gateways of the erosion of state sovereignty and provides a critical account of several variants of post-sovereignty. Techno-libertarianism and anarchism, the most radical variant, stipulates that market exchange based on voluntary contracts finally will prevail and replace the received distinction between private and public domain. Techno-liberals are on the contrary convinced that distributional problems get aggravated in the digitalised world and have to be taken care of by public authorities. Sturn insists that re-inventing sovereignty and redrawing the private-public demarcation are amongst the most difficult and pressing tasks facing the digitalised world.

In Chapter 21, "Antitrust law and digital markets" the legal scholar Viktoria H.S.E. Robertson provides an overview of the challenges that data-driven digital markets pose for competition law. She focuses on five key areas of competition law, market definition, market power assessments, anti-competitive agreements, abusive conduct and merger control, and discusses for each one the main issues associated with digital markets, how European competition law has started or intends to attend to these issues, and how some of these areas need to be adapted to the digital age. For example, one main challenge lies in the difficulties regarding the assessment of market power in digital markets and platforms, which should ideally take into account the effects that lead to the emergence of market dominance, such as network effects, economies of scale, switching barriers for users, etc.

Shifting the focus to the issue of power in the platform economy in Chapter 22, titled "Platform regulation: coordination of markets and curation of sociality on the internet", Ulrich Dolata offers a techno-sociological perspective of the mechanisms through which the leading Internet companies – with their networked platforms – fulfil their role as structure-forming, rule-setting and action-coordinating key players in today's web. The author considers them to be far more than just infrastructure providers; they are rather "differentiated societal systems with a distinct institutional foundation". The chapter provides a detailed discussion of two main areas of regulation: first, the private-sector coordination of markets and, second, the technically mediated structuring and curation of sociality. The author remarks that power is highly asymmetrically distributed in the platform economy and reflects on rooms for manoeuvre to deal with this asymmetry at the two levels of social and political intervention. He finds that "the political regulatory approaches, to date, are not suitable for substantially correcting or controlling the regulatory sovereignty of the platform operators". He proposes the unbundling of those large corporations and setting up public supervisory and regulatory bodies to cope with these problems.

Apart from these dimensions of the governance and institutional framework of the digital transformation, another relevant dimension concerns that of the design and distinct features of policy programmes targeting innovation in smart technologies.

Focusing on current European Union innovation policy, in Chapter 23 on "New mission-oriented innovation policy in the digital era: how policy-based social technologies fuel the development of smart technologies", Marlies Schütz and Rita Strohmaier conceptualise the endogeneity and dynamics of mission-oriented innovation policy. Based on evolutionary economic thinking together with borrowings from the field of ethology with regard to three key characteristics of a technology (viz., purpose, function and benefit) and the core aspect of reflexivity, the authors work out the analytical framework of "policy-based social technologies". This allows evaluating mission-oriented innovation policies from an activity-centred and process-oriented perspective. By means of this concept greater analytical rigour is applied to mission-oriented innovation policy initiatives and this is validated in a comprehensive case study of ECSEL JU, a policy initiative of the European Union in the field of electronic components and systems. The authors further demonstrate the relevance of the proposed analytical framework for mapping mission-oriented innovation policy as organisational innovation that offers solutions to "grand societal challenges" – a topic that will be further addressed in Part 5 of the Handbook.

After discussing the policy perspective and how innovation policy shapes the development of smart technologies, the next two chapters focus on the reverse causal link, namely how smart technologies mould institutions and what their socio-organisational effects are.

In Chapter 24 on "Crypto assets", Tobias Eibinger, Ernst Brudna and Beat Weber ask in what way crypto assets – as a decentralised system of digital money – may impact on transaction costs and affect the "need of trusting others" in financial transactions. After having introduced the reader to the techno-economic features of crypto assets and blockchain technology, the authors contextualise cryptocurrencies with the centralised institutions of the monetary and financial system in a modern capitalist market economy and discuss their potential to subvert the current organisation of the monetary and financial sector. Apart from a detailed account of cryptocurrencies, they also explain the concepts of "tokenisation" and "decentralised finance" as two alternative, non-conventional modern financial instruments. One of their main findings is that "digital currencies" will at present most probably not compete with traditional currencies, hence their impact on the monetary policy of public institutions can be expected to remain small. The authors close the chapter by reflecting on the challenges that crypto assets may pose to the socio-economic sphere, in particular with regard to their potential to shape the "background conditions" of a society.

Along these lines, in Chapter 25, Brendan Markey-Towler provides a general account of "Blockchain and the 'smart-ification' of governance: the last 'building block' in the smart economy". The author takes an evolutionary-institutional perspective on the economy, where the latter is perceived as a complex, ever-evolving web of rules operating at different levels (micro, meso, macro) and guiding production, exchange and governance. Within such an economy, smart technologies like artificial intelligence, platforms or the Internet of Things eventually enable completely decentralised, distributed and automated networks of production and exchange that are based on interconnected cybernetic control systems. This inevitably requires adjustments in the institutional authority. In this context, Markey-Towler argues that blockchain – as an institutional technology – facilitates a new mode of smart governance for Internet-based socio-economic platforms that is itself decentralised, distributed, automated and cybernetically controlled. As such, it represents "the last 'building block' in the smart economy". This will require a new reading and theory of economics, likely emerging from the field of institutional cryptoeconomics, to deal with the "fuzzy blend" of decentralised structures and strong cybernetic control that characterise the new smart economy.

In view of the rapid developments in the field of smart technologies, the question arises as to the place of the human being in the new social order and what an insatiable drive for progress implies for the planetary boundaries. Part 5 on "Smart technologies and grand societal challenges", including five chapters, is devoted to this tightrope act between disruptive technological change and environmentally sustainable and socially responsible transformation. Against the backdrop of digitalisation, it discusses some of the most pressing problems of our time, as articulated by the European Commission in the context of the Horizon 2020 framework: food security and sustainable agriculture, health and well-being, secure and efficient energy, as well as secure societies.

As the chapters introduced so far show, digitalisation bears an enormous potential of change for economy and society but may equally represent one of the greatest threats to our planet and to the life as we know it: overall, the digital revolution can contribute to dematerialisation and increased resource efficiency, reduce emissions, achieve cost savings in education and production, and enable people to participate more fully in society. However, there are also major dangers: increasing energy demand and electronic waste, rebound effects from increased consumption, geopolitical power struggles and, last but not least, the exacerbation of existing inequalities within a society and between different world regions.

In Chapter 26, "'Back to the future': smart technologies and the sustainable development goals", J. Carlos Domínguez, Claudia Ortiz Chao and Simone Lucatello set the stage for a broad discussion of how smart technologies enable or prevent opportunities for the sustainability transformation, as set out in the 2030 Agenda of the United Nations. Can smart technologies support efforts regarding poverty reduction and the eradication of hunger without exacerbating existing structural inequalities? Can they help achieving healthier lifestyles without trapping individuals into continuous surveillance and social control? Can they enable sustainable cities and communities without making them overly dependent on technology? Can they stimulate the decarbonisation of the economy without further fuelling energy demand? These and other questions are critically discussed in this chapter. The authors demand caution when it comes to the potential of smart technologies: "they might be significant enablers, but they are not magic bullets to solve all development problems". With this conclusion, the chapter leads over to the following contributions that focus on specific instances of the aforementioned tightrope act.

In the digital Anthropocene, Hans Jonas's "dialectic of progress" – creating new problems while solving those it generated in the past – is omnipresent. The only technologies that are risk-free are in fact those that are never built. What is needed regarding the digital transformation is strong governance by nation states as well as by the international community that not only incentivises but also defines the direction of transformation. An entrepreneurial state envisioning sustainability not only minimises the risks involved in technology *development*; it also has the supreme responsibility to carefully weigh the opportunities and unintended consequences that the *use* of these technologies has on a global scale.

This is particularly crucial when it comes to securing the survival of humankind: the world's population is expected to grow to almost 10 billion by 2050. To feed them, agricultural yields will need to rise significantly. Precision agriculture is as important as biotechnology in achieving these necessary efficiency gains. Both fields have benefited enormously from the digitisation trend. In Chapter 27, Andreas Stamm deals precisely with this topic. In "North-South divide in research and innovation and the challenges of global technology assessment" he discusses "the case of smart technologies in agriculture" and raises three important points: (1) technology assessment must be reconceptualised in the light of global challenges; (2) consequently, technology assessment needs to be carried out on an international stage; (3) the Global North – bearing the greatest responsibility for anthropogenic climate change – has an obligation to support the development of technologies that potentially boost economic growth and social well-being in developing countries and help them tackle their environmental challenges. As regards digital agriculture, the author argues that smart technologies may lead to decreasing job opportunities in the Global South. The case of the plant breeding technology CRISPR-Cas9 shows how legal rulings in the Global North delay and negatively impact the development of a technology potentially essential for combating hunger and securing livelihoods in developing countries. New forms of science, technology and innovation partnerships are necessary to warrant more inclusive, just and sustainable technological and global change.

In the following three chapters, the perspective shifts from a rather geopolitical to a societal level. Chapter 28 by Dan van der Horst is on "Smart technologies, energy demand and vulnerable groups; the scope for 'just' metering?" It sheds light on the extent to which vulnerable groups benefit from ongoing technological advances in smart metering and remote monitoring. With a focus on the roll-out of smart metering in the UK, van der Horst juxtaposes the claims of potential benefits of the private sector and the population at large with those of vulnerable consumers, such as poor, physically and mentally challenged or elderly citizens. In this regard the author maintains that digital and remote metering and monitoring could fulfil important functions that go beyond efficient domestic consumption of energy but also affect environmental conditions in the home and the health conditions of its inhabitants. Thus, to fully capitalise on the advantages of smart energy services for vulnerable groups, they need to be connected to three interrelated trends: assisted living, digital health as well as social networks and peer support. Thus, the government has a key role to play in encouraging and facilitating innovations that feed into a more systemic and inclusive narrative of the smart society while also tackling issues of increased surveillance and data control.

Connecting smart energy with digital health and assisted living, the chapter also bridges to the upcoming topic in this part, viz. the impact of smart technologies on health and well-being: innovations such as wearable health monitoring systems or biosensors for diagnostic purposes as well as new medical tools (such as 3D printing of implants, pacemakers or even body parts) create a vision of the "transparent" body, whose condition can be shared on Internet platforms. In addition, telemedicine provides medical care for people living in remote areas and supports the health sector in the Covid-19 pandemic. Aspects like these are discussed in Chapter 29 on "Smart Health", where the health economists Thomas Czypionka and Susanne Drexler introduce the reader to the digital transformation of the healthcare system that causes disruptive changes at the different stakeholder levels (service providers, insurers, patients and the research and innovation system). The authors distinguish telemedicine from related concepts such as e-health, telehealth and teleconsultation, and consider the role of artificial intelligence in healthcare. To give a small taster of the future potential of smart technologies in the health sector, these topics are subsequently illustrated by a number of technical solutions and applications; among others, EPItect, an in-ear sensor to detect epileptic seizures; MySugr, an app that supports diabetes patients; and e-health platforms in Denmark and the UK. To address the intricacies involved in this topic, the chapter closes with a brief resumé of the socio-economic implications and the obstacles that inhibit the vast opportunities of smart health at present.

The final chapter of Part 5 touches upon one further aspect of smart technologies: taking a gambit with digitalisation, individuals and society must not be sacrificed. This in turn requires a new ethical framework that reflects the full scope of our drive for innovation. The same technologies that enable the digital transformation also pose its greatest threat: they facilitate illegal access to and misuse of personal data, sabotage of industrial plants and critical infrastructure, such as water and energy systems, and turn into deadly weapons in cyberwars. In Chapter 30, devoted to "Cybersecurity and ethics: a uncommon yet indispensable combination of issues", Karsten Weber stresses that cybersecurity does not only entail technical challenges but since the emergence of this topic in the late 1960s has also nurtured paramount institutional, economic and organisational issues that have shaped the discussion up until now. The reluctance of public institutions to safeguard the society and the environment from the immense damage caused by cyberattacks have eroded the trust of citizens in the government, further exacerbating the problem. Besides the economic dimension, the subject raises serious ethical questions, as truly secure computer systems would also restrict our freedom, flexibility and privacy. Describing the various challenges related to such fundamental moral aims and values, Weber lays out potential routes to balancing this tightrope act between cybersecurity and individual and societal autonomy. These routes ought to help manoeuvre society through the dangerous waters of cybercrime and cyberwar towards a society empowered and not held captive by smart technologies.

Finally, Part 6, "Smart technologies: case studies", widens the scope of the discussion and in three chapters presents a number of case studies on the transformative capacity of smart technologies. With a focus on the digital transformation in Asia, selected nation states' policy initiatives are studied. These shed light on how the public sector has engaged in sparking the development of smart technologies and how their absorption has already transformed the socio-economic system and will continue to do so in the near future.

Aimed at preparing the Japanese economy and society for the digital transformation, Japan in 2016 has launched "Society 5.0", which constitutes a holistic policy framework that seeks to establish a completely networked Japanese society, tackling especially the demographic challenge of an (over-) ageing population. Addressing this topic from a policy perspective, Yuko Harayama and René Carraz in Chapter 31 present a detailed case study of "A digital society for an ageing population: the Japanese experience". After having introduced the reader to the age distribution and the composition of Japan's population, past policy approaches are reviewed. Then, the authors provide an in-depth study of the policy path that led to the initiation of the policy model Society 5.0, its salient features and the Japanese experience of the deployment of smart technologies for raising the quality of life, especially for the elderly people.

In Chapter 32 on "Digitalisation and development in India: an overview", Syed Mohib Ali Ahmed discusses the concerted efforts of the Indian state to promote structural transformation through smart technologies and simultaneously bridge the digital divide and promote inclusive growth. The author focuses on two particular strategies the government pursues on its path towards "digital India": the digitalisation of services and digital payments that should foster digital and financial inclusion. The author argues that "in a structurally retrogressive and informal economy like India", one-sided policy activities to support the digital transformation only feed further into a regime of jobless economic growth, since the sharp rise of services in national value-added and the hitherto largely unrealised productivity gains from digitalisation in the manufacturing sector cannot absorb the surplus labour from low productivity sectors like agriculture. At the same time, the lack in digital literacy makes it difficult for citizens to access new financial services and e-governance. In addition to growth of aggregate demand, a supportive infrastructure is therefore necessary to enhance the digital skills and competences of the workforce and in this way create complementary job opportunities in the production sector.

With a special focus on "Industry 4.0 in China", Han Li and Wei Zhang in Chapter 33 explore the digitalisation of production in China from an economic perspective. The chapter starts with a brief overview of China's economic history and on past industrial policies since the foundation of the People's Republic in 1949. The authors then discuss the current Chinese policy framework of the digital transformation and present empirical evidence on China's current status with regard to the adoption and diffusion of the relevant smart technologies, such as industrial robots, big data or artificial intelligence. The authors address also socio-economic effects accompanying the diffusion of smart technologies as well as their potential for upgrading China's industry landscape. The chapter closes with a reflection on the future potential of Industry 4.0 and identifies a list of challenges ahead. Finally, the authors derive some policy implications, including, first, the need to replace existing industrial policies by competition policies, second, the support of small- and medium-sized enterprises and, third, setting incentives for human capital formation.

Acknowledgements

This Handbook is an outcome of a research project on the digital transformation and the socio-economic effects of smart machines, which was carried out by the editors at the Graz Schumpeter Centre, University of Graz between 2017 and 2021. The research project was funded by the Austrian Science Fund (FWF) under grant number P 30434-G27. The editors gratefully acknowledge the financial support of the Austrian Science Fund (FWF).

Each of the chapters was reviewed by the editors and by external reviewers. We are most grateful to the following colleagues who kindly mustered time and energy to support us with the review process:

- Wiljan van den Berge, School of Economics, University of Utrecht, The Netherlands
- Susana Borrás, Department of Organization, Copenhagen Business School, Denmark
- José van Dijck, Institute for Cultural Inquiry, University of Utrecht, The Netherlands
- Christian Fuchs, Westminster School of Media and Communication, University of Westminster, UK
- Manfred Füllsack, Institute of Systems Sciences, Innovation and Sustainability Research, University of Graz, Austria
- Harald Hagemann, Department of Economics, University of Hohenheim, Germany
- Gudrun Haindlmaier, Department of Geography and Regional Research, University of Vienna, Austria
- Tom Hargreaves, School of Environmental Sciences, University of East Anglia, UK
- Illona Horwath, Institute for Lightweight Design with Hybrid Systems, University of Paderborn, Germany
- Christian Lager, Institute of Economics, University of Graz, Austria
- Bernhard Mahlberg, Institute for Industrial Research, Vienna, Austria
- Shah Nitin, Vice President of Nokia-Siemens, New Delhi, India
- Marc Pilkington, Department of Economics, University of Bourgogne, France
- David Rigby, Institute of the Environment and Sustainability, Department of Geography, University of California, Los Angeles, United States
- Marcel Smolka, International Institute of Management and Economic Education, Department of International and Institutional Economics, University of Flensburg, Germany
- Petra Sauer, (LIS)^{2}ER Tony Atkinson Research Fellow, Luxembourg Institute of Socio-economic Research (LISER) & Luxembourg Income Study (LIS), Luxembourg
- Gabriel Schaaf, Institute of Crop Science and Resource Conservation, University of Bonn, Germany
- Carolyn Snell, Social Policy and Social Work, University of York, UK
- Manoj Singh, Innovation Consultant, IT and Telecom Sectors, Anisha Global, Pune, India
- Dirk Nicolas Wagner, Strategic Management, Karlshochschule International University, Karlsruhe, Germany
- Beat Weber, The Oesterreichische Nationalbank (OeNB), Vienna, Austria
- Karsten Weber, OTH Regensburg, Germany

Our special thanks go to Julia Wurzinger from the Graz Schumpeter Centre, University of Graz, who helped us with proofreading and excellent editorial assistance.

PART 1

Disruptive technological change

Historical record, economic analysis, methods and tools

1
IS TECHNOLOGICAL PROGRESS INEVITABLE?

Robert Skidelsky

Introduction

Since the Industrial Revolution, technological innovation has subjected economies to what the economist Joseph Schumpeter called "gales of creative destruction" (Schumpeter 1994: 82–83). This is a novel experience in human affairs. Between the invention of the plough 10,000 years ago and about 1700 there was technology, but no technological momentum. Technologies came and went; some were known in some places and unknown in others; some were forgotten, others frozen; they remained primitive; there was no basic improvement in 'man's estate' for thousands of years. As John Maynard Keynes noted:

> From the earliest times of which we have record – back, say to two thousand years before Christ – down to the beginning of the eighteenth century, there was no very great change in the standard of life of the average man living in the civilised countries of the earth. Ups and downs certainly. Visitations of plague, famine, and war. Golden intervals. But no progressive, violent change. . . . This slow rate of progress, or lack of progress, was due to two reasons – to the remarkable absence of important technological improvements and to the failure of capital to accumulate.
>
> *(Keynes 1978a: 323)*

The "remarkable absence of important technological improvements" should alert us to the possibility that the technological dynamism of the last 250 years, while being a crucial episode in the ascent of our species, is destined to fizzle out as it exhausts its contribution to human well-being. It was the historian Arnold Toynbee who pointed out that automation, by simplifying the apparatus of living, produced a "consequent transfer of energy from some lower sphere of being or action to a higher one" (Toynbee 1974: 198). Keynes himself said the same thing in his essay cited previously: the solution to the problem of production would free humans for the first time in their history to devote their energies to non-economic purposes (Keynes 1978a: 326). Technology, like Mephistopheles, having done its work, could retire from the scene. It is difficult to embrace this view *tout court*, but we should at least expect a much greater degree of discrimination between the development and uses of different types of technology, with medical technology, for example, continuing to advance, but the pace of work and consumer technology slowing.

DOI: 10.4324/9780429351921-3

The two absences, or failures, mentioned by Keynes – stagnation of technology and capital – are linked. Technology failed to improve because capital failed to accumulate: economists would say it was 'consumed' rather than being invested in new tools and machines.

Capitalism was the first social organisation in history to make capital accumulation its object. Its three essential properties were the drive for riches, competition for profit in markets, and concentration of capital ownership in a class of capitalists. Although Karl Marx saw the concentration of capital ownership as the *unique* property of capitalism, the emergence of capitalist civilisation preceded the concentration of capital and is thus a part of the story of capital accumulation. But capitalism was not solely responsible for the acceleration of technology. Parallel to its emergence as the dominant socio-economic form was the emergence of a scientific outlook, which reoriented intellectual life to the production of useful things. It was this which made a sustained improvement in the tools of production seem feasible and desirable, thus making competition between capitalists for money a social instrument of technological improvement. Capitalism and science jointly reversed the "remarkable absence of important technological improvements" to which Keynes referred.

Capitalism has a strong claim to be humanity's greatest social invention, by virtue of the fact that it solved for the human species the problem which has plagued all living things: the imbalance between population and resources. From its start, the human population has waxed and waned round relatively fixed means to support it. The Reverend Thomas Malthus made a law out of these historical cycles: owing to its almost limitless fertility, population tended to expand faster than the land available to feed it. The problem of excess mouths to feed was met in one of two ways: starvation or migration. But the cycle then started up again. Fewer mouths meant more food per mouth, and with more food came more children, and so on.

Technology offered a non-tragic solution to the problem by increasing the yield of natural resources. If productivity could keep pace with population, the Malthusian problem was overcome. The technological dynamism of capitalism made this possible for the first time in human history. In fact, capitalist civilisation did better than that: it generated social mechanisms which caused population growth to fall relatively to productivity growth, enabling the world to feed ever greater numbers at a higher standard of living.

However, by making the economy dynamic, capitalism made it disruptive. Since the 19th century the economy itself has been the main source of social disruption. Keynes rightly mentioned the "visitations of plague, famine, and war". War has been a declining factor in human affairs as economic competition replaced military competition. Plagues and famines continue, but on an incomparably lesser scale than in the past. Instead we have periodic "crises of capitalism". Capitalism made disruption *endogenous*. Since the 19th century capitalism has been in the business of destroying old and creating new ways and forms of life. Periodic redundancy of jobs has replaced periodic redundancy of population.

Capitalism and science are yoked together, not in a base-superstructure relation, as Marxists have claimed, but in a complicated civilisational matrix which it will be the purpose of this essay to investigate. How did this unique civilisation arise?

The discussion will be divided as follows: lucky and unlucky geography, attitudes to work, the key institutions, and from religion to science.

Lucky and unlucky geography

Although humans are creators of the tools they use, the kind of tools they use and the pace of their adoption has always depended on their location: by location meaning the natural endowments of a given region, including climate. The claim is that different locations give rise to different needs and habits of thought and these give rise to different mental outlooks and institutions. These mentalities and institutions persist, rendering the societies which have them 'path-dependent' long after the geographic impetus is exhausted.

The geographic hypothesis was invented in the 18th century to account for the perception, of which western Europe had by then become very much aware, that it was forging ahead in wealth from the rest of the world; and it has been extensively used from Montesquieu onwards to explain (and celebrate) the inventive energy of the west in contrast to the supposed lethargy of everyone else.

Jared Diamond's theory of the 'lucky location', in his books *The Third Chimpanzee* (1991) and *Guns, Germs, and Steel* (1997), is the most celebrated recent example of the geographic hypothesis. Rejecting the idea that dominant peoples come from 'superior' genetic stock, he argues that those who came to dominate others did so because of advantages in their local environment. His principal claim is that, because of similarities in climate, technology diffuses more readily latitudinally than longitudinally. That is why technologies spread along Eurasia, but not up and down Africa and the Americas.

The most famous historical rendering of the geographical theory is Karl Wittfogel's *Oriental Despotism* (1957). Wittfogel believed that the way people earn their living determines their institutions, and their institutions in turn determine their receptivity to innovation. The river-based delta agricultures of the Middle East, India, and China favoured 'despotism'; the rain-based, geographically fractured, farming of Europe favoured warring states and free institutions.

The geographic hypothesis has done sterling work in explaining some aspects of the technological record as revealed by archaeology and history. The ancient empires of the 'Fertile Crescent' (Mesopotamia) around 5000BC to 500BC, together with those of India and China at roughly the same time, have been called 'hydraulic civilisations'. Friedrich Klemm writes of the Fertile Crescent:

> The technology of Mesopotamia and of Egypt was to a considerable extent determined by the three great rivers, the Euphrates, the Tigris and the Nile. Irrigation, the building of dykes and canals, the control and utilization of flood waters, were major engineering undertakings on a scale that necessitated organization by the State.
>
> *(Klemm 1964: 18)*

The ability to control river flooding by dykes and earthworks could increase grain production tenfold (Headrick 2009: 18). In China, a major 'hydraulic' economy, the first flood control system along the banks of the Yellow River dates from 2200 BCE; the Grand Canal, a waterway of over 1000 kilometres, linking the Yellow River to the Yangtze, dates from about 500 BCE. After a thousand years of hydraulic engineering the country was covered with a network of waterways, which banished serious floods and droughts for 22 centuries. The first Indian population centres from around 6000–5000 BCE were along the banks of the Indus River (in modern Pakistan) and the Ganges basin. The Mesopotamian empires have long since disappeared, but 'hydraulic' empires persisted into modern times in China.

All the institutions and practices of the hydraulic economy were built around the need to control water. Not only could these civilisations support far more people than their Neolithic ancestors, "but they also built monuments and cities, invented writing, mathematics, and calendars, and created elaborate religions, literatures, philosophies, and other forms of culture" (Headrick 2009: 17). David Graeber has described how early currencies grew out of the need for a convertible and portable way to pay standing armies (Graeber 2011: 225–228).

Geography can also help explain the original 'retardation' of the Mediterranean lands compared with those of the Fertile Crescent. The neoliths of Greece and Italy, who emigrated from Anatolia, "avoided becoming 'civilised' as long as possible", because there were always new lands for foraging and settlement. Not till the first millennium BCE – that is 9,000 years after the Asian settlements – did inhabitants of Greece and Italy come to occupy all the land available and found themselves in the same predicament as others: they had to civilise themselves or perish (Headrick 2009: 34). However, their climatic conditions then started to bring economic and civilisational benefits.

The usefulness of geography as an explanatory tool is not exhausted by considering only ancient economies. Immanuel Wallerstein does not dismiss the factors of "climate, epidemiology, soil conditions" in explaining the "crisis of feudalism" in 14th- and 15th-century Europe, and the search for new land in the west and east to replace depleted seigneurial revenues (Wallerstein 1974). Geography helps explain why the steam engine was invented in Britain and not China, and why it was first applied to railways in Britain and paddle-ships in the USA.[1] A frontier or 'edge' culture' always differs from an interior one, an urban culture from a rural one. The 'frontier thesis' of Frederick Jackson Turner helps explain the enduring mindset of the American mid-west (Turner 2003).

The assertion that today's 'virtual networks' have abolished the importance of location is a claim too far. States, which exist in geographical spaces, and are, to some extent their product, can close down global networks, and have already done so in the name of national security. Yuval Noah Harari's conclusion is the safest: "Geographical, biological and economic forces create constraints. Yet these constraints leave ample room for surprising developments, which do not seem bound by any deterministic laws" (Harari 2014: 197).

Attitudes to work

Technology is labour-saving. As economists tell it, technological innovation is a function of the balance between capital and labour. When labour is scarce (expensive) there is an incentive to substitute capital for labour; with plentiful (cheap) labour this incentive disappears. However, historically, conquest not technical innovation has been the main method of overcoming labour scarcity. Ancient economies were slave economies. Slavery was the fruit of conquest, which brought both more land and more hands to work it. But it was also the product of an attitude of mind which regarded manual work as degrading and 'fit only for slaves'. The thought that one day this work might be done by 'mechanical slaves' occurred to Aristotle, but in his own time there were enough human slaves to make labour-saving irrelevant.

Paganism and the contempt for work

Slavery is a property system in which some human beings are owned by others, to be used and disposed of at the will of the owners, just like any tool. With such cheap and pliant human tools at their disposal, property owners lack any incentive to invest in mechanical tools. Slaves made up about one-third of the population of the Roman Empire; slaves were also common at this time in China and India. They were the mobile capital of the classical world, sufficiently cheap to make innovation unnecessary, especially as the installation costs of slaves (conquest) were

1 Great Britain had "ample deposits of coal, iron, and tin and an indented coastline and navigable rivers that made transportation easy" (Headrick 2009: 92).

not individually attributable. They enabled the Romans to ignore such productivity-improving devices as the water mill and reaping machine, both invented in the 1st century CE.

The classical ideal of the good life, a mixture of contemplation, politics, and gracious living, was shaped by the fact that the noble class did not have to 'work' for its living. In the grandiose cities of the ancient world with their imposing monuments philosophers discussed the principles of good government and the good life; aristocrats "enjoyed their baths, their theatres, and their villas" (Gibbon 1996). In his peerless *History of the Decline and Fall of the Roman Empire*, Edward Gibbon fixes the years 96–180 CE as the period in which "the condition of the human race was most happy and prosperous" (Gibbon 1996). This view from the top explains the enduring appeal of the aristocratic ideal of leisure. Gibbon says nothing about the happiness of the slaves. The happiness of the few was made possible by the toil of the many. "Technical work" as Klemm puts it, "was the affair of slaves and the free citizen usually despised manual work and even activity devoted to the furtherance of discovery" (Klemm 1964: 42).

Christianity and the duty to work

The end of slavery is conventionally explained by "diminution in the supply of slaves" (Klemm 1964: 62). But it must have, in part, been a cultural choice to abandon a way of life which might hope to replenish the slave supply. Larry Siedentop has drawn attention to the importance of the "impinging of Christian moral norms on social roles" (Siedentop 2014: 165). This happened in two ways. First, the Church's doctrine of the equality of all in the eyes of God (only God could 'own' souls) undermined the classical assumption of human inequality. The ground for the later concept of individual rights was also laid in the church doctrine that the individual rather than lordship was the true subject of the law (Siedentop 2014: 184). Second was the Church's claim that secular laws had no legitimacy if they contravened laws of God. Thus "Christian beliefs provided grounds for an appeal against injustice that had not been available in the ancient world" (Siedentop 2014: 176). Peasant uprisings started to be not merely in defence of 'customary' law, but in defence of a wider concept of justice. Hegel's powerful attack on the revived slavery of the 19th century was in the name of denying the humanity of both master and slave (Smith 1992: 97–124).

The theological assault on the classical contempt for work pervades Christian apologetics. Early Christian writers often imagined God as the supreme artisan of the natural world, and the marginal status of early Christians "tended to generate among [them] a toleration of labour and laborers not found in the Greco-Roman world" (Ovitt 1986: 490). "Happy is he who earns his bread by the work of his hands" declared John Chrysostom in the second half of 4th century. Disgusted by the decadence of his time, the 2nd-century monk Theophilus of Antioch inveighed against "sloth and levity. repugnant in the eyes of God" (Q in Klemm 1964: 66). The historian Edward Gibbon paid a back-handed compliment to the power of such ideas when he blamed Christianity for the decline in the martial vigour which sustained the Roman Empire. Echoing Gibbon, Nietzsche would dub Christianity the 'slaves' religion'.

The Christian 'duty to work' drew inspiration from the Old Testament: in Genesis, God created the world in six days, and having worked hard, on the seventh he rested. This was a very different god to the Greek gods, who, while they often had craft skills, but except for Hephaestus, did not toil. Christ was the son of a carpenter, although he gave up work to preach. As a result of these influences, Christianity never had the same disdain for work as the ancient Greek citizens. On the contrary, Christian theology emphasised the value of work and of serving others. Work was a defence against idleness, and tool for charity and expiation.

Because of the ascetic streak in Christianity, work was an important element of monastic life from the beginning. It was institutionalised by John Cassian in the *Institutes* (c. 400AD), and then St Benedict in his *Regula* (c. 500AD). The *Regula* of St Benedict not only laid down the rules for organising and governing a monastery, but also laid out strictly what activities should be done at any time of the day or night. In the *Rule of Saint Benedict*, we read that "Leisure is the enemy of the soul, and for this reason the brothers must spend a certain amount of time in doing manual work as well as the time spent in divine reading" (Ovitt 1986: 498). The Benedictine emphasis on the close relationship between manual and spiritual work is summed up in the injunction "*ora et labora*", or "pray and work". The routine labours of daily life, the very fact of their tedium, was spiritualised in the early middle ages.

Paracelsus, a 16th-century alchemist, put the Christian attitude like this: we were not originally constituted for labour. Since the Curse, outside of Paradise, it was ordained that we should work, a course laid on us through the Angel who spoke these words: "In the sweat of thy brow shalt thou eat bread" (Genesis 3:19). Properly understood God's punishment was not to make people work, but to make work painful, as one would expect for any expiation of sin. No comparable 'duty to work' is found in Asian societies. In China, work is not valued in itself, as in Christianity. Physical work, in particular, is condemned as menial. The ultimate aim in life is to reach a position in which one can order others to work rather than work oneself.

The injunction to work carried with it no injunction to innovate. True enough, the so-called 'Dark Ages' which followed the collapse of Roman power produced three innovations which 'set the stage for the modern world': the mouldboard plough, three-field rotation, and the breast strap for horses and draft animals, the last imported from China (Headrick 2009: 56–58). All three very slowly raised agricultural productivity and made smaller landholdings economically viable.

However, the main response to the problem of labour shortage created by the collapse of the Roman world was a rearrangement of property and political relationships to which the general name 'feudalism' is given.

Feudalism is a form of property ownership, in which ownership is formally conditional on fulfilment of obligations and workers are attached to landed estates. It plays a key part in the institutional theory of European technology, both as explaining the initial retardation of technology and its eventual flowering.

Labour shortage was met by attaching cultivators to the land. This was 'serfdom', a system of modified slavery in which peasants were guaranteed land in return for labour services: the bottom rung of the system of land in return for services which ran all the way up to the monarch. Serfdom, not technical innovation, was the favoured means of provisioning a shrunken political estate. There was a parallel flight to the monasteries, which acquired economic functions as strongholds of literacy, craftsmanship, and cultivation.

The key institutions

It is generally agreed that capitalism started in Europe. The story told by Jean Baechler is severely institutional: states in this region, and especially England, were the first to remove the institutional obstacles to economic efficiency (Baechler 1975). The enabling background was the failure in Europe to re-establish the Roman Empire, and the consequent separation of church and state and the division of the European world into competing states. As we shall see, the institutional view fails to explain why modern capitalism specifically originated in north-western Europe. For this we need a moral hypothesis.

From the perspective of the 18th-century Enlightenment, feudalism was part of the 'Dark Ages' between the collapse of Rome and the onset of modernity. By contrast modern institutional history sees its structure of weak monarchies, armed aristocracies with juridical functions, 'free' cities each with its own rights and privileges, and above all competing of states within a single religious polity (Christianity) as seedbeds of the constitutional state and the private property system. Some historians trace the transformation of feudal Europe into capitalist Europe to the Black Death of 1348, which removed between a third and a half of Europe's population: a classic apocalypse which forced humans to rethink and redo their political and social relations, their past and future. To the Dutch historian Johan Huizinga (1987), the plague started the replacement of the universal Christian culture of Europe by secular nation states.

The constitutional state

A crucial difference between the European system and that of Asian countries was the fracturing of unified Roman state authority into the separate institutions of Church and State, whereas the Islamic, Indian, and Chinese states remained theocratic. In principle, the institutional separation of religious and secular authority was a functional dispute within a single Christianity polity. In time, though it broadened into a political argument concerning the proper grounds of authority and coercion. The Church's claim to exercise a co-equal authority with the (largely defunct) Empire ("Render unto Caesar the things which are Caesar's and unto God the things that are God's") was a crucial contribution to the idea of a contractual state.

The Reformation accelerated the separation between church and state. The Protestant revolt against the ecumenical claims of Catholicism brought religious reformers like Martin Luther, Philip Melanchthon, Huldrych Zwingli, and John Calvin the patronage of various European monarchs who saw in Protestantism an ideological basis for nation-state building.

The Reformation thus fractured Papal authority in Europe, splintering and decentring it. It was only in post-Reformation Europe, where even largely Catholic countries like France were riven with deep religious conflicts, that ideas about politics, science, and philosophy as independent from divine revelation could truly take root. If there already existed a significant rift between church and state power in early modern Europe, then the scientific revolution was the wedge to drive the two finally apart.

One important consequence of the Black Death was the gradual disarmament of the nobility and cities and the concentration of 'law and order' in the hands of the state. The technical means which made this possible was the application of gunpowder to warfare, perhaps as a response to infantry shortage. Before the development of effective artillery, castles were a huge obstacle to the centralisation of royal power. Gunpowder was invented in China, but not used for military purposes. Its application to warfare in Europe shifted the comparative advantage in violence to monarchies by allowing economies of scale in fighting. By the 15th century it had cancelled the advantages of castles and walled cities: the fall of the heavily fortified Constantinople in 1453 to Ottoman artillery signalled not just the closure of the Silk Road linking the Mediterranean to Eurasia, but the end of the dispersed feudal power system. Europe was gradually consolidated into competitive, centralised monarchies; these monarchies established national markets by breaking down feudal protection of local markets and weakening the hold of the guild system.

To summarise: historians now regard the plurality of feudal institutions as marking a fundamental divergence from the 'Asiatic' type of despotic rule, in which the feudal tenure system was weak or non-existent, where religious and secular authority was fused,[2] and where imperial, not

2 The fusion took two opposite forms: in China the state took over the church; in Islam, the church took over the state in the form of the caliphate.

aristocratic, government was the norm. That notable social liberal Montesquieu (2002) likened Europe's intermediate feudal institutions to the 'weeds and pebbles' that lie scattered along the shore obstructing the otherwise overflowing tide of autocracy.

Private property

Private property in the sense we know it was born from the collapse of the institution of serfdom. Here again the Black Death played a crucial role. According to David Landes (1988: 18), it "*compelled* [my italics] the propertied classes to offer substantial inducements to attract and hold the manpower needed to work their estates". As a result, feudal obligations were gradually commuted into cash rentals, landlords obtained sole ownership of their estates, and a class of free peasants, owing no obligations in cash or kind, but relying for their income on the sale of their products in local markets, came into existence.

A remoter effect of the Black Death was the search for new lands to colonise and fresh supplies of colonial labour. As Wallerstein (1974: 51) explains, "Europe needed a larger land base . . . which could compensate for the critical decline in seigniorial revenues . . . which the crisis of feudalism implied". Germanic peoples migrated eastwards to the Slav lands and western Europeans to the Americas. The western 'voyages of discovery', starting with Vasco Da Gama and Christopher Columbus, were key events in the history of technology: the moment when western Europe broke out of its peninsular confinement on the edge of the Eurasian heartland and found a 'New World' for itself. Utilising improved naval technology, Holland, England, and France closely followed Portugal and Spain in grabbing land in the Americas in a modern explosion of 'slash and burn', and followed Portugal in establishing mercantile empires in the Pacific. The opening up of oceanic trade routes to replace the now closed Silk Road shifted the location of wealth and power from the Italian city-states of southern Europe to the 'emerging nations' of western Europe. The profits of commerce and slavery strengthened monarchies and merchants at the expense of the feudal lords. The ocean opened up an unlimited frontier for 'new men' to seek their fortunes.

In the institutional story it is the vesting of property rights in individuals rather than corporations which made innovation profitable. For the first time in history, property in both land and labour was freed up to be bought and sold by individual owners. This was the institutional basis of capitalism, in which the three 'factors' of production, land, capital, and labour were subjected to market exchange. This released and prioritised lust for gain and quest for economic efficiency. Competition between 'capitalists' incentivised them to install labour-saving machinery to force down labour costs. Karl Marx identified the Darwinian competition between capitalist firms for profit as the taproot of technological dynamism.

The neoclassical economist tells essentially the same story in the language of scientific neutrality. The institutional change most pertinent to technological innovation was the establishment of 'intellectual property rights', normally secured by patents, by which governments grant inventors temporary monopoly rights over the products, processes, and techniques which they have invented. This is the only way of avoiding that great disincentive to innovation known as the 'free rider problem', the inability of an inventor to profit personally from an invention because knowledge is a free good. In Britain, guarantee of private property in inventions dates from the Statute of Monopolies in 1624, when it was established that the Crown might grant patents (temporary monopolies) only to the first inventors of new manufacturers or processes 'useful to the state' and not to anyone it wanted (Klemm 1964: 172). The guarantee provided an incentive to innovate in cost-saving technology. But it was not the only possible stimulus. Between the mid-18th and mid-19th centuries Britain's Royal Society of Arts awarded over

2000 innovation 'prizes'. Whether by way of patents or prizes the requirement was that the 'private rate of return' from an innovation equalled 'the social rate of return'.[3]

The economic historian Douglass North compares Britain and Spain to illustrate the power of this hypothesis. The modernisation of property rights in Britain set it on its growth path, by making it profitable for 'improving' landlords to capture the profits of their improvements. By contrast, in Spain, the Crown failed to curtail the right of the *Mesta* (the shepherds' guild) to drive their sheep across the land wherever they wanted. "A landlord who carefully prepared and grew a crop might expect at any moment to have it eaten or trampled by flocks of migrating sheep".[4] North's handy formula can be applied generally: the predations of the state can have the same effect as those of the *Mesta* in retarding technological innovation.[5]

Protection of private property rights, the rule of law, independent central banks, and the reliance of rulers on parliaments to supply them funds made up a quadrant of related institutions which together freed up capital to create more capital in a property-regime which guaranteed freedom of possession and rights of inheritance.

Historians have explained how countries in western Europe gradually acquired the balance between state power needed to guarantee security of private property against predators and independent legal and power centres needed to prevent the state from becoming predator-in-chief.

This is a fertile explanatory thread. Most historians now trace capitalism's pathway from the commercial republics of the 15th-century Italian city-states, through the 16th-century voyages of discovery and first overseas European empires, to the almost continuous dynastic warfare between states from between the 16th and 18th centuries. These were the centuries of Europe's breakout from peninsular to world position: the escape of the whale from the embrace of the bear. Over the same period, the Asian empires stagnated, crippled, as westerners saw it, by their despotic, bureaucratic systems.

From religion to science

The weakness in all institutional accounts of the emergence of capitalism is the failure to link the growth of capital accumulation to the growth of the scientific outlook. They grew together like Siamese twins, but each with its separate personality. The intense competition between states created the free spaces which the scientific outlook needed to gain its territorial hold; while the scientific world view directed the form which capitalist competition took. Together they explain the technological acceleration which took place in the 19th century. The connecting link is Protestantism.

3 But see Klaus Muhlhahn, *Making China Modern: From the Great Qing to Xi Jinping* (2019: 60). Muhlhahn points out that Qing China had 'large transaction markets for land', even though 'ownership of a single plot could be vested in parties endowed with separate rights over the surface and subsurface. Rights could be sold, leased, or used as collateral. He admits though that property rights in imperial China depended on politics rather than the law (2019: 68). The absence till recently of patent protection in China has been used to explain its technological retardation. The US has seven of the largest ten IT companies, China two, by market value. China is nowhere in pharmaceuticals, whereas the US has six of the top ten.

4 Douglass C. North and Robert Paul Thomas, *The Rise of the Western World: A New Economic History* (1973: 4); cf. Daron Acemoglu, James A. Robinson, and Simon Johnson, 'The Rise of Europe: Atlantic Trade, Institutional Change, and Economic Growth', in *The American Economic Review*, Vol. 95, No. 3, (June, 2005), who explain the success of Britain and Holland after 1500 together with failure of Spain and Portugal through institutional constraints on executive power.

5 Neoclassical economists consider 'enclosure' the rational solution to the problem of 'over-grazing', or 'tragedy of the commons'. Commonly owned property will be over-grazed as each flock owner takes advantage of the 'free land' to increase the size of his flock, leading to exhaustion of the land.

This 'culture of capitalism' is commonly ascribed to three things: the emergence of Protestantism, the individualistic outlook, and the changed view of nature associated with the Enlightenment. With Protestantism came the view that wealth not poverty gave the best assurance of salvation; with Individualism, the belief in individual freedom as the basis of progress; with the Enlightenment science, for the first time, was harnessed to the discovery of 'useful' truths. These pieces of cosmological invention produced a cast of mind in the outstanding thinkers and actors of the time and a pattern of activities and wants in populations which may be seen as the 'ultimate cause' of the scientific and technological trajectory.

The Protestant Ethic

The German sociologist Max Weber crucially distinguished between the 'spirit of capitalism' and the institutions of capitalist society. In *The Protestant Ethic and the Spirit of Capitalism*, Weber (1930) claimed that the "spirit of capitalism", by which he meant the spirit of gain, or love of money, far from being inherent in man, entered history at a particular time (16th century) and place (north-western Europe). It was an unintended consequence of the Calvinist doctrine of predestination.[6]

Calvinists believed that God had divided people into the saved and damned, and there was nothing they could do to influence His selection. Weber's argument was that they responded to the anxiety produced by this uncertainty with a redoubling of their efforts in order to convince themselves – if not the Almighty – that they were among the saved. Crucially, success in worldly affairs was taken as a 'sign' or 'proof' of grace. In this way the love of money which had hitherto been regarded as a vice, was turned into a virtue; and the active life privileged over the contemplative one. Calvinism was the most radical expression of the link between salvation and 'industriousness'; the 'industrious revolution' was the moral and ideational foundation of the Industrial Revolution. With this brilliant hypothesis, Weber simultaneously explained both why the spirit of capitalism flowered in Protestant Europe and failed to take root in Asia. It also warns against locating the origins of capitalism in the Renaissance. Renaissance Europe saw a splendid flowering of thought and art. But capital did not accumulate in its centres and there was little technical progress.

Individualism

Liberal political theorists, from the 17th century onwards, were anxious to map out an area of individual freedom immune from external interference. Social contract theories of the state had this as their object. Individuals should be protected from the arbitrary acts of a political tyrant as a matter of natural right. But the functional benefit of individual freedom was also in the mind of political philosophers and economists. David Hume (1741) was quite clear about this: free thought required a limited state. "An unlimited despotism", he wrote, "effectively puts a stop to all improvement, and keeps men from attaining knowledge".[7] For John Stuart Mill (2003: 76) a century later, the danger was seen to arise not so much from political as from social tyranny: "the tyranny of prevailing opinion and feeling . . . the tendency of society to impose,

6 Baechler denies this: he sees nothing new about the love of gain, only an expanded institutional means for its expression.

7 A Chinese scholar of the 1st century CE had long anticipated Hume when he wrote of the first Ch'in emperor: "The resources of the empire were exhausted in supplying [Shihuang's] government, and yet were insufficient to satisfy his desires" (Q. in Goodrich 1948).

by other means than civil penalties, its own ideas . . . on those who dissent from them". It was in order to protect the genius, the original thinker, the eccentric, in short those persons who advance thought and civilisation, from the despotism of social convention that Mill postulated absolute freedom of thought and discussion and endeavoured to separate 'that part of a person's life which concerns only himself' from that part which is of legitimate concern to society (Mill 2003: 143). It would seem that Mill's chief concern was to ensure conditions of freedom in which great thinkers can flourish, for, as he observed, original natures cannot be fitted "without hurtful compression into any of the small number of moulds which society provide in order to save its members of the trouble of forming their own character" (Mill 2003: 129). Individual freedom could be justified on utilitarian grounds as a pathway to knowledge. Despite hesitations, Mill subscribed to the scientific idea of knowledge, since for him, as for all Enlightenment thinkers, custom and religion remained the egregious sources of error.

The scientific outlook

Modern science started out as sceptical of theology and has ended up as a new theology, claiming it could read the secrets of the universe better than the old theology. Its claim to God-like knowledge had been foreshadowed by the 13th-century Franciscan friar Roger Bacon who made the case for experimental science.

> For there are two modes of acquiring knowledge, namely by reasoning and experience. Reasoning draws a conclusion and makes us grant the conclusion, but does not make the conclusions certain of truth, unless the mind discovers it by path of experience.
>
> *(Q. in Klemm 1964: 94)*

His 17th-century namesake Francis Bacon attributed the advance in knowledge to the new empirical method of discovering truth. Theology, in the scientific conception, was like magic because its 'truths' couldn't be validated or refuted by experiment: they corresponded only to Roger Bacon's first principle, drawing a conclusion by reasoning. The question was posed: what if the two methods diverged?

Despite its strong backward-looking bias ("It's all in Aristotle"), the Renaissance was a bridge between medieval scholasticism and scientific empiricism. Renaissance thought was experimental. In addition to the truth revealed by scholastic reasoning, there was truth to be discovered. Whereas Copernicus was a model-builder, the astronomer Johannes Kepler (1571–1630) set out to verify Copernicus's heliocentric hypothesis by observation. Kepler's observations inspired Isaac Newton (1643–1727) to produce a general law which could not be reconciled with religious teaching. Newton was the first purely secular law maker: his general law of motion was the first genuinely scientific law, which applied the force of gravity to both falling apples and orbiting planets. In Newton the man, science, and magic were not yet wholly disentangled, but with 'Newtonism' the process of what Weber called the 'disenchantment' of nature became unstoppable (Keynes 1978b: 363–374). Nature came to be seen as a machine, subject to physical laws. The subjugation of nature could, in turn, be used to improve the material and intellectual condition of humanity. The idea of a human nature sufficiently plastic to be moulded by science and technology was a logical culmination of this line of advance.

At first this changed world view had little effect on technology. The knowledge embodied in tools was local, 'tacit' knowledge, what the Greeks called *phronesis* or skill. The technologies which resulted were local and scattered, not part of a general and advancing body of knowledge.

Unsupported by science, they rose and fell with the particular material and social conditions which produced them. Thus, many technologies simply disappeared with the fall of the Roman Empire.

But from the 18th century onwards, scientific curiosity was stimulated to look for an explanation of why tools worked; and the tool-makers started to look to science to improve their tinkering. The mathematical models of the universe made by scientists could be applied to engineering, so technology in the modern sense was born. Scientists and engineers met and read papers in the learned societies which sprang up in Britain and France in the 18th centuries. Their ideas spread throughout the educated classes, creating a 'culture of innovation'. Inventors had a market for their wares which ramified further than the traditional public demand for improved military technology. Stimulating the curiosity of the scientist and the ingenuity of the engineer lay an expanded private demand for the comforts and luxuries they promised.

Why Britain?

According to David Landes (1988: 69), Britain was the first society 'that interposed relatively few institutional barriers' to fundamental technical innovation. This, crucially, enabled Britain to become what Peter Mathias (2001) called The First Industrial Nation.

Why Britain? In David Starkey's (2020) words it was a

> product of a not yet fully understood . . . concatenation of circumstances: of the wealth generated by empire and slavery abroad and of freedoms – political, legal, economic and intellectual at home, all spiced by consumerism and a precocious mass market for luxury goods.

The traditional story starts with a cluster of technical inventions which multiplied the power of human labour. As a schoolboy once wrote: "About 1760 a wave of gadgets swept over England". The 'new economic history' downplays the gadget theory. There was no sudden 'take-off' into a technological future. Rather the mis-named 'industrial revolution' was a result of a long historical and institutional development unique to Britain. Although 'enclosure' of the commons was common in Europe from the 16th century onwards (leading to peasant revolts) Britain developed capitalist forms of agriculture much sooner in a long process of concentrating ownership in the hands of large landlords, going back to the Black Death. This concentration of ownership and the improvements in agricultural productivity which it enabled, forced redundant labour into the towns, where it was cheaply available for manufacture. Long before the Industrial Revolution, England was much more urbanised than France and other Continental countries (Crafts 1985: 37). The English revolution of 1649 had also clipped the wings of its monarchy, forcing it to bargain with landlords and merchants for revenue, and initiating a naval competition with the Dutch for commercial mastery which led the first British empire overseas in North America.

The account given by Nicholas Crafts in *British Economic Growth during the Industrial Revolution* is consonant with that of Douglass North and the new institutional history. Crafts himself mentions the development of the following institutions efficient for innovation: (1) specialist insurance and banking services like Lloyd's underwriters which reduced the uncertainties and costs of foreign trade, (2) improvements in domestic credit through the development of internal bills of exchange, and reorganisation of government finances in the 1730s which reduced interest rates, encouraged trading in liquid assets, and strengthened the credit basis of country banks, (3) new forms of business enterprises for raising capital and limiting risk: joint stock companies, limited liability companies, stock and commodity markets, and (4) the replacement of the family as the unit of production by 'capitalistic' market-oriented enterprises, starting in agriculture and spreading to industrial like textiles.

Both the strengths and weaknesses of the institutional approach are on display here. It cannot be denied that institutions had to be permissive, even supportive, of technological innovation, but as Joel Mokyr (2017: 5) points out, they do not explain the 'surge' of 'technological creativity'. These were the result of changes in attitudes to wealth and work associated with Protestantism and the changes in attitude to Nature associated with the Enlightenment. What we see is a re-engineering of institutions to make possible what powerful groups in society wanted to do. The scientific-technological idea impelled the creation of institutions needed to give it effect.

Previous societies, too, had been very wealthy: why did they not innovate the institutions needed to put their wealth to productive or different use? The modern answer is that the monarchs, priests, and nobles, power holders, of the pre-modern world preferred to *consume* the wealth created by producers in wars, church-building, and conspicuous displays rather than *invest* it in new machines. The gold and silver pouring in from the New World was spent by Spain on both God and Empire. Starting in north-west Europe – Holland, France, Britain – capital was 'put to work' for the first time in history. It was reorientation of human thought and activity to pursuit of wealth which finally overcame the Malthusian trap.

Conclusion

The capitalist civilisation which created today's technological momentum was thus the result of a complex interplay between institutional development and the development of ideas. Neither provides a complete explanation on its own. However, it is the thesis of this essay that technological innovation is both enabled and disabled but not caused by institutions. Innovation is caused by an attitude of mind which makes people spend their energies and money on one thing rather than another. It depends, that is, on a culture of innovation, which is the obverse of the religious culture.

One needs to be clear about this in thinking about the future of technology. The view that technology determines the future rests on the belief that it generates an unchallengeable commentary on itself. If, on the other hand, the technology we use is shaped by reflection about the place of technology in the great scheme of things then the future opens up. Humans choose their own destinies; they are not imprisoned by them.

References

Acemoglu, D., Robinson, J. A. and Johnson, S. (2005): The Rise of Europe: Atlantic Trade, Institutional Change, and Economic Growth, *The American Economic Review* 95(3), 546–579.

Baechler, J. (1975): *The Origins of Capitalism*. Oxford: Basil Blackwell.

Crafts, N. (1985): *British Economic Growth During the Industrial Revolution*. Oxford: Clarendon Press.

Gibbon, E. (1996): *The History of the Decline and Fall of the Roman Empire*. London: Penguin Books.

Goodrich, L. C. (1948): *A Short History of the Chinese People*. Sydney: Allen and Unwin.

Graeber, D. (2011): *Debt: The First 5,000 Years*. New York: Melville House.

Harari, Y. N. (2014): *Sapiens: A Brief History of Humankind*. Canada: Signal.

Headrick, D. R. (2009): *Technology: A World History*. Oxford: Oxford University Press.

Huizinga, J. H. (1987): *The Waning of the Middle Ages: A Study of the Forms of Life, Thought, and Art in France and the Netherlands in the Fourteenth and Fifteenth Centuries*. London: Penguin Books.

Hume, D. (1741): Of Liberty and Despotism, in *Essays, Moral and Political*. Edinburgh: A. Kincaid, 173–187.

Keynes, J. M. (1978a): Economic Possibilities for our Grandchildren, in Johnson, E. and Moggridge, D. (eds.): *The Collected Writings of John Maynard Keynes: Vol. 9*. London: Royal Economic Society, 321–332.

Keynes, J. M. (1978b): Newton, the Man, in Johnson, E. and Moggridge, D. (eds.): *The Collected Writings of John Maynard Keynes: Vol. 10*. London: Royal Economic Society, 363–374.
Klemm, F. (1964): *A History of Western Technology*. Cambridge, MA: MIT Press.
Landes, D. S. (1988): *The Unbound Prometheus: Technological Change and Industrial Development in Western Europe from 1750 to the Present*. Cambridge: Cambridge University Press.
Mathias, P. (2001): *The First Industrial Nation: The Economic History of Britain, 1700–1914*. Abingdon: Routledge.
Mill, J. S. (2003): *On Liberty*. Edited by Bromwich, D. and Kateb, G. New Haven: Yale University Press.
Mokyr, J. (2017): *A Culture of Growth: The Origins of the Modern Economy*. The Graz Schumpeter Lectures. Princeton: Princeton University Press.
Montesquieu (2002): *The Spirit of the Laws*. Edited by Cohler, A. M., Miller, B. C. and Stone, H. S. Cambridge: Cambridge University Press.
Muhlhahn, K. (2019): *Making China Modern: From the Great Qing to Xi Jinping*. Cambridge, MA: Harvard University Press.
North, D. C. and Thomas, R. P. (1973): *The Rise of the Western World: A New Economic History*. New York: Cambridge University Press.
Ovitt, Jr., G. (1986): The Cultural Context of Western Technology: Early Christian Attitudes toward Manual Labor, *Technology and Culture* 27(3), 477–500.
Schumpeter, J. A. (1994): *Capitalism, Socialism and Democracy*. London: Routledge.
Siedentop, L. (2014): *Inventing the Individual: The Origins of Western Liberalism*. Cambridge, MA: The Belknap Press.
Smith, S. B. (1992): Hegel on Slavery and Domination, *The Review of Metaphysics* 46(1), 97–124.
Starkey, D. (2020): A Perversion of Puritanism, *The Critic*, 22. June. Available at: https://thecritic.co.uk/a-perversion-of-puritanism-which-aims-to-trash-our-history/ [28.01.2021]
Toynbee, A. J. (1974): *A Study of History: Abridgement of Volumes I-VI*, D.C. Oxford: Oxford University Press.
Turner, F. J. (2003): *The Significance of the Frontier in American History*. Tucson: The University of Arizona Press.
Wallerstein, I. (1974): *The Modern World-System: Capitalist Agriculture and the Origins of the European World-Economy in the Sixteenth Century*. New York: Academic Press.
Weber, M. (1930): *The Protestant Ethic and the Spirit of Capitalism*. London: George Allen & Unwin Ltd.
Wittfogel, K. (1957): *Oriental Despotism. A Comparative Study of Total Power*. New Haven: Yale University Press.

2

DISRUPTIVE TECHNOLOGICAL CHANGE IN RECENT ECONOMIC HISTORY

Werner Plumpe

Introduction

Technological change is a continuous process that results primarily from the pragmatic use of technical problem solutions under changing (economic) conditions. Joseph Schumpeter's distinction between adaptive and destructive change can be made, although this distinction is conceptually easier to make than can be shown in economic history in practice because economic change phenomena often merged almost inseparably. It is true that, as was often the case in the older history of technology, the invention of new technologies can be dated quite accurately (e.g. the invention of the steam engine, the automobile, the first numerical calculating machine, etc.) (Radkau 2008); but whether these inventions led to disruptive technological change cannot be deduced from their existence alone. Even the mere adaptation of technical solutions to problems, as such expressed in shifts in relative prices, can lead to major technological advances; one need only think of the small-scale technological adaptations identified by Joel Mokyr in the core of the British Industrial Revolution for pragmatic purposes, which were more important for the implementation of the modern factory system as a whole than the mere fact of certain technical discoveries (Mokyr 2002, 2009).

In the analysis of technological change, it is therefore very important how this change is conceptualised and what perspectives on the subject matter result from this. Numerous technical innovations that have contributed significantly to an accelerated increase in productivity since the 18th century have been known for a long time, many of them since antiquity. Whether a technical invention actually became an economic innovation obviously depended less on the existence of the new idea than on the conditions under which this innovation was technically used, how quickly it could spread in the subsequent development and become a kind of technical-economic common good. Whether the respective innovation had what it took to trigger revolutionary changes or whether it led to the broad stream of technical adaptation processes that are always taking place anyway, was only clear when its use had long been decided. Whether it was a basic innovation, as Gerhard Mensch (1975: 54) called it, and if so, what its scope was, was therefore completely open in advance, as a reason for decision, so to speak. As a rule, decisions on use were based on much more complex, but in essence mostly pragmatic problem-solving calculations and corresponding current pressure to act or associated expectations for the future. It is thus obvious that it was not only the economic potential of technical innovations

 DOI: 10.4324/9780429351921-4

that decided on their use, but also a complex structure of practical everyday problems, institutional possibilities and restrictions, and (semantic) ideas of appropriate action (Plumpe 2009). Technical-economic change, especially the acceleration of corresponding changes, cannot be adequately explained from a technical and economic point of view alone, even if this is the obvious core of every argument. Older and more recent economic historiography, but not least also the history of technology, have chosen correspondingly more comprehensive approaches, whereby a combination of different perspectives now prevails, namely of a technical, knowledge and economic-historical nature, executed in an exemplary manner – above all in the works of Joel Mokyr (2002, 2009), who can certainly fall back on a relatively extensive tradition of relevant works, the beginning of which in the strict sense probably lies in David S. Landes's study on the unleashed Prometheus (Landes 1969). Landes later elaborated on his ideas about the success of technical change under Western liberal conditions (Landes 1998), which was criticised as an unjustified apotheosis of a kind of Western supremacy that underestimated the dynamics of non-Western societies as well as the violent character of Western expansion (Pomeranz 2000). The particular dynamics of Western expansion was, therefore, in the perspective of critique, logically rather attributed to a favourable factor endowment (coal deposits), geographical advantages and the willingness to acquire colonial resources by force in case of doubt (Beckert 2014, among others), while the actual question of an increased productivity development due to changing technology use receded into the background. Despite indisputable individual empirical findings and a certain conceptual attractiveness, this criticism has not been able to assert itself in the academic debate, because the reasons for the disparity in development in Northwest Europe and Asia since the 18th century could not be clarified in this way (Vries 2013): Northwest Europe's withdrawal from the global static was not a consequence of asymmetrical power distribution and use, which itself should have been explained. Rather, there are numerous endogenous change factors that ultimately favoured a market-based organisation of the economy, which developed an astonishing elasticity of supply in relation to population growth and technical improvements. In the unique co-evolution of privately constituted decentralisation, a more or less dominant market selection with considerable competitive pressure and an institutional framework, that at least tends to favour this, lies the real "secret" of Northwest Europe and later North America. The individual moments of this co-evolution were by no means a unique feature of these spaces; they can be found in many, sometimes more advanced forms in other parts of the world. But the co-evolutionary constellation was unique, at least in this period (Jones 1981; Mitterauer 2003; Plumpe 2019). Numerous individual studies on technical change in Western Europe have made precisely clear: the actual dynamics of change resulted only or even predominantly from a kind of internal cascade of mutually reinforcing technical problem-solving steps, which corresponded above all to population growth, urbanisation, changing energy use and certain institutional conditions (Wrigley 2010), in the wake of which "capitalist practices" first proved possible and then particularly productive, first in Holland and later in parts of Great Britain. Nor did their spread follow any logic of violence, even though violence and exploitation were always a moment of European colonial expansion, which were carried primarily (but by no means alone!) by the centres of the early capitalist transformation (Reinhard 2016). With a targeted export of capitalism, if it had been attempted at all, the colonial powers failed not only in India and China but also in Africa, where they largely refused a capitalist reorganisation of the work and production processes, which was typically diagnosed by the colonial masters as a deficit of civilisation. And the contribution of colonial exploitation to European-American economic and technological development, for all its significance in detail (Inikori 2002), was not the reason why capitalist forms of mass production and mass consumption first established themselves here. They were helpful, but not the cause; their quantitative volume alone was too small for this, quite apart from the fact that in the chronology of capitalism the mass demand for colonial raw materials and consumer goods was a consequence of, rather than a prerequisite for, the establishment of capitalism. It is, therefore, necessary to clarify the question of why the disruptive technological change took place primarily in Northwest Europe and North America, from where it gradually spread globally.

The temporal access to the economic-technological change is in principle open, but from the perspective outlined here, it is advisable to limit the period under consideration to the time since the emergence of modern capitalism approximately in the 17th century (Plumpe 2019). This makes sense not only in terms of the matter in hand (accumulation of disruptive phenomena); the economic dynamics have also changed fundamentally since this period. Since then, in parallel with the increase in productivity, there has been an increase in the intensity of technological change, which in turn has become a driver of productivity, a co-evolution that was indeed new in world history and marked a clear break with the older times, but also with developments in the non-European world. Admittedly, research today is far from underestimating the older technological change, as has long been the case. However, a more objective view of the processes of change in the pre-modern era should not give the impression that these are basically at best marginal differences, which is what this is all about. At the latest with the implementation of capitalist procedures, the economic use of new technological opportunities for action took on a changed dynamic that was simply unknown in the previous world (Fossier 2008), because the interrelation of fundamental technological innovations and small-scale technological adaptations enabled a kind of continuous experimentation that, under the conditions of competition and competition for market opportunities, generated an astonishing intensity of technological change that has lasted from the 18th century to the present. There are good reasons why this can bė seen as the core of the emergence of the modern economy (Mokyr 2018); the data of Angus Maddison (2007), as controversial as they may be in detail, leave no reasonable doubt about such a periodisation. The annual rates of change in national product in the main Western European countries accelerated from a low 0.4% in the period before 1800 to 1.75% in the first two-thirds of the 19th century and to over 2% in the period before the First World War, and – in the event of setbacks due to war – have remained at this level until the present day, while Asia experienced a severe setback during this period, from which some parts of the continent have only been slowly recovering since the end of the 19th century. In price-adjusted data, the Western European national product per capita of the population rose from USD 907 (price basis 1992) in 1600 to USD 3,688 in 1913, while Chinese per capita income fell during the same period and Indian per capita income virtually stagnated (Maddison 2007: 382). This opening of a growing gap, which has been described as a Great Divergence, is very closely connected with the technological upheavals mentioned earlier and their economic exploitation, which of course did not result in a kind of continuous growth process, but rather initiated, enabled and digested a development characterised by recurring growth and crisis phenomena (Plumpe 2017), which clearly points to the special character of this dynamic, which is characterised by intensive structural change. The technological change associated with this was therefore not a one-off event, but rather initiated a phase of its regular recurrence, which determines the core of the dynamics of the modern economy.

Schumpeter's "long waves"

The modern economy is therefore characterised by two things, namely firstly by a hitherto unknown growth dynamic, which then takes place not in simple upward movements but in a sequence of upswing and stagnation phases, which admittedly oscillate around a stable, upward trend. Such long-wave phases of change have been addressed in various contexts; in terms of economic history, building on other preliminary work, Joseph Schumpeter, taking up thoughts of Nikolai Kondratjew (1926) and Arthur Spiethoff (Schumpeter 1933), was the first to define them precisely in 1939 (Schumpeter 1939) and in doing so also established a certain temporal pattern of "long waves of economic development" resulting from approximately 25 to 30 years of upswing and equally long downturn phases. This is not set in stone, especially since the waves can be plausibly observed empirically rather than necessarily shown theoretically, and more recent studies have therefore also established other time periods. Behind this wave-like structural change lies innovative thrusts which, at least up to now, have enabled longer upswings and downswings with a certain regularity. These fundamental innovative thrusts, which Gerhard Mensch called "basic innovations", were always associated with technological disruptions, in other words, with what Joseph Schumpeter called "creative destruction". Such disruptive changes are therefore not a continuous moment of economic change, but rather appear in a certain cyclical nature, the duration and extent of which is disputed in detail, but whose existence can hardly be plausibly denied (Petzina and van Roon 1981). Such basal upheavals are by no means the only technological disruptions in economic structural change, which is rather carried by small-scale and, in a certain sense, continuous changes, as recent economic history research has shown that the major economic upheavals essentially consist of an accumulation of small-scale change and adjustment processes. It is generally true that innovations in the production process or product design always displace older procedures or products, and even the implementation of basic innovations is ultimately based on a large number of small and, in detail, insignificant changes in everyday economic behaviour (fundamentally Mokyr 2002, 2009). Therefore, if in the following we are primarily concerned with the major technological-economic thrusts, the significance of small-scale change phenomena is by no means negated or relativised. On the contrary, the longer-term economic structural change is so difficult to break down, especially since the small-scale changes as a whole have made revolutionary changes possible, which in turn can be easily grasped with certain breaks in time. But at the heart of them are always those basic innovations which, to put it again, are individual technical changes. In the style of Thomas Kuhn, these trigger a kind of technical-economic paradigm shift in everyday economic activity, whereby a basal change ultimately leads to a wealth of change processes, which in turn can also be disruptive in their individual aspects, but are in fact fundamental in their cumulative effect. Such basic changes are firstly found in the context of the factory use of machine tools, whereby the previous production processes, which were based on the direct production effectiveness of human labour, were placed under massive change potentials, then in the transport revolution, which started with the use of steam power to enable mobility and led to a hitherto unimaginable market standardisation, in the implementation of scientific knowledge in production processes, which not only loosened and in some cases finally completely replaced the link between production and natural or nature-identical substances, but also made completely new products and production processes possible, then in the individual mobilisation by the automobile, in the change in information processes by microelectronics, which have now reached an enormous degree of diffusion. At present, a possible new "long wave" is emerging, for which keywords such as Artificial Intelligence, Internet of Things and Industry 4.0 stand, but such phenomena are historical phenomena and not the expression of a lawful change that would allow reliable forecasts. Where the technical development will lead is not completely open, but it is not determined either, especially since it is connected with an explosive power-political situation that contains its very own logic of escalation. Each of these disruptive, innovative thrusts will be made transparent in the following with the help of examples.

The aim is not to explain such processes of change theoretically, but to grasp them historically and empirically and to demonstrate their respective significance.

Historical phases

If one takes advantage of the intensity of the economic use of technical changes, a clear caesura can indeed be identified, which can be dated to the turn of the 17th to 18th century, even if it is illusory to believe that technical change can be fixed with dates. Incidentally, the issue here is not technical change per se, but rather its economic use and – within this framework – pragmatic further development in accordance with the economic opportunities for action associated with it. And here the period of the 17th/18th century is of downright drastic importance because the framework conditions of everyday economic activity have indeed changed dramatically. The French historian Jean Gimpel spoke as early as the 1970s of the "Industrial Revolution" of the Middle Ages (Gimpel 1980), which he identified above all in the interplay between population growth, urbanisation, the beginning of mass production, changed energy use, mining, textile production and new technical processes, especially in architecture and the building industry, but also in the field of precision mechanics and optics, especially in the area of measuring and generally quantifying procedures, an observation that Alfred W. Crosby shares in his book on *The Measure of Reality* (Crosby 1997).

Technical knowledge was highly advanced in the High and Late Middle Ages, and even after that, if one takes the engineering achievements of the Renaissance as a yardstick, it did not decline (Gille 1968). What changed dramatically in the 14th century, however, was the demand for agricultural and commercial services, which collapsed dramatically in the centres of Europe with the "Black Death" of the 1340s. Depending on the region and specific conditions, between a third and a half of the population was lost as a result of the plague trains of the time; in some places, such as Florence, a large part of the population died, while other regions were spared (Bergdolt 2011; Fouquet and Zeilinger 2011). This slump in the population put an end, one could say with Gimpel, to the "Industrial Revolution of the Middle Ages", because for a longer period the mass markets literally collapsed, and with them the technical constraints of efficiently managing mass consumption phenomena.

To put it bluntly, the plague trains destroyed or blocked the interaction between technical knowledge and its economic use, with the question of mass demand for commercial and agricultural goods being the most decisive factor. The conditions after the end of the plague and the gradual stabilisation of the population situation were by no means bad for the use of technical opportunities for action, but they became relevant above all from the point of view of satisfying the demand of the upper class, which was also connected with the fact that relative prices shifted in the wake of the plague. With a general decline in demand, the prices of agricultural goods in particular fell; although mass demand for industrial goods also fell, since urban labour was scarce, wages rose and working conditions improved in favour of the urban labour force. Consequently, commercial goods became more expensive, at least in comparison to agricultural goods. The upturn in urban patricians in many cities of Upper Germany or Upper Italy reflects precisely this situation, in which urban luxury trades produced comparatively expensive luxury goods for a relatively narrow demand segment in the courtyard and upper classes (Kriedte 1980). Long-distance trade, which at that time ran in particular via Genoa and Venice, was also specialised in this market segment since the high costs alone prohibited the mass transport of simple goods (Braudel 1986). Therefore, while in the textile, mining and food production sectors the older techniques continued to be used, but only had to be adapted selectively to meet new mass demand, the demand for certain high-priced luxury goods increased, but this was not a mass phenomenon in itself and could be satisfied by traditional forms of individual handicraft production, even if in some cases on a large scale. Recent research on the history of consumption largely confirms this finding; the wave of consumption in the 15th century, in particular, was very different from the process of change in demand that began in the 17th century, in which the focus was now on cheaper goods for everyday use (de Vries 2008).

The war and the "modernisation" of arms production

However, it would be too short-sighted to attribute the technical changes solely to the pragmatic adaptation of conventional production methods to the changing (mass) demand, even if current economic-historical research agrees that it was above all population growth and urbanisation that created the decisive conditions for the enforcement and establishment of the modern capitalist economy. But in this context, technological change cannot be explained in isolation, especially, since it is primarily a matter of its use by economic actors. Technical innovations are not necessarily phenomena of economic origin; the great changes in technical knowledge since the 16th century have much to do with the military conflicts of the time and a "military revolution" (Parker 1988) taking place within this framework, namely the conversion to firearms, especially artillery, which substantially changed military tactics. Siege artillery and fortress construction, on the one hand, conversion of infantry tactics and introduction of mobile field artillery, on the other hand, not only increased the financial expenditure enormously, but they also resulted in considerable economic expansions, since the mass production of corresponding guns, rifles and equipment was now required, which far exceeded the production possibilities of conventional weapon and equipment production. The rise of correspondingly privileged gun, rifle and uniform manufacturers can be observed throughout Europe in the early modern period. However, the use of new techniques and materials (iron casting) did not follow the increase in "private" demand, but rather the needs of the corresponding authorities, so that there were no forms of supply in conformity with the market, but rather companies privileged by the authorities, whose dynamics of change depended less on the exploitation of market opportunities, but initially on the requirements and expectations of the authorities.

The effects of the sharp rise in demand for such materials can, of course, hardly be overestimated. The consequences are likely to have been particularly great for the economic improvement of the iron smelting process in Great Britain, where bronze production was limited due to a shortage of raw materials, especially since the material iron, later steel, has an extremely wide range of applications which are by no means limited to military use. The experimentation with ore processing in greater heat, the use of coal and finally coke, and the improvement of furnace technology (introduction of the blast furnace) by the Darby industrialist family during the 18th century (Hammond and Hammond 2005: 136f), the introduction and continuous improvement of steel production, firstly by the crucible steel process in Sheffield (Beyer 2007: 41), then, in terms of mass production, puddling (Paulinyi 1987), enabled ever larger fields of application for the new material steel, which finally became a central moment of industrialisation. Without it, the fundamental transformation of mechanical engineering in the second half of the 18th century would have been as little possible as the introduction of the railway as a fundamental change in the entire transport system.

The upturn in Dutch maritime trade and shipbuilding

Of outstanding importance was undoubtedly the upturn in maritime trade, which would hardly have been conceivable without a corresponding expansion and improvement in shipbuilding techniques (Braudel 1986). Here too, the 17th century saw significant technical improvements which did not revolutionise shipbuilding in general, but which changed it to a large extent with new types of ships and, above all, rational manufacturing processes. Pioneering functions were assumed by the Dutch shipyards, which specialised in the series production of simple but technically efficient and extremely economical ship types, the best-known type being the so-called "fluyte", from which the developing Dutch maritime trade benefited greatly (North 2001: 28f). This is because this maritime trade took place mainly in the North Sea and the Baltic Sea. Although it was not as spectacular as the Atlantic and Asian trade, which had to be handled by large merchants, it was of much greater economic importance. About 80% of Dutch maritime trade took place in the North Sea and the Baltic Sea; the Baltic trade was soon called the "moedercommercie" because here the great mass of ships was mainly used in the timber and grain trade, but also in the transport of beer and salt as well as in general cargo trade. The Dutch ship types, which were both efficient and inexpensive, were a decisive factor in the competitiveness of Dutch maritime trade, which for a long time succeeded in largely ousting the Hanseatic and British competition (Braudel 1986: 143ff; North 2014). What was new in Dutch shipbuilding technology was not spectacular in detail; it consisted mainly in improved process technology and division of labour, which was made possible by the formation of a regional cluster. This led to the creation of a large number of interrelated companies with a division of labour, each specialising in certain parts of the ship or its equipment. Favourable energy utilisation costs (windmill technology), high specialisation (masts, ropes, sails, etc.) and short transport routes made it possible to move on to series production, although the essential material remained wood, as the use of peat as an energy source did not allow the high temperatures necessary for the rational production of iron. In the context of wood technology, however, Dutch shipbuilding was almost legendary; this is still evidenced today by the famous voyage of the Russian Tsar Peter, who anonymously practised the profession of ship carpenter at a Dutch shipyard at the end of the 17th century, at least for a short time.

The big cities, the building industry and the breweries

At the same time, maritime trade was an essential factor in the rise of the Dutch economy, which rightly deserves the title of having been the first modern economy ever (de Vries and van der Woude 1997). Efficient maritime trade enabled two phenomena to develop simultaneously, namely wealth creation and urbanisation, for which the northern Netherlands became the epitome in the 17th century, its "Golden Age" (Schama 1988). The population of Amsterdam and the Dutch and Zeeland cities "exploded" in the 17th century, and accordingly, the building industry and utilities grew in importance, while the necessary supply of goods (grain, wood, other industrial products) was ensured by trade, the income from which in turn made a major contribution to wealth creation. In this way, the brickworks became an outstanding economic sector, which in Holland became downright industrialised. At the same time, enormous brewing complexes were created in the larger cities, which increased the number of conventional brewing processes to previously unimaginable levels (Meußdoerffer and Zarnkow 2014). Technical inventions were seldom connected with this in a basic sense; innovations were mostly related to the organisation of the production processes and the size of the production processes. However, the cumulative effects of these changes should not be underestimated. Jan de Vries (2008) speaks in this context of an "Industrious Revolution", which preceded the Industrial Revolution, challenging it in a certain way, because the increase in consumption opportunities changed people's working behaviour, increased their productivity and thus increased the supply elasticity of the economy. This consumption-based change was of particular importance because it was no longer limited to the luxury consumption of a narrow, affluent upper class, but was primarily a phenomenon of the emerging and rapidly expanding urban middle classes, and even of craftsmen who were generally relatively well off in economic terms. The interaction of urbanisation and changing consumption structures have therefore been of central importance for economic change, which would hardly have been possible even without the rapid changes in shipbuilding technology. The changing demand, some of which was only partly created in the first place, therefore affected not only the demand for raw materials and materials or capital goods but above all objects of daily life, the quantity and quality of which were now changing significantly. Starting with clothing and household items, this soon affected the entire field of furniture and home decoration; in the 17th century, Dutch painting thus became a veritable "industry" with mass production phenomena, as it almost became a status compulsion to furnish one's city apartment with appropriate works of art and furniture, not to mention tableware, cutlery and table linen (North 2001; Schama 1988).

The Netherlands, led by Amsterdam, were innovative in one area, in particular, namely the financial sector (Hart 1993). It is true that most financial and trade finance techniques had been known for a long time and had been used to a lesser extent in other places, namely in the northern Italian cities and later in Antwerp. However, Amsterdam became the real origin of a phenomenon that would later be called the "financial revolution" and which ranged from the mass use and mathematical processing of commercial bills, the establishment of exchange and credit banks and regular stock exchange events, to the establishment of the first major public limited companies, which soon attracted trade and the (wealthy) public throughout Europe.

This change culminated in the establishment and consolidation of a transparent public financial economy with regular state debt management, transparent tax requirements and a spending policy that was obviously oriented towards the benefit of the own economy and guaranteed by patrician governance. In this way, Amsterdam gained an almost legendary reputation as Europe's first financial centre and was – for this very reason – in a position to finance the costly war against Spain, which lasted until 1648, at low interest rates without going bankrupt, while the potentially disproportionately richer Spain not only lost the war, but also had to declare national bankruptcy on several occasions and dragged part of Europe's high finance down with it.

The mass demand for textile goods, in particular for cotton, and the industrialisation of the cotton trade

Due to the growing population in the 17th and 18th centuries (population decline in Germany due to the 30-year war was an exception, not the rule), the demand for commercial goods increased, especially in the cities, which now increasingly met with an increased supply, also because the upswing in maritime trade made "colonial" goods more readily available and, above all, cheaper (Trentmann 2016). Coffee and tea, tobacco and chocolate thus gradually became everyday goods, and the supply and consumption of sugar, which made some of these goods edible in the first place (Menninger 2008), also increased dramatically. The American or Asian plantation economy, which was developing in the background, was of little importance; its role for economic development was that it was able to satisfy the hunger for raw materials of the developing capitalist economies of the European Northwest at comparatively low prices, whereby the prices of cotton, for example, depended not only on production costs but also on global demand. After the only major technical innovation, the introduction of the cotton engine machine, the so-called "cotton gin", invented in 1793, it was possible to greatly expand the cultivation of cotton (Beckert 2014); however, the actual work in the fields remained mainly manual labour carried out by slaves (Inikori 2002). The rising demand for cotton then caused cotton prices to rise in the 19th century, which in turn increased the demand for slaves, so that overall the relatively low technical level was profitable for a long time, at least when cotton prices were stable and high (Fogel and Engerman 1974; Fogel 2003). Fundamental technical innovations did not end the phase of manual labour until much later, which continued for the time being even after the end of slavery (Beckert 2014).

Behind the strong increase in demand for cotton was an almost explosive growth in demand in Europe for the new fabrics, which were initially successful as fashion items but then quickly became mass products due to their quality, wearing comfort and price (Berg 2005). The strong demand not only resulted in a considerable increase in the demand for raw materials; it also put the production of cotton fabrics on a completely different level. Cotton was not unknown in Europe, but in terms of wool and linen it remained a niche raw material until the end of the 18th century. Increased cotton imports from India, however, met with a demand in Europe that was driven by fashion preferences, which gradually gained in importance and constituted a market that opened up attractive opportunities for action, which also increased due to the discrimination against imports of Indian finished cotton fabrics as a result of navigational acts. With the traditional technical possibilities (home spinning, home weaving), a considerable expansion of the supply was possible and took place in the putting-out system (Kriedte et al. 1977). This bottleneck, as David Landes precisely described a long time ago (Landes 1969), now provoked technical experiments that led to mechanisation and later factory organisation of the spinning mill, which was reflected in the first mechanical spinning mill ever, the "Spinnerei", founded by Richard Arkwright in 1771. This spinning mill was technically still very traditional: it simply combined several spinning rollers driven by water. However, the continuous improvement of spinning machines and the use of fossil fuels (steam engine) led to a process of change, the cumulative effect of which was to rapidly increase the supply of industrial yarns, whose quality and homogeneity were far superior to those of previous spinning products. This process of technical change gradually shifted the bottleneck into the field of weaving, which was no longer able to cope with the rapidly growing quantities of yarn conventionally; in any case, the question of its mechanisation also arose, which found an initial answer as early as the 1780s with the "Power Loom", the first mechanical loom. In contrast to machine spinning, however, the mechanical loom of Edmund Cartwright, which at the same time bundled or took up technical changes that had already been made in the first half of the century ("Schnellschütze"), met with massive social protests, as weavers working according to conventional methods rightly feared for their income opportunities. After a transitional phase marked by machine storming and self-exploitation of the weavers, the machine loom and thus the modern textile factory did not become established until the first half of the 19th century, a process of mechanisation that was to last much longer in the area of the long-dominant raw materials linen and wool.

The significant technical changes, which are still closely linked to the idea of the Industrial Revolution today, were not particularly impressive in this sense, nor should their quantitative significance be overestimated. Technically, the mechanisation of cotton processing was a process of change consisting of many small steps without a single revolutionary technical act, but in its cumulative effect it was indeed a revolutionary force (Mokyr 2009). The role of the cotton industry has also long been overestimated in economic terms. However, it was by no means triggered by a kind of comprehensive industrialisation, and even explosively increasing growth rates were by no means associated with its modernisation (Crafts 2004). However, its "collateral effects", which, moreover, absorbed or reinforced further moments, ultimately led to the quite correct idea of a radical break in the previous use of economic opportunities for action. On the one hand, the "modern cotton industry" successively displaced its older competitors, indeed contributed significantly to the massive increase in the relevant competitive phenomena, which was one of the main reasons for the global success of the English textile industry, especially in the first half of the 19th century, which was able to offer large quantities of high-quality textiles at such unbeatably low prices that the decline of Indian textile production can be easily explained by this price competition alone. On the other hand, the mechanisation of textile production itself became an important moment of technological change, as the modern textile factories themselves strongly stimulated the demand for energy, materials and above all machine tools. The use of coal had long been commonplace in the relatively wood-poor United Kingdom, as there were easily mineable coal deposits and the sea route favoured their transport (Wrigley 2010). The technical possibilities of coal firing, which was clearly superior to the use of peat in the Netherlands, for example, were gradually emerging, in particular the possibility of generating large process heat that could be used to significantly improve metallurgical processes.

The coke blast furnace has already been mentioned, as well as the subsequent iron processing and steelmaking techniques. The associated demand for coal, in turn, necessitated more complex mining methods, in particular improved mine drainage, for which simple steam engines were used at an early stage and have now been continuously improved. The combined effects of all this can hardly be overestimated, even if each individual expansion step seemed to have little significance on its own (Allen 2009).

Energy, materials and machinery

The expansion of coal mining, which was itself a major consumer of energy, machinery and equipment, was already underway due to the use of coal for domestic heating but was greatly accelerated by the needs of the emerging textile industry and other industries (brewing was mentioned, but pottery played no less important role), especially the rise of the iron and steel industry, which itself became the main supplier of materials for modern production techniques. Apart from the use of steam power for draining the shafts and extracting coal, mining also remained tied to traditional manual work for a long time, so the increase in production was mainly dependent on the employment of large groups of miners. Although the capital outlay per shaft rose enormously with increasing depths – especially the large underground mines on the European continent, which were built from the middle of the 19th century onwards, required considerable capital advances – their use, however, depended very traditionally on manual mining work, which was only gradually replaced by mechanical mining and extraction techniques from the 1920s onwards (Brüggemeier 1983; Burghardt 1995). Nevertheless, it was possible to achieve a huge increase in production before that time; the price to be paid for this was the emergence of large, densely populated coal-mining areas, which formed a very special profile of modern, technically driven urbanisation and which were found in an exemplary fashion in northern England, Wallonia or the Ruhr. The coal mined by the hundreds of thousands of miners was, however, a complex material, the mass use of which brought with it ever new possibilities. The coking of coal released large quantities of coal gas and coal tar, which gradually enabled the development of a gaslighting industry, as well as the use of coal tar derivatives, which became the starting point for the modern chemical industry of organic compounds through various representations of substances (tar dyes) in the 1840s and 1850s. Above all, however, the mass production of coke made it possible to accelerate the smelting processes and facilitated the further processing of pig iron into steel, which in turn was the starting product of a large number of other capital goods, which in turn fundamentally changed industrial production. Steel became the material of industry par excellence, but above all of the modern (tool) engineering, whose products made it possible for the first time to substitute human with machine labour on a large scale and thus to increase labour productivity decisively.

As I said, the Industrial Revolution was, therefore – if one takes the already-mentioned phenomena of industrial mass production together, not the result or the sequence of individual disruptive technological events – the consequence of a cumulative process of technological change, which in its result showed extremely disruptive effects, which were perhaps most clearly reflected in the futile attempts of traditional producers to stop this change by force (machine storm). In this respect, the period from the beginning of the factory age to the end of the machine storm, between the 1770s and the 1820s, marks a clear break. The destruction of wool and cotton spinning mills reached its peak in the years around 1812, when civil war-like riots broke out in the emerging industrial centres of Great Britain, which were violently suppressed by the military. Many leaders and participants of the so-called Luddism were condemned, executed or deported; the resistance weakened afterwards and finally ceased altogether; even if there were still individual fights, for example in the Aachen area, the advantages of the new production methods were too obvious. For, as has been shown, they did not remain limited to the field of mechanisation of textile production, but over time changed the entire field of production of raw materials, capital and investment goods. In this context, Joseph Schumpeter speaks of a first long wave of economic activity, the upswing of which between 1770 and 1810 was indeed marked by the cotton industry, which he therefore rightly identifies as the technological and economic leading sector of this upswing (Schumpeter 1939). The subsequent downturn phase in the 1810s and 1820s, which did not end until the 1830s, was therefore also the actual background to the Europe-wide poverty crisis of that time, so-called pauperism, since, among other things, the cotton companies reacted to the economic problems by keeping costs down and by forced technical rationalisation, which was reflected in correspondingly precarious social conditions, which provided Friedrich Engels with the occasion and material for his drastic description of the situation of the working class in England (Engels 1845/1970). The social crisis was exacerbated by contingent factors such as the Indonesian volcanic eruption and the ensuing global agricultural crisis, the consequences of the collapse of Napoleon's continental system and the huge national debt resulting from the wars, but in essence, it was a cyclical phenomenon resulting from the implementation of major technological innovations.

The revolution in transport

However, the great economic crises of the 1810s, 1820s and 1830s were by no means the expression of a kind of final agony of the still young capitalism; rather, a new upswing set in at the same time in these years, the extent of which was to far exceed the expansion of the first upswing phase since the 1770s. At its core was once again a fundamental innovation surge, which cannot be attributed to individual technical innovations, but rather combined various technical innovations into a kind of breakthrough technology, which then radically changed the entire economic landscape with enormous consequences. There is talk of the railway, and in a somewhat more limited sense also of steamship construction, whose mass spread was indeed associated with something like a "transport revolution" (Schivelbusch 1977), after which almost everything changed economically. This was initially because space was "shrinking"; the speeds reduced the time spent between European cities, which had previously been many hours, if not days, to a few hours. The high speeds also presupposed a sufficiently well-functioning signalling system; the co-evolution of rail transport and telegraphy is obvious, although the latter could also be used independently of the railway. The major supra-regional and finally transatlantic cable links that were established between the 1840s and 1870s, and with which, for example, the upswing of the House of Siemens is directly linked, further unified the world; economic news from New York could now spread globally very quickly, one of the factors that contributed significantly to the fact that regional economic crises could become global phenomena in the blink of an eye, something that also gave the founding crash of 1873 its special dynamic (Wenzlhümer 2013).

Space shrank, the globe unified, time adapted: only now did it become necessary to have uniform time structures, without which timetables or long-distance connections would hardly have been possible, whereas previously local time had sufficed to orientate oneself adequately. By far the greatest effect, however, came from the possibility of now being able to transport heavy and bulky goods over long distances at low cost. The complex significance of railway construction, which very quickly became the leading sector of the second long wave of economic development, based solely on the enormous sums of investment it attracted, was demonstrated above all by Alfred D. Chandler (1962, 1990) and, using the German example, Rainer Fremdling (1985). It begins with the technical homogenisation of market structures. Mass markets open up opportunities for action and thus provide technical and economic leeway that would not exist without them; in this respect alone, the cut caused by the drastic decline in transport prices and transport times can hardly be overestimated. Above all, however, the construction and operation of railways themselves had complex consequences. From the demand for construction and operating materials, through the entire technical apparatus (locomotives, wagons, rails, stations, telegraphy, maintenance) to the large numbers of employees, a leading technical sector developed which, between the 1840s and the 1880s, attracted the bulk of non-agricultural investment and eclipsed actual industrial investment. Finally, it was the capital-intensive operation of large-scale corporate structures which, from financing to a day-to-day organisation, heralded a fundamental change in the form of enterprise, which eventually became standard practice in similar enterprises. From the separation of ownership and control to functional and regional differentiation and modern financing structures, railway companies became the type of modern, management-run enterprise par excellence. The relative loss of importance of the textile industry, the upswing in the iron and steel industry, and the strong expansion of the capital and capital goods industry itself: everything is causally or significantly related to the expansion of the railway system, which, apart from telegraphy, whose technical perfection fell in the middle of the 19th century, was itself largely rooted in older technological knowledge. Steam engine construction and steel production were children of the 18th century, as was rail transport, even though artificial channels for the transport of bulky goods had played a greater role at this time. The use of steam engines to enable movement in space was an obvious development, nothing revolutionary new. But the complex effects were serious, not least the complete transformation of the city through its technical opening. Neither the immense migration processes would have been possible without the railways (and thus also the progressive urbanisation), nor modern urban spatial design and development would have been possible without the corresponding local transport and its space-structuring functions (Reulecke 1985; Marschalck 1984; Schivelbusch 1977; Reulecke 1997).

The railway boom affected the different spaces in a very peculiar way. Great Britain was the pioneer in this area for a long time, and since the 1840s the European continent followed, with Germany soon surpassing Great Britain in terms of area network and amount of investment, just as it had become self-sufficient in rolling stock and railway construction material since the 1850s, which previously had to be obtained from Great Britain. In the other European countries, the spread of the system was slower, and in some cases, it occurred later (in Russia in particular). The real hotspot, however, had not been in Europe since the 1840s anyway; the USA became the country of the railways' par excellence, which developed an unusual dynamic here. In view of the size of the still undeveloped country, the financing through capital and technology imports from Europe and the often-speculative nature of the exploration and development projects, the railway companies there combined all the decisive characteristics of this new branch of industry in an almost paradigmatic way. Their significance was such that the European public seeking investment flocked to the US in search of corresponding US securities, which consequently became the values in which speculative waves and crises have been brutally felt again and again since the 1850s. But none of this outweighed the fascination that the railways exuded in the course of the country's migration to the West.

Of similar importance for the development of what is today called "the world economy" were the breaks in maritime transport technology. The opportunities offered by rapid and cheap transport on the new, constantly improved steamships since the second half of the century were decisive for the change in relative prices for certain product groups (grain, meat, raw materials). Since the 1870s and 1880s, American wheat, whether from the USA or Argentina, has been able to enter European markets at unbeatable prices, and has even plunged Europe's less productive agricultural sector into a creeping structural crisis. It was not even the large quantities of grain or the mass of frozen meat that drove up trade volumes. While certain groups of goods even such as sugar, which had long dominated the market, or textile products grew only slowly, the transport volumes of coal, metal products and raw materials, which now came from the colonial areas or the countries of Latin America, provided the fuel for the industries in the USA and Europe on a large scale (Fischer 1979).

Not only did the merchant fleets change their technical appearance; the sailing ship, which was still of great importance in the second third of the century, gradually disappeared, albeit not abruptly, from the large trading ports; the ever-larger steam-powered merchant ship determined the picture and at the same time demanded a completely different structure of port facilities than had previously been customary. The fact that transport capacities virtually exploded in the last decades before the outbreak of the First World War was of decisive importance for this (Osterhammel and Peterson 2003). The British merchant navy, the largest in the world, doubled its volume between the 1880s and 1914, but its relative growth eclipsed that of other countries. The clear winner of the race in those years was Japan, whose merchant fleet grew by more than 4,000%, and Germany's merchant fleet, the second largest in world trade after Britain, grew more than three times as fast. The capacity of the merchant fleet of the 15 largest countries increased from 13 million NRT to about 30 million NRT between the 1880s and the beginning of the First World War (Sartorius von Waltershausen 1931). Together with the revolution in land transport, shipping was a major driver of the so-called first globalisation; in economic terms, this contributed above all to commodity market integration, that is it enabled market integration in terms of price, which of course made the other competitive factors more prominent. The different regions of the world economy thus in fact entered into an open competition which was no longer hidden by the weight of transport prices (Findlay and O'Rourke 2007: 402ff).

By the 1870s at the latest, the disruptive significance of the railway was over; now it was an established economic practice, still dynamic and promising, but increasingly without the aura and fascination of the new. Not least because of this fading away of the formerly comprehensive railway euphoria, but certainly, also because of other factors such as the lack of gold discoveries, the international growth rate slowed down in the 1870s and 1880s after the founding crash of 1873; a period of repeated setbacks and intensified competition set in, which also put an end to the long dominant economic liberalism. Under the leadership of the USA, protectionism spread internationally, which also took hold of the European continent (Torp 2014). Only the Netherlands, already a hub of world trade at that time, and Great Britain, which was able to afford its European balance of trade deficit due to the revenues from colonial business, maintained extensive free trade. Otherwise, moderate customs duties prevailed in Europe, which were partially prohibitive in the USA – a period of "neo-mercantilism" (Joseph A. Schumpeter) began (Findlay and O'Rourke 2007: 387–411).

The "Second Industrial Revolution"

Nevertheless, the "Great Depression" was over at the beginning of the 1890s. Gold discoveries in Alaska and South Africa put an end to the deflationary mood of the previous years, but what contributed to the renewed dynamism of the world economy was something else. The literature speaks of a Second Industrial Revolution, this time a scientific and technological revolution, which definitely brought the era of the first industrialisation to an end (North 1981). Its old industries, from textiles to coal and steel, remained important, but largely lost their avant-garde role to so-called new industries, whose basis consisted in the use of scientific discoveries and material representations for everyday industrial life. From the 1840s onwards, there had been a significant increase in the stock of technical knowledge (constituted and supported by universities), with the conversion of research to experimental work (Liebig principle) playing a major role and university institutes and research laboratories organised in this way occupying a prominent position (Heuss 1949). This did not mean that ingenious inventors or lone wolves had become meaningless; but they themselves or the implementation of their representations were increasingly dependent on a scientific environment, a scientific infrastructure, which a single "tinkerer" could hardly represent. At the latest since the turn of the century, but in some cases already years before, numerous companies followed this research and development path, in which they built up their own research and development capacities, often in imitation of university institutes, the design of which soon exceeded the scope of the rather small university institutes (Murmann 2003). Without wanting to give comprehensive details, broad areas of the electrotechnical and chemical, pharmaceutical and optical/precision mechanical industries owed their phoenix-like rise to this constellation, which, in turn, radiated into other areas and thus triggered the momentum of an economic wave that lasted at least until the outbreak of the World War. Once again, it is impossible to place individual technical events in a kind of causal chain at the beginning, but there are inventions that are good examples of this indeed disruptive change. If one wants to isolate individual inventions, there are many candidates, but hardly any is as suitable as the microscope produced according to scientific rules by Carl Zeiss in Jena, which was based on the physical and optical calculations of the university physicist Ernst Abbe. While Joseph von Fraunhofer's inventions had already made a significant breakthrough in macroscopy, albeit still based on the work of an autodidact (Plumpe 2020), Ernst Abbe's calculations enabled the serial production of qualitatively homogeneous, high-performance microscopes; the simultaneous establishment of a corresponding glass production facility by Schott'sche Glaswerke thus made possible the serial production of high-quality measuring instruments for the first time, which in turn became the central prerequisite for intensifying the corresponding research processes, for example in the field of chemical and pharmaceutical research or materials science (Auerbach 1918). The establishment of research laboratories in the chemical industry, which began in the 1880s, would have been just as impossible without these instruments as the establishment of corresponding research capacities at universities, whose representations of substances and their economic use as dyes or medicines in turn became a prerequisite for a corresponding intensification of research. The enabling of science-based innovations ultimately became a decisive criterion of competitiveness here, and accordingly, the entire complex of chemical-pharmaceutical production was already turning over more and more rapidly before the First World War, and only those companies that had the appropriate capabilities in this area were able to hold their own. Consequently, the large companies in the corresponding industries became veritable patent machines, starting with the metallurgical and electrotechnical industries, through the chemical industry, to pharmaceuticals and precision engineering/optics (Schneider 2017; Abelshauser et al. 2004).

The disruptive character of the new type of production resulted primarily from the fact that it was now possible to substitute a great many natural substances with human creations and at the same time to free oneself from the rigid specifications of coal and steam in the area of energy use. Thus, natural dyes quickly lost their importance; with the indigo synthesis, which became possible in a profitable way since 1900, the last natural mass dye lost its market almost overnight (Engel 2009). At the same time, chemotherapy, the curing of diseases by using artificially produced chemical substances, started from simple active substances, by-products of colour production. Aspirin, the further development of an intermediate product of colour production, became the most well-known drug, and a completely new branch of the pharmaceutical industry began (Wimmer 1994). While weak-current electrics had revolutionised telecommunications in particular, since the presentation of the dynamo principle it has been possible to develop strong-current electrical apparatus, starting with lighting and including the operation of small electric motors, which made the once large transmission mechanisms of steam power obsolete. The electric motor and the electric lighting, in turn, required enormous amounts of electricity and a corresponding transmission technology; power stations, overhead lines and power grids were the inevitable consequence, which in turn stimulated the corresponding electrotechnical apparatus construction and the development of a corresponding infrastructure, extremely capital-intensive innovations whose radiating effect cannot be overestimated (Pohl 1988; Pinner 1918). These new possibilities (including local public transport) were linked to a further intensification of urbanisation, which at the same time provoked a kind of communal provision of general interest, which in turn depended to a large extent on the very technical services that were now becoming possible (lighting, canalisation, freshwater supply, development of a medical infrastructure, etc.). This type of technical-economic change was not yet primarily geared to mass consumption, even though it gained importance in the years before the First World War. It was still primarily a matter of building up a modern capital stock and developing a corresponding infrastructure. Starting from the USA, however, the direction of technological change was now to change significantly; mass consumption, which itself was becoming the driving force behind technological change, came more and more to the fore (Grazia 2005).

Cars and mass consumption: the Fordist age

One of the technical innovations that had a similar broad effect as the scientification of the industry since the 1880s is a product that was invented in those years, but whose broad economic impact was only felt selectively before the First World War. Its significance cannot be overestimated, however, because the automobile (Feldenkirchen 2003), which we are now talking about, triggered a wave similar to that associated with the modern industries of the 1880s. This was not so much due to the individual mobilisation itself, which had long been known and was now technically improved. It was primarily due to how the new automobile was technically developed and thus became usable as a mass product. The first automobiles were technically difficult to master; anyone who could afford such an expensive vehicle usually had the means to employ a technically skilled chauffeur. Only with the successive simplification of the so-called control elements of the automobile did it become possible to operate such a vehicle comparatively easily by oneself, thus potentially increasing the possibilities of use infinitely. But for this to happen, it was necessary to be able to manufacture and sell the car at a price that could be mass marketable. And this is where Henry Ford's American automobile company comes in, which not without good reason gave its stamp and name to the subsequent phase of industrial development: the Fordist era. The car became the nucleus of a "production regime", which ranged from serial car production at low average costs with correspondingly organised production processes to the mass sale of the products to the company's employees; in short, the workers in the Ford factories were to earn enough to afford their own products. The assembly line, which, in turn, depended on the efficient use of decentralised drive units (electric motors!), became the symbol of Fordism; the simple, easy-to-use and relatively inexpensive Model T, the famous "Blechliesel" (in English: "tin lizzy"), became the world's first car produced in millions. During this production period between 1908 and 1927, 15 million units were built. Henry Ford became a legend during his lifetime, and mass motorisation became a reality in the USA as early as the 1920s (Ford 1924). This coupling of mass production and mass sales (Beynon and Nichols 2006) initially remained an American speciality, including relatively high wages, advertising and the development of a supply infrastructure, which was guaranteed relatively smoothly by the American oil companies.

This complex, although technically not particularly sophisticated, was not so easy to establish in Europe, which was relatively impoverished as a result of the war; the purchasing power of demand, vulgo: the relatively low wages did not allow this and, in turn, in limited markets, the construction of large production capacities, which are only profitable when capacity utilisation is high, makes little sense. The beginnings of an automotive technological complex can therefore be observed everywhere; in the vast majority of cases, the European automotive companies still in existence today have their beginnings before the First World War. But companies that could have rivalled the US giants of the time, namely Ford and General Motors, were at best found in Britain and France, where the automobile was somewhat widespread in the interwar period due to the relatively wealthy middle classes, while Germany's data lagged far behind, although the automobile already exerted its fascination here as well. The political project of a "Volkswagen" took up this fascination and the constituent idea of Fordism, without, of course, becoming reality before 1945; the corresponding capacities were still reserved for military purposes in Germany. That the Fordist model would also be of resounding importance in Europe was shown by the meteoric rise of the Volkswagen Group since the 1950s, which had parallels in the major European industrialised countries, for example with Fiat in Italy (Wellhöner 1996).

The impact of the expanding automotive industry was enormous, comparable only to the railways since the 1840s. Similarly, the mass use of the automobile changed the structure of urbanised spaces, making the functional division between living, working and leisure time only practically possible, thus making the concepts of contemporary architecture feasible at all, albeit at the cost of intensifying traffic (Gartman 2009). The car-friendly city already emerged in the USA in the 1920s, which was favoured in this respect due to the lack of historical "burdens"; since the 1950s it has become a worldwide reality, including the explosion of the road network and the strong expansion of the supply infrastructure (petrol stations, service stations, repair shops). Despite the obvious ecological consequences and the accident risks associated with motorisation, the car has spread throughout the world and is now making its triumphant advance even where there is still room for it. At the same time, automobile production or technically driven production clusters have become the largest production units in many countries, and for a long time they have clearly outperformed other industrial sectors in terms of employment. Finally, the automotive industry is of considerable importance as a pull factor for upstream and downstream industries, be it the steel industry, be it the entire complex of supplier companies, be it finally the industrial provision of consumables. Since the product was constantly changing at the same time, car manufacture itself became a major driver of technological change, even if the disruptive significance of small-scale change is now low. It is difficult to judge whether the current politically forced conversion of automobiles from internal combustion engines to electric drives has any special technological significance, especially since this transformation would not exist without its political subsidisation.

The wave of technical and economic change triggered by the car experienced its global peak in the 1950s and 1960s, when the car became the epitome of the assertive mass consumer society. The new forms in which mass consumption society was reflected, namely the green-field shopping centre, mass tourism and the so-called "occupational commuting", which found their more or less definitive structure in the 1970s, were now also associated with it and its use. At the same time, the spread of the automobile reached saturation levels that allowed the wave to run out, indeed to tip over into a veritable crisis, as the simple, inexpensive and technically relatively undemanding car models were no longer accepted in markets characterised by rising incomes. This self-fulfilment of a wave had already hit Ford in the 1920s, when it became clear that the policy of various models and brands with which the competitor General Motors responded to the large and differentiated demand would be more successful. Similarly, the successful model of many European companies to conquer large market shares with simple models led to a veritable crisis in the late 1960s and then in the 1970s, as these models proved increasingly difficult to sell and demand demanded a product variety that was hardly compatible with the companies' previous expansion strategies (Diez 2012; Tilly and Triebel 2013).

The saturation of mass markets, new competition and diversification

The end of the "Fordist" wave was reached in the 1970s worldwide when at least the large developed markets were saturated to a certain extent; although there was still potential where there was considerable pent-up demand, this did not pose any major technical challenges. The temporarily emerging superiority of East Asian, especially Japanese, manufacturers in the automotive, consumer electronics or optical industries was less the result of innovative technical achievements than of a better organisation of production processes and a more efficient organisation of work, which European and US companies were temporarily unable to counter in a saturated market without great technical dynamism (Kleinschmidt 2002; Raphael 2019). Whether the 1970s can therefore be described globally as a special decade of crisis is not conclusively clarified in economic history research; these years were crisis-ridden above all for the centres of the "Fordist" age, but here too with remarkable differences (Ferguson et al. 2010). The decisive factor was above all the technological exhaustion after the long years of boom, which now spread to mass manufacturers such as VW, or the limited ability to correct one's own organisational and technological path dependencies. Instead, an entirely new literary genre emerged that addressed precisely this bottleneck, the lack of competitiveness of modern Western companies. However, the resulting conclusions about how to regain success were usually as unsuccessful as the attempts to explain Japanese success. Books such as the legendary McKinsey publication by the authors Peters and Watermann on the 100 best-managed US companies quickly attracted ridicule, as many of these supposedly well-managed companies had already gone under themselves in one of the numerous new editions of the book (Plumpe 2008). The relevant literature (Nicolai and Kieser 2002) was certainly not without effect, especially since it often sang a joint chorus with management consultancies, which had become increasingly influential in recent years. They consistently recommended a variation of conventional organisational structures, corporate strategies and management styles, to which they prescribed more flexibility and a greater contemporary orientation away from the technocratically managed group (Boltanski and Chiapello 2005). As a consequence, a process of reorganisation and simultaneous inflation of many companies began to intensify from the 1970s onwards, which sought to cope with their growing size through organisational diversification until it was discovered in the 1980s that these "fashions" were not very successful. Since then, a renewed process of reorganisation began, which threw the strategy of diversified size formation overboard and focused on market-related streamlining of companies while concentrating on profitable core businesses; the incipient process of corporate reorganisation was handled and financed by complex capital market transactions, which in turn fuelled investment banking (Plumpe et al. 2020; Paul et al. 2020). The shift towards a financial market capitalism that has since been haunting the social science literature was therefore not a phenomenon of a technically driven change in the traditional capitalist business model, but its restructuring under the conditions of global competition, with financial transactions gaining importance for another reason. With the Eurodollars, dollar loans outside of US regulation, which had constituted a kind of free money market, especially in Europe, and the petrodollars that strengthened this market (Merki 2005), the globally floating money supply grew rapidly. Due to the lack of good investment opportunities, the money supply sought profitable investment opportunities in the weakening economies of old capitalism and found them in the still growing economies of the so-called Third World, especially in Latin America (Ferguson et al. 2010). The consequences, namely the sharp increase in debt and the varying debt crises since then, are well known. In 1980, interest rates in the USA shot up after the Federal Reserve, under its new president Paul Volcker, had switched to a restrictive money supply policy in 1979; the illusion of simple and above all economically beneficial debt burst as it were overnight; numerous debtors were no longer able to meet their obligations and many international banks were sitting on virtually uncollectible loans. This "debt crisis" now proved to be an innovative impulse in that strategies were sought to continue to make it possible to grant loans (at the most attractive conditions possible) and at the same time to effectively limit their risks (Sattler 2019). The securitisation, valuation and subsequent tradability of loans on the stock exchange, which caused the risks to disappear from the banks' books (starting with the so-called "Brady Plan"), combined with the beginning policy of "quantitative easing" – a strong expansion of the money supply and simultaneous keeping of interest rates low, which had been pursued since the 1980s and reached its peak in the years after the world financial crisis and since then continuing through the euro crisis to the current coronavirus pandemic – was the consequence of the debt crisis, even if this type of risk structuring is today essentially regarded as the cause of the financial crisis of 2008 (Admati and Hellwig 2013). However, the risk management of banks did not change solely as a result of the new form of securitisation and "securitisation" of credit instruments; the new possibilities of the digital organisation of banking operations, which arose at the same time, allowed a comparatively smooth expansion of the business model and – in retrospect – a strong inflation of bank balance sheets, which could be disguised relatively well to the outside world, among other things by the establishment of special purpose vehicles (McMillan 2014). Thus, under the conditions of the new information technologies and at the same time a policy of "cheap money", a specific form of coping with the debt crisis became a powerful driver for the inflation of the financial sector, the other side of which became the increasing indebtedness of companies, private households and – above all – states. This then in fact included a kind of "disruptive break", which was reflected above all in the increasing dependence of states on the financial markets, whose guarantee therefore itself became a kind of condition of existence for state budgets capable of being financed (Streeck 2014). However, the economic effects of this inflation of a financial sector that was essentially guaranteed by government action and cheap money/low interest rates were mixed; while the states were kept liquid and did not collapse under their debt burden, productivity development stagnated, even regressed in numerous states, not least because now few productive projects found financing opportunities, but above all, because the financial sector itself was absorbing even larger amounts of funds that were not invested in the so-called "real economy".

The 1970s and 1980s are therefore still considered to be years of technological stagnation, which favoured the shift of the ever-growing financial masses into a so-called financial capitalism, which, however, was less associated with a surge in productivity than with a new, less productive way of generating profits. For the economy itself, therefore, this period was primarily determined by shifts in location and reorganisation and restructuring phenomena, which further intensified the drift in favour of the financial and capital markets. However, this drift also benefited from new technical possibilities, which gained ever clearer contours in the 1980s and have since matured into a technological push without precedent. We are talking about the microelectronic or information revolution.

The microelectronic and information revolution

The computer, the execution of complex arithmetical operations on the basis of the digitalisation of data, is a comparatively old phenomenon; even the old Jacqard looms of the Lyon weavers of the 18th century had moments of this technology, which is indispensable today. However, the use, networking and acceleration of information processes only gradually developed, but experienced a huge boost with the miniaturisation of the devices in the 1980s and their gradual, first military, then general networking, since it was now possible to link an infinitely large number of decentralised computer capacities with one another and thus not only optimise computer performance, but also enable completely new information and control processes. In view of the fact that the first decentralised computers were still regarded as better typewriters and were therefore completely underestimated, the revolutionary change that had been gradually taking hold since the 1990s was only all the more serious. As memory and processor capacities evolved, each of these decentralised, networked units became more and more powerful, with the result that computing and control capacities literally exploded, while at the same time gradually minimising their costs. At present, billions of smartphones worldwide, each of which has a performance many times greater than that of the large computers of the 1970s and 1980s, are in interactive use and make a central contribution to shaping everyday human life by almost completely replacing conventional information paths, and at conditions and prices that have made the cost of individual pieces of information negligible. The significance of this revolution is also comprehensive because, very similar to the railway, it is a process that triggers comprehensive impulses: the development and production of devices, their operating and application systems and their networking is already a significant economic impulse, which, entirely in the Schumpeterian sense, led to the creative destruction first of all of the conventional methods of text production and text control (typewriter), then of almost all text-based production processes (printing and publishing) and, in addition, soon covered the entire information base of economic transactions. The control and sequence of economic transactions could now be successively "automated", and the machine tools known since the Industrial Revolution became "intelligent" in that their human fine-tuning became increasingly superfluous. Instead, the machines could be networked with one another in such a way (Industry 4.0) that largely autonomous, self-correcting production processes became possible, which in turn made it possible to redesign the entire upstream and downstream areas of industrial production, from supply to packaging and sales. The productivity surges in the industrial sector were, at least initially, enormous; they enabled the significant decline in industrial employment rates, even if their sharp decline in formerly industrialised countries such as the USA or Great Britain was mainly the result of the regional relocation of industrial production.

Moreover, the new information technologies were not limited to individual moments of industrial action; they covered the entire field of human activity, indeed, certain areas were opened up for economic activity in the first place (e.g. today's social networks) by transforming human communication to a large extent into an economically tradable good, which in turn gained importance as a "data complex", since it makes behaviour and processes modellable and plannable, at least in statistical aggregation, and thus itself becomes the basis of something for which the term "Artificial Intelligence" has become established, the dimension of which is becoming apparent, although it has not yet been realised in a comprehensive sense. This is currently the "breaking point" of technological change, which is of such significance that the AI question is highly politicised since the disposal of AI seems to be an important factor of future economic significance and political-military power. Although still in its infancy, its fundamental importance is already becoming apparent, whether in the area of the reorganisation of the financial sector and monetary systems, the shaping of a supra-regional and global division of labour (communications sector) or the regrouping of regional specialisations. Thus, it is now quite conceivable that the trend towards relocating industrial production in line with the respective regional labour costs will be halted or even reversed, at least in some areas, given that labour costs play an increasingly minor role in modern production.

This economic change as a result of the use of complex control and information technologies has, of course, another side, which, as "technological unemployment" or underemployment, represents the other side of every major innovative thrust. This phenomenon was evident in machine storming and pauperism, in the mass unemployment of the interwar period or the "rationalisation of work" since the 1960s. Today, too, clear, technology-induced shifts in the world of work can be seen, less in the sense that "work" is running out, but nevertheless in so far as the labour market shows clear gaps between highly demanded, usually highly qualified workers and a cheap, even precarious segment of low-qualified employment, the latter not only increasing significantly because the middle range of industrial skilled work is becoming smaller. Also, and above all, migration processes are causing a strong increase in this area, whose economic use is usually not very productive. Whether this is related to the stagnation in productivity development that has been observed for some time cannot be discussed here; it probably is. But this is inevitably part of the dialectic of disruptive technological change, namely being "creative destruction".

The information revolution is also so effective in its significance because it is to be seen both as a factor and an indicator of the second wave of globalisation, which has increasingly determined the structures of the world economy since the 1970s at the latest. With China's capitalist transformation and the demise of the Soviet system, the political blockades of globalisation initially gradually disappeared. Since the 1990s, capitalism and an institutionally liberal world economy have been a global phenomenon in which the exchange of capital, goods and people can take place increasingly freely, since, parallel to the institutional opening up of the former communist economies, the global level of protection also declined significantly for a long time. Besides, comparable to the last third of the 19th century, transport costs continued to fall as the large container fleets gained ground, so that sea transport from Asia to Europe and America reached a historically low level – with the well-known consequences for global relocation, which have become more and more of a political problem since the 2010s at the latest. Just as globalisation fired up competition between the great powers at the end of the 19th century and was at least one of the causes of the First World War, globalisation does not diminish the major conflicts in any way, but rather contributes to their more pronounced nature.

Conclusion

In the economic history of modern capitalism, disruptive technological change has been a comparatively regular, recurring phenomenon, at least up to the present, even if no legality or compelling regularity is hidden behind this wave-shaped economic structural change. Moreover, these disruptive processes were by no means due to simple events that would have had a revolutionary effect. Presenting change in this way has long been popular (the invention of the steam engine, the railway, the car, the computer, etc.), and there is little to be said against symbolically labelling this disruptive change with individual inventions. However, its probability does not depend on individual inventions, but rather on the extent to which these innovations are (or can be) used economically and what consequences they have. It is not the invention itself that determines its utilisation environment; many more factors play a decisive role, some of which could be mentioned earlier, but most of which had to be assumed. The ability of an economy to make efficient use of technological innovations and thus to trigger comprehensive transformation effects is a highly presuppositional phenomenon; this ability is undoubtedly the secret of the upswing that started in northwestern Europe and has since spread globally under the name of capitalism (Plumpe 2019).

The disruptive moment is therefore not the individual technical event, but the ability to transform such events into fundamental economic changes. This peculiar dialectic of technical innovation and processual use with a simultaneous progressive adaptation of the technical constellation to the conditions of efficient production and a, thoroughly volatile, consumer demand was comprehensively demonstrated by Joel Mokyr for the English Industrial Revolution; however, it is already found in the revolution of shipbuilding and consumer goods production in the Netherlands in the 17th century as well as later in the use of steam power as a universal means of propulsion for machines and transport vehicles. Here again, it was not a matter of individual inventions; the first simple steam engines of the early 18th century had little to do with the complex apparatuses that eventually drove locomotives and steamships. It was their further development, adaptation, selective improvement, in other words, a continuous process of use, which in turn provided the experience for further adaptations and improvements, until finally, at the end of the 19th century, oil and its derivatives made new propulsion technologies technically possible, which then underwent a similar process of change as steam power in the automotive industry of the 20th century and had the same complex effects on production, transport and consumption phenomena. Here, too, we find again the peculiar wave-like nature of change, which has been sufficiently described by Joseph Schumpeter, and which is characterised by the term's invention, innovation and diffusion, as well as made transparent. The end of the "Fordist" cycle in the 1960s and 1970s was not associated with the disappearance of this type of economy; it lost its pioneering function and was consequently no longer the area that followed or only significantly inspired the other areas of the economy. In this respect, the 1970s almost logically triggered a kind of technological pessimism (Mensch 1975), which provided the debates until the early 1980s with a peculiar undertone, but which then disappeared very quickly in the context of the unfolding information economy, not least because the "digital world" emerged at an almost breath-taking pace as a new key technology whose innovative potential was so enormous that it was able to overcome the exaggerations of the so-called dot.com bubble, which burst at the beginning of the new millennium.

A phenomenon of change not originally connected with this made the contextual dependence of disruptive change quite obvious during this period, namely the beginning rises of the Chinese economy and the associated slow closing of the so-called "Great Divergence" (Pomeranz 2000), which had become a characteristic of the processes of global economic change since the 18th century. Ultimately, the results of the Chinese reform processes were and are unmistakable in this respect (Deaton 2013), the institutional framework conditions that decisively determine the possibilities of exploiting economic opportunities for action (Hodgson 2015), above all whether they allow a process of open variation and market-based selection of economic activity, whereby its dynamics are significantly increased, albeit at the price of the necessarily associated failure. This renewed dynamism, which has become characteristic of the world economy since the 1990s, also had to do with globalisation, that is the opening and establishment of opportunities for supra-regional division of labour, the technical exploitation of which became increasingly economically attractive in view of increasing and more efficient transport methods. Since then, the connection between technical-economic opportunities for action and the framework conditions for their use has been obvious. However, even this evolutionary complex cannot completely emancipate itself from the technological possibilities, although China's race to catch up was initially less in the form of innovative steps, but more in the form of comprehensible steps. This did not stop the decline in technological dynamics in the context of the diffusion of microelectronic technical possibilities (Gordon 2017); instead, the emergence of Chinese suppliers in a generally rather stagnant global economic environment intensified a drastic increase in politically mediated economic competition and disputes, the core of which is not quite coincidentally the question of which world region will be the carrier of future innovative thrusts, the use of which will then also enable corresponding economic success and political weight. This is just as obvious as the more than coincidental correlation between global economic opening and economic prosperity, the connection between economic success and political capacity to act. The question of the future enabling and shaping of technological change phenomena is thus in central respect also a question of the respective institutional orders and political contexts, which can, of course, set contradictory, or at least different, impulses. While an open global economy based on the division of labour not only favours the variety of economic action and a corresponding market-based selection of variants promises relatively high efficiency, the political design of such action tends to be restrictive in order to correct by force the possibly undesirable regional manifestations of this change, which in turn favour the individual regions differently. This does not stop technical change as a whole, but subjects its design to considerable restrictions and generates escalation potentials, which can also have economic downturns in their wake or, as can be seen very clearly in the 20th century in the context of the Great Depression of 1929, can significantly intensify them. And in the same way, the dialectic of this change, its creative destructive power, is a permanent sting that does not lose its meaning under capitalist conditions, which is rather renewed with every fundamental technological break – in an admittedly equally new form. In this dialectic, a disruptive change appears both as a solution to the problem and as a cause of the problem; one cannot be obtained without the other. Perhaps the underestimation of this dialectic is also responsible for the naivety of many critics who hope for solutions to problems that do not turn into causes of problems. But this is illusionary.

References

Abelshauser, W. et al. (eds.) (2004): *German Industry and Global Enterprise. BASF: The History of a Company.* Cambridge: Cambridge University Press.

Admati, A. and Hellwig, M. (2013): *The Banker's New Clothes: What's Wrong with Banking and What to Do about it.* Princeton: Princeton University Press.

Allen, R. C. (2009): *The British Industrial Revolution in Global Perspective.* Cambridge: Cambridge University Press.

Auerbach, F. (1918): *Ernst Abbe, sein Leben, sein Wirken, seine Persönlichkeit, nach den Quellen und aus eigener Erfahrung geschildert.* Leipzig: Rockstuhl.

Beckert, S. (2014): *Empire of Cotton: A Global History*. New York: Alfred A. Knopf.
Berg, M. (2005): *Luxury and Pleasure in Eighteenth Century Britain*. Oxford: Oxford University Press.
Bergdolt, K. (2011): *Der schwarze Tod in Europa, Die große Pest und das Ende des Mittelalters*. München: C. H. Beck.
Beyer, B. (2007): *Vom Tiegelstahl zum Kruppstahl. Technik- und Unternehmensgeschichte der Gussstahlfabrik von Friedrich Krupp in der ersten Hälfte des 19. Jahrhunderts*. Essen: Klartext.
Beynon, H. and Nichols, T. (eds.) (2006): *The Fordism of Ford and Modern Management: Fordism and Postfordism*. Cheltenham: Edward Elgar.
Boltanski, L. and Chiapello, E. (2005): *The New Spirit of Capitalism*. London: Verso.
Braudel, F. (1986): *Sozialgeschichte des 15. – 18. Jahrhunderts: Der Handel*. München: Kindler.
Brüggemeier, F.-J. (1983): *Leben vor Ort. Ruhrbergleute und Ruhrbergbau 1889 bis 1919*. München: C. H. Beck.
Burghardt, U. (1995): *Die Mechanisierung des Ruhrbergbaus 1890–1930*. München: C. H. Beck.
Chandler, A. D. (1962): *Strategy and Structure: Chapters in the History of the Industrial Enterprise*. Cambridge: Beard Books.
Chandler, A. D. (1990): *Scale and Scope: The Dynamics of Industrial Capitalism*. Cambridge, MA: Harvard University Press.
Crafts, N. (2004): Productivity Growth in the Industrial Revolution. A New Growth Accounting Perspective, *The Journal of Economic History* 64, 521–535.
Crosby, A. W. (1997): *The Measure of Reality. Quantification and Western Society, 1250–1600*. Cambridge: Cambridge University Press.
Deaton, A. (2013): *The Great Escape: Health, Wealth and the Origins of Inequality*. Princeton: Princeton University Press.
Diez, W. (2012): *Die internationale Wettbewerbsfähigkeit der deutschen Automobilindustrie. Herausforderungen und Perspektiven*. München: Oldenbourg.
Engel, A. (2009): *Farben der Globalisierung. Die Entstehung moderner Märkte für Farbstoffe 1500–1900*. Frankfurt am Main: Campus.
Engels, F. (1845/1970): Die Lage der arbeitenden Klasse in England. Nach eigener Anschauung und authentischen Quellen (1845), in *Marx/Engels-Werke (MEW)*, Bd. 2. Berlin: Dietz, 225–506.
Feldenkirchen, W. (2003): *Vom Guten das Beste: Von Daimler und Benz zur DaimlerChrysler AG*. München: F.A. Herbig.
Ferguson, N. et al. (eds.) (2010): *The Shock of the Global: The 1970s in Perspective*. Cambridge, MA: Harvard University Press.
Findlay, R. and O'Rourke, K. H. (2007): *Power and Plenty. Trade, War, and the World Economy in the Second Millennium*. Princeton: Princeton University Press.
Fischer, W. (1979): *Die Weltwirtschaft im 20. Jahrhundert*. Göttingen: Vandenhoeck & Ruprecht.
Fogel, R. W. (2003): *The Slavery Debates 1952–1990. A Retrospective*. Baton Rouge: Louisiana State University Press.
Fogel, R. W. and Engerman, S. L. (1974): *Time on the Cross: The Economics of American Negro Slavery*. Boston: Little, Brown & Company.
Ford, H. (1924): *My Life and Work*. London: Doubleday, Page.
Fossier, R. (2008): *Das Leben im Mittelalter*. München: Piper.
Fouquet, G. and Zeilinger, G. (2011): *Katastrophen im Spätmittelalter*. Darmstadt: Philipp von Zabern.
Fremdling, R. (1985): *Eisenbahnen und deutsches Wirtschaftswachstum 1840–1879. Ein Beitrag zur Entwicklungstheorie und zur Theorie der Infrastruktur*. 2. erw. Aufl. Dortmund: Gesellschaft für westfälische Wirtschaftsgeschichte.
Gartman, D. (2009): *From Autos to Architecture: Fordism and Architectural Aesthetics in the 20th Century*. New York: Princeton Architectural Press.
Gille, B. (1968): *Ingenieure der Renaissance*. Düsseldorf: Econ.
Gimpel, J. (1980): *Die industrielle Revolution des Mittelalters*. Zürich, München: Artemis.
Gordon, R. J. (2017): *The Rise and Fall of American Growth. The US-Standard of Living since the Civil War*. Princeton: Princeton University Press.
Grazia, V. de (2005): *Irresistible Empire: America's Advance through 20th Century Europe*. Cambridge, MA: Harvard University Press.
Hammond, J. L. and Hammond, B. (2005): *The Rise of Modern Industry*. London: Routledge.
Hart, M. C 't (1993): *The Making of a Bourgeois State: War, Politics, and Finance during the Dutch Revolt*. Manchester: Manchester University Press.

Heuss, T. (1949): *Justus von Liebig. Vom Genius der Forschung*. Hamburg: Hoffmann und Campe.
Hodgson, G. M. (2015): *Conceptualizing Capitalism. Institutions, Evolution, Future*. Chicago: Chicago University Press.
Inikori, J. E. (2002): *Africans and the Industrial Revolution in England. A Study in International Trade and Economic Development*. Cambridge: Cambridge University Press.
Jones, E. L. (1981): *The European Miracle: Environments, Economies and Geopolitics in the History of Europe and Asia*. Cambridge: Cambridge University Press.
Kleinschmidt, C. (2002): *Der produktive Blick. Wahrnehmung amerikanischer und japanischer Management- und Produktionsmethoden durch deutsche Unternehmer 1950–1985*. Berlin: Akademie Verlag.
Kondratjew, N. (1926): Die langen Wellen der Konjunktur, *Archiv für Sozialwissenschaft und Sozialpolitik* 56, 573–609.
Kriedte, P. (1980): *Spätfeudalismus und Handelskapital. Grundlinien der europäischen Wirtschaftsgeschichte vom 16. bis zum Ausgang des 18. Jahrhunderts*. Göttingen: Vandenhoeck & Ruprecht.
Kriedte, P., Medick, H. and Schlumbohm, J. (1977): *Industrialisierung vor der Industrialisierung: Gewerbliche Warenproduktion auf dem Land in der Formationsperiode des Kapitalismus*. Göttingen: Vandenhoeck & Ruprecht.
Landes, D. S. (1969): *The Unbound Prometheus. Technical Change and Industrial Development in Western Europe from 1750 to the Present*. Cambridge: Cambridge University Press.
Landes, D. S. (1998): *The Wealth and Poverty of Nations. Why Some are so Rich and Some so Poor*. New York: W.W. Norton & Co.
Maddison, A. (2007): *Contours of the World Economy 1–2030 AD. Essays in Macro-Economic History*. Oxford: Oxford University Press.
Marschalck, P. (1984): *Bevölkerungsgeschichte Deutschlands im 19. und 20. Jahrhundert*. Frankfurt am Main: Suhrkamp.
McMillan, J. (2014): *The End of Banking. Money, Credit, and the Digital Revolution*. Zürich: Zero/One Economics.
Menninger, A. (2008): *Genuss im kulturellen Wandel: Tabak, Kaffee, Tee und Schokolade in Europa (16.-19. Jahrhundert)*. Stuttgart: Franz Steiner.
Mensch, G. (1975): *Das technologische Patt. Innovationen überwinden die Depression*. Frankfurt am Main: Umschau.
Merki, C. M. (ed.) (2005): *Europas Finanzzentren. Geschichte und Bedeutung im 20. Jahrhundert*. Frankfurt am Main: Campus.
Meußdoerffer, F. and Zarnkow, M. (2014): *Bier. Eine Geschichte von Hopfen und Malz*. München: C. H. Beck.
Mitterauer, M. (2003): *Warum Europa? Mittelalterliche Grundlagen eines Sonderwegs*. München: C. H. Beck.
Mokyr, J. (2002): *The Gifts of Athena. Historical Origins of the Knowledge Economy*. Princeton: Princeton University Press.
Mokyr, J. (2009): *The Enlightened Economy. An Economic History of Britain 1700–1850*. New Haven: Yale University Press.
Mokyr, J. (2018): *A Culture of Growth: The Origins of the Modern Economy*. 5. Aufl. Princeton: Princeton University Press.
Murmann, J. P. (2003): *Knowledge and Competitive Advantage. The Coevolution of Firms, Technology and National Institutions*. Cambridge: Cambridge University Press.
Nicolai, A. and Kieser, A. (2002): Trotz eklatanter Erfolglosigkeit: Die Erfolgsfaktorenforschung weiter auf Erfolgskurs, *Die Betriebswirtschaft* 62(6), 579–596.
North, D. C. (1981): *Structure and Change in Economic History*. New York: W.W. Norton & Co.
North, M. (2001): *Das Goldene Zeitalter. Kunst und Kommerz in der niederländischen Malerei des 17. Jahrhunderts*. Köln: Böhlau.
North, M. (2014): *Kommunikation, Handel, Geld und Banken in der frühen Neuzeit*. München: De Gruyter Oldenbourg.
Osterhammel, J. and Peterson, N. (2003): *Geschichte der Globalisierung*. München: C.H. Beck.
Parker, G. (1988): *The Military Revolution. Military Innovation and the Rise of the West, 1500–1800*. Cambridge: Cambridge University Press.
Paul, S., Sattler, F. and Ziegler, D. (2020): *Hundertfünfzig Jahre Commerzbank 1870–2020*. München: Siedler.
Paulinyi, A. (1987): *Das Puddeln. Ein Kapitel aus der Geschichte des Eisens in der Industriellen Revolution*. München: Oldenbourg.
Petzina, D. and van Roon, G. (eds.) (1981): *Konjunktur, Krise, Gesellschaft: Wirtschaftliche Wechsellagen und soziale Entwicklung im 19. und 20. Jahrhundert*. Stuttgart: Klett-Cotta.

Pinner, F. (1918): *Emil Rathenau und das elektrische Zeitalter.* Leipzig: Akademische Verlagsgesellschaft.

Plumpe, W. (2008): Nützliche Fiktionen. Der Wandel der Unternehmen und die Literatur der Berater, in Reitmayer, M. and Rosenberger, R. (eds.): *Unternehmen am Ende des „goldenen Zeitalters". Die 1970er Jahre in unternehmens- und wirtschaftshistorischer Perspektive.* Essen: Klartext, 251–270.

Plumpe, W. (2009): Ökonomisches Denken und wirtschaftliche Entwicklung. Zum Zusammenhang von Wirtschaftsgeschichte und historischer Semantik der Ökonomie, *Jahrbuch für Wirtschaftsgeschichte* 50(1), 27–52.

Plumpe, W. (2017): *Wirtschaftskrisen. Geschichte und Gegenwart.* 5. Aufl. München: C.H. Beck.

Plumpe, W. (2019): *Das Kalte Herz. Kapitalismus: Geschichte einer andauernden Revolution.* Berlin: Rowohlt.

Plumpe, W. (2020): 1805: Fernsicht aus Benediktbeuern, in Fahrmeir, A. (ed.): *Deutschland: Globalgeschichte einer Nation.* München: C.H. Beck, 313–317.

Plumpe, W., Nützenadel, A. and Schenk, C. R. (2020): *Deutsche Bank 1870–2020: The Global Hausbank,* London: Bloomsbury.

Pohl, M. (1988): *Emil Rathenau und die AEG.* Mainz: Hase & Koehler.

Pomeranz, K. (2000): *The Great Divergence. China, Europe, and the Making of the Modern World Economy.* Princeton: Princeton University Press.

Radkau, J. (2008): *Technik in Deutschland. Vom 18. Jahrhundert bis heute.* Frankfurt am Main: Campus.

Raphael, L. (2019): *Jenseits von Kohle und Stahl. Eine Gesellschaftsgeschichte Westeuropas nach dem Boom.* Berlin: Suhrkamp.

Reinhard, W. (2016): *Die Unterwerfung der Welt. Globalgeschichte der europäischen Expansion 1415–2015.* München: C.H. Beck.

Reulecke, J. (1985): *Geschichte der Urbanisierung in Deutschland.* Frankfurt am Main: Suhrkamp.

Reulecke, J. (ed.) (1997): *Geschichte des Wohnens.* Bd.3: Das bürgerliche Zeitalter 1800–1918. Stuttgart: DVA.

Sartorius von Waltershausen, A. (1931): *Die Entstehung der Weltwirtschaft. Geschichte des zwischenstaatlichen Wirtschaftslebens vom letzten Viertel des 18. Jahrhunderts bis 1914.* Jena: G. Fischer.

Sattler, F. (2019): *Herrhausen: Banker, Querdenker, Global Player. Ein deutsches Leben.* München: Siedler Verlag.

Schama, S. (1988): *The Embarrassment of Riches. An Interpretation of Dutch Culture in the Golden Age.* New York: University of California Press.

Schivelbusch, W. (1977): *Geschichte der Eisenbahnreise. Zur Industrialisierung von Raum und Zeit im 19. Jahrhundert.* München: Hanser.

Schneider, M. C. (2017): Das wissenschaftliche Unternehmen. Zur chemisch-pharmazeutischen Forschung bei E. Merck, Darmstadt, ca. 1900 bis 1930, *Zeitschrift für Unternehmensgeschichte* 62, 163–203.

Schumpeter, J. A. (ed.) (1933): *Der Stand und die nächste Zukunft der Konjunkturforschung.* Festschrift für Arthur Spiethoff. München: Duncker and Humblot.

Schumpeter, J. A. (1939): *Business Cycles. A Theoretical, Historical and Statistical Analysis of the Capitalist Process.* New York: McGraw-Hill Book Company (dt. 1961).

Streeck, W. (2014): *Buying Time: The Delayed Crisis of Democratic Capitalism.* London: Verso.

Tilly, S. And Triebel, F. (eds.) (2013): *Automobilindustrie 1945–2000. Eine Schlüsselindustrie zwischen Boom und Krise.* München: Oldenbourg.

Torp, C. (2014): *The Challenges of Globalization. Economy and Politics in Germany, 1860–1914.* New York: Berghahn Books.

Trentmann, F. (2016): *Empire of Things. How We became a World of Consumers from the Fifteenth Century to the Twenty-first.* London: Allen Lane.

Vries, J. de (2008): *The Industrious Revolution. Consumer Behavior and the Household Economy, 1650 to the Present.* Cambridge: Cambridge University Press.

Vries, J. de and van der Woude, A. (1997): *The First Modern Economy. Success, Failure and Perseverance of the Dutch Economy, 1500–1815.* Cambridge: Cambridge University Press.

Vries, P. (2013): *Escaping Poverty: The Origins of Modern Economic Growth.* Göttingen: V&R Unipress.

Wellhöner, V. (1996): *„Wirtschaftswunder", Weltmarkt, westdeutscher Fordismus: Der Fall Volkswagen.* Münster: Westfälisches Dampfboot.

Wenzlhümer, R. (2013): *Connecting the Nineteenth Century World: The Telegraph and Globalization.* Cambridge: Cambridge University Press.

Wimmer, W. (1994): *"Wir haben fast immer was Neues": Gesundheitswesen und Innovationen der Pharma-Industrie in Deutschland 1880–1935.* Berlin: Duncker & Humblot.

Wrigley, E. A. (2010): *Energy and the English Industrial Revolution.* Cambridge: Cambridge University Press.

3

ON MACHINE AGES

Causes, forms and effects of technological change*

Heinz D. Kurz

Man is a tool-making animal.
(Benjamin Franklin)

Introduction

Adam Smith defines the objects of political economy, "considered as a branch of the science of a statesman or legislator", to consist in the following:

> first, to provide a plentiful revenue or subsistence for the people, or more properly to enable them to provide such a revenue or subsistence for themselves; and secondly, to supply the state or commonwealth with a revenue sufficient for the publick services. It proposes to enrich both the people and the sovereign.
>
> *(Smith [1776] 1976, WN IV.1)*

The growth of both kinds of revenue is taken to be furthered by a deepening of the "social division of labour", Smith's catch-all term for technological and organisational progress. He insists that the "quantity of science" available to a nation and put to productive use decides about her productivity and wealth (WN I.i.9).

According to Karl Marx, in capitalism labour productivity increases "as in a greenhouse" (MEGA II/5: 505): it is growing "geometrically" with no upper boundary in sight. Innovations, Joseph A. Schumpeter insists, "are the overwhelming fact in the economic history of the capitalist society" (1912: 159): they involve a sequence of thrusts of "creative destruction". While in the long run they typically tend to benefit many people in terms of more and better goods, medical services, life expectancy and living conditions at large, they do not only have winners, but also losers. Innovations help to satisfy needs and wants of people and solve pressing problems of various kinds, but they are often the source of entirely new problems, caused inter alia by the by-products that accompany them, such as, for example, pollution, environmental

* I am grateful to Marlies Schütz for useful comments and to Julia Wurzinger for help with the graphical illustrations. For an in-depth discussion of some of the issues raised in this chapter, see Kurz et al. (2018).

DOI: 10.4324/9780429351921-5

degradation or the difficulty of disposing of nuclear waste. Technological change, we might say without too much of an exaggeration, feeds upon itself and the results it produces. It affects deeply the socio-economic metabolism and gives rise to changing systems of production, consumption and disposal. These changes may be more or less smooth, but some may be disruptive and force the system to quick, painful and spasmodic adjustments in an attempt to absorb the new and eliminate the old.

Whoever wishes to understand what Marx and Schumpeter called the "law of motion" of modern society ought to study its dynamism from within and its apparently insatiable thirst for novelty and change. What are the sources of the system's restlessness, the forms of the technological and organisational progress it generates and the effects these have on the properties of the system?

This chapter provides a brief account of how a handful of political economists dealt with these issues. Interestingly, systematic and general economic analysis began broadly at a time when Northwest Europe took off from a state of relative stagnation – the "Malthusian era", for short – and gradually embarked on a path of sustained growth of labour productivity, real social product per capita and population. Therefore, our account also starts in this period. We set the stage by first looking briefly at the time prior to the Industrial Revolution, its antechamber, so to speak. This was not a time without some noteworthy technological improvements and phases of modest economic growth. What was still missing, however, was that these phenomena had become a sort of new normal. The main part of the chapter deals with the time beginning with the Industrial Revolution and how major political economists and students of technology and the economy reflected upon what was happening – the causes, forms and effects of technological change. It was the time of the First Machine Age: machine power replaced progressively labour and animal power, but machines did not yet learn themselves, although there were early attempts pointing in this direction. The authors dealt with briefly include Adam Smith (1723–1790), David Ricardo (1772–1823), Charles Babbage (1791–1871), and Karl Marx (1818–1883). The Second Machine Age (Brynjolfsson and McAfee 2014) is instead characterised by machines learning human skills such as perception, cognition and communication. We experience a new type of dynamically increasing returns and the rise of a new type of monopolies – "superstar firms" (Autor et al. 2020). These will be illustrated in terms of some of the concepts and tools the classical economists elaborated. In order to avoid duplications, important contributions by major economists then and now that are covered in other chapters of this Handbook will not be dealt with here. These concern inter alia Joseph A. Schumpeter's (1883–1950) contribution and the theory of long waves.

In the antechamber of the Industrial Revolution

What caused the fundamental change of the socio-economic regime that brought about a "culture of growth" (Mokyr 2017) in Western Europe and the "great divergence" (Pomeranz 2000) between it and the rest of the world? Which empirical evidence reflects what happened? How does it affect the social order and the distribution of power amongst various groups of society, the circulation of elites, the might and glory of a nation and its rulers? A book originally written in the late 1670s, published posthumously in 1690, is titled *Political Arithmetick*. Its author, William Petty (1623–1687), advocates a "physician's outlook" on economic problems and decides to express himself

> in terms of *Number, Weight*, or *Measure* . . . and to consider only such Cases, as have visible Foundations in Nature, leaving those that depend upon the mutable

> Minds, Opinions, Appetites, and Passions of particular Men, to the Consideration of others.
>
> *(Petty 1986: 244)*

Petty sees himself as the founder of a new science that was both empirical and analytical, concerned with providing a quantitative analysis of the productive capacity of a nation and the means and ways to increase it. Marx ([1867] 1954: 259) calls him "the father of Political Economy, and to some extent the founder of Statistics". However, Petty is not only a philosopher of science and political economist, working for some time as the personal secretary of Thomas Hobbes. He is also an inventor: in 1647 he files a patent application for a printing machine and in 1663 he presents a "double-bottom ship", a sort of catamaran, in the harbour of London, which, alas, sinks on this occasion – a paradigmatic example of the misfortune of a would-be pioneer. Petty's inventions were incremental, elaborating on existing ideas. For instance, while as a boy working on a merchant ship and later in the service of the Royal Navy he in all probability came across outrigger boats in East India, which served as a paragon for his double-bottom ship.

This and similar cases illustrate a widely held view at the time that trade was a major source of technological improvements. According to David Hume (1711–1776) trade is a medium of "open communication" between nations and of the diffusion of economically useful knowledge across the globe. He goes as far as to argue that "every improvement" which Great Britain has made in the arts both of agriculture and manufactures "has arisen from our imitation of foreigners; and we ought so far to esteem it happy, that they had previously made advances in arts and ingenuity" (Hume [1777] 1985: 328). In this perspective, socio-economic development was not the result of significantly moving the frontier of economically useful knowledge outwards, but rather of gradually approaching the existing frontier worldwide: it was not so much a result of innovation as of imitation.

In Volume I of *Capital*, published in 1867, Karl Marx distinguishes between the following two extreme forms of generating surplus value and profits:

> The surplus-value produced by prolongation of the working-day, I call *absolute surplus-value*. On the other hand, the surplus-value arising from the curtailment of the necessary labour-time, and from the corresponding alteration in the respective lengths of the two components of the working-day, I call *relative surplus-value*.
>
> *(Marx [1867] 1954: 299)*

In the former case the productivity of labour is fairly constant, because technical conditions of production are. Surplus labour time and therefore surplus value and profits are increased by increasing the length of the working day or by increasing the intensity of labour, given the amount of labour needed to reproduce the commodities constituting the given real wages, which Marx called "necessary labour". In the latter case, the length of the working day and real wages are basically constant, but due to improved methods of production labour productivity is increased: the commodities constituting real wages can be reproduced in a shorter time, which curtails necessary labour time and increases surplus labour time.

Historically, the production of absolute surplus value dominated early phases of capitalist development, especially in regions and countries that after the Reformation came under the spell of radical protestant and puritan sects. William Petty explains the economic success of the "Hollanders" in the following way: "Dissenters of this kind [i.e. Calvinists and Baptists], are for the most part, thinking, sober, and patient Men, and such as believe that Labour and Industry is their Duty towards God" (Petty 1986: 262). Marx observes: "Protestantism, by changing almost all the traditional holidays into workdays, plays an important part in the genesis of capital" ([1867] 1954: 262, n.2). A lengthening of the working day, an abolition of holidays and an intensification of labour at fairly constant real wages per worker implied a decrease of the hourly wage rate and correspondingly an increase in the rate of profits. For a given propensity to accumulate capital out of profits, this implies a higher rate of accumulation and a higher rate of economic growth. As Max Weber ([1904/1905] 1930) put it, the "Spirit of Capitalism", fuelled by the "Protestant Ethic", led to an acceleration of economic growth without substantial innovations and technological change.[1] Weber therefore confirmed Marx's concept of the production of absolute surplus value.

However, socio-economic expansion based on the production of absolute surplus value had its severe limitations: the length of the working day and the intensity of labour were bounded from above, the lowering of real wages from below. Speeding up what was essentially only extensive growth requested a parallel speeding up also of population growth. This was not possible, however, at constant or even deteriorating living and working conditions of the labouring classes. Labour power tended to become scarce (also in the sense of a falling proportion of workers in good health and physical and mental strength) and firms had to seek other ways of surviving the competitive struggle and safeguarding and increasing their profits. This ushered in a new era, based on the production of relative surplus value: development and growth were henceforth driven by new, labour-saving methods of production. These methods, Marx argues, replenished time and again the "reserve army of the unemployed", which kept the aspirations of workers at bay and rendered a continual economic expansion possible.

In the second half of the eighteenth century, the (first) Industrial Revolution gradually gained momentum, reflecting a fundamental change of regime: the Malthusian era gave way to a culture of innovation and growth. Important elements of this change included: the strengthening of the view that humankind can understand nature with the help of new experimental methods; the elaboration of new technologies and new goods by putting this understanding to productive use; the confirmation of Francis Bacon's conviction that people can significantly improve their living conditions by putting especially the natural and engineering sciences at the service of agriculture, manufacturing and commerce; and, as an upshot of all this, the emergence of a virtuous circle that turns technological change and economic development from a sporadic phenomenon into a sustained process. Tools and machines quickly gained in importance, gradually redefining and reshaping the labour process and controlling large parts of the work force rather than the other way round. A new era was dawning – the "First Machine Age".

Adam Smith on the social division of labour

In the *Wealth of Nations*, first published in 1776, Adam Smith insists that the social division of labour decides about the "skill, dexterity, and judgment" with which a society's labour is applied (WN I.3). Its deepening improves the productive powers of labour and impacts on the distribution of the product amongst workers, capitalists and landlords (see WN I.5 and II.iii.32). While no single consideration of Smith in this regard is truly novel, their combination is and has

1 The claim that the acceleration of economic growth was the result of human capital formation associated with higher literacy rates of protestants who were obliged to read the Bible by themselves (see, for example, Becker and Woessmann 2009) gets little empirical support; see, for example, Kelly et al. (2020).

become a locus classicus in the discussion of technological change and economic development (see Eltis 1984; Aspromourgos 2009; Kurz 2010).

Smith deals both with the causes, forms and consequences of technological change. As regards the causes, he opines: "the desire of bettering our conditions . . . comes with us from the womb, and never leaves us till we go into the grave" (WN II.iii.28). Private and public opulence derive originally from this principle, which governs the productive efforts of humans. While bad government may fetter the desire, it "is frequently powerful enough to maintain the natural progress of things toward improvement" (WN II.iii.31). Another "propensity in human nature", namely "to truck, barter and exchange one thing for another" (WN I.ii.1) – probably "the necessary consequences of the faculties of reason and speech" (WN I.ii.2) – is said to give occasion to the social division of labour and socio-economic progress. (We will see later how in the Second Machine Age machines gradually acquire certain human faculties, such as to perceive, identify, speak and learn.) Smith's concept of the social division of labour encompasses all aspects and varieties of technological and organisational change and learning processes, including learning by doing and learning by using. However, there is no sharp specification of different forms of technological progress and the specific consequences these engender. He stresses that while some of the consequences are intended, others are not, a fact that contributes to the remarkable difficulty of the analysis.

The social division of labour, Smith emphasises, depends on the "extent of the market" (WN I.iii.1). According to Allyn Young (1928: 529) this is Smith's central "theorem" and "one of the most illuminating and fruitful generalisations" in economics: it is a cornerstone of the doctrine of dynamically increasing returns. Smith stresses that the division of labour and the technological change accompanying it propagate themselves in a cumulative way so that change becomes progressive and persistent. He sees essentially three channels through which the division of labour increases productivity:

1 Improvements in the dexterity of workers as a gain from specialisation;
2 A better use of workers and the plant and equipment they operate by saving the time lost in passing from one kind of work to another; and
3 Invention of machines that replace labour power by machine power (see WN I.i.6–8).

While items 1 and 2 face rather narrow boundaries, item 3 does not, but Smith is not aware of the productive potential of this fact. Ironically, he opines that the manufacturing sector provides much more ample opportunities for the division of labour than any other sector in the economy, especially agriculture. At the same time he misses the role of manufactures as the "engine of growth" during the Industrial Revolution by claiming that the sector produces essentially only trinkets and luxury goods for the propertied classes, but not investment goods – tools and machinery – for the economy as a whole. He in fact considers "corn" (wheat), the representative agricultural product, as the only (composite) product that enters directly or indirectly in the production of all products, including itself (see WN IV.v.a.23). Manufacturing products may enter in their own production, but do not enter in the production of other products (i.e. are means of production of limited purpose and scope or simply pure consumption goods).[2]

According to Smith, in the course of a deepening of the social division of labour, and as a natural by-product of it, there emerges what today is called a sector specialising in Research & Development. The reference is to

2 In terms of Sraffa's distinction (1960: 7–8), corn is a basic product, whereas manufactures are non-basics.

> philosophers or men of speculation [that is, scientists], whose trade it is, not to do anything, but to observe every thing; and who, upon that account, are often capable of *combining together the powers of the most distant and dissimilar objects*.

He continues: "In the progress of society, philosophy or speculation becomes, like every other employment, the principal or sole trade and occupation of a particular class of citizens" (WN I.i.9; emphasis added). Reconfiguring and recombining existing particles of knowledge may generate new particles: technological change is necessarily path-dependent. The combinatorial metaphor recurs, inter alia, in the concept of "new combinations" in Marx ([1890] 1959: 255) and Schumpeter (1912), and recently in Weitzman's (1998) concept of "recombinant growth".

At the time of Smith, technological change had become fairly ubiquitous and regular, systematically generated by institutions concerned with the production of new, economically valuable knowledge. Huge extra profits awaited the successful pioneer, but failure was frequent:

> These profits sometimes are very great, and sometimes, more frequently, perhaps, they are quite otherwise; but in general they bear no regular proportion to those of the other old trades in the neighbourhood. If the project succeeds, they are commonly at first very high. When the trade or practice becomes thoroughly established and well known, the competition reduces them to the level of other trades.
>
> *(WN I.x.b.43)*

In a few lines Smith summarises the complex process triggered by the introduction of a new method of production or a new commodity that disturbs the current situation and forces the economic system to leave the trodden path. The gradual diffusion and eventual generalisation of the new and the concurrent elimination of the old involves a transition from a given system of production to a new one. In the latter the extraordinary (or monopoly) profits of the pioneer are competed away and a tendency towards a uniform rate of profits establishes itself again. Comparing the two systems of production or fully adjusted positions of the economy with one another informs about the effects of technological change on the most important state variables of the economy, such as output, employment, income distribution and so on.

Smith can be said to have sensed an important fact, properly established only by David Ricardo, namely that to a technologically fully specified system of production corresponds a constraint binding hypothetical changes in the distributional variables, especially the general rate of profits (r) and the real wage rate (w), or $w - r$ relationship (see Aspromourgos 2009: 100 and 180; see also Kurz and Salvadori 1995: chap 4).[3] This concept, and its movement over time in $w - r$ space caused by technological change (and other factors at work, such as the increasing scarcity of certain natural resources), played (at least implicitly) an important role in Smith's vision of the dynamical properties of the economy. In Figure 3.1 T represents the "old" system of production and D the "new" one, reflecting a deeper social division of labour. The new system of production has a larger labour productivity and therefore a larger maximum level of the real wage rate corresponding to a zero rate of profits, W; that is, $W_D > W_T$. This follows directly from elements 1 and 2 and indirectly from element 3 of Smith's previous characterisation of the division of labour. A counteracting tendency results from the fact, stressed by Smith, that in factories a larger number of overseers and people employed in monitoring and enforcing work discipline are needed.

3 In marginalist or neoclassical analyses the commonly used term is "factor price frontier", proposed by John Hicks. See, for example, the use of the concept in Korinek and Stiglitz (2021: 6).

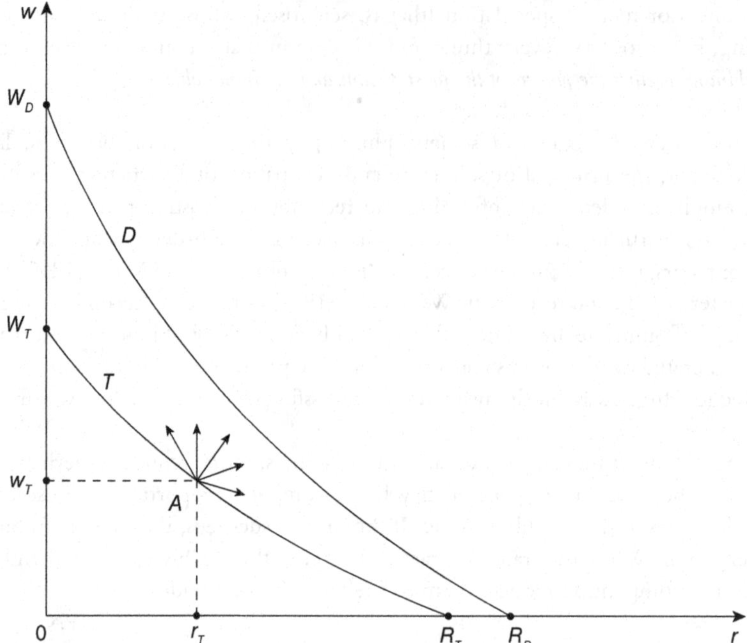

Figure 3.1 Division of labour

As regards the maximum rate of profits of the system, *R*, corresponding to hypothetically vanishing wages, Smith's argument may be translated as implying a rise or at least no fall in it, $R_D \geq R_T$. This follows directly from element 2 earlier, which implies a better utilisation of the existing plant and equipment and thus a lower capital-output ratio. However, there is also element 3, the replacement of labour power by machine power, which together with the other elements may go in either direction. Here I assume that Smith's view can be represented by a slight increase of the maximum rate of profits. Accordingly, the $w - r$ relationship tends to move outwards and get steeper over time. A deeper division of labour is cost reducing, which is reflected by the fact that for a given *r*, *w* will be larger.

The movement of the $w - r$ relationship may explain Smith's extreme optimism as to the blessings of the "system of natural liberty" and its capacity to provide growing revenues of the people and the state. Starting from point *A* in Figure 3.1, the development of the wealth of a nation may lead to rising levels of real wages without necessitating a fall in the general rate of profits. Despite his optimistic outlook Smith was nevertheless convinced that the general rate of profits was bound to fall in the course of time. He argued that the accumulation of capital intensified the competition amongst individual capitals that bid down the rate of return on capital. However, he did not reason correctly: since he had assumed *free competition* from the beginning, that is the absence of any barriers to entry in or exit from markets, there was no room for an intensification of competition.

David Ricardo on improved machinery

We owe David Ricardo the discovery of what is known as the "fundamental law of income distribution", given the system of production in use: "The greater the portion of the result of labour that is given to the labourer, the smaller must be the rate of profits, and vice versa" (Ricardo 1951–1973, *Works* VIII: 194). The levels of the two distributional variables are inversely related to each other. With a change in technology, the $w - r$ relationship changes too.

In several respects Ricardo enhances considerably our knowledge regarding the subject under consideration and corrects some of Smith's misleading propositions. He does so especially in his *Principles of Political Economy*, first published in 1817. Most importantly, and directly connected with our main theme, Smith had explained the rents of land and the alleged superior productivity of agriculture with reference to the labour that nature performs in this sector for free. This "free gift" is seen to distinguish agriculture from all other sectors, with rent reflecting the "generosity" of nature. Ricardo strongly opposes this doctrine and argues instead that rent expresses nature's "niggardliness", because if land of the best quality was available in abundance there could be no rent, since nobody would pay for a service that is available in excess supply. However, his criticism goes deeper and questions the basis of Smith's entire reasoning. Ricardo asks:

> Does nature nothing for man in manufactures? Are the powers of wind and water, which move our machinery, and assist navigation, nothing? . . . There is not a manufacture which can be mentioned, in which nature does not give her assistance to man, and gives it too, generously and gratuitously.
>
> *(Works I: 76, fn)*

Smith's view was fundamentally mistaken: his false perception of the origin of rent spoilt his theory of value and distribution, his view of the role of the manufacturing sector and his vision of the economy's technological dynamism. In Ricardo's perspective, technological progress consists in increasing humankind's control over the powers of nature. Since all sectors in the economy benefit from these powers, they all benefit from improvements in the knowledge about them.

Ricardo is frequently lumped together with his intellectual sparring partner Thomas Robert Malthus (1766–1834) and qualified as a horseman of the apocalypse. This is utterly misleading. According to Malthus, it was the destiny of the overwhelming part of humankind to live in misery and distress. The Anglican clergyman saw the human race under the spell of two immutable laws – diminishing returns in agriculture and a "law of population". These tend to swiftly annihilate any improvement in productivity due to accidental inventions. Ricardo is strongly opposed to this gloomy perspective, which derived from Malthus's gross underestimation of the pace of technological change. He repeatedly expresses the conviction that "we are yet at a great distance from the end of our resources" (*Works* IV: 34), and that "we are happily yet in the progressive state, and may look forward with confidence for a long course of prosperity" (*Works* VII: 17). The widespread misconception that Ricardo saw the stationary state lurking around the corner mistakes his method of counterfactual reasoning for a proposition concerning the actual trend of the economy. In fact he asks: what would be the course of things in the purely hypothetical case in which capital accumulates and the population grows, but there is *no* further technological progress? In this construed case the general rate of profits is bound to fall, which suffocates economic dynamics. In the actual economy there *is* however technological progress (see, e.g., *Works* I: 120 and V: 125–6).

As early as the second chapter of his *Principles*, Ricardo distinguishes between different forms of technological change – in today's verbiage: (direct) labour saving, capital (i.e. indirect labour) saving and land saving – and stresses that these typically have different effects. He insists, inter alia, that "improvements in agriculture" are of two kinds; "those which increase the productive powers of the land, and those which enable us, by improving our machinery, to obtain its produce with less labour" (*Works* I: 80). He is clear that land-saving progress is detrimental to the interests of the landed gentry, because it tends to diminish their rents. This is also the reason why in economic history there were cases in which landlords tried to prevent the introduction of new methods cultivating the land.[4]

Ricardo stresses repeatedly that different forms of technological progress generally affect different classes of society differently. While in the course of the mechanisation of production workers will typically get "more liberally rewarded" (*Works* I: 48), there are cases of improved machinery in which they will suffer, at least temporarily, from (increased) unemployment and falling wages. In response to the Luddite movement, Ricardo reconsiders the doctrine of automatic compensation of any displacement of workers that his close follower John Ramsay McCulloch had put forward. According to it

> the application of machinery to any branch of production, as should have the effect of saving labour, was a general good, accompanied only with that portion of inconvenience which in most cases attends the removal of capital and labour from one employment to another.
>
> *(Works I: 386)*

In the new chapter "On machinery", added to the third edition of the *Principles* of 1821, Ricardo recants this view and stresses that certain forms of improved machinery are harmful to the interests of workers and give rise to what was later dubbed "technological unemployment". He stresses: "I am convinced, that the substitution of machinery for human labour, is often very injurious to the interests of the class of labourers" (*Works* I: 388).

The case he has in mind can be illustrated by means of Figure 3.2. The $w - r$ relationship T represents the technique prior to the introduction and diffusion of improved machinery and the one indicated by M the technique thereupon. The latter exhibits a higher labour productivity and therefore a higher maximum real wage rate, and a higher ratio of fixed capital (machinery) to gross output, or capital-output ratio, and therefore a lower maximum rate of profits, $R_M < R_T$. In this case, mechanising the economy requests that some labour be withdrawn from the production of consumption goods ("necessaries") and employed in the production of improved machinery. This implies a reduction in consumption output and therefore of wage goods, so that the same number of workers can no longer be employed at the given real wage rate. Some workers will lose their jobs and as a consequence "gross produce" will shrink. Employment will only rise again when economic growth picks up as a consequence of increased profitability and accelerated capital accumulation. This may take some time, and subsequent cases of labour-saving technological change may further retard the adjustment process towards higher levels of employment.

The form of technological progress Ricardo contemplates here involves an increase of "net income" of society, profits, and a decrease of "gross income", which in his labour-value accounting equals the amount of productive labour employed during the year. Therefore, "the same cause which may increase the net revenue of the country, may at the same time render the population redundant, and deteriorate the condition of the labourer" (*Works* I: 388).

4 For statements that the landed gentry frequently tried to suppress agricultural improvements, see, for example, *Works* IV: 41, and more than a century before Ricardo William Petty (1986: 249–50).

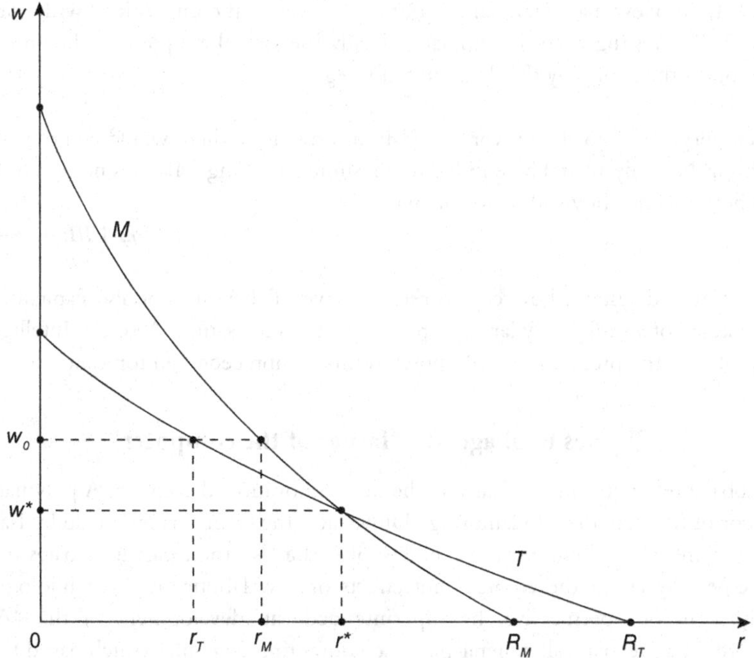

Figure 3.2 Gross-produce-reducing machinery

Two final observations regarding Figure 3.2 are appropriate. First, the fact that the $w-r$ relationships pertaining to techniques T and M intersect at $r = r^*$ (and $w = w^*$), the switch-point between the two, means that the invention of improved machinery does not ipso facto lead to its adoption and thus an innovation: it can profitably be employed if and only if the actual real wage rate in the economy is larger than w^*. Hence, whether a new method of production will be adopted depends not only on its own physical characteristics, but also on the economic characteristics of the world into which it is born. If the wage rate happens to be equal to w_0, technique M will replace technique T. The innovator will initially pocket extra profits, which competition will later erode. For a given and constant real wage rate, w_0, the general rate of profits will rise from r_T to r_M. (If the general rate of profits happens to remain constant, the corresponding real wage rate will be higher.) A rising capital-output ratio and falling maximum rate of profits are obviously compatible with a rising actual rate. (We will come back to this finding in the section dealing with Marx.) Technical change is typically accompanied by a change in relative ("natural") prices. The prices of commodities in whose production technical change occurs will fall relative to the prices of the other commodities. Secondly, the adoption of a new technique may not be profitable at the given prices and wages. However, if for whichever reason these will change, it may become profitable and get adopted. This is the case of what was later called an "induced innovation".

Adam Smith had been perfectly aware of the fact that technological progress typically affects the distribution of power between workers, capitalists and landlords in the "dispute" over the distribution of the product. Ricardo confirms Smith's view by speaking of machines as the "mute agents" of production (*Works* I: 42), which unlike workers do not ask for higher wages or go on strike. In his view machinery and labour are "in constant competition" with one another (*Works* I: 395). Interestingly, he contemplates already the virtual endpoint of the process of the mechanisation of the economy that had only just begun:

> If machinery could do all the work that labour now does, there would be no demand for labour. Nobody would be entitled to consume any thing who was not a capitalist, and who could not buy or hire a machine.
>
> *(Works VIII: 399–400)*

In a fully automated system, how can workers survive, if they are not also capitalists or land owners, possessed of a sufficiently large property? With digitalisation, Artificial Intelligence and smart technologies the problem Ricardo poses assumes unprecedented topicality.

Charles Babbage: the "father of the computer"

Charles Babbage represents in an ideal way the new, innovation-driven era. A polymath, Lucasian Professor of Mathematics at Cambridge University (the chair formerly held by Isaac Newton), inventor, mechanical engineer and philosopher, he also took part in debates in political economy, especially about the employment effects of new labour-saving technologies. Most importantly from our perspective, he superintended the development of the "Analytical Engine" in 1837, a mechanical, general-purpose computing machine, which paved the way to the computer.[5] The device was designed to calculate polynomial functions and solve linear systems of simultaneous equations.[6] Babbage was therefore not without some justification called the "father of the computer" (Copeland 2020).

In 1832 he published *Of the Economy of Machinery and of Manufactures*; 1835 saw the fourth edition. The book was swiftly translated into several languages and was widely read. Interestingly, Babbage stresses that the treatise is "one of the consequences that have resulted from the Calculating-Engine" ([1835] 1986: iii) – a precursor of the Analytical Engine, which provided him with "principles of generalization" applied when systematising the various forms of the mechanical arts. The book is subdivided in two sections: Section I informs about the most important mechanical principles that regulate the application of machinery to arts and manufactures; Section II deals with the political economy of the subject.

Machinery is applied in order "to supersede the skill and power of the human arm" (ibid.: 1). Babbage distinguishes between two types of machines: (1) those that generate power and (2) those that transmit force and execute work (ibid.: 16). In the literature on the Industrial Revolution, James Watt's steam engine and Sir Richard Arkwright's spinning machine are typically taken to have been the major inventions driving the development. While there is no doubt that the two inventions are of great importance, things are invariably more complex. As Babbage emphasises, there is a whole set of technological breakthroughs that are complementary, each enhancing the effectiveness of the other. Better machines generating power are crucial, but so are machines executing work. The perfecting of machine tools admitted a self-reproducing,

5 In this project was also involved the mathematician Ada Lovelace, the daughter of the poet Lord Byron. (The programming language ADA is named after her.) While the machine was never built, Ada Lovelace elaborated the first program for it.

6 Knowing how to solve linear systems of equations would have been of great help to the classical economists, who struggled with the problem that the prices of commodities were interdependent because, as the title of Piero Sraffa's book (1960) indicates, commodities are produced by means of commodities.

quasi-autonomous industrial development. Especially the automatic lathe – the slide rest – allowed the construction of more powerful machines, possessed of higher precision and greater efficacy. Babbage stresses the overwhelming importance of machine tools and working machines for industrial development: the former were a precondition of the latter and enabled their production in large numbers, which accelerated the mechanisation of the industrial sector. The slide rest permitted several machine operations to be performed simultaneously and implied a substantial reduction in production costs.

The use of machinery increases labour productivity and may reverse absolute advantages in trade. A most dramatic case to which Babbage draws the attention is superior British machinery outcompeting coarser Indian machinery and giving the British cotton industry a competitive edge despite Indian wages being only one-seventh of the English: higher productivity may easily more than compensate higher labour costs. Babbage sees three kinds of advantages associated with machinery and manufactures: (1) the addition which they make to human power; (2) the economy they produce of human time; and (3) the conversion of common and worthless substances into valuable products (ibid.: 6). Advantage (1) refers to forces deriving from wind, water and steam that add to human power in operations of lifting, carrying, pulling, pushing and rotating objects. Advantage (2) refers to the economy of human time: it may be seen to embrace, Babbage surmises, all the other advantages under a single heading, that of the genuinely *labour-saving character of innovations*.[7] He illustrates it in terms of numerous examples from all kinds of industries. Advantage (3) is on a different level. It refers to the investigation of the useful properties of things and therefore the rise of the material sciences, and contains as a subcase the use of by-products of production processes. The latter includes the recycling of materials: Babbage mentions inter alia the melting of the steel of obsolete machines and its re-use in building improved ones.

As regards the political economy of machinery and manufactures, Babbage stresses that "Perhaps the most important principle on which the economy of a manufacture depends, is the *division of labour* amongst the persons who perform the work" (ibid.: 169). Against the background of Adam Smith's discussion of the division of labour he observes: "the most important and influential cause has been altogether unnoticed". This cause is at the heart of what became known as the "Babbage principle":

> *That the master manufacturer, by dividing the work to be executed into different processes, each requiring different degrees of skill or of force, can purchase exactly that precise quantity of both which is necessary for each process; whereas if the whole work were executed by one workman, that person must possess sufficient skill to perform the most difficult, and sufficient strength to execute the most laborious, of the operations into which the art is divided.*
>
> *(Ibid.: 175–6; Babbage's emphasis)*[8]

The principle applies to all kinds of work. Its importance is accentuated by three factors. First, according to Babbage the division of labour in the First Machine Age is on the whole upskilling rather than deskilling, as Adam Smith had maintained. He refers to the "increased intelligence

7 A most important advantage that may derive from the use of machinery, Babbage stresses, "is from the check which it affords against inattention, the idleness, or the dishonesty of human agents" (ibid.: 54). Machines can monitor and dictate the speed of the labour process and increase work discipline.

8 The principle played an important role in the literature on "scientific management" and especially in Frederick Winslow Taylor's (1856–1915) designing of factories and allocating workers possessed of different skills to different tasks.

amongst the working classes" (ibid.: 340), which improves their ability to foresee the implications of technological change on their own profession and take precautionary measures that "mitigate the privations which arise from the fluctuations in the value of labour", that is, wages (ibid.: 341).

Secondly, the division of labour "can be applied with equal success to mental as to mechanical operations" and "ensures in both the same economy of time" (ibid.: 191). To this Babbage adds a remark that reads like a clairvoyant vision of a central element of the Second Machine Age. The arrangements, he writes, which "ought to regulate the interior economy of a manufactury, are founded on *principles of deeper root* . . . and are capable of being usefully employed in *preparing the road to some of the sublimest investigations of the human mind*" (ibid.: emphasis added). As a result of going down this road, today the concept of artificial neural networks as a branch of AI come quickly to one's mind.

Third, the division of labour typically necessitates larger factories that allow one to exploit increasing returns to scale because of indivisibilities and fixed cost degression (ibid.: chap. XXII). Babbage is optimistic that lower unit costs will lead to lower prices, although the case he describes is one in which there is a tendency towards a natural monopoly and no reason is given why monopolies would pass on reductions in cost to their customers.

Babbage is on the whole optimistic that the introduction and diffusion of labour-saving machines will not lead to significant or lasting technological unemployment of workers performing "hand-labour". He begins his reasoning by recalling the doctrine of automatic compensation. Surprisingly, he does not enter into a discussion of Ricardo's criticism of the doctrine, which he in all probability knew. He rather insists: "the ultimate consequence of improvements in machinery is almost invariable to cause a greater demand for labour" (ibid.: 335). He mentions, however, frictional losses of employment in case the new labour requires higher skills than the old.

Babbage asks what is less detrimental to workers – a rapid or a slow introduction of machinery that replaces hand-labour. In case the labour needed to operate the machines requests higher skills, quickly worsening living conditions of workers provide a strong incentive to learn the requested new skills (ibid.: 336), so that the adjustment process can be expected to be more intense, but also more rapid. But he admits that there are cases in which the problems will not go away quickly. This is so especially when workers with lower skills than those they replace can operate improved machinery. One is reminded of child labour substituted for adult labour. In this case, "competition amongst the working classes themselves" will intensify (ibid.: 336). Such conflicts will depress wages and worsen working conditions.

A great concern of Babbage's is the view that the interest of workers and that of their employers may be "at variance" (ibid.: 250). That this is repeatedly the case can hardly be denied as the case of the Luddite movement illustrates. Babbage therefore advocates the principle "that every person employed should derive advantage from the success of the whole" (ibid.: 251). The instalment of a scheme of profit sharing would allow workers to benefit from the prosperity of the business without affecting their daily wages. He lays out in some detail such a scheme, but reckons with the stiff opposition from capital owners. Such a scheme would get workers directly interested in the success of their firms and spur their diligence and inventiveness.

Karl Marx: technological progress – societally destructive and creative

Karl Marx comes to political economy via Friedrich Engels's maiden paper "Umrisse zu einer Kritik der Nationalökonomie" (Outlines of a Critique of Political Economy), published in 1844 (Engels 1844). In it Engels polemicises against the "liberal economists", including Smith,

Ricardo, Jean-Baptiste Say and Malthus, whom he accuses of being "hypocrites" by presenting the socio-economic order based on private property, competition and class antagonism as ethically and economically superior to an order based on collective ownership. Receiving a commercial training in one of his father's factories in Manchester, young Engels experiences first-hand the factory system and gets to know major technological and organisational innovations. He is full of praise for the achievements of the natural and technical sciences that generate economically useful knowledge at an accelerating speed and ridicule the predominant concern of economists with diminishing returns in agriculture. Thanks to the sciences, productive powers were growing "immeasurably". Engels refers to leading inventors like Arkwright, Berthollet, Cartwright, Crompton, Davy, Hargreaves, von Liebig and Watt. Their inventions raise the productivity of the soil and of labour "infinitely" and allow "the cause of misery, poverty and crime" to be eliminated by a social transformation that abolishes private property, competition and class conflict. Technological change does not only make a moral society possible, it also makes it necessary. Revolution, Engels is convinced, is imminent.

Marx shares the thrust of Engels's argument and throws himself into political economy. Technological progress is apparently of key importance for an understanding of the law of motion of modern society. Which form of technological progress is congenial to the capitalist mode of production and how does it shape its path? Marx is keen to derive this form from the characteristic features of capitalism and especially from the intensifying antagonism between capital and labour. This necessitates revising Engels's analysis. In particular, the core element of "scientific socialism" has to be a proof that the bias of technological change in capitalism leads inevitably to a falling general rate of profits. This is taken to show that capitalism, just like previous modes of production, is transient and will eventually be replaced by socialism.

Which form of technological progress is "just another expression peculiar to the capitalist mode of production" (1959: 213)? The conflict between capitalists and workers, Marx argues, directs technological change towards a rising "organic composition of capital" (in price terms, a rising capital-output ratio): the ratio of "constant capital" representing "dead labour", embodied in the means of production (tools, machinery, etc.), to "living labour", the labour commanded by means of "variable capital" (wages). Capitalists replace the element they cannot fully control and discipline, the worker, by the element they can, the tool and the machine. However, by leading dead labour into the battle against living labour, capitalists make the source from which surplus value and profits spring, living labour, progressively redundant. Ironically, the replacement of labour power by machine power, while in the short run allowing individual capitalists to reduce costs and increase profits and prevail against their "inimical brothers", their competitors, in the long run it undermines the position of the capitalist class as a whole. Marx apparently counts upon a ruse of history that turns individual rationality into collective irrationality, reflected in a falling general rate of profits.

Interestingly, the form of technological progress that according to Marx seals the fate of capitalism is precisely the one Ricardo had identified as most detrimental, not to the class of capitalists, but to that of workers: see its illustration in Figure 3.2. Marx is in fact keen to put Ricardo's reasoning upside down. The gist of his argument can be put in the following way. The inverse of the organic composition of capital is the ratio of living labour, L, employed during a given year, and the labour "embodied" in the plant and equipment available in the economy, C. Yet L/C in Marx's labour-value accounting gives the maximum rate of profits that can be obtained in a given system of production in the hypothetical case of zero wages. An organic composition that increases over time therefore implies a falling maximum rate of profits. If the organic composition rises without limit, the maximum rate of profits vanishes. In this case the actual rate of profits cannot forever escape a downward trend: capitalism will fall, because profitability falls.

There is no need to enter into a detailed discussion of why Marx's argument lacks cogency (see, for example, Kurz 2018: 1205–18). A few observations must suffice. Empirically, the capital-output ratio, $v = K/Y$, increases in certain periods of time, including roughly the one in which Marx wrote *Capital*, and decreases in others. We do not observe, however, a permanently rising capital-output ratio. And even if it should rise all the time, but happens to be constrained from above, the actual rate of profit need not fall. Theoretically, Marx's argument is incoherent, because it does not adequately take into account the necessary fall in the (values and) prices of the means of production and the wage goods (in terms of some invariant standard), consequent upon the increase in labour productivity. The former limits the rise in the capital-output ratio, whereas the latter implies a rise in the share of profits in national income, $\pi = P/Y$, given the real wage rate. Since $r = \pi/v$, it is not immediately clear what happens to the general rate of profits. However, as Ricardo had already conjectured, and Sraffa (1960) shown convincingly, cost-minimising behaviour cannot lead to a falling general rate of profits, *if* the real wage rate is taken to be constant and there are no scarce natural resources (see Schefold 1976, Kurz and Salvadori 1995 and Kurz 2010). If there are such resources, the general rate of profits is bound to fall, unless there is sufficient technological progress that counteracts the fall, as Ricardo had argued.

Writing several decades after Ricardo, Marx had witnessed a great deal more of the Industrial Revolution unfolding before his eyes. In the British Library he studied books on technology and its history, but appears to have relied on sources that were not up to date, such as Andrew Ure's *Philosophy of Manufactures* (1835) (see Paulinyi 1997 and Illner 2020). While he perused cursorily Babbage's book, he missed some of its main messages, including the one on the extraordinary importance of the slide rest. However, in two regards he went beyond some of the previous contributions to the subject. These concerned, first, a still more refined distinction between different forms of technological progress than Ricardo's, including an anticipation of what later became known as Harrod-neutral progress; secondly, a characterisation of different stages of socio-economic history, each with its own prevalent pattern of technological and structural change: cooperation; division of labour and manufacture; machinery and modern industry (see Marx [1867] 1954: part IV).

An interim period

While technological change, economic development and growth were at the centre of interest of the classical economists and Marx, the rise to predominance of the marginalist school of thought in the period from around 1870 up until 1939 moved them largely to the periphery (see Rostow 1990). Problems of the allocation of given resources to alternative ends within a static framework of the analysis assumed centre stage, whereas the accumulation of resources and their employment with respect to changing ends in a dynamic framework were largely removed from sight. There were, of course, notable exceptions to the rule. These included Schumpeter's theory of innovation, but also Wassily Leontief's input-output analysis to the extent to which it was concerned with economic change and development (see especially Leontief 1928 and 1985). Finally there is Piero Sraffa's (1960) resumption and reformulation of the classical surplus approach to the problem of value and distribution, which triggered a huge theoretical and empirical literature, a branch of which was concerned with studying technological change and economic growth. It deserves to be noted that Schumpeter, Leontief and Sraffa all elaborated their analyses in a critical confrontation with the classical economists (including François Quesnay) and Marx (see Kurz 2011).

Here a brief summary account of some contributions must suffice. The impact of different forms of technical progress on the constraint binding changes in the distributional variables has been investigated in numerous studies either in a macroeconomic framework or using input-output data. See, for example, Foley and Marquetti (1999), who conclude in their macro study that technological progress in several countries up until the mid 1990s showed an increase in labour productivity accompanied by a slight increase in the capital-output ratio; in the period studied they maintain that technical progress exhibited a "Marx-bias". For studies using input-output data in investigating the movement of the $w - r$ relationship of numerous countries over time, see, for example, Mariolis and Tsoulfidis (2016) and Shaikh (2016). Kurz (2020) provides a critical account of the scope and limits of input-output analysis in the regard under consideration.

There have been numerous contributions using dynamic input-output models to study the impact of technological progress on the time profile of labour employment and its skill composition, the growth patterns of different sectors of the economy and structural change and so on. See, in particular, Leontief and Duchin (1986) and Kalmbach and Kurz (1991, 1992). These studies dealt with robotics and automation, which need the computer, but not yet with what characterises the Second Machine Age – self-learning systems and AI.

The new machine age

Innovations are typically seen to be associated with increases in labour productivity and at least temporary monopoly positions of the successful pioneers that allow them to reap potentially huge extra profits. While there is some controversy about the (expected) magnitude of productivity increases, there is widespread agreement that the current wave of technological progress favours the emergence of huge companies possessed of some kind of monopoly power that are difficult to contest and can be expected to persist for a long time. Then there is the question whether digitalisation, AI and so on will revive the spectre of "technological unemployment". In this regard, some major commentators are rather pessimistic. We deal with these issues in turn.

There is still significant uncertainty about the productivity-enhancing potential of machine learning and AI. While some observers see it as the source of an unprecedented acceleration of technological progress and rise in labour productivity, others draw the attention to productivity statistics, which, at least up until now, do not support the occasionally exuberant expectations – on the contrary. There is talk about a new productivity "puzzle" or "paradox", following Robert Solow's 1987 remark that the computer age is to be seen everywhere except in the productivity statistics. In fact, measured productivity growth has slowed significantly in several countries in recent times. Advocates of the view that the new technologies are possessed of a huge transformative capacity consider this perplexing phenomenon as reflecting the proverbial calm before the storm (see, e.g., Brynjolfsson et al. 2019). Staying with the metaphor, the storm will break out as soon as complementary innovations will enhance the effectiveness of the new technologies and the necessary infrastructures and organisations have been implemented. This includes the formation of numerous intangible assets that are difficult to measure but indispensable (new kinds of software, developed datasets, new skills, novel business practices and the like). The fact that the learning capacity of machines is swiftly improving is bound to have a significant impact on the economy and society. Most types of human work require the skills of perception, cognition and communication. In all three regards there have been substantial advances in machine learning: machines are able to recognise objects; they use the data they process cumulatively to make reliable decisions and predictions; they learn the language of humans, communicate with them and amongst themselves; they are able to solve autonomously a growing array of problems. All these developments speak in favour of a surging transformative impact of new technologies that will before long also be reflected in productivity numbers.

There remains the intricate problem of mismeasurement due to the inadequacy of the received statistical practices (data collection, processing and interpretation) in a world of bits and bytes.

With respect to the market power that comes with successful product and process innovations, Adam Smith was of the opinion that it will typically erode unless the bearers of such power manage to ward off the onslaught of competitors or get protected by the government. Is this still a largely valid description of what happens today, or will the new forms of technological change give rise to monopolies that show much greater resilience than in the past? According to some commentators they do: information and communication technologies and especially AI apparently favour the rise of natural monopolies or "superstar firms" (Autor et al. 2020) and a "winner-takes-all dynamics" (Korinek and Stiglitz 2021). The "Big Five" (Apple, Google, Facebook, Amazon, Microsoft) are cases in point with respect to the United States, but a comparable list could be put together with respect to China. There appears to be a wide consensus that monopolies based on machine learning, platforms and networks benefit from new forms of dynamically increasing returns and are possessed of a high degree of stability and perseverance. Once ahead of their competitors, the lead of such firms can be expected to increase over time and eventually force competitors to leave the market. The built-in engine of self-optimising machines in "data capitalism" involves a persistent and in all probability increasing profitability spread between superstar firms, on the one hand, and firms in the competitive segment, on the other.

In terms of a version of the distributional scheme elaborated on earlier, we may illustrate the situation for a given state of technological knowledge and its distribution amongst the two segments in the following way (for a formal analysis, see, inter alia, Yoshihara and Veneziani 2019). The monopolistic segment yields an average rate of profits r_m, the competitive one a rate r_c, where $r_m > r_c > 0$. Assume in addition that only two kinds of labour are employed, skilled and unskilled, and are paid different wage rates w_s and w_u, respectively, where $w_s > w_u > 0$. Then the relationship between the four distributional variables may be written in implicit form as

$$\Phi(w_s, w_u, r_m, r_c) = 0$$

In case the wage differential $\delta > 0$ is given, that is, $w = w_s = (1 + \delta)w_u$, $w_u > 0$, the distribution frontier can be illustrated in three dimensions: w, r_m and r_c. On the further assumption that factors such as joint production and the like are absent, any two of the distributional variables are inversely related to one another for any (non-negative) level of the third variable, given the system of production in use (and assuming that the inequality of r_m and r_c is preserved). That is, $\partial r_m/\partial w < 0$, $\partial r_c/\partial w < 0$ and $\partial r_c/\partial r_m < 0$. In Figure 3.3 the three-dimensional constraint binding changes in the distributional variables A illustrates such a system. For a given wage rate $w = w^0$ and a given profit rate differential $\gamma > 0$, we have $r_m = (1 + \gamma)r_c$. The point $P\ (w^0, r_m^0, r_c^0)$ on the constraint then gives the constellation prior to technological progress.

Assume now that there has been technological progress and that the system is already fully adjusted to a particular phase in the development of the new technological regime. This can obviously be described again in terms of a constraint relating the three distributional variables. In the literature there is almost unanimous agreement that the new technology is labour saving, which means that the point of intersection of the constraint with the w-axis will be further removed from the origin. The conviction is also widespread that it is on the whole resource saving, which means that the other points of intersection are also moved outwards (see, e.g., Korinek and Stiglitz 2021). Some commentators see also a strong capital-saving bias, which would amplify this tendency, while others argue that it is capital-using, which would counteract it. In Figure 3.3, which serves only an illustrative purpose, we have assumed that the

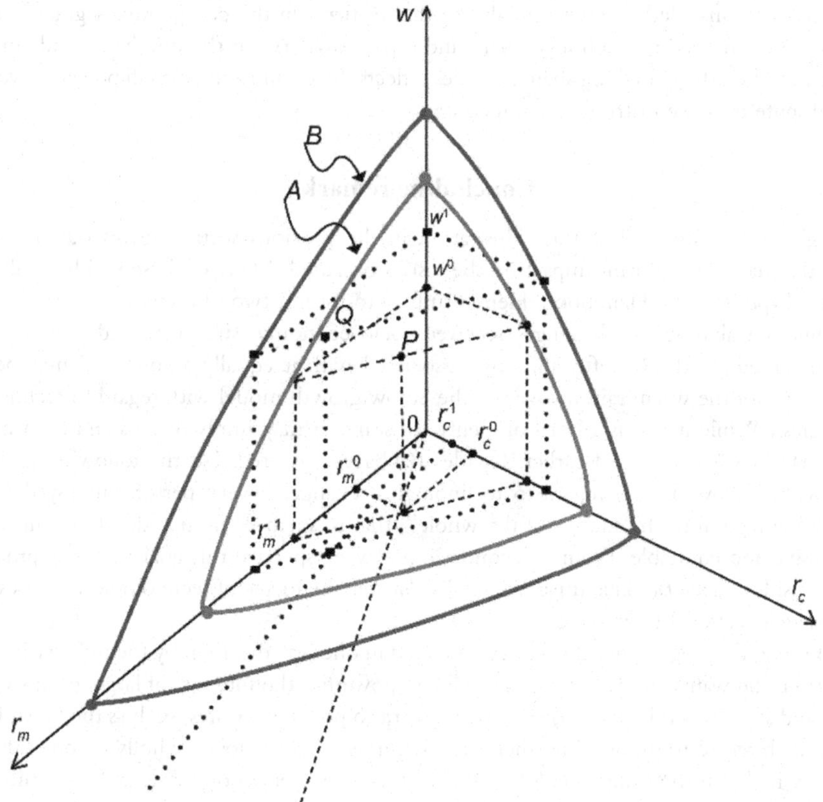

Figure 3.3 Machine learning and superstar firms

constraint corresponding to the new technology is given by *B* and is both labour-, resource- and capital-saving. Therefore, *B* lies above *A* for all non-negative values of the three variables: the new technology is unequivocally superior to the old one and will be adopted by firms irrespective of the wage level.

Obviously, income distribution is bound to change vis-à-vis the new situation. The changes may concern the wage rates for unskilled and skilled labour, and therefore the wage differential δ, and the profit rates of the competitive and the monopolistic segment, and therefore the profitability differential γ. Because of the expected net displacement of unskilled workers their bargaining position will deteriorate, whereas that of skilled workers is seen to increase. Hence the wage differential will become larger. Here we assume that w in the new situation will be larger and equal to w^1. A higher average wage rate is, of course, perfectly compatible with (1) a lower wage rate for unskilled labour and (2) a lower share of wages in national income, provided the average rate increases by less than the average labour productivity. Both phenomena have been observed in several countries in recent times. We assume also that the profitability differential increases, reflecting the growing market power of superstar firms. Point Q (w^1, $r_m^{\ 1}$, r_c^1) on constraint *B* reflects the new situation. While the profit rate in the monopolistic segment has risen, in the competitive segment it has fallen.

Summarising the case illustrated: the new wave of technological progress benefits first and foremost the monopolistic segment and to some extent also skilled workers. It is detrimental to the interests of unskilled workers and also to that of firms in the competitive segment. There is a double segmentation – amongst firms and capital owners, on the one hand, and amongst workers, on the other. A falling share of wages reflects the diminishing overall power of workers in the dispute over the distribution of income.

Concluding remark

The highly schematic and abstract illustration in the previous section leaves out important aspects that have a significant impact on the results obtained. Many of these will be dealt with in other chapters of the Handbook. Here it suffices to list just two of them.

Technological progress challenges received socio-economic structures and conditions. It does not distribute the benefits and costs associated with it equally amongst all members of society, it is not the warm rain invoked in the Solow growth model with regard to technological progress. While it opens a horn of plenty for some, it may imply distress and even misery for others. As we have heard earlier, Charles Babbage suggested that the following principle ought to be followed with regard to technological change: "every person employed should derive advantage from the success of the whole", that is, socio-economic development should be inclusive and equitable. From an economic policy perspective this goal can be approached from two sides – *ante factum* and *post factum*: by *directing* the forms of technological progress and by *distributing* its benefits and costs.

Technological progress has always been directed in one way or other, by the military interests of a nation, the wants and wishes of the rich and powerful, the interests of large groups within society and so on. As Ricardo was clear with regard to policy measures, such as the Corn Laws, but also with regard to innovations, there are cases in which "the loss is wholly on one side, and the gain wholly on the other", and "the loss . . . is far greater on one side, than the gain is on the other" (*Works* I: 336). In such cases, those who gain could not even in principle compensate those who lose. Such innovations ought in general to be avoided. Innovations should rather be steered in directions that are favourable to the living conditions of humankind and other species at large and ward off the dangers threatening their survival.

Schumpeter referred to innovations aptly as processes of "creative destruction". People typically admire the successful innovator for the creative part of his or her activity. They tend to ignore the destructive part of it in terms of lost jobs and skills, bankrupt firms or the disappearance of entire industries. Clearly both parts have to be imputed to the innovator. This leads immediately to the question of how to involve the innovator in compensating the losers. In the absence of ideal risk markets and confronted with fundamental uncertainty, the costs and benefits of innovations and their distribution amongst agents will only become clear as time goes by. This excludes the possibility of compensating the losers ex ante and necessitates compensating them ex post. Hence a scheme of redistributing income and wealth appears to be indispensable and cannot be rejected with reference to the efficient functioning of ideal markets, because in the turmoil caused by the process of creative destruction there are no such markets.

References

Aspromourgos, T. (2009): *The Science of Wealth. Adam Smith and the Framing of Political Economy*. London: Routledge.

Autor, D., Dorn, D., Katz, L. F., Patterson, C. and Van Reenen, J. (2020): The Fall of the Labor Share and the Rise of Superstar firms, *Quarterly Journal of Economics* 135(2), 645–709.

Babbage, C. (1835): *On the Economy of Machinery and Manufactures*. 1st ed. 1832, 4th enlarged edn 1835. London: Charles Knight. Reprint: A. M. Kelley, Fairfield NJ, 1986.

Becker, S. O. and Woessmann, L. (2009): Was Weber Wrong? A Human Capital Theory of Protestant Economic History, *Quarterly Journal of Economics* 124(2), 531–596.

Brynjolfsson, E. and McAfee, A. (2014): *The Second Machine Age. Work, Progress, and Prosperity in a Time of Brilliant Technologies*. New York and London: W. W. Norton & Company.

Brynjolfsson, E., Rock, D. and Syverson, C. (2019): Artificial Intelligence and the Modern Productivity Paradox: A Clash of Expectations and Statistics, in Agrawal, A., Gans, J. and Goldfarb, A. (eds.): *The Economics of Artificial Intelligence: An Agenda*. Chicago: Chicago University Press, 23–57.

Copeland, B. J. (2020): The Modern History of Computing, in *Stanford Encyclopedia of Philosophy* (Winter 2020 Edition), Edward N. Zalta (ed.). Available at: https://plato.stanford.edu/archives/win2020/entries/computing-history/ [30.04.2020]

Eltis, W. (1984): *The Classical Theory of Economic Growth*. London: Palgrave Macmillan.

Engels, F. (1844): Umrisse zu einer Kritik der Nationalökonomie. *Deutsch-Französische Jahrbücher*, 1./2. Lieferung, Paris. Reprinted in MEGA I/2, 467–494.

Foley, D. K. and Marquetti, A. (1999): Productivity, Employment and Growth in European Integration, *Metroeconomica* 50(3), 277–300.

Hume, D. ([1777] 1985): *Essays. Moral, Political, and Literary*. Edited by Miller, E. F. Indianapolis: Liberty Fund.

Illner, E. (2020): Mensch und Maschine. Technikvorstellungen bei Friedrich Engels, Karl Marx und Ernst Knapp, in Illner, E., Frambach, H. and Koubek, N. (eds.): *Friedrich Engels. Das rot-schwarze Chamäleon*. Darmstadt: wbg (Wissenschaftliche Buchgesellschaft), 104–139 and 408–411.

Kalmbach, P. and Kurz, H. D. (1991): Microelectronics and Employment: A Dynamic Input-output Study of the West German Economy, *Structural Change and Economic Dynamics* 2, 171–189.

Kalmbach, P. and Kurz, H. D. (1992): Chips und Jobs, in *Zu den Beschäftigungswirkungen programmgesteuerter Arbeitsmittel*. Marburg: Metropolis.

Kelly, M., Ó Gráda, C. and Mokyr, J. (2020): The Mechanics of the Industrial Revolution, UCD Centre for Economic Research. *Working Paper No. 20/16*.

Korinek, A. and Stiglitz, J. E. (2021): Artificial Intelligence, Globalization, and Strategies for Economic Development. Institute of New Economic Thinking. *Working Paper No. 146*.

Kurz, H. D. (2010): Technical Progress, Capital Accumulation and Income Distribution in Classical Economics: Adam Smith, David Ricardo and Karl Marx, *European Journal of the History of Economic Thought* 17(5), 1183–1222.

Kurz, H. D. (2011): Who is Going to Kiss Sleeping Beauty? On the "Classical" Analytical Origins and Perspectives of Input-output Analysis, *Review of Political Economy* 23(1), 25–47.

Kurz, H. D. (2018): Hin zu Marx und über ihn hinaus: Zum 200. Geburtstag eines deutschen politischen Ökonomen von historischem Rang, *Perspektiven der Wirtschaftspolitik* 13(3), 245–265.

Kurz, H. D. (2020): The Theory of Value and Distribution and the Problem of Capital, *European Journal of Economics and Economic Policies: Intervention* 17(2), 241–264.

Kurz, H. D. and Salvadori, N. (1995): *Theory of Production. A Long-Period Analysis*. Cambridge: Cambridge University Press. Revised paperback edn 1997.

Kurz, H. D., Schütz, M, Strohmaier, R. and Zilian, S. (2018): Riding a New Wave of Innovations. A Long-term View at the Current Process of Creative Destruction, *Wirtschaft und Gesellschaft* 44(4), 545–583.

Leontief, W. (1928): Die Wirtschaft als Kreislauf, *Archiv für Sozialwissenschaft und Sozialpolitik* 60, 577–623. Translation of parts as (1991) The economy as a circular flow. *Structural Change and Economic Dynamics* 2, 177–212.

Leontief, W. (1985): The Choice of Technology, *Scientific American* 252(6), 25–33.

Leontief, W. and Duchin, F. (1986): *The Future Impact of Automation on Workers*. New York: Oxford University Press.

Mariolis, T. and Tsoulfidis, L. (2016): *Modern Classical Economics and Reality: A Spectral Analysis of the Theory of Value and Distribution*. Tokyo: Springer.

Marx, K. ([1867] 1954): *Capital*. Vol. I. London: Lawrence and Wishart.

Marx, K. ([1890] 1959): *Capital*. Vol. III. London: Progress Publishers.

MEGA (1975 ff.): *Karl Marx/Friedrich Engels, Gesamtausgabe*. Edited by the International Marx-Engels Society (IMES). Berlin: De Gruyter.

Mokyr, J. (2017): *A Culture of Growth: The Origins of the Modern Economy*, in *The Graz Schumpeter Lectures*. Princeton and Oxford: Princeton University Press.

Paulinyi, A. (1997): *Karl Marx und die Technik seiner Zeit*, Mannheim: Landesmuseum für Technik und Arbeit, *LTA-Forschung* 26.

Petty, W. (1986): *The Economic Writings of Sir William Petty*. Edited by Hull, C. H. Vols I and II, originally published in 1899. Cambridge: Cambridge University Press. Reprinted in one volume 1986. New York: Augustus M. Kelley.

Pomeranz, K. (2000): *Great Divergence: China, Europe, and the Making of the Modern World Economy*. Princeton, Oxford: Princeton University Press.

Ricardo, D. (1951–1973): *The Works and Correspondence of David Ricardo*, 11 vols. Edited by Piero Sraffa with the collaboration of Maurice H. Dobb. Cambridge: Cambridge University Press. In the text referred to as *Works* followed by the volume number.

Rostow, W. W. (1990): *Theories of Economic Growth from David Hume to the Present*. New York: Oxford University Press.

Schefold, B. (1976): Different Forms of Technical Progress, *Economic Journal* 86, 806–819.

Schumpeter, J. A. (1912): *Theorie der wirtschaftlichen Entwicklung*. Berlin: Duncker & Humblot.

Shaikh, A. (2016): *Capitalism: Competition, Conflict, Crises*. Oxford: Oxford University Press.

Smith, A. ([1776] 1976): An Inquiry into the Nature and Causes of the Wealth of Nations, 2 vols, in Campbell, R. H. and Skinner, A. S. (eds.): *The Glasgow Edition of the Works and Correspondence of Adam Smith*. Oxford: Oxford University Press. In the text referred to as WN.

Sraffa, P. (1960): *Production of Commodities by Means of Commodities*. Cambridge: Cambridge University Press.

Ure, A. (1835): *The Philosophy of Manufactures: Or, an Exposition of the Scientific, Moral, and Commercial Economy of the Factory System of Great-Britain*. London: Charles Knight.

Weber, M. ([1904/1905] 1930): *The Protestant Ethic and the Spirit of Capitalism*. Translation from German by Talcott Parsons. Reprint 2001. London: Routledge.

Weitzman, M. (1998): Recombinant Growth, *Quarterly Journal of Economics* 113, 331–360.

Yoshihara, N. and Veneziani, R. (2019): Technical Progress, Capital Accumulation and Distribution. Queen Mary University of London, School of Economics and Finance. *Working Paper No. 899*.

Young, A. (1928): Increasing Returns and Economic Progress, *Economic Journal* 38(152), 527–542.

4

TOOLS AND CONCEPTS FOR UNDERSTANDING DISRUPTIVE TECHNOLOGICAL CHANGE AFTER SCHUMPETER

Mark Knell and Simone Vannuccini

Introduction

This chapter is about radical innovation and disruptive technological change. Discovering the sources, nature, and mechanisms of disruptive technological change can help to understand the long-run dynamics of innovation and map profound transformation in socio-economic systems. Schumpeter (1912) argued that innovation comes from within the economic system, will happen discontinuously, displace old equilibria, and create radically new conditions over extended periods. Innovative entrepreneurs and enterprises propel the evolution of the socio-economic system by carrying out "'new combinations' of already existing ideas" (Kurz 2012: 872). These result in new products, new methods of production, new markets, new materials and resources, and new forms of business organisation (including new business models and new types of networks). Schumpeter continued this line of inquiry in his book *Business Cycles* where he explains how technology and technological revolutions shape exceptionally long-run economic dynamics. These revolutionary disturbances can be enormous in the sense that they "will disrupt the existing system and enforce a distinct process of adaptation" (Schumpeter 1939: 101). Schumpeter (1942: 83, emphasis in original) then described how the "process of industrial mutation . . . incessantly revolutionizes the economic structure *from within*, incessantly destroying the old one, incessantly creating a new one".

Radical innovation, technological revolutions, and *creative destruction* are tools and concepts essential for understanding disruptive technological change and economic fluctuations from a Schumpeterian perspective. A central idea of Schumpeter (1935: 233) was that innovations are not "evenly distributed through time" but "appear, if at all, discontinuously in groups or swarms", often in close geographic proximity with other innovations. The resulting *swarms* of innovations created new and rapidly growing sectors, led to considerable structural transformation, a narrative Schumpeter (1942) referred to as the "perennial gale of creative destruction". A vast literature appeared from these ideas, which now forms the post-Schumpeterian narrative that overlaps with the evolutionary approach to economics (Nelson et al. 2018), adaptive growth (Metcalfe et al. 2006), and transformational growth (Nell 1998).

 DOI: 10.4324/9780429351921-6

Technological discontinuities were a main concern of Schumpeter, but few economists use the same vocabulary today. This may be because economists typically look for the cumulative effects of the radical innovation, and not the radical innovation itself (Harris 1991). Yet, the very idea that technological discontinuities occur stem from a broader philosophical debate on the nature of change and evolution. Mokyr (1990) compares technological discontinuities (macroinventions) with biological speciation – the radical appearance of a new species that directly contradicts Darwin's claim that *nature does not make leaps*. He characterises technological change by punctuated equilibria, with prolonged periods of incremental improvements succeeded by rapid bursts of radical novelties. From this perspective, technological change displays non-linear dynamics, and to have an idea whether this non-linearity shows some identifiable pattern is important, as that would help the prediction of long-run outcomes.

Arthur (2009: 17) suggested that evolutionary models cannot account for radically novel technologies. Innovative technologies are supposedly "radical" when the functions they perform are novel and when, at the same time, they give rise to entirely new configurations of known functions (Langlois 2017). The invention of the microprocessor in 1971 is a good example of a radical technological discontinuity that became the core technology of the digital revolution. It was truly transformational. Technological discontinuities appeared throughout history, and include the factory system, steam engine, electricity, steel, and the Internet. Social scientists hardly notice them at first, but over time they transformed the way we live and work, enabled new business models, and the creation of new firms and industries. To derive an ideal type of technological discontinuities is not a trivial task, especially because of the recursive and systemic nature of technologies allows for the coexistence of discontinuities and incremental improvements at the same time and at different scales.[1]

Schumpeter (1935, 1939) employed the idea of long waves to describe an Industrial Revolution or Kondratiev long cycle. The Industrial Revolution was a specific kind of technology revolution that began with the introduction of the water-frame spinning machine introduced by Arkwright in the early 1770s. The idea of Industrial Revolution was initially part of the continental economic discourse in the mid-19th century when Friedrich Engels ([1845] 1975: 307) used the term to describe how "the invention of the steam-engine and of machinery for working cotton" gave rise "to an industrial revolution, a revolution which altered the whole civil society". Toynbee (1884) introduced the expression into the English discourse through a series of lectures on the Industrial Revolution, published posthumously.

The literature on the economics of technical change has only recently become aware of the heterogeneity characterising technologies. The idea of general purpose technology (GPT) takes this into account (Helpman 1998). A similar dynamic appears in the business and management literature, that represents major technological innovations as radical, breakthrough, discontinuous, or disruptive innovations. All these constructs rely on similar foundations, but rarely refer to Schumpeter. Business Studies stress the importance of discontinuities, individual inventors, innovation networks, or sudden breakthroughs that allow firms, industries, and markets to appear, transform, or disappear (see Tushman and Anderson 1986, Christensen 1997 and Tushman and O'Reilly 1996). Christensen (1997) pioneered the term "disruptive innovation", which takes a more product and market-oriented perspective on the issue of technological discontinuity. Disruptive innovations need not be radical but can produce a sizeable turbulence in the markets in which they appear. Technological discontinuities underlying the business cycle

1 Following Schumpeter, Arthur (2009: 18–19) argues that the evolution of technology is due to combinatorial evolution or fresh combinations of what already exists. Arthur also highlights the *problem-solving* nature of technological innovation.

often have disruptive effects on the structure of an industry, and revolutionary effects on the fabric of society.

Finally, there is the issue of emerging technologies. Rotolo et al. (2015) define emerging technologies by five attributes: radical novelty, fast growth, coherence, prominent impact, and uncertainty and ambiguity. They use scientometric techniques to map emerging technologies in terms of growth, radical novelty, and coherence, and *future-oriented technology analysis* to select emerging technologies, including their impact, uncertainty, and ambiguity. At the end of this chapter, we consider possible emerging technologies including nanotechnology, biotechnology, quantum computing, and artificial intelligence (AI), and what role they might play in the next technology revolution (or Kondratiev long wave).

In this chapter, we review four concepts essential for understanding radical and disruptive technological change: long waves, techno-economic paradigms, general purpose technologies, and disruptive technologies. This chapter draws its inspiration from Schumpeter's vision of long waves (1939) developed further by Knell (2010, 2013, 2021) and Cantner and Vannuccini (2012). Our focus is on technology, but we rule out a reading based on technological determinism. In line with Heilbroner (1967), we are aware that technology influences but does not exclusively decide the socio-economic order. The following section introduces early long wave theory and the Kondratiev long cycle. The next section summarises Schumpeter's three cycle model, emphasising the Kondratiev long wave. We summarise long waves modelling methods after Schumpeter after that. In the following section, we describe the theory behind the techno-economic paradigm and present a brief historical narrative supporting the theory in the section afterwards. We then cover GPTs and disruptive technologies. After that, we consider the range between discontinuous and continuous innovation. The concluding section covers the emerging technologies.

On the origin of long waves

Schumpeter (1939: 252) regarded Kondratiev as the first economist to provide a "reasonably clear statistical description" of the long wave. Several economists had noticed them previously, namely Hyde Clarke, W.S. Jevons, and Clement Juglar in the 19th century, and John Bates Clark, Knut Wicksell, Vilfredo Pareto, and Mikhail Tugan-Baranovsky in the early 20th century.[2] Using long-term price data, they found climate, demographics, land speculation, sunspots and famines, secular stagnation, and the possibility of crisis were all plausible explanations of the long cycle. Marx and Engels were not direct precursors to long wave theory; though, they had experienced frequent periods of unrest, especially the global crisis of 1857–58, secular stagnation, and extreme inequality in their lifetime.

An active participant in the *Soviet Industrialisation Policy Debate* of the 1920s, Nikolai Kondratiev published several papers on long wave cycle before his death in 1938 during the Great Purge. Two English translations had an immediate impact followed by two all-inclusive versions published posthumously in 1979 and 1984 as well as his collected works in 1999. Inspiration originated in Marshall's approach to competitive equilibrium, Marx's periodic renewal of fixed capital, and Tugan-Baranovsky's endogenous theory of the business cycle.[3] Kondratiev used the

2 See Schumpeter (1954), Barr (1979), van Duijn (1983, chapter 4), Barnett (1998), and Freeman and Louçã (2001, part 1).

3 Tugan-Baranovsky ([1894] 1954) found long-term patterns underlying the normal business cycle, related to prolong periods of rising and falling commodity prices and the expansion and contraction of trade that fluctuates at 50-year intervals (Barnett 2001).

term long wave cycle to stress their cyclical character (Garvy 1943; Louçã 1999), but others named them long waves, long cycles, or long swings (Tylecote 1992). These papers received much criticism in his lifetime for his use of statistical and econometric techniques. Kondratiev (1925, 1928) addressed some of these concerns in a methodological paper on eversible dynamic processes. Trotsky (1973: 273) believed that the downward phase of a Kondratiev cycle implied the decline of capitalism, harshly criticising Kondratiev's interpretation of equilibrium and self-reinforcing mechanisms in economic development.[4]

Kondratiev (1979, 1984) named three phases in the cycle, expansion, stagnation, and recession. Using statistics on price behaviour, including wages, interest rates, raw material (wholesale) prices, production and consumption of coal and pig iron, foreign trade, and bank deposits, Kondratiev dated two long cycles of expansion and decline, one upswing from 1789 to 1814 and a downward swing from 1814 to 1849 (first Industrial Revolution), and a second wave including one upswing from 1849 to 1873 and a second downswing from 1873 to 1896 (age of railroads, steam, and steel). A third wave began in 1896 and ended in 1920 (electricity, chemistry, and motors). He applied the method of least squares (linear trend and third-order non-linear trend) to estimate the theoretical curve.

In his model, Kondratiev distinguished between normal cycles of fixed capital replacement and extraordinary waves of investment in infrastructure, including large plants and buildings, railways, canals, etc., that could last 50 years.[5] His main belief was that long swings reflect spurts in the reinvestment of fixed capital. Kondratiev modified Marx's idea that periodic reinvestment of fixed capital produced business cycles by introducing diverse types of capital (investment) goods with different degrees of durability (Garvy 1943). There is an endogenous mechanism in the theory, but not all variables are endogenous to the system, interpreting them in consequence rather than a cause of its rhythm (Rostow 1975). Radical technological change, wars and revolutions, and changes in gold production and monetary circulation appear exogenous to the long cycle (Tylecote 1992). Yet Kondratiev believed that technological changes happened as a response to endogenous forces within capitalism (Rosenberg and Frischtak 1984). He recognised that profound changes in society take place before the beginning of a long upswing.

Schumpeter on long waves

Schumpeter (1935, 1939) developed a theory of long waves within his three-cycle scheme or model. Bunching or clustering of radical innovations propelled the endogenous long-investment cycle in the model (Silverberg 2003). The model became a "convenient descriptive device" with its main purpose to explain cyclical behaviour over time (Schumpeter (1939: 177). Modelling of this kind built on oscillations generated by the capital goods, investment bunching, and self-reinforcing behaviour. His historical narrative focused on cotton textiles, railroads, steel, automobiles, and electric power in the United States, Britain, and Germany from 1790 to 1920 (McCraw 2007).[6] In this narrative, there was an uneven distribution of innovative activi-

4 On June 23, 1921, Trotsky (1945: 227) wrote, Capitalist equilibrium is an extremely complex phenomenon. Capitalism produces this equilibrium, disrupts it, restores it anew in order to disrupt it anew, concurrently extending the limits of its domination. In the economic sphere these constant disruptions and restorations of the equilibrium take the shape of crises and booms.

5 The length of a cycle depends on the economic lifespan of durable fixed capital (machinery and factories).

6 In *Economic Development* Schumpeter (1934: 229) writes, "every normal boom starts in one or a few branches of industry (railway building, electrical, and chemical industries, and so forth), and that it derives its character from the innovations in the industry where it begins".

ties across countries, industries, and time. The broadly defined technology that characterised the period dated the cycle.

Schumpeter named the three cycles for prominent business-cycle theorists: a short-term 'Kitchin' inventory cycle of about 40 months' duration that captured information asymmetries, a medium-term investment or 'Juglar' cycle that captured the dynamics of both the monetary or financial markets over an eight-to-ten-year period,[7] and a long-term 'Kondratiev' cycle capturing the evolution of major technological innovations over a 50-to-60-year period. Integral to each long wave are shorter cycles: six Juglars to each Kondratiev and three Kitchins to each Juglar. Both the Juglar and Kitchin cycles could generate financial crisis and economic recessions at regular intervals, depending on the timing of each cycle. Schumpeter believed that the demand for capital and credit would generate cyclical fluctuations as Juglar reasoned, but he also understood them to be a consequence of innovative behaviour rather than excessive speculative behaviour.

Business Cycles laid the foundation for a post-Schumpeterian cyclical approach to radical innovation, clusters, long waves, and technological revolutions. Schumpeter (1939: 31) called these the "tools of analysis". The scheme developed by Schumpeter (1935, 1939) provided the building blocks to define, date, and weight inventions and innovations, as well as a sound theoretical basis for modelling long waves with different underlying mechanisms and different periodicity (Kurz et al. 2018).

Long waves after Schumpeter

After a hiatus of academic interest, long waves returned to the theoretical discourse in the 1980s. Mensch (1979) started the discussion by arguing that basic innovations occur in groups or swarms that appear toward the end of the long wave downswing, resembling a "depression-induced accelerator", which then stimulates inventive activity and kick-starts a new long wave. Clark et al. (1981) rejected the depression trigger hypothesis and found weak evidence for a bunching of innovation. They also questioned the timing of the innovations in the long wave, giving diffusion a more prominent role in the cycle. Tinbergen (1981) noticed this idea, suggesting that the system dynamics perspective was ideal for modelling long waves. Forrester (1977) developed a two-sector system dynamics model with capital accumulation and depreciation that simulates successive long waves of overexpansion and collapse in the capital-producing sector. While the long wave tends to be a self-sustaining cycle, innovation and clustering are not fully endogenous in the model. This spawned a discussion and debate on the nature and existence of long waves that included institutes such as Science Policy Research Unit (SPRU) (Freeman et al. 1982; Freeman 1984; Freeman and Perez 1988), International Institute for Applied Systems Analysis (IIASA) (Bianchi 1985; Vasko 1987; Di Matteo et al. 1989; Vasko et al. 1990), and several heterodox economists.

The revival of classical and Marxian political economy gave impetus to long wave theory. Mandel (1964) sought early on to combine Marx's crisis theory with long waves. Mandel (1980) viewed long waves as a unity of endogenous economic mechanisms and exogenous social and political factors which combine to define the path of capitalist development. The falling tendency in the profit rate appeared as a mirror image of a technological revolution and

7 Juglar's analysis of business cycles based on easy credit and speculative behaviour was different from Schumpeter's so-called Juglar cycle, driven by investment and technological innovation (Legrand and Hagemann 2007). Schumpeter described the Juglar cycle as a fixed investment cycles that average a decade.

where the starting point and turning point appear in the cycle.[8] Mandel (1980), as well as Wallerstein (1984) and Shaikh (1992), accepted that profitability drives accumulation and that long cycles are waves of capital accumulation, not only price waves, but innovation appears absent in the approach. Gordon (1991) introduced the social structure of accumulation (SSA) approach to explain tensions between the forces of production, relations of production, institutional arrangements, and the technical superstructure.[9] In the model, the long-swing upturn depends on a complex process of institutional transformation including "exogenous" forces and stage-theoretic considerations.

A conference session on long waves and economic growth organised for the *American Economic Association* in December 1982 was pivotal in the understanding the debate on long waves.[10] In this session, Mansfield (1983: 141) found, "There are enormous difficulties in defining, dating, and weighting inventions and innovations, . . . [which limit] the extent to which they cluster together and whether these clusters (if they exist) occur about forty to sixty years apart". Rosenberg and Frischtak (1983: 150) supposed long waves

> repeat over time, either because the wave-generating factors in the form of innovation clusters are themselves cyclical (or at least recur with a certain regularity), or because there is an endogenous mechanism in the economic system which necessarily and regularly brings a succession of turning points.

In their view, *causality*, *timing*, *economy-wide repercussions*, and *recurrence* are necessary conditions to generate long waves. Gordon et al. (1983) applied the SSA approach to long swing expansion (reproductive cycle) and crisis (nonreproductive cycle) to stimulate profitability, investment, and growth.

There are many different descriptions of long waves in more recent debates. Nelson (2001) called long waves "a succession of eras (Hobsbawm 1968), phases (Maddison 1991), accumulation regimes . . . [and] a succession of socio-institutional systems built on a cluster of technologies". Gordon (1991: 271) named them different tendencies or schools, "bunched investments, traditional Schumpeterian, neo-Schumpeterian, modified Trotskyist, traditional Marxist, world-systems, economic/warfare interactions, and French regulationist approach". Freeman and Louçã (2001: 97) named three modelling methodologies: (1) traditional statistical methods, (2) simulation of formal models, and (3) historical narratives. Kondratiev and Schumpeter used a historical and narrative approach and early econometric techniques. Gordon (1991), Freeman and Louçã (2001), Reati and Toporowski (2009), Silverberg (2003), and Scherrer (2016) review some of the models and tools used to conceptualise long waves.

Freeman and Louçã (2001), Louçã (1997, 2007, 2019), and Metz (2011) raised serious concerns about the statistical and econometric analyses of long waves. Strong and unrealistic

8 Shaikh (1992: 189) argues,

> The falling tendency in the rate of profit chokes off the initial acceleration in the mass of profit, which then decelerates and eventually stagnates. The point of stagnation in the mass of profit, which Marx called the "point of absolute overaccumulation", signals the turning point in the long wave.

Shaikh (2016) suggests wholesale price index divided (normalised) by the price of gold displays long fluctuations.

9 The SSA approach is about the institutional arrangements "that regulate the accumulation process and establish the conditions for profitability" (Gordon et al. 1983). It is closely related to the French Regulation School.

10 Papers and Proceedings of the Ninety-Fifth Annual Meeting of the American Economic Association, New York, New York, December 29–30, 1982. Session on Long Waves and Economic Growth: A Critical Appraisal. In the *American Economic Review*, May 1983, Vol. 73, No. 2.

assumptions underlying the trend-cycle make it difficult to separate growth and fluctuations in the general equilibrium theory. Researchers used a wide variety of statistical tools to study long waves. Some, including Kondratiev (1979, 1984), Kuznets (1940),[11] and van Duijn (1983) used moving-average smoothing techniques and the trend-deviation computation to obtain the trend-cycle.[12] Others, such as Goldstein (1988) and Solomou (1988) used correlation analysis. Another group, including Mandel (1980), Gordon (1991), Kleinknecht (1987), and Kleinknecht et al. (1992), transform the growth rate and then separate the analysis of long fluctuations from the behaviour of shorter ones. A third group, including van Ewijk (1981, 1982), Beenstock (1983), Kleinknecht (1992, part 1), and Reijnders (1990), used spectral and cross-spectral analysis, but found little or no empirical support for the Kondratiev cycle, while Korotayev and Tsirel (2010), Grinin et al. (2016), and Ozouni et al. (2018) found support for it. Van Duijn (1983) found compelling evidence of a global long cycle. Diebolt (2014) detected Kuznets cycles and no Kondratiev cycles. Limitations of spectral methods led Gallegati et al. (2017) and Gallegati (2019) to develop a wavelet method for dating long waves based on the phase difference between a representation in growth rates, which is consistent with the findings of Korotayev and Tsirel (2010). Using a similar method, Staccioli and Virgillito (2020) confirm Freeman and Louçã's (2001) dating of the cycle. Silverberg and Verspagen (2003) suggest non-deterministic processes limit the possibilities for forecasting and the explanatory power of the long wave theory.

Contributions from the field of complexity and chaos theory offer the most sophisticated models of long waves. Goodwin (1990) extended Schumpeter's three-cycle scheme to show how investment in innovative activities produces both the short and the long wave, and implicitly assumes that the two are independent. While there is no obvious solution to the problem, Goodwin integrates effective demand and complexity (self-reinforcing mechanism) to produce non-linear dynamics in his theory. This results in limit cycle behaviour with an unstable Juglar cycle and stable Kondratiev cycle.[13] The self-reinforcing mechanism in the model, however, only oscillates within a given long wave and requires an exogenous shock to move between revolutions. Fatás-Villafranca et al. (2012) develop a Goodwin model that generates endogenous cyclical growth as a disequilibrium process that partly captures the succession of long waves.[14]

Strohmaier et al. (2019) propose a new way to capture the systemic transformations induced by technological discontinuities by tracking the spread of technological change through networked socio-economic structures. Their approach is not explicitly based on long waves but builds on the structuralist-evolutionary framework of Lipsey et al. (2005). A two-mode network connects a variety of indicators to the macro-components of an economy to map its structure. Indicators' values change over time, and the authors use this information to measure economies' evolution given their structure. They interpret the resulting mapping as an absorbing Markov chain. This allows to use measurements from network theory to assess how performance indicators (e.g. productivity or employment growth) "absorb" shocks initiated in any part of the economy.

Freeman's et al. (1982) commentary of Mensch (1979) is the starting point for the post-Schumpeterian theory of long waves. This study led to several publications on long waves,

11 Kuznets (1930) found infrastructural investment cycles, of between 15 and 25 years, before Schumpeter.

12 Evidence of long waves appeared when the average growth rate over a lengthy period was either above or below the long-term trend.

13 See Goodwin (1990: 44) and Goodwin (1989). Goodwin (1987) also introduced non-linear dissipative systems into Schumpeter's long wave, which acts like a 'frequency converter'. The model produces a non-linear, long-term evolution of the economy.

14 Using a Goodwin model, Silverberg and Lehnert (1993) showed that clustering was not necessary to produce long-period cyclical behaviour.

which we discuss in the following two sections. Freeman (1995) adopts a "reasoned history" approach to long waves. Nelson (2020) defined the method to include "qualitative verbal descriptions and narratives, spliced by numbers, and structured in a 'reasoned' but not formal way" (Nelson 2020) and resembles the "appreciative theory" approach of Nelson and Winter (1982). In a reasoned history approach, radical innovation depends on the interplay between institutional set-ups and technical change.

Long waves, technological revolutions, techno-economic paradigms, and great surges of development

One shortcoming in Schumpeter (1939) was his "use of an a-historical theory of entrepreneurship in place of theories of the firm and of technical change" (Perez 2015: 70). This limitation inspired Freeman and Perez to introduce the techno-economic paradigm as a conceptual tool in the 1980s. Perez (1983) and Freeman and Perez (1988) focused on (macro) techno-economic paradigms that connect cyclical theories of technological evolution with path dependence and structural and institutional changes (Freeman 1994). Prior to this, Dosi (1982, 1988) proposed an analogy between Kuhnian scientific paradigms and (micro) technological paradigms, where the evolution of technologies follows identifiable trajectories. The evolution of specific technological trajectories underlying (micro) technological paradigm must be consistent with a (macro) techno-economic paradigm. Perez (1985: 443) defines a techno-economic paradigm as "a set of common-sense guidelines for technological and investment decisions" which guides the actor's choices. Freeman (2019) grounded this idea in the historical approach, which recognises the importance of "institutional diversity and of path dependence for firms, industries, national economies and technologies". By taking this approach, Schumpeter's long waves appear as a succession of techno-economic paradigms. Freeman and Louçã (2001) called them long waves, while Perez (2002, 2015) called them great surges of development.[15]

Perez (1983, 1985) argued that each great surge has its own all-pervasive low-cost key or leading sector(s).[16] These usually include a new source of energy or a new material, technologies, products and processes, and unique organisation (Perez 2010; Knell 2013). A key sector in a techno-economic paradigm must satisfy four conditions: (1) low and declining relative cost; (2) unlimited supply; (3) pervasiveness of low-cost core input; and (4) reduction of the cost and improvement of the quality of labour, capital equipment, and intermediate products. There is an increasing mismatch between the techno-economic and the socio-institutional spheres during the first half of the techno-economic paradigm (Perez 1983, 2002).[17]

Freeman and Perez (1988) associated the Kondratiev long wave with technological revolutions. They use the term *techno-economic paradigm* to describe the systemic transformation following a technological revolution. Techno-economic paradigms introduce a more complex perspective into the long wave debate through the interplay between technology, the economic structure, management, social institutions, and the way people relate to the

15 Perez (2015: 74) stressed that long waves are about statistical testing of long time series, while great surges are a historical narrative.

16 These "key" or "leading" sectors of the economy or "basic industries" as Sraffa (1960) put it, are commodities that enter directly or indirectly in the production of all commodities, which then spread novelty throughout the economy. Pasinetti (1981) extends the multi-sectoral conceptual framework to handle leading sectors and creative destruction.

17 This mismatch also appears between physical and social technology Beinhocker (2006).

technology.[18] Perez (2002: 8) defined a technological revolution as "a powerful and highly visible cluster of new and dynamic technologies, products and industries, capable of bringing about an upheaval in the whole fabric of the economy". It is also "a set of interrelated radical breakthroughs, forming a major constellation of interdependent technologies", namely "a cluster of clusters or a system of systems" (Perez 2010: 189).

Every revolution has its own logic that shapes a specific techno-economic paradigm. This involves not only the appearance and evolution of certain core inputs and physical technologies, but also a change in the way people relate to the technology (Nelson 2005). The microprocessor, for example, not only transformed the way economic system generates economic growth, but also, through its applications, how people interact. The same holds with mechanisation of production, the emergence of the chemical industry, and mass production, just to name a few of the 'big bang' innovation kick-starting technological revolutions. There will be one or more core technologies in each revolution. A core technology may appear as a cluster of interrelated radical breakthroughs and form a core constellation of interdependent technologies that will drive down the cost of production over time. Long-term economic growth is not only causally related to the major technological innovations, but also their diffusion throughout the economy.

Economists already understood the idea of S-shaped (sigmoid function) growth curves (Nelson and Winter 1982). The diffusion of radical technologies appears S-shaped as they spawn novel ideas and new combinations until exhaustion of the paradigm. Perez (2002) interpreted the growth curve as a S-shaped diffusion curve or technology life cycle to describe a technology revolution or great surge. Diffusion involves a learning process in which there are leaders and followers that evolve over the course of the techno-economic paradigm.[19] Once a particular paradigm becomes dominant, the breakthrough technology will define the evolution of the technology life cycle. Figure 4.1 shows the four phases in each techno-economic paradigm: (1) irruption phase, when the latest technology supplants old technology; (2) frenzy phase, or the period of intense exploration; (3) synergy phase, when the new technology is diffused throughout the economy; and (4) maturity phase, as the diffusion process becomes complete. There is also a gestation period at the beginning of the technology life cycle, in which a laboratory-invention phase, with prototypes, patents, and early applications will develop. This phase challenges the dominant technology system, and can last for decades, culminating in the new paradigm.

Freeman and Louçã (2001) name six phases in each long wave, implying two technology systems may coexist, with the laboratory phase foreseeing the new paradigm. In this instance, the diffusion of each revolution may last more than a century. They take a systemic perspective on long waves, focusing on macro processes of technological diffusion. By contrast, Perez (2002, 2015) emphasised the diffusion process of each technological revolution and its transformative effects on all aspects of the economy and society. But Perez then asserts, "great surges are about the rhythm and path of assimilation of each technological revolution and its paradigm". This changed the focus of analysis and Schumpeter's dating of the cycle (called 'great surge' in Table 4.1). The diffusion of the mass production revolution or the digital revolution are good examples.

18 The *techno-economic paradigm* is more than the sum of its parts – it is a coordination mechanism that evolves as a *conditio sine qua non* alongside a new technological regime. This theory goes from single innovations and their technological trajectories and builds up hierarchically to the whole technological 'envelope' of an economy.

19 Diffusion is not prominent in *Economic Development* and only appears as a process by which firms copy, imitate, and gradually improve on the original innovation, or what he described as *induced innovations*.

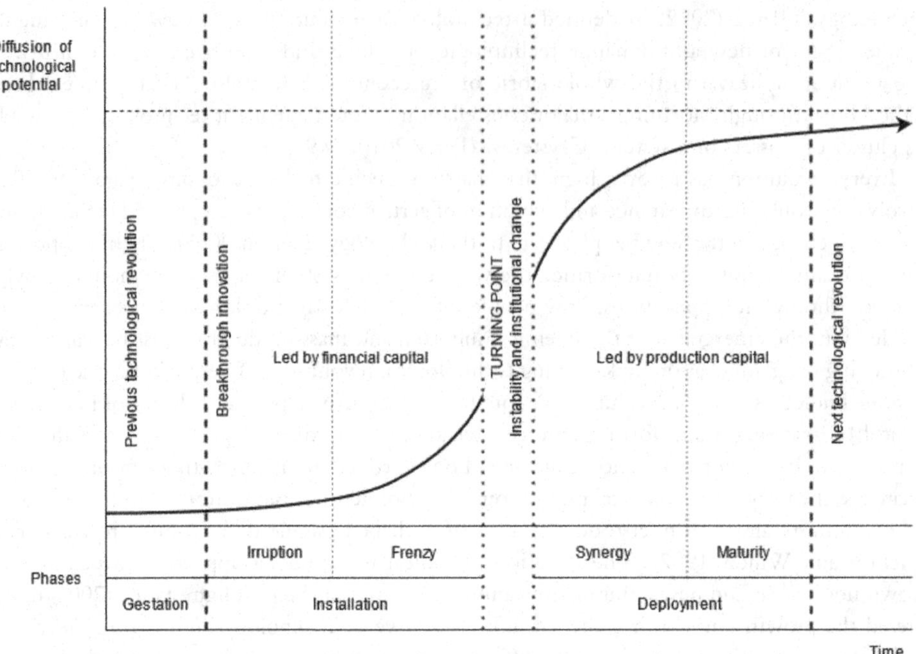

Figure 4.1 Phases of the techno-economic paradigm

Source: Based on Perez (2002: 48)

A novel feature of Perez (2002) is that there is a turning point in the middle of each technological revolution where production capital supplants financial capital.[20] This idea follows from Schumpeter's recognition that the entrepreneur and financier are two independent economic agents that drive the innovation process.[21] Perez then argues that the financier dominates in the first two phases of the cycle, and the entrepreneur dominates in the second two phases.[22] There can be significant and turbulent changes to industrial structure and the regulatory regime during the first phase. Increasing financial instability will culminate in frenzy and one or more major technology (financial) bubbles might appear as the phase progresses. Financial bubbles are most common just before the turning point as confidence in the financial system becomes more turbulent but support for the latest technology gains momentum. Substantial political unrest can also take place near the turning point. The instability should be powerful enough to weaken the power of finance and restore the long-term interests of production capital and the regulatory power of the State (Perez 2013). Dynamic expansion, economies of scale and diffusion are most common during the synergy phase, when producers tend to dominate, and economic growth is

20 At the turning point, stagnation combined with financial fragility will become an acute problem, similar to what happened during the canal mania, when Britain invested huge sums in canal building in the 1790s, during the speculative frenzy or railway mania in the mid-1840s in the UK, the panic of 1893 in the US, the great US stock market crash of 1929, and finally the internet mania of 2000 and financial collapse of 2007 in the US.

21 Schumpeter (1912) emphasised the role of banks and bank credit in financing investment and radical innovation (see also Knell 2015).

22 This suggests that there is a break in the great surge over two periods.

Table 4.1 Summary of great surges (technological revolutions)

	Key energy source	*Initial breakthrough*	*Physical technologies*	*Social technologies*	*Networks*
First (1771)	Water-powered industry	Arkwright's water-frame	Iron, raw cotton, coal, machine tools	Factory system vs. domestic system, shift-work, entrepreneurs, interchangeable parts	Canals, roads, sailing ships, local networks
Second (1829)	Steam-powered industry	Manchester-Liverpool railway, Brunel's steam ship	Iron, coal, railways, steamships	Factory system and capital process, worker loss of control, professional management, joint stock companies	Railways, steam ships, telegraph networks, national power networks
Third (1875)	Electrification of industry	The Carnegie Bessemer steel plant, Edison's electric power	Steel, copper, metal alloys, internal combustion engines, electricity	American system, key machine tools, vertical integration, horizontal integration, management, R&D lab	Transport, communication, and electricity networks
Fourth (1908)	Fossil fuel–driven industry	Ford moving assembly line; Burton oil cracking process	Oil, gas, synthetic materials, automobiles, aerospace	Mass production, managerial control (scientific management), multinational corporations	Production networks, radio, television, motorways, airports
Fifth (1971)	Digitalisation of energy systems	Intel microprocessor	Integrated circuits, computers, robots, software, AI, e-services, satellites	Factory and workplace automation, flexible specialisation, integrative management	Global innovation networks, internet, mobile connectivity networks

Source: Own summary based on Freeman and Perez (1988), Perez (2002), and Tunzelmann (1995)

balanced. In the last phase, complacency appears as the technology reaches maturity and diffuses through the economy, and diminishing returns begins to set in.

From the Industrial Revolution to the digital revolution

Freeman and Perez (1988: 50) give a brief overview of the main characteristic of successive long waves or techno-economic paradigms. In their sketch, there have been five technological revolutions since Richard Arkwright introduced the first mechanical spinning machine, or water-frame, in a Cromford water-powered mill in 1771. Precise dating of the first Industrial Revolution may be arguable, but Schumpeter (1939), Freeman and Louçã (2001), and Perez (2002) recognise Richard Arkwright's patent on the water-frame in 1769. Table 4.1 summarises the five technological revolutions and follows the dating of Perez (2002). The first column names the key energy source in each revolution. Renewable energy sources become crucial for the digitalisation of energy systems in the last half of the fifth techno-economic paradigm. Column two shows the first breakthrough in each revolution. Columns three and four condense the key physical and social technologies in each paradigm, following the broad distinction made by Nelson (2005). Physical technologies include the core technologies and leading (or carrier) industries (e.g. Arthur 2009) and social technologies are the institutional arrangements of the economy, which include "changing modes of division of labour, and new ways or organising and governing work" (Nelson 2005: 195, e.g. Tunzelmann 1995). Finally, the fifth column names the key networks in each revolution.

Innovation in the machine tool industry and in precision engineering gradually led to the design of smaller high-pressure steam engines. The Rainhill Locomotive trials and the opening of the Liverpool-Manchester railway between 1829 and 1831 marked the *Second Technological Revolution*. As resource prices fell, steam-powered mechanisation of industry became the key technology (Chandler 1977). Agglomeration, standardisation, and specialisation accelerated productivity growth during this time. Railways, telegraphs, transatlantic steamship navigation, and a universal postal service also made it possible to network the economy. Timesaving management and specialisation in tasks within the enterprise drove labour productivity during this paradigm.[23]

The *Third Technological Revolution* started the electrification of industry, transport, and a new electrical equipment industry from 1875. Innovative technology also made steel inexpensive. Steel, chemicals, copper, and other metal alloys were key for the development of shipping, railways, bridges, tunnels, and other large civil engineering projects. Thomas Edison created the first professional research and development (R&D) laboratory, with the explicit aim to provide the market with steady stream of new products, which later included the phonograph, microphones, electric lighting, and a system for electrical distribution, as well as other goods (Isreal 1998). This made it easier to recombine old ideas in new ways and to encourage the creation of social networks both within and outside the laboratory.

Declining oil prices combined with the moving assembly line (to produce Ford's Model T) in 1913 made it possible to produce inexpensive motorised vehicles. This application of mass production techniques was essential to the *Fourth Technological Revolution*, including making use of machines and presses to stamp out parts and ensure interchangeability, leading to the relative cheapness of large-scale production and the emergence of mass consumption (Hounshell 1984). Many industries appeared using mass production techniques, including

23 Large joint stock companies appeared during this time, creating the need for principled management. Repression of the working class and inequality was extreme at the time (Tunzelmann 1995).

automotive components, tractors, aircraft, consumer durables, and synthetic materials, and to ensure that mass consumption continued (Freeman and Louçã 2001). After the Second World War, the US and Europe experienced extended periods of high growth, innovative product design, and intensive positive feedback effects. This led to the creation of even larger corporations that needed new ways of managing diverse operations (Chandler 1977).

The digital revolution was bubbling underneath the Fourth Industrial Revolution. Advances in the vacuum-tube (1935) and transistor (1947), followed by many prototypes, patents, and early applications at Bell Labs, anticipated the digital revolution. A vibrant electronic cluster then appeared in the Santa Clara (Silicon) Valley (Lécuyer 2006). Intel announced plans to make the first commercially practical microprocessor (known as the Intel 4004) out of silicon in 1969 (realised in November 1971), the same year the US department of defence installed the first computers on the Advanced Research Projects Agency Network (ARPANET) (which became the internet). This microprocessor made it possible to incorporate all the functions of a central processing unit (CPU) onto a single integrated circuit. The Apollo Guidance Computer was the first silicon integrated circuit-based computer, which led to the microprocessor. Nevertheless, news about Apollo 11 mission completely drowned out the announcement by Intel (Isaacson 2014).

The microprocessor was a revolutionary breakthrough. Advances in photolithography (and planar) techniques increased the number of transistors in an integrated circuit, doubling about every two years (the so-called Moore's Law).[24] Soon afterwards, several new technological trajectories appeared within the digital techno-economic paradigm, which evolved into clusters of new and dynamic technologies, products, and industries that rippled through the entire economy and society. New enterprises appeared and interacted with each other in complex networks. This led to the development of a global digital telecommunications network and the internet, together with electronic mail and other e-services.

We are now in the second phase of the digital revolution (Perez 2013; Reati and Toporowski 2009). The financial collapse of 2008 marked the turning point, as power shifted away from financiers to entrepreneurs and enterprises, and currently we should be in the period of high economic growth, or the golden age. Here diffusion is key, as is the complete digitalisation of everyday objects and activities, or what we might refer to as ubiquitous computing (Weiser 1991). The blending of the digital and real world has opened room to the creation of digital twins, as well as for the appearance of cyber-physical systems (Lombardi and Vannuccini 2021). A wide array of new products dependent on computing devices have appeared in recent decades, including situated robotics, artificial intelligence (AI) technologies, and smart energy networks (Perez 2013; Knell 2013). Modern robots can be autonomous or semi-autonomous, appearing human-like at times, but most often they are just a sophisticated industrial (and numerical control) machine with little guiding intelligence. These included industrial robots, warehouse robots, agricultural robots, autonomous vehicles, caring robots, medical robots, and robots in education. A few robots embed AI in the sense of (statistical) learning algorithms, but these are narrow and limited by the inability to deal with common sense solutions to everyday problems (Mitchell 2019).

General purpose technologies

General purpose technologies (GPTs) are a particular type of radical technology. They are a valuable tool for understanding technological change. Bresnahan and Trajtenberg (1995: 84) define

24 The original Intel 4004 microprocessor had 2,250 transistors at the 10,000 nm node. As of 2021, the Apple M1 Max has 57 billion transistors at the 5 nm node.

GPTs as key technologies that have "the potential for pervasive use in a wide range of sectors and by their technological dynamism". Rosenberg and Trajtenberg (2004) called them "epochal innovations".[25] Analyses of GPTs originally focused on industries' interdependence in innovation,[26] and are narrower in scope than the techno-economic paradigm. However, industrial change at scale results in economic growth. Economists have since embedded GPTs into endogenous growth models, as their diffusion process offered fruitful way to explain growth cycles (Cantner and Vannuccini 2012; Faccarello and Kurz 2016).

Lipsey et al. (2005: 98) defined a GPT as "a single generic technology, recognisable as such over its whole lifetime, that initially has much scope for improvement and eventually comes to be widely used, to have many uses, and to have many spillover effects". GPTs are a distinct class of technologies by virtue of three features characterising them. For Bresnahan and Trajtenberg (1995), these are "pervasiveness, inherent potential for technical improvements, and 'innovational complementarities', giving rise to increasing returns-to-scale" (see also Bresnahan 2010; Cantner and Vannuccini 2012). General applicability suggests that a technology is pervasive in scale and scope, that is, economic actors use the technology in a large array of different economic activities (scope) and in sizeable quantity (scale). Technological dynamism captures the very steep learning curve of these technologies. Complementarities in innovative activity describe the push that a GPT gives to innovation in user sectors (and vice versa): adopting this radical technology makes innovation downstream easier by raising the returns of investing in R&D (downstream R&D and GPT quality are supermodular).

Mainstream economists tried to model long waves (Jovanovic and Rob 1990) and technological revolutions (Shy 1996). But GPTs strip the historical nuances out of long wave theory while keeping the central abstract mechanism stimulating the diffusion of radical technologies. The approach is a response to the critical view of long waves as "ex-post rationalization of historic events" (Lipsey et al. 2005). Dosi and Nuvolari (2020) called them "long waves for neoclassical economists". This explains why few economists link the two concepts, although the mechanisms they aim to capture are similar (Nuvolari 2019). Bresnahan (2010: 763) summarised the goals of GPT research programme:

> One goal lies in growth macroeconomics, to provide an explanation of the close link between whole eras of economic growth and the innovative application of certain technologies, called GPTs, such as the steam engine, the electric motor, or computers. Another goal is in the microeconomics of technical change and proceeds by differentiating between innovations of different types.

From the growth macroeconomics perspective, GPT-driven growth theory builds on the endogenous growth models with expanding product variety (Romer 1990; Grossman and Helpman 1991) and the Schumpeterian growth model of Aghion and Howitt (1992, 2009). GPT models also extend the models of Lucas (1988) and Romer (1986, 1990) by explicitly considering the

25 Rosenberg and Trajtenberg (2004) show how the adoption of the Corliss steam engine produced what they call "a relaxation of geographical constraints". Adopting the new technology, economic activities were not anymore bound to water as source of power and could relocate, boosting agglomeration and urbanisation processes as other social transformations. This historical account of GPTs demonstrates their far-reaching impact and shows the similarities with the literature on techno-economic paradigms.

26 Bresnahan and Trajtenberg (1995) introduced GPTs after they realised that their research agendas on digitalisation in the workplace were dependent on technological advances in and the wide availability of common input technologies (Bresnahan 2010).

interaction between technology breakthroughs and complementary technologies (Faccarello and Kurz 2016).[27] Helpman (2004) suggests a significant difference between GPTs and endogenous growth models is that radical innovation drives the former, while incremental innovation drives the latter.

Long cycles derived from GPT-driven growth models are different from long wave theory. In GPT models, radical innovation can appear deterministically or stochastically but is usually exogenous to the system (Helpman 1998; Cantner and Vannuccini 2012). It also sparks a growth phase in long wave models while GPT-driven growth models initially display a slump (in productivity and output) because of the reallocation of resources needed to guarantee the implementation of the novel technology. The economy shifts to a new, higher, equilibrium growth path only after implementation lags. Shocks produced by radical innovations such as those kick-starting a new technological revolution should appear in the patterns of industrial growth. This produces what Harberger (1998) labelled a synchronous "yeast" process, as compared to the "mushroom" process of idiosyncratic, localised growth. In practice, the elasticity of growth to technological shocks can vary across sectors of economic activities because of a complex array of reasons that range from the very microeconomic (e.g. firms incentives and competition) to meso- and macroeconomic one (e.g. the structure of inter-industry demand of intermediate goods, or retardation effects in industrial growth that are not uniformly distribution across economic activities). Even a synchronous technological shock can produce asynchronous effect, and more problematic, non-revolutionary asynchronous and localised shocks can average out pre-existing sectoral differences. This makes it exceedingly difficult to show empirically the pushes deriving purely by radical innovations such as GPTs (Napoletano et al. 2006).

From the microeconomics of technical change perspective, GPTs set in motion a key mechanism that Bresnahan and Trajtenberg (1995) label "dual inducement". The dual inducement captures the interdependence in innovation incentives between GPT-producing and GPT-using sectors. When a dispersed set of application (downstream) sectors use a GPT as a core technology in their innovative activities, the improvement in the core technology is a function of user sectors' innovation decision. Yet, innovation in GPT-user sectors depends on the GPT-producer's decision about its own innovation intensity. The underlying implication of the incentive structure is not the generation of long wave patterns, but a less ambitious (though more abstract) coordination problem. Adoption may not happen when a radical innovation is not 'good enough'. The coordination failure associated with GPTs provide room for technology policy intervention. Public procurement policies aimed at creating enough critical mass of downstream demand to start a dual inducement feedback loop and, thus, the diffusion of a GPT throughout the economy. Without this, a potential GPT might never become pervasive (Cantner and Vannuccini 2021).

The idea of GPT appears at the intersection of three major trajectories of research in the economics of technological change. It features in models of industrial organisation (e.g. Bresnahan and Trajtenberg 1995), endogenous growth (e.g. Helpman 1998), and economic history (e.g. Rosenberg and Trajtenberg 2004). And it is the cornerstone of recent industrial and innovation policy designs such as smart specialisation (Foray 2015, 2018). The essential hub-and-spoke network structure between a core, radical technology and its user application domains that underly the GPT concept is an agile representation of the interdependencies at work in a system or cluster of technologies (Bresnahan and Yin 2010). Furthermore, the mechanism

27 A GPT appears as a productivity parameter in the model and excludes technology and innovation. Emphasis on new technical (efficient) solutions of the GPT implies a simple view of the complex process of adoption of a radically different technology across heterogeneous agents (see Faccarello and Kurz 2016: 549).

captured by GPTs approximates the dynamics that Rosenberg (1963) explored in his study of the innovation and industry dynamics at work in the machine tool industry at the turn of the 20th century. Rosenberg uncovered the process of "technological convergence", namely the emergence of a new upstream industry as the result of a growing similarity in the knowledge base between different downstream industries. The novel industry that takes over the task of supplying technologies for a large set of user sectors, is likely to rely on a GPT.

Lipsey and colleagues (Lipsey et al. 2005; Bekar et al. 2018) link the GPT approach used in industrial organisation and endogenous growth models with the historical, appreciative, structuralist, and evolutionary approaches. They developed several multi-sectoral growth models that introduce uncertainty on the arrival of a new GPT as well as inter-industry spillovers (Carlaw and Lipsey 2011). Many economists consider the non-linear path of GPTs diffusion and their lengthy implementation lags as an explanation for productivity slowdowns and booms (Jovanovic and Rousseau 2005; Brynjolfsson et al. 2021). Lipsey and co-authors (Carlaw and Lipsey 2002; Bekar et al. 2018) stress, instead, how technological revolutions of sort like those induced by the adoption of GPTs can occur through forms of technological replacement that fundamentally change the fabric, working logic and infrastructure of economic systems without necessarily displaying reverberations on macroeconomic performance such as productivity.

GPT models rely on diverse assumptions about the nature of technological and economic change. As a conceptual tool, GPTs have become the object of several academic disputes as the research programme builds on many conflicting ideas (Bekar et al. 2018). Bresnahan and Yin (2010) expanded the list of GPTs to include the succession of different computer platforms. Nevertheless, economic historians willing to capture the nature of technological revolutions with a more operational conceptual tool have been disappointed by the abstraction. For example, Field (2008) characterised GPTs as a "meme" in the field of economic history, while David and Wright (1999) suggest that the concept "may be getting out of hand" given its ambiguous use. Because of their heuristic power, GPTs can be the bridging concept between economic history and the microeconomics of technological change (Cantner and Vannuccini 2012). For this reason, they remain one of the most important tools for understanding radical technological change after Schumpeter.

Disruptive technologies

Disruptive innovation is a predictive tool that helps understand the impact of innovative technology. It is a theory about why businesses fail, but it does not explain technological change (Lepore 2014). So, it quickly became a catchphrase and rhetorical narrative in the management and strategy literature, often misused and misunderstood. Bower and Christensen (1995) introduced the term *disruptive technology*, which Christensen (1997) developed further in "The Innovator's Dilemma". Christensen and Raynor (2003) then used the term *disruptive innovations* to emphasise the idea that technologies only become disruptive once they are part of a company's business model. Although they changed the terminology, its meaning has remained the same, that is, to explain a paradoxical pattern of industrial evolution where smaller competitors that supply low-range niche products come to replace leading companies. Christensen et al. (2015) defined disruption as "a process whereby a smaller company with fewer resources is able to successfully challenge established incumbent businesses" (see Christensen et al. 2018). The innovator dilemma is the one that hits incumbent firms. By doing precisely what has made them successful, dominant firms overlook market segments from which disruptors can enter and progressively conquer the market.

Christensen and Raynor (2003) found this pattern repeated in industries, such as hard disks, steel mills, and earthmovers and began to investigate whether this pattern appeared because the industry leaders did not innovate or respond to customers' needs. On the contrary, they discovered that being innovative and sensitive to their customers' needs were the reason these industry leaders failed. They found that by catering to the needs of the mainstream customers, the leading companies tend to ignore downmarket opportunities for cheaper and lower performing products. These markets tended to be smaller and less profitable and often left to newcomers which supplied products that were cheap and tawdry, but often offered an added advantage, such as "portability" (for the first transistor-based radios) or smaller size (for successive generations of hard disks).

Gans (2016) distinguishes between demand side and supply side disruption, with the former being the 'classic' process. Disruption can appear on the supply side, when "new innovations are especially difficult for incumbents to adopt and offer competitively because they involve changes in the entire architecture of a product (that is, how components link together) rather than in the components themselves". Gans associates this type of disruption to the theory of architectural innovation (Henderson and Clark 1990). Here a reconfiguration in the structure of interdependence among the components of a product or technology presents added challenges to incumbent companies that can count less on the complementary assets they accumulated before the architectural innovation arrives.

Utterback and Acee (2005) provide a useful extension that links disruptive innovation with Schumpeterian 'destruction'. They place disruption in context by highlighting how the "attack from below" typical of Christensen accounts is just one type of discontinuous innovation. Different drivers can trigger discontinuous change, not just niche demand. Most of the sources of change tend to come from the outside of a given industry. Still, they suggest the true (societal) impact of discontinuous types of technological change may not displace incumbents or established technologies but lie in the market expansion they generate.

Both the supply and demand side theories of disruption offer a useful tool to understand the impact of technological change. However, compared with the grand view of long waves and technological revolutions, or with the economic mechanisms at work in the diffusion of GPTs, disruptive innovation describes a specific path to market dominance and offers insights for strategic management of enterprises. Firms exploiting disruptive innovation to gain market leadership can certainly have a broader societal impact, stretching out until the working of the whole economy (the role played by current technical giants is a case in point). Yet the level of analysis they influence is that of the organisational and market structure, rather than the whole socio-technical "envelope".

Between discontinuous and continuous innovation

Schumpeter and post-Schumpeterians characterise innovations as either being radical or incremental. But many other possibilities lie between these two extremes. This requires defining the extent of innovation according to the extent of change or degree of novelty. Real novelty and increasing complexity were of interest to Schumpeter (1934, 1939), but he never fully embraced these concepts. In a paper written in 1932, Schumpeter (2005: 115) defined novelty as the "transition from one norm of the economic system to another norm in such a way that this transition cannot be decomposed into infinitesimal steps". One consequence of genuine novelty is increasing societal complexity. The most novel innovations appear as discontinuities through time, while the less novel innovations appear as imitations and gradual improvements on the original innovation.

Novelty also is originality, and as such can distinguish between the creation of new knowledge and the adoption and use of existing knowledge. Originality often involves greater risk and uncertainty (Rosenberg 1976), but it can also lead to greater diffusion as other enterprises try to replicate the innovation. Diffusion does not play a vital role in Schumpeter's theory, he does recognise the importance of copying, imitating, and gradually improving on the original "induced" innovation. Some enterprises follow an offensive strategy and introduce highly novel products onto the market, while others follow a defensive and imitative strategy and introduce fewer novel innovations (Freeman and Soete 1997). The commodity, enterprise, and entrepreneur are the units of analysis that are important for Schumpeter, which makes it possible to discuss some of the issues important to the degree of novelty.

Every commercial innovation displays a degree of novelty (and risk). The degree of novelty (degree of newness) is a conceptual and measurement problem. It shows whether the innovation is "new for the industry" or "new for the firm" (reflecting adoption rather than innovation in the narrow sense) or "new to the world". These distinctions suggest that replication matters for the degree of novelty to redesign and adapt existing products (Winter and Szulanski 2001; Becker et al. 2006). Replication is complex, requiring adaptive behaviour as it is not possible to replicate new knowledge, but discover it. The OECD/Eurostat (2018: 78–79) stresses that disruptive or radical innovations are difficult to find and measure. They assert a world-first product innovation implies a qualitatively greater degree of novelty than a new-to-market innovation.[28]

The discrete view on innovation is mostly a representational issue. Various approaches to the study of technology understand innovations exist in a spectrum that has radical and incremental as extremes. For example, the multi-level perspective on socio-technical transitions and the more recent theory of deep transitions (Schot and Kanger 2018) stress how innovation occurs within a multi-layered landscape that goes from single rules to meta-rules, regimes, meta-regimes, and socio-technical systems. The complexity approach to technology highlighted the recursive nature of technology that "builds itself out of itself" (Arthur and Polak 2006). Radical innovation is a complex process that requires a series of smaller complementary advances that makes the final innovation possible, which Silverberg and Verspagen (2005) approximate as a percolation process through a technology lattice. In sum, the radicalness of innovation is a matter of degree rather than clearly separable categories.

Emerging technologies in the sixth techno-economic paradigm

Today it is fashionable to write about Industry 4.0 and the Fourth Industrial Revolution. The idea of Industry 4.0 began as a marketing-style promotion during a press conference at the Hannover Messe, a fair in Germany, in 2011. Institutions took up the concept and two years later, the Federal Academy of Science and Engineering, with the support of the Federal Ministry of Research (BMBF), presented a research agenda and implementation recommendations (Kagermann et al. 2013). This agenda became part of a specialised discourse meant to promote the digitalisation of production and innovation networks that integrate machinery, warehousing systems, and production facilities through cyber-physical systems (Pfeiffer 2017). Since then, the term Industry 4.0 has stimulated lively, ongoing debates about the future of work and society.

The motto of the 2016 World Economic Forum meeting in Davos was "Mastering the Fourth Industrial Revolution". Schwab (2016) associated it with the diffusion of digital technologies and emerging breakthrough technologies such as biotechnology, nanotechnology,

28 Dahlin and Behrens (2005) suggest a measure of radicalness for patents.

artificial intelligence, robotics, 3D printing, the Internet of Things, and quantum computing. He named four revolutions, broadly corresponding to long waves: (1) water- and steam-powered mechanical manufacturing (first two long waves); (2) electricity-powered mass production (second two long waves); (3) digitalisation (fifth long wave); and (4) cyber-physical systems (sixth long wave). However, in this framework there is no discussion of nor reference to long waves, technological revolutions, and techno-economic paradigms. The discussion resembles the crafting of a strategic narrative rather than a solid conceptual scheme to interpret techno-economic dynamics. Schiølin (2020: 4) labels the Fourth Industrial Revolution as an act of "Future essentialism", or the

> discourses, narratives or visions that . . . produce and promote an imaginary of a fixed and scripted, indeed inevitable, future, and that can be desirable if harnessed in an appropriate and timely fashion but is likewise dangerous if humanity fails to grasp its dynamics.

Compared to this, the concepts and tools we highlight in this chapter are not 'imaginaries', but patterns derived from historical or analytical analysis.

Similar discourses on digitalisation and automation technologies have taken place over the past decade, often re-igniting decades-old debates between techno-optimists and techno-pessimists (Mokyr et al. 2015). Our concepts and tools will help to understand these discussions and debates. For example, Rifkin (2011) claims "green" technologies, such as wind, solar photovoltaic, hydropower, and wave and tidal power, will encourage a new Industrial Revolution. Brynjolfsson and McAfee (2014) believe we are in the second machine age where productivity gains from intangible assets and digitalisation of the economy intensified changes in skills, organisations, and institutions. They believe earlier technology-driven revolutions, such as the steam engine and electrification, took longer to diffuse through the economy. Schot and Kanger (2018) grouped the first four technological revolutions into what they call "industrial modernity", or the first deep transition. Their framework builds on the concepts we describe, as for them Deep Transitions describe "an evolutionary multi-level model of surges". Labour and machines were complementary during the first transition. The second deep transition began with the digital revolution and involves the automation of cognitive tasks (especially symbol processing) that make humans and software-driven machines substitutes, rather than complements. Brynjolfsson et al. (2021) suggested that productivity growth may have "paused" in the first decades of the digital revolution but might surge again as the novel wave of digital technologies deploy its effect, resembling the way ICTs pushed productivity growth in the nineties after the puzzling slowdown in the eighties. These narratives confirm Perez's (2013) idea that we are now in the golden age of the digital revolution.

Can we envisage the future and the sixth technological revolution through the tools and concepts we outlined? Possible emerging technologies include nanotechnology, biotechnology, quantum computing, and AI. This is where the physical, digital, and biological worlds could converge in what Lombardi and Vannuccini (2021) call a "cyber-physical universe". Inspiration for the idea originates in part in Richard Feynman's 1959 lecture, "There's Plenty of Room at the Bottom". Here Feynman described a process in which scientists would be able to manipulate and control individual atoms and DNA molecules. The idea of the transistor and the microprocessor triggered the process of miniaturisation. Feynman anticipated nanoscience and nanotechnology, which influenced other science fields, including chemistry, biology, physics, materials science, and engineering. We might call Feynman's idea the Quantum Technological Revolution, which would blend the digital revolution and its informational basis and the sixth technological revolution, much like Brynjolfsson and McAfee (2014) and Schot and Kanger (2018) might suggest.

It is possible that two or three independent technological systems could converge into one system (or a system of systems, to use Perez's expression), which would then trigger the explosive take-off into the sixth technological revolution (Roco and Bainbridge 2003; Knell 2010). Atoms, DNA, bits, and synapses will supply the basic elements and foundational tools that will make it possible to integrate several emerging technologies, including nanotechnology, biotechnology, quantum biology, information technology, and the latest cognitive technologies, into multifunctional systems. It is a long way from Kondratiev and Schumpeter's contributions, but the tools we illustrated remain the cornerstone of a useful theory of innovation and change even in our current complex socio-technical landscape.

References

Aghion, P. and Howitt, P. (1992): A Model of Growth Through Creative Destruction, *Econometrica* 60(2), 323–351.

Aghion, P. and Howitt, P. (2009): *The Economics of Growth*. Cambridge, MA: MIT press.

Arthur, W. B. (2009): *The Nature of Technology: What it is and How it Evolves*. London: Allen Lane.

Arthur, W. B. and Polak, W. (2006): The Evolution of Technology within a Simple Computer Model, *Complexity* 11(5), 23–31.

Barnett, V. (1998): *Kondratiev and the Dynamics of Economic Development: Long Cycles and Industrial Growth in Historical Context*. London: Palgrave Macmillan.

Barnett, V. (2001): Tugan-Baranovsky as a Pioneer of Trade Cycle Analysis, *Journal of the History of Economic Thought* 23(4), 443–466.

Barr, K. (1979): Long Waves: A Selective, Annotated Bibliography, *Review* 2, 675–718.

Becker, M. C., Knudsen, T. and March, J. G. (2006): Schumpeter, Winter, and the Sources of Novelty, *Industry and Corporate Change* 15, 353–371.

Beenstock, M. (1983): *The World Economy in Transition*. London: George Allen and Unwin.

Beinhocker, E. D. (2006): *The Origin of Wealth: Evolution, Complexity, and the Radical Remaking of Economics*. Boston, MA: Harvard Business School Press.

Bekar, C., Carlaw, K. and Lipsey, R. (2018): General Purpose Technologies in Theory, Application and Controversy: A Review, *Journal of Evolutionary Economics* 28(5), 1005–1033.

Bianchi, G., Bruckmann, G., Delbeke, J. And Vasko, T. (eds.) (1985): *Long Waves, Depression, and Innovation: Implications for National and Regional Economic Policy*. IIASA Collaborative Paper, Laxenburg, Austria.

Bower, J. and Christensen, C. M. (1995): Disruptive Technologies: Catching the Wave, *Harvard Business Review* 73(1), 43–53.

Bresnahan, T. F. (2010): General Purpose Technologies, in Hall, B. and Rosenberg, N. (eds.): *Handbook of the Economics of Innovation*. Amsterdam: North-Holland, 761–791.

Bresnahan, T. F. and Trajtenberg, M. (1995): General Purpose Technologies 'Engines of Growth'? *Journal of Econometrics* 65(1), 83–108.

Bresnahan, T. F. and Yin, P. L. (2010): Reallocating Innovative Resources around Growth Bottlenecks, *Industrial and Corporate Change* 19(5), 1589–1627.

Brynjolfsson, E. and McAfee, A. (2014): *The Second Machine Age: Work, Progress and Prosperity in a Time of Brilliant Technologies*. New York: W.W. Norton.

Brynjolfsson, E., Rock, D. and Syverson, C. (2021): The Productivity J-curve: How Intangibles Complement General Purpose Technologies, *American Economic Journal: Macroeconomics* 13(1), 333–372.

Cantner, U. and Vannuccini, S. (2012): A New View of General Purpose Technologies, in Wagner, A. and Heilemann, U. (eds.): *Empirische Makroökonomik und mehr: Festschrift zum 80. Geburtstag von Karl Heinrich Oppenländer*. Berlin: De Gruyter, 71–96.

Cantner, U. and Vannuccini, S. (2021): Pervasive Technologies and Industrial Linkages: Modeling Acquired Purposes. *Structural Change and Economic Dynamics* 56, 386–399. https://doi.org/10.1016/j.strueco.2017.11.002

Carlaw, K. I. and Lipsey, R. G. (2002): Externalities, Technological Complementarities and Sustained Economic Growth, *Research Policy* 31(8–9), 1305–1315.

Carlaw, K. I. and Lipsey, R. G. (2011): Sustained Endogenous Growth Driven by Structured and Evolving General Purpose Technologies, *Journal of Evolutionary Economics* 21(4), 563–593.

Chandler, A. D. Jr (1977): *The Visible Hand*. Cambridge, MA: Harvard University Press.

Christensen, C. M. (1997): *The Innovator's Dilemma: When New Technologies cause Great Firms to Fail*. Boston, MA: Harvard Business School Press.

Christensen, C. M., McDonald, R. Altman, E. J. and Palmer, J. E. (2018): Disruptive Innovation: An Intellectual History and Directions for Future Research, *Journal of Management Studies* 55(7), 1053–1078.

Christensen, C. M. and Raynor, M. E. (2003): *The Innovator's Solution*. Boston, MA: Harvard Business School Press.

Christensen, C. M., Raynor, M. E. and Rory McDonald, R. (2015): What is Disruptive Innovation? *Harvard Business Review*, 44–53.

Clark, J., Freeman, C. and Soete, L. (1981): Long Waves, Inventions, and Innovations, *Futures* 4, 308–322.

Dahlin, K. B. and Behrens, D. M. (2005): When is an Invention Really Radical? Defining and Measuring Technological Radicalness, *Research Policy* 34, 717–737.

David, P. A. and Wright, G. (1999): General Purpose Technologies and Surges in Productivity: Historical Reflections on the Future of the ICT Revolution, in David, P. A. and Thomas, M. (eds.): *The Economic Future in Historical Perspective*. Oxford: Oxford University Press.

Diebolt, C. (2014): Kuznets versus Kondratieff: An Essay in Historical Macroeconometrics, *Cahiers d'économie politique* 67, 81–117.

Di Matteo, M., Goodwin, R. M. and Vercelli, A. (eds.) (1989): *Technological and Social Factors in Long Term Fluctuations*. Berlin, New York: Springer.

Dosi, G. (1982): Technological Paradigms and Technological Trajectories, *Research Policy* 11, 147–162.

Dosi, G. (1988): Sources, Procedures, and Microeconomic Effects of Innovation, *Journal of Economic Literature*, 1120–1171.

Dosi, G. and Nuvolari, A. (2020): Introduction: Chris Freeman's "History, Co-Evolution and Economic Growth": An Affectionate Reappraisal, *Industrial and Corporate Change* 29(4), 1021–1034.

Engels, F. ([1845] 1975): The Condition of the Working-Class in England, in Marx, K. and Engels, F. (eds.): *Collected Works*. Vol. 4. London: Lawrence and Wishart.

Faccarello, G. and Kurz, H. D. (eds.) (2016): *Handbook on the History of Economic Analysis, Vol. III: Developments in Major Fields of Economics*. Cheltenham, UK: Edward Elgar.

Fatás-Villafranca, F., Jarne, G. and Sánchez-Chóliz, J. (2012): Innovation, Cycles and Growth, *Journal Evolutionary Economics* 22, 207–233.

Field, A. J. (2008): *Does Economic History Need GPTs?* SSRN Library 1275023.

Foray, D. (2015): *Smart Specialisation: Opportunities and Challenges for Regional Innovation Policy*. London: Routledge.

Foray, D. (2018): Smart Specialization Strategies as a Case of Mission-oriented policy – A Case Study on the Emergence of New Policy Practices, *Industrial and Corporate Change* 27(5), 817–832.

Forrester, J. W. (1977): Growth Cycles, *De Economist* 125, 525–543.

Freeman, C. (ed.) (1984): *Long Waves in the World Economy*. London: Frances Pinter.

Freeman, C. (1994): The Economics of Technical Change, *Cambridge Journal of Economics*, 18(5), 463–514.

Freeman, C. [1995] (2019): History, Co-Evolution and Economic Growth, *Industrial and Corporate Change* 28(1), 1–44.

Freeman, C., Clark, J. and Soete, L. (1982): *Unemployment and Technical Innovation: A Study of Long Waves and Economic Development*. London: Frances Pinter.

Freeman, C. and Louçã, F. (2001): *As Time Goes By. From the Industrial Revolution to the Information Revolution*. Oxford: Oxford University Press.

Freeman, C. and Perez, C. (1988): Structural Crisis of Adjustment, Business Cycles and Investment Behaviour, in Dosi, G. et al. (eds.): *Technical Change and Economic Theory*. London: Pinter, 38–66.

Freeman, C. and Soete, L. (1997): *The Economics of Industrial Innovation*. 3rd ed. London: Pinter.

Gallegati, M. (2019): A System for Dating Long Wave Phases in Economic Development, *Journal of Evolutionary Economics* 29(3), 803–822.

Gallegati, M., Gallegati, M., Ramsey, J. B. and Semmler, W. (2017): Long Waves in Prices: New Evidence from Wavelet Analysis, *Cliometrica* 11, 127–151.

Gans, J. (2016): *The Disruption Dilemma*. Cambridge, MA: MIT Press.

Garvy, G. (1943): Kondratiev's Theory of Long Cycles, *Review of Economic Statistics* 25, 203–220.

Goldstein, J. (1988): *Long Cycles: Prosperity and War in the Modern Age*. London and New Haven: Yale University Press.

Goodwin, R. M. (1987): The Economy as an Evolutionary Pulsator. In *The Long-Wave Debate* (pp. 27–34). Heidelberg: Springer.

Goodwin, R. M. (1989): Towards a Theory of Long Waves, in Di Matteo, M., et al. (eds.): *Technological and Social Factors in Long Term Fluctuations*. Heidelberg: Springer.

Goodwin, R. M. (1990): *Chaotic Economic Dynamics*. Oxford: Oxford University Press.

Gordon, D. M. (1991): Inside and Outside the Long Swing: The Endogeneity/Exogeneity Debate and the Social Structures of Accumulation Approach, *Review* (Fernand Braudel Center) 14(2), 263–312.

Gordon, D. M. Weisskopf, T. W. and Bowles, S. (1983): Long Swings and the Nonreproductive Cycle, *American Economic Review* 73(2), 146–151.

Grinin, L. Korotayev, A. and Tausch, A. et al. (2016): *Economic Cycles, Crises, and the Global Periphery*. Cham, Switzerland: Springer.

Grossman, G. and Helpman, E. (1991): *Innovation and Growth in the Global Economy*. Cambridge: MIT Press.

Harberger, A. C. (1998): A Vision of the Growth Process, *The American Economic Review* 88(1), 1–32.

Harris, D. J. (1991): Equilibrium and Stability in Classical Theory, in Nell, E. J. and Semmler, W. (eds.): *Nicholas Kaldor and Mainstream Economics*. London: Palgrave Macmillan.

Heilbroner, R. L. (1967): Do Machines Make History? *Technology and Culture* 8(3), 335–345.

Helpman, E. (ed.) (1998): *General Purpose Technologies and Economic Growth*. Cambridge, MA: MIT Press.

Helpman, E. (2004): *The Mystery of Economic Growth*. Cambridge, MA: Harvard University Press.

Henderson, R. M. and Clark, K. B. (1990): Architectural Innovation: The Reconfiguration of Existing Product Technologies and the Failure of Established Firms, *Administrative Science Quarterly* 35, 9–30.

Hobsbawm, E. (1968): *Industry and Empire: From 1750 to the Present Day*. London: Pelican.

Hounshell, D. A. (1984): *From the American System to Mass Production, 1800–1932: The Development of Manufacturing Technology in the United States*. Baltimore: Johns Hopkins University Press.

Isaacson, W. (2014): *The Innovators: How a Group of Hackers, Geniuses, and Geeks Created the Digital Revolution*. New York: Simon and Schuster.

Isreal, P. (1998): *Edison: A Life of Invention*. New York: John Wiley.

Jovanovic, B. and Rob, R. (1990): Long Waves and Short Waves: Growth Through Intensive and Extensive Search, *Econometrica* 58(6), 1391–1409

Jovanovic, B. and Rousseau, P. L. (2005): General Purpose Technologies, in Aghion, P. and Durlauf, S. (eds.): *Handbook of Economic Growth*. Vol. 1. Amsterdam: Elsevier, 1181–1224.

Kagermann, H., Wahlster, W. and Helbig, J. (2013): *Recommendations for Implementing the Strategic Initiative INDUSTRIE 4.0*, final report of the Industrie 4.0 Working Group, April 2013.

Kleinknecht, A. (1987): *Innovation Patterns in Crisis and Prosperity: Schumpeter's Long Cycle Reconsidered*. London: Macmillan.

Kleinknecht, A., Mandel, E. and Wallerstein, I. (eds.) (1992): *New Findings in Long-Wave Research*. London: Macmillan.

Knell, M. (2010): Nanotechnology and the Sixth Technological Revolution, in Cozzens, S. and Wetmore, J. (eds.): *Nanotechnology and the Challenges of Equity, Equality and Development*. Heidelberg: Springer.

Knell, M. (2013): Multi-source Energy Networks and the ICT Revolution, *European Planning* 21, 1838–1852.

Knell, M. (2015): Schumpeter, Minsky and the Financial Instability Hypothesis, *Journal of Evolutionary Economics* 25, 293–310.

Knell, M. (2021): The Digital Revolution and Digitalized Network Society, *Review of Evolutionary Political Economy*, forthcoming.

Kondratiev, N. D. (1925): The Static and the Dynamic View of Economics, *The Quarterly Journal of Economics* 39(4), 575–583.

Kondratiev, N. D. ([1925] (1979): Long Business Cycles, *Voprosy kon'iunktury* I, 28–79. Translated as: Long Waves of Economic Life, in *Review (Fernand Braudel Center)* 2, 519–562. Shorter version in *The Review of Economic Statistics* 17(6), 105–115 (1935).

Kondratiev, N. D. ([1926] 1984): *The Long Wave Cycle*. New York: Richardson and Snyder.

Kondratiev, N. D. (1928): *Bol'shie tsikly kon'iunktury (Major Economic Cycles)*. Moscow: Krasnaia Presnia.

Korotayev, A. V. and Tsirel, S. V. (2010): A Spectral Analysis of World GDP Dynamics: Kondratiev Waves, Kuznets Swings, Juglar and Kitchin Cycles in Global Economic Development, and the 2008–2009 Economic Crisis, *Structure and Dynamics* 4, 3–57.

Kurz, H. D. (2012): Schumpeter's New Combinations, *Journal of Evolutionary Economics* 22(5), 871–899.

Kurz, H. D., Schütz, M., Strohmaier, R. and Zilian, S. (2018): Riding a New Wave of Innovations: A Long-term View at the Current Process of Creative Destruction, *Wirtschaft und Gesellschaft* 44(4), 545–583.

Kuznets, S. S. (1930): *Secular Movements in Production and Prices: Their Nature and their Bearing Upon Cyclical Fluctuations*. Boston: Houghton Mifflin Company.

Kuznets, S. S. (1940): Schumpeter's Business Cycles, *The American Economic Review* 30, 257–271.

Langlois, R. N. (2017): Fission, Forking, and Fine Tuning, *Journal of Institutional Economics*, 1–22.

Lécuyer, C. (2006): *Making Silicon Valley: Innovation and the Growth of High Tech, 1930–1970*. Cambridge: MIT Press.

Legrand, M. D.-P. and Hagemann, H. (2007): Business Cycles in Juglar and Schumpeter, *The History of Economic Thought* 49(1), 1–18.

Lepore, J. (2014): The Disruption Machine, *The New Yorker* 23, 30–36.

Lipsey, R. G., Carlaw, K. I. and Bekar, C. T. (2005): *Economic Transformations: General Purpose Technologies and Long Term Economic Growth*. Oxford: Oxford University Press.

Lombardi, M. and Vannuccini, S. (2021): *A Paradigm Shift for Decision-making in an Era of Deep and Extended Changes*. SSRN Library 3807948.

Louçã, F. (1997): *Turbulence in Economies: An Evolutionary Appraisal of Cycles and Complexity in Historical Processes*. Cheltenham: Edward Elgar.

Louçã, F. (1999): Nikolai Kondratiev and the Early Consensus and Dissensions about History and Statistics, *History of Political Economy* 31, 169–205.

Louçã, F. (2007): *The Years of High Econometrics: A Short History of the Generation that Reinvented Economics*. New York: Routledge.

Louçã, F. (2019): As Time Went By – Long Waves in the Light of Evolving Evolutionary Economics. *SPRU Working Paper Series* SWPS 2019-05.

Lucas, R. E. (1988): On the Mechanics of Economic Development, *Journal of Monetary Economics* 22(1), 3–42.

Maddison, A. (1991): *Dynamic Forces in Capitalist Development*. Oxford: Oxford University Press.

Mandel, E. (1964): The Economics of Neo-Capitalism, *Socialist Register* 1, 56–67.

Mandel, E. (1980): *Long Waves of Capitalist Development: The Marxist Interpretation*. New York: Cambridge University Press.

Mansfield, E. (1983): Long Waves and Technological Innovation, *American Economic Review* 73(2), 146–151.

McCraw, T. K. (2007): *Prophet of innovation: Joseph Schumpeter and Creative Destruction*. Cambridge, MA: Harvard University Press.

Mensch, G. ([1975] 1979): *Stalemate in Technology. Innovations Overcome Depression*. Cambridge, MA: Ballinger.

Metcalfe, J. S., Foster, J. and Ramlogan, R. (2006): Adaptive Economic Growth, *Cambridge Journal of Economics* 30(1), 7–32.

Metz, R. (2011): Do Kondratiev Waves Exist? How Time Series Techniques Can Help to Solve the Problem, *Cliometrica* 5, 205–238.

Mitchell, M. (2019): *Artificial Intelligence: A Guide for Thinking Humans*. London: Pelican Books.

Mokyr, J. (1990): Punctuated Equilibria and Technological Progress, *The American Economic Review* 80(2), 350–354.

Mokyr, J., Vickers, C. and Ziebarth, N. L. (2015): The History of Technological Anxiety and the Future of Economic Growth: Is this Time Different? *Journal of Economic Perspectives* 29(3), 31–50.

Napoletano, M., Roventini, A. and Sapio, S. (2006): Modelling Smooth and Uneven Cross-sectoral Growth Patterns: An Identification Problem, *Economic Bulletin* 15(6), 1–8.

Nell, E. J. (1998): *The General Theory of Transformational Growth*. Cambridge: Cambridge University Press.

Nelson, R. R. (2001): Foreword, in Freeman, C. and Louçã, F. (2001): *As Time Goes By. From the Industrial Revolution to the Information Revolution*. Oxford: Oxford University Press.

Nelson, R. R. (2005): *Technology, Institutions, and Economic Growth*. Cambridge, MA: Harvard University Press.

Nelson, R. R. (2020): On "Reasoned History", *Industrial and Corporate Change* 29/4, 1035–1036.

Nelson, R. R., Dosi, G., Helfat, C., Pyka, A., Saviotti, P., Lee, K., . . . Winter, S. (2018): *Modern Evolutionary Economics: An Overview*. Cambridge: Cambridge University Press. doi:10.1017/9781108661928

Nelson, R. R. and Winter, S. G. (1982): *An Evolutionary Theory of Economic Change*. Cambridge, MA: Harvard University Press.

Nuvolari, A. (2019): Understanding Successive Industrial Revolutions: A "Development Block" Approach, *Environmental Innovation and Societal Transitions* 32, 33–44.

OECD/Eurostat (2018): *Oslo Manual 2018: Guidelines for Collecting, Reporting and Using Data on Innovation*, 4th Edition. Paris: OECD Publishing.

Ozouni, E., Katrakylidis, C. and Zarotiadis, G. (2018): Technology Evolution and Long Waves: Investigating their Relation with Spectral and Cross-spectral Analysis, *Journal of Applied Economics* 21(1), 160–174.

Pasinetti, L. L. (1981): *Structural Change and Economic Growth: A Theoretical Essay on the Dynamics of the Wealth of Nations*. Cambridge: Cambridge University Press.

Perez, C. (1983): Structural Change and Assimilation of New Technologies in the Economic and Social Systems, *Futures* 15(5), 357–375.

Perez, C. (1985): Microelectronics, Long Waves and World Structural Change: New Perspectives for Developing Countries, *World Development* 13(3), 441–463.

Perez, C. (2002): *Technological Revolutions and Finance Capital: The Dynamics of Bubbles and GOLDEN Ages*. Cheltenham: Edward Elgar.

Perez, C. (2010): Technological Revolutions and Techno-economic Paradigms, *Cambridge Journal of Economics* 34, 185–202

Perez, C. (2013): Unleashing a Golden Age after the Financial Collapse: Drawing Lessons from History, *Environmental Innovation and Societal Transitions* 6, 9–23.

Perez, C. (2015): From Long Waves to Great Surges, *European Journal of Economic Social Systems* 27(1–2), 69–79.

Pfeiffer, S. (2017): The Vision of Industrie 4.0 in the Making – a Case of Future Told, Tamed, and Traded, *Nanoethics* 11, 107–121.

Reati, A. and Toporowski, J. (2009): An Economic Policy for the Fifth Long Wave, *PSL Quarterly Review* 62, 147–190.

Reijnders, J. P. G. (1990): *Long Waves in Economic Development*. Aldershot: Edward Elgar.

Rifkin, J. (2011): *The Third Industrial Revolution: How Lateral Power Is Transforming Energy, the Economy, and the World*. New York: Palgrave Macmillan.

Roco, M. C. and Bainbridge, W. S. (2003): *Converging Technologies for Improving Human Performance: Nanotechnology, Biotechnology, Information Technology, and Cognitive Science*. Dordrecht: Springer.

Romer, P. (1986): Increasing Returns and Long-run Growth, *Journal of Political Economy* 94(5), 1002–1037.

Romer, P. (1990): Endogenous Technological Change, *Journal of Political Economy* 98(5), 71–102.

Rosenberg, N. (1963): Technological Change in the Machine Tool Industry, 1840–1910, *Journal of Economic History* 23(4), 414–443.

Rosenberg, N. (1976): On Technological Expectations, *The Economic Journal* 86(343), 523–535.

Rosenberg, N. and Frischtak, C. R. (1983): Long Waves and Economic Growth: A Critical Appraisal, *American Economic Review* 73(2), 146–151.

Rosenberg, N. and Frischtak, C. R. (1984): Technological Innovation and Long Waves, *Cambridge Journal of Economics* 8(1), 7–24.

Rosenberg, N. and Trajtenberg, M. (2004): A General-Purpose Technology at Work: The Corliss Steam Engine in the Late-Nineteenth-Century United States, *Journal of Economic History* 64(01), 61–99.

Rostow, W. W. (1975): Kondratiev, Schumpeter, and Kuznets: Trend Periods Revisited, *The Journal of Economic History* 35(4), 719–753.

Rotolo, D., Hicks, D. and Martin, B. R. (2015): What is an Emerging Technology? *Research Policy* 44, 1827–1843

Scherrer, W. (2016): Technology and Socio-Economic Development in the Long Run, in Hilpert, U. (ed.): *Routledge Handbook of Politics and Technology*. New York: Routledge, 50–64.

Schiølin, K. (2020): Revolutionary Dreams: Future Essentialism and the Sociotechnical Imaginary of the Fourth Industrial Revolution in Denmark, *Social Studies of Science* 50(4), 542–566.

Schot, J. and Kanger, L. (2018): Deep Transitions: Emergence, Acceleration, Stabilization, and Directionality, *Research Policy* 47, 1045–1059.

Schumpeter, J. A. ([1912] 1934): *Theorie der wirtschaftlichen Entwicklung*, Leipzig, Verlag von Duncker & Humblot. Third edition translated by R. Opie as *The Theory of Economic Development*. Cambridge, MA: Harvard University Press.

Schumpeter, J. A. ([1932] 2005): *Development*, translated by M.C. Becker and T. Knudsen, *Journal of Economic Literature* XLIII, 108–120.

Schumpeter, J. A. (1935): The Analysis of Economic Change, *The Review of Economics and Statistics* 17(4), 2–10

Schumpeter, J. A. (1939): *Business Cycles: A Theoretical, Historical, and Statistical Analysis of the Capitalist Process*. New York: McGraw-Hill.

Schumpeter, J. A. (1942): *Capitalism, Socialism and Democracy*. New York: Harper and Brothers.

Schumpeter, J. A. (1954): *History of Economic Analysis*. New York: Oxford University Press.

Schwab, K. (2016): *The Fourth Industrial Revolution*. London, Penguin.
Shaikh, A. (1992): The Falling Rate of Profit as the Cause of Long Waves: Theory and Empirical Evidence, in Klienknecht, A. et al. (eds.): *New Findings in Long-Wave Research*. London: Palgrave Macmillan, 174–202.
Shaikh, A. M. (2016): *Capitalism. Competition, Conflict, Crises*. Oxford: Oxford University Press.
Shy, O. (1996): Technology Revolutions in the Presence of Network Externalities. *International Journal of Industrial Organization* 14(6), 785–800.
Silverberg, G. (2003): Long Waves: Conceptual, Empirical and Modelling Issues, in Hanuch, H. and Pyka, A. (eds.): *Elgar Companion to neo-Schumpeterian Economics*. Aldershot: Edward Elgar.
Silverberg, G. and Lehnert, D. (1993): Long Waves and Evolutionary Chaos in a Simple Schumpeterian Model of Embodied Technical Change, *Structural Change and Economic Dynamics* 4, 9–37.
Silverberg, G. and Verspagen, B. (2003): Breaking the Waves: A Poisson Regression Approach to Schumpeterian Clustering of Basic Innovations, *Cambridge Journal of Economics* 27(5), 671–693.
Silverberg, G. and Verspagen, B. (2005): A Percolation Model of Innovation in Complex Technology Spaces, *Journal of Economic Dynamics and Control* 29(1–2), 225–244.
Solomou, S. (1988): *Phases of Economic Growth, 1850–1973: Kondratiev Waves and Kuznets Swings*. Cambridge: Cambridge University Press.
Sraffa, P. (1960): *Production of Commodities By Means of Commodities. Prelude to a Critique of Economic Theory*. Cambridge: Cambridge University Press.
Staccioli, J. and Virgillito, M. E. (2020): Back to the Past: The Historical Roots of Labour-saving Automation, *GLO Discussion Paper, No. 721*, Global Labour Organization (GLO), Essen.
Strohmaier, R., Schuetz, M. and Vannuccini, S. (2019): A Systemic Perspective on Socioeconomic Transformation in the Digital Age, *Journal of Industrial and Business Economics* 46(3), 361–378.
Tinbergen, J. (1981): Kondratiev Cycles and So-called Long Waves: The Early Research, *Futures*, 258–263.
Toynbee, A. (1884): *Lectures on the Industrial Revolution in England*. London: Rivingtons.
Trotsky, L. ([1945] 1972): *The First Five Years of the Communist International*, 2 vols. New York: Pathfinder.
Trotsky, L. (1973): The Curve of Capitalist Development, in *Problems of Everyday Life and Other Writings on Culture and Science*. New York: Monad, 273–280.
Tugan-Baranovsky, M. ([1894] 1954): Periodic Industrial Crises, *Annals of the Ukrainian Academy of Arts and Sciences in the United States*, 745–802.
Tunzelmann, G. N. von (1995): *Technology and Industrial Progress: The Foundations of Economic Growth*. Aldershot: Edward Elgar.
Tushman, M. L. and Anderson, P. (1986): Technological Discontinuities and Organizational Environments, *Administrative Science Quarterly* 31(3), 439–465.
Tushman, M. L. and O'Reilly, C. A. (1996): The Ambidextrous Organization: Managing Evolutionary and Revolutionary Change, *California Management Review* 38, 1–23.
Tylecote, A. (1992): *The Long Wave in the World Economy: The Current Crisis in Historical Perspective*. London: Routledge.
Utterback, J. M. and Acee, H. J. (2005): Disruptive Technologies: An Expanded View, *International Journal of Innovation Management* 9(01), 1–17.
van Duijn, J. J. (1983): *The Long Wave in Economic Life*. London: Allen and Unwin.
van Ewijk, C. (1981): The Long Wave – A Real Phenomenon? *De Economist* 129(3), 324–372.
van Ewijk, C. (1982): A Spectral Analysis of the Kondratieff Cycle, *Kyklos* 35, 468–499.
Vasko, T. (ed.) (1987): *The Long-Wave Debate; Selected Papers from an IIASA International Meeting*. Weimar, GDR, June 10–14, 1985. Heidelberg: Springer.
Vasko, T., Ayres, R. U. and Fontvielle, L. (eds.) (1990): *Long Cycles and Long Waves*. Berlin: Springer.
Wallerstein, I. (1984): Long Waves as Capitalist Process, *Review* 7(4), 559–575.
Weiser, M. (1991): The Computer for the 21st Century, *Scientific American* 265(3), 94–104 (September).
Winter, S. G. and Szulanski, G. (2001): Replication as Strategy, *Organizational Science* 12, 730–743.

5

ENTREPRENEURSHIP AND INDUSTRIAL ORGANISATION

Uwe Cantner and Thomas Grebel

Introduction

The ongoing digital transformation manifests itself in a number of new smart technologies. These include robotics, artificial intelligence, autonomous systems, platforms, and new digital business models. They create new market opportunities. Both supply-side and demand-side factors determine the respective development paths, the type of players involved, but also the form and intensity of innovation activity. In these processes of market or industry development, the new smart technologies mentioned are still at an early stage and their further development is still open. Nevertheless, possible development patterns and their driving factors can be indicated by considering the literature on industrial organisation (IO) and industrial dynamics. And that is precisely the aim of this chapter, with a special focus on the role of entrepreneurs and start-ups, their opportunities for, and obstacles to success.

Overview

The following sections address the role of entrepreneurship against the background of industrial organisation. The impetus of economic change roots in the entrepreneurial behaviour of individual actors. They shape the industry structure and its evolution. Entrepreneurial behaviour, or more general, entrepreneurship constitutes an endogenous element in industrial dynamics. In a reciprocal fashion, entrepreneurial behaviour is also influenced by industry pattern and related characteristics. It is two sides of the same coin. Entrepreneurship and industrial organisation theory should therefore be close friends.

Whereas the focus in the entrepreneurship literature is rather put on understanding the spur and nature of the successful entrepreneur or successful entrepreneurial behaviour, the literature on industrial organisation brings the role of industry structure, actor conduct, and actor performance to the fore.

One of the early predecessors in entrepreneurship research, Joseph Alois Schumpeter, conceived the entrepreneur as the main driver of economic change.[1] In his seminal contribution *The*

1 There are many concepts of the entrepreneur. Hébert and Link (1988) identify 12 themes, each of which emphasising a specific aspect such as the entrepreneur as a risk taker, a financer, a decision-maker, or most prominently the entrepreneur as an innovator. Compare also Grebel (2004).

DOI: 10.4324/9780429351921-7

Theory of Economic Development, Schumpeter (1912/1934) portrays the entrepreneur as a leader who is willing to seize business opportunities and carry out *new combinations*. In other words, Schumpeter describes the entrepreneur as innovator that destroys existing (industry) structures while creating new ones. In later contributions, Schumpeter (1912/1934 1942) also realised that it is larger firms playing a major role in boosting technical change through innovation.

At the onset of industrial organisation studies (Bain 1956, 1959), the vantage point was to investigate the impact of industry structures on the economic performance of actors. Out of the early structural approaches, the Structure-Conduct-Performance (SCP) approach arose, in which the interplay between the conduct of firms (i.e. profit maximisation) and the industry leads to a specific performance. Looking through the lens of the SCP approach, Schumpeter (1912/1934, 1942) offers an implicit story of industry dynamics with his entrepreneurship concept. The heroic Schumpeter entrepreneur (Schumpeter 1912/1934) who is not satisfied with the economic performance of competitive markets (structure) tries to innovate (conduct) in order to destroy existing industry structures and to achieve monopoly rents (performance) at best.

Both strands ofliterature, entrepreneurship and industrial organisation, have meanwhile developed a large body of complementary theories and empirical studies, some of which we briefly sketch in this chapter. We start with the classical discussion about the impact of market structure on innovative activities motivated by the Structure-Conduct-Performance approach in the 1970s, followed by tournament and non-tournament models in "New Industrial Economics" – including an empirical perspective – that endogenised the link between entrepreneurship/innovation and market structure in the 1980s. Recognising that innovation is tantamount to saying that entrepreneurship always includes the production of knowledge/technology. The successive section is dedicated to the impact of appropriability conditions within markets and industries on entrepreneurial behaviour. Thereafter, the dynamic perspective of entrepreneurship and industry life cycle theory is addressed, followed by a section on the concept of entrepreneurial ecosystems emphasising the fact that entrepreneurship and thus endogenous change of industry structures is a restless, perpetual economic process. Demand-side-related aspects to industrial dynamics and their relation to entrepreneurship and entrepreneurial opportunities are addressed in the section before the conclusion; in the latter, we suggest some research desiderata.

Market structure, innovation, and entrepreneurship – from Structure-Conduct-Performance to New Industrial Economics

As pointed out, combining one of the most basic research questions in entrepreneurship research – "What drives entrepreneurial/innovative behaviour?" – with the industrial organisation perspective – "What role does market structure play in economic performance?" – the immediate complementary research question is: "What drives entrepreneurial/innovative behaviour, given a certain market structure?" Early studies on innovation activities and entrepreneurship focused on the relationship of structural conditions of industries and innovation-related profit opportunities. Consequently, absolute firm size on the one hand and industry structure and hence relative firm size, on the other, were dominating the discussion at the time: what firm size or market structure, respectively, is more conducive to entrepreneurial activity?

Structure-Conduct-Performance approach

A first systematic treatment of the role of industry structure and economic performance was offered by the Structure-Conduct-Performance approach (SCP) (Bain 1956, 1959). The gist of its argument was that firm behaviour (conduct) and the economic/industrial structure they are embedded in determine overall economic performance. For simplicity, market structure, be it competitive or monopolistic, or anything in between, was interpreted as the structural dimension of an industry, and the conduct of firms as profit-maximising behaviour.

Evidently, market structure and relative firm size go hand in hand. In general, a monopolist is large, while, under perfect competition, individual firms tend to be negligible in size. By and large, this is tantamount to comparing entrepreneurs (small firms) with large firms, a distinction that has become one of the core topics in economics of innovation.

First and foremost, it was Schumpeter himself pointing at this distinction. In contrast to *The Theory of Economic Development* (1912) where he identified the entrepreneur as the main driver of innovative and economic change, Schumpeter proposed large firms with deliberate research efforts as *deus ex machina* in his 1942 book *Socialism, Capitalism and Democracy*. This distinction entailed a long-lasting discussion on context factors to what extent they favour either small or large firms as main innovators. From an industrial organisation viewpoint, this boils down to the question about the most conducive market structure: monopoly, perfect competition, or oligopoly. Hence, the so-called Neo-Schumpeterian hypotheses were born, claiming that large firms or more monopolistic market structures respectively favour innovation activities. The difference between the two viewpoints is simply absolute versus relative firm size.

In a seminal paper, Arrow (1962) argues by constructing a model with exogenous innovation that firms in a competitive environment (i.e. small firms) should have a higher incentive to innovate than a monopolist has. A competitive small firm, earning zero extra profits before innovation, will earn a monopoly profit afterwards. So would a monopolist, but the monopolist would lose the profit he or she already had before innovation. Thus, the notion of the *cannibalisation effect* was born with the conclusion that a small, entrepreneurial firm should have a comparatively higher incentive to innovate than an established large firm. Reformulated in terms of market structure, a more competitive market structure should appear to be more conducive to innovation activities than a monopolistic one. Hence, the questions related to Neo-Schumpeterian hypotheses were answered in favour of small firms and competitive market structures.

New Industrial Economics

Arrow's model suffers from several drawbacks; for example, it conceives innovative behaviour in isolation and treats market structure as exogenous. In the aftermath, new types of models were developed that take a game-theoretic approach, of which some deal with dynamic aspects. This strand of research has become known as the so-called "New Industrial Economics" approach (NIE). It introduced a further impetus to the discussion of innovation and entrepreneurship in an industry context.

Non-tournament models and endogenous market structure

Most prominent and path breaking in this context is the work by Dasgupta and Stiglitz (1982). They design a model in which both R&D investment of firms and market structure are endogenous. They criticise Arrow's (1962) assumption that innovation would not affect the market structure and innovations were exogenous. The model by Dasgupta-Stiglitz has become a standard modelling element (Levin and Reiss 1984; Klepper 1996), which meanwhile is labelled a non-tournament model.

The basic assumption is that all firms invest in R&D and decide on the quantity of a (homogenous) output in a profit-maximising way. According to the results of the Dasgupta-Stiglitz model, it rather is an oligopolistic market structure with a few firms that is conducive to innovation than a competitive market structure, or small entrepreneurial firms. In the vein of Dasgupta-Stiglitz's model, a firm's willingness to invest into R&D, proxied by the share of R&D expenditures in sales, the so-called R&D intensity, should be lower for small firms. This, in turn, contrasts the conclusions of Arrow (1962); Dasgupta and Stiglitz (1982) propose the opposite: for firms to strive for innovation, the market structure needs to differ from perfect competition.

Though Dasgupta and Stiglitz formulate their theory as a static model, it suggests an implicit dynamic driven by profit opportunities. New entrepreneurial firms enter the market as long as profit opportunities are positive. With an increasing number of firms entering the market, these opportunities eventually vanish and so does firm entry. In other words, market structure, interpreted as the final number of firms in the market, is endogenous in the model. By construction, the equilibrium number of firms in the market is always greater than one and smaller than infinity, which leads to the conclusion that an oligopolistic market structure is most conducive to innovation.

Tournament models and "the winner takes it all"

In contrast to non-tournament models – where all firms innovate successfully – tournament models, or synonymously, patent races assume that "the winner takes it all". Some of these models feature innovation as a sequence of innovations. In each innovation round a monopolist is determined, be that an incumbent or an entering (entrepreneurial) firm. Consequently, market or industry development reflect a sequence of monopolies. With respect to the incentive to innovate, as Reinganum (1983, 1984) as well as Harris and Vickers (1985) show, firms or entrepreneurs entering the market have a higher incentive to invest in R&D than incumbent firms do because the latter always run the risk to cannibalise existing profits – entrepreneurs do not. This leads to the equivalent conclusion that a potential entrepreneurial entrant should be more willing to invest in innovation than an established firm does.

Empirics

Looking at the empirical evidence, the validation of the Neo-Schumpeterian hypotheses (see Cohen 2010 for a splendid overview) shows a mixed picture. At the most, a robust result is that in absolute terms large firms spend more on R&D than small ones (entrepreneurs); in relative terms, however, no significant difference is detectable. Only with respect to R&D productivity, measured by the number of patents or innovations per unit of R&D, small firms appear superior.

However, many empirical studies identify an inverted u-shape relationship between market concentration and R&D investment, its explanatory power remains weak. Including further explanatory variables such as technological opportunity and profit appropriation conditions reduce the significance of market concentration as an explanatory variable drastically. Only 1.5–4% of innovativeness can be attributed to market concentration, whereas industry type, demand structure, technological opportunities, and appropriability conditions explain up to 32–50%.

Overall, the evidence for large firms and concentrated markets sustaining and promoting higher innovative activities is weak and hence the validation of the Neo-Schumpeter hypotheses fails. Remaining variables such as technological opportunities, average cost of R&D projects, continuity and predictability of technological development, or learning curve effects appear to have more explanatory power. Further neglected aspects in empirical studies are demand-side effects such as product differentiation, or the intensity of price competition. Moreover, the majority of empirical studies use cross-sectional data, which do not allow any conclusions on causality and the dynamic nature of innovation within industries.

Contextual factors concerning technology, demand, or other economic aspects need not only be considered each distinctively, they also have to be put into the timely context of an industry's development. As the following sections show, with changing industry conditions, the type of innovation and the incentive for start-ups to enter the market change, too.

Innovation structures, appropriation regimes, and entrepreneurship – industry/technology classifications

In view of the mixed empirical evidence concerning the Neo-Schumpeterian hypotheses and the observation that it is industry- or technology-specific factors rather than market concentration explaining entrepreneurial activities and innovation within a market/industry, a new stream of literature started to develop in the 1980s. The primary objective was to understand industries in terms of their ability to foster and promote certain types of innovation activities (i.e. internal, external, and cooperative R&D; product or process innovation). Thus, the prospects on innovation rents and the role of appropriation conditions came to the fore. In these mostly empirical studies, the focus was put on an inductively generated, broad account of empirical regularities, which eventually led to the so-called OACK approach (opportunities, appropriability conditions, cumulativeness of technological change, and the specific nature of knowledge).

In various studies, Malerba and Orsenigo (1993, 1997) attempted to classify industries into Schumpeter Mark I and Schumpeter Mark II industries, the former reflecting the ideas of Schumpeter (1912), the latter the elaborated concept of Schumpeter (1942). Using patent data to measure innovativeness allows them to do so, where mainly IPC classes represent technologies, sometimes coinciding with industries. Malerba and Orsenigo discover, in what they call Schumpeter Mark I technologies (i.e. technologies in which small firms and innovative start-ups dominate innovative activities), that the concentration of innovators is quite low and innovator rankings regularly change over time. Since new or entering innovators constantly destroy or at least damage the economic value generated by previous innovators, Malerba and Orsenigo conclude that *creative destruction* is prevalent in these technologies. The story in Schumpeter Mark II technologies looks different. Large firms dominate innovative activities. Innovator concentration is high and, on the contrary, the ranking of innovators remains stable over time. Additionally, the manner of innovation is different. Innovators primarily introduce improvements of previous innovations. For this reason, Malerba and Orsenigo conceive innovation as a process of "creative accumulation" in those technologies. With little surprise, this also has an impact on firm entry and exit. Technologies of the Schumpeter Mark I type experience a high rate of innovative entrepreneurial entry and exit, whereas, in Schumpeter Mark II technologies, entry and exit remain low.

One of the main insights of Malerba and Orsenigo is that there is a significant positive correlation between firm size, innovation concentration, stability in performance rank-orders, on the one hand, and a significant negative correlation of these technology characteristics with the rate of innovative entry, on the other. At least for the technologies in the major industrialised countries, these findings appear to be robust.

The two classes by Malerba and Orsenigo, Schumpeter Mark I and Schumpeter Mark II, suggest that, aside from profit opportunities, the innovative performance of industries also depends on appropriation conditions, the specific characteristics of knowledge creation and its cumulativeness as well as the kind and scope of innovation. These dimensions interact in a specific way and characterise the two classes, Schumpeter Mark I and Schumpeter Mark II. First, the opportunity dimension simply takes account of whether the industry under consideration offers relevant and exploitable innovation opportunities. In case this condition does not hold, neither small entrepreneurial firms, nor innovative start-ups, nor established innovators will engage in innovative activities. A second factor, appropriation conditions, which directly refer to the characteristics of knowledge created during the innovation process, determines the actual degree of appropriation. To the extent to which knowledge represents a public good, (potential) competitors will make use of it. Given sufficient absorptive capacities (Cohen and Levinthal 1990; Criscuolo and Narula 2008), they will imitate the innovator's technology. In case newly created knowledge is generic, codified, well-structured, and self-containing, imitators will be able to exploit the new knowledge. In case knowledge is highly specific, tacit, rather complex, and systemic, its imitation/exploitation comes at cost, which might be unsurmountable for potential imitators, due to the lack of absorptive capacity. The cumulativeness of knowledge is one aspect that will aggravate its exploitability by external actors. It contributes to its systemic instead of independent character as well as to its complex instead of well-structured dimension and partly to its tacit versus codified nature. Hence, cumulativeness and appropriability are more to the benefit of established firms than to entrepreneurial firms or innovative start-ups. The latter might rather be inclined to innovate new products and services, whereas established firms with a large accumulated stock of knowledge tend to engage in process and organisational innovation.

Along this line of reasoning, the two classes, Schumpeter Mark I and Schumpeter Mark II, have become labels for two regimes in innovation activities. The entrepreneurial regime is based on Schumpeter Mark I, characterised by a low degree of appropriability due to the specific type of knowledge, low cumulativeness in new knowledge generation, and a higher inclination for product innovation, which, in addition, will motivate small firms and entrepreneurial start-ups to enter the market. Conversely, the Schumpeter Mark II regime represents a routinised regime in which large firms with a high degree of cumulated knowledge focus on process and organisational innovation, since the cumulativeness of knowledge improves appropriation conditions because small entrepreneurial firms lack absorptive capacities and often the required financial scope. In other words, the path dependence of knowledge creates barriers to entry and provides large incumbent firms with a competitive advantage in innovation.

Entrepreneurship over the industry life cycle

The inconclusiveness of the empirical evidence on the Neo-Schumpeterian hypotheses, whether it is *small* versus *large* firms being the primary innovators, revealed that mainly industry-specific factors (i.e. profit opportunities, the specificities in knowledge creation, and appropriation conditions) are factors that explain entrepreneurial behaviour and innovative activities. Moreover, with the classification of innovation regimes and the insight that industry conditions change over time, a large body of literature has developed, called industrial dynamics (e.g. Dosi et al. 1995), which deals with the following basic pattern. Firms usually enter a market or industry as small firms with an innovation, some of which grow large over time by further innovating; some have to exit and this altogether determines the dynamics of an industry.

An interesting elaboration of industrial dynamics is the industry life cycle approach (ILC). Through radical innovations, new markets/industries are born experiencing certain patterns of evolution and thus shape an industry life cycle. This approach offers to track the dynamics of industries from cradle to grave, although the analyses run in principle only until maturity. Evidently, entrepreneurial behaviour, entrepreneurship, and innovative activities play an important role in the narrative of an industry life cycle. However, the kind and quality of these phenomena change along the industry life cycle. First, it is entrepreneurs/innovators who create an industry or set the initial seed for an industry. Secondly, entrepreneurs constantly keep the competitive pressure up within an industry. Thirdly, entrepreneurs/innovators shape the evolving structure of an industry in all phases of the ILC. Vice versa, it is the ILC pattern and its dynamics that feed back on entrepreneurial behaviour and the type of innovation regime.

Industry life cycle

To give an example, the industry life cycle of the German car industry was born in 1886 when Carl Benz invented the first automobile. Benz, and shortly afterwards Daimler, founded new firms to market automobiles. After that, many new firms emerged. Figure 5.1 graphs the development of this emerging industry showing the number of car companies over the years.

The development follows a pattern that comes close to a stereotypical ILC, as indicated by the bold line indicating the number of firms for each year. In a first phase, only a few firms exist in the industry (here until about 1897); then, in a second phase, a rapid take-up follows until a peak is reached (here 1924). A third phase follows, the shake-out phase, which manifests a sharp drop in the number of firms (here ending in 1930). In a last, fourth phase, the number of incumbent firms stabilises and entry-exit turbulences fade.

The second phase shows two sub-phases, a decline in the number of firms between 1908 and 1918 followed by a sharp increase between 1918 and 1924. Hypothesising about the effect of World War I, it seems plausible to assume that business prospects before the economic shock started to deteriorate; after the shock, entrepreneurial opportunities improved rapidly, possibly accelerating firm entry even more as it might have been the case without such shock (Grebel 2004; Grebel et al. 2003). If no shock had occurred, more firms would have entered the industry and the industry life cycle may have taken the shape of the dashed line (stylised ILC 1).

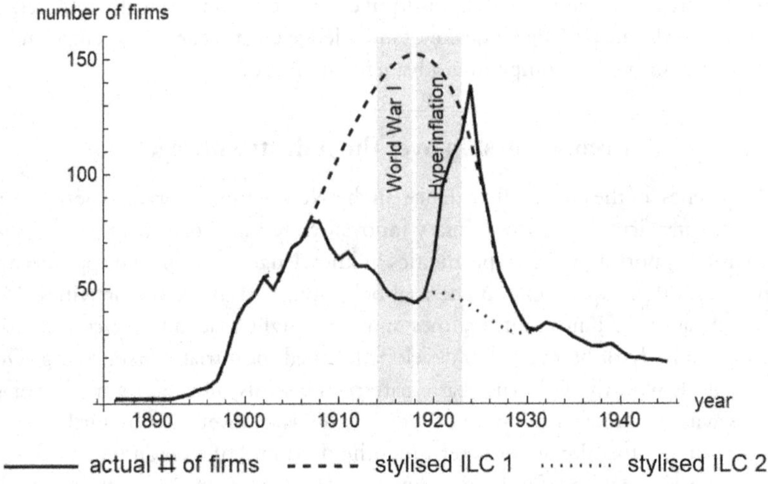

Figure 5.1 Development of the German automobile industry from 1886 to 1945

Source: Based on Cantner et al. (2009)

Note: There were two shocks to the ILC: World War I and hyperinflation. Both shocks affected the economy as a whole and the car industry in particular. The hypothesis that World War I inhibited entrepreneurial behaviour in the beginning and boosted it later seems self-evident. Assuming no such shock, the dashed line as hypothetical stylised ILC would have been conceivable. The second shock, hyperinflation, leading to a run into real investment possibly had entrepreneurial entry overshooting. Assuming this shock away, the dotted line marks another conceivable, counterfactual path of the ILC.

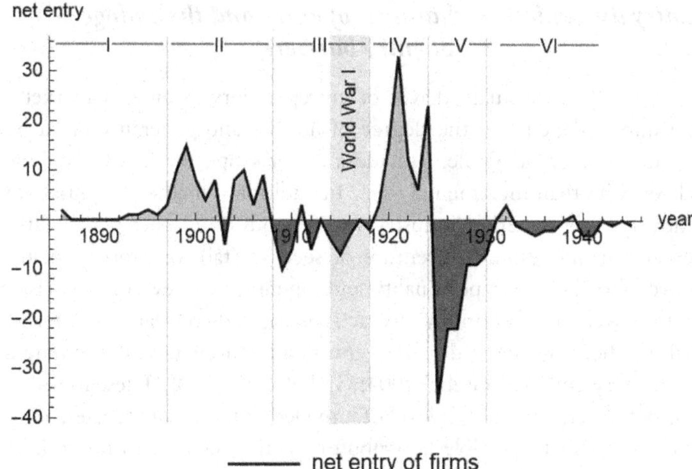

Figure 5.2 Development of entry and exit in the German automobile industry from 1886 to 1945

Source: Based on Cantner et al. (2009)

Alternatively, the sharp peak can be explained by the effect of the hyperinflation period (1918–1923). The induced run into real investment had firm entry overshoot aggravating the shakeout phase. In this scenario, the industry life cycle of the automobile industry might have taken the path of the dotted line (stylised ILC 2) given no hyperinflation.

The pattern of an ILC results from two underlying dynamics, entry and exit. Figure 5.2 displays the net effect of the two dynamics, which allows to detect several phases in the ILC. The phases are distinguished by whether net entry is positive, negative, or zero. By this criterion we distinguish phases I to VI, characterised as in Figure 5.2.

The bold line represents net entry as the difference between the number of entering and exiting firms. From 1886 to 1896 entry is moderate and exit did not occur, our phase I. From then on until 1907, phase II, the entry dynamics dominate the exit dynamics leading to positive net entry. The reverse holds for phase III, 1908 to 1917, thereafter, in phase IV, net entry is positive again until 1924, followed by a harsh shake-out phase V until 1930 with a strongly negative net entry. From then on, in phase VI, the winnowing of the industry slows down and net entry becomes very small, with slightly more exits than entries.

The type of entry in this development has different sources. It contains newly founded firms but also lateral entry, when established firms of other industries enter the automobile industry. With regard to the exit dynamics, observed exits comprise several ways of leaving the market such as bankruptcy, lateral exits, and mergers and acquisitions.

Overall, the ILC describes the change of an industry's organisation as well as a change in the type of entrepreneurial behaviour and innovation. Both, entrepreneurial behaviour and the industry structure, follow a mutually influencing pattern of change. Entrepreneurs found firms, create new markets, or enter a market to seize perceived profit opportunities. In the early phase of an ILC, it primarily is innovative start-ups driving the entry dynamics, when *creative destruction* constantly reshuffles innovator rankings. The shake-out phase winnows successful from unsuccessful innovators before the industry eventually consolidates and established firms start dominating the industry dynamics in an innovation regime labelled *creative accumulation*. This, however, already anticipates the discussion of entrepreneurial ecosystems in the following section.

Entry dynamics: mechanisms of entry and the endogeneity of entry barriers

Making profits certainly is the main driver of entrepreneurs to enter a market. The potential profit opportunities depend on the degree of novelty and generality of an initial innovation. Radical innovations, as a whole, provide a larger scope for follow-up innovations and entrepreneurial activities than incremental ones. In addition, whether potential entrepreneurs/innovators eventually enter a market/industry also depends on his/her personality, knowledge, expertise, access to various resources (venture or seed capital), or properly trained staff, and beyond and above, also his or her personality and aspiration to newness and risk-taking. The actual decision to entrepreneurial/innovative actions depends on various determinants. Only if this individual threshold is exceeded market entry and a subsequent entrepreneurship-driven evolution of an industry emerges (Grebel 2004; Grebel et al. 2003). These lower bounds represent a kind of entry barrier, a minimum level of knowledge and competencies, a minimum level of venture capital or access to possible contributors to the venture. In the course of industry development, these entry barriers are not fixed but may change over time and tend to become binding.

No innovation and fixed entry barriers

In the model by Klepper and Graddy (1990), such kind of entry barrier is modelled as a fixed monetary amount *e*. Only if this fixed cost of entry can be covered by the first period's returns, entrepreneurs decide to enter the market. Each potential entrant or entrepreneur *i* is characterised by firm-specific per unit production cost c_i, which is randomly assigned and drawn from a half-open interval with a given lower bound c^*. After the random assignment of unit costs c_i in the first period, they remain constant for each firm.[2] Whenever a potential entrant's per unit profit, measured as the difference between the market clearing price p and unit costs c_i, covers entry costs, $p - c_i \geq e$, an entrepreneur decides to enter – potential entrepreneurs with unit cost such that $p - c_i < e$ decide to remain outside the market.

As the industry evolves, new firms enter and established firms expand their total output. Given a conventional market demand function, the market-clearing price p_t constantly declines over time and so do per unit profits $p_t - c_i$ for an incumbent firm and $p_t - c_i - e$ for an entrepreneur that just entered.

For potential entrepreneurs outside an industry, a declining p_t implies declining chances for a firm entry. Given fixed entry cost *e* the positive margin of market entry shrinks over time; as for a potentially entering firm it must hold that $p_t - e \geq c_i \geq c^*$. Once p_t has reached a certain level so that $p_t < c^* + e$, entrepreneurial entry peters out. For firm performance, measured by c_i, this means that entry cohort by entry cohort, the average of the unit cost of the entering firms decrease. Hence, the steadily improving performance of cohorts finally approach c^*.

Looking at the contribution of entrepreneurship to the industry dynamics, the following can be summarised. The fundamental mechanism of industry development, aside from firm growth, is entrepreneurship and entrepreneurial entry. The higher the entry barrier to an industry, the less important entrepreneurial entry – in terms of number of entering firm per period and in terms of the number of periods in which entry takes place at all. In the course of an ILC, the

2 In the model of Klepper and Graddy (1990) unit costs in the first period can randomly be changed once but then stay fixed.

productive performance of entering firms increases, or put differently, in later phases of the industry's development, successful entry requires a better productive performance.

Innovation and endogenous entry barriers

In his 1996 contribution, Klepper extends the previous model by Klepper and Graddy (1990). Klepper (1996) introduces product and process innovation and endogenises the entry barrier.

He considers industries as driven by incumbents, entrepreneurial entries, exits (of incumbents), and firm growth. In each period incumbents and entrepreneurs pursue two types of innovation activities, product and process innovation. In addition, incumbents decide about expanding their production capacity. Introducing a new product yields a one-period innovation rent to the innovator. The expected return depends on the innovative expertise s_i of firm i, which is assumed to be a random number $s_i \in [0; s^*]$ exogenously assigned to firm i, where s^* is the upper bound. In the subsequent period, the product innovation becomes an integral part of the standard product. In order for firms to be able to imitate all product innovations of the previous period and integrate them into the standard product, they all have to screen the market that incurs cost F in every period. F is assumed to be constant and to exceed the innovation rent of product innovation.[3] This implies that each firm, established or new, needs to produce the standard product and invest in process R&D to reduce unit production cost in order to cover the losses arising from screening the market for imitation due to the assumption that the return to product innovation never covers those costs. Nonetheless, it is the innovation expertise s_i and the level of optimal product R&D that influence the final firm profit.

The optimal level of process R&D depends, inter alia, on firm size, the expansion of the market size – since the market-clearing price continuously falls and the corresponding firm output Q_{it} of the standard product increases. Because of the assumption that incumbents steadily increase their production of the standard product in the course of time, older firms grow large and manage to generate higher unit cost reductions in the standard product than smaller (younger) ones. In principle, the smaller (younger) a firm, the lower the price of the standard product, the less it invests in process R&D to reduce unit production cost c, and the lower the profit out of the standard product.

Putting the focus on our core interest, that is, entrepreneurial entry and exit, the crucial condition is the expected firm profit $E(\pi_{it}) = f(s_i, Q_{it}, p_t)$ of an incumbent or a new firm i in period t that determines firm entry and exit. The pivotal parameters in the model are the innovative expertise s_i of firms and the moment in time they enter the market. In the course of the ILC, the price p_t steadily falls and therewith the individual firm margin $(p_t - c)$ of the standard product.

In each period t, a minimum level of innovative expertise $\tilde{s}_{et}$ in $\left[0; s^*\right]$ is required to ensure expected profit to be non-negative, $E(\pi_{it}) = f(s_i, Q_{it}, p_t)\big|_{s_i \geq \tilde{s}_{et}} \geq 0$. This holds when the expected profit is at least as high as to cover the market screening cost minus the innovation rent from a new product plus the profit from the standard product. The minimum level $\tilde{s}_{et}$ is specific for every entry cohort in period t. Only potential entrepreneurs of a certain cohort that meet or better exceed $\tilde{s}_{et}$ have the expertise and knowledge to generate a product innovation with a high enough probability for a non-negative expected total profit. Only if this is satisfied, entrepreneurs decide to enter the industry.

3 By model construction, optimal product R&D, the resulting product innovation rent, and the screening cost are identical and constant for all firms, both incumbents and entrants.

As the industry develops, the minimum level of expertise $\tilde{s}_{et}$ required for entry increases from cohort to cohort. Due to a steadily declining market-clearing price for the standard product p_t, the prospective gains from process R&D shrink. The declining price lowers the return a firm can reap from process innovation and constrains the scope to cover expected losses when product-oriented R&D activities fail. Hence, for potential entrepreneurs of different cohorts with an equivalent s, the expected profit $E(\pi_{it}) = f(s_i, Q_{it}, p_t) > E(\pi_{it+n}) = f(s_i, Q_{it+n}, p_{t+n})$ with $Q_{it} \cong Q_{it+n}$ and $p_t > p_{t+n}$ $(n \geq 1)$. The higher the entrepreneur's innovative expertise s_i, the higher her expected profits. The later firms enter the market, the higher the minimum level $\tilde{s}_{et}$ required to enter. Eventually, firm entry comes to a halt, when $\tilde{s}_{et}$ has reached the upper bound s^*.

This is how Klepper (1996) endogenises the entry barrier: it is arising from the need to cover expected losses in product innovation plus imitation activities, develop endogenously and get tighter over time. Regarding the number of entries, whether they go up or down over time depends on the random assignment of s to the potential entrants. However, the average quality level of entrants measured by innovation competence s increases from cohort to cohort. Eventually, at a certain point T in the industry development the required competence level $\tilde{s}_T$ gets higher than s^*, the upper bound. Then, entry dynamics come to an end.

Summarising, industry dynamics in terms of entrepreneurship is characterised by the following entry dynamics. Over time, the quality of newly founded entrepreneurial firms, as indicated by their innovative expertise, increases. Although the rate of entry may fluctuate, the average expertise of entering firms increases along the industry life cycle. Entry eventually fades out when the entry barrier has reached the upper bound s^* and potential entrants are no longer able to meet the minimal innovation expertise required for a non-negative expected profit.

Exit dynamics: entrepreneurial entry and survival

Not all entrepreneurs entering a market survive. If expected profits, when staying in the market, are negative, firms will exit the market. In Klepper and Graddy (1990) the negative profit arises from the difference between the firm-specific (constant) level of unit cost in production and the market-clearing price, which gradually declines over time as the number of firms in the market and the total industry production increases. In each period, incumbents stay only in the market if $p_t - c_i \geq 0$, otherwise they exit. The exit dynamics stop when the price has declined sufficiently enough so that $p_t = c^*$. When comparing the average unit cost of firms in exit cohorts, we see a steady decline in the average of unit cost between consecutive exit cohorts. This means that, over time, more and more better performing incumbents are forced to exit.

In the Klepper (1996) model, as for the entry dynamics, the required minimum innovative expertise also drives the exit decision. The fact that the expected return to product innovation never fully covers the imitation cost F, it must be the return from process innovation and from selling the standard product to compensate for that loss in order to have positive expected total profits. Age also is a relevant factor that increases the chance of positive expected profits. The younger (smaller) a firm, the lower is the expected profit from process R&D. This profit might be too small to compensate for expected losses arising from product innovation plus imitation. For the latter it again holds true that the expected loss is the smaller, the higher the innovation competence s of a firm.

There again exists a minimum required competence $\tilde{s}_{et}$ at which the expected loss out of product innovation plus imitation is low enough to get just compensated for by the profits out of process R&D. For a small (young) firm $\tilde{s}_{et}$ is relatively large and may even be higher than s^*, the maximum possible innovation competence. For larger (older) firms, however, $\tilde{s}_{et}$ will be relatively low.

Along with industry development, the minimum innovative competence $\tilde{s}_{et}$ required to stay in an industry increases over time. The reason for that is a gradually falling market clearing price for the standard product. Forces behind this are process innovations that decrease production costs and the increase in total production in the industry due to entry and due to expansion of established firms. With a declining market price, the per unit profit out of the standard product declines and with it the degree by which a loss out of product innovation plus imitation can be covered. For that larger firms have more financial means than smaller ones. Consequently, for two firms in period t that are equal in innovation expertise s, the one which is smaller (younger) will have a higher minimum required innovation competence $\tilde{s}$ and thus is more likely to exit. For two firms that are of equal size in period t, both face the same $\tilde{s}$ and the one with the lower s_i is more likely to exit. Hence, in terms of comparing profits, it holds true that

$$E(\pi_{it}) = f(s_i, Q_{it}, p_t) > E(\pi_{jt}) = f(s_j, Q_{jt}, p_t)$$

with either $Q_{it} > Q_{jt}$ and $s_i = s_j$, or $Q_{it} = Q_{jt}$ and $s_i > s_j$, or proper combinations of the two.

In view of these relationships, a clear-cut time pattern of exit dynamics cannot be derived. Certainly, over time, younger and thus smaller firms are more likely to exit than older and larger firms as incumbents have the advantage of a higher profit out of the standard product. However, these dynamics interfere with the dynamics of entering entrepreneurs that steadily exhibit better terms of innovation competence, often even better than incumbents. This may lead to situations in which larger firms are forced to exit before smaller firms.

Empirics

In this section, we discuss some empirical results on the aforementioned entry and exit dynamics of entrepreneurial firms during the industry life cycle. The studies we draw on refer to the German automobile industry from 1986 to 1945, as introduced earlier.[4]

For the purpose of this chapter, we consider the number of entering firms including their innovation expertise on the one hand and, on the other, the number of exiting firms including their respective innovation expertise. For innovation expertise we take two alternative indicators; first a dummy variable indicating patenting of a firm and secondly a firm patenting dummy combined with its duration in the industry. Firms that enter without a patent are considered imitators.

We demarcate certain phases during the time span 1886 to 1945 (see Figure 5.2). An early phase with a moderate rate of entry is 1886–1896; a first phase of high entry is 1897–1907; 1908–1917 shows a negative net entry rate; for 1918–1924 a sharp increase in entry is observed; 1925–1930 follows with an industry shake-out; a slightly negative or rather zero net entry rate follows in 1931–1945.

Entry dynamics

From 1886 to 1945 the total number of entering firms is 351 of which 333 exited the industry again. Of those entering, 128 held patents, of those exiting 113. The average duration in the industry is about 7 years with a maximum of 60 and a minimum of 1 year. Some of the exits occurred shortly after entry. We run an alternative analysis on those entries that lived for 5 years

4 The data and the way they were collected is described in Cantner et al. (2006, 2009, 2011) and Dressler (2006).

Table 5.1 Entry and entry performance – patenting versus not-patenting entries

Period	*All firms*			*Firms with survival of at least 5 years*				
	entries	*patenting entries*	*patenting entries in all entries*	*entries*	*patenting entries*	*patenting entry in all entries*	*avg survival patenting entries*	*avg survival non-patenting entries*
	[1]	*[2]*	*[2]/[1] = [3]*	*[4]*	*[5]*	*[5]/[4] = [6]*	*[7]*	*[8]*
1886–1896	8	6	0,75	7	6	0,86	28,50	6,50
1897–1907	101	43	0,43	65	35	0,54	17,12	5,95
1908–1917	46	20	0,43	25	17	0,68	17,15	2,54
1918–1924	101	29	0,29	31	14	0,45	3,83	2,15
1925–1930	69	24	0,35	12	10	0,83	18,96	1,53
1931–1945	26	6	0,23	7	3	0,43	21,00	4,10
sums	351	128		147	85			

and more. That leaves us with 147 entering and 129 exiting firms, which held 85 and 70 patents, respectively.

As to entry dynamics, the proposition is that the entry barrier in terms of product innovation expertise is increasing over time so that fewer and fewer firms are able to meet it. A proper test for this would be a comparison of entering and non-entering firms in their innovation expertise. However, data on potential entrants that did not enter are unavailable. As a kind of auxiliary analysis, we simply track innovation driven entries over time. We look at all entries and additionally on those that lived at least for 5 years.

Table 5.1 delivers the result of our analysis. First, looking at the full-time span from 1886 to 1945, for firms with and without innovation expertise there is early on an increase followed by a drop in the number of entries – which follows a life cycle pattern. Secondly, in terms of the share in innovation-based entries in all entries, for the full sample, this share starts with 75% in 1886–1896, then drops gradually to 23% in phase 1931–1945. For the sample of entry surviving at least 5 years, the drop is less severe: from 86% in 1886–1896 it falls to 43% in 1931–1945. In terms of innovative entry, fewer and fewer firms are observed in absolute as well as in relative terms. Conversely, imitation, for which firms without patents may strive for, seems to become comparatively easier over time.

Thirdly, in order to qualify patenting, we use a further measure, the average survival time of patenting versus non-patenting entries – where higher survival indicates higher innovative quality. Looking at the subsample of survivors of at least 5 years, patenting firms start in 1886–1896 with an average duration of about 28 years. This is about four times higher than an average non-patenting firm. The duration drops in the next sub-periods to 17 years and even 4 years (in World War I times). From 1925 on, it rises up to 21 years in 1931–1945, which is about five times as much as an average non-patenting firm. Hence, except for the starting sub-period and during wartime, the quality of patenting entries increases absolutely as well as in comparison to non-patenting (imitating) firms. An interpretation of this result is that the entry barrier, in terms of a minimum innovation expertise, increased over time. In this sense, the core result of our discussion of entrepreneurial entry dynamics over the industry life cycle finds some empirical support.

Table 5.2 Exit – patenting versus non-patenting

Period	*All firms*			*Firms with survival of at least 5 years*				
	exits	*patenting exits*	*patenting exits in all exits*	*exits*	*patenting exits*	*patenting exits in all exits*	*avg survival patenting exits*	*avg survival non-patenting exits*
	[1]	*[2]*	*[2]/[1] = [3]*	*[4]*	*[5]*	*[5]/[4] = [6]*	*[7]*	*[8]*
1886–1896	1	0		0	0			
1897–1907	43	17	0,40	15	9	0,60	3,53	5,83
1908–1917	68	18	0,26	39	15	0,38	7,39	8,21
1918–1924	42	9	0,21	4	1	0,25	1,44	18,33
1925–1930	140	50	0,36	52	30	0,58	9,98	10,55
1931–1945	39	19	0,49	19	15	0,79	17,74	7,75
sums	333	113		129	70			

Exit dynamics

As to the exit dynamics, we start with an analysis equivalent to the entry dynamics in order to find an indication for an increasing minimum innovation expertise. Table 5.2 reports our results.

First, for the whole sample as well as for the subsample of firm survival of at least 5 years, exit activities start low, reach a peak, and then decline. This holds for both firms with and without patenting. For the whole sample, twice as many non-patenting firms exit as patenting firms; for the reduced sample, it is rather balanced. Secondly, in terms of the share of exiting firms that patent in all exits, for both samples, this ratio increases over time. This is due to non-patenting firms exiting more in earlier phases and only when the required minimal innovation expertise increased high enough, more and more patenting firms are forced to exit. Thirdly, as to the average survival time of exiting firms, patenting dropouts compared to non-patenting ones live shorter on average – in all phases before 1931–1945. Average survival is increasing from 3,5 years in 1897–1907 to about 18 years in 1931–1945. This indicates that in earlier phases of the ILC recent entries and hence smaller (younger) firms are forced to exit; in later phases, this shifts larger (older) companies.

Additional results on ILC and entrepreneurs

Besides these descriptive results on required innovation expertise to stay in the industry, survival analyses deliver insights on the type of entrepreneurial firms that are more likely to survive (longer) in an industry. The following results are of interest (see Cantner et al. 2006, 2009, 2011): first, entrepreneurial firms that entered an industry early/earlier and survived for some time (i.e. larger and thus older firms), are less likely to exit. Consequently, entrepreneurs entering later have a harder time surviving. Secondly, entrepreneurial firms that entered an industry with an innovation are less likely to exit than those entrepreneurs that entered without innovation. Moreover, those that entered earlier and with an innovation are less likely to exit than entrepreneurs that entered later without innovation. Interestingly, being late can be compensated by being innovative, given a start-up meets the minimum innovation expertise. Thirdly, entrepreneurial start-ups that entered with more experience they have accumulated elsewhere and which contributes to their innovation expertise are less likely to exit than those without that kind of expertise.

The results presented in this subsection largely offer evidence on the interaction of entrepreneurial action and context. In view of this chapter, the conditions for entrepreneurial entry and firm exit change endogenously; entry and exit rates shape the pattern of industry dynamics that in turn feed back on entry and exit conditions. Obviously, the fit of the theoretical model(s) and the empirical evidence needs to be improved on both sides. There is potential for further progress.

Entrepreneurship and the dynamics of entrepreneurial ecosystems

Entrepreneurship and new firms take a substantial impact on industries along their life cycle. However, entrepreneurship plays its main role in early phases of the industry life cycle. By the degree to which industries mature, this impact vanishes. In the final phase of the industry life cycle, when the market structure stabilises, entrepreneurial activities are of no relevance anymore concerning the concurrent main technology.

However, a conceptual lacuna opens up here, the question about what comes after an industry has matured and what kind of further or different developments may follow. For the latter, new firms may be the constituting factor. For instance, in a mature industry characterised by an established core technology, further improvements of this technology become more difficult to achieve. This makes innovation activities more and more expensive. Instead of trying to improve the state-of-the-art technology any further, the attempt to aim at radical innovation may appear more promising to sustain profits. This radical change may take place via substituting the former key technology with a new one. For that type of change, new firms are more likely to be the promotors. Established firms, often having huge production capacities and a high stock of human capital related to the old technology, are less likely to succeed in exploiting such economic opportunities. Hence, with the maturity of an industry and an increasingly costly-to-improve key technology, the context conditions call for new entrepreneurial activities – new directions for innovation, new key technologies, and hence new technological and economic opportunities.

Here, the concept of entrepreneurial ecosystems comes into play. It serves as a framework to explain entrepreneurial activities within regions and industrial sectors. Cantner et al. (2020) who show how innovation activities shift between entrepreneurs, intrapreneurs, or incumbents in an industry or a region, have suggested a dynamic version of this approach. Entrepreneurs are predominantly active in start-ups in the sense of the ILC approach. The latter fits rather well into the ILC setting because in principle they are the established firms that run innovation projects, corporate or internal ones.

In an entrepreneurial ecosystem, the relation between the two types of actors can be considered as follows: in an early phase, phase 1, entrepreneurs and their innovation activities dominate the pattern of development – phase 2, the *growth phase*, in Cantner et al. (2020). In phase 3 – the *stabilisation and maturity phase* – intrapreneurs pursue more and more innovation activities such as scaling up former start-ups or acquiring small firms with their innovations. Dominance by intrapreneurs is achieved in phase 4 – the *phase of decline*. Entrepreneurial activities decline to lead to a downturn of the entrepreneurial ecosystem. So far, the similarity to an ILC model is obvious: the model of a dynamic entrepreneurial ecosystem resembles this mechanism of shifting innovation activities from entrepreneurs towards intrapreneurs as time goes on.

In addition, entrepreneurial ecosystems conceptualise the shift of innovation activities away from intrapreneurs back to entrepreneurs. It is this shifting, back and forth, of innovation activities between entrepreneurs and intrapreneurs that is at the core of a dynamic entrepreneurial ecosystem approach. To accomplish this, phase 1 – the *birth phase* of an entrepreneurial ecosystem – is connected to phase 5, the *reemergence phase*. In phase 5, incumbent firms offer a pool of not commercialised ideas for entrepreneurs. By the degree to which these entrepreneurs use these opportunities, which are often quite different from the fundamentals of current prevailing technologies, a new cycle of an entrepreneurial ecosystem takes off and another phase 1 is initiated.

Hence, on this conceptual basis, the role of entrepreneurship is not only to get a new technology established – just as in the ILC approach – but also to take the lead in radical change. Established firms are quite reluctant to redirect their daily business. They rather try to avoid cannibalisation effects and, due to lock-in effects, they often lack the proper expertise. New entrepreneurial firms are flexible enough, not too much imprinted by the past in order to go into new directions and pursue radically new technologies. This is what the dynamic version of the entrepreneurial ecosystem approach substantiates. In this sense, it describes a continuous reloading of an entrepreneurial ecosystem cycle. Notwithstanding this, an empirical corroboration of concept is still pending – but on the agenda.

Demand-side aspects on entrepreneurial entry

The analysis in the previous sections focuses on the supply-side analysis of innovation decisions in general and on start-up decisions by entrepreneurs in particular. The demand side was considered here either via the assumptions of a given fixed demand function or exogenously growing demand. In this section, we will neglect the less specific consideration of the demand side and take a closer look at its importance for innovative entrepreneurs.

A connection between market size – and thus demand – and innovation was first established by Adam Smith (Smith 1776). As market size increases, the possibilities for the division of labour increase, and the resulting, more efficient use of labour is due, among other things, to "the invention of a great number of machines, which facilitate and shorten labour, and enable one man to do the work of many" (Smith 1776, chapter 3). This explicit link between market size and innovation has rarely been discussed further.

A notable exception is the work of the sociologist of invention Gilfillan (1935a, 1935b), who not only echoed Adam Smith's idea but also proposed an additional role for the demand side in the innovation process. First, he suggested that the pace of technology should be faster in industries where the number of potential adopters, and thus the incentive for firms to innovate, is higher. Second, he suggested that demand not only creates incentives, but also alerts to new needs that must be met by the supply side. And both factors, when viewed in this way, create increased opportunities for entrepreneurs and innovative start-ups. However, this relationship requires at least two qualifications.

First, in a dynamic context with growing demand, this correlation holds only true whenever the growth of demand is higher than the growth of established companies. This relationship can be explained by the development of the number of companies in the German automobile industry, as shown in Figure 5.1. This development takes place in a context in which the demand for automobiles grows continuously from year to year. In years with an increasing number of companies, the incumbents grow less than demand; in years with a decreasing number of companies, demand growth is lower than incumbent growth. A major reason for strong growth of firms is returns to scale (i.e. static and especially dynamic returns to scale). The latter can be found in Klepper's (1996) model that we presented previously, where dynamic returns to scale occur in process innovations. Seen this way, the exploitation of new opportunities, opened up by an increasing demand, can be inhibited by a too small scale of companies. Therefore, it will be incumbents rather than entrepreneurial entrants that take advantage of these opportunities.

Second, in addition to dynamic returns to scale on the supply side, demand-side dynamic returns to scale can break up the positive relation between demand-side growth and entrepreneurial entry. Dynamic returns to scale on the demand side, also called network effects, are based on the phenomenon that the utility a user or adopter of a technology can enjoy depends on the number of other users or adopters of this technology (see e.g. Arthur 1989). The development of an industry where such effects are at work shows a trend toward a reduction in the variety of alternative technological approaches that eventually may lead to a dominant technological design. Such a reduction, in turn, diminishes the economic opportunities for entrepreneurs and hence entry rates.

In view of these findings and relationships, also the demand side has ambiguous effects on innovative entrepreneurial entry into an industry. Higher demand potentially opens up more opportunities for entrepreneurs to enter an industry with an innovation. This may be counteracted by demand side–based network effects as well as with dynamic economies of scale on the supply side.

Conclusion

Entrepreneurship and industrial organisation

This chapter combines two strands of literature, entrepreneurship and industrial organisation, and it shows that both are inextricably intertwined. Entrepreneurial behaviour and entrepreneurship give impetus to economic change as put forward by the work of Joseph A. Schumpeter. It creates new industries and shapes their economic evolution. The structural approach found in the industrial organisation literature theorises about the consequential economic behaviour idiosyncratic industry structures induce, such as asking whether it is a monopoly structure that offers higher incentives to perform innovative activities than perfectly competitive markets. The Structure-Conduct-Performance approach provides mixed propositions. Arrow claimed that perfect markets have a higher incentive to innovate than monopolies. The literature on the New Industrial Economics sees oligopoly markets as main drivers of innovation.

Likewise, empirical evidence does not solve the puzzle satisfactorily, either. Whether large or small firms are the main innovators, as Schumpeter reasoned, there is no significant difference in R&D efforts according to firm size. It rather is industry-specific characteristics such as technology, knowledge, or absorptive capacity that entrepreneurs/firms need to be endowed with in order to explain industrial dynamics. Industry life cycle theory, taking up this dynamic perspective, vividly documents that entrepreneurial behaviour, entrepreneurship, innovation as much as technological, economic, and behavioural change are time- and context-dependent as well as reciprocal dynamic phenomena asking for an integral approach. This necessarily needs to include demand-side factors as these may on the one hand push entrepreneurship and show features that reduce entrepreneurial opportunities, on the other.

The theoretical approaches as well as the empirical studies addressed in this chapter underline the important role of entrepreneurship and entrepreneurs. They seek and seize new technological opportunities for innovative activities. As long as profits can be reaped, entrepreneurs will stay active; they do not stop activities before profit opportunities are exploited or entry barriers are too high. Then, entrepreneurial dynamics slow down. In the entrepreneurial ecosystem approach, this slowing down is seen as a shift of innovation activities from entrepreneurs towards intrapreneurs (as located in established firms). The story usually ends, here. However, economists should be able to say more about what comes after. Entrepreneurs not only exploit given opportunities but sometimes they are the ones that create new ones. This process is much less understood since the pattern of radical change and radical innovation are not well understood. To look deeper into entrepreneurial behaviour that goes against the odds or the mainstream and to relate that to radical structural change and even the birth of new industries is an avenue to follow in future research. In our current situation with radical changes in many industries and technologies, with transformative tendencies and with the emergence of new actors on a broad scale, the object of analysis is clear: modelling and empiricism are waiting to be pursued and aligned further.

Relevance for digital transformation

Referring back to the introduction, the literature-based discussion of the relation between entrepreneurship and industrial organisation earlier is meant to inform about potential developments in prominent new digital technologies and their respective industries and markets. Aspects of small versus big firms, followers versus leaders, appropriability conditions, the whole range of life cycle pattern and dynamics as well as demand-side factors are to be expected to play their role herein too, although not necessarily with equal importance. Especially the role of scale and network effects appear to be of far more and higher importance in digital technologies than in "traditional industries". The combination of supply-side dynamic scale economies – mainly due to rather negligible marginal costs of additional users – and demand-side network externalities may be a reason for a faster decline of economic opportunities for additional and new entrepreneurs and their innovative ideas. Entry costs rise in a relatively short time and prevent even good, new ideas from falling on fertile ground.

In addition, digital technologies have functions such as easy matching of demand and supply and they operate in contexts that were not considered in industrial dynamics literature. Platforms, for instance, are such a new form of a market in virtual space which are particularly relevant for new, digital business models. On the one hand, platforms significantly strengthen the market performance of platform members, as both sides of the market reinforce each other's attractiveness, due to easier accessibility and contractibility. For outsiders, such a constellation tends to reduce the economic opportunities. These, be they entrepreneurs or established companies, find it increasingly difficult to compete with an evolving platform and to innovate. Setting up a new, competitive platform instead will become less probable over time. On the other hand, a platform itself can be particularly attractive for entrepreneurs to become members and pursue innovation activities on it. Here the benefits of a platform arise from the fact that entrepreneurs can reach the demand faster and more efficiently. Research in this respect concerning industry patterns and dynamics is still pending.

References

Arrow, K. J. (1962): Economic Welfare and the Allocation of Resources to Invention, in Nelson, R. R. (ed.): *The Rate and Direction of Economic Activity*. New York: Princeton University Press, 609–626.

Arthur, W. B. (1989): Competing Technologies, Increasing Returns, and Lock-in by Historical Events, *Economic Journal* 99(394), 116–131.

Bain, J. S. (1956): *Barriers to New Competition: The Character and Consequences in Manufacturing Industries*. Cambridge, MA: Harvard University Press.

Bain, J. S. (1959): *Industrial Organization*. New York: John Wiley & Sons, Inc.

Cantner, U., Cunningham, J. A., Lehmann, E. E. and Menter, M. (2020): Entrepreneurial Ecosystems: A Dynamic Lifecycle Model, *Small Business Economics*. https://doi.org/10.1007/s11187-020-00316-0

Cantner, U., Dressler, K. and Krüger, J. (2006): Firm Survival in the German Automobile Industry, *Empirica* 33, 49–60.

Cantner, U., Krüger, J. and v. Rhein, K. (2009): Knowledge and Creative Destruction over the Industry Life Cycle – The Case of the German Automobile Industry, *Economica* 76(301), 132–148.

Cantner, U., Krüger, J. and v. Rhein, K. (2011): Knowledge Compensation in the German Automobile Industry, *Applied Economics* 43(22), 2941–2951.

Cohen, W. M. (2010): Chapter 4 – Fifty Years of Empirical Studies of Innovative Activity and Performance, in Hall, B. H. and Rosenberg, N. (eds.): *Handbook of the Economics of Innovation*. Vol. 1. Amsterdam, the Netherlands, Oxford: North-Holland (Elsevier), 129–213.

Cohen, W. M. and Levinthal, D. A. (1990): Absorptive Capacity: A New Perspective on Learning and Innovation, *Administrative Science Quarterly* 35, 128–152.

Criscuolo, P. and Narula, R. (2008): A Novel Approach to National Technological Accumulation and Absorptive Capacity: Aggregating Cohen and Levinthal, *The European Journal of Development Research* 20(1), 56–73.

Dasgupta, P. and Stiglitz, J. E. (1982): Industrial Structure and the Nature of Innovation Activity, *Economic Journal* 90, 266–293.

Dosi, G., Marsili, O., Orsenigo, L. and Salvatore, R. (1995): Learning, Market Selection and The Evolution of Industrial Structures, *Small Business Economics* 7, 411–436.

Dressler, K. (2006): *Der Lebenszyklus der deutschen Automobilindustrie – Know-how und Überleben von Unternehmen 1886–1939*. Köln, Lohmar: Josef Eul.

Gilfillan, S. C. (1935a): *The Sociology of Invention*. Chicago: Follett Publishing Company.

Gilfillan, S. C. (1935b): *Inventing the Ship*. Chicago: Follett Publishing Company.

Grebel, T. (2004): *Entrepreneurship*. London: Routledge.

Grebel, T., Pyka, A. and Hanusch, H. (2003): An Evolutionary Approach to the Theory of Entrepreneurship, *Industry & Innovation* 10, 493–514.

Harris, C. and Vickers, J. (1985): Perfect Equilibrium in a Model of a Race, *Review of Economic Studies* 52(2), 193–209.

Hébert, R. F. and Link, A. N. (1988): *The Entrepreneur: Mainstream Views & Radical Critiques*. New York: Praeger Publishers.

Klepper, S. (1996): Entry, Exit, Growth, and Innovation over the Product Life Cycle, *American Economic Review* 86, 562–583.

Klepper, S. and Graddy, E. (1990): The Evolution of New Industries and the Determinants of Market Structure, *Rand Journal of Economics* 21, 27–44.

Levin, R. and Reiss, P. (1984): Tests of a Schumpeterian Model of R&D and Market Structure, in Griliches, Z. (Ed.): *R&D, Patents, and Productivity*. Chicago: University of Chicago Press, 175–208.

Malerba, F. and Orsenigo, L. (1993): Technological Regimes and Firm Behaviour, *Industrial and Corporate Change* 2, 45–72.

Malerba, F. and Orsenigo, L. (1997): Technological Regimes and Sectoral Patterns of Innovative Activities, *Industrial and Corporate Change* 6, 83–118.

Reinganum, J. R. (1983): Uncertain Innovation and the Persistence of Monopoly, *American Economic Review* 73(4), 741–748.

Reinganum, J. R. (1984): Practical Implications of Game Theoretic Models of R&D, *American Economic Review Papers and Proceedings* 74(2), 61–66.

Schumpeter, J. A. (1912/1934): *The Theory of Economic Development*. Harvard: Harvard University Press.

Schumpeter, J. A. ([1987] 1942): *Socialism, Capitalism and Democracy*, reprint. London: George Allen & Unwin.

Smith, A. ([1776] 1976): *An Inquiry into the Nature and Causes of the Wealth of Nations*. Chicago: University of Chicago Press.

6

IS THIS TIME DIFFERENT? A NOTE ON AUTOMATION AND LABOUR IN THE FOURTH INDUSTRIAL REVOLUTION

Luigi Marengo

Introduction

Nearly 90 years ago John Maynard Keynes published a short and visionary article where he tried to imagine economic life in two generations. He envisaged a sharp decline of labour demand due to technological substitution that he called "technological unemployment":

> We are being afflicted with a new disease of which some readers may not yet have heard the name, but of which they will hear a great deal in the years to come – namely, *technological unemployment*. This means unemployment due to our discovery of means of economising the use of labour outrunning the pace at which we can find new uses for labour.
>
> *(Keynes 1930: 360, emphasis in original)*

Technological unemployment is indeed an unavoidable and ubiquitous consequence of technological change based on the introduction of new capital goods. At the outset of the first Industrial Revolution Luddites destroyed textile machinery because it was stealing their jobs and wages. Indeed, we can now say that Luddites were right in the short run but utterly wrong in the long run. The machines which the first Industrial Revolution was based on did destroy jobs of the workers they replaced, but they hugely increased the productivity of labour and increased its value and remuneration and, especially, lots of new jobs were created by the new organisation of production. Machines had to be produced by the rising machine sector. The factory system was "delegating" many simple routine production tasks to machines but also creating opportunities for new less simple and routine tasks. For instance, the factory system was a complex organisation which required coordination jobs which simply did not exist before. The internal division of labour needed coordination, the factory had to work on a continuous basis

 DOI: 10.4324/9780429351921-8

and input and output markets had to be constantly monitored and secured in order to avoid disruptions, financial management also became far more complex. It was the birth of the industrial and service sectors' white collar class.

Today we are entering the fourth Industrial Revolution,[1] accompanied by a new wave of substitution between labour and capital and we can even encounter new forms of Luddism (Jones 2006). Can we expect that, like the previous Industrial Revolutions, job losses will only characterise a temporary short run phase and that new jobs will be finally created in a number at least equal to those that will be lost? And that the new jobs will be, on the whole, of better quality than the lost ones – less routine, more interesting, more creative, more productive and therefore better paid?

An "optimistic" vision tends to answer yes to such questions, claiming that roughly the same virtuous adjustments that took place in the previous Industrial Revolutions will operate also in the present one. Such an adjustment may be painful and require one or more generational turnovers and profound institutional changes in the education system and labour markets, but it will finally happen. According to this view, technological unemployment is a temporary frictional phenomenon, and governments, trade unions and other institutions, especially those devoted to education, should operate to facilitate such transition and to help make it socially acceptable by providing support to the "losers" of the transition.

But an opposite view claims that this time it is different. The fourth Industrial Revolution has some common characteristics with the previous ones, but also some important peculiarities that make its nature and in particular its long-term impact on the labour market different from what we observed in the past revolutions. In particular, past mechanisations could only substitute routine manual work, while today's automation is likely to concern jobs which involve more and more sophisticated cognitive skills and even learning. According to this view, digital technologies, AI, machine learning, big data, internet of things have some important features which will mark a sharp discontinuity with previous Industrial Revolutions. Persistent technological unemployment and growing inequality will characterise a long historical phase and the consequences on the organisation of production and society will go well beyond frictional adjustments.

Keynes himself in his 1930 essay gave a fundamentally optimistic view and considered technological unemployment as "a temporary phase of maladjustment". If we are going towards a world in which production requires little labour, then "[a]ll this means in the long run *that mankind is solving its economic problem*" (ibid., emphasis in original). However, his optimism was founded on a rather radical change in the organisation of society (i.e. massive reduction of working hours per week and massive redistribution programs) funded by high taxes on capital income.

In this short note I will briefly review the main arguments of the two visions. Personally, I am more convinced by those put forward by the "pessimistic" view, mainly because I think that the underlying "microeconomic" features of the new technologies are profoundly different from those of the previous Industrial Revolutions. I will argue that some of virtuous circles which have been activated in the past are unlikely to operate this time. Needless to say, I am absolutely aware that long-term forecasts should always be taken cautiously and that reality often surprises us. New kinds of virtuous circles may well arise in the future that we do not envisage

1 The fourth Industrial Revolution is based upon robotics, artificial intelligence, big data, internet of things, biotech, nanotech (Schwab 2016). Some authors consider it as a development of the third Industrial Revolution (based on computers and internet) rather than a new one. Regardless how we label it, it is a wave of technological change which is just beginning, and we can hardly imagine the precise directions it will take.

today. Thus, my position is that yes, this time is different and we are going to face some serious societal challenges caused by the new technologies, but of course we do not know what the long-term consequences will be.

Automation and codification

Let me begin with an extremely sketchy and somehow rough view of the first Industrial Revolution. It was essentially based upon two interrelated processes: codification of knowledge and division of labour. The former was part of a long-lasting effort to have a "scientific" approach to productive knowledge. Formerly tacit knowledge, embedded in the minds, hands and actions of artisans was codified, reported in textual records, analysed and taught. This process of codification and understanding also allowed a parallel process of decomposition of such knowledge into simpler sub-tasks. This division of labour allowed specialisation, the use of a less qualified workforce and vast opportunities for the development of capital goods which could perform some of these extremely simple routine tasks with higher strength, continuity, reliability and efficiency than human workers.

This process of codification of tacit knowledge was and is still today pushed by strong economic forces. Tacit knowledge suffers in fact from dramatic organisational inefficiencies. Tacit knowledge can be apprehended only by a long and painstaking period of practice which consists of working side by side with someone who possesses this knowledge. This was reflected in the masters-journeymen-apprentices organisation of work typical of pre-industrial guilds. Codified and divided productive knowledge consists instead either of simple tasks that can be learned very quickly (those of the shop floor worker) or of highly specialised codified tasks that are learned in the formal training of engineers, chemists and other highly qualified technicians. However, typically, the latter training is mainly paid not by firms but by society as a whole and/or by the trainee.

Codification was a precondition for automation: machines can only perform perfectly codified tasks, and when machines could be introduced, they quickly outperformed human workers and made them redundant. In this sense Luddites were right. But this was not the end of the story, since many compensating factors counterbalanced the reduction of work per unit product. Just to summarise the main ones, lots of new jobs were created in the machine-producing sectors, in activities different from mere shop floor jobs (clerical, management and all sorts of "coordinating" tasks), and because of the general decrease of prices and increase of quantities thanks to technological change, innovation itself fuelled economic growth and increase of income (see, e.g., Piva and Vivarelli 2017 for a critical survey).

The overall outcome was an increase of employment, wages, and also of "quality" of work, though the process also involved a lot of social suffering and unrest and major societal and institutional changes.

Today, the technologies of the third and fourth Industrial Revolution are pushing the codification of knowledge to unprecedented levels. Such technologies are nothing else than codification technologies, as they are based upon digitisation, that is the translation of any kind of information and knowledge into a digital code that can be understood and automatically processed by a computer. In addition, the new "machines" are not bound to executing simple, repetitive, physical tasks, but they are more and more capable of performing highly sophisticated and complex procedures, involving change, adaptation, cognition and even learning. Are the same "virtuous" compensation mechanisms going to operate also for these new technologies or not? In the next two paragraphs I will briefly review the main arguments for a "yes" or "no" answer.

The optimistic view

The "optimistic" view acknowledges that new technologies will cause the loss of many jobs but foresees that in the long run compensation effects will occur, new jobs will be created and the labour markets will adjust. Some of these compensation effects will be very similar to those that have been operating in the previous Industrial Revolutions: new jobs will be created in the machine-producing industries, in human tasks which complement such machines, and in all sectors which, though experiencing a decrease of the labour input per unit product, will benefit from a considerable growth of demand due to the decreasing prices and higher efficiency enabled by automation itself.

Other compensation effects will instead be specific to the fourth Industrial Revolution. Many analysts envisage in the next decades a shift of demand from manufacturing goods to services in which the human component is central and cannot be substituted: healthcare, personal care, relational goods, creative goods, etc.

Of course, also the optimistic view acknowledges that compensation effects will take time, important changes in the educational institutions and in the labour markets and, not least, in the very mentality and culture of people. A few generational turnovers may be needed before the adaptation is completed and in the meantime, technological unemployment can considerably rise and create social unrest. New jobs will be very different from the old ones and they will require new competences that the dismissed workers will probably lack. Public policies will play a key role by both providing some safety net to the victims of the new technologies and by promoting the necessary institutional changes. As to the former, various measures can be envisaged from unemployment benefits to forms of basic income and other poverty reduction measures. As to the latter, public policies should promote a quick and smooth transition. On one side, industrial policies should favour the adoption of new technologies; on the other side, they should promote the necessary changes in the education system and in the labour markets. The education system plays a particularly central role: Goldin and Katz (2008) provide a formidable account of the interaction between education and technology and show that the supply of highly qualified workers with the right qualifications is key in boosting income growth and equality of its distribution.

The pessimistic view

The pessimistic view claims that that the fourth Industrial Revolution has some important features which cast serious doubts on the predictions that also this time compensation effects will be able to initiate and feed the virtuous circles we observed in the past. Such novel features have to do with the nature of the technologies at the core of the fourth Industrial Revolution but also with the socio-economic transformations that are accompanying the current wave of technological change.

The starting point is the observation that digital technologies are pushing the process of codification of the human know-how to an unprecedented level. The new technologies are nothing else than codification technologies: they codify more and more complex tasks, and with artificial intelligence they even promise to codify complex cognitive task and learning. In the previous Industrial Revolutions machines were only able to perform simple repetitive tasks. Their advantage over human workers was limited to a higher force, higher precision and lower exposure to fatigue and boredom. But whenever information processing, flexibility, adaptation, learning, creativity, etc. were involved, machines had no chance to substitute humans, but at most to complement them.

Computer-based technologies on the contrary far exceed human capabilities in information processing and are almost inevitably going to substitute humans in all the tasks based on information processing, regardless of their complexity. Technology scholars were citing only a couple of decades ago such tasks as driving a car as a typical task that could never be automatised. They were saying that driving a car involves lots of tacit knowledge and learning by doing (it is in fact a task that we learn through practice), in which a broad repertoire of simple tasks, such as braking, pressing the clutch, changing gear, etc. must be combined in a specific and often novel way to cope with a potentially infinite set of situations whose detection requires the processing of a big amount of sensory data. Indeed, it was said, some of the simple tasks (such as gear changing) could be automatised, but not their creative, complex and heavily information-dependent combinations. Today self-driving cars are real and will presumably become a dominant technology quite soon. Interestingly they do not mimic human tacit knowledge but exploit the huge computational and information processing power of modern computers and algorithms.

The increase of information processing power and the advancement in learning algorithms are likely to generate a wave of technological innovations which will expand the domain of machines to non-routine, complex and cognitive tasks. Algorithms can now take complex decisions in financial markets, diagnose illnesses, make intelligent searches of legal databases, all tasks which were before performed by highly qualified workers with college degrees. Far from being relegated to shop floor tasks, machines are substituting clerical and even professional specialist labour.

Beaudry et al. (2013) find a "great reversal" occurring since the beginning of the new millennium: the demand for highly qualified cognitive jobs in the US is declining, while US universities keep increasing the number of graduates. The outcome is the increase of unemployment and underemployment and the decrease of wages for many categories of college graduates.

Frey and Osborne (2017) have considered 702 jobs and estimated their probability to be taken over by computers. They conclude that most of the jobs in the manufacturing and in clerical work will disappear, while those which have the lower risk of disappearing are those which involve human interaction, such as most of the jobs in health and education, political and managerial skills or highly specialised technical and scientific competencies. Filippi and Trento (2019) have recently carried out a similar inquiry on the Italian case and found also a relevant, though weaker, impact of the new technologies.

But in addition to this standard labour substitution argument, new technologies may have a negative impact on employment for some additional and less obvious reasons, which I will discuss in the next section.

Employment and productivity in automation technologies

As already mentioned, one of the fundamental compensation effects in an Industrial Revolution is the creation of new jobs in the industries which produce the capital goods which are at the core of the new technologies. In the first and second Industrial Revolution a considerable amount of jobs were created in the industries producing engines, electricity, steel, machine tools, etc. The third and fourth Industrial Revolutions are based on computers and digital goods – software and any other kind of file which can be processed by a computer. Although by far not the only constituent element, digital technologies and digital goods are at the core of the third and fourth Industrial Revolution, as the steam engine, steel and electricity were at the core of the first two Industrial Revolutions. The production technologies for digital goods such as software and any other kind of files which can be processed by a computer are quite different from the production of standard manufacturing goods and services. Once the first unit of a digital good has been produced, duplicating it has almost zero cost (Quah 2003). Thus, the costs of production are essentially fixed costs whereas marginal costs are negligible and the labour input goes almost entirely into the production of the first unit. This cost structure has some important consequences (Guellec and Paunov 2017). First, the link between quantity produced and employment is rather weak as quantity can be increased with little labour input. This compensation effect was due to the increase of production which was perhaps the most important virtuous circle operating in the previous Industrial Revolutions: the labour input per unit product decreased but output increased much faster, producing a positive net effect on employment. In digital technologies instead, the effect on employment of increasing output tends to be limited.

Second, this cost structure with almost zero marginal costs resembles one of natural monopolies and leads therefore naturally to high concentration, especially in the presence of network economies that often characterise digital technologies.

A second important phenomenon is the impact that internet and digital technologies have on the production services. It is well known that production of services was normally subject to a major source of inefficiency as they required contemporaneity and co-presence of production and consumption. If today I want to buy a car in Rome I can buy one which was produced weeks ago in a factory located thousands of miles away. If instead I need a medical service, the physician must have spare production capacity here and now and devote it specifically to me. This makes the production of services highly inefficient because it can serve only a small local market, cannot exploit economies of scale and cannot produce or buffer inventories. But, on the other hand, services have been a source of employment, often of high quality, which was inevitably much more geographically diffused then employment in manufacturing, which tends to be much more concentrated. Information technologies and in perspective AI are partly delinking production and consumption in many services: with internet banking, we can operate with a branch of our bank located on the other side of the earth, or simply located in the "cyberspace". The potential for a massive process of "industrialisation" in the production services with economies of scale and agglomeration, creation of large production units serving the large markets and opportunities for division of labour and specialisation is there in many other service sectors: from distribution to financial, medical and educational services.

Third, it is well known and largely debated that the current wave of technological innovation is not paralleled by a general increase of labour productivity. There are many possible explanations of this apparent paradox (Brynjolfsson 1993; Gordon 2000; Acemoglu et al. 2014) but one of them suggests that an important part of the current wave of innovation is not enhancing the productivity of workers but simply substituting them even in tasks in which machines are not performing much better than humans. Moreover, also the technology-driven creation of new jobs is disappointing:

> if all we do is continue down the path of automation, with no counterbalancing innovations to generate new tasks, the implications for labour are depressing. It will not be the end of work anytime soon, but the trend towards lower labour share and anaemic growth in labour demand will continue – with potentially disastrous consequences for income inequality and social cohesion.
>
> *(Acemoglu and Restrepo 2019: 5)*

Conclusion: automation and inequality

I argued in this short note that the age of technological unemployment that Keynes envisaged 90 years ago might be approaching, though much later than Keynes foresaw. The fourth

Industrial Revolution is at our doorstep and promises to change radically the organisation of production and the role of labour. There is little doubt that many jobs will become obsolete and technological unemployment will rise. The long-term issue is whether this unemployment will be absorbed by a parallel creation of new and mostly better jobs. It is hard to give a precise answer to such a question, but I argued that some features of the new technologies and the new organisational forms of production cast serious doubts on the possibility that also this time the virtuous circles that in the past produced a positive long-run effect on labour will also operate in the next decades.

In the meantime, we are already observing clear signs of a general decrease of the labour share (Karabarbounis and Neiman 2014) and increase of job polarisation (Autor et al. 2006). These two phenomena are among the leading causes of the general increase of income inequality we have been experiencing in the last decades in all Western advanced economies (Piketty 2013; Milanovic 2016). Roughly stable for a long time, the share of the produced value which goes to labour has been declining since the 1980s to the advantage of the share which goes to capital. This is a clear sign of a macro phenomenon of substitution of labour with capital. On the other hand, the labour markets in the advanced economies are becoming more and more polarised, in the sense that we observe a sharp decline of employment in the "middle" range of skills and wages, while employment is stable or increasing at the extremes (i.e. for high skill and generously paid jobs on one side and for low skill and low pay jobs on the other). The former are jobs requiring very high qualifications, often linked to the creation, design and management of the new technologies. The latter are on the contrary jobs with very low qualification but in which the human component cannot be substituted or is so cheap that substitution is not economically viable. Migrations from poorer countries are often an additional factor that keeps wages for these jobs particularly low.

Last but not least, globalisation is accelerating many of the phenomena described. The few remaining labour-intensive tasks in production are often relocated in developing countries where wages and labour regulations allow major savings in labour costs. The global extensions of digital markets generate global monopolies that we could hardly observe in the past. The industrialisation of services is also accompanied by a relocation of production and jobs on the global market. Finally, migrations have, on one hand, created a kind of global *élite* of workers with top qualifications and offering labour on a global market and, on the other hand, of the spectrum, a global *proletariat* which exerts downward pressure on wages for the low-skill jobs.

References

Acemoglu, D., Autor, D., Dorn, D., Hanson, G. and Price, B. (2014): Return of the Solow Paradox? IT, Productivity, and Employment in US Manufacturing, *American Economic Review* 104*(5)*, 394–399.

Acemoglu, D. and Restrepo, P. (2019): The Wrong Kind of AI? Artificial Intelligence and the Future of Labor Demand. *NBER Working Paper 25682.*

Autor, D., Katz, L. and Kearney, M. (2006): The Polarization of the Labor Market, *American Economic Review* 96(2), 189–194.

Beaudry, P., Green, D. and Sand, B. (2013): The Great Reversal in the Demand for Skill and Cognitive Tasks, Cambridge MA. *NBER Working Paper No. 18901.*

Brynjolfsson, E. (1993): The Productivity Paradox of Information Technology, *Communications of the ACM* 36(12), 66–77.

Filippi, E. and Trento, S. (2019): *The Probability of Automation of Occupations in Italy, Department of Economics and Management.* University of Trento, mimeo.

Frey, C. and Osborne, M. (2017): The Future of Employment: How Susceptible are Jobs To Computerisation? *Technological Forecasting and Social Change 114(C)*, 254–280.

Goldin, C. and Katz, L. (2008): *The Race Between Education and Technology*. Cambridge, MA: Harvard University Press.

Gordon, R. J. (2000): Does the 'New Economy' Measure Up to the Great Inventions of the Past? *Journal of Economic Perspectives* 14(4), 49–74.

Guellec, D. and Paunov, C. (2017): Digital Innovation and the Distribution of Income. *NBER Working Paper No. 23987.*

Jones, S. (2006): *Against Technology*. London: Routledge.

Karabarbounis, L. and Neiman, B. (2014): The Global Decline of the Labor Share, *Quarterly Journal of Economics* 129(1), 61–103

Keynes, J. M. (1930): Economic Possibilities for our Grandchildren, reprinted in *Essays in Persuasion (1963).* New York: W. W. Norton & Co.

Milanovic, B. (2016): *Global Inequality: A New Approach for the Age of Globalization*. Cambridge, MA: The Belknap Press of Harvard University Press.

Piketty, T. (2013): *Le Capital au 21ᵉ siècle*. Paris: Ed. du Seuil.

Piva, M. and Vivarelli, M. (2017): Technological Change and Employment: Were Ricardo and Marx Right? *IZA Discussion Paper 10471.*

Quah, D. (2003): Digital Goods and the New Economy, in Jones, D. C. (ed.): *The New Economy Handbook.* Amsterdam; London: Academic Press/San Diego, US: Elsevier Science, 289–321.

Schwab, K. (2016): *The Fourth Industrial Revolution*. New York: Penguin.

PART 2

Smart technologies and work

7

SMART TECHNOLOGIES AND THE CHANGING SKILLS LANDSCAPE IN DEVELOPING COUNTRIES

Karishma Banga

Introduction

Developing countries currently face a critical challenge in terms of creating productive employment and skills-development. For instance, 18 million new and productive jobs will need to be created each year through to 2035 to keep up with demographic challenges of youth influx into the sub-Saharan labour market (IMF 2015). Moreover, the Fourth Industrial Revolution is raising concerns of 'jobless growth', poor quality informal work and in general a lack of 'decent work' – over 1.5 billion people remain in vulnerable employment (ILO 2016). Furthermore, the COVID-19 pandemic has given a rapid boost to the use of smart technologies, with a rise in tele-working, tele-medicines, tele-education, e-commerce and online gaming, etc. This has increased the importance of digital skills in building resilience to shocks such as the pandemic. A survey of 68,574 people in the age group 16–35 from six countries in ASEAN, for instance, reveals that 87% of the respondents increased their usage of at least one digital tool during the pandemic and 42% picked up at least one new digital tool (WEF-ASEAN survey 2020). The number of users on African Development Bank's Coding for Employment platform (CfE) e-learning platform also witnessed a sharp spike in online learning; the number of users rose by 38.5% to 9,000 within one week (AfDB 2020).

To prepare the youth for productive employment, and for changes in the labour market, developing economies need to adapt to the changing landscape of skills. In an increasingly digitalised economy, employability is being increasingly determined by 'new skills for the future' – mainly digital and soft skills. These can be developed through government policies on education and skills-development as well as private sector investment into targeted skills-development. It is also key to note that the relationship between policies and digitalisation is two-way; while policies can boost skills for the digital economy, digitalisation can increase viability and efficiency of policy solutions, such as through online portals for skills-development, mass online courses, EdTech, etc.

This paper proceeds as follows; first, the evidence on the changing landscape of labour markets is discussed at the country level, followed by an examination of the sectoral shifts in employment. Afterwards, a tasks-based framework is presented, which is followed by a discussion on the implications of this framework for labour markets in developed and developing countries. The next section focuses on smart technologies and future skill-needs. The last section concludes and discusses policy implications.

 DOI: 10.4324/9780429351921-10

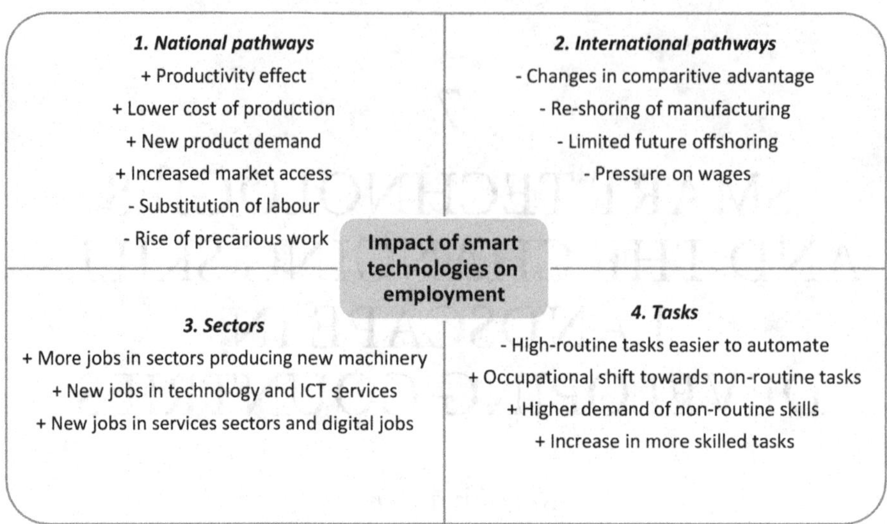

Figure 7.1 Matrix for analysing the impact of digitalisation on employment

Source: Author

Smart technologies and country-level employment changes

Analysing the implications of digital technologies on labour markets is a complex issue, since digital technologies can affect employment in a country through national and international pathways, with its impact differing across type of technology considered, sectors, industries and type of tasks (see Figure 7.1). The majority of the evidence on employment effects of automation is from developed countries, with a nascent body of literature on developing countries.

Smart technologies and job creation

The potential of digitalisation to create new jobs should not be underestimated. Digital technologies can boost the supply of employment opportunities through a number of channels, including increase in productivity leading to higher output and exports; lower cost of production leading to higher profits which can be re-invested into existing and new product-lines; rise in existing product demand through expansion in customer base using online e-commerce platforms; and lower barriers to entry in the export market – summarised as pathways under Block 1 (Figure 7.1).

Some studies document a rise in jobs due to automation but these are mainly based on data from developed countries. For instance, Gregory et al. (2016) find that computerisation between 1999 and 2010 in the European Union (EU) led to the creation of 11.6 million jobs, not due, as one might expect, to the absence of labour replacing capital but to an increase in product demand and spillover effects in the non-tradable sectors. Similarly, Booz and Company's (2012) estimates show that digitisation created 19 million jobs in the global economy between 2009 to 2010. Muro and Andes (2015) confirm that developed countries which invest more in robots lose fewer manufacturing jobs than countries which do not.

Even if robotics and automation are substituting labour in certain manufacturing industries, these job losses can be offset by job growth in other complementary manufacturing industries, such as electronics and ICT hardware (WTO 2017), as well as job growth in services sectors

(Dauth et al. 2017) which are producing and managing these technologies, such as ICT and ICT-enabled services. Even with a manufacturing firm, increasing automation may displace workers performing repetitive tasks in the factory but can create new jobs for more skilled labour in the firm, as well as jobs in the service of automation tools and machinery. However, Autor and Salomons's (2018) analysis, using country-industry data for 18 developed countries, finds that while automation in one industry indeed displaces employment in that industry, there is lack of evidence for own-industry employment losses being recovered in other sectors. Direct employment losses are offset to some extent by indirect employment gains in customer industries and through increases in aggregate demand (Autor and Salomons 2018). A key drawback in some of the studies that examine the impact of automation on employment is that the labour market is assumed to be frictionless (see, for instance, Autor et al. 2003; Zeira 1998). This implies that a firm's decision to automate is solely based on the relative price of labour versus technology. Recent evidence reveals that due to frictions in the labour market, workers find it difficult to adjust to industry-level import shocks, leading to high costs of job losses and unemployment, particularly for those with low initial wage levels (Autor et al. 2014). Labour market frictions also differ across countries; labour mobility costs are much higher in developing countries as compared to developed countries (Hollweg et al. 2014).

Smart technologies and job destruction

Some studies find automation to have a labour-substituting effect – that is, automation can displace jobs and substitute labour, affecting overall employment negatively (see, for instance, Frey and Osborne 2017; Bowles 2014; Acemoglu and Restrepo 2017). Examining the impact of computerisation on employment, Frey and Osborne (2017) find that 47% of the jobs in US are at risk of being automated. Using the same methodology, Bowles (2014) finds that 40–60% of the workforce in EU can be displaced by technological changes, with particularly strong effects in labour markets of Romania, Portugal, Greece and Bulgaria, while 57% of jobs in the OECD are at risk of being automated (Oxford Martin School 2016). Some studies, such as Frey and Rahbari (2016) and World Bank (2016) also report that jobs in developing countries are at risk, with particular threat to the manufacturing sector (Hallward-Driemeier and Nayyar 2017; Schlogl and Sumner 2020). Estimates of jobs losses are as high as 69% in India and 77% in China (Oxford Martin School 2016).

A serious drawback of some of these studies is that they assume that occupations as a whole can be automated away. For instance, Frey and Osborne (2017) use a binary categorisation of occupations in their paper, automatable occupations and non-automatable occupations. There is a significant variability in tasks within each occupation, with certain tasks and activities in a job susceptible to automation, while others are not, implying that automatability of jobs is not a binary characteristic (Autor and Handel 2013; Willcocks 2020). A better approach when examining the employment impact of digitalisation is therefore to analyse the task content of individual jobs. On breaking down occupation into tasks that have different levels of automatability, the share of jobs that can be automated in the OECD countries falls to 9% in the US and between 6 to 12% in the OECD, with significant differences across countries depending on regulation, behaviour of users, economic and technical feasibility and labour market dynamics (Arntz et al. 2016).

More recently, McKinsey (2017) examines the potential impact of automation in roughly 2,000 work activities for more than 800 occupations, across developed and developing countries (based on the US labour market data). Assessing the ability of robots in 'sensory perception, cognitive capabilities, natural language processing, social and emotional capabilities, and physical capabilities', it is found that only a small percentage (less than 5%) of occupations can be fully automated, although roughly 50% of work activities in almost all occupations can be automated using current digital technologies. Building on this, Manyika et al. (2017) estimate that by 2030, roughly 3–14% of the global workforce will need to switch occupational categories. While some work activities across all skills-levels can be automated, automation is likely to affect lower-skilled workers more, particularly those involved in the manufacturing sector (McKinsey 2017). Sectoral composition of economic activity is key to understanding economic development (Herrendorf et al. 2014). The national and international pathways discussed earlier are likely to affect employment differently across sectors, industries and tasks (Blocks 3 and 4). For instance, while the manufacturing sector overall is the most susceptible to automation, the rate of automation will differ across industries, depending on technological and economic feasibility. This is discussed further in the section to follow.

In conclusion, there is no consensus that emerges from the literature on automation and employment. Studies use different types of data and methodologies across different regions, with some studies focusing more explicitly on employment impact, acknowledging that estimation of probability/potential of technical automation does not equate with job losses (Parschau and Hauge 2020).

Smart technologies; sectoral shifts in employment

Compared to developed countries, developing economies have a larger agricultural sector, lower shares of employment and value-added in industry and manufacturing, as well as a large informal service sector (Schlogl and Sumner 2020). Smart technologies in agriculture can play a critical role in boosting job creation through increased productivity, value-addition and diversification of functions, increased regional trade, formalisation of work and increased opportunities for women and youth (Krishnan et al. 2020). For instance, use of digital platforms in agricultural value chain and collection of data histories can help in keeping records of creditworthiness of farmers, youth and women who previously may not have been eligible for working capital or personal loans, bringing traditionally marginalised sections of the society into more formal work setting (ibid.). However, it is critical to note that a persistent gendered divide in access to digital technologies can exacerbate the gendered divide in employment. A large proportion of women in developing countries work in the agricultural sector – for instance, 76% in Kenya and 84% in Rwanda – but continue to be marginalised as a result of socio-cultural norms that curb their basic rights and entitlements (such as land ownership), lack of access to the internet, basic skills and education (Commonwealth Secretariat 2020).

In addition to a gendered digital divide in 'access' to smart technologies, some studies report a digital divide between men and women in the 'use' of these technologies. For instance, a survey of 821 Ugandan farmers shows that female farmers have lower access to digital platforms than men, but even on these platforms, female farmers are lagging in internet access, access to credit and formal work opportunities and access to productivity-enhancing services (ODI 2020). This is linked to lower ownership of mobile phones, lower education and lower digital skills-level in women compared to men (ibid.). In fact, the gendered digital skills-divide is quite stark between developed and developing countries; for instance, while roughly 60% of the female population in the UK can conduct the basic function of copying or moving a file or a folder on the computer, this falls down to as low as 2.9% of the female population in Pakistan who can carry out this basic ICT function (Commonwealth Secretariat 2020). Women also have a lower access to financial services, particularly through mobile technology (Hunt and Samman 2016). Women are 14% less likely to own a mobile phone as compared to men, which translates into 200 million fewer women than men owning mobile phones in low- and middle-income countries (GSMA 2016).

Moving labour and other resources away from lower productivity sectors, such as agriculture, to higher productivity sectors, such as manufacturing and services, has been labelled as structural change or transformation, including through adoption of new technologies (Mc Millan et al. 2017). Manufacturing-led development has been historically used as a stepping-stone for job creation in developing countries. However, barring a small group of Asian economies, the share of manufacturing value-added in GDP has been declining in developing economies, as a result of structural changes, changing global demand and technological progress – also known as 'premature de-industrialisation' (Rodrik 2013). Compared to the services and agriculture sectors, the manufacturing sector is more intensive in 'routine' tasks (OECD 2016), which are more easily codifiable and therefore easier to automate. In line with this, the World Economic Forum (WEF) Job Survey (2016) predicts an overall decline of 1.6% in manufacturing and production employment in the period 2015–2020, largely driven by labour-substituting technologies (across developed and developing countries), with 3D printing expected to reduce employment by 3.5%, followed by the changing nature of work (−3%), new energy supplies and robotics. However, these estimates need to be treated with caution since they do not account for labour-complementing productivity improvements through digital technologies, such as robotics. A survey by Capgemini Worldwide[1] (2017) in the United States, the United Kingdom, France, Germany, Italy, Sweden, China and India finds that 76% of manufacturers already have an ongoing 'smart factory' initiative or are working on one. These smart factories operate on Industrial Internet of Things (IoT), a digital technology that is expected to have a major impact on the landscape of manufacturing. It allows scaling up of interconnected manufacturing, where machinery and equipment communicate with each other through the internet, without a human operator. Such IoT-based manufacturing will require transmission of data across the entire production chain, indicating the increasingly important role of ICT services in the manufacturing processes. This also highlights the growing role of data processing services and need for advanced data analytics.

Within the manufacturing sector, investment in automation is taking place at different rates across industries, depending both on technological and economic feasibility. For instance, the deployment of robots in the garments sector, particularly for sewing operations, has been a challenge due to limpness of the fabric and frequent slipping of materials, indicating that the garment industry could be one of the last to be fully digitally automated (Altenburg et al. 2020). Robots in this sector would require more dexterity as compared to other sectors. At the same time, average wages in the garments sector are relatively low compared to the automotive sector, indicating that automation may not be economical.

Comparing the distribution of robot sales across the main industries of selected countries in Figure 7.2, one can observe that robot deployment in the US, UK and China is much more diversified. Robots are being used in all of the following industries: automotive, electrical, chemical, metal products, food and beverage and other which includes apparels, footwear, etc. In contrast, in developing economies of India, South Africa and Mexico, roughly more than 60% of total robot deployment is concentrated just in the automotive sector. Similarly, in Malaysia, 64% of robot deployment is in the electrical/electronical industries.

Focusing on developing Asia's use of industrial robots, AfDB et al. (2018) confirm that robot deployment is concentrated mainly in capital-intensive manufacturing – which has relatively

1 A Paris-based multinational information technology consulting corporation.

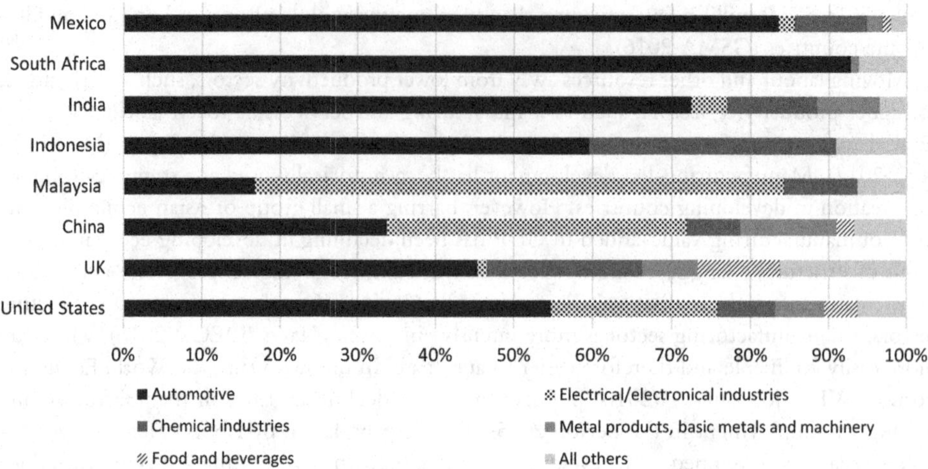

Figure 7.2 Distribution of robots across different industries

Source: Author, constructed from IFR (2017)

low employment levels to begin with. As per their report, while the electrical/electronics sector and automotive sector each accounted for 39% of total robot use in 2015, these sectors accounted for only 9.2% and 4.2% of total manufacturing employment, respectively. In contrast, textiles, apparel and leather combined accounted for only 0.1% of robot usage in 2015, but 19.2% of total manufacturing employment. Given that robot usage is concentrated in sectors with relatively low employment, AfDB et al. (2018) point out that concerns about robots replacing workers may be overstated. Furthermore, labour displacement in developing economies is restricted by the low elasticity of factor substitution (Dao et al. 2017). Globally, industries of paper and paper products, wood and wood products, basic metals, food, beverage and tobacco, and textiles and garments are less affected by global technological changes (Hallward-Driemeier and Nayyar 2017). Developing countries, which are still at nascent stages of installing smart technologies, therefore have a window of opportunity in these sectors, which will still require workforce with traditional industrial skills.

But how long will this window of opportunity last? Although it might take longer compared to developed economies, the relative price of investment goods is eventually likely to fall in developing economies due to continuous and significant advances in technical feasibility of digital technologies and falling capital costs. Banga and Velde (2018a) demonstrate this for the case of the furniture industry; the authors find that while a robot in the US will become cheaper than labour in the US furniture manufacturing industry by 2023, a robot in the Kenyan furniture manufacturing will become cheaper than Kenyan labour more than a decade later – in 2034. Moreover, by 2033, operating a robot in the US becomes cheaper than Kenyan labour, signalling the possible re-shoring of furniture manufacturing tasks to the US[2] around this time.

Smart technologies and shifting patterns of globalisation question the potential of manufacturing-led development for job creation, particularly for developing countries. A strand of literature has subsequently emerged examining the role of services-led economic transformation.

2 These estimates are based on relative prices of operating a robot and per unit cost of labour. Labour productivity increases over time are accounted for but do not include other factors such as transportation and time to market costs.

Smart technologies are opening up new avenues of value-added in the services sector, including in IT or IT-enabled services, such as business and financial services (Newfarmer et al. 2019; Gollin 2018). In African countries, for instance, mobile telephony and mobile banking have already achieved some successes, and other services sectors of potential include social, education and healthcare sectors, as well as tourism and infrastructure construction.

However, a services-led development model may not be employment-intensive; highly productive and tradable services, such as IT services, require highly skilled workers, while non-tradeable services, such as social care and other personal services, are neither highly value-adding in nature nor scalable (Schlogl and Sumner 2020). Moreover, in some services sectors as well, digital technologies are found to be employment reducing. For instance, mobile internet and cloud technology is predicted to reduce employment by roughly 3.9% in installation and maintenance, and by 5.82% in office and administrative work (WEF 2016). In office and administrative work, virtual assistants are rapidly changing the role or tasks provided by secretaries, particularly in small- and medium-sized enterprises. While traditionally, secretaries have worked in an office setting, virtual assistants are independent contractors or employees who can remotely offer many secretarial services, including office services such as calendar management, documents and filing support, as well as technical services such as social media support, email marketing, transcription, etc. Tasks in financial services, such as data entry and accounting and filing claims as well as manual, clerical, logistical tasks in transportation and storage, inventory management and back-office processing face high exposure to routinisation.

Overall, services-led development tends to require relatively high skills, particularly IT services, which further needs long-term, steady investments in education, infrastructure, institutions and governance. A more balanced approach in developing countries, which incorporates services, natural resources, agriculture and some manufacturing, may therefore be important. This seems to be particularly relevant for Africa, where 60% of the population works in agriculture; natural resources are in abundance; and digital services are probably only realistic for the better-educated segments of the labour market.

Towards a tasks-based framework for skills-analysis

From the previous sections, an important point emerges around shifting the focus from 'automatability of jobs' to 'automatability of tasks' for understanding future skill-needs in developing countries. The underlying assumption in some studies that occupations as a whole can be 'automated away may result in overestimating job losses, since there is a great variability in the tasks within each occupation, which is not accounted for in studies of Frey and Osborne and others who rely on the same methodology (Autor and Handel 2013).

Skill requirements across types of tasks can be understood using the skills-tasks matrix, developed by Banga and te Velde (2018b) and presented in Figure 7.3. Tasks can be classified as routine or non-routine and manual or cognitive. Different combinations of these tasks' categories determine the associated skill-needs:

- The first combination is *routine-manual tasks*; physical tasks which follow explicit rules and are easier to codify. Examples include operating machinery, carpentry and constructions, performed mainly by middle-skilled workers with physical skills. These tasks are repetitive and do not involve much analytical thinking, and therefore are being increasingly automated. Some routine manual tasks require hand dexterity, such as stitching in garments, which cannot yet be easily performed by robots.

	Manual	*Cognitive*
Routine	**Physical skills** such as controlling and operating machinery, assembling parts, construction, carpentry.	**Enterprise skills** such as bookkeeping, accountancy, pay-roll processing, cashier and clerk work
Non-routine	**Physical skills** of operating vehicles, industrial truck operations, **interpersonal skills and socio-emotional and empathetic skills** for sales occupations and service occupations such as protecting, caring, janitorial services.	**Job-neutral** hard skills such as basic and intermediate digital skills, proficiency in foreign language. **Job-neutral** soft skills such as interpersonal skills, managerial skills, analytical and critical thinking, problem-solving and creative and design skills. **Job-specific** hard skills such as legal skills, pharmacy skills, advanced digital skills

Figure 7.3 Skills-tasks matrix

Source: Banga and te Velde (2018b)

- The second combination is *routine-cognitive tasks*. Compared to routine-manual tasks, these tasks are associated with higher education levels, reading and writing capabilities, as well as more developed cognitive abilities, such as analytical and critical thinking, logical and mathematical reasoning and managing. Routine-cognitive tasks are also performed mostly by middle-skilled workers such as bookkeepers, secretaries, bank tellers and clerks, and are being automated at a fast rate. For example, deployment of self-checkout counters instead of cashiers in stores, and free business management software such as Money Manager EX, TurboCASH and so on.
- *Non-routine manual tasks* are those that either involve agile physical skills (such as operating a vehicle) or 'soft' skills (such as interpersonal skills, socio-emotional skills, or empathetic skills). These tasks relate more to service occupations such as nursing, childcare, janitorial work, security work, etc. These tasks, generally performed by low-skilled workers, do not follow any set pattern of rules and require tacit knowledge or personal interactions which are hard to automate.
- *Non-routine cognitive tasks* require job-neutral 'hard' skills (measurable skills) such as digital skills (collecting information from online sources, data analysis, etc.) and job-neutral 'soft' skills, such as interpersonal skills, managerial skills, analytical thinking and creative-thinking skills. They can also involve job-specific skills, such as legal writing, coding and programming, healthcare, education and training, professional and technical services.

With the rise of smart technologies, the demand for labour in non-routine cognitive tasks is likely to increase. The demand for workers performing non-routine tasks has been on an increasing trend in a number of economies; since the 2000s, the employment share of occupations intensive in non-routine cognitive skills (such as analytical and critical thinking) and socio-emotional skills has increased from 19 to 23% in emerging economies, and from 33 to 41% in advanced economies (World Development Report 2019). Examining 13 major developed and emerging economies, WEF (2016) finds that the percentage of jobs requiring cognitive abilities as a core skill is expected to rise by 15% between 2015–2020. Furthermore, among all the jobs requiring cognitive abilities as part of their core skill sets, 52% of them do not have such requirements now but are expected to experience increasing demand for cognitive abilities by 2020 (WEF 2016).

Smart technologies and 'polarisation' of skills?

From the previous section, it is clear that there is no consensus in the literature on the impact of automation on employment. Several studies have, however, established an increase in demand for labour in non-routine tasks linked to rise in smart technologies. A strand of literature on developed countries argues that automation has led to 'labour market polarisation' or 'hollowing out' of the middle-skilled workforce, with increasing demand for high-skilled and low-skilled labour relative to middle-skilled workers (see Autor et al. 2006; Goos and Manning 2007; Autor and Dorn 2013; Goos et al. 2014; Beaudry et al. 2016). The main reason for this polarisation is 'routinisation'; middle-skilled workers[3] are engaged in occupations that consist of routine tasks that can be more easily automated. In contrast, high-skilled workers perform non-routine tasks, complementary to new technologies, such as research and development, managing, designing (Beaudry et al. 2016), and low-skilled workers perform non-routine manual tasks that are hard to automate, such as nursing and childcare. Highly paid skilled work has in turn raised the demand for low-paid services, reinforcing polarisation of occupations into 'lovely' and 'lousy' jobs in developed countries (Goos and Manning 2007). These low-skilled services include non-routine tasks such as catering, construction, cleaning and childcare, which do not follow precise procedures and are therefore much harder to automate (Autor and Dorn 2013). Many of these services, however, remain the least educated and least paid categories of employment.

Job polarisation due to routinisation of tasks has been mainly documented for developed economies (see, for instance, Spitz-Oener 2006; Autor and Dorn 2013; Michaels et al. 2014; Goos et al. 2014; Ikenaga and Kamibayashi 2016), with the degree of dislocation of middle-skilled workers varying significantly. For developing countries, the literature on job polarisation is limited and there appears to be no single picture that has emerged. There is evidence of job polarisation in Chile during the 2000s (Messina et al. 2016) and in Brazil and Mexico (Maloney and Molina 2016). Some studies, such as World Bank (2016) and Reijnders and De Vries's (2017) find declining shares of middle-skilled workers in some developing countries, and conclude that as being indicative of job polarisation. As per World Bank (2016), some countries, such as Malaysia and Uganda, witnessed declining share of middle-skilled workers in the period 1995–2012, but the rate of decline is lower in developing countries – at 0.39% as opposed to 0.59% a year in developed countries (World Bank 2016). Focusing on Asian economies, AfDB et al. (2018) find that between 2005–2015, the share of jobs intensive in non-routine[4] tasks increased, while routine-intensive jobs declined. ADB's (2018) analysis of employment trends in five developing Asian economies (India, Indonesia, the Philippines, Thailand, and Viet Nam) shows that over the past decade, annual expansion of employment in jobs intensive in non-routine cognitive tasks, social interactions and the use of ICT was 2.6% faster than total employment. Figure 7.4 shows employment growth by task-intensity and type across the

3 Middle-skilled occupations are those that are intensive in routine cognitive and manual skills, such as clerks, crafts and related workers, plant and machine operators. In contrast, high-skilled occupations are intensive in non-routine cognitive and interpersonal skills such as technicians and professionals, while low-skilled occupations are intensive in non-routine manual skills such as sales and service workers.

4 Classification into routine and non-routine is based on Autor and Dorn (2013) and excludes agricultural occupations.

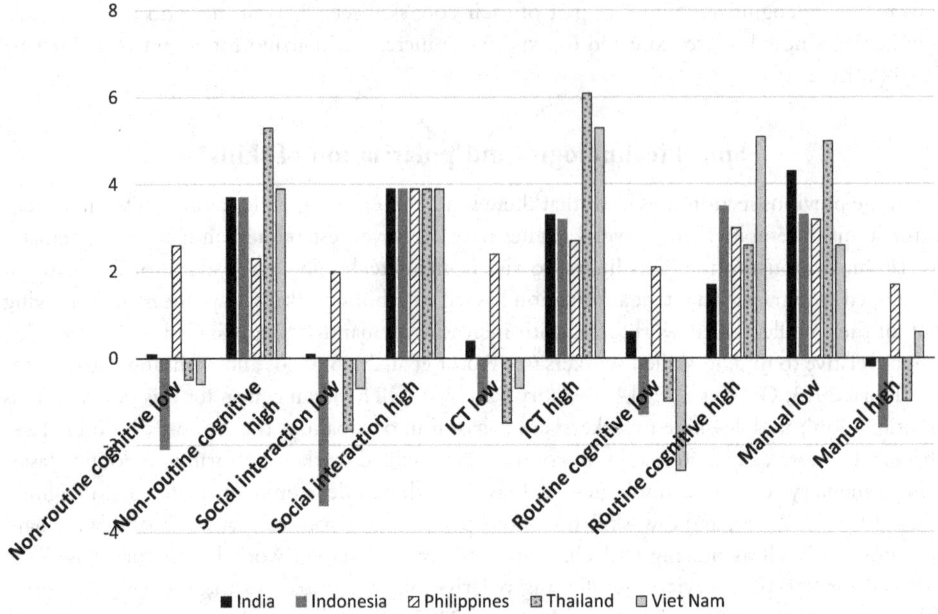

Figure 7.4 Annual employment growth (in %) in jobs, by task-intensity and type

Source: Reproduced from ADB (2018)

Note: The time frames vary across countries; Viet Nam (2007–2015), Thailand (2000–2010), India (2000–2012), the Philippines (2001–2013) and Indonesia (2000–2014). Jobs are classified in terms of whether they are high-intensive or low-intensive with respective to the five task category.

five Asian economies; employment growth is observed to be positive in cognitive tasks (both routine and non-routine) and tasks which are intensive in social interaction and ICT. Rapid digital progress in some sectors accompanied by a proliferation of low-paid service jobs has also been documented in Turner (2018). Detailed analysis of occupation titles in three Asian countries – Malaysia, Philippines and India[5] – by the ADB (2018) further reveals that new job titles have emerged to handle new technologies, particularly in occupations of ICT operations and user support technicians, architect and designers, software application developers and analysts, medical and pharmaceutical technicians, electronics and telecommunications installers, sales and marketing professionals as well as database and network professionals and other health associate professionals. In contrast, manual jobs have experienced limited employment and wage growth, contributing to rising inequality (ibid.).

Studies that use declining shares of middle-skilled workers to show job polarisation, such as World Bank (2016), suffer from the serious drawback of assuming broad groups of middle-skilled occupations being routine-intensive (Das and Hilgenstock 2018). Within occupations, there is heterogeneity in routine-intensity of tasks, which needs to be accounted for in order to avoid overestimation of the decline in middle-skilled occupations (ibid.). Moreover, for routinisation to result in polarisation, a significant share of the economy has to be engaged in middle-skilled occupations (Das and Hilgenstock 2018), which is not the case for developing

5 Data for Malaysia is for the year 2008, India, 2015; Philippines, 2012.

economies, where jobs tend to be more concentrated in industries with low susceptibility to automation (ILO 2014; Maloney and Molina 2016). Roughly, 40% of the workforce in developing countries continues to be employed in the primary sector (ILO 2014). Furthermore, a key driver of routinisation-induced polarisation is the decline in the relative price of investment goods, which is more of a developed country phenomenon (Dao et al. 2017). In contrast to the evidence on job polarisation in developed countries, there is some evidence of smart technologies shifting employment towards less-skilled workers in developing countries. For example, digital technologies in Chilean manufacturing in the period 2007–2013 shifted employment structure towards less skilled workers performing routine and manual tasks (Almeida et al. 2017). In middle-income countries in Europe, such as in Bulgaria and Romania, while the demand for non-routine cognitive and interpersonal work is increasing, there is no increase in demand for low-skilled non-routine manual work (Gorka et al. 2017).

A zoom-in on skills for the future

In the age of smart technologies, there is likely to be growth in non-routine cognitive tasks with developed countries experiencing job-polarisation, explained further in the following sections. While the evidence of job-polarisation in the case of developing countries is mixed, it is clear that smart technologies will increasingly require digital and soft skills. We explore these categories, and how developing economies are faring in these skills, in more detail.

As demonstrated in Figure 7.3, under the block of non-routine cognitive tasks, digital skills can be categorised as job-neutral or job-specific. These skills range from basic to intermediate levels. Basic job-neutral skills are those that are neutral to the industry of the worker and are important for workers to function at a minimum level in a digital economy. These skills can be further broken down into hardware skills – the skills to operate a computer and touch-screen smartphones, as well as software/online skills such as emailing and searching the internet which can help gain access to information and people (ITU 2018). Intermediate job-neutral digital skills refer to skills for engaging with internet and digital technologies in a more productive manner that can be used across a wide range of digital tasks. Examples include skills to use computer technology (for instance, use of Microsoft® Office and PowerPoint), digital design (User Interface design, Photoshop, etc.), digital marketing (use of social media and electronic platforms) and data analytics and storage, as well as secure use of the internet to carry out such tasks (ITU 2018). Job-specific digital skills are advanced or specialised technical skills, such as computer programming, network management, coding, big data analytics, cryptography, and so on. Such skills are mainly acquired through advanced formal education but can also be learnt from other options such as incubators, boot camps, etc. (ITU 2018). Banga and te Velde (2018b) show that the share of ICT professionals – such as web programmers, coders etc. – in total employment is much lower in developing economies compared to the US; the share of ICT professionals in total employment is around 2.4% in the US, but as low as 0.18 and 0.04% in developing economies of Thailand and Indonesia respectively.

Table 7.1 presents ITU data on the percentage of population across a range of low-, middle- and high-income countries, across types of skills. It is observed that while over 60% of the population in high-income countries of UK and Germany have basic digital skills of copying/moving a file or folder or using copy/paste tools, only 3–5% of the populations in Pakistan, Zimbabwe and Sudan have basic digital skills. Similarly, over 30% of the population in Germany, Singapore, UK and Italy can use basic arithmetic formulas on a spreadsheet but this falls to below 10% in countries of Cambodia, Niger, Zimbabwe, Sudan, Pakistan, Jamaica and Tunisia. A wide variation is also noticed in terms of intermediate digital skills across income

Table 7.1 ICT skills, by type of skill (% of population)

	Copying or moving a file or folder	*Using copy and paste tools to duplicate or move information within a document*	*Using basic arithmetic formula in a spreadsheet*	*Sending emails with attached files*	*Connecting and installing new devices*	*Finding, downloading, installing and configuring software*	*Creating electronic presentations with presentation software*	*Transferring files between a computer and other devices*	*Writing a computer program using a specialised programming language*
Low-income countries									
Niger	13	7.7	1.1	6.2	4.6	3.7	0.9	5.4	0.9
Sudan	4.3	3.6	1.8	2.4	2.8	2.4	1.9	2.9	1.6
Middle-income countries									
Zimbabwe	5.1	3.5	1.8	4.3	1.7	1.8	1.7	3.2	0.6
Botswana	34	31.3	19.7	19.9	19.2		11.6	23.6	4.8
Brazil	23.5	20.8	11.8	21.4	9.5	16.6	9.9	17.7	2.9
Cambodia	27.8	26.8	9	37.6	1.5	0.6	2.5	20.6	0.1
Indonesia	49.5	49.5	7.9	7.1	21.8	12.5	7.9	56.6	
Jamaica	15.4	14	5.3	15.3	7.1	7.1	3.1	7.9	
Malaysia	55.5	55.2	24.8	43	48.4	33.7	20.2	42.4	7.5
Pakistan	5.5	4	2.1	3.7		3.4	1.7	2.4	1.5
Russian Federation	56.8	46.8	24.5	33.5	10.5	3.7	9.8	29.6	1.3
Thailand		5.1		17.2		7.4	27.6	25.8	
Tunisia	23	11.6	5.8	8.7	4.6	3.2	2.7	3.2	1.9
High-income countries									
France	52.1						33.6	54.7	5.6
Germany	68.2	62	37.7			63.2	39.1	61.7	5.6
Italy	49	42	30.7			34.9	31	43.2	4.3
Spain	56					51.9	37.7	53.4	6.4
United Kingdom	63.7	62.1	47.4			61.3	47.2	55.1	8.5

Source: ITU (2018)

status. In terms of advanced digital skills, it is observed that 6–8% of the population in Spain, UK, Singapore and Malaysia are using the internet for writing a computer programme using specialised programming language, but this falls to as low as 0.6% in Zimbabwe, 0.9% in Nigeria and 0.1% in Cambodia.

Soft skills refer to interpersonal skills, managerial skills, analytical, critical thinking and problem-solving skills as well as creative skills and adaptive skills. Employers in Benin, Liberia, Malawi and Zambia suggest technical skills, teamwork, communication and problem-solving skills are the most important set of skills for the future (World Bank 2019). Even advanced digital technologies can require soft skills. For instance, 3D printing involves the knowledge of computer-aided design and 'additive manufacturing', advanced digital skills of 3D modelling but also soft skills of problem-solving, critical thinking and creative designing. Developing countries continue to lag behind in soft skills compared to developed countries (Banga and te Velde 2018b).

Conclusion and policy implications

A number of important insights have emerged from this study. First, automation may create new employment opportunities, replace labour in certain tasks and also change the role of workers in certain occupations. The rate of automation differs across countries and sectors, depending on technological and economic feasibility; the impact of automation on the labour market also differs across countries, depending on a number of factors, including globalisation, new product demand, occupational structure in employment, labour frictions, institutional differences affecting relative wages and so on. Routinisation has led to labour market polarisation in developed economies. But there is less evidence of declining employment shares of middle-skilled workers in developing economies, possibly owing to labour being more concentrated in low-skilled and low-routine occupations as well as the slower decline in the relative price of investment. The rate of polarisation also varies across developing countries.

However, given the fast pace of technical advances and the falling cost of capital equipment, polarisation is a possible scenario for developing economies in the future. Given that digitalisation affects both employment and employment structures, it is important to identify what skills are likely to be in demand in the future. In the context of the digital economy, the study identifies core skills that can directly increase the competitiveness of the workforce: (1) job-neutral digital skills; (2) job-specific digital skills; and (3) job-neutral soft skills such as communication, management, analytical and critical thinking and creativity.

Complementary actions at different levels are needed for developing countries to close or reduce the digital and soft-skills gap with regards to developed countries. This will require addressing both supply-side and demand-side challenges to skills-development. On the supply side, there is a need to incorporate digital literacy and basic ICT skills at the primary and lower-secondary level of education, and increasing Technical and Vocational Education and Training (TVET) enrolment at the upper-secondary and tertiary levels to increase provision of intermediate to advanced digital skills and soft skills (Banga and te Velde 2019). Non-formal TVET can be extended through apprenticeships and Recognition of Prior Learning (RPL). With increasing deployment of smart technologies in production, employed-led training is emerging as an important channel of skills-supply, due to inability of TVET curricula to catch up with the fast pace of technological change (Wolter and Ryan 2011) and lack of effective trainers in TVET with expertise/knowledge of new technologies (ACET-MCF 2019).

Policies that create the right incentives for firms to demand technology and invest in skills-development are also important. This includes policies that facilitate competitive domestic market, R&D culture, technology transfer, foreign direct investment, upgrading in global value chains, etc. Improving national co-ordinating mechanisms to address systemic failures between the supply and demand for technology and skills is also important. This can be done through collaborations across educational and research institutes, training providers, enterprises and fostering skills diffusion. Some examples of these intermediate institutions include industry associations, R&D consortia and technology hubs or parks, in addition to online portals and platforms that can help in reduce information gaps across the demand and supply of skills (Banga and te Velde 2019).

With limited budgets, countries may need to make trade-offs. An important trade-off is between today's workers – a large percentage of whom lack basic digital and soft skills – and the future labour force for which early investments into digital and soft skills should be prioritised given the relatively higher returns (World Bank 2019). Ultimately, the relative weight placed on education policy priorities will depend on a country's cultural, political and economic context. For instance, countries which are more digitalised may choose to direct their resources in reorienting secondary and tertiary TVET towards digital skills, while those still at nascent stages may choose to focus on expanding basic digital literacy in the population. Coherence between education, skills development and other policies will further require solid institutional structures and clear mapping of responsibilities, better dialogue and effective co-ordination.

Identifying future skill-needs and channels of skills-development have become more important now than ever. This chapter showed that increasing use of smart technologies has significantly changed employment structures across countries and skills landscape. The COVID-19 pandemic is likely to accelerate the use of smart technologies across the globe, with businesses shifting online, people working from home, and students learning remotely. However, the pandemic also threatens to exacerbate the existing digital skills-divide across countries. Key areas of future research there include identifying how the pandemic is affecting deployment of smart technologies in global production; its subsequent effects on employment patterns and skills structures; examining whether it is exacerbating or creating job polarisation in developing countries; and how skills-development policies can contribute to inclusive post-crisis recovery.

References

Acemoglu, D. and Restrepo, P. (2017): *Robots and Jobs: Evidence from US Labor Markets.* Cambridge, MA: National Bureau of Economic Research (NBER).

ACET-MCF (2019): *The Future of Work in Africa: Implications for Secondary Education and TVET Systems.* Ghana, Africa: African Center for Economic Transformation (ACET). Available at: https://mastercardfdn.org/wp-content/uploads/2019/05/Future-of-Work-and-Implications-for-Secondary-Education-and-TVET-FINAL.pdf [02.04.2021]

AfDB (2020): The Relevance of Digital Skills in the COVID-19 Era. Available at: www.afdb.org/en/news-and-events/relevance-digital-skills-covid-19-era-36244 [30.03.2021]

African Development Bank (AfDB), Asian Development Bank (ADB), European Bank for Reconstruction and Development (EBRD), Inter-American Development Bank (IDB) (2018): The Future of Work: Regional Perspectives. Available at: https://publications.iadb.org/handle/11319/8840 [30.03.2021]

Almeida, R., Fernandes, A. M. and Viollaz, M. (2017): Does the Adoption of Complex Software Impact Employment Composition and the Skill Content of Occupations? Evidence from Chilean Firms. *Policy Research Working Paper 8110.* Washington, DC: World Bank.

Altenburg, T., Chen, X., Lütkenhorst, W., Staritz, C. and Whitfield, L. (2020): *Exporting Out of China or Out of Africa? Automation versus Relocation in the Global Clothing Industry* (No. 1/2020). Discussion Paper.

Arntz, M., Gregory, T. and Zierahn, U. (2016): The Risk of Automation for Jobs in OECD Countries: A Comparative Analysis. *OECD Social, Employment, and Migration Working Paper 189: 0–1.* Paris: OECD.

Asian Development Bank (ADB) (2018): *Asian Development Outlook: How Technology Affects Jobs.* Manila: Asian Development Bank. http://dx.doi.org/10.22617/FLS189310-3

Autor, D., Katz, L. F. and Kearney, M. (2006): May. Measuring and Interpreting Trends in Economic Inequality, *AEA Papers and Proceedings* 96(2), 189–194.

Autor, D. and Salomons, A. (2018): Is Automation Labor-displacing? Productivity Growth, Employment, and the Labor Share. *Brookings Papers on Economic Activity*.

Autor, D. H. and Dorn, D. (2013): The Growth of Low-Skill Service Jobs and the Polarization of the US Labor Market, *American Economic Review* 103(5), 1553–1597.

Autor, D. H., Dorn, D., Hanson, G. H. and Song, J. (2014): Trade Adjustment: Worker-level Evidence, *The Quarterly Journal of Economics* 129(4), 1799–1860.

Autor, D. H. and Handel, M. J. (2013): Putting Tasks to the Test: Human Capital, Job Tasks, and Wages, *Journal of Labor Economics* 31(1), S59–S96.

Autor, D. H., Levy, F. and Murnane, R. J. (2003): The Skill Content of Recent Technological Change: An Empirical Exploration, *Quarterly Journal of Economics* 118(4), 1279–1333.

Banga, K. and te Velde, D. W. (2018a): *Digitalisation and the Future of Manufacturing in Africa. Supporting Economic Transformation*. London: ODI.

Banga, K. and te Velde, D. W. (2018b): Skill Needs for the Future. *Background Paper, 10*. Pathways for Prosperity. Oxford University.

Banga, K. and te Velde, D. W. (2019): Overseas Development Institute, *Pathways for Prosperity Commission Background Paper Series no. 29 Oxford*. United Kingdom.

Beaudry, P., Green, D. A. and Sand, B. M. (2016): The Great Reversal in the Demand for Skill and Cognitive Tasks, *Journal of Labor Economics* 34(S1), S199–S247.

Booz and Company (2012): *Maximizing the Impact of Digitization*. Amsterdam et al: PwC.

Bowles, J. (2014): *The Computerisation of European Jobs*. Brussels: Bruegel.

Capgemini Worldwide (2017): Smart Factories and the Modern Manufacturer. Available at: www.capgemini.com/service/smart-factories-and-the-modern-manufacturer [30.03.2021]

Commonwealth Secretariat (2020): The State of the Digital Economy in the Commonwealth. Available at: https://thecommonwealth.org/sites/default/files/inline/Digital%20Connectivity%20Report_low%20res_.pdf [30.03.2021]

Dao, M. C., Das, M. M., Koczan, Z. and Lian, W. (2017): Why is Labor Receiving a Smaller Share of Global Income? Theory and Empirical Evidence. *International Monetary Fund Working Paper No. 17/169*.

Das, M. M. and Hilgenstock, B. (2018): The Exposure to Routinization: Labor Market Implications for Developed and Developing Economies. *International Monetary Fund Working Paper No. 18/135*.

Dauth, W., Findeisen, S., Südekum, J. and Woessner, N. (2017): German Robots – the Impact of Industrial Robots on Workers, *CEPR Discussion Paper No. DP12306*.

Frey, C. B. and Osborne, M. A. (2017): The Future of Employment: How Susceptible are Jobs to Computerisation? *Technological Forecasting and Social Change* 114, 254–280.

Frey, C. B. and Rahbari, E. (2016): Do Labor-saving Technologies Spell the Death of Jobs in the Developing World? Prepared for the 2016 Brookings Blum Roundtable. Available at: www.brookings.edu/wp-content/uploads/2016/07/Global_20160720_Blum_FreyRahbari.pdf [30.03.2021]

Gollin, D. (2018): *Structural Transformation and Growth without Industrialisation*. Background Paper. Oxford: Pathways for Prosperity Commission.

Goos, M. and Manning, A. (2007): Lousy and Lovely Jobs: The Rising Polarization of Work in Britain, *The Review of Economics and Statistics* 89(1), 118–133.

Goos, M., Manning, A. and Salomons, A. (2014): Explaining Job Polarization: Routine-biased Technological Change and Offshoring, *American Economic Review* 104(8), 2509–2526.

Gorka, S., Hardy, W., Keister, R. and Lewandowski, P. (2017): *Tasks and Skills in European Labor Markets*. IBS Research Report 03/2017. Warsaw: Institute for Structural Research.

Gregory, T., Salomons, A. and Zierahn, U. (2016): Racing with or against the Machine? Evidence from Europe. *Centre for European Economic Research Discussion Paper No. 16–053*.

GSMA (2016): Bridging the Gender Gap: Mobile Access and Usage in Low and Middle-income Countries. Available at: www.gsma.com/mobilefordevelopment/wp-content/uploads/2016/02/Connected-Women-Gender-Gap.pdf [30.03.2021]

Hallward-Driemeier, M. and Nayyar, G. (2017): *Trouble in the Making? The Future of Manufacturing-led Development*. Washington, DC: The World Bank.

Herrendorf, B., Rogerson, R. and Valentinyi, A. (2014): Growth and Structural Transformation. *NBER Working Paper Series No. 18996*. Cambridge, MA: NBER. Available at: www.nber.org/papers/w18996 [30.03.2021]

Hollweg, C. H., Lederman, D., Rojas, D. and Ruppert Bulmer, E. (2014): *Sticky Feet: How Labor Market Frictions Shape the Impact of International Trade on Jobs and Wages*. Washington, DC: The World Bank.

Hunt, A. and Samman, E. (2016): *Women's Economic Empowerment: Navigating Enablers and Constraints*. London: Overseas Development Institute. Available at: www.odi.org/sites/odi.org.uk/files/resource-documents/10683.pdf [30.03.2021]

Ikenaga, T. and Kamibayashi, R. (2016): Task Polarization in the Japanese Labor Market: Evidence of a Long-Term Trend, *Industrial Relations* 55, 267–293.

ILO (2016): Where are Workers More at Risk of Being in Vulnerable Employment? World Employment and Social Outlook – Trends 2016. Available at: www.ilo.org/global/about-the-ilo/multimedia/maps-and-charts/WCMS_443264/lang – en/index.htm [30.03.2021]

IMF (2015): *Africa Economic Outlook, April*. Washington, DC: International Monetary Fund.

International Labour Organization (ILO) (2014): *World of Work Report 2014: Developing with Jobs*. Geneva: ILO.

International Telecommunication Union (ITU) (2018): Digital Skills Toolkit. Available at: www. itu.int/pub/D-PHCB-CAP_BLD.02–2018 [30.03.2021]

The International Federation of Robotics (IFR) database, 2017. Available at https://ifr.org/

Krishnan, A., Banga, K. and Mendez-Parra, M. (2020): Disruptive Technologies in Agricultural Value Chains, *ODI Working Paper 576*. Available at: https://cdn.odi.org/media/documents/disruptive_agritech_-_5_mar_2020_-_final_draft.pdf [30.03.2021]

Maloney, W. and Molina, C. (2016): Are Automation and Trade Polarizing Developing Country Labor Markets, Too? *World Bank Policy Research Working Paper 7922*. Washington, DC: World Bank.

Manyika, J. Lund, S., Chui, M., Bughin, J., Woetzel, J., Batra, P., Ko, R. and Sanghvi, S. (2017): Jobs Lost, Jobs Gained: Workforce Transitions in a Time of Automation. McKinsey Global Institute. Available at: www.mckinsey.com/~/media/mckinsey/industries/public%20and%20social%20sector/our%20insights/what%20the%20future%20of%20work%20will%20mean%20for%20jobs%20skills%20and%20wages/mgi%20jobs%20lost-jobs%20gained_report_december%202017.pdf [30.03.2021]

McKinsey Global Institute (2017): A Future That Works: Automation, Employment, and Productivity. Available at: www.mckinsey.com/~/media/mckinsey/featured%20insights/Digital%20Disruption/Harnessing%20automation%20for%20a%20future%20that%20works/MGI-A-future-that-works-Executive-summary.ashx [30.03.2021]

Mc Millan, M., Rodrik, D. and Sepulveda, C. (2017): *Structural Change, Fundamentals, and Growth: A Framework and Case Studies*. Washington, DC: The World Bank.

Messina, J., Oviedo, A. M. and Pica, G. (2016): *Job Polarization in Latin America*. Mimeo. Washington, DC: World Bank.

Michaels, G., Natraj, A. and Van Reenen, J. (2014): Has ICT Polarized Skill Demand? Evidence from Eleven Countries over Twenty-five Years, *Review of Economics and Statistics* 96(1), 60–77.

Muro, M. and Andes, S. (2015): Robots Seem to be Improving Productivity, Not Costing Jobs, *Harvard Business Review*. Available at: https://hbr.org/2015/06/robots-seem-to-be-improvingproductivity-not-costing-jobs [30.03.2021]

Newfarmer, R., Page, J. and Tarp, F. (eds.) (2019): *Industries without Smokestacks: Industrialization in Africa Reconsidered*. Oxford, UK: Oxford University Press.

ODI (2020): *Ag platforms as Disruptors in Value-chains: Evidence from Uganda*. ODI Report. London: Overseas Development Institute.

OECD (2016): Routine Jobs, Employment and Technological Innovation in Global Value Chains. Available at: www.oecd.org/sti/ind/GVC-Jobs-Routine-Content-Occupations.pdf [30.03.2021]

Oxford Martin School (2016): Technology at Work: The Future Is Not What It Used To Be. Available at: www.oxfordmartin.ox.ac.uk/downloads/reports/Citi_GPS_Technology_Work_2.pdf [30.03.2021]

Parschau, C. and Hauge, J. (2020): Is Automation Stealing Manufacturing Jobs? Evidence from South Africa's Apparel Industry, *Geoforum* 115, 120–131.

Reijnders, L. and de Vries, G. (2017): Job Polarization in Advanced and Emerging Countries. *University of Groningen Working Paper*.

Rodrik, D. (2013): The Perils of premature Deindustrialization, *Project Syndicate* 11(10). Available at: www.project-syndicate.org/commentary/dani-rodrikdeveloping-economies–missing-manufacturing?barrier=accesspaylog [30.04.2021]

Schlogl, L. and Sumner, A. (2020): Automation and Structural Transformation in Developing Countries, in Schlogl, L. and Summer, A. (eds.): *Disrupted Development and the Future of Inequality in the Age of Automation*. Hampshire: Palgrave Pivot, 51–78.

Spitz-Oener, A. (2006): Technical Change, Job Tasks, and Rising Educational Demands: Looking Outside the Wage Structure, *Journal of Labor Economics* 24(2), 235–270.

Turner, A. (2018): Capitalism in the Age of Robots: Work, Income and Wealth in the 21st-Century. Available at: www.ineteconomics.org/research/research-papers/capitalism-in-the-age-ofrobots-work-income-and-wealth-in-the-21st-century [30.03.2021]

Wolter, S. C. and Ryan, P. (2011): Apprenticeship, in Hanushek, E., Machin, S. and Woessmann, L. (eds.): *Handbook of the Economics of Education*. Vol. 3. Amsterdam, Netherlands: Elsevier, 521–576.

World Economic Forum (WEF) (2016): The Future of Jobs: Employment, Skills and Workforce Strategy for the Fourth Industrial Revolution. Available at: www3.weforum.org/docs/WEF_Future_of_Jobs.pdf [30.03.2021]

WEF-ASEAN Survey (2020): COVID-19 – The True Test of ASEAN Youth's Resilience and Adaptability Impact of Social Distancing on ASEAN Youth. ASEAN YOUTH SURVEY 2020 EDITION. Available at: http://www3.weforum.org/docs/WEF_ASEAN_Youth_Survey_2020_Report.pdf [30.03.2021]

Willcocks, L. (2020): Robo-Apocalypse Cancelled? Reframing the Automation and Future of Work Debate, *Journal of Information Technology* 35(4), 286–302.

World Bank (2016): *World Development Report 2016: Digital Dividends*. Washington, DC: The World Bank.

World Bank (2019): World Development Report 2019: *The Changing Nature of Work*. Washington, DC: The World Bank.

World Trade Organisation (WTO) (2017): *Trade, Technology and Jobs*. New York: WTO.

Zeira, J. (1998): Workers, Machines, and Economic Growth, *The Quarterly Journal of Economics* 113(4), 1091–1117.

8
THE IMPACT OF DISRUPTIVE TECHNOLOGIES ON WORK AND EMPLOYMENT

Irene Mandl, Ricardo Rodriguez Contreras, Eleonora Peruffo and Martina Bisello

Introduction

Europe – like other world regions – is affected by several 'mega-trends'. Next to globalisation, the transition to a carbon-neutral economy, demographic developments and also digitalisation are changing our economy, labour market and society. Technological innovation is not a new phenomenon but a structural feature of the economy, and a necessary one to remain competitive. However, while some innovations are incremental, there are others that are more disruptive or 'game-changing' by changing ways of production and service provision, working and living. Since recently, there seems to be an increasing number of innovations that have the potential to result in deep technological and social transformation. While some of them, such as medical exoskeletons, are already used in practice, for others, such as plant communication, it is expected to take a long time before they will be operationally implemented. Nevertheless, from a policy perspective, it is recommended to start exploring also the potentials of these technologies as well as their possible impacts in order to be better prepared for the future and to consider relevant interventions.

Disruptive technologies can result in a 'win-win situation' for employers and employees if companies sense the potential of technological improvements, strategically use them to realise a competitive advantage, while at the same time adjusting work organisation to maintain or improve working conditions and job quality for their staff. However, if awareness about the potential impact of game-changing technologies on work and employment is lacking, or implementation is not realised by considering the potential disadvantages for the workforce, the work and employment standards developed over the last decades in Europe might be at risk.

Against this background, this article aims to highlight first indications of potential positive and negative effects of disruptive technologies on work and employment. Its objective is to increase awareness among scholars, policymakers and practitioners to make them consider the best possible use of modern technologies for the benefit of all.

The next section defines the concepts and technologies used in this article. The sections 'Employment and labour market impact' and 'Impact on working conditions and job equality' provide an overview of the main identified effects of selected disruptive technologies on the labour market (macro perspective) and working conditions/job quality (micro perspective). Afterwards, the impact of game-changing technologies on industrial relations and social

DOI: 10.4324/9780429351921-11

dialogue will be discussed. Finally, the conclusion indicates some policy pointers derived from the analysis.

Concepts and definitions

Digitalisation affects employment, working conditions and social dialogue; but talking about digitalisation as such might not be precise enough to understand the phenomenon. The digital age results from the development and re-combination of technologies and their application to new areas (Pérez 2003). Some of these technologies can be defined as 'game-changing' because of the potential disruptive effect they could have on employment, working and living conditions and labour market institutions (Eurofound 2018a).

This article follows Eurofound's conceptualisation on digitalisation (Fernández-Macías 2018) which distinguishes technologies into three drivers:

- Automation: the substitution of human input by machine input
- Digitisation: the transformation of physical objects and documents into bits (and vice versa)
- Coordination by platforms: the use of digital networks to organise economic transactions in an algorithmic way

These three drivers should be understood as an analytical tool to help better identify work and employment implications; in practice, these drivers are often combined in operational applications (Eurofound 2018a) and examples will be given throughout the chapter.

Under each of these drivers, there fall several specific technologies. In what follows, a selection of technologies is analysed as regards their impact on work and employment. Under the automation driver, technologies such as advanced robotics, automated software (including Artificial Intelligence (AI)[1]) and autonomous vehicles can be listed. The digitisation driver includes additive manufacturing (3D printing at industrial level), Internet of Things (IoT and its industrial applications), virtual reality/augmented reality (VR/AR) and wearables. Coordination by platforms focuses on platform work, understood as the matching of supply and demand for paid labour through an online platform or an app (Eurofound 2015, 2018f) (Table 8.1).

Employment and labour market impact

Automation

Automation has particularly strong implications for the evolution of the types of human input necessary for the production and service provision processes, and therefore the structure of employment by occupation and sector, as well as the skill levels required. This might explain why in recent years, along with interest in studying the effects of automation, concerns have also grown. According to Eurobarometer 2017, around 70% of respondents think that robots will steal our jobs and that they destroy more jobs than they create.

However, when analysing the impact of automation on the labour market, one can take two different theoretical perspectives: a job/occupation perspective and a so-called *task* perspective, where tasks are the smallest distinct units of labour input (specific actions of transformation or combination carried out by human operators within the production and service provision processes).

1 Machine learning and deep learning.

Table 8.1 Classification of driver and specific technologies discussed in this chapter

Drivers	*Technologies*
Automation	Advanced robotics
	Automated software (including Artificial Intelligence (AI))
	Autonomous vehicles
Digitisation	Additive manufacturing (3D printing at industrial level)
	Internet of Things (IoT and its industrial applications)
	Virtual/augmented reality (VR/AR) and wearables
Coordination by platforms	Online platforms or apps

Conceptualising jobs as 'bundles of tasks' allows for the possibility that only parts of a job – rather than all of it – may be substituted by a specific automation technology (Acemoglu and Autor 2011).[2]

Estimates, such as those from the OECD, that do take into account the variety of workers' tasks within occupations when estimating the impact of automation (Arntz et al. 2016) are substantially lower than those provided by the pioneering work of Frey and Osborne (2013, 2017) for the US where the entire occupation was considered as main unit of analysis.[3]

Next to job loss, the replacement of specific tasks within jobs due to automation technologies may also lead to a transformation (upskilling) of job profiles, enhanced productivity and potentially employment growth induced by demand effects – depending among other things on the distribution of productivity gains (Eurofound 2018a; Bessen 2018). The emergence of new products and services due to the introduction of new technologies is also a possible outcome, although more difficult to assess in terms of potential job creation.

Looking more specifically at the two automation technologies considered in this article, the potential substitution of human input is found in both services and manufacturing, but its extent varies across industries. *Advanced industrial robotics* (AIR), which encompasses the use of digitally enabled robots working within industrial environments that are equipped with advanced functionality allowing them to deal with less structured applications and in many cases to collaborate with humans, could have both negative and positive effects. On the one hand, AIR could lead to direct labour saving and therefore have a negative impact on the traditional manufacturing job profile – that of the non- or semi-skilled, blue-collar, production line worker[4] – and therefore mostly affect sectors such as vehicle, machinery and

2 Beyond the purely technical feasibility of replacing human input for the performance of specific tasks, the way work is organised has a significant influence on which specific tasks can be automated. If work is organised in a way that reduces the importance of key human labour attributes by centralising, standardising and breaking tasks down, the possibilities for its automation may significantly increase (Bisello et al. 2019).

3 On average across 21 of the OECD countries, only 9% of jobs are found to be automatable, compared with around 47% in the US as found by Frey and Osborne (2013). More recently, Nedelkoska and Quintini (2018) estimate that 14% of jobs in 32 OECD countries are highly automatable (having a likelihood of automation of over 70%) and a further 32% have a 50–70% likelihood of automation. See also Eurofound (2019e) for an analysis of three different automation scenarios; the objective of the study is to determine which tasks are most susceptible to automation and then to determine in which jobs these tasks predominate.

4 Employment in this type of manufacturing job has been shrinking for many years in developed economies and these production line jobs have been the fastest declining. While trade is also a factor contributing to such developments, namely in terms of offshoring of predictable work processes that are easy to replicate and displace, there is a consensus that technology has been the dominant vector of manufacturing job loss (Eurofound 2018a).

consumer goods manufacturing (Eurofound 2018a, 2018d). This trend is likely to intensify in the future and to have a higher disruptive potential if more advanced technologies, such as robots enhanced by AI, will be deployed at a significant scale (for a review of the literature and a discussion of employment effects of robots in the manufacturing sector, see Klenert et al. 2020).

On the other hand, the use of AIR can have compensatory positive employment effects related to the emergence of new job profiles in manufacturing which would lead to increased demand for specialised, highly digitally skilled workers. Such workers will likely include data scientists and mechatronics engineers. Apart from specialised technical skills, these employees will need to have highly developed social and communication skills in order to collaborate effectively with departments and teams of other disciplines (Eurofound 2020) (also see the following section about 'Impact on working conditions and job quality' under 'Automation').

Finally, the impact of automation in the services sector will greatly depend on the extent to which advanced robots are able to safely interact with humans or their ability to navigate in unstructured environments. It is expected that specific areas which involve engaging with robots, supervising or developing automating technologies will be potentially more affected (for example, mobile robots used in factories or warehouses). Robots have also started to replace humans in dangerous occupations and in environments that humans cannot access, enabling new service functions (such as in the areas of emergency and rescue).

As regards *autonomous vehicles*, considerable job loss in the transport and storage sector could be triggered if the transportation of goods and people were to be fully realised by self-driving cars or drones. While autonomous vehicles may provide some relief in those countries experiencing a labour shortage in terms of professional truck drivers, the overall negative impact in terms of job loss also for other categories, such as bus or taxi drivers, could offset the benefits (Eurofound 2019d). However, the horizon for mass adoption of fully autonomous vehicles is assessed to be longer than ten years and it is unclear exactly how many jobs could be lost and when this transition will occur. At the same time, the deployment of autonomous transport devices could also create new jobs directly related to transport services, such as autonomous transport planners, analysts, fleet managers and supply chain managers. Other services could also be positively impacted, although indirectly, by the changes in the transport sector: for example, the insurance industry could need new types of experts who can understand driving algorithms and interpret the dynamics of accidents.

Overall, the two analysed automation technologies show some potential for both job creation and job destruction. However, assessing the net outcome is a challenging exercise (see Autor 2015 for a discussion on the future of workplace automation). In general, it can be reasonably expected that the demand for higher-skilled occupations such as specialised and technical professions will grow, while low-skilled routine jobs will be mostly negatively affected – especially in manufacturing.

Digitisation

During the past decade the digitisation of processes has intensified in the manufacturing sector with the so-called industry 4.0 where not only advanced robotics but also *additive manufacturing, sensors* and the use of *AI* (machine learning and deep learning) are used. The analysis of data produced by these machines contributes to the creation of new production processes and production environments. Beyond manufacturing, some of these technologies are also applied in services, for example virtual reality for training rescue emergency workers or augmented reality in logistics. In logistics, augmented reality smart-glasses can be used by pickers to guide them to pick up articles instead of using a handheld label reading device (Heust 2017; Eurofound 2019a), thus freeing their movements.

As of today, the aggregate effects of digitisation technologies on employment in terms of job creation and destruction are not clear. Studies on the deployment of these technologies (Eurofound 2018b, 2019a, 2019b, 2019c) indicate rather a shift in tasks and the transformation of existing job profiles as well as the creation of new ones.

In manufacturing, so called cyber-physical or industry 4.0 factories (Peruffo et al. 2017) combine sensors (*Industrial Internet of Things*) and machines, including advanced robotics, into new work processes which can be supervised digitally. When highly digitised processes are grouped together a 'virtual twin' of the factory can be built, that is a digital copy of the factory. Virtual twins are also applied in airplanes monitoring and there are applications for digital copies of hospitals and even of cities. The digitisation of processes entails a change in workers' tasks, that is an emphasis on monitoring and thus the skill to manage digital information and sensors' data. In terms of job profiles, this translates into demand for data analysis experts as well as software engineers and other profiles which work on interfaces between machinery and data. But the most important implication embedded in the digitisation of processes is the enabling of remote working. If the virtual twin contains all the information about the process, the worker does not have to be in situ: the task can be performed anywhere, provided that there is a fast and reliable internet connection and interoperability across applications as well as cybersecurity is provided.

Another shift in work processes in industry 4.0 is the adoption of *additive manufacturing* which enables fast prototyping and the production of new shapes and textures not previously achievable with traditional methods and thus requiring fewer assembling tasks. Some new tasks have been observed; tasks such as additive manufacturing equipment loading/unloading (printing material) and cleaning tasks, for instance the residue and dust of the 3D printing. These tasks might be integrated into existing job profiles or be bundled into a new job depending on how companies decide to organise work.

Wearables, which are the application of devices with sensors to people and which could be called the 'internet of people', are being used in everyday life (for example to monitor one's exercise) but can be used in different work settings. For example, monitoring devices have an important application in healthcare allowing patients' conditions supervision remotely. This type of applications, as for the industrial IoT, are likely to increase demand of ICT professionals and data scientists who can help manage and analyse data flows.

VR, as of 2019, has begun to be integrated into the workplace as a training tool. Initially used by the defence forces, the use of VR has been extended to rescue emergency workers such as firefighters and surgeons. Through VR, the training can take place in a setting where the hazard has been taken away. In certain cases, like in the case of firefighters training, this avoids the need, and the cost, of burning down a building. Moreover, the possibility of repeating the training allows the worker to learn from mistakes and improve performance (Eurofound 2019a). In manufacturing, apart from the afore-mentioned virtual twins, VR can be used in health and safety training as well. In healthcare, VR can support surgical operations preparation by allowing the surgeon to see a virtual model of the patient's anatomy.

AR has been piloted in surgical operations for training purposes, in logistics, in urban planning and in the tourism sector. In urban planning, AR tools have the potential to increase the understanding of a proposed urban change by superimposing the new features to the real environment, thus increasing citizens' understanding of the proposed plan (Eurofound 2019**a**). For what concerns the tourism sector, the use of AR 'guides' which can display a 'what it looked like' scenario instead of the narrative provided by a human guide could diminish the need of human guides, although in many museums an automation/digitisation of tours has already taken place with the use of audio-guides.

To sum up, while the quantitative employment impact of digitisation is still difficult to assess, there is a high likelihood that job profiles will change, and new ones will be created. This mainly refers to a shift towards higher-skilled specialist occupations, and those taking on a role in supervising the technologies.

Platform work

Platform work is still a marginal phenomenon in Europe, with less than 2% of the population doing it as a main job (Pesole et al. forthcoming). However, there is wide agreement among experts that it has been dynamically growing for the last few years, and will continue to expand.

Platform work can be particularly advantageous for disadvantaged groups on the labour market, as it offers easy and unbureaucratic access to work and income, notably when considering small, low-skilled tasks (Eurofound 2019i). However, an important aspect that needs to be considered is whether platform work offers sustainable career options, acts as a stepping stone into more traditional employment, or results in a situation in which the workers are locked into an employment form perceived as unfavourable by them and/or could result in labour market segmentation. For the time being, no information about such effects, nor about the potential of platform work to crowd out traditional employment, is available.

Those types of platform work that are related to moderately to higher-skilled tasks, strategically conducted by the workers and providing them with high discretion as regards work organisation can contribute to foster entrepreneurial spirit and self-employment (Eurofound 2019i). They provide the workers with an easy and low-risk opportunity to try whether they have the required entrepreneurial skills (like self-organisation or dealing with clients) or to stabilise or expand their activity if they are already working as self-employed or freelancer.

An important labour market risk inherent to platform work is the potential misclassification of workers as self-employed, while in practice they are subject to subordination like employees and have limited discretion to balance the increased higher risk related to self-employment. If this happens, employment rights and working conditions standards are passed by, which might result in a race to the bottom which might also affect other types of employment in the longer run.

Related to the employment status is the potentially disruptive effect of platform work on legalising undeclared work. The fact that collecting data is a key element of the business models and mechanisms in the platform economy provides opportunities to make work more transparent compared to similar tasks conducted in the traditional economy, hence contributing to declaring work that is traditionally done in the shadow economy. At the same time, due to the fragmented and potentially international character of platform work, there are some assumptions that such work is not properly taxed or registered with social insurance authorities, hence potentially increasing the level of undeclared work.

To conclude, numerical labour market effects of platform work cannot yet be assessed. From a qualitative perspective, platform work is deemed to positively contribute to labour market integration and income generation, and to foster self-employment in Europe. On the negative side, it has the potential to disrupt established employment rights and labour standards, and to contribute to undeclared work and labour market segmentation.

Impact on working conditions and job quality

Automation

As indicated in the previous section, automation technologies have the potential to transform the tasks performed within jobs, and therefore the content of the job itself and the competencies needed to perform it. This has in turn strong implications, for instance, in terms of *skills use* and *development*. Overall, there is a tendency observed among the analysed automation technologies of a decrease in manual tasks and an increase in the need for intellectual skills. This is clear in the case of the adoption of advanced robots in manufacturing, which may drive employment demand for jobs involving tasks such as engaging with, supervising or developing automating technologies. This shift towards monitoring, programming and machine-control tasks, which is considered to result in better job quality, would also require existing workers to upskill (Eurofound 2018c). With regards to autonomous vehicles instead, these could considerably challenge professional drivers who would need training to develop new skills for tasks that are not fully pre-defined. Indeed, while the main task of driving would be replaced, autonomous vehicles might still require 'hosts' or 'conductors' to act as neutral parties in the shared, enclosed environment of the vehicle. This could require more creative and social skills, related to controlling passengers' behaviour and reacting to unforeseen circumstances, for instance. At the same time, it could also mean a shift towards more manual tasks such as loading and unloading, which would not necessarily provide an opportunity to obtain better quality jobs.

The adoption of automation technologies will also necessitate changes to *work organisation* and *work environment*. A better understanding of how the robotic system relates to the social and qualitative aspects of the overall work system will be needed. In particular, the management of human-robots interaction will be a key aspect to consider because of the complexities related to it: people management is expected to be significantly affected, especially in terms of supervision of robots (and possibly supervision by robots). Work environment will necessarily be impacted too, due to the technological limitation of current robot technology to work in unstructured environments and respond to unexpected scenarios. In services where the 'working space' is generally not as clearly defined as the manufacturing production line, this could either limit the application of robots (confining them to fixed locations or restricted areas of motion) or necessitate a redesign of service delivery areas. For autonomous vehicles, work organisation seems to be oriented in two directions according to the type of transport involved; if transporting goods, the organisation of work might shift to remote supervision and less necessity to move location for workers. If transporting passengers, the scenario is still unclear since it might range from remote supervision to unsupervised and fully autonomous vehicles, and for what concerns workers, they might be needed on board or not (helping passengers with luggage, cleaning, supervising peoples' behaviour). The use of autonomous vehicles might open up possibilities to use time spent driving and paying attention to the road with time used to complete work tasks, although this could depend on the length of the trip.

In terms of *working time*, advanced robotics has the potential to shorten working hours due to reduced workload. In order to maximise return on capital investments, production facilities are likely to operate 24/7, hence some specialised staff may be required to be constantly on call and work at unsocial hours in order to deal, for instance, with machine failure. In general, traditional and predictable working time schedules could be eroded. On the contrary, the deployment of autonomous vehicles could lead to more regular working hours and more consistent work schedules, which would have not only a positive effect on the experience of professional drivers at work but also on their quality of life. Autonomous transport devices could also reduce the commuting time for workers between their place of employment and their homes; it would also allow for the possibility to use the time not spent driving for work activities (e.g. checking emails, making phone calls, etc.) to be added to the total daily hours worked.

Finally, the impact on *occupational health and safety* is also very relevant. In general, automation technologies are assumed to result in less physical strain: in the case of advance robotics, physically burdensome tasks can be conducted by machines; in the case of autonomous vehicles, traditional challenges related to posture problems and need for break times in the transport sector could be overcome. This in turn also implies reduced risks of accidents and injuries. However, it is acknowledged by experts that such positive developments are more likely to occur when employers and employees are informed about specificities and potential dangers related to working with such technologies and are instructed on how to react to potential accidents. More generally, there is a need to invest in the safety of processes in situations where robots cooperate with human workers: this is especially crucial in work environments that are not particularly structured, like it is the case for the services sector. Similarly, while autonomous transport devices could improve safety in workplaces where they are used – by reducing injuries or fatalities that occur as a result of collisions caused by human error – they could also result in new risks in the early stage of deployment related to supervision of such vehicles and interactions with them. Interestingly, automation technologies such as advanced robotics are expected to bring about a greater risk of negative psychosocial effects due to machine control of work processes and increasingly secondary role of human intervention.

Digitisation

Among the technologies with a prevalence of digitisation, the *skills requirements* point towards a shift from manual to intellectual tasks or, in already digitised roles, to an intensification of intellectual tasks. Roles likely to be increased are those where ICT and data skills are required; these skills can help companies, both in the manufacturing and in the services sectors, to collect, store and make sense of data to improve their processes or products. Emphasis is also placed on the ability to communicate among disciplines: for instance the analysis of data flows, be it coming from production processes (IIoT) or from people (wearables), will need to be supported by workers who are familiar with software development, data analysis including AI, cybersecurity and data quality management.

Also, digitisation of work processes offers the possibility of radical changes to *work organisation*. The way in which each application is used and interacts with other technologies and the workers will shape not only work and production processes but also the other aspects of job quality (working time, work intensity, autonomy, flexibility, control and health and safety). Work organisation choices that a company can make include how much autonomy can be given to workers, for example if the pace of work is dictated by the rhythm of the machine or which type of tasks, very simple or very complex, are carried out by technology and which by humans. Of course, this type of choice is not entirely up to companies but is intertwined with the type of product or service being created. A job with a high social interaction component might not be suitable for digitisation even though it is technically feasible (for example in some healthcare contexts). For two technologies, wearables and VR/AR, an indication of changes seems to be more easily identifiable: in case of wearables workflows will be more data driven while for VR/AR there is a risk of task-driven organisation resulting in workers becoming passive recipients of task assignments.

In terms of *working time* the digitisation of work could on the one hand allow for remote working and a better work-life balance (although the benefits of teleworking are not clear cut

(Eurofound and the International Labour Office 2017); on the other hand, the supervision of highly automated and digitised processes means that these are potentially running 24/7, implying that a malfunction could occur at unsociable hours and require human intervention.

In general, the effects of digitisation technology on *autonomy*, *flexibility* and *control* are strictly related with sector of activity and jobs. In those jobs where wearables are used there is a tendency for less autonomy due to the potential for increased monitoring through the technology. On the other hand, notably where digitisation enhances remote working, workers are expected to experience a higher degree of autonomy and discretion in their decisions on how to structure their work and when to realise individual tasks.

As regards *health and safety*, in general, similarly to automation technologies, digitisation technologies reduce physical risks either by transferring a number of operations to the online or virtual space (IoT, VR/AR) or by providing new effective ways of risk-free training (VR). VR could also improve employees' interaction and or empathy and perceptiveness. Wearables also have a risk reducing effect when, for example, workers' physical or workspace conditions are monitored to get potential danger alerts. Potential health and safety risks have been identified around additive manufacturing, as some of the powders and pastes used in these processes are new materials and their safety in the long term has still to be proven (for example, release of toxic particles). Also, additive manufacturing machines have moving parts and high temperature nozzles from which workers should be sheltered. Some hints of negative psychosocial effects have been noted with the use of wearables since they increase the amount of control over workers.

Platform work

The algorithmic matching of supply and demand is a key characteristic of platform work. Beyond matching, algorithms can also be used to exert *control* over the worker (Vandaele 2018), thereby limiting their *autonomy* and *flexibility* which is generally promoted as one of the biggest advantages for platform workers.

Flexibility over *working time* is generally given in those types of platform work related to higher skills requirements and less influence of the platform in the work organisation. However, in online tasks – even if higher-skilled – global competition tends to create tight deadlines, which in turn results in high work intensity and stress levels. Additionally, some types of platform work are related to unsocial working hours (that is, beyond core working hours, like on evenings or weekends). Nevertheless, this is to be attributed to the task requirements rather than the specific characteristics of platform work, and hence not different from similar tasks in the traditional economy (e.g. food delivery, taxi services, emergency home maintenance tasks). In some types of platform work the mechanisms of the business model can incentivise workers to realise high work intensity. Examples are pay-by-task models (motivating workers to work at high speed to maximise their income in a given period) or systems with non-transparent task assignment algorithms or client decisions which tend to create insecurity among workers due to the unpredictability of work.

Platform work is often criticised for its low *earning* opportunities. This is widely based on early surveys flagging the low rates of individual tasks and the low overall income workers achieve in platform work, often limited to online micro-tasks (Ipeirotis 2010; Berg 2016; Leimeister et al. 2016). More nuanced research (Eurofound 2019i) shows that for other types of platform work, notably those delivered on location and where workers have discretion to set payment rates, income corresponds to market rates in the traditional economy. Predictability of earnings is rather good in those types where the platform goes beyond matching by determining the work organisation of tasks delivered on location. In contrast, it is rather non-existent in most online tasks, where also a high incidence of unpaid working time is observed (searching and bidding for tasks which might not be assigned to the worker at the end).

The physical environment in platform work does not differ from similar situations in the traditional tasks. Online workers are confronted with aspects related to office/computer work (for example, posture and ergonomics, eye strain) while on-location workers potentially have to face physical hazards (like road accidents in transport-related or handicraft tasks, exposure to chemicals in cleaning tasks). However, *health and safety* responsibilities (e.g. provision of adequate equipment) are not clear in the platform economy. This can be problematic if the mechanisms of the platform (pay-by-task) and the young age of the worker (lower level of awareness of risks and precaution measures, and interest in investing in such) incentivises workers to prioritise earnings over safety.

Impact on labour market institutions

The impact of disruptive technologies on industrial relations

This impact of disruptive technologies on collective employment relations depends on the nature and degree of deployment of each specific technology as well as the economic activity considered. In general, the impact of the implementation of game-changing technologies, especially automation-related ones, have been even stronger in well-established manufacturing activities. To a lesser extent, the effects can be noticed in the services sector, even though they are also quite remarkable in some sectors such as logistics and transport.

Industrial relations are challenged when disruptive technologies impact employment levels or the job profiles within sectors. The potential job losses or changes in production and work organisation condition considerably the collective employment relations raising the workers' representatives and unions' needs to adapt to a changing work environment as well as workforce, and a requirement to come up with responses which might need to deviate from traditional approaches. This is even more challenging when considering that the impact of the implementation of disruptive technologies takes place in a wider context of change and especially higher competition in specific business activities, for example, the automobile, finance and banking, and the telecom sectors. As far as these alterations in production and service provision require significant changes in work organisation and skills needed, workers' representation and unions should also be requested to have a say.

Throughout the history of industrial revolutions, automation processes were *finally* integrated in the framework of national industrial relations systems. It took place in various ways and with difficulty over a long period of time. In the end, throughout the last century, organised workers in unions made it possible to embed the subsequent changes in work processes and working conditions mainly through collective bargaining at different levels. Therefore, labour and capital found an institutionalised way to organise their relations and settle their disputes derived from the periods of technological disruption.

Depending on each technology and sector, this time the impact of new technologies may lead to more disruptive potential in social dialogue at the company and sectoral level, particularly when used in combination of various forms of AI. The higher the impact of implementing each technology on the nature of work, workforce and working conditions, the more likely are the implications for collective employment relations. In this regard, the impact of disruptive

Table 8.2 Implications of changes caused by disruptive technologies for collective employment relations

Changes stemming from the implementation of disruptive technologies	*Implications for collective employment relations*
Transformations in production processes and work organisation	
Outsourcing and subcontracting of digital processes or tasks: blurring company boundaries	Potential disruption in collective workers' representation as the reduction in the number of employees may make it harder to reach thresholds to set up representation bodies – work councils
Digitised (mainly IoT-supported devices and VR) processes and tasks enabling remote control of the machinery and equipment	Increasing difficulties for unions to represent and organise geographically dispersed workers
Use of platform services and platform workers	Disruption of union solidarity
Transformations in working conditions	
New digital skills needed in production processes and tasks	Labour demand tends to favour higher-skilled profiles where the level of unionisation is lower
Emerging occupational health risks, rather psychosocial than physical ones	New field to be researched, monitored and negotiated by workers' representatives and unions in specific or general bodies
Constant data management process and monitoring production tools	Risk to fully or partially individualised wage setting and/or supplementary remuneration, undermining collective bargaining
Applying centralised management algorithms	Concern for the autonomy and privacy of employees

technologies on employment relations may be considered indirect as long as the technologies condition core aspects of workers' representation such as membership and organisational capacity of trade unions as well as influence collective bargaining and power relations with employers. A number of indirect effects of game-changing technologies and overall technological change in collective employment relations can be stressed in Table 8.2.

Automation blurs company boundaries and creates more complex organisational forms of production, both of which impact collective employment relations. In combination with developments related to work organisation – for example, increased remote working (Johnston and Land-Kazlauskas 2018) – these shifts resulting from automation could compromise the minimum threshold for consultation rights and thus the ability of employee representatives to engage in negotiations (IBA GEI 2017). Overall, these tendencies will diminish opportunities for workers to organise into trade unions and, consequently, reduce collective bargaining rights and workers' participation in the decision-making processes that affect working conditions.

The earlier discussed effects of disruptive technologies of setting more complex organisational forms of production may impact workers' representation and unions' representativeness at the workplace level. Even though there would be different approaches to the implementation of technology at the company level, it is quite likely that growing physical distance as a result of remote working can render unionisation less easy to organise.

Nevertheless, some of the technologies discussed can contribute to improve the organisation of workers. For example, it has been pointed out that IoT-supported social media and other digital platforms can also be used for labour organisation, with platforms enabling crowdfunding efforts to fund union initiatives or opening up alternative methods of collaboration between workers' representatives and new forms in worker representation (BDA 2015).

The role of social dialogue and collective bargaining

Available literature and studies have described how tripartite social dialogue has dealt with technological change (Eurofound 2016, 2017; EESC 2018). It becomes evident that this issue has become part of the employers' and unions' agendas, likely driven by the German industry 4.0-like strategies resulting in first national tripartite statements, joint declarations and different types of 'digitalisation' agreements showing the governments' willingness to involve the social partners.

This joint approach has also been adopted at the sectoral level between the social partners themselves. Beyond the cross/industry statement on digitalisation produced by the EU social partners (BusinessEurope, ETUC, CEEP and SMEunited) in 2016, over the past years the EU sectoral social partners have reached various declarations, joint statements and other instruments stressing the pathway to adapt businesses and workforce to technological developments. Particularly the need to adapt skills policies as well as labour market regulations and institutions has been highlighted.

In most EU countries, collective agreements at both sectoral and company levels are just mentioning the importance of technological change and digital transformation committing the signatory parties to continue discussing this challenge and above all, promoting training to facilitate technological adaptation of employees. Just isolated examples of collective agreements can be found regulating the right to disconnect in relation to working time as well as fragmented measures related to technical adaptation or privacy at the workplace.

This lack of specific measures may be due to the following reasons:

- Technologies, and game-changing technologies in particular, or the combination thereof, evolve fast, while collective bargaining is a 'heavy' institution reacting slower to digital developments. For example, most collective agreements last more than two years before being renegotiated or renewed.
- Employers and unions have wage setting and working time 'over-the-top' in their negotiation agendas when facing collective bargaining. Impact of technological change in production is considered a discretionary power of the employer, and it is assumed that these issues should not be subject to negotiation as such, unless they have substantial implications for work organisation (work shifts, for example).

However, the effects of technological transformation at company and workplace levels are taking place step by step in collective bargaining mainly through a combination of working time measures and skills policies. In this regard, some experiences are leading the way to manage the effects of automation: carmakers in Germany such as Daimler, Volkswagen, BMW, Audi and Bosch, as well as automotive parts manufacturers like Continental, have introduced ambitious remote working programmes for hundreds of thousands of employees (where, for instance, robots can be managed remotely). Following the same track, SAP's 22,000 employees in Germany also have the same right to work wherever they want in the country.

In other sectors, collective agreements in large companies as TIM (formerly Telecom Italia) or Unilever (industry 4.0 agreement) intend to deploy massive training opportunities and re-qualifications programmes for the entire staff. Interestingly, in line with the recommendations regarding the redistribution of the 'value of data ownership' of the High-Level Expert group on the Impact of the Digital Transformation on EU Labour Markets (European Commission 2019), the collective agreement in Lamborghini has agreed to establish a bilateral commission to analyse to whom the data produced by the company's IT systems will be available, either the workers or the company only.[5]

It is in the financial sector where the examples of managing the impact of digitalisation using collective bargaining are more frequent and explicit. In the Italian banking sector, the collective agreement (280,000 workers) includes the creation of a joint national committee to analyse the impact of new forms of technology and digitalisation aimed at identifying the skills required in the future.[6] In the same sector in Germany (near 200,000), the collective agreement set up specific programmes to assess the readiness of workers for the digital transition.[7] Similarly, the collective agreement in the banking sector in Belgium has established pathways for workers whose jobs may be under threat due to digital transformation, providing them with training and coaching in a digital platform.[8]

These examples do not hide that in the first decade of the century there were relevant cases of economic activities and sectors such as banking, media and post mail services dramatically restructured due to digital transformation. In these cases, the whole industrial relations system (strong social partner representativeness and well-established social dialogue) played a crucial role to smoothly manage technological redundancies with a combination of early retirement schemes and training, amongst other social and labour measures.

More recently, the increasing emergence of platform work, with its blurring impact on employment statuses and less favourable conditions of employment for workers in terms of job stability, income security and predictability as well as working hours, has many potential implications for industrial relations and social dialogue. Platform work has been extensively analysed over the past years (Eurofound 2017, 2018e, 2018f, 2019f, 2019g, 2019h, 2019i). The specific characteristics of platform work and the unclear employment status of workers challenge the ability of platform workers to have their interests represented. As they are widely considered self-employed, at least in some member states traditional trade unions do not have a mandate to represent them, and competition regulation may not permit them to organise through other means. In this regard, the ILO Global Commission on the Future of Work (ILO 2019) recommends 'the development of an international governance system for digital labour platforms that sets and requires platforms (and their clients) to respect certain minimum rights and protections'.

Platform economy actors are at an early stage of engaging in forms of social dialogue or collective bargaining. Some initiatives to organise and mobilise platform workers are emerging in several EU Member States, driven by trade unions or by grassroots organisations (Vandaele 2018; Eurofound 2019i) as the Belgian Collectif des Courier-e-s, a self-organised collective of food-delivery riders.

As an example of institutionalised collective voice, recent legislation in France provided platform workers with the right to set up or join a trade union and to organise or participate in a strike without negative consequences for their contractual relationship. There are also some examples of collective agreements signed by institutional trade unions workers' autonomous collectives and the management of a platform firm, such as the food-delivery company

5 Planet Labor, 25 July 2019: www.planetlabor.com/en/industrial-relations-en/national-industrial-relations/italy-new-collective-agreement-for-lamborghini-provides-for-greater-participation-of-staff-representatives-on-industry-4-0-and-data-collection/

6 Planet Labor, 20 December 2019: www.planetlabor.com/en/industrial-relations-en/national-industrial-relations/italy-banking-sector-collective-agreement-renewed/

7 Planet Labor, 5 July 2019: www.planetlabor.com/en/industrial-relations-en/national-industrial-relations/germany-agreement-in-banking-sector-for-4-pay-rise-and-measures-in-view-of-digital-transition/

8 Planet Labor, 3 October 2019: www.planetlabor.com/en/industrial-relations-en/national-industrial-relations/belgium-a-banking-sector-agreement-that-focuses-on-training-and-teleworking/

Sgnam-MyMenu in Italy. The agreement set a fixed hourly rate in line with the sector's minimum wage, compensation for overtime, holidays, bad weather and bicycle maintenance compensation (Aloisi 2019). Similarly, and widely promoted as the first collective agreement on platform work, the agreement between Hilfr (a platform company providing cleaning services) and the largest Danish trade union, 3F, established a minimum wage and workers' entitlement to contributions to pensions, holiday pay and sickness benefits (Eurofound 2019i).

Interestingly, flexibility in collective bargaining may offer capacity to frame platform work through extension mechanisms. Thus, the agreement signed by Bzzt (a platform of personal transport by tuck tuck) covers its drivers under the taxi agreement (Swedish Transport Workers' Union). Another example is the coverage provided by the collective agreement for temporary agency work for Instajobs workers (a platform for student work) signed between the platform and the largest Swedish trade union Unionen. The same 'temporary agency work window' has been followed by Gigstr (low-skilled tasks) in Sweden (Jesnes et al. 2019).

An interesting example – likely the first one at the sectoral level – of collective agreement covering food delivery riders was signed in the Italian logistics sector in 2017. The agreement covers working time, the requirement for notice and compensation for changes in working schedules and compensation in case of illness.

It is quite likely that more examples of collective bargaining or other forms of social dialogue will follow. The European Commission announced the discussion on challenges related to platform work and possible solutions as an essential commitment in the Road Map presented in January 2020. Priority issues like employment status, working conditions and access to social protection of platform workers, access to collective representation and bargaining have been discussed during 2021 through different policy documents and public consultations such the one on 'Collective bargaining agreements for self-employed – scope of application EU competition rules'. It is expected that the European Commission publishes a proposal addressed to improving the working conditions of platform workers by the end of 2021, including a legislative initiative, an impact assessment and further consultation of the social partners under Article 153 TFEU.

Conclusions and policy pointers

Role for EU investment and policies

Despite accounting for only about 7% of the world's population, Europe accounts for 20% of global R&D investment. However, statistics relating to the number of patents and other indicators suggest that the EU is lagging behind other regions – notably the US and China – in technological development, including those technologies that are considered potentially more disruptive.

EU policy has started to address this gap, for example by financially supporting the development of disruptive technologies such as AI, although further efforts and more holistic approaches might be needed. Efforts are also required to close the financing gap between R&D grants and private investment.

EU regulations supporting disruptive technologies

While many policies support the development and adoption of technologies, legislative action should also be taken to integrate automated processes in the workplace. Regulatory frameworks, especially in the area of standardisation and interoperability, need to be established and fine-tuned both for companies (e.g. related to IoT) and workers (e.g. to prevent potential hazards related to additive manufacturing or advanced robotics).

Even though GDPR protects personal data, the types of data produced during working time need to be defined more specifically. The ownership of workers' data, in particular related to performance and behaviour, and in blurred employment relationships like platform work, should be specifically regulated. It should also be recognised that workers' and consumers' data are used to increase a firm's value, and they should be compensated accordingly.

Strategic use of technologies for labour market purposes

The new capabilities of disruptive technologies should be exploited to improve labour markets. Technologies falling under the digitisation cluster (i.e. IoT, additive manufacturing and VR/AR) can foster flexibility through remote working. VR/AR can also provide alternative training methods for hazardous occupations.

Advanced robotics, or the use of exoskeletons, can reduce physical strain and hazardous tasks, which could enable the integration of disabled or older workers into the workplace. Furthermore, there should be a focus on research into hazardous tasks that can be carried out by machines in order to make workplaces safer.

Platform work could be used as a strategic tool to foster labour market integration of disadvantaged groups, to extend working life or to legalise undeclared work. However, before deciding on such a pathway, more needs to be learned about potential unintended side effects, such as crowding out of traditional employment with more favourable working conditions or its contribution to labour market segmentation.

A European vision for technology and digitalisation

Digitalisation entails more than solely implementing new technologies in work processes, as it is interrelated with other mega-trends like internationalisation and the transition to a low carbon economy. Accordingly, it requires a long-term and holistic strategic vision, and that the implications of digitalisation for both human capital and the value chain be taken into consideration. Usually, digitalisation will involve reorganisation at the company level, driven by competition in an increasingly technological and globalised environment.

The way in which technology is adopted and implemented is highly significant and therefore prompts the need for effective human resources management approaches, which should involve worker representatives or trade unions. Human resources management policies are increasingly expected to be able to ease technological change through people management.

Technological transformation does not take place overnight, and the pace of change should be tailored to help manage the transition period. Organisations need to identify early on the strategic vision driving the adoption of technology and, consequently, the overall effects for the organisation and for the staff. This is particularly important for those organisations adopting robotisation in combination with other general-purpose technologies such as IoT and AI, which are usually run from the cloud. Such combinations of technologies enable the implementation of smart factories or digitised networked workplaces, with profound impacts on production and work organisation.

EU companies have significant experience in dealing with transformative automation, and lessons learned should be applied to the current technological transition. A narrow approach to technological change, which mostly takes advantage of automation to reduce the workforce and labour costs, should be avoided. Instead, labour-friendly strategies, which seek to reskill the workforce in line with efficiency and productivity gains, should be applied.

Employment and skills

After the first wave of near-dystopian predictions about the impact of automation on job losses, more recent estimates confirm the limited quantitative effects of automation on employment. It is widely accepted that a job is composed of a bundle of tasks to which workers apply their skills in exchange for wage; some of these tasks can be automated, but others cannot. Presently, estimates suggest that automation will thus replace only very few jobs completely (automating all task content in a given job), while impacting the job profile of others. The latter is of importance as it alters skills requirements to which employers, workers and the education systems need to adapt.

The increase of remote working enabled by some of the analysed disruptive technologies could pose serious challenges to policymakers and workers alike. If jobs can be done remotely, they may not be performed – and therefore paid – at the local level. Technological change should be viewed as an opportunity to adjust work processes and reorganise the production process and the provision of services to improve, rather than replace, workforce capacity. As more tasks become automated over time, tasks that are not replaced by machines are likely to change and could increase in value, while other new creative tasks requiring human talent may be created. In this transition period, further public and private research and monitoring is needed to ensure that workers and machines work together effectively.

Significant investment in skills at all levels is essential in this technological transition. As future jobs in the technological sphere will be constantly evolving, continuous and lifelong training should be at the core of the EU strategy to address technological change. To stay ahead of global competition, the EU needs to solve two issues in particular: how to equip the new generation of workers with the right skills, both transferable and specific, and how to help workers whose jobs might be lost or substantially changed through the adoption of disruptive technologies.

Training policies should be based on effective tools anticipating future skills needs and close collaboration and synergies between business, education providers, social partners and governments. Apprenticeship systems and vocational education and training need to be reoriented in line with these changes, and collective bargaining at all levels can play an important role in anticipating and managing change by establishing agreed training and skills policies.

Social dialogue and employment relations

The more disruptive the technology, the greater the impact it has on the labour force and on working conditions, and thus on social dialogue. Companies focusing exclusively on technical change, rather than on the people working with the technologies, are pursuing a short-term strategy that is unlikely to foster workforce cohesion. Social dialogue involving workers affected by disruptive technologies – ideally going beyond standard employees but also focusing on non-standard work including self-employment – therefore becomes a crucial way to create new, more appropriate agendas for negotiation and identify new areas for employer–worker cooperation.

Contributing to manage the impact of technological change and the effects of game-changing technologies is already on the cross-industry and sectoral policy agenda of employer organisations and trade unions, both at the EU and national level. The social partners have addressed these challenges at the macroeconomic level (the potential employment effects and the need to refocus the skills gaps) and at the sectoral and company level.

Collective bargaining contributes to effectively manage transition periods supporting companies and sectors to restructure and adapt accordingly, dealing with the risks of increasing wage inequality and polarisation in working conditions. A new wave of collective agreements managing digital transformation and digital organisation of work through working time flexibility, reskilling and upskilling, work-life balance and working conditions such as workers' health and safety, including psychosocial risks, are already leading the way for updating collective bargaining.

In this regard, despite the existing innovative experiences, collective bargaining coverage of platform workers is still at a very early stage. However, social partners and particularly trade unions are making efforts to reach agreements addressing social protection and other working conditions standards stemming from the new form of work in digital platforms. Furthermore, the use of digital means and platforms could contribute to better organise the activities and influence of the social partners themselves as well as to monitor and extend their capacities and increase their membership.

References

Acemoglu, D. and Autor, D. (2011): Skills, Tasks and Technologies: Implications for Employment and Earnings, in Card, D. and Ashenfelter, O. (eds.): *Handbook of Labor Economics 4(B)*. Amsterdam: Elsevier, 1043–1171.

Aloisi, A. (2019): At the Table, Not on the Menu: Non-Standard Workers and Collective Bargaining in the Platform Economy. *EUIdeas*. 25 June 2019. https://euideas.eui.eu/2019/06/25/at-the-table-not-on-the-menu-non-standard-workers-and-collective-bargaining-in-the-platform-economy/

Arntz, M., Gregory, T. and Zierahn, U. (2016): The Risk of Automation for Jobs in OECD Countries: A Comparative Analysis. *Working Paper No. 189*. Paris: OECD Publishing.

Autor, D. H. (2015): Why Are There Still So Many Jobs? The History and Future of Workplace Automation, *Journal of Economic Perspectives* 29(3), 3–30.

BDA (2015): *Seize the Opportunities of Digitisation, BDA Position on the Digitisation of Business and the Working World*. May 2015.

Berg, J. (2016): *Income Security in the On-demand Economy: Findings and Policy Lessons from a Survey of Crowdworkers*. Conditions of Work and Employment Series, No. 74. Geneva, Switzerland: ILO (International Labour Organisation).

Bessen, J. (2018): AI and Jobs: The Role of Demand. *NBER Working Paper No. 24235*. https://doi.org/10.3386/w24235

Bisello, M., Peruffo, E., Fernandez-Macias, E. and Rinaldi, R. (2019): *How Computerisation is Transforming Jobs: Evidence from the Eurofound's European Working Conditions Survey*. Brussels: European Commission.

EESC (2018): *Overview of the National Strategies on Work 4.0: A Coherent Analysis of the Role of the Social Partners. European Economic and Social Committee*. Published September 14.

Eurofound (2015): *New Forms of Employment*. Luxembourg: Publications Office of the European Union.

Eurofound (2016): *Foundation Seminar Series 2016: The Impact of Digitalisation on Work*. Dublin: Eurofound.

Eurofound (2017): *Addressing Digital and Technological Change through Social Dialogue*. Dublin: Eurofound.

Eurofound (2018a): *Game-changing Technologies: Exploring the Impact on Production Processes and Work*. Luxembourg: Publications Office of the European Union.

Eurofound (2018b): Additive Manufacturing: A Layered Revolution. *Working Paper*. Dublin: Eurofound.

Eurofound (2018c): *New Tasks in Old Jobs: Drivers of Change and Implications for Job Quality*. Luxembourg: Publications Office of the European Union.

Eurofound (2018d): The Impact of Advanced Industrial Robotics on European Manufacturing: Taking Human – Robot Collaboration to the Next Level. *Working Paper*. Dublin: Eurofound.

Eurofound (2018e): *Employment and Working Conditions of Selected Types of Platform Work*. Luxembourg: Publications Office of the European Union.

Eurofound (2018f): *Platform Work: Types and Implications for Work and Employment – Literature Review*. Dublin: Eurofound.

Eurofound (2019a): Virtual and Augmented Reality: Implications of Game-changing Technologies in the Services Sector in Europe. *Working Paper*. Dublin: Eurofound.

Eurofound (2019b): Wearable Devices: Implications of Game-changing Technologies in Services in Europe. *Working Paper*. Dublin: Eurofound.

Eurofound (2019c): Blockchain: Implications of Game-changing Technologies in the Services Sector in Europe. *Working Paper*. Dublin: Eurofound.

Eurofound (2019d): Autonomous Transport Devices: Implications of Game-changing Technologies in the Services Sector in Europe. *Working Paper*. Dublin: Eurofound.

Eurofound (2019e): *Technology Scenario: Employment Implications of Radical Automation*. Luxembourg: Publications Office of the European Union.

Eurofound (2019f): *Literature Review – Online Moderately Skilled Click-work: Employment and Working Conditions*. Dublin: Eurofound.

Eurofound (2019g): *On-location Client-determined Moderately Skilled Platform Work: Employment and Working Conditions*. Dublin: Eurofound.

Eurofound (2019h): *Mapping the Contours of the Platform Economy*. Dublin: Eurofound.

Eurofound (2019i): *Platform Work: Maximising the Potential While Safeguarding Standards?* Luxembourg: Publications Office of the European Union.

Eurofound (2020): *Game-changing Technologies: Transforming Production and Employment in Europe*. Luxembourg: Publications Office of the European Union.

Eurofound and the International Labour Office (2017): *Working Anytime, Anywhere: The Effects on the World of Work*. Geneva: Publications Office of the European Union, Luxembourg, and the International Labour Office, Geneva. Labour Office.

European Commission (2019): *The Impact of the Digital Transformation on EU Labour Markets. High-Level Expert Group*. Brussels: European Commission.

Fernández-Macías, E. (2018): *Automation, Digitisation and Platforms: Implications for Work and Employment*. Dublin: Eurofound.

Frey, C. B. and Osborne, M. (2013): *The Future of Employment: How Susceptible are Jobs to Computerisation?* Oxford: Oxford Martin Programme on Technology and Employment.

Frey, C. B. and Osborne, M. (2017): The Future of Employment: How Susceptible are Jobs to Computerisation? *Technological Forecasting and Social Change* 114(C), 254–280.

Heust, P. (2017): *DHL Experiments with Augmented Reality, HesaMag, #1 Autumn-Winter 2017*. Brussels: ETUI.

IBA GEI (International Bar Association Global Employment Institute) (2017): *Artificial Intelligence and Robotics and Their Impact on the Workplace*. London: IBA Global Employment Institute.

ILO (2019): *Work for a Brighter Future – Global Commission on the Future of Work*. Geneva: International Labour Office.

Ipeirotis, P. G. (2010): Analyzing the Amazon Mechanical Turk Marketplace, *XRDS* 17(2), 16–21.

Jesnes, K., Ilsøe, A. and Hotvedt, M. J. (2019): *Collective Agreements for Platform Workers? Examples from the Nordic Countries*. Nordic future of work Brief 3, FAFO. March 2019.

Johnston, H. and Land-Kazlauskas, C. (2018): *Organizing On-demand: Representation, Voice, and Collective Bargaining in the Gig Economy*. Geneva: International Labour Office.

Klenert, D., Fernández-Macías, E. and Antón, J. I. (2020): Don't Blame it on the Machines: Robots and Employment in Europe. Available at: https://voxeu.org/article/dont-blame-it-machines-robots-and-employment-europe [30.03.2021]

Leimeister, J. M., Durward, D. and Zogaj, S. (2016): *Crowd Worker in Deutschland: Eine empirische Studie zum Arbeitsumfeld auf externen Crowdsourcing-Plattformen*. No. 323. Düsseldorf: Hans Böckler Stiftung.

Nedelkoska, L. and Quintini, G. (2018): Automation, Skills Use and Training. *Working Paper No. 202*. Paris: OECD Publishing.

Pérez, C. (2003): *Technological Revolutions and Financial Capital*. Cheltenham, UK; Northampton, MA: Edward Elgar Publishing.

Peruffo, E., Rodriguez Contreras, R., Molinuevo, D. and Schmidlechner, L. (2017): Digitisation of Processes Literature review. *Working Paper*. Dublin: Eurofound.

Pesole, A., Brancati, U. and Fernández-Macías, E. (forthcoming): *Platform Workers in Europe: Evidence from the COLLEEM II Survey*. Luxembourg: Publications Office of the European Union.

Vandaele, K. (2018): Will Trade Unions Survive in the Platform Economy? Emerging Patterns of Platform Workers' Collective Voice and Representation in Europe. *Working Paper 2018.05*. Brussels: European Trade Union Institute.

9

THE FOURTH INDUSTRIAL REVOLUTION AND THE DISTRIBUTION OF INCOME

Stella S. Zilian and Laura S. Zilian

Introduction

The recent stream of technological innovations, especially in the field of cognitive computing technologies, is expected to lead to significant structural changes in the economy and society – comparable to periods of technological upheaval in the past. While technology is commonly seen as a main driver of economic growth, phases of major technological change have always been accompanied by anxieties and fears associated with the possible negative effects technology may have on employment as machines have increasingly become able to substitute for human labour (Mokyr et al. 2015). Thus far, the predictions of high technological unemployment have never been fulfilled. On the contrary, innovation and new technologies have usually given rise to the creation of new jobs and industries as labour productivity and productive capacities increased, living standards improved, and new products were developed.

However, even though previous technological revolutions have had positive long-run effects, they were accompanied by significant disruptions such as rising income inequality. In fact, technological change has long been identified as one of the main drivers of income inequality (Atkinson 2015). The more recent phase of technological development, which is often referred to as the fourth industrial revolution, is no exception and might therefore exacerbate income inequalities, which are currently at historically high levels (Piketty 2014). Income inequality encompasses two different, but interconnected aspects: the personal income distribution and the functional income distribution. The former describes the distribution of wages on the individual or household level, while the latter refers to the distribution of income between the owners of the factors of production.

In economic theory new technologies are treated either as substitutes or complements of human labour depending on the skill-content of jobs or tasks. Hence, innovations affect the wage distribution between different skill groups by increasing the relative productivity of some (the high-skilled workers) while decreasing the relative productivity of others (the low-skilled workers). Based on the assumption that wages are paid according to the marginal productivity of labour, this implies that technological change increases wage dispersion. The hypothesis of *skill-biased technological change* and the newer concept of *routine-biased technological change* have been studied intensively (Acemoglu 2002; Acemoglu and Autor 2011; Autor 2013; Autor et al.

DOI: 10.4324/9780429351921-12

2003, 2008; Card and DiNardo 2002; Goos et al. 2009, 2014; Goos and Manning 2003, 2007) but need to be reconsidered, nonetheless. Given the progress in fields such as Artificial Intelligence, machines can perform a rising number of tasks of increasing complexity, such as driving or speech recognition (Brynjolfsson and McAfee 2014). Hence, to remain competitive, people will need to acquire new skills for the upcoming era of the fourth industrial revolution (Acemoglu and Restrepo 2020; Furman and Seamans 2019). One set of new skills can be subsumed under the term digital skills, which refer to abilities related to the use of digital technologies, encompassing a broad range of skills such as creating and editing digital media, using online services, or protecting privacy and data. It is well-known from the literature in sociology and media and communication studies on digital inequality that these skills are unequally distributed along the usual dimensions of social inequality, such as age, gender, race, or socio-economic status (DiMaggio et al. 2004; Drabowicz 2014; Robinson et al. 2015; van Deursen et al. 2017). The unequal distribution of digital skills, which can be explained by theories of social stratification, might therefore perpetuate existing wage inequalities as the demand for these skills grows in the labour market.

But technical progress may also affect the income distribution by changing market structures. The recent wave of technological innovations is characterised by the growing importance of knowledge-based capital (KBC), which gives rise to economies of scale. Due to these increasing returns to scale, KBC-intensive sectors tend to be highly concentrated and dominated by few incumbent firms that may achieve global market power and benefit from monopoly or oligopoly rents and high capital returns. These oligopolistic market structures as well as the high innovative capacity in KBC-intensive sectors increase the risk for competitors to be successful in such markets. Consequently, the risk premium is higher for investment in KBC, increasing the return on capital for successful investors. These effects may lead to rising capital shares in knowledge-based economies and thus, lower wage shares.

The goal of this chapter is to present an overview of the most popular explanations to relate income inequality with technological change in the economics literature and to discuss what this relationship may look like in the era of increasingly smart technologies, which "enable[e] intelligence, processing, communication, and networking capabilities in all products, systems, and processes, influencing all parts of society" (Beernaert and Fribourg-Blanc 2017: 567). This does by no means mean that new technologies are the only explanation for rising income inequality. Indeed, there are several important (and usually intertwined) factors explaining income inequality, for example, globalisation, redistribution policies, financialisation or the reduced bargaining power of labour. Hence, our account provides one puzzle piece for explaining social and economic inequality.

The chapter is structured as follows. First, we discuss technological revolutions from a historical perspective. In the following section, we present the key mechanisms how digitalisation may affect income inequality. Then, we discuss how product and process innovation affect labour demand and consequently the earnings distribution. Afterwards, we review recent research on the impact of digital innovations on market concentration, the functional income distribution, and the rise of the top 1% income share. The last section concludes.

Transformative technologies: a historical perspective

Adding a historical perspective helps to inform our understanding of the ways in which current technological changes may affect the economy and society. The use of the term fourth industrial revolution, coined by Schwab (2016), suggests that recent technological developments, often summarised under the catchword "digitalisation",[1] have a similarly transformative character as technological revolutions of the past. This is also reflected in the conceptualisation provided by Lee et al. (2018: 7) who describe the fourth industrial revolution as "the broad changes in industries as well as society that are affected by the disruptive technological changes in artificial intelligence, automation, and hyper-connectivity".

The first great industrial revolution refers to the onset of the mechanisation of production with the help of steam and waterpower at the end of the 18th century and marks the transition from an agricultural society to an industrial society – at least in those countries that are considered the most highly developed economies today. The transformative character of the general purpose technology[2] (GPT) that drove the first industrial revolution, namely steam power, can be illustrated by two examples: firstly, the invention of the steam locomotive made it possible to bridge geographical distances at a speed never seen before. Secondly, the concept of wage labour emerged with the implementation of the industrial mode of production and, conversely, a class of capitalists was formed while for the first time in history, the workplace was separated from home. These two examples illustrate an important characteristic of a general purpose technology (GPT): they do not only change the economy by increasing productivity, but they change society as a whole (Lipsey et al. 2005). The social changes brought about by the first industrial revolution in Britain are well-studied (see for example Feinstein 1998) and the general conclusion that is drawn distinguishes between long-run and short-run effects: despite the recurring concerns that machines will make human labour redundant (see Mokyr et al. 2015) the mechanisation of production did not lead to technological unemployment at a large scale in the long run because sustained productivity growth and the emergence of new industries and occupations were able to absorb the displaced workers. However, the authors emphasise that these developments took time and, estimations suggest, longer than an average working life. In fact, the transition phase was characterised by substantial disruptions such as the worsening of working conditions as production shifted from artisan shops to factory mass production, and rising income inequality (Mokyr et al. 2015).

The second industrial revolution is commonly dated back to the late 19th and the early 20th century when mass production caught on with the help of the GPT of that time, electricity. One example illustrating the transformative character of electricity is the electric telegraph, which was as revolutionary as the Internet in accelerating the flow of information and thus facilitating trade and commerce, among others (Mokyr 1999; Wajcman 2013). Furthermore, the electrification of the assembly line was crucial for the mass production of goods, turning former niche or luxury markets to mass markets. A classic example is the market for the automobile: the high productive efficiency translated to a massive fall in prices for cars from an average of 2,126 US-$ in 1908 to 317 US-$ (in 1908 dollars) in 1923. This drop in prices is reflected in a massive growth of annual sales from 64,000 in 1908 to 3.6 million in 1923 (Wajcman 2013). Furthermore, the organisational requirements of mass production, especially the need to plan both the product as well as the work process in advance, contributed to the emergence of new bureaucratic corporation structures, which resulted in a growing demand for white-collar workers (Lauder et al. 2018). The mass market does not only require mass production, but also mass consumers – the emergence of consumerism is therefore driven by the early industrial revolutions. Indeed, between 1870 and 1914 real wages increased considerably and standards of

1 There is a variety of terminologies and characterisations of the on-going phase of technological change; see for example Kurz et al. (2018).

2 A general purpose technology is characterised by the following features: they are not restricted to individual industries, they are evolving and reduce costs for users, and they ease the diffusion of innovations.

living, for example in terms of social insurance, working hours, nutrition, or housing, started to improve in Western industrialised countries (Mokyr 1999).

The third industrial revolution, which started in the 1970s, is characterised by the increasing usage of IT and electronics to automate production processes even further. It marks the beginning transition from the industrial to the information society. At the same time, this period is characterised by an acceleration of globalisation trends, which is closely connected to the development of information and communication technologies. The most important GPTs of this period are computers and the Internet, whose global spread increased rapidly between 1970 and 2000. While there is disagreement to what extent these technologies affected productivity, it is widely agreed upon that computerisation has altered the nature of work and the structure of employment, for example through its impact on the task composition within industries and occupations with considerable shifts towards analytical and interactive non-routine tasks (Spitz-Oener 2006; Black and Spitz-Oener 2010).

ICT and the Internet still play a dominant role for the fourth industrial revolution, but it is difficult to foresee what the next GPT will be. One candidate is Artificial Intelligence (AI), which relies heavily on the availability of enormous data stocks (big data) to improve pattern recognition and analysis (OECD 2020). Pratt (2015) estimates that global data storage is roughly equivalent to the capacity of ten million human brains. Of course, this should by no means be equated with the capacity of ten million human brains, but it is a good indicator of the scale of global information storage. Since AI benefits greatly from large amounts of data and the implementation and availability of local wireless communication as well as the increasing spread and performance of the Internet, the foundations for AI-intensification in the production of goods and services seem to have been laid. Moreover, given the efforts in various countries to promote variants of the German "Industry 4.0", a concept of industrial production in which smart machines, production parts, and storage systems communicate with each other by autonomously exchanging data and thus automatically initiating the execution of work tasks (Brödner 2018), it is not surprising that David Ricardo's "machinery question" has re-returned in recent years. In general, the historical evidence suggests that the impact of technological revolutions on the economy and society is positive for economic and social welfare. However, even optimistic accounts, such as the one of Mokyr et al. (2015), recognise that there are significant detrimental impacts on some parts of the society in the short and medium term, thereby increasing social inequality. Hence, it is important to study the consequences of these disruptions that will hit society as the fourth industrial revolution evolves. Especially during the first industrial revolution no measures were taken to alleviate the negative impacts of mechanisation, but the ongoing fourth industrial revolution can be shaped so that risks and opportunities are more evenly shared. "Technology is not destiny" (Brynjolfsson and McAfee 2014: 257) and neither is the impact of a technological revolution on society.

The key relationships between the fourth industrial revolution and income distribution

In the following sections we discuss the key mechanisms through which the fourth industrial revolution affects the income distribution illustrated in the conceptual diagram presented in Figure 9.1. The diagram provides a summary of the main arguments encountered in the literature on the impact of the recent wave of technological change on income inequality and guides the structure of this section. First, we describe the general macroeconomic effects of technological innovation on labour demand, illustrated as channel (1), using the compensation theory as presented in Vivarelli (1995, 2014). We then discuss the consequence of changing

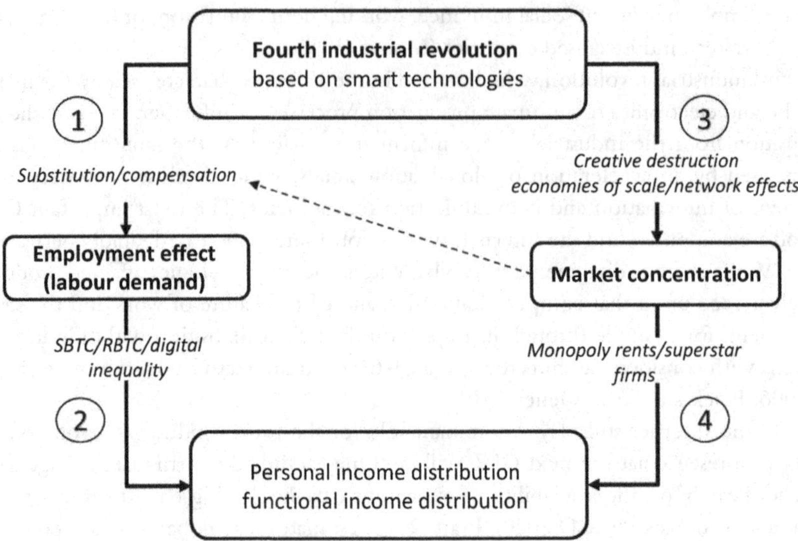

Figure 9.1 Conceptual diagram to illustrate the effect of technological innovations triggering the fourth industrial revolution and the income distribution

Source: Author's illustration

Note: In the diagram we see the effects of new technologies on the income distribution. (1) describes the substitution and compensation effect of technological innovations on labour demand. (2) relates the labour demand effect to the personal income distribution via the skill-biased technological change hypothesis, the routine-biased technological change hypothesis and digital inequality. (3) describes the impact of the fourth industrial revolution on market concentration. (4) relates effects on market concentration with the functional and personal income distribution via the distribution of monopoly rents and the impact of superstar firms. The dashed line indicates that the degree of market concentration affects the size of the compensation effects.

skill requirements due to the employment effect induced by process and product innovations for the personal income distribution, illustrated as channel (2), based on the commonly used explanations in the economic literature, skill-biased technological change and routine-biased technological change (e.g., Acemoglu and Autor 2011). In addition, we relate the demand for "new" digital skills to wage inequality based on the digital inequality literature and the stratification hypothesis. Next, we turn to the potential effects, which the fourth industrial revolution might have on market concentration due to economies of scale, network effects, and the process of creative destruction, illustrated in channel (3). Finally, we describe how rising market concentration may affect the distribution of income due to the distribution of monopoly rents and potential superstar effects illustrated as channel (4).

The fourth industrial revolution, employment, and the distribution of wage income

Macroeconomic effects of process and product innovations on labour demand

For a simplified analysis of macroeconomic employment effects of technological change, one can distinguish between process innovations and product innovations. While the former are generally assumed to increase labour productivity by improving the production process (labour saving technological change), the latter are assumed to enlarge product markets and, thus, tend to have positive effects on employment and aggregate income (Vivarelli 1995). Smart technologies, which play a key role in the fourth industrial revolution, are characterised by both types of innovations. On the one hand they are used in industrial production, for example in the form of robots, where they displace more costly human labour in the production process. On the other hand, some smart technologies are products that can be seen to (indirectly) expand the market while others are merely substitutes for existing goods. For example, smartphones exhibit both properties: they substitute for standard mobile phones and at the same time, they create new markets complementary to smartphones, such as the market for apps.

The productivity gains associated with process innovations can trigger compensation effects in quite distinct parts of the economic system that might mitigate or even offset their negative displacement effects (Acemoglu and Restrepo 2020; Vivarelli 1995, 2014). The following description of the compensation effects follows Vivarelli (1995: 26ff.) and Vivarelli (2014: 125ff.) who takes into account institutional and market mechanisms. One commonly used argument in favour of compensating effects assumes that, in perfectly competitive markets, falling production costs lead to lower prices for consumers which stimulates demand for products and, thus, labour demand. The effectiveness of this mechanism of compensation via *decrease in prices* crucially depends on competitive market structures and on the price elasticity of demand: the lower the price elasticity of demand, the weaker the compensation effect. If markets are incomplete and if there is imperfect competition, cost saving does not necessarily lead to lower product prices for consumers which weakens the compensation effect due to lower prices.

If market power prevails, cost reductions might in fact result in higher profits for companies and/or higher wages for those employed in those innovating firms. Higher profits, in turn, can be used in different ways: they can be distributed as dividends to shareholders, reinvested in the company (for example in R&D to foster further innovation), or invested in the financial market. In principle, profits that are shared with employees in the form of higher wages or if they are shared as dividends with shareholders, this may induce a compensation effect via *increases in income* by generating additional demand. The scale of this compensation effect via increases in income depends on factors related to bargaining power, but also the propensity of households to consume or propensity of firms to invest. This demand-driven compensation effect in the Kaldorian and Keynesian tradition stands in contrast to the neoclassical compensation effect via decrease in real wages: although it is of course possible that firms hire more workers when wages decrease, as suggested by partial labour market theory approaches, this may be offset due to potential negative impacts of lower income on effective demand.

The compensation effects described thus far emerge due to process innovations. But an important aspect of technological progress and innovation is the invention and development of new products and services that may lead to the emergence of entirely new markets and sectors with new employment opportunities. The magnitude of this effect, *compensation via new products*, depends on the substitutability of new and old products as well as on the labour intensity of the production of new products. This effect is the most difficult to foresee or speculate about because it impossible to know what goods or services people will want or need in the future, but it has clearly been a highly effective compensation mechanism in the past. Just think about the vast number of jobs in all kinds of sectors of the economy created around the invention of the automobile, the computer or the Internet. Related to these effects caused by product innovation, is the creation of new tasks or the alteration of "old" tasks due to the development of new technologies that complement labour (Acemoglu and Restrepo 2018). One good example for a new task created by the Internet is web development while tasks related to social media management can be seen as a mix of new and traditional marketing tasks. Thus, if humans continue to have a comparative advantage in performing certain tasks and if R&D efforts are directed towards labour-friendly innovations, for example assistive technologies in care, even sophisticated computer technologies will not lead to persistently high technological unemployment (Autor 2015; Acemoglu and Restrepo 2018). The interaction of the substitution and compensation effects determine the net impact of process and product innovation on the demand for labour.[3] Regarding the technologies of the fourth industrial revolution, much of the literature focuses on quantifying substitution effects, or more precisely substitution potentials, without assessing possible compensation effects. Overall, these studies predict that new technologies can substitute for an increasing number of tasks and consequently jobs that rely heavily on automatable tasks. For example, Dengler and Matthes (2020) show in a recent analysis that 34% of male and 15% of female employees in Germany work in jobs with a high potential of substitution (i.e., more than 70% of the tasks could be automated). The gender differences are explained by the gender-specific segregation of the German labour market: the risk of substitution is high in male-dominated occupations, while it is low in social and cultural service jobs, which are dominated by women. They further find that employment growth between 2013 and 2016 is significantly weaker in occupations with a higher potential for substitution. This might indicate that these occupations are subject to on-going automation processes.

While Dengler and Matthes (2020) and similar research, such as Frey and Osborne (2017) or Arntz et al. (2017), study the technological potential for substitution, they leave aside the potential for compensation. So far history has shown that compensation effects always more than offset the substitution effects on the aggregate, due to sustained productivity growth and the expansion of markets following the introduction of new products as well as the creation of new tasks, as described in Acemoglu and Restrepo (2018). Moreover, technological unemployment was also avoided by dividing work between a greater number of people thereby reducing the total numbers of hours worked (Leontief 1982). For example, the quantity of annual hours worked per employees halved between 1870 and 1998 in highly industrialised Western economies (Mokyr et al. 2015).[4] Nevertheless, current developments, such as the slowdown of productivity growth and in particular its decoupling from employment growth in recent years (Brynjolfsson and McAfee 2013), suggest that for the fourth industrial revolution, some compensation effects might just not be as strong as for previous industrial revolutions which might be a consequence of the different nature of the capital used in the production of goods and services today: the fourth industrial revolution is characterised by the growing importance of so-called knowledge-based capital (i.e., intangible capital goods such as software, data, but also Research & Development [R&D] and intellectual property rights). Due to the non-rivalry property of knowledge, KBC-intensive industries are characterised by increasing returns to scale giving rise to growing market concentration (Guellec and Paunov 2017). This may reduce the size of the compensation effects significantly if innovating firms do not share innovation rents with their employees or invest them to expand production, but rather engage in rent-seeking, or use the rents to invest in (unproductive) financial capital.

3 This is discussed in a similar way in Acemoglu and Restrepo (2018, 2020). However, the account of Vivarelli (2014) presented here gives a more general overview as it also discusses substitution and compensation effects in terms of how institutional and market structures can affect the overall labour demand outcomes.

4 Even though it is reasonable to assume that humankind will never actually run out of work, the unequal distribution of work will remain one of the most pressing issues for modern societies. For example, in recent years a polarisation of hours worked can be observed; the share of people working many hours and the share of people working no or few hours have both been increasing, which contributes to the polarisation of employment (Wajcman 2013).

Labour demand and wage inequality: skill-biased technological change, routine-biased technological change, and digital inequality

The dominant view in economics is that technological change is expected to increase income inequality because new technologies complement high-skilled workers while they substitute for low-skilled workers and thus raise the demand for high skills relative to the demand for low skills (Atkinson 2015). The argument is rooted in human capital theory and it relies on the assumption that wages reflect a worker's marginal productivity of labour. Consequently, according to the skill-biased technological change hypothesis, new technologies increase the productivity of high-skilled workers who receive higher wages as they are remunerated according to their marginal product of labour. An early contribution fuelling the idea of biased technological change is the paper of Atkinson and Stiglitz (1969) in which they argue that technological change does not affect workers' productivity in general but is often specific to certain production activities. Atkinson and Stiglitz (1969) deviate from the then-orthodoxy by recognising that technological change is not neutral, but that the localisation of technological change in the work process matters. Ideas of the model proposed by Atkinson and Stiglitz (1969) found application in the literature on skill-biased technological change but even more so in the literature on task-biased technological change. In this section we discuss both, as they are still the dominant explanations of rising wage inequality due to technological change.

SKILL-BIASED AND ROUTINE-BIASED TECHNOLOGICAL CHANGE

The textbook model relating wage dispersion to technological change in economics draws on the hypothesis that modern technologies are essentially skill-biased in favour of highly skilled workers (Acemoglu and Autor 2011). According to the hypothesis of skill-biased technological change (SBTC), technological change affects the wage distribution between different skill groups by increasing the productivity of high-skilled workers relative to that of low-skilled workers. This was extensively studied regarding the computer revolution: at least since Krueger (1993) economists argue that the use of computers increases the productivity of employees and they are thus rewarded higher for their computer skills.[5] The computer wage premium therefore contributes to rising earnings inequality between those who have these skills and those who do not.

While the idea that new technologies are pre-dominantly upskilling seems to receive empirical support,[6] they fail to explain more recent developments such as the polarisation of labour markets with growing employment shares at the top and at the bottom of the skill or income distribution at the cost of middle-income jobs, or the rising share of top 1% income earners. While the former has been addressed by economists through shifting the focus from skills, or rather formal education, to tasks (Autor et al. 2003), the latter, and we will return to this in

5 Kristal and Edler (2019) briefly discuss status distinction as an alternative reason for why computer use at work is rewarded. The general idea is that computer use at work can help to "generate resources that workers can use in making claims on starting salary and pay raises" (4). This goes back to DiMaggio and Bonikowski (2008) who argue that computer savviness may signal competence or greater skills and employers might therefore be inclined to pay them higher wages. Furthermore, apt Internet users can access a greater range of labour market information, increasing their bargaining power, and they can expand their social network and may therefore benefit from wage effects due to greater social capital.

6 See for instance Acemoglu (2002) for a review and Card and DiNardo (2002), Lauder et al. (2018), and Bogliacino (2014) for critical discussions of SBTC.

the following section, is related to "superstar" economics and rising returns to capital and the unequal distribution of capital ownership.

To explain the phenomenon of polarisation of labour markets, Autor et al. (2003) develop a task model which recognises that jobs consist of different tasks that can be categorised along the dimension's routine/non-routine and manual/cognitive tasks. The distinction of skill and task is the most important difference between the canonical model and the task model (Acemoglu and Autor 2011): while a task is defined as a unit of work activity which produces output, skills refer to a worker's general stock of capabilities to perform different tasks. This implies that workers of a given skill level, which has been acquired by formal education and experience, can reallocate their skills towards different sets of tasks in response to needs and wants. In other words, within the defined skill groups workers are essentially free to change the set of tasks they perform and consequently, they can adapt to changing labour market conditions. The main argument put forward in Autor et al. (2003) is that technologies can replace specific, codifiable routine tasks while they cannot perform non-routine manual and cognitive tasks. Whether a job is prone to substitution depends on the mix of codifiable routine tasks and more complex, non-routine tasks. The routine-intensive jobs do not necessarily have to be low-skill and low-paying jobs. In fact, many low-paying jobs in personal services are non-routine intensive while some typical middle-paying occupations, such as administrative work or bank tellers, are routine intensive (Autor 2013, 2015).

Since the introduction of the task model, numerous empirical studies confirm its prediction of a shift away from jobs with a high-routine task content towards jobs with a low-routine task content and corresponding wage dispersion (for example Autor 2015; Autor et al. 2008; Goos et al. 2014) or put differently, the past decades were characterised by the dual growth of what Goos and Manning (2003) call good MacJobs and lousy McJobs (Wajcman 2013). In addition, Black and Spitz-Oener (2010) show that the task content *within* occupations changed considerably because of computerisation between 1979 and 1999. They observe a significant increase of the share of non-routine analytical tasks.[7]

Considering the fourth industrial revolution the question arises to what extent new smart technologies such as AI will change the demand for skills and therefore wage differentials. It is difficult to foresee which tasks will become routine in the future – for example, the initial work of Autor et al. (2003) considered driving a car as a non-routine task. Nowadays autonomous driving is not fiction anymore from a technical point of view. This highlights how difficult it is to predict which skills will be needed to perform which tasks in the future. While high skills are needed to develop, design, and programme smart technologies, medium skills are needed to supervise, maintain, and manage the production processes (in manufacturing as well as in services). Low skills to perform, for example, non-routine manual tasks that require responsive reactions, such as the tasks carried out by click-workers, will also be demanded in the future. Deming (2017), on the other hand, highlights the growing importance of social and interpersonal skills since the 1980s. One explanation, following Autor (2015) is that tasks which require such skills cannot easily be carried out by machines – not even by very smart ones. Nevertheless, although precise estimates of future skill demand are not feasible, there is broad consensus that the task content of occupations has already been subject to significant changes and that the demand for (more or less) new digital skills and other technology-complementing skills will

7 This development was particularly pronounced for women and Black and Spitz-Oener (2010) argue that this was a main driver of the declining gender wage gap. While we believe that gender equality in the fourth industrial revolution is a very important topic, the literature is too large to discuss it here but Howcroft and Rubery (2018) provide an informative overview of gender equality issues in the fourth industrial revolution.

increase (Deming 2017; Ferrari 2013; Furman and Seamans 2019). Furman and Seamans (2019) specifically point to machine-learning skills for the era of AI while Deming (2017) emphasises social and so-called "soft" skills as important complementary skills. Moreover, Acemoglu and Restrepo (2018) highlight the problem of workers acquiring the "wrong" set of skills for the upcoming era of technologies such as AI. They further argue that a shortage of certain skills might have significant negative effects with far-reaching implications for inequality. Since economic research shows that there is a wage premium for digital skills (Falck et al. 2016; Grundke et al. 2018; Hanushek et al. 2015), rising demand for digital skills will affect wage inequality between those who possess these skills and those who do not.

DIGITAL SKILLS

Digital skills are understood as skills which are necessary to perform activities and meet the requirements of digital technologies. The range of activities that require a certain level of digital skills is manifold and differ regarding their levels of complexity: using a smartphone, using a self-service cash register, maintaining an industrial robot, using spreadsheets to prepare budget plans and various forms of Internet use for research purposes, programming a search engine algorithm, etc. In short, digital technologies already permeate all areas of life and digital skills are therefore not only becoming increasingly important for (finding and keeping) work, but also increasingly determine the extent of social participation in everyday life (e.g., to use online services provided by the government).

It is often assumed that as the fourth industrial revolution progresses, relative demand for (new) digital skills will continue to rise (Berger and Frey 2016) and consequently their value, which results in higher wages for jobs requiring those skills intensively. However, these jobs will most likely be filled by people who already possess digital skills since it is more costly to acquire new skills than to "update" skills that already exist. As a result, the gap between people who are more comfortable using digital technologies and those who are not will widen. The crucial question is what causes this gap. From an economist's perspective, people possess different digital skills, a specific form of human capital, due to a variety of individual factors such as personal characteristics (e.g., innate abilities), acquired knowledge through formal and informal learning processes, and selection mechanisms following considerations of comparative advantage (Rosen 2008). Human capital formation is essentially assumed to follow the same logic as any other investment decision, but socio-economic inequalities mediate the returns to these investments. This is not too different from research in sociology that emphasises how existing structural social inequalities determine and shape the patterns of digital inequality and the distribution of digital skills (Robinson et al. 2015). However, scholars of social stratification and social exclusion are particularly interested in understanding the interplay of social structures (e.g., class, status group, or power) and digital inequality (Ragnedda 2017).

In this section we switch to the sociological perspective and discuss how wage inequality in the digital age is related to digital inequality, which can be described as the unequal access to and the differentiated use of digital technologies across different social strata of the society (e.g., Helsper 2012; DiMaggio et al. 2004; Robinson et al. 2015). The theoretical basis for digital inequality research is provided by Helsper (2012) who develops a corresponding field model to explain differences in ICT use as a complex interplay of economic, cultural, social, and personal factors in the offline and the digital world, which reinforce each other. In this model, social inclusion is restricted by economic constraints (e.g., lack of income, joblessness), cultural constraints (e.g., gender or other stereotypes), social constraints (e.g., networks of friends and family), or personal constraints (e.g., physical/mental well-being or psychological traits). Helsper (2012) argues that exclusion from these offline fields corresponds to exclusion in digital economic, cultural, social, and personal fields – and vice versa. In the context of this model, digital skills are an important mediator of the influence of offline social exclusion on digital exclusion (Helsper 2012). Consequently, their distribution determines to what extent offline social inequality is reproduced, amplified, or might even be reduced. However, the vast body of literature on the digital divide and digital inequality indicates that the distribution of those skills is characterised by existing patterns of social inequality and reveals the structural differences in the access to digital technologies (first-level divide) and in the use of digital technologies (second-level divide) along the usual dimensions of social inequality, such as gender, socio-economic background, age, ethnicity, etc. (DiMaggio et al. 2004; Robinson et al. 2015; van Deursen and van Dijk 2014; Zilian and Zilian 2020; Zillien and Hargittai 2009). Over the past years, the concept of the third-level digital divide, which focuses on inequalities in outcomes related to ICT use, gained attention (Ragnedda 2017; van Deursen et al. 2017).

When the digital divide research first appeared in the 1990s, it was mainly concerned with differences in Internet access, both in a comparative and intra-country context (van Dijk 2005; Zillien and Hargittai 2009) and many studies confirm the hypothesis that socio-economic inequalities are reflected in access to the Internet (for overviews, see DiMaggio et al. 2004; van Dijk 2005; Robinson et al. 2015). With the on-going process of digitalisation of an increasing number of different aspects of economic, social, and political life, the first-level digital divide lays the foundation for further digital inequalities. This idea is well summarised in van Dijk (2005: 15) who puts forward that the digital divide is driven by the unequal distribution of resources leading to "unequal access to digital technology [that] brings about unequal participation in society".

Thus, as digital technologies and the Internet have become more widespread, the focus of research has shifted from access to the Internet, to analysing differences in the way people use computer technologies and how this affects their success in different life domains (Ragnedda 2017). For example, Zillien and Hargittai (2009) find that people with a higher socio-economic status are more likely to use the Internet for informing themselves about politics, economics, or about stock markets than people with a lower socio-economic status. More recently, van Deursen et al. (2017) study the sequential and compound pattern of digital inequality based on a representative survey of Dutch Internet users. They find that digital exclusion starts with differences in skills (not only technical but also social and creative skills), followed by differences regarding the purpose of Internet use and finally differences in tangible outcomes in the offline world. The authors therefore provide evidence for the hypothesis that the Internet exacerbates offline inequalities. For Austria, Zilian and Zilian (2020) demonstrate that being female and having a low socio-economic background have a negative impact on digital problem-solving skills, as measured by the PIAAC[8] survey in 2011/12. However, the degree of ICT engagement in everyday life absorbs these differences suggesting that the acquisition of digital problem-solving skills is characterised by "learning-by-doing", which is in line with van Dijk (2005). Hence, enabling the integration of ICT use in everyday life may help to reduce digital inequality but to what extent does not only depend on the quantity of ICT use, but ultimately on the quality – and this is primarily a question of the distribution of financial resources: despite the shifting focus towards the second-level and third-level digital divide, the first-level divide regarding access to the Internet and hardware still matters, even in developed countries. A notebook provides different features and possibilities to develop digital skills than a smartphone or a tablet. An old and slow computer or a bad and slow Internet connection may negatively affect

8 The OECD Programme for the International Assessment of Adult Competencies.

the user experience compared to a high-end product or a stable and fast Internet connection. Hence, efforts to reduce digital inequality through, for example, providing more digital education is not sufficient if social inequalities are not reduced. The problem of digital inequality can therefore not be solved at the individual level, but must be addressed at a higher, societal level (van Deursen et al. 2017).

In summary, the fourth industrial revolution affects wage income inequality mainly because of its impact on skill requirements. The literature on routine-biased technological change shows that significant changes of employment and occupational structures have been taking place since the beginning of the 1990s and polarisation patterns can be observed in many Western industrialised countries. It is reasonable to expect that technology-driven transformations of the task content and organisation of work will continue. While it is impossible to predict which skills will be demanded in the future, there is broad consensus that the importance of digital skills but also of social and interpersonal skills will rise. But as the extensive, mainly sociological, literature on digital inequality shows, the distribution of digital skills is characterised by social stratification. This may well lead to rising income inequality between the digitally skilled and the digitally unskilled or even to the digital exclusion of those people, who are already socially excluded.

The fourth industrial revolution, market concentration, and the functional income distribution

At a general level, two important developments can be observed regarding the distribution of income between the production factors labour and capital: firstly, the importance of the factor capital in relation to gross domestic product (GDP) has increased significantly over the past three decades (Piketty 2014; Atkinson 2015) and a persistent decline in the wage share has been recorded in (almost) all OECD countries since the end of the 1980s (Autor et al. 2020; Karabarbounis and Neiman 2014; OECD 2011). Technological progress is assumed to be important for this development because it induces firms to switch to more capital-intensive production processes as the relative price for capital decreases, which in turn reduces the labour share (Dao et al. 2017; Karabarbounis and Neiman 2014). Secondly, increasing evidence indicates that rising market concentration, the emergence of "winner-takes-all markets" and "superstar firms" is closely connected to digitalisation, which has not only contributed to a fall of the labour share but also to the rise of income concentration at the very top (Autor et al. 2020; Furman and Orszag 2018; Guellec and Paunov 2017).

In this section we start by discussing the connection between the growing importance of intangible capital goods, which lie at the heart of smart technologies and the fourth industrial revolution, and rising market concentration before turning to the distributional effects associated with these changing market structures.

Digitalisation and market concentration

The literature on the impact of the fourth industrial revolution, or rather its driving technological innovations, on market concentration often focuses on two counteracting mechanisms, which are both related to the nature of knowledge-based capital that has been becoming increasingly important as a factor of production and should not be equated with the physical capital stock (OECD 2013). Firstly, digital innovations enable the incumbents to strengthen and defend their market power due to economies of scale, perpetuated by network and lock-in effects, which are associated with knowledge-based capital. Secondly, there is significant potential for high business dynamism due to "creative destruction" processes since entry barriers are low in

KBC-intensive markets if the process of imitation and diffusion is unrestricted. Knowledge-based capital subsumes intangible capital goods such as R&D, software, databases, algorithms, or intellectual property rights. The non-rivalrous nature of knowledge is associated with increasing returns to scale – once knowledge has been produced, mostly through costly R&D, it can be used and reproduced at (almost) no costs (Guellec and Paunov 2017). Thus, economies of scale arise due to the possibility to produce at low to zero marginal costs; for example, large profits can be generated through the licensing of software products without imposing any additional production costs. The economies of scale are reinforced by network and lock-in effects (Guellec and Paunov 2017; Krämer 2019). Network effects, well known from the literature on public goods, occur when the benefit of a good increases with the number of users, which is the case for social networks or digital platforms that offer services, such as Uber or Airbnb (Allen 2017). Lock-in effects are related to network effects and they arise as it becomes more costly for consumers to switch between different products. However, incumbents can use lock-in effects, for example by restricting compatibility with similar products from competitors (e.g., Apple), to defend their market position (Krämer 2019). These mechanisms are well described in Shapiro and Varian (1999: 77) who identify the positive feedback effects related to network effects as a main determinant for the emergence of monopolistic market structures in the knowledge economy and may, in its "most extreme form" lead to the establishment of a "winner-takes-all-market, in which a single firm or technology vanquishes all others". As prominent examples, the authors cite the triumph of Microsoft and Intel in the PC sector and the penetration of VHS over Betamax. Nowadays, the importance of network effects is even more evident. Facebook and the acquired services WhatsApp and Instagram, which connect billions of users worldwide, dominate social media and competitors can hardly assert themselves or, as in the case of WhatsApp, strong start-ups are bought up (Makridakis 2017).

At the same time, the low or zero marginal cost property of KBC in combination with trends such as rapidly falling prices of, for example, cloud computing, implies that the growing importance of KBC reduces entry barriers to many markets by lowering innovation costs thereby driving the process of "creative destruction" (Guellec and Paunov 2017). Creative destruction in the sense of Schumpeter (1912) describes the dynamic development of a capitalist economy in which the "old" is permanently displaced by the "new". The "new" is defined as innovations produced by entrepreneurs who are eager to experiment. Schumpeter is often interpreted in the sense that for innovation-driven growth, entrepreneurs must have monopoly power that allows them to generate innovation rents to compensate them for bearing the risks associated with innovation activities. The time to skim off these Schumpeterian rents is limited: as soon as innovations diffuse due to imitation by competitors, they cannot be generated anymore. Consequently, the establishment of persistent dominant monopoly positions is assumed to be restricted by the continuous process of imitation and diffusion.

Even though there seems to be great potential for processes of creative destruction in the fourth industrial revolution, the empirical literature on recent developments, especially on the US economy, points towards a slowdown of business dynamism and the emergence of monopolistic or oligopolistic market structures (Autor et al. 2020; Council of Economic Advisers 2016; Decker et al. 2014). For example, Decker et al. (2014) show that the annual start-up rate, measured as the share of new companies in all companies, has fallen in the US from an average of 12% at the end of the 1980s to an average of 10.6% just before the financial crisis, only to fall to below 8% during the recession. They further show that while the survival rate of start-ups is low, those that do survive are characterised by high growth and they contribute disproportionately to job creation in the US. Furthermore, recent evidence suggests that large players tend to acquire small start-ups: Google, Apple, Amazon, and Facebook acquired more than 400 companies up to 2016 (Makridakis 2017). According to a report by the Council of Economic Advisors (CEA) from 2016, rising market concentration and decreasing intensity of competition in many sectors of the US economy is driven by a sharp rise in Merger & Acquisitions. The report also shows that the profit rate between firms has become much more heterogeneous since 1995, which is reflected in the observation that the profit rates of the top companies have increasingly moved away from the average. This heterogeneity between firms is largely traced back to differences in productivity driven by managerial and technological capabilities (Decker et al. 2017; Van Reenen 2018; Autor et al. 2020). Digital innovations are expected to facilitate the emergence of "superstar" firms, which are characterised by particularly large market shares and high rates of return. For example, Autor et al. (2020) argue that the most productive firms can attract most of the revenues in their sectors because of scale effects associated with the growth of KBC, growing platform competition, and greater possibilities for consumers to compare prices due to the Internet and search engines.

To sum up, the documented decline in start-up rates in the US indicates that the role of entrepreneurs has been becoming less important. Moreover, empirical evidence suggests a positive link between digitalisation and market concentration and the rise of superstar firms. Makridakis (2017) illustrates this with the four big American IT giants Apple, Google, Amazon, and Facebook, for whose success the efficient use of the Internet to offer their services and products is a decisive factor. In addition, the network effects already mentioned contribute significantly to the fact that digital companies can capture and defend high market shares. Finally, especially regarding the development and use of one of the most important technologies, namely AI, the digital superstars already have a considerable head start over their competitors.

Market concentration and the distribution of income

As discussed in the previous section, the tendency towards monopolisation and concentration of market power may weaken some of the compensation effects associated with new technologies such as the mechanism of compensation via decrease in prices (monopolistic firms can act as price-setters), and the effect via increase in income as the latter depends crucially on the rent-sharing behaviour of firms and the balance of power between employers and workers. Hence, monopolistic market structures and the rise of superstar firms have several potential implications for the income distribution. Firstly, the low labour intensity and low wage share in superstar firms affects the functional income distribution by pushing down the aggregate wage share (Autor et al. 2020). Secondly, the growing heterogeneity between firms with superstars extracting super-normal returns on capital due to their market dominance directly affects the personal income distribution since capital ownership is highly concentrated at the top of the income distribution (Atkinson 2015; Dao et al. 2017; Piketty 2014). Thirdly, the distribution of monopoly rents can theoretically increase or decrease income inequality depending on with whom the rents are shared or, if invested, on the type of investment. However, recent research suggest that monopoly rents are largely shared with shareholders, investors, and the management, thus increasing income inequality between the top-income earners and the average workers (Furman and Orszag 2018; Guellec and Paunov 2017).

SUPERSTAR FIRMS

In their influential paper, Autor et al. (2020) examine, both theoretically and empirically, the implications of the phenomenon of superstars for the development of the wage share. They show that the worldwide fall in the wage share (i.e., the reduced share of labour income in total income) since the 1970s can be explained by the rising dominance of superstar firms as they are characterised by low within-firm wage shares. Due to their size, superstar firms have lower fixed costs per output and because of their market power, they can achieve higher profit margins and therefore generate higher profits – both factors contribute to the reduction of the wage share in favour of the profit share. As superstar firms increase their share in the total economy, they disproportionately affect the overall development of the wage share. Autor et al. (2020) show for the US that there has been a concentration of turnover within many industries (excluding the public sector) over the last 30 years. At the same time, there has been a decline in the wage share within all industries, which has been particularly pronounced in industries that have experienced a relatively high increase in market concentration. From the shift in turnover between firms within sectors it can be concluded that superstar firms are partly responsible for this development: the extent of the fall in the wage share within firms within a sector varied considerably but since superstar firms with lower wage shares gained more weight within their sectors, the sector-specific wage shares decreased.

Another contributing factor to the fall in the wage share relates to the textbook notion of capital-enhancing technological change. In particular, production processes of digital or KBC-intensive superstar firms are less labour intensive than those of traditional firms. For example, Karabarbounis and Neiman (2014) identify the low labour intensity of production in ICT-intensive industries as one key factor driving the global decline in the wage share since 1975. They argue that technological change in the form of improvements of ICT led to falling capital user costs and this induced firms to increasingly shift away from labour as a factor of production. Indeed, comparing employment and profitability figures of digital and traditional firms reveals significant differences. While the digital superstars Apple, Google, Amazon, and Facebook employed a total of 458,000 people in 2016, the traditional superstar companies Walmart, Johnson and Johnson, Berkshire, and Toyota employ a total of 3.004 million people. Unsurprisingly this reflects in other key figures such as market capitalisation per employee ($4.13 million in digital vs. $0.38 million in traditional firms) and revenue per employee ($950,000 vs. $310,000) (Makridakis 2017).

The growing productivity dispersion between firms can also partly explain the growing earnings dispersion between employees. For instance, Song et al. (2019) show that there is an agglomeration of higher-paid workers in highly productive firms that pay better than other firms, or as they put it, one can observe a sorting of the labour force. In addition, high-paid workers often work together with other high-paid workers and low-paid workers work together with other low-paid workers. The latter phenomenon of segregation between high-paid and low-paid workers is partly due to the increasing possibilities of outsourcing a higher proportion of services or tasks. This often concerns low-paid workers, such as those providing cleaning services, but also higher-paid accounting and IT services are increasingly being outsourced, which is facilitated by digitalisation. This results in rising inequalities both within and between industries. Moreover, Song et al. (2019) demonstrate that the part of rising inequality due to within-firm inequality can primarily be traced back to growing within-firm earnings dispersion in very large companies with more than 1,000 employees, and even more so, the growing concentration of top incomes at mega companies with more than 10,000 employees. Considering the findings of Autor et al. (2020), the evidence provided by Song et al. (2019) could be interpreted as an indication that superstar firms do not only drive inequality at the aggregate, but also at the individual level.

Finally, recent research on monopsony in the labour market – a situation in which employers can exert considerable market power over workers because workers cannot instantaneously find or switch jobs – indicates that the impact of firm heterogeneity on wage inequality is increased in the presence of asymmetric power relations (Manning 2021). Monopsony power often prevails in low-wage segments of the labour markets but in recent years different forms of online work have attracted attention. For example, Dube et al. (2020) provide a thorough analysis of Amazon Mechanical Turk (AMT), a platform where mostly low-skilled human intelligence tasks are posted by requesters (employers) who set the wage ex ante and workers can then decide to do the task or not. The requesters then review the work but only pay workers whose submitted work they approve of. Dube et al. (2020) find evidence for significant monopsony power in terms of a low elasticity of labour supply. Although some research in labour economics suggests that monopsony power contributes to rising inequality, the evidence is still inconclusive (see Manning 2021 for a recent literature review) and there is significant scope for future research – especially as new forms of work organisation, such as platform work, become more widespread.

Distribution of monopoly rents

An economic rent is defined as "the return to a factor of production in excess of what would be needed to keep it in the market" (Furman and Orszag 2018: 2). In a Schumpeterian framework temporary innovation rents are necessary to incite further innovative activities. Moreover, the process of creative destruction increases social mobility as "new" market entrants may become the next winners while the "old" incumbents might become the future losers (Guellec and Paunov 2017). The impact of creative destruction and innovation on top income inequality on an aggregate level is examined in Aghion et al. (2019) for the US. Their empirical findings confirm the predictions of their Schumpeterian model: the rate of innovation is positively correlated with the top 1% income share – reflecting innovation rents – while no such relationship can be established for broader inequality measures, such as the Gini coefficient. In addition, they find a positive correlation between the rate of innovation and social mobility between 1996 and 2010 which is primarily driven by new market entrants. This is interpreted as the confirmation of the effect of creative destruction on social mobility.

While Aghion et al. (2019) argue that the innovation rents are only temporary because the positive correlation weakens when lagged effects are considered, the discussion in this chapter has shown that the accumulation of KBC facilitates the emergence of highly concentrated industries. It follows that digital innovations are associated with more permanent monopoly rents that go beyond the Schumpeterian innovation rents.[9] This ultimately leads to the question of the distribution of these rents which is determined by the bargaining power of various actors, such as inventors, shareholders, investors, different workers, and management (Atkinson 2015; Guellec and Paunov 2017). Empirical evidence suggest that the economic rents are disproportionately shared with capital owners, management, and shareholders, hence contributing to the fall of the labour share as well as to the rise of income concentration at the very top (Furman and Orszag 2018; Guellec and Paunov 2017). Furman and Orszag (2018) argue that returns on invested capital (ROIC), reflecting rents accruing to capital owners, differ greatly between firms; especially firms in the IT and health sector enjoyed very high ROIC of more than 45% on average between 2010 and 2014. Similarly, according to Guellec and Paunov (2017), the observed diverging movement of rising corporate profits and decreasing interest rates indicate

9 Aghion et al. (2019) slightly touch on this issue by introducing the distinction between productive and defensive innovations. While the former increase productivity and are carried out by new market entrants, the latter are used by incumbents to raise entry barriers and defend their market power. Since the share of defensive innovations are rising, the imitation and diffusion part of the creative destruction process may be distorted leading to the establishment of more persistent monopoly positions and consequently monopoly rents.

that rents are to large extents captured by investors and/or capital owners. Moreover, the growth of profits strongly correlates with the growth of top 1% incomes but not with the growth of the middling 40% of incomes between 1992 and 2013 thereby increasing income inequality.

Guellec and Paunov (2017) further emphasise the increasing market instability associated with digital innovations, which leads to rising risk premia for investors and entrepreneurs: winner-takes-all markets are not only characterised by higher market concentration, but also by a high degree of volatility since successful innovations can easily result in gaining very large market shares almost instantaneously[10] and consequently investors, business owners, and executives are compensated for taking this risk.

To sum up, digital innovations seem to foster higher market concentration, superstar dynamics, and the emergence of winner-takes-all structures, which potentially contribute to rising income inequality in terms of the functional as well as the personal income distribution. While the presented evidence does not necessarily imply that these trends of the recent past will continue as the fourth industrial revolution develops, it nevertheless sheds light on the potential risks associated with an economy that increasingly relies on KBC as a factor of production as well as a consumption good. Since many empirical studies use US data, future research should focus more on international comparisons which may help to inform about the role institutional settings play in mediating the described effects of digital innovation on market concentration and income inequality.

Conclusion

In this chapter we highlighted potential effects of technological change on the personal and functional income distribution in the context of the fourth industrial revolution. To help inform our understanding of these effects, we started by summarising how previous industrial revolutions affected economy and society, where negative effects were counterbalanced by positive effects. In particular, the recurring fears of persistent technological unemployment were unfounded, as the long-run effects on employment, income, and standards-of-living were positive. However, the short run, especially in the first industrial revolution, was characterised by rising inequality and deterioration of working conditions. The key insight from the historical perspective is therefore that processes of rapid technological change and their implications for the society need to be continuously studied to inform policy makers about potential measures that can be taken to counterbalance the negative effects from the short-run disruptions associated with industrial revolutions. Therefore, we present the main mechanisms identified in the recent economics literature relating income inequality and technological change against the background of the innovations, which drive the fourth industrial revolution.

We used compensation theory to discuss macroeconomic employment effects of technological change. The main idea of the compensation theory is that unemployment arising from process innovations which lead to the substitution of human labour by machines, is offset by several key mechanisms related to indirect effects on real income and investment, among others. Furthermore, product innovations induce the emergence of new markets, such as the app economy, increasing the demand for goods and creating new employment opportunities. In the history of economic thought, the compensation effects, especially those related to new products, were usually underestimated while the displacements effects were overestimated (Mokyr et al. 2015).

10 For example, the market for mobile phones was dominated by Nokia until Apple introduced its iPhone and Samsung followed suit with its Samsung Galaxy. In recent years, Huawei and Xiaomi have established themselves as serious competitors.

However, due to certain characteristics of the innovations driving the fourth industrial revolution related to their negative impact on market competition, we argue that the magnitude of compensation effects might be weakened as the balance of power between employers and workers gets distorted. Our discussion shows that digital innovations, which rely heavily on intangible knowledge-based capital, seem to drive increased market concentration, enable superstar dynamics, and contribute to the emergence of winner-takes-all structures. We identify several explanations discussed in the recent economics literature for how these technology-driven monopolistic market structures might affect the distribution of income. Firstly, since KBC-intensive superstar firms are characterised by low labour intensities of production and low labour shares, their rising importance in the economy affects the distribution between labour and capital by pushing down the aggregate wage share (Autor et al. 2020). Secondly, Furman and Orszag (2018) show that these superstar firms can also extract super-normal returns on capital and given that capital ownership is highly concentrated at the top of the income distribution (Atkinson 2015) this contributes to rising income inequality between the rich and the poor. Thirdly, growing heterogeneity between firms in terms of productivity and the agglomeration of high-skilled workers in high-productivity firms and vice versa increases income inequality between people working in highly and lowly productive firms since firms act as wage-setters rather than wage-takers (Song et al. 2019). Finally, the results from recent empirical research suggests that monopoly rents are mostly shared with shareholders, investors, and the management rather than with the average employee. This increases income inequality between the top-income earners and the average workers (Guellec and Paunov 2017). While the presented evidence does not necessarily imply that these trends of the recent past will continue as the fourth industrial revolution develops, it nevertheless sheds light on the potential risks associated with an economy that increasingly relies on KBC as a factor of production as well as a consumption good. However, as many empirical studies use US data, there is significant scope for future research to place a focus on international comparisons. This may help to inform about the role of institutional settings in mediating the described effects of digital innovation on market concentration and income inequality.

While the discussed economic research on the relationship between digital technologies, market concentration and the distribution of income has grown over the past five years, the impact of digital technologies on wage inequality has been studied extensively since the third industrial revolution. Economists usually assume that technological change in general increases wage income inequality due to the skill-bias or routine-bias of new technologies. The fourth industrial revolution and the technologies driving it are also expected to considerably increase the demand for high skills and more specifically, for digital skills while at the same time social and interpersonal skills become increasingly important. Since economic research shows that there is a wage premium for these "new" digital skills (Falck et al. 2016; Hanushek et al. 2015), rising relative demand for them will affect wage inequality between those who possess digital skills and those who do not.

The popularity and persistency of the hypotheses of skill-biased technological and routine-biased technological change may partly be attributed to their implicit message that wage inequality arising from technological change simply is a logical consequence of the technology's skill-biased or routine-biased nature. This is convenient for policy makers as the strategy to battle this form of inequality is to turn to the panacea of more and better education – a political request that will hardly ever encounter any resistance. However, a mere quantitative increase of formal education is unlikely to help fighting wage inequality that may arise due to increasingly smart technologies if the distribution of the skills of the future, digital skills, and competencies, is itself shaped by existing social stratification. The unequal distribution of digital skills is well-studied in the literature on the digital divide and, more recently, digital inequality and many studies confirm that existing inequalities along socio-demographic characteristics such as age, ethnicity, gender, educational background, or social status influence the use of digital technologies and the development of digital skills (van Dijk 2005, Helsper 2012; DiMaggio et al. 2004; Drabowicz 2014; Robinson et al. 2015; Hargittai 2010; Zillien and Hargittai 2009; van Deursen et al. 2017; Zilian and Zilian 2020). So even though digital inequality seems to be a new phenomenon, it in fact simply mirrors the well-known structural inequalities with the inherent danger of exacerbating them as the fourth industrial revolution progresses. Research on the relationship between digital competencies, which captures a variety of skills beyond the mere usage of ICT at work, and wage differentials is still scarce but due to the increasing availability of new and internationally comparable data, such as the OECD Programme for the International Assessment of Adult Competencies, digital inequality and its consequence for wage income inequality provides a fruitful path for future research.

Finally, the role of (labour market) institutions is neglected in most of the literature on SBTC and RBTC and it is assumed that the impact of technological change on wage inequality is more or less the same across countries. An exception is Kristal and Edler (2019) who show that there are significant differences regarding the size of the computer wage premium and consequently the computer wage gap across different varieties of capitalism. A higher degree of central market coordination, which is characteristic for Nordic coordinated countries, is associated with lower computer wage gaps, ergo lower wage inequality, compared to liberal countries. Moreover, as the recent research on platform work shows, this new form of work organisation is characterised by considerable monopsony power and may affect income inequality. Thus, the research agenda on wage inequality in the fourth industrial revolution should be expanded by considering explicitly the interaction of politics and technology and how they affect the wage distribution. To fight inequality arising from digital innovations, traditional redistributive policies through labour taxation is stretched to its limits. Although it remains a suitable instrument to reduce inequality arising from between-firm differences in average wages, it does not affect income derived from capital ownership or other sources of income aside from labour. To address top income inequality, Atkinson (2015) proposes to introduce a progressive lifetime capital receipts tax, similar to the Capital Acquisitions Tax in the Republic of Ireland, to tax receipts of inheritance and gifts. In addition, he argues in favour of an increase of the top personal income tax rate (e.g., 65% for the UK). Concerning inequality arising from technological change, he supports Mazzucato (2013: 119) and her endorsement of a more active role of the state to guide the direction of technological change by "encouraging innovation in a form that increases the employability of workers and emphasises the human dimension of service provision".

The active role of the state is particularly important considering the rising influence of superstar firms. While the superstar firms are undoubtedly among the most successful companies in their respective industries, they also use very successfully so-called "dark" business practices such as excessive tax avoidance and political influence via lobbying activities to further entrench their own supremacy (The Economist 2016). Nobel laureate Joseph E. Stiglitz estimates the extent of tax avoidance in a contribution for *The Nation*:

> Just five American firms, Apple, Microsoft, Google, Cisco, and Oracle, collectively have more than a half trillion dollars stashed abroad as they achieve tax rates in some cases well under 1% of profits. We can debate what a "fair share" of taxes is, but what these companies pay is below any reasonable standard.
>
> *(Stiglitz 2017)*

Declining wage shares in combination with successful tax avoidance practices, rising earnings dispersion, and political influence exercised by superstars are questionable in terms of democracy, the future of the welfare state, and, consequently, social cohesion. Hence, restoring the distorted balance of power must be addressed by public policy and with regard to rising market concentration, one means to this end carved out by Atkinson (2015) seems to be of particular urgency: competition policy needs to explicitly take into account the distributional dimension of monopolistic and oligopolistic market structures.

References

Acemoglu, D. (2002): Technical Change, Inequality, and the Labor Market, *Journal of Economic Literature* 40(1), 7–72. https://doi.org/10.1257/jel.40.1.7

Acemoglu, D. and Autor, D. H. (2011): Skills, Tasks and Technologies: Implications for Employment and Earnings, in Ashenfelter, O. and Card, D. (eds.): *Handbook of Labor Economics*. Amsterdam: North-Holland, 1043–1171.

Acemoglu, D. and Restrepo, P. (2018): The Race between Man and Machine: Implications of Technology for Growth, Factor Shares, and Employment. *American Economic Review* 108(6), 1488–1542.

Acemoglu, D. and Restrepo, P. (2020): Robots and Jobs: Evidence from US Labor Markets, *Journal of Political Economy* 128(6), 2188–2244. https://doi.org/10.1086/705716

Aghion, P., Akcigit, U., Bergeaud, A., Blundell, R. and Hemous, D. (2019): Innovation and Top Income Inequality, *Review of Economic Studies* 86(1), 1–45. https://doi.org/10.1093/restud/rdy027

Allen, J. P. (2017): *Technology and Inequality*. Cham: Palgrave Macmillan.

Arntz, M, Terry Gregory, T. and Zierahn, U. (2017): Revisiting the Risk of Automation, *Economics Letters* 159(2017), 157–160

Atkinson, A. B. (2015): *Inequality – What Can Be Done?* Cambridge, MA and London: Harvard University Press.

Atkinson, A. B. and Stiglitz, J. E. (1969): A New View of Technological Change, *The Economic Journal* 79(315), 573–578.

Autor, D. H. (2013): The "Task Approach" to Labor Markets: An Overview, *Journal for Labour Market Research* 46(3), 185–199. https://doi.org/10.1007/s12651-013-0128-z

Autor, D. H. (2015): Why are There still so Many Jobs? The History and Future of Workplace Automation, *The Journal of Economic Perspectives* 29(3), 3–30.

Autor, D. H., Dorn, D., Katz, L. F., Patterson, C. and van Reenen, J. (2020): The Fall of the Labor Share and the Rise of Superstar Firms, *The Quarterly Journal of Economics* 135(2), 645–709. https://doi.org/10.1093/qje/qjaa004.Advance

Autor, D. H., Katz, L. F. and Kearney, M. S. (2008): Trends in U.S. Wage Inequality: Revising the Revisionists, *Review of Economics and Statistics* 90(2), 300–323. https://doi.org/10.1162/rest.90.2.300

Autor, D. H., Levy, F. and Murnane, R. J. (2003): The Skill Content of Recent Technological Change: An Empirical Exploration, *The Quarterly Journal of Economics* 118(4), 1279–1333.

Beernaert, D. and Fribourg-Blanc, E. (2017): Thirty Years of Cooperative Research and Innovation in Europe: The Case for Micro- and Nanoelectronics and Smart Systems Integration, in Van de Voorde, M., Puers, R., Baldi, L. and van Nooten, S. E. (eds.): *Nanoelectronics: Materials, Devices, Applications*. Weinheim: Wiley-VCH Verlag GmbH & Co. KGaA, 567–593.

Berger, T. and Frey, C. (2016): Structural Transformation in the OECD: Digitalisation, Deindustrialisation and the Future of Work. *OECD Social, Employment and Migration Working Papers, No. 193*. Paris: OECD Publishing. https://doi.org/10.1787/5jlr068802f7-en.

Black, S. E. and Spitz-Oener, A. (2010): Explaining Women's Success: Technological Change and the Skill Content of Women's Work, *The Review of Economics and Statistics* 92(1), 187–194.

Bogliacino, F. (2014): A Critical Review of the Technology-inequality Debate, *Suma de Negocios* 5(12), 124–135. https://doi.org/10.1016/s2215-910x(14)70034-5

Brödner, P. (2018): Industrie 4.0 und Big Data – wirklich ein neuer Technologieschub? in Hirsch-Kreinsen, H., Ittermann, P. and Niehaus, J. (eds.): *Digitalisierung industrieller Arbeit*. Baden-Baden: Nomos Verlagsgesellschaft mbH & Co. KG, 323–346. https://doi.org/10.5771/9783845283340-322

Brynjolfsson, E. and McAfee, A. (2013): The Great Decoupling, *New Perspectives Quarterly* 30(1), 61–63.

Brynjolfsson, E. and McAfee, A. (2014): *The Second Machine Age: Work, Progress, and Prosperity in a Time of Brilliant Technologies*. Boston: WW Norton & Company.

Card, D. and DiNardo, J. E. (2002): Skill-Biased Technological Change and Rising Wage Inequality, *Journal of Labor Economics* 20(4), 733–783.

Council of Economic Advisers. (2016): Benefits of Competition and Indicators, in *Issue Brief*. Available at: https://escudrojo.files.wordpress.com/2016/11/counseleconadvisor-onmarketconditions.pdf [30.04.2021]

Dao, M. C., Das, M. M., Koczan, Z. and Lian, W. (2017): Why is Labor Receiving a Smaller Share of Global Income? Theory and Empirical Evidence. *International Monetary Fund. IMF Working Paper.*

Decker, R. A., Haltiwanger, J., Jarmin, R. S. and Miranda, J. (2014): The Role of Entrepreneurship in US Job Creation and Economic Dynamism, *Journal of Economic Perspectives* 28(3), 3–24. https://doi.org/10.1257/jep.28.3.3

Decker, R. A., Haltiwanger, J., Jarmin, R. S. and Miranda, J. (2017): Declining Dynamism, Allocative Efficiency, and the Productivity Slowdown, *American Economic Review* 107(5), 322–326.

Deming, D. J. (2017): The Growing Importance of Social Skills in the Labor Market, *The Quarterly Journal of Economics* 132(4), 1593–1640.

Dengler, K. and Matthes, B. (2020): *Substituierbarkeitspotenziale von Berufen und die möglichen Folgen für die Gleichstellung auf dem Arbeitsmarkt*. Expertise für den Dritten Gleichstellungsbericht der Bundesregierung. Available at: www.dritter-gleichstellungsbericht.de [21.04.2021]

DiMaggio, P. and Bonikowski, B. (2008): Make Money Surfing the Web? The Impact of Internet Use on the Earnings of U.S. Workers, *American Sociological Review* 73, 227–250.

DiMaggio, P., Hargittai, E., Celeste, C. and Shafer, S. (2004): Digital Inequality: From Unequal Access to Differentiated Use, in Neckerman, K. M. (ed.): *Social Inequality*. New York: Russell Sage Foundation, 355–400.

Drabowicz, T. (2014): Gender and Digital Usage Inequality among Adolescents: A Comparative Study of 39 Countries, *Computers & Education* 74, 98–111.

Dube, A., Jacobs, S., Naidu, S. and Suri, S. (2020): Monopsony in Online Labor Markets, *American Economic Review: Insights* 2(1), 33–46.

Falck, O., Heimisch, A. and Wiederhold, S. (2016): Returns to ICT Skills. *OECD Education Working Papers, 134*, 0_1.

Feinstein, C. H. (1998): Pessimism Perpetuated: Real Wages and the Standard of Living in Britain during and after the Industrial Revolution, *Journal of Economic History* 58(3), 625–658. https://doi.org/10.1017/S0022050700021100

Ferrari, A. (2013): *DIGCOMP: A Framework for Developing and Understanding Digital Competence in Europe*. Vol. 26035. Publications Office of the European Union.

Frey, C. B. and Osborne, M. A. (2017): The Future of Employment: How Susceptible are Jobs to Computerisation? *Technological Forecasting and Social Change* 114(C), 254–280.

Furman, J. and Orszag, P. (2018): A Firm-level Perspective on the Role of Rents in the Rise in Inequality, in Guzman, M. (ed.): *Toward a Just Society: Joseph Stiglitz and Twenty-First Century Economics*. New York: Columbia University Press, 19–46.

Furman, J. and Seamans, R. (2019): AI and the Economy, *Innovation Policy and the Economy* 19(1), 161–191.

Goos, M. and Manning, A. (2003): McJobs and MacJobs: The Growing Polarisation of Jobs in the UK, in Dickens, R., Gregg, P. and Wadsworth, J. (eds.): *The Labour Market Under New Labour*. London: Palgrave Macmillan UK, 70–85. https://doi.org/10.1057/9780230598454_6

Goos, M. and Manning, A. (2007): Lousy and Lovely Jobs: The Rising Polarization of Work in Britain, *The Review of Economics and Statistics* 89(1), 118–133.

Goos, M., Manning, A. and Salomons, A. (2009): Job polarization in Europe, *The American Economic Review* 99(2), 58–63.

Goos, M., Manning, A. and Salomons, A. (2014): Explaining Job Polarization: Routine-biased Technological Change and Offshoring, *The American Economic Review* 104(8), 2509–2526.

Grundke, R., Marcolin, L., Squicciarini, M. and others. (2018): Which Skills for the Digital Era? Returns to Skills Analysis. *OECD Science, Technology and Industry Working Papers* 2018(9).

Guellec, D. and Paunov, C. (2017): Digital Innovation and the Distribution of Income. *NBER Working Paper, 23987*.

Hanushek, E. A., Schwerdt, G., Wiederhold, S. and Woessmann, L. (2015): Returns to Skills around the World: Evidence from PIAAC, *European Economic Review* 73, 103–130.

Hargittai, E. (2010): Digital Na(t)ives? Variation in Internet Skills and Uses among Members of the "Net Generation". *Sociological Inquiry* 80(1), 92–113. https://doi.org/10.1111/j.1475-682X.2009.00317.x

Helsper, E. J. (2012): A Corresponding Fields Model for the Links Between Social and Digital Exclusion, *Communication Theory*, 22(4), 403–426. https://doi.org/10.1111/j.1468-2885.2012.01416.x

Howcroft, D. and Rubery, J. (2018): Gender Equality Prospects and the Fourth Industrial Revolution, in Neufeind, M., O'Reilly, J. and Ranft, F. (eds.): *Work in the Digital Age: Challenges of the Fourth Industrial Revolution*. New York: Rowman & Littlefield International, 63–74.

Karabarbounis, L. and Neiman, B. (2014): The Global Decline of the Labor Share, *Quarterly Journal of Economics* 129(1), 61–103.

Krämer, H. (2019): Digitalisierung, Monopolbildung und wirtschaftliche Ungleichheit, *Wirtschaftsdienst* 99(1), 47–52.

Kristal, T. and Edler, S. (2019): Computers Meet Politics at Wage Structure: An Analysis of the Computer Wage Premium across Rich Countries, *Socio-Economic Review* 0(0), 1–32.

Krueger, A. B. (1993): How Computers have Changed the Wage Structure, *The Quarterly Journal of Economics* 108(1), 33–60.

Kurz, H. D., Schütz, M., Strohmaier, R. and Zilian, S. (2018): Riding a New Wave of Innovations: A Long-term View at the Current Process of Creative Destruction, *Wirtschaft Und Gesellschaft: Wirtschaftspolitische Zeitschrift Der Kammer Für Arbeiter Und Angestellte Für Wien* 44(4), 545–583.

Lauder, H., Brown, P. and Cheung, S. Y. (2018): Fractures in the Education-economy Relationship: The End of the Skill Bias Technological Change Research Programme? *Oxford Review of Economic Policy* 34(3), 495–515.

Lee, M., Yun, J., Pyka, A., Won, D., Kodama, F., Schiuma, G., Park, H., Jeon, J., Park, K., Jung, K., Yan, M.-R., Lee, S. and Zhao, X. (2018): How to Respond to the Fourth Industrial Revolution, or the Second Information Technology Revolution? Dynamic New Combinations between Technology, Market, and Society through Open Innovation, *Journal of Open Innovation: Technology, Market, and Complexity* 4(3), 21. https://doi.org/10.3390/joitmc4030021

Leontief, W. W. (1982): The Distribution of Work and Income, *Scientific American* 247(3), 188–204. https://doi.org/10.1038/scientificamerican0982-188

Lipsey, R. G., Bekar, C. and Carlaw, K. (2005): *Economic Transformations: General Purpose Technologies and Long-term Economic Growth*. Oxford: Oxford University Press.

Makridakis, S. (2017): The Forthcoming Artificial Intelligence (AI) Revolution: Its Impact on Society and Firms, *Futures* 90, 46–60.

Manning, A. (2021): Monopsony in Labor Markets: A Review, *ILR Review* 74(1), 3–26.

Mazzucato, M. (2013): *The Entrepreneurial State: Debunking Public vs. Private Sector Myths*. London: Anthem Press.

Mokyr, J. (1999): The Second Industrial Revolution, 1870–1914 Joel, in Castronovo, V. (ed.): *In Valerio Castronovo, ed., Storia dell'economia Mondiale*. Rome: Laterza Publishing, 219–245.

Mokyr, J., Vickers, C. and Ziebarth, N. L. (2015): The History of Technological Anxiety and the Future of Economic Growth: Is This Time Different? *Journal of Economic Perspectives* 29(3), 31–50. https://doi.org/10.1257/jep.29.3.31

OECD (2011): *Divided We Stand: Why Inequality Keeps Rising – Country Note Germany*. OECD Publishing. Available at: www.oecd.org/germany/49177659.pdf [21.04.2021]

OECD (2013): *Supporting Investment in Knowledge Capital, Growth and Innovation*. OECD Publishing. https://doi.org/10.1787/9789264193307-en

OECD (2020): *OECD Digital Economy Outlook 2020*. Paris: OECD Publishing. https://doi.org/10.1787/bb167041-en.

Piketty, T. (2014): *Capital in the 21st Century*. Cambridge, MA, London: Harvard University Press.

Pratt, G. A. (2015): Is a Cambrian Explosion Coming for Robotics? *Journal of Economic Perspectives* 29(3), 51–60.

Ragnedda, M. (2017): *The Third Digital Divide: A Weberian Approach to Digital Inequalities*. Oxford: Routledge.

Robinson, L., Cotten, S. R., Ono, H., Quan-Haase, A., Mesch, G., Chen, W., Schulz, J., Hale, T. M. and Stern, M. J. (2015): Digital Inequalities and Why They Matter, *Information, Communication & Society* 18(5), 569–582. https://doi.org/10.1080/1369118X.2015.1012532

Rosen, S. (2008): Human Capital, in Durlauf, S. N. and Blume, L. E. (eds.): *The New Palgrave Dictionary of Economics*. London: Palgrave Macmillan, 2816–2827.

Schumpeter, J. (1912): *Theorie der wirtschaftlichen Entwicklung. Eine Untersuchung über Unternehmergewinn, Kapital, Kredit, Zins und den Konjunkturzyklus*. Leipzig: Duncker & Humblot.

Schwab, K. (2016): *The Fourth Industrial Revolution*. Geneva: World Economic Forum.

Shapiro, C. and Varian, H. (1999): *Information Rules: A Strategic Guide to the Network Economy*. Boston: Harvard Business School Press.

Song, J., Price, D. J., Guvenen, F., Bloom, N. and von Wachter, T. (2019): Firming Up Inequality, *The Quarterly Journal of Economics* 134(1), 1–50. https://doi.org/10.1093/qje/qjs020.Advance

Spitz-Oener, A. (2006): Technical Change, Job Tasks, and Rising Educational Demand: Looking outside the Wage Structure, *Journal of Labor Economics* 24(2), 235–270.

Stiglitz, J. (2017): America Has a Monopoly Problem – and It's Huge. *The Nation,* 23 October 2017. Available at: www.thenation.com/article/archive/america-has-a-monopoly-problem-and-its-huge/ [06.05.2021]

The Economist (2016): *What Goes around*. Available at: www.economist.com/node/21707051/print [30.04.2021]

van Deursen, A. J. A. M., Helsper, E. J., Eynon, R. and van Dijk, J. A. G. M. (2017): The Compoundness and Sequentiality of Digital Inequality, *International Journal of Communication* 11, 452–473. http://eprints.lse.ac.uk/68921/

van Deursen, A. J. A. M. and van Dijk, J. A. G. M. (2014): The Digital Divide Shifts to Differences in Usage, *New Media and Society* 16(3), 507–526. https://doi.org/10.1177/1461444813487959

van Dijk, J. (2005): *The Deepening Divide: Inequality in the Information Society*. London: Sage Publications.

van Reenen, J. (2018): Increasing Differences Between Firms: Market Power and the Macro-Economy. *CEP Discussion Paper, 1576.*

Vivarelli, M. (1995): *The Economics of Technology and Employment: Theory and Empirical Evidence*. Aldershot: Edward Elgar Publishing.

Vivarelli, M. (2014): Innovation, Employment and Skills in Advanced and Developing Countries: A Survey of Economic Literature, *Journal of Economic Issues* 48(1), 123–154.

Wajcman, J. (2013): *Pressed for Time: The Acceleration of Life in Digital Capitalism*. Chicago: University of Chicago Press.

Zilian, S. S. and Zilian, L. S. (2020): Digital Inequality in Austria: Empirical Evidence from the Survey of the OECD "Programme for the International Assessment of Adult Competencies", *Technology in Society* 63(September), 101397. https://doi.org/10.1016/j.techsoc.2020.101397

Zillien, N. and Hargittai, E. (2009): Digital Distinction: Status-specific Types of Internet Usage, *Social Science Quarterly* 90(2), 274–291. https://doi.org/10.1111/j.1540-6237.2009.00617.x

10
THE LEGAL PROTECTION OF PLATFORM WORKERS

Jeremias Adams-Prassl and Martin Gruber-Risak

Introduction

In its 2016 Communication 'A European agenda for the collaborative economy', the European Commission highlighted how:

> The collaborative economy creates new opportunities for consumers and entrepreneurs. The Commission considers that it can therefore make an important contribution to jobs and growth in the European Union, if encouraged and developed in a responsible manner. Driven by innovation, new business models have a significant potential to contribute to competitiveness and growth. The success of collaborative platforms are at times challenging for existing market operators and practices, but by enabling individual citizens to offer services, they also promote new employment opportunities, flexible working arrangements and new sources of income. For consumers, the collaborative economy can provide benefits through new services, an extended supply, and lower prices. It can also encourage more asset-sharing and more efficient use of resources, which can contribute to the EU's sustainability agenda and to the transition to the circular economy.
>
> *(European Commission 2016: 2)*

The opportunities are great – but so are the potential downsides: many platform workers struggle with low pay, long hours, and precarious conditions.[1] In this contribution, we focus on the challenges this form of work organisation in the 'on-demand' or 'gig economy' pose to traditional labour market regulation, and explain how some of these might be met.

1 In this chapter we use the terminology suggested by Eurofound, *Employment and working conditions of selected types of platform work* (Publications Office of the European Union 2018: 9): "Platform work is a form of employment that uses an online platform to enable organisations or individuals to access other organisations or individuals to solve problems or to provide services in exchange for payment".The manuskript was completed in December 2020, developments after this date like the proposal of the European Commission for a Directive on improving working conditions in platform work as presented in December 2021 are therefore not considered.

 DOI: 10.4324/9780429351921-13

This is a timely endeavour as the political guidelines of the present President of European Commission Ursula Von der Leyen for the period 2019–2024 includes a passage touching on exactly this issue:

> Digital transformation brings fast change that affects our labour markets. I will look at ways of improving the labour conditions of platform workers, notably by focusing on skills and education.
>
> *(Von der Leyen 2019a: 10)*

This call is particularised in the mission letter to the Commissioner for Jobs, Nicolas Schmit, which puts an explicit emphasis on the improvement of working conditions of platform workers:

> Dignified, transparent and predictable working conditions are essential to our economic model. I want you to closely monitor and enforce existing EU law in this area and to look at ways to improve the labour conditions of platform workers.
>
> *(Von der Leyen 2019b: 5)*

In this chapter we explore different options to achieve this aim – ways to improve labour conditions for platform workers. To this end, the chapter is structured as follows. The first subsections briefly introduce platform work, highlighting salient aspects for subsequent debate: there is a broad spectrum of platforms, from physical service provision to exclusively digital work. Platform workers' experiences are similarly varied, ranging from successful entrepreneurs to those stuck with monotonous tasks and long hours, remunerated significantly below minimum wage rates. The successive section then outlines the regulatory challenges arising from platform work: one of the very purposes of employment and labour law is to draw a distinction between the genuinely self-employed and those requiring protection, and to bring the latter within its protective scope. The multiplicity of contractual relationships and competing legal characterisations in the arrangements between platforms, workers, and customers, on the other hand, sits uneasily with the traditional binary divide. It is this mismatch which lies at the core of classification problems in the gig economy, and the resulting exclusion of platform workers from even the most basic labour standards.

In the following sections we develop different approaches for addressing this problem. They deal with interpretative approaches to the notion of employee and employer in an attempt to enlarge (or restore) the scope of employment law to include those working in the virtual realm. First, an approach based on Prassl's functional-typological concept of the employer is outlined, developed on the basis of a catalogue of five employer functions (Prassl 2015). Then another re-interpretation of the employee is proposed, emphasising economic arguments over organisational ones. Both approaches have the advantage of requiring little legislative activity and may therefore be the most easily applicable, especially as judges are increasingly asked to adjudicate upon employment status in platform work.[2]

Another (much-disputed) approach to regulating work in the on-demand economy is based on the idea that an intermediate legal category, situated between the employee and the self-employed, might be the most apt to deal with the legal issues arising from platform work. The

2 E.g. the London Central Employment Tribunal 28.10.2016, 2202551/2015 & Others, Aslam, Farrar & Others v Uber B.V., Uber London Ltd. & Uber Britannia Ltd, www.judiciary.gov.uk/judgments/mr-y-aslam-mr-j-farrar-and-others-v-uber/.

successive section looks at existing models and recent litigation in Austria, Germany, and the United Kingdom to demonstrate the potentially different effects of such an approach.

The third and maybe most obvious way to deal with platform work discussed here is to introduce special legislation for platform work. This approach would follow established patterns of regulation for other forms of atypical work especially temporary agency work that is also construed as a multi-party relationship. On the European as well as national level, special legislative provisions have been enacted to deal with the specific problems arising from the multiplicity of contracts and contractual partners found in outsourcing and agency work. The next section reflects on this avenue and points out possible advantages of this approach, followed by a section that assesses the importance of coherent legal protection through the lens of the COVID-19 pandemic, which swept European and international labour markets in the spring of 2020.

Whilst the avenues to be chosen are thus potentially manifold, one thing is clear: there is an urgent need for legislators and practitioners to address the often vulnerable situation of persons working in the gig economy, offering maximum flexibility but getting very little security in return. This must be changed, while keeping in mind that any proposed solution or mix of solutions must be able to respond flexibly to changing economic and organisational models, but at the same time must offer conceptual coherence in the face of factual complexity.

Platform work

Developments in information and communication technology ('ICT') and the Internet have made it easier than ever before to match demand and supply in real time, both locally and globally. This has led not only to fundamental changes in traditional work relationships, but also to the emergence of new forms of employment located in the grey and often unchartered territory between employment contracts and freelance work (Eurofound 2015; Aloisi 2016; De Stefano 2016a; Prassl 2018: Chapter 5). A particularly salient instance of this phenomenon is platform work, a relatively recent model also known as crowdsourcing of labour, gig work, on-demand work, or crowd employment. These notions describe an ICT-based form of organising the outsourcing of tasks to a large pool of workers. The work (ranging from transportation services and cleaning to digital transcription or programming tasks) is referred to in a variety of ways, including 'gigs', 'rides', or 'tasks', and is offered to a large number of people (the 'crowd') by means of an Internet-based 'crowdsourcing platform'.[3] This organisational model forms part of a larger set of processes known as 'crowdsourcing';[4] with customers (or indeed employers) referred to as 'crowdsourcers'. The resulting contractual relationships are manifold and complex: whilst the work is usually managed through an intermediary (the crowdsourcing platform), some will insist on direct contractual relationships between crowdsourcer clients and platform workers, whereas others will opt for tripartite contractual structures, akin to traditional models of agency work and labour outsourcing (Kittur et al. 2012; Leimeister et al. 2014b).

There is a nearly unlimited factual variety that characterises the emergence of online platforms, both in terms of crowdsourcing in general (crowdfunding,[5] allocation of non-labour resources such as accommodation[6]), and crowdsourcing of labour, or – as it is also called –

3 Such as, notably, Amazon's Mechanical Turk (www.mturk.com) see Strube (2014); for the German platform Clickworker see Lutz (2017).

4 This term derives from a combination of the words 'outsourcing' and 'crowd', and was used by Jeff Howe for the first time (Howe 2006).

5 www.KickStarter.com.

6 www.Airbnb.com.

crowdwork, in particular. It is therefore not useful, or indeed feasible, to construct an overall taxonomy of crowdsourcing platforms. For present purposes, however, a few fundamental distinctions may be drawn.

Crowdsourcing, first, can take place internally or externally, depending on whether the crowd comprises a company's internal workforce or simply any number of individuals on a given platform. With external crowdsourcing, the crowdsourcer generally uses crowdsourcing platforms that already have an active crowd of registered workers. In this contribution, we will look solely at external crowdsourcing, as internal crowdsourcing is generally arranged within the context of existing employment relationships, and therefore poses fewer fundamental legal problems (Klebe and Neugebauer 2014: 4), regardless of whether the platform is operated by an independent enterprise or by the company itself (Eurofound 2015: 110).

Work crowdsourced to an external crowd, on the other hand, can be seen as clustered along a spectrum of services and arrangements.[7] On the one end, we find physical services to be undertaken in the 'real' (offline) world, where the platform worker comes into direct contact with the customer. Examples include transportation delivered via apps such as Uber, domestic services (cleaning, repair work, etc.) delivered via platforms such as Helpling,[8] and clerical work (e.g. customer service or accounting) provided by platforms like UpWork.[9] Uber customers, for example, use an app on their smartphones to request rides from a specific location, information that is instantly broadcast to drivers in the area.[10] Once the request has been accepted by a driver, she is directed to the passenger and onwards to the required destination, through her version of the Uber app. Payment is taken automatically from the customer by the platform, and after the taking of a commission of between 20 and 30%,[11] passed on to the driver. Customers and drivers rate each other anonymously following each journey; the resulting scores are displayed to passengers and operators respectively before the next trip commences. Helpling operates in a similar way, even though the physical work takes place in the client's home or business premises – customers log in on a platform or app, type in their postcode and when cleaning is required. They are then offered profiles of workers available in their vicinity, with further information about each individual, and an online facility to complete bookings. Payment is processed via the platform, and after completion of their tasks, all cleaners are rated by their customers, with the resulting score displayed online in order to inform future customers.

On the other end of the spectrum, there is digital work delivered in the virtual world, usually via an interface provided by the platform. The tasks involved here are often very simple, repetitive activities involving low pay and highly standardised or automated processes. These 'microtasks' include digital labelling and the creation of image descriptions, categorising data and products, and the translation or proofreading of short texts, with larger tasks often broken down into smaller subtasks to be worked on independently. These microtasks are then posted on platforms, where platform workers can find and complete them. The leading platforms for this kind of 'cognitive piece work' (Schmidt 2014: 378) or 'Neo-Taylorism' (Leimeister et al. 2014a: 32)

7 For a typology of platform work Eurofound (2018: 53 et seq.), Platform work: Types and implications for work and employment – Literature review.

8 www.helpling.com – in the UK, this platform operates under a different brand, www.hassle.com.

9 www.upwork.com.

10 www.uber.com/features.

11 Though the platform continues to experiment with varying the amount of commission payable, which is particularly harmful to part-time workers, as the percentage retained falls with the number of trips offered: www.cnet.com/news/uber-tests-30-commission-for-new-drivers-in-san-francisco/.

include Amazon's Mechanical Turk[12] and Clickworker.[13] Survey research has shown that 25% of the tasks offered on Amazon Mechanical Turk are valued at $0.01, 70% offer $0.05 or less and 90% pay less than $0.10 per completed task; thus equalling an average wage of about $2 per hour (Eurofound 2015: 115).

This segment of the labour market is also referred to as the 'gig economy', characterised by the prevalence of short-term contracts or freelance work, as opposed to permanent jobs. Workers earn their livelihood not from one open-ended full-time employment contract but, like musicians, from a series of 'gigs'. For some, this is a working environment that offers flexibility with regard to employment hours, a diversity of different jobs, and the opportunity to earn additional income; to others, it represents precarious, low-paid work offering little legal protection.

Working in the platform economy

Historically, the main advantage of hierarchical employment relationships over contracts with independent contractors was understood to be the entrepreneur's degree of control, and the resulting decrease in transaction costs, whether in the search, selection, and training of workers, or the employer's tight control over the production process (Coase 1973). An increasing desire for labour flexibility, on the other hand, was the driver behind the more recent creation of different forms of atypical work, including agency work, part-time work, and fixed-term employment (Eurofound 2001; European Parliament 1998).

Platform work is a rather novel combination of these factors, insofar as platforms attempt to increase flexibility for the employer or customer and to reduce the costs of 'empty' or unproductive moments, whilst at the same time maintaining full control over the production process in order to keep transaction costs to a minimum. In order to meet these seemingly contradictory goals, two preconditions must be met: first, the crowd must be large enough in order to always have individuals available when needed, and to maintain enough competition between platform workers to keep prices low. This is usually achieved through platforms' large and active crowds, with different platforms specialising in different segments of the crowdsourcing market.

Secondly, instead of the command-and-control systems inherent in 'traditional' employment relationships, crowdsourcers and platforms rely on 'digital reputation' mechanisms to guide the selection of platform workers and to ensure efficient performance control. Individual models vary, but the fundamental approach is consistent: platform workers are awarded points, stars, or other symbols of status by the crowdsourcer or customer after completing a task.[14] Quality control itself can thus be crowdsourced by the platform to its customers or other crowdsourcers, tapping the 'wisdom of the crowd' (Surowiecki 2004) in order to determine the performance levels of each individual platform worker.

The potential upsides of this emerging model for firms and workers alike should not be underestimated.[15] Through the use of platforms, businesses ranging from restaurants to IT service providers can draw on a large crowd of flexible workers to reduce or even eliminate the cost of unproductive time at work, and rely on reputation mechanisms to maintain full control over the production process or service delivery. The resulting competition between platform

12 www.mturk.com/mturk/welcome.

13 www.clickworker.com.

14 On some platforms, including Uber, customers are rated by crowdworkers in turn: cf. Langlois (2015).

15 www.telegraph.co.uk/technology/uber/12086500/In-praise-of-the-gig-economy.html

workers will ensure that quality remains high whilst wages are low. As Lukas Biewald, then CEO of the platform Crowdflower, put it bluntly in 2010,

> Before the Internet, it would be really difficult to find someone, sit them down for ten minutes and get them to work for you, and then fire them after those ten minutes. But with technology, you can actually find them, pay them the tiny amount of money, and then get rid of them when you don't need them anymore.
>
> *(Marvit 2014)*

Platform work similarly offers significant potential upsides for (at least some of its) workers, first and foremost, in terms of flexibility: platform workers can decide when to work, where to work, and what kinds of tasks to accept. Platform work might therefore be more compatible with other duties, such as childcare. The flexibility and potentially limited nature of individual engagements can also help the underemployed, providing additional income to their regular earnings,[16] and (at least through virtual work platform work) allowing those excluded from regular labour markets due to disabilities or other factors to find opportunities for gainful employment (Zyskowski et al. 2015). Finally, there are an increasing number of genuinely successful small entrepreneurs, focused on particular niches or offering special skills, for whom platform work has become a very profitable source of new business.

At the same time, however, it is important to note that working conditions for the vast majority of platform workers appear to be poor, irrespective of the work being delivered (Marvit 2014). The lack of unions or organising powers, the oligopoly of but a few platforms offering certain kinds of tasks, and constant economic and legal insecurity result in a massive imbalance of bargaining power, noticeable primarily in low wage-rates and heavily slanted terms and conditions in platform use agreements. In the case of virtual platform work, global competition and dislocated physical workplaces further aggravate these problems, as a lack of regulation leads to what some have called 'digital slaves' (Rosenblum 2013) working away in their 'virtual sweatshops'.[17]

Two problems in particular are repeatedly highlighted: low wages, and workers' dependence on their ratings with a particular platform. As regards the former problem, for example, some reports suggest that the average wage on Amazon's Mechanical Turk is less than $2 per hour (Ross et al. 2010), considerably below the US minimum wage.[18] A related aspect is insecurity as regards payment: in accordance with the general terms and conditions of microtasking-platforms, crowdsourcers have the right to reject the work without having to give a reason or providing payment, whilst still receiving the fruits of a worker's labour.[19]

Various systems of 'digital reputation', or rating mechanisms, which form one of the core elements of platform work, raise a second set of difficult questions:[20] a customer-input based system of stars or points not only puts platform workers in a state of permanent probation, but also infringes their mobility as it ties them to particular platforms. As the more attractive and

16 www.dailyworth.com/posts/3410-flexible-side-gigs-to-bring-in-extra-income/4.

17 See so-called 'gold farming' (professional online gaming to collect virtual money in games like World of Warcraft): Dibbel (2016).

18 In the US there are different minimum wages, depending on the Federal State. Cf. www.dol.gov/whd/minwage/america.htm.

19 For an illustration in the context of Mechanical Turk, see Martin et al. (2014).

20 There are also increasing reports of discrimination and bias hampering the operation of rating systems: see www.bostonglobe.com/news/science/2013/08/08/the-pitfalls-crowdsourcing-online-ratings-vulnerable-bias/

better paid tasks are only offered and assigned to those who have the best reputation, and as a worker's digital reputation is not transferable between individual platforms, a change of platforms will be difficult – a fact which also further impairs the bargaining position of platform workers.

As far as the volume of platform work is concerned, 173 platforms organising services are active throughout Europe (Fabo et al. 2017). Around 2% of the working age population (16–74 years) in 14 member states have platform work as their main occupation. Around 6% derive a substantial income from platform work (at least 25% of their average income from a 40-hour working week) and almost 8% perform tasks on digital platforms at least once a month (Pesole et al. 2018). Like agency work, platform work therefore is not that widespread but a relevant phenomenon nevertheless. This is the case not only because of the numbers, but also because it challenges the standard employment relationship at its core which will be pointed out in the following section.

Regulatory challenges

Working conditions in the platform economy

It cannot be denied that platform work offers significant potential benefits for (at least some of) its workers, first and foremost, in terms of flexibility and in the creation of new working and business opportunities. At the same time, however, working conditions for the vast majority of platform workers appear to be poor, irrespective of the work being delivered. As this has been laid out earlier it suffices to highlight the two most pressing problems platform workers themselves raise: low wages, unstable work, and workers' dependence on their ratings with a particular platform that hinders workers' mobility. The first is mostly a result of the legal uncertainty of the legal status of the platform workers as well of the lack of worker representation in this industry.

The underlying (legal) problems

One of the very purposes of employment and labour law is to draw a distinction between the genuinely self-employed and those requiring protection against many of the problems already outlined, bringing the latter group within its protective scope. Most jurisdictions have developed a more or less elaborate legal framework regulating the employment relationship based on the idea of the existence of an imbalance of bargaining power when negotiating pay and conditions of work (Freedland and Davies 1983: 14, 69). This usually includes the right to organise, to bargain collectively and to take collective action. Self-employed persons, on the other hand, do not enjoy any of these rights, including minimum wages, sick pay, or protection against unfair dismissal. Indeed, they may even be forbidden from coming to mutual arrangements over basic terms, such as minimum payments, as this might contravene competition or anti-trust laws.[21]

It is therefore important to analyse where the line is drawn between the status of an employee and a self-employed person or independent contractor. As we have pointed out elsewhere (Prassl and Risak 2016: 633), this becomes very hard when more than two parties are involved, as the

21 Cf. European Court of Justice Case C-413/13 *FNV Kunsten Informatie en Media v Staat der Nederlanden* [2014] ECLI:EU:C:2014:2411.

received analytical approach was developed in the context of bilateral employment relationships. Employment law thus struggles with the crowdsourcing of labour given the involvement of an intermediary or platform in addition to the platform workers and crowdsourcers. A traditional analysis would split the three-party arrangements underlying platform work scenarios into a series of bilateral contractual relationships, and attempt to classify each relationship separately. The economic situation of platform workers, however, is not accurately reflected in the sum of these fragments of contracts. Looking only at individual relationships at a time, without also considering their interwoven nature because of the crowdsourcing platform, is akin to determining the nature of a cloth by looking only at its differently coloured threads of wool, without taking into account the knitting pattern. The received analytical approach tends to ignore complex multi-party relationships, and analyses the resulting fragments without reference to the broader context and economic effects of platform work. This, then, is at the core of its shortcomings when faced with multiple parties: there is little analysis of contractual relationships as an interdependent net of contracts that only makes sense as a whole.

Possible solutions

In the following sections, we will point out four different ways to deal with the regulatory challenges, starting with the one that is the least 'intrusive' (i.e. requiring the least changes in labour regulation and jurisprudence) through to the one requiring the most detailed legislative activity. We start out with an approach that focuses on who is the employer based on a functional concept asking who can best meet the responsibilities deriving from the employer functions (see following section). Another approach is the widening of the notion of the employee, which up to now (at least in some jurisdictions) has been primarily based on organisational criteria and less on the economic dependency on a single or few contractual partners. Another approach might be the introduction of an intermediate category or – where it already exists – its application to virtual platform workers. The last regulatory avenue explored is the one of special statutory regulation of platform work, similar to temporary agency work. This will be further discussed in the following sections.

The different ways of dealing with the issues of virtual platform workers are complementary rather than mutually exclusive to one another. They also do not solve the problems to a different extent: while an extension of the notion of the employee will bring platform workers (or at least some of them) into the protective scope of employment law, this solution does not clearly solve those issues connected with multiple-party work relationships. And of course, the different paths for reform do very much depend on the status quo and general approach to labour law in any given jurisdiction; where employment regulation is primarily based on collective bargaining, for example, the extension of the possibilities to do so will be the focus, while in those systems with close-knit statutory protection, the extension of the scope of application of key protective norms will be more crucial.

A functional concept of the employer

As *The Concept of the Employer* (Prassl 2015) suggests, in order to restore congruence to the application of employment law norms, the very definition of the employer must carefully be reconceptualised as a more openly functional one, whether through judicial recognition of that notion in litigation, or through legislative action. Present space limitations prohibit an extensive rehearsal of the development of that notion; two crucial steps can nonetheless be highlighted. First, the argument that the traditional unitary analysis of the employer has long been accompanied by functional elements: employment law identifies, at least indirectly, a series of five employer functions – from hiring workers to setting their rates of pay – and regulates them in one or several areas – from anti-discrimination law to minimum wage provisions.

For purposes of this analysis, a 'function' of being an employer is one of the various actions employers are entitled or obliged to take as part of the bundle of rights and duties falling within the scope of the open-ended contract of service. These functions are rarely set out explicitly: indeed, in most jurisdictions, the definition of the employer is seen as an afterthought in determining the scope of worker-protective norms. Upon closer inspection, however, it quickly appears that the concept implicitly mirrors the definition of the employee or worker, allowing for a 'reverse-engineering' of employer functions out of factors defining the employee (Prassl 2015: 24–25).

In trawling the established tests of employment status, such as control, economic dependence, or mutuality of obligation for these employer functions, there are endless possible mutations of different fact scenarios, rendering categorisation purely on the basis of past decisions of limited assistance.[22] The result of this analysis of concepts underlying different fact patterns, rather than the actual results on a case-by-case basis, is the following set of functions, with the presence or absence of individual factors becoming less relevant than the specific role they play in any given context. Individual elements can vary from situation to situation, as long as they fulfil the same function when looked at as a whole.[23]

The *five main functions* and their functional underpinnings of the employer are:[24]

1 Inception and termination of the employment relationship

This category includes all powers of the employer over the very existence of its relationship with the employee, from the 'power of selection' to the right to dismiss.

2 Receiving labour and its fruits

Duties owed by the employee to the employer, specifically to provide his or her labour and the results thereof, as well as rights incidental to it.

3 Providing work and pay

The employer's obligations towards its employees, such as, for example, the payment of wages.

4 Managing the enterprise-internal market

Coordination through control over all factors of production, up to and including the power to require both how and what is to be done.

5 Managing the enterprise-external market

Undertaking economic activity in return for potential profit, whilst also being exposed to any losses that may result from the enterprise.

Key to this concept of the *multi-functional* employer is the fact that not one function mentioned is relevant in and of itself. Rather, it is the *ensemble* of the five functions that matters: each of them covers one of the facets necessary to create, maintain, and commercially exploit employment relationships, thus coming together to make up the received legal concept of employing workers or acting as an employer – and being subject to the appropriate range of employee-protective norms.

22 Whilst subsequent examples are drawn primarily from common law jurisdictions, we suggest that the approach is capable of being similarly developed in civilian jurisdictions.

23 The 'equipollency principle' (*Äquivalenzprinzip*): Nogler (2009).

24 For earlier attempts at such lists see, e.g., Freedland (2003: 40).

A functional conceptualisation of the employer, then, is one in which the contractual identification of the employer is replaced by an emphasis on the exercise of each function – whether by a single entity, as demonstrated immediately following, or in situations where different functions may be exercised from more than one *locus* of control.[25] Indeed, in the platform work context, one particular challenge arises from the fact that functions may sometimes be jointly exercised by platforms, customers, and potentially even the platform worker herself. The shared exercise between two or more entities, or one where functions are parcelled out between different parties, arise where platform work arrangements lead to a fragmented exercise of employer functions – it is in those scenarios that the functional model of the employer will now be put to the test: there may be elements of genuine self-employment, platforms performing employer roles, and even customers potentially becoming subject to regulatory obligations.

In order to reconcile these contradictions, and ensure a consistent application of employment law in the face of factual complexity, our conceptualisation of the concept of the employer needs to move from the current rigidly formalistic approach to a flexible, *functional* concept. In more concrete terms, the following working definition has been offered by Prassl (2015: 155): the functional concept of the employer should come to mean

> the entity, or combination of entities, playing a decisive role in the exercise of relational employing functions, and regulated or controlled as such in each particular domain of employment law.

Calling for a functional definition of the employer is not a completely novel approach to the problems arising from multilateral employment arrangements. Judy Fudge (2003: 636 et seq.), for example, has long noted the "need to go beyond contract and the corporate form, and adopt a relational and functional approach to ascribing employment-related responsibilities in situations involving multilateral work arrangements in employing enterprises" (Deakin 2001: 79 et seq.).

In order to embrace a functional approach, however, the law's underlying methods of reasoning need to evolve in part. The present sub-section thus sets out to consider the meaning of 'functional' in the proposed functional concept on a more abstract level, in the hope that this will allow for a clearer account of that approach. It further aims to develop functional typology as a richer concept than simply a contrast to the perceived formalism of the current bilateral-contractual approach (Fudge 2006), thus avoiding at least some of the dangers of the 'transcendental nonsense' which can result from the indiscriminate use of the 'functional approach' as a panacea to various analytical problems, "often . . . with as little meaning as any of the magical legal concepts against which it is directed" (Cohen 1935).

The key idea of this *functional approach* is to focus on the specific role different elements play in the relevant context, instead of looking at the mere absence or presence of predetermined factors. The presence of a contract of employment (or other contract) can thus be an important indicator in particular fields (for example the obligation to pay wages), but it is by no means the only one. A functional concept of the employer is one where the employing entity or entities are defined not via the absence or presence of a particular factor, but via the exercise of specific functions. This exercise of specific functions extends to include a decisive role in their exercise, in order to take account of the judicial recognition in existing cases that as regards employer

25 The term *locus* of control is designed to avoid additional complexities arising out of the fact, noted inter alia by Freedland (2003), that even in traditional companies without external influence, management control is often exercised by more than one person amongst a group of relatively senior executives.

functions the right to play a decisive role in a particular function is as relevant as the actual exercise thereof.

We have applied this concept to existing platform models in a paper and reached the following conclusions (Prassl and Risak 2016: 636 et seq.): an examination of the transportation service Uber's business model demonstrated that, where a platform exercises all employer functions, it can easily be identified as an employer, with drivers consequently to be seen as workers, rather than independent contractors. Most platforms, on the other hand, lead to a fragmentation of employer functions as demonstrated in the case of TaskRabbit which provides household services. We concluded that, just as different functions may be exercised by various parties, concomitant responsibility should be ascribed to whichever entity – or combination of entities – has exercised the relevant function. As a result, multiple entities may come to be seen as employers for different purposes; the model is able at the same time to recognise elements of (genuine) self-employment, as the concluding examples have demonstrated.

Redefining the notion of the employee

Two of the core questions of labour law relate to the scope and justification of employment protection; put differently: who is protected, and why? The scope of employment should extend to those in need of protection because of their unique situation. This leads us to the second question, namely what makes the employment relationship so special and the employee in need of special protection. One of the most frequently cited underlying rationales of labour law is the twofold economic dependence of the employee. This refers, first, to the fact that resources (e.g. materials, machines, or an organisation) are typically needed to perform the work and that employees have, at least historically, depended on the employer to provide them. Secondly, it implies dependence of the employee on 'selling' his or her labour in exchange for remuneration from the employment relationship to sustain his or her living. Some legal orders, however, do not refer to these economic arguments, focusing instead on the way the work is actually performed.[26] Especially the second aspect (dependence on salary to earn a living) is considered impractical, as employers often have no means of ascertaining whether their contractual partners actually have other sources of income or their reasons for working more generally.

The European Court of Justice applies a similar approach. It is settled case law that the essential feature of the employment relationship is that for a certain period of time one person performs services for and under the direction of another person in return for which she receives remuneration.[27] It is of major importance that a person acts under the direction of his or her employer as regards, in particular, the freedom to choose the time, place, and content of the work,[28] that the employee does not share in the employer's commercial risks,[29] and, for the duration of that relationship, forms an integral part of that employer's undertaking, so forming an economic unit with that undertaking.[30] In its first decision on platform work of 22 April 2020 dealing *in concretu* with the applicability of the Working Time Directive 2003/88/EC to 'neighbourhood parcel delivery couriers' organised via an online platform, the European Court of

26 For Austria cf. Risak (2010: 36); Brodil and Risak (2020: 14); for Germany cf. Weiss and Schmidt (2008: 45).

27 ECJ in *N.*, C-46/12, EU:C:2013:97, para. 40 and the case-law cited, and ECJ in *Haralambidis*, C-270/13, EU:C:2014:2185, para. 28.

28 ECJ in *Allonby*, EU:C:2004:18, para. 72.

29 ECJ in *Agegate*, C-3/87, EU:C:1989:650, para. 36.

30 ECJ in *Becu and Others*, C-22/98, EU:C:1999:419, para. 26.

Justice again applied the traditional criteria mentioned earlier.[31] It was not even discussed in this order of the court to adapt them to the platform work environment let alone develop them further. There is good reason to do so as we will argue in the following.

For decades this organisational approach focusing on the restricted self-determination when working on the one hand delivered satisfactory results and on the other was practical and relatively easy to apply. This was based on the fact that only those having enough resources were able to become self-employed and that they were able to negotiate for pay that satisfied their needs. On the other hand, those working under the close supervision of another person often did not have enough bargaining power when negotiating pay and conditions of work (Freedland and Davies 1983: 14, 69). In those circumstances, it is rather unproblematic to equal organisational and economic dependency in the past. This picture, however, has changed due to a number of factors and has led to the emergence of a growing number of self-employed: advances in digital technologies, the widespread availability of handheld devices, and ever-increasing high-speed connectivity have combined with the realities presented by several cycles of economic downturn, shifts in lifestyle, and generational preferences (Lobel 2016: 2). These new 'solo-entrepreneurs' and freelancers are very different from the ones in the past, where 'liberal professions' such as lawyers, architects, and other high-skilled professionals had the power to bargain for high remuneration and controlled their own working conditions. Platform workers active in the virtual realms of the gig economy today resemble the workers of the 19th century who did not have any other alternatives than to sell their labour in a highly competitive market. They compete with a large reserve army of virtual labour unlike those self-employed in liberal professions. They are also similar to traditional employees as they do work in person and thereby sell their labour and not an end product. Finally, they are also vulnerable as they earn their livelihood by doing this vis-à-vis only one or a very limited number of immediate contractual partners (viz., the platforms). The only difference between them and traditional employees is the fact that they are formally free to work on what and when they choose – but this freedom may often be no more than formal, due to an economic situation which does not leave them a lot of alternatives to selling their labour in a certain way to certain contractual partners (Risak and Lutz 2017: 358).

Against this background it makes sense to open up a range of employment rights, not least the rights to organise, to bargain collectively, and to take collective action, to this group of vulnerable self-employed. At first glance, this might be in conflict with European Union competition and anti-trust law, as Art. 101 TFEU forbids all agreements and concerted practices which have as their object or effect the prevention, restriction, or distortion of competition: collective agreements could be characterised as a restriction on competition between employees, thus contravening that provision. The European Court of Justice has held, however, that agreements entered into within the framework of collective bargaining between employers and employees and intended to improve employment and working conditions must, by virtue of their nature and purpose, be regarded as not falling within the scope of Art. 101(1) TFEU.[32] In our view, it is therefore crucial to either redefine the notion of the employee or take specific legislative initiatives in order to open up collective bargaining to this group of self-employed with limited bargaining powers. In December 2015, for example, the Seattle City Council unanimously

31 ECJ in *Yodel Delivery Network Ltd*, ECLI:EU:C:2020:288.

32 ECJ in *Albany*, EU:C:1999:430, para. 60; *Brentjens'*, EU:C:1999:434, para. 57; *Drijvende Bokken*, EU:C:1999:437, para. 47; *Pavlov* and Others, C-180/98 to C-184/98, EU:C:2000:428, para. 67; *van der Woude*, EU:C:2000:475, para. 22; *AG2R Prévoyance*, C-437/09, EU:C:2011:112, para. 29; *FNV Kunsten Informatie en Media v Staat der Nederlanden*, ECLI:EU:C:2014:2411, para. 23.

enacted legislation granting the city's drivers "a voice on the job and the opportunity to negotiate for improved working conditions at their companies".[33]

Redefining the notion of the employee, or specifically including the self-employed within the scope of certain employment law norms, would also widen the scope of application of individual labour law (i.e. the set of rules granting individual rights and entitlements and therefore protecting employees from unfair und unhealthy working conditions). This body of laws usually encompasses, among others, minimum wages, working time restrictions, right to paid sick leave, and holidays, as well as protection against dismissals. If the economic situation of the employee is the reason why these rights and entitlements have been developed in the first place, it is hard to argue why not to extend the scope of their application to persons in the same situation only based on the argument that they are not formally integrated enough into the business of their contractual partners.

That redefining the concept of worker is not an easy endeavour became evident by the fierce opposition to even include a very traditional definition of 'worker' into recent EU legislation, the Directive (EU) 2019/1152 on transparent and predictable working conditions. It has to be seen as a success that in Recital 8 it is mentioned that – provided that they fulfil the criterion of subordination – platform workers among others could fall within the scope of this Directive.

Introduction or extension of an intermediate category

Another option to protect virtual platform workers that has been mooted is a suggestion that the law might recognise an intermediate category of worker between employee and independent contractor (Lobel 2016: 10; Harris and Krueger 2015). In this way, the argument runs, the level of protection may be graded, and the fact that the personal integration of some platform workers is less intense and that they enjoy a certain level of flexibility and freedom can actually be used to their advantage.

The examples are numerous: in Canada, jurisprudence has developed the category of dependent contractor for cases in which a contractor has worked exclusively or largely exclusively for one client for an extended period. They are then deemed a dependent contractor for purposes of termination notification and representation.[34] In Italy, *lavoro etero-organizzato* relationships enjoy some level of statutory protection,[35] and in Germany and Austria some employment regulations are to be applied also to employee-like (*arbeitnehmerähnliche*) persons. In Austria, these persons are defined as persons who perform work/services by order of and on account of another person without being in an employment relationship, but who may be considered employee-like due to their economic dependence. Only some provisions of labour law apply to those employee-like persons, for example, those on the competence of the labour courts,[36] agency work,[37] employee liability,[38] and anti-discrimination.[39] In Germany, the intermediate category is defined similarly, and is also covered by the Act on Collective Agreements (*Tarifvertragsgesetz*), and may therefore

33 www.seattle.gov/council/issues/giving-drivers-a-voice.

34 Cf. Superior Court of Justice, 14.8.2014, *Wyman v. Kadlec*, 2014 ONSC 4710 (CanLII), http://canlii.ca/t/g8lnv (26.19.2016); Court of Appel for Ontario, 23.12.2009, *McKee v. Reid's Heritage Homes Ltd.*, 2009 ONCA 916 (CanLII), http://canlii.ca/t/27551.

35 Cf. De Stefano (2016b: 20).

36 The Labour and Social Courts Act s 51 (3) 2.

37 The Act on Agency Work s 3.

38 The Employees' Liability Act s 1 (2).

39 The Equal Treatment Act ss 1 (3) 2 and 16 (3) 2.

conclude collective agreements with normative effect. In the United Kingdom the extension of employee rights beyond the employment contract seems to be the furthest developed, as discussed in our analysis of the 2016 Uber decision, immediately following.

Instead of building on these specific domestic experiences, Harris and Krueger (2015) have instead argued in favour of the statutory introduction of a novel third, intermediate category to capture gig economy workers: their 'independent worker' status would be entitled to some protection, including collective bargaining and elements of social security provision, whilst being denied recourse to basic standards such as wage and hours protection. This approach differs from existing models, insofar as platforms would immediately be relieved of some of employers' most costly obligations – whilst continuing to litigate over independent contractor status.

As an exasperated US District Judge famously noted, the task of determining worker status is often akin to being "handed a square peg and asked to choose between two round holes".[40] Adding a third round hole is therefore unlikely to solve any classification problems. Indeed, it appears that even those jurisdictions that have recognised a third category have done so without resolving any of the fundamental classificatory problems. If anything, more confusion is introduced, as became evident during recent UK litigation against Uber, with legal arguments focused on the third category recognised in English employment law.

In *Aslam v. Uber BV* the Central London employment tribunal ruled on 28 October 2016 that Uber drivers were workers for purposes of s. 230(3)(b) of the Employment Rights Act 1996, rather than independent contractors, as the company had long maintained. In a clear and powerful judgement, the tribunal found that the company's "resorting in its documentation to fictions, twisted language and even brand new terminology" merited "a degree of scepticism" (at paragraph [87]), and found that drivers were workers. As a result, Uber's drivers will now be entitled (subject to a series of pending appeals) to a small number of core rights attached to worker status, including, importantly, the National Minimum Wage Act 1998 and the Working Time Regulations 1998.[41]

Such basic protection will overcome some of the worst problems faced by Uber drivers – not least because the tribunal (rightly) ruled that a driver is 'working' for the entire time that his (the vast majority being male, as noted in the decision) Uber drivers' app is switched on, and he is able and willing to accept rides, not just when he has a passenger in his car. In the longer run, however, Uber drivers – even when classified as workers – will face many of the problems encountered by zero-hours workers across the United Kingdom (Adams et al. 2015): from low income to struggling with unpredictable shifts due to a lack of guaranteed work. This, then, is the fundamental problem with the creation of a novel third status category: not only would it fail to alleviate the uncertainty and classificatory problems identified previously; it would provide platform workers with a lower degree of protection even though, as previous discussion has shown, they might often be amongst the most vulnerable participants in the labour market.

Beyond the United Kingdom, the experience with this intermediate category is similarly varied. While its introduction does not, at first glance, appear to change anything to the disadvantage of traditional employees because of employers moving over to this now-legitimate group, the Italian example seems to indicate otherwise. De Stefano (2016b: 20) points out that the workers that would qualify for full protection as employees under the traditional legal

40 United States District Court, Northern District of California, Order of March 11, 2015 Denying Cross-Motion for Summary Judgment (Case No. 13-cv-04065-VC) 19, Cotter et al. v. Lyft Inc.

41 The Employment Tribunal's findings were upheld in the Employment Appeal Tribunal and by a majority (Underhill LJ dissenting) in the Court of Appeal. The UK Supreme Court dismissed Uber's final appeal on 19 February 2021 [2021] UKSC 5.

tests would likely become deprived of many rights if they were crammed into an 'intermediate bucket'. He warns that regulating dependent self-employment as a distinct group is no panacea for addressing the changes in business and work organisation driven by the disintegration of vertical firms. Some argue, on the other hand, that as existing law no longer protects a growing number of persons who once would have enjoyed the status of employees and who are now slipping out of the protective scope of labour law due to their increased formal freedom and flexibility, there is nonetheless the need for such intervention. It is arguable, for example, that the lack of any intermediate status effectively provides greater incentives for employers to reclassify their workers as independent contractors and that an intermediary category may well provide them with those rights they actually need (Lobel 2016: 12). In our view, however, current proposals are flawed insofar as they do not even recognise the full set of 'basic' employment rights, including the right to organise and to bargain collectively as well as the application of minimum wage legislation, and as they would lead to little additional clarity or faster dispute resolution.

From the point of view of EU-law, a further difficulty arises from the issue of the applicable law and the choice of law in situations where platforms operate across multiple jurisdictions. In cases concerning cross-border contractual relationships, Regulation (EC) Number 593/2008 of the European Parliament and of the Council on the law applicable to contractual obligations (Rome I Regulation)[42] applies, according to which there is freedom of choice regarding the applicable law (Article 3). However, this is limited when it comes to employment contracts (Article 8). In these cases, the level of protection cannot fall below that which would be provided in the absence of choice. Platform workers who are not considered to be employees thus lack significant protection, not only as regards the application of statutory protection is concerned, but also insofar as they – or better their contractual partners – may choose any law without any restrictions. It is very likely that platforms will include the choice of a legal order that is favourable for them in their terms and conditions and thereby make it harder again for the platform workers to enforce even the limited number of rights they have. An extension of the limitation of the choice of law provisions for employees to the intermediate category therefore seems of the essence.

Special legislation

The last option to be highlighted in this contribution is the creation of a special legislative act dealing with the issues involved with platform work, as has been done in many European countries with temporary agency work in the transposition of the Temporary Agency Work Directive 2008/104/EC. This is the most complicated solution as it must take into account that the platform economy is very diverse and that a one-size-fits-all-approach will hardly work. We can therefore only sketch in very rough strokes what such an act might look like.

The aim would be to ensure the protection of platform workers and to improve the quality of platform work. It should also take into account that platform work may contribute to the creation of jobs and to the development of flexible forms of working by introducing creative and innovative business models but also keep in mind that there is nothing innovative about precarious work. The primary goal thus would be the creation of a level playing field for those platforms that endorse an approach that is platform worker friendly, rather than one based on low labour costs and value extraction.

42 OJ L 177, 4 July 2008, pp. 6–16; cf. Kozak (2017).

The heart of such an Act should be a *rebuttable legal assumption* that the underlying contractual relationship is an employment contract between the platform worker and the platform. Attempts to classify the legal relationships underlying the gig economy have shown that it is very hard to get an insight into how the work is actually organised by the platform and the mechanisms behind it. This knowledge is of significant importance to prove before a court of law that an employment contract has been concluded – but, as the platform worker has no means of getting to the information necessary to do so, this often amounts to the impossibility of providing the court with the necessary evidence. The proposed legal assumption would recalibrate the massive imbalance of information and thereby justify a departure from the otherwise existing contribution of the burden of proof (Risak and Lutz 2017: 356).

A core provision might thus be – as in the case of agency work (Art. 5 of Directive 2008/104/EC) – a *principle of equal treatment* with a corporate customer's existing workforce, to ensure that jobs are not crowdsourced just for the sake of contravening minimum wage and other employment provisions. The basic working and employment conditions of platform workers shall therefore be at least those that would apply if they had been recruited directly by the crowdsourcer to occupy the same job for the duration of working on tasks or actively looking for them – if the general availability is part of the business model, as is the case with most platforms.[43] This would also establish the equal treatment of temporary agency workers and avoid the circumvention of the laws protecting them by switching over to platform work.

It should be noted, however, that this equal-treatment-approach very likely only works in cases where the platform worker is actually working for a business that would otherwise employ an employee and that actually crowdsources labour. In cases where the contractual partner is a consumer and the alternative is contracting directly with a self-employed person (e.g. with a cleaner) avoiding the intermediary (e.g. the platform TaskRabbit), the equal treatment principle cannot apply. In these cases, no crowdsourcing of employment contracts takes place and a host of other issues arise, not least as regards the application of minimum wages to those platform workers.

Other topics that seem to be important might be the *prohibition of certain clauses* in contracts with platform workers and the terms and conditions of the platforms. This can refer to the notorious clauses that enable the contractual partner to refuse to accept a completed task without having to give a reason and refrain from paying the advertised remuneration or provisions that the result may be kept even in those cases. Other possible issues are non-compete clauses as well as clauses that restrict the hiring of platform workers by crowdsourcers (Risak and Lutz 2017: 357). Finally, workers should also be permitted to port their ratings across different platforms to ensure that their expertise and experience is adequately recognised.

The very tricky question for legislation will be to draw the role and the responsibility of the crowdsourcing platforms in a transparent way in order to give platform workers certainty of their legal position in this set-up, without, however, suffocating those crowdsourcing models that are based on genuine self-employment (and thus not necessarily in need of statutory protection). This final concern, however, should not be – in our view – a hindrance to, or excuse from, protecting those genuinely in need of protection. Finally, it should also be noted that any platform work-specific legislation ought not to fall into the trap of technological exceptionalism, and recognise that, fundamentally, platform work should be regulated as work first and foremost.

Some, but by far not all of these issues are also addressed in the so-called 'Platform to Business' (P2B) Regulation (EU) 1150/2019, which applies only to self-employed platform

43 Cf. Employment Tribunals 28.10.2016, 2202551/2015 & Others, Aslam, Farrar & Others v Uber B.V., Uber London Ltd. & Uber Britannia Ltd, www.judiciary.gov.uk/judgments/mr-y-aslam-mr-j-farrar-and-others-v-uber/.

workers. This concerns in particular the restriction, suspension, and termination of platform access (Article 4), restrictions of competition (Article 10), and internal dispute resolution (Article 11). The provisions on rankings (Article 5), on the other hand, concern the appearance of platform workers on the platform and not customer ratings. An extension of the scope of application to platform employees who qualify as employees would therefore not do justice to their interests. For this reason, specific legislation on platform work should be given preference on the EU level if the P2B Regulation is not to be massively expanded in terms of content.

Platform work in times of crisis

The COVID-19 pandemic has served as a stark reminder of just how far the gig economy's risk has been shifted towards a breaking point: medical staff apart, many of the jobs which have become subsumed by gig models are amongst those facing high coronavirus risks.[44] Official advice to workers[45] is clear: self-isolate if you experience symptoms. Work from home, even if you don't. Only head out to work equipped with suitable protective kit, from face masks and shields to gloves and hand sanitiser. And maintain social distancing at all times.

For gig workers, that's easier said than done: self-isolation is rarely financially viable; working from home impossible; and social distancing tough to implement in practice. Take self-isolation, first. As we have seen, employees are entitled to receive pay, even when they're off sick. They'll be protected from layoffs. And employers in many countries will receive government support to finance these schemes. It's easy, in other words, to self-isolate when you don't have to worry about income continuity – a luxury not enjoyed by independent contractors.

The same is true for home working: by definition, it's impossible to work from home when your job involves picking up goods in a warehouse or dashing around town delivering food. (There are some purely digital gigs, of course – but earnings quickly take a hit when workers face regular interruptions,[46] such as home-schooling kids where schools are closed). Those on low incomes are particularly hard hit. A recent survey by economists in Oxford, Zurich, and Cambridge[47] shows just how close the correlation between earnings and the ability to work from home is in times of COVID-19: workers on less than US$ 20k are half as likely to be able to do their job from home when compared with those on incomes of more than US$ 40k.

When gig workers head to work, finally, they are exposed to a much higher frequency of high-risk contacts – whether its door handles and turnstiles, or customers and fellow workers. Protective equipment is scarce – and expensive, even if you can find it. In the early summer of 2020, a court ruling[48] in France exposed a long list of health- and safety-related shortcomings in Amazon's warehouses, leading to their temporary closure.

A high-risk gamble – for all of us

Put together, the absence of employment rights quickly leads to a double whammy for gig workers: not only is it impossible to stay at home, whether sick or not – but when they are forced to go out, it's into a world of frequent high-risk exposure.

44 www.nytimes.com/interactive/2020/03/15/business/economy/coronavirus-worker-risk.html

45 www.who.int/emergencies/diseases/novel-coronavirus-2019/advice-for-public

46 https://cepr.org/active/publications/discussion_papers/dp.php?dpno=14294

47 https://abiadams.com/wp-content/uploads/2020/04/US_Inequality_Briefing.pdf

48 www.bbc.com/news/world-europe-52301446

And the risks are by no means limited to workers: customer health is also at stake. A lack of employment protection leaves gig workers who suspect that they might have come into contact with COVID-19 with an invidious choice. Go to work, and risk spreading the infection – or stay at home, unable to make rent and pay for necessities. No wonder, then, that workers around the world report having to go to work,[49] even when public safety advice dictates that they should stay at home.

Changing tides

How can we make gig work safe, for workers and customers alike? As the pandemic drags on, a number of options have emerged: some platforms are rolling out contactless deliveries,[50] and stepping in to provide their workers with financial assistance[51] and protective equipment – even if not always voluntarily.[52] Several countries have begun to offer financial protection to the self-employed.[53]

Those are promising fixes in the short term. But their longer-term sustainability is questionable: platforms' offerings have come under criticism for being 'not enough',[54] and workers have walked out[55] over lack of safety precautions. Government support schemes for the self-employed can take far too long[56] to make payments to workers in dire financial straits, have the potential to create serious moral hazard, and will be impossible to finance for extended periods of time.

In the long run, the only sustainable and efficient[57] solution lies in risk diversification – which is precisely what the different options explored in this paper offer. Recognising that gig workers are employees is now more important than ever. Even before the advent of the crisis, the tide had already begun to turn. The past few months have seen a flurry of legal activity across the world, whether it's the enactment of AB5[58] in California, or a series of rulings by senior courts in Europe.[59] Given the tight control exercised by platforms over all elements of product and service delivery, judges and legislators consistently fail to be impressed by independent contractor claims.

The assertion that employment law somehow leads to rigid, inflexible work, finally, is simply untrue. Legally, there is nothing at all in any employment law system we have ever studied to stop employers from giving workers full employment rights, as well as unlimited flexibility, from flexible shift arrangements to unmeasured working time. Any suggestion that employment rights are inherently incompatible with flexibility is a myth – and in times of COVID-19, a dangerous one for us all.

49 www.ft.com/content/48e4a311-6a7c-4f80-b541-7893a5f97ea4

50 https://au.deliveroo.news/news/contactless.html

51 www.uber.com/en-BH/blog/update-covid-19-financial/

52 www.studiolegalecarozza.it/news/obbligo-di-dispositivi-di-protezione-contro-il-rischio-covid-19-e-tutela-urgente.htm

53 www.gov.uk/guidance/claim-a-grant-through-the-coronavirus-covid-19-self-employment-income-support-scheme

54 https://therideshareguy.com/uber-driver-coronavirus-compensation/

55 www.cnet.com/news/instacart-workers-strike-amid-covid-19-fears-call-company-response-a-sick-joke/

56 www.theguardian.com/world/2020/mar/25/almost-500000-people-in-uk-apply-for-universal-credit-in-nine-days

57 https://econpapers.repec.org/article/aeaaecrev/v_3a79_3ay_3a1989_3ai_3a2_3ap_3a177-83.htm

58 https://leginfo.legislature.ca.gov/faces/billTextClient.xhtml?bill_id=201920200AB5

59 https://twitter.com/JeremiasPrassl/status/1235238405468041217

Conclusion

In this contribution, we set out to explore a series of potential legal solutions to the problem that platform workers in the gig economy offer a lot of flexibility but get little security in return. In concluding, it is important to note that whilst the phenomenon of 'gigs' and 'platforms' is indeed a novel one, the legal implications – particularly as regards employment law – are much less so. Seen from a historical labour law perspective, platform work is but the most recent threat to emerge to the law's quest for underlying coherence in the scope of protective norms in the face of dramatic changes in the labour market: online platforms or 'apps' act as intermediaries in a spot-market for labour, providing clients with workers for a wide range of jobs referred to as 'gigs', 'rides', or 'tasks' that are, from a legal perspective, not all that different from traditional outsourcing and agency relationships, or the more recent phenomenon of zero-hours contracts in the United Kingdom.

At first glance, the advantages for business, customers, and workers resulting from the 'gig economy' are immense: platform work does away with many of the regulatory costs traditionally associated with employing individuals; customers can receive a nearly infinite number of services at cut-price rates; and workers can find flexible work to suit their schedules and income needs. Upon closer inspection, however, a series of problems arising from this fragmentation of traditional work arrangements quickly emerges – in particular for workers, who often find themselves outside the scope of employment protective norms as a result of digital platforms' business models, thus suffering low pay, no job security, and challenging working conditions.

Each of the models scrutinised has its peculiar advantages and drawbacks. Present space limitations do not permit for a detailed summary, but three points may nonetheless be made. First, the importance of recognising that whichever regulatory solutions are adopted, we should be careful of reinventing the wheel: many of the problems we encounter are not novel, so efforts should be made to fit platform work into existing regulatory structures, with only partial additions as and when required. Second, new regulatory measures, if adopted, should not lead to the dilution of workers' rights, as might be the case with some 'third status' proposals, in particular. Finally, and perhaps most importantly, given the vast heterogeneity of platforms, users, and working conditions, it is unlikely that an easy solution could be found: platform work can cater to the needs of successful entrepreneurs, but it can also become a low-wage trap. Only a sophisticated and responsive approach will be able to address the vast range of problems identified.

References

Adams, A., Freedland, M. and Prassl, J. (2015): The "Zero-Hours Contract": Regulating Casual Work, or Legitimating Precarity? *Giornale di Diritto del Lavoro e di Relazioni Industriali* 147, 529–556.

Aloisi, A. (2016): Commoditized Workers – The Rise of On-Demand Work, a Case Study Research on a Set of Online Platforms and Apps', CLLPJ *Comp. Labor Law & Policy Journal* 37, 653–690.

Brodil, W. and Risak, M. (2020): *Arbeitsrecht in Grundzügen*. 10th ed. Austria: LexisNexis.

Coase, R. (1973): The Nature of the Firm, *Economica* 4(16), 386–405.

Cohen, F. (1935): Transcendental Nonsense and the Functional Approach, *Columbia Law Review* 35, 814–817.

Deakin, S. (2001): The Changing Concept of the "Employer" in Labour Law, *Industrial Law Journal* 72, 72–84.

De Stefano, V. (2016a): The Rise of the "Just-in-time Workforce": On-demand Work, Crowd Work and Labour Protection in the "Gig Economy", *Comparative Labor Law & Policy Journal* 37, 471–503

De Stefano, V. (2016b) *The Rise of the "Just-in-time Workforce": On-demand Work, Crowdwork and Labour Protection in the "Gig-economy"*. ILO Working Paper. Available at: https://EconPapers.repec.org/RePEc:ilo:ilowps:994899823402676 [28.01.2021]

Dibbell, J. (2003): The Unreal Estate Boom, *Wired Magazine* 1/2003. Available at: www.wired.com/2003/01/gaming-2 [28.01.2021]

Dibbell, J. (2016): The Life of the Chinese Gold Farmer, *NYT Magazine* 17. June. Available at: www.nytimes.com/2007/06/17/magazine/17lootfarmers-t.html?_r=2&oref=slogin [28.01.2021]

Eurofound (2001): *Third European Survey on Working Conditions*. Luxembourg: Office for Official Publications of the European Communities.

Eurofound (2015): *New Forms of Employment*. Luxembourg: Publications Office of the European Union.

Eurofound (2018): *Employment and Working Conditions of Selected Types of Platform Work*. Luxembourg: Publications Office of the European Union.

European Commission (2016): *A European Agenda for the Collaborative Economy*. COM 356 final

European Parliament (1998): *Atypical Work in the EU* (SOCI106EN).

Fabo, B., Beblavý, M., Kilhoffer, Z. and Lenaerts, K. (2017): *Overview of European Platforms: Scope and Business Models, Study Performed for JRC*. Luxembourg: Publications Office of the European Union.

Felstiner, A. (2011): Working the Crowd: Employment and Labour Law in the Crowdsourcing Industry, *Berkeley Journal of Employment and Labor Law* 32(1), 143–203.

Freedland, M. (2003): *The Personal Employment Contract*. Oxford: Oxford University Press.

Freedland, M. and Davies, P. (1983): *Kahn Freund's Labour and the Law*. London: Stevens & Sons.

Fudge, J. (2003): Fragmenting Work and Fragmenting Organizations: The Contract of Employment and the Scope of Labour Regulation, *Osgoode Hall Law Journal* 44(4), 609–648.

Fudge, J. (2006): The Legal Boundaries of the Employer, Precarious Workers, and Labour Protection, in Davidov, G. and Langile, B. (eds.): *Boundaries and Frontiers of Labour Law*. Oxford and Portland: Hart, 310–313.

Harris, D. and Krueger, A. (2015): A Proposal for Modernizing Labor Laws for Twenty-First Century Work: The "Independent Worker". Hamilton Project. *Discussion Paper 2015–10*.

Howe, J. (2006): The Rise of Crowdsourcing, in *Wired Mag* 14. June. Available at: www.wired.com/2006/06/crowds/ [29.01.2021]

Kittur, A. et al. (2012): *The Future of Crowd Work*. Paper presented at 16th ACM Conference on Computer Supported Cooperative Work. Available at: www.lri.fr/~mbl/ENS/CSCW/2012/papers/Kittur-CSCW13.pdf [29.01.2021]

Klebe, T. and Neugebauer, J. (2014): Crowdsourcing: Für eine handvoll Dollar oder Workers of the crowd unite? *Arbeit und Recht* (1), 4–7.

Kozak, W. (2017): Crowdwork mit Auslandsbezug, in Lutz, D. and Risak, M. (eds.): *Arbeit in der Gig-Economy*. Wien: ÖGB, 304–319.

Langlois, S. (2015): Don't Tip Your Uber Driver? It Could Cost You a 5-star Rating, in *Market Watch* 12. August. Available at: www.marketwatch.com/story/dont-tip-your-uber-driver-it-could-cost-you-a-5-star-rating-2015-08-12 [29.01.2021].

Leimeister, J. et al. (2014a): Crowdwork – Digitale Wertschöpfung in der Wolke, in Benner, C. (ed.): *Crowdwork – zurück in die Zukunft*. Frankfurt am Main: Bund, 9–42.

Leimeister, J. et al. (2014b): Crowdwork – digitale Wertschöpfung in der Wolke, in Brenner, W. and Hess, T. (eds.): *Wirtschaftsinformatik in Wissenschaft und Praxis*. Berlin, Heidelberg: Springer, 51–64.

Lobel, O. (2016): The Gig Economy & The Future of Employment and Labor Law. *USD Legal Studies Research Paper Series*, Research Paper No. 16–223.

Lutz, D. (2017): Virtuelles Crowdwork: Clickworker, in Lutz, D. and Risak, M. (eds.): *Arbeit in der Gig Economy*. Wien: ÖGB, 62–105.

Martin, D. et al. (2014): Being a Turker, in *CSCW'14*. Proceedings of the 17th ACM Conference on Computer Supported Cooperative Work & Social Computing. Available at: http://dl.acm.org/citation.cfm?id=2531602 [28.01.2021]

Marvit (2014): How Crowdworkers Became the Ghosts in the Digital Machine, *The Nation* 5. February. Available at: www.thenation.com/article/how-crowdworkers-became-ghosts-digital-machine/ [28.01.2021]

Nogler, L. (2009): Die Typologisch-Funktionale Methode am Beispiel des Arbeitnehmerbegriffs, *ZESAR* 11, 461–469.

Pesole, A., Urzì Brancati, C., Fernández-Macías, E., Biagia, F. and González Vázquez, I. (2018): *Platform Workers in Europe: Evidence from the COLLEEM Survey*. Luxembourg: Publications Office of the European Union.

Prassl, J. (2015): *The Concept of the Employer*. Oxford: Oxford University Press.

Prassl, J. (2018): *Humans as a Service: the Promise and Perils of Work in the Gig Economy*. Oxford: Oxford University Press.

Prassl, J. and Risak, M. (2016): Uber, Taskrabbit, and Co.: Platforms as Employers? Rethinking the Legal Analysis of Crowdwork, *Comparative Labor Law & Policy Journal* 37, 619–651

Prassl, J. and Risak, M. (2017): The Legal Protection of Crowdworkers – Four Avenues for +-Workers' Rights in the Virtual Realm, in Meil, P. and Kirov, V. (eds.): *Policy Implications of Virtual Work*. Cham, Switzerland: Palgrave, 273–295.

Risak, M. (2010): Austria, in *International Encyclopaedia for Labour Law and Industrial Relations*. Wolters Kluwer.

Risak, M. and Lutz, D. (2017): Gute Arbeitsbedingungen in der Gig-Economy – was tun? in Lutz, D. and Risak, M. (eds.): *Arbeit in der Gig-Economy*. Wien: ÖGB, 352–363.

Rosenblum, M. (2013): The Digital Slave – That would be You, *Huffington Post* 5 June. Available at: www.huffingtonpost.com/michael-rosenblum/the-digital-slave-that-wo_b_3222785.html [28.01.2021]

Ross, J. et al. (2010): Who are the Crowdworkers? Shifting Demographics in Amazon Mechanical Turk, Paper prepared for Chi. Available at: http://dl.acm.org/citation.cfm?doid=1753846.1753873 [28.01.2021]

Schmidt, F. A. (2014): The Good the Bad and the Ugly, in Benner, C. (ed.): *Crowdwork – zurück in die Zukunft*. Frankfurt am Main: Bund, 367–386.

Strube, S. (2014): Vom Outsourcing zum Crowdsourcing – Wie Amazons Mechanical Turk funktioniert, in Benner, C. (ed.): *Crowdwork – Zurück in die Zukunft?* Frankfurt am Main: Bund, 75–92.

Surowiecki, J. (2004): *The Wisdom of the Crowds*. New York: Doubleday.

Weiss, M. and Schmidt, M. (2008): Germany (Fed.Rep.), in *International Encyclopaedia for Labour Law and Industrial Relations*. Wolters Kluwer.

Von der Leyen, U. (2019a): A Union that Strives for More – My Agenda for Europe. Available at: https://ec.europa.eu/info/sites/info/files/political-guidelines-next-commission_en_0.pdf [28.01.2021]

Von der Leyen, U. (2019b): Mission Letter to Nicolas Schmit, Commissioner-designated for Jobs. Available at: https://ec.europa.eu/info/sites/info/files/mission-letter-nicolas-schmit_en.pdf [28.01.2021]

Zyskowski, K. et al. (2015): Accessible Crowdwork? Understanding the Value in and Challenge of Microtask Employment for People with Disabilities. Available at: http://research.microsoft.com/pubs/228714/crowdwork_and_disability.pdf [28.01.2021]

11

SMART TECHNOLOGIES AND GENDER: A NEVER-ENDING STORY

Knut H. Sørensen and Vivian Anette Lagesen

A world without women

Many of the discourses related to smart technologies and the ideas of widespread digitalisation of human societies are promissory. They express attractive sociotechnical imaginaries where access to electronic resources of communication, problem solving, information, and entertainment is ubiquitous; it should result in progress for everybody. However, the promises are ambiguous in the sense that the developments represent both opportunities and challenges, not the least with respect to artificial intelligence and robotics. At the core of these promissory performances is the issue of understanding a variety of human practices and how they may be supported or changed. Thus, it is important to ask who are engaged in such explorations of smart technologies and whose understandings count when decisions about design and deployment are made? This chapter engages with these questions with a focus on gender issues.

We can illustrate the importance of pursuing such issues by recalling the promissory discourses related to the slogan of 'the paperless office', which emerged in the late 1970s. Computers were claimed to facilitate office automation to the extent that around 40% of the workforce consisting mainly of women would be redundant. However, the effects of the digitalisation of office work turned out to be much less dramatic, in part because the claim of severe cuts was based on a lack of knowledge about the work performed by women office workers and the skills needed to perform them. Their tasks were more comprehensive and complex than the men in the computer industry assumed and thus not so easy to automate (Webster 1996). What happened was a gross misjudgement based on gendered stereotypes – women's work requires little skills – and no effort had been made to investigate this assumption.

Today, the widespread belief is that the situation with respect to gender and equal opportunities for men and women has changed fundamentally since the 1980s. However, as we shall see when we delve into the issue of gender and smart technologies, the state of affairs is not fundamentally different from what it was 40 years ago. The computer industry is still numerically dominated by men, and the assumptions underlying the design of smart technologies seems to rely to a surprising degree singularly on the experiences and the tastes of men designers, with little interest in analysing women's needs and practices. The consequences of the lack of gender balance in the industry are palpable, and the unreflexive use of gender stereotypes in design is quite remarkable.

DOI: 10.4324/9780429351921-14

In an editorial on November 23, 2019, the *Economist* complains that "Silicon Valley is bad at making products that suit women. This is a missed opportunity".[1] What is at stake are the entrenched practices of design, in line with the observation of Greek philosopher Protagoras that "man is the measure of all things". Thus, the complaint is that smart technologies are constructed with inherent gendered biases, such as virtual-reality headsets that do not fit women whose pupils are close together than men's and smartphones that are too big to fit comfortably in the hands of the average woman.

The editorial sees the problem as emerging from the dominant position of men in the IT industry as a entanglements of gender biases in design and the gender imbalance among those constructing smart technologies. The problem is even framed as a business case: "Women are 50% of the population and make 70–80% of the world's consumer-spending decisions. That means they control the deployment of more than $40trn a year". In line with this observation, the editorial rhetorically asks about the industry, "What is holding them back?".

Unfortunately, there are some rather obvious answers. Liza Mundy's (2017) aptly titled article "Why is Silicon Valley so awful to women?" suggests that the gender balance problem of the industry is deeply entrenched and difficult to change. She and other analysts (e.g., Emily Chang 2018) describe the culture of Silicon Valley as deeply alienating to and harassing of women, as a "Brotopia", to use Chang's expression. Even if the companies offer attractive wages and interesting work, the work environment is sexist and even more so than in most other industries. The perseverance of the subtle micro-aggression directed at women is particularly a problem because it is hard to recognise and even harder to complain about. As we will show, there are discursive practices that render women invisible and unwanted with respect to smart technologies. The high level of aggression occasionally expressed (and which is hard to believe) is evident from the discourses articulated by men engaged with computer games in the incident known as 'gamergate'.[2]

Thus, in this chapter, we will discuss the issue of smart technologies and gender with a view both to the position of women in IT and the gender biases inherent in the designs of the IT industry, including some of the consequences of such biases. However, it seems pertinent to start by inquiring into the CONCEPT of smart technologies; what happens when we try to unpack it? Rommetveit et al. (2017) show that there are many uses of 'smart' that includes reference to characteristics of the technology, processes of modernising, and professional achievement. Oxford English Dictionary offers among several other definitions that 'smart' may be said to be: "Of a device or machine: appearing to have a degree of intelligence; able to react or respond to differing requirements, varying situations, or past events; programmed so as to be capable of some independent action".[3] The Wikipedia article on 'smart devices' mentions the following examples: smartphones, smart cars, smart thermostats, smart doorbells, smart locks, smart refrigerators, phablets and tablets, smartwatches, smart bands, smart key chains, and smart speakers.[4]

These references do not address gender issues directly as they use terms that we at face value consider to be gender neutral. However, Rommetveit et al. (2017: 8–9) suggest the importance of addressing

1 www.economist.com/leaders/2019/11/21/silicon-valley-is-bad-at-making-products-that-suit-women-that-is-a-missed-opportunity (accessed May 14, 2020)

2 See www.theguardian.com/commentisfree/2019/aug/20/the-guardian-view-on-gamergate-when-hatred-escaped (accessed May 14, 2020).

3 www.oed.com/view/Entry/182448#eid22356150 (accessed May 14, 2020)

4 https://en.wikipedia.org/wiki/Smart_device (accessed May 14, 2020)

> the gendered and elitist imaginaries of technology use, who the 'citizen' is and the ways in which citizens are seen as actively engaged, empowered, rational, calculating, and so on. This we foreground . . . as issues of inclusion/exclusion, especially in scenarios and other depictions of lifeworlds that appear to be populated for the most part by able-bodied Western males and over-simplified stereotyping of groups such as the family and the elderly.

The Wikipedia list of smart devices seemingly connotes more to the lifeworlds of men than that of women, but this needs further inquiry.

In the following, we first present some prominent approaches that are helpful in analysing the issues of this chapter, often labelled as feminist technoscience. We continue by addressing the gender imbalance problems related to computer science and engineering, presenting some of the relevant research in this area. Then, we discuss some features of gender biases in the construction and design of smart technologies, with a focus on what we consider as the mismeasure of women. Finally, we provide some suggestions about how to deal with gender biases and the lack of women in computer science and engineering.

Backdrop: some important feminist technoscience approaches

As a social science concern, the gender and technology issue is basically a post-1980 phenomenon. The early feminist critique of technoscience emerged from observations that new technology was not a neutral force of production, but rather a tool that could be employed to undermine women's position in the labour force or as a basis for differential treatment of men and women (Cockburn 1983, 1985). This tradition of research developed a topical interest in how the introduction of computers would shape and be shaped by the gender division of labour in workplaces (for an overview, see Webster 1996). Later, there came a growing concern about the comparatively low and declining numbers of women in computer science and engineering (e.g., Cohoon and Aspray 2006; Lagesen 2007; Fox et al. 2009).

Further, the image of computers was transformed from highlighting computing machines to identifying them as the core of information and communication technologies (ICT). This reflected that the technologies became common household goods as well as standard workplace tools. Accordingly, and in line with the observations made in the introduction, feminist technoscience put two distinct problem areas on the agenda (see Faulkner 2001). First, the initial concern about women *in* ICT: why so few? Second, the issue of women *and* ICTs: women's situation as users and non-users of the new technologies. Women seemed to be on the wrong side of a digital divide between those with and without access to and competence in using and making the new ICTs.

Thus, the dominant concern of early feminist technoscience work was the exclusion of women as users as well as professionals with respect to ICTs. While some scholars studied the relations between men and technology and the role of technology as constituent of modern masculinity (see, e.g., Hacker 1989, 1990; Mellström 1995, 2003; Turkle 1984), the exclusion of women remained the major focus (see, e.g., Abbate 2012; Ahuja 2002; Archibald et al. 2005; Barker and Aspray 2006; Cronin and Roger 1999). Two main exclusion accounts relevant to women and/in ICT emerged in the feminist technoscience literature (Sørensen et al. 2011). The first may be called 'A world without women', focusing on absolute exclusion and the ICT arenas as men's worlds. A second account, 'A chilly culture', told about the problems of retaining women when they are a minority as users or designers of ICT.

The account of exclusion as a mechanism upholding 'a world without women' is based on the argument that throughout history men scientists and engineers have made explicit efforts to keep women out of technoscientific arenas (e.g., Noble 1992; Merchant 1980). Ruth Oldenziel (1999) shows that the modern profession of engineering in the USA was established through conscious efforts to exclude women (as well as lower class and ethnic minority men), including the making of masculine symbolic representations of 'technology' and defining the emerging field of engineering as belonging to white middle-class men. Thus, technoscience as 'a world without women' is described as a culture of science and technology where women and femininity appear as matter out of place. Seemingly, this legitimises sexism and harassment of women (Chang 2018).

What is produced and reproduced, according to this first account of exclusion, is an outspoken gender-based division of labour in relation to science and technology, combined with a gendered construction and appreciation of skills that renders women as less competent and less relevant than men. Gender-related divisions and differences have also been observed with respect to use, from children's toys to men's and women's different relationship to artefacts and activities in everyday life (Horowitz 2001; Lie and Sørensen 1996; Kleif and Faulkner 2003). Some scholars have pointed to (some) men's fascination with technology as an outcome of erotic relationships imbued with pleasures of exercising power (Hacker 1989, 1990); as a flight from the challenges of social relations especially with women (Turkle 1984); and as a way of reproducing a kind of brotherhood around technology (Mellström 1995).

The second account of exclusion that describes the ICT culture that women experience as chilly, came from widespread observations of women choosing not to study computer science and engineering as well as the phenomenon metaphorically described as 'the leaking pipeline'. Such metaphors are widely used to explain the lack of women in science and engineering (Blickenstaff 2005; Husu 2001; Moratti 2020). The underlying idea of 'the leaking pipeline' is to consider educational or career paths as pipelines. The argument is that when the share of women is reduced, for example from the PhD to the full professor stage, this is due to women opting or being coerced out of the academic career path; thus, the career pipeline is leaking. Overall, proportionately more women than men were seen to have left specialist ICT work throughout their career trajectory (Millar and Jagger 2001). Accordingly, the chilliness account was invoked in two ways. It was employed to capture how women often experience ICT education and workplaces as unwelcoming and to describe the problem of recruitment: why many women appear reluctant to become involved with ICT. In this latter context, the chilly ICT culture account emphasised the symbolic meaning of this technology as a domain for men as well as the alienating aspect of the privileging of hacker or geek practices that many women experienced.

The making of computer science and engineering as a domain for men was a result of historical circumstances. Early on, computers seemed to be interpreted as a fairly gender-neutral technology (Sørensen and Berg 1987), possibly due to the important role of women as programmers in the pioneering period of computers (Ensmenger 2010; Misa 2010). During the 1990s, however, the initial promise of gender neutrality was seen to be corrupted because, increasingly, the use of computers appeared to be dominated by young men, while young women felt alienated (Rasmussen and Håpnes 2003; Misa 2010; Abbate 2012). A centrepiece of this argument was the figure of the hacker/nerd/geek, first identified by Joseph Weizenbaum (1976). They were predominantly young men pre-occupied with computers, seen to be the embodiment of the chilling or excluding aspects of the culture of computing (Gansmo et al. 2003a; 2003b). Sherry Turkle (1984) in her study of hackers at MIT added a gender dimension, describing the ambivalent attitudes towards them. The young men were admired by other students for their enthusiasm and skills but pitied because of their lack of social competence and their social seclusion. Turkle (1988) also found evidence of capable women opting out of computing because they saw the asocial aspect of the hackers as rendering computing, in Faulkner's terminology (2009), a "gender inauthentic" option for them. Even women computer enthusiasts appeared to diminish their own abilities compared to men co-enthusiasts (Nordli 2003).

In the feminist technoscience literature, we also find a counter storyline that Sørensen et al. (2011) name "the woman communicator". It represents a more optimistic focus on what might make computers attractive to women, emphasising the coming of the Internet as a positive turn for women's relationship with ICT. This belief – also called cyberfeminism (Bell and Kennedy 2000; Hawthorne and Klein 1999) – maintained that the Internet signified increased emphasis on communication compared to programming and calculation. For example, Sadie Plant (1997) argues that women, computers, virtual reality, and cyberspace are linked together in dispersed and distributed connections with an inherently feminine character. More commonly, scholars highlighted how ICT came to represent a transgression of the idea that computers are made mainly for calculation and management of information; this area of technology is not the least interwoven with communication. The latter quality was assumed to be beneficial for women's careers. They were expected to profit from their comparatively better communication skills (Rasmussen and Håpnes 2003; Spender 1995). More generally, cyberfeminism celebrated digital technologies, above all the Internet, as potentially liberating for women and contesting men's dominance with respect to computers (Everett 2004).

However, the optimism was challenged. Van Zoonen (2002) claims that the Internet has been shaped by men, and that even women's everyday use of the Internet frequently was embedded in constructions made by men-dominated communities of designers. A crucial weakness was the tendency to reproduce an essentialist belief in 'woman, the great communicator' (Sørensen et al. 2011). Thus, such an argument based on a discourse on 'cyberfeminism' (Plant 1997) invoked a traditional dichotomous thinking about gender but based on a reversal of the usual status hierarchy; women knowing communication were more important than men knowing programming.

We may still observe all the three accounts present in technofeminism in action, also with respect to smart technologies. However, the observations presented in the introduction suggest a diagnosis of a continued predominance of the chilly culture, not the least with respect to the experiences of women working in the ICT industry. Regarding women as users of ICT, a lot of previous research was concerned about women being on the wrong side of the digital divide (Sørensen et al. 2011). The digital divide – the split between those who could or could not use ICT – was mainly considered as a matter of access to computers and the Internet and of having the skills needed. Policy initiatives aimed to fix the situation by providing training and affordable online connections. Still, such problems remain because, as Pippa Norris (2001: 91–92) puts it: "(T)he heart of the problem lies in broader patterns of social stratification that shape not just access to the virtual world, but also full participation in other common forms of information and communication technologies".

As indicated in the introduction, we also see an emerging narrative of what Caroline Criado Perez (2019) calls "invisible women". This account of mismeasures and neglect of women's characteristics and needs in the design of smart technologies adds to the previous ones by emphasising the need to be concerned with the shape and the performance of these technologies. Cynthia Cockburn (1983) made an early observation of the phenomenon in her study of the design of typesetting machines; they were constructed to keep women skilled in typewriting out of the printing industry. However, for a long time, the focus was on the symbolic rather than the practical aspects of computers and the Internet, emphasising the perception that these were masculine technologies. What Perez and others show is not so much explicit efforts to keep women from using smart technologies as a long-standing tradition of not actively considering women as users and investigating their needs and preferences, but just using (white) men as 'the measure of all things' and not pursuing diversity as input to design efforts.

These accounts are useful as tools to diagnose problems. What more can feminist technoscience offer to explore the issues related to gender and smart technologies?

Beyond exclusion and neglect: the cyborg, co-production, and assemblage optics as feminist technoscience analytics

The idea that gender and technology interact and reciprocally interfere is basic to feminist technoscience. In principle, this framework embraces and combines non-essentialist and non-binary understandings of gender with constructivist understandings of technology, to provide an inherently dynamic perspective on both objects of inquiry (Cockburn and Ormrod 1993; Cockburn and Fürst-Dilic 1994; Faulkner 2001; Wajcman 2004). This idea may be developed in different ways. Here, we draw on Sheila Jasanoff's (2004) proposed idiom of co-production, noticing that when somebody does gender s/he also does technology. Thus, we need to study actors. At the heart of invoking the idiom to study gender technology issues is the understanding that stability with respect to performing gender or smart technologies should be analysed as achievements and not be taken as a given. If, for example, the gendering of smart technologies appears to be stable in a given context, it is because it is made stable. Jasanoff's framework helps us to observe how actors help to produce stability through (1) the making and re-making of identities of people designing or using technology (as for example competent and authentic), (2) the making of institutions, for example R&D units, that reproduce ways of designing technologies, (3) the making of discourses that upholds certain views on gender and technology (in for example consumer tests or textbooks), and (4) the making of representations, such as use metrics, that uphold stable practices because metrics may make practices seem 'natural'. A key advantage of adopting a co-production framework is that, when we investigate the production of stability and repetition with respect to gender and smart technologies, we may at the same time observe sources of instability and change.

The latter point is rooted in the cyborg metaphor, introduced by Donna Haraway (1991). She argues that the cyborg figure represents an implosion of human and machine, which seems a pertinent perspective given the invasive character of smart technologies into bodies and human life. Haraway is particularly interested in disruptions and ambiguities in the relation between gender and technoscience. She advocates the exploration of the relationships between men, women, and technology and the need to pay attention to the complexities and contradictions of these relationships. A significant point is that cyborgs are infused with what Haraway calls trickster qualities, a mix of seductive features and surprising outcomes. This seems a highly relevant perspective on smart technologies, given how they are entrenched in promissory discourses about progress.

Some have also advocated that a version of actor-network theory (ANT) is useful to study gender and technology (Lagesen 2012; Singleton 1995). These efforts share some features with the cyborg approach, not the least the focus on the interactions between human and non-human elements of an object of study, but ANT emphasises the process of assembling the elements that constitute the object. A main tenet of ANT is that society is an achievement of actors (human and non-human) engaged in producing a variety of associations among human and non-human elements. Researchers should trace these associations by following the actors engaged in making them (Latour 2005). Hence, we should see the hybrids of gender and smart technologies discussed in this chapter as outcomes of processes of reassembling associations among human and non-human elements. From this perspective, when a woman engages with a smartphone or a smartwatch, she might end up doing gender as well organising her everyday life in a different way than before, with potentially variable outcomes.

The *Economist* editorial we quoted in the introduction should serve as a reminder that the making of smart technologies means business, that the endgame is profit. This is of course an important aspect of the processes of assembling for example a smartphone. The many instances listed by Perez (2019) suggest that often women are not included in the assembling efforts; they are not part of the equation. This does not mean that women never are in focus. For example, as Cassidy (2001) shows, during the 1990s, the US computer industry tried to advertise the personal computer with a kind of feminine identity. These advertisements promoted the idea that personal computers were an important work tool as well as an instrument of developing family life. One of the most highly profiled initiatives to include women as users of ICT – in this case girls and computer games – was the development of the 'Barbie Fashion Designer', a video game that has been described as "a beachhead in the dynamic dialogue between girls and computers". This interpretation offers a glimpse into what Sherry Turkle (2011) calls "the mirror of the machine", the ability of computers to highlight the possibilities in using ICT to foreground different and possibly evolving images of femininity to be experienced, also with regard to computer games (Kafai et al. 2008).

The Barbie fashion designer game has also been heralded as an example of what Cassell and Jenkins (1998: 14f) calls entrepreneurial feminism, an assembly strategy that includes feminist ideas in the targeting of new markets consisting of women. Hendrik Spilker and Knut H. Sørensen (2000, 2002) analysed two other examples of entrepreneurial feminism. The first was the design and marketing of a CD-ROM meant to help young women discover the world of personal computers and the Internet. They labelled this kind of inclusion effort women-in-particular because the design sought to attract young women specifically by guiding the assembly efforts to include elements thought to be particularly interesting to this group. The second example was found in one media consortium's effort to make their web page attractive to all 'normal' users, women as well as men. This women-and-everybody-else strategy was based on efforts to make women part of the hybrid of web technology and active users of the Internet. The web design was guided by the idea that there existed a standardised, unisex mode for utilising the web page. Making the web page accordingly was assumed to make it particularly attractive to women.

In the following, we return to the topical issues identified in the introduction. First, we discuss the argument of women's invisibility in the making of smart technologies and some of the consequences of this. Then we move 'upstream' to review some of the more recent literature on women in computer science and engineering, reflecting on what the continued underrepresentation of women convey about the sociotechnical imaginaries related to smart technologies. In the conclusion, we discuss the possible co-production of using men as the measure of most (all?) smart technologies and the lack of women engaged in the design of these technologies.

In the image of men

The argument of women's invisibility implies a particular gendering of smart technologies, where women's needs and tastes are not considered. By gendering we mean a process of assembling human and non-human elements where gender biases influence the process, intentionally or not. Often, biases are unconscious reflections, but they may also result from ideological conviction or explicit endorsement of gender stereotypes. Importantly, the gendering of objects may not be stable. For example, in 2004 Epson launched a printer for women, designed by an all-women team. The Epson E-100 printer was shaped like a beauty-box and came out of an 'All Women, For Women' programme through an effort to make a printer that was 'easy for women to use'. The printer was designed and marketed in Japan. Interestingly, the same printer was marketed in the United States. The only noticeable difference was the colour, which was changed to grey metallic. In the US, this printer was advertised as a printer for both men and women (Sørensen and Lagesen 2005).

The example of the printer for women illustrates the difficulty with translating gender unambiguously into physical form. As Sørensen (1992) points out, it is generally hard to explicitly implement masculine (or feminine) values in the design of technologies. The problem partly stems from the non-linearity of design processes, but also in assigning gender to physical form. When engineering students in Sørensen and Berg's (1987) study aligned big, noisy, and dirty technologies to men and small, silent, and clean technologies to women, this reflected stereotypical perceptions of men and women's work. The students' gendering of technologies was based on gendered interpretations of different kinds of work.

Still, technologies may be seen as gendered in a more literal sense, with reference to physiological differences. Perez (2019: 159f) points to the size of smartphones as an example of the gendering of smart technologies, arguing that the relatively large screens of the most advanced mobiles on the average fit men's hands better than those of women. Voice recognition is another example. Such features play an increasing role in the design of smart technologies. However, such systems tend to recognise men's voices better than women's, even if women have higher speech intelligibility. Yet another of Perez's examples is the implicit gender bias in the algorithms of many artificial intelligence systems, such as those used to assess CVs of job applicants in some companies. These biases may not be intended but just a result of unreflected application of gender stereotypes.

There is a deep irony in that a society where it is commonly believed that the differences between men and women have a biological origin, technoscientists do biomedical and psychological research and design technologies as if there were no such dissimilarities. The idea seems to be that it is sufficient to study men or depart from men's practices, even if the outcomes are supposed to serve everybody. This contrasts even design strategies where the aim is to serve 'women and everybody else' (Spilker and Sørensen 2002). As we read Perez, gender bias with respect to smart technologies refers to at least three different practices:

- Unreflexively adapting technologies to men's anatomy, such as the size of hands or the sound of voices
- Generalising from studies of men
- Neglecting behavioural differences

An example of the third kind of bias is the design of health apps for mobiles based on the assumption that the phone always is carried on the body. While men tend to have their mobiles in their pockets, which produces somewhat reliable data of movements, women tend to have their mobiles in their handbags. Arguably, smartphones contribute to the cyborg features of today's people, but from a gender equality perspective, their trickster features are evident.

To a considerable extent, such gender biases are the result of sloppy and unreflexive design where user involvement is, at best, weak and limited. This is aptly illustrated by the development of so-called smart homes. Smart homes are heterogeneous assemblages of a variety of smart technologies, buildings, appliances, work, consumption, entertainment, emotions, aesthetics, etc. (Maalsen 2020; Wilson et al. 2015). The phenomenon also referred to as home automation dates back to at least the 1960s, promising protection, productivity, and pleasure (Strengers et al. 2019). When Anne-Jorunn Berg (1994) studied efforts of constructing smart homes in the US, she found that the underlying ideas reflected the interests and fascinations of the men involved in the design. Housework, however, received little attention and was something of which the designers knew little. The gender bias was evident but not in a binary fashion. The smart home was not designed to cater to the interests of all men but to men with an outspoken fascination for technology, probably also with considerable technical competence.

Yolande Strengers (2013, 2014) uses the label 'the Resource Man' to signify such men as those implicitly or explicitly targeted by the smart home designers. She developed this concept from studies of one aspect of smart homes – energy consumption. The Resource Man is a person, usually a man, who "embodies a unified vision for the smart energy consumer. . . . In his ultimate state, Resource Man is interested in his own energy data, understands it, and wants to use it to change the way he uses energy" (Strengers 2014: 26). The Resource Man is an assemblage of gender, competence, interest, and probably also income, since many smart home technologies are quite expensive. He may also be considered a digital housekeeper (Kennedy et al. 2015), the person who manages the digital technologies of a household, such as broadband connections, routers, computers, and digital media. This is a concept that highlights the gendered distribution of expertise and engagement with smart technologies.

From a feminist perspective, it is interesting to note a trickster quality of smart home technologies that affect men. The enactment of the role of the Resource Man or the digital housekeeper is time consuming. Thus, as Strengers and Nicholls (2018) note, it means 'more work for father', playing with Ruth Schwartz Cowan's (1983) classic book title *More work for mother*. It aptly summarises Cowan's findings from her analysis of the consequences of the development of household technologies for white, middle-class women in the US. Thus, the gendering of smart home technologies may be experienced as ambiguous. 'More work for father' probably also means that the Resource Man exercises greater influence regarding the acquisition and the domestication of these technologies but otherwise that smart homes mainly retain the gendered division of labour well known from 'normal' households.

Robots developed for domestic use, such as vacuum cleaners, lawnmowers, and window cleaners, may also involve 'more work for father'. However, Fortunati (2018) argues that the domestic sphere now is replacing the workplace as the most important area for innovation in robotics, which should lead to the design of robots that can take part in other kinds of housework and care. How this will affect the gendered division of labour in households is still an open issue. It is reasonable to believe that housework and care work will be transformed, but it is too early to speculate if this will lead to less work for both men and women or if Cowan's history will be repeated, that more tasks with higher demands on quality will be performed in the household of the future.

Even if some developers claim that their robots are gender-neutral, robots are easily gendered (Søraa 2017). Not the least, the shape and the sound of the voice of a robot tend to be interpreted as either masculine or feminine. This perception may affect the interaction between humans and robots, but the gendering of such interactions appears to be shifting and complex (Nomura 2017). Moreover, gendering of robots may happen because

> Much of what roboticists take for granted in their own gendered socialization and quotidian lives is reproduced and reified in the robots they design (. . .). How robot-makers gender their humanoids is a tangible manifestation of their tacit understanding of femininity in relation to masculinity, and vice versa.
>
> *(Robertson 2010: 4)*

However, the gendering of robots may also be quite explicit, for example in the design of sex robots (Strengers and Kennedy 2020).

Thus, there is a considerable risk that robots will be gendered because their design departs from established practices and entrenched gender bias. A similar phenomenon has been observed with respect to artificial intelligence and machine learning. To train machines, huge amounts of data are accessed from existing datasets, and these datasets tend to reflect existing, gendered practices (Zou and Schiebinger 2018; Schiebinger and Ogawa 2018). To avoid such tacit gendering will require concerted action.

This argument is augmented by Strengers and Kennedy (2020). They argue that main parts of the present development in robotics and AI may be interpreted as the making of what they call 'smart wives' – digital assistants that are feminised to appear as friendly and sometimes flirty, docile, and efficient. The aim of the development, Strengers and Kennedy claim, is to design smart technologies by digitally reproducing the stereotypical housewife of the 1950s: a white, middle-class, and heteronormative housekeeper with high standards for cleanliness and personal services. In this way, there is considerable risk that developments in robotics and AI contribute to uphold outdated stereotypes, even reversing progress with respect to gender, sexual, and ethnic equality. Consequently, Strengers and Kennedy ask for a reboot of 'the smart wife'.

These critical observations are also relevant with respect to 'smart cities', which represent another sociotechnical imaginary that is based on smart technologies, above all the so-called Internet of Things (Talari et al. 2017; Silva et al. 2018). The European Union presents its version of the imaginary:

> A smart city is a place where traditional networks and services are made more efficient with the use of digital and telecommunication technologies for the benefit of its inhabitants and business. A smart city goes beyond the use of information and communication technologies (ICT) for better resource use and less emissions. It means smarter urban transport networks, upgraded water supply and waste disposal facilities and more efficient ways to light and heat buildings. It also means a more interactive and responsive city administration, safer public spaces and meeting the needs of an ageing population.[5]

The smart city imaginaries such as this one quoted tend to be without any concern for gender (or ethnicity for that matter); the underlying idea is progress for everybody. Such assemblages of technologies are implicitly assumed to be able to cater flexibly to all sorts of needs, without any outspoken consideration of priorities, differences, and conflicts among the public. Thus, arguably, smart cities are gendered through the neglect of gender issues. An example is mobility, where needs and travel patterns definitively are gendered (Uteng 2019). The neglect may be due to the fact that smart city technologies largely are tools of governance, particularly through various forms of surveillance – sensors, cameras, etc. – and the use of models and algorithms to analyse and make sense of the large amount of data that is produced. The number crunching allows for indicators of pollution, traffic, energy use, water supply, and so on. Maybe the Resource Man will find these indicators useful, but who else besides city bureaucrats? On the other hand, digital surveillance has potentially harmful consequences for women and people of colour emanating from the employment of stereotypes in designs and calculations rather than neglect (Dubrofsky and Magnet 2015).

Oudshoorn et al. (2016) take these arguments further by suggesting that diversity gets lost in the design of smart technologies. It is not just gender that is neglected. Age, class, and ethnicity

5 https://ec.europa.eu/info/eu-regional-and-urban-development/topics/cities-and-urban-development/city-initiatives/smart-cities_en (accessed May 14, 2020).

also tend to be overlooked. This is a bit surprising, given an increasing concern for diversity in many societies. However, the communities engaging in the development of smart technologies may have been oblivious to such concerns, maybe because of the widespread tradition of considering *the* user in singular in design discourses (Woolgar 1990; Akrich 1995). Another issue may be the lack of diversity within these communities, such as their gender imbalance. Thus, we need to look further at why the ICT professions are so resilient with respect to improving the gender diversity of the industry.

A world of and for men

For 25 years, researchers have engaged with the issue of why there are so few women studying and working with computer science and engineering, finding many explanations as shown earlier. The state of affairs has remained the same, despite numerous attempts to make amendments. Thus, the field of computer science seems to be and to remain gendered to the effect that women continue to be a minority in the field. This situation could be interpreted as the outcome of a co-production of gender and technology (Vitores and Gil-Juárez 2016). However, some care should be exercised. Women are not a minority in computer science everywhere (Johnson et al. 2019). One such example is Malaysia, where women constitute a majority of computer science students and computer science is considered 'suitable' for women (Lagesen 2008). Clearly, the gendering of computer science depends on cultural context. Similarly, in the US, women dominated as programmers in the 1940s, because programming was interpreted as secretarial work (Ensmenger 2010).

Still, in countries such as the US, which has been in the forefront of the development of smart technology, men dominate the industry. The situation seems similar in China and in countries in Western Europe. The gender balance was considerably better until the late 1980s, as Nathan Ensmenger observes in his book (2010) *The computer boys take over*. He explains the gendering process as related to the recruitment practices of the computer industry:

> The primary selection mechanism used by the industry selected for antisocial, mathematically inclined males, and therefore antisocial, mathematically inclined males were overrepresented in the programmer population; this in turn reinforced the popular perception that programmers ought to be antisocial and mathematically inclined (and therefore male), and so on ad infinitum. Combined with the often-explicit association of programming personnel with beards, sandals, and scruffiness, it is no wonder that women felt increasingly excluded from the center of the computing community.
>
> *(Ensmenger 2010: 78–79)*

Thus, the gendering process provided for a subgroup of men with special qualities, qualities that were reflected in recruitment tests, etc. The outcome seems to follow Phelps's (1972) statistical theory of racism and sexism. The computer industry did not search for women but not for all kinds of men either. As Ensmenger shows, programming was made into something mysterious that required special skills and attitudes. In turn, this contributed to an image of the computer industry that reinforced the limited recruitment, providing for a rather homogeneous set of employees with a homogeneous culture. The result has been the development of what Chang (2018) calls a "Brotopia", a technology-focused, promissory culture of men that features aggressiveness, misogyny, and workaholism. The few women that succeed clearly struggle.[6]

6 https://fortune.com/2020/10/09/sheryl-sandberg-lean-in-rachel-thomas-women-leaving-workforce/ (accessed February 3, 2021).

Chang's information stems from Silicon Valley companies. They may not be representative of the industry more generally, but there is little doubt that companies such as Google and Facebook are at the front of developing smart technologies. This makes the "Brotopia" culture an important context of smart technologies. Moreover, the Silicon Valley companies may represent an extreme version of an exclusive, homogeneous culture of men computer specialists, but even more moderate versions seem just as dominated by men. The mechanisms of exclusion may be different but no less effective in keeping women (and many men) out. Only a subset of men with special interests and personal qualities is engaged in the making of smart technologies. The culture seems akin to the hacker cultures, described as dominated by men, androcentric, stuck in discourses of meritocracy that mask inequity and lack of diversity and allowing for sexual harassment and exclusion of women (Steinmetz et al. 2020).

It is worrisome that the culture underpinning the development of smart technologies has proved to be so resilient to improve the diversity of the workforce. Numerous attempts have been made to recruit women students to computer science and engineering. Some of these have been successful (Frieze and Quesenberry 2019; Lagesen 2007; Margolis and Fisher 2002) but less so in the long run. This contrasts remarkably to the substantial changes in the gender balance that have taken place in formerly men-dominated professions such as medicine and law. As we observed previously, explanations of the lack of women in computer science and engineering abound. However, we know much less about the mechanisms that also seem to keep many men out of the making of smart technologies, even though Ensmenger's research provides useful indications. His observations of the industry's emphasis on antisocial and mathematical inclination when they recruit, is also a sign of warning regarding the gendering of smart technologies. The resulting culture seems a problematic point of departure for design of inclusive smart technologies as well as technologies that address a wider range of needs than those of the small subgroup of men employed in the industry.

Thus, we suspect that smart technologies may be gendered, often intentionally, to serve a subset of men such as 'the Resource Man' (Strengers 2013, 2014), although 'the smart wife' may attract wider popularity (Strengers and Kennedy 2020). We know that women use a lot of the technologies we call smart and that women encounter many such technologies that supervise or control them as well as men, such as we find in 'smart cities'. Despite the striking examples highlighted in Perez (2019), the most important problem is not the exclusion of women but the feeble efforts to include them (Sørensen et al. 2011).

Conclusion: the gendering of smart technologies through the co-production of the lack of women and the lack of interest in women

We argue that the dual gendering of designers of smart technologies as well as the content/shape of these technologies remain stable and resilient to change. Sheila Jasanoff's idiom of co-production offers some important clues about why this dual gendering seems so entrenched, also in the case of a rather dynamic development of technology. As previously mentioned, she invites us to consider the following four ordering instruments: making identities, making institutions, making discourses, and making representations. These ordering instruments have helped to stabilise the gender and computer technologies as we have demonstrated. For example, the identity of ICT professionals has been gendered in a quite stable way. The predominant elements include being a man with little social interest, a competitive orientation, and good mathematical skills. According to Ensmenger (2010), this identity is reproduced through the recruitment policy of the ICT industry and it provides for boundary work to keep the industry – at least the technology professionals – employing only an exclusive group of people (Lagesen and Sørensen 2009). Thus, the ICT industry works as an institution that is a stable repository of knowledge and power, despite the discourse of disruptive innovations that has been prevalent for quite some time. Moreover, there is no disruption in the gendered discourse of expertise, which place women on the outside of the world of designing smart technologies. This discourse has mobilised and continues to mobilise a representation of the profession of computer science and engineering as gender inauthentic to women (Faulkner 2009). In addition, the ICT industry is – as we have shown – experienced by women as a chilly and sometimes also as a hostile environment that is not particularly attractive as a place of work.

In turn, we have identified a co-production of the exclusion of women from the design of smart technologies and the lack of concern for gender diversity when engaging with such design. The invisibility of women in this context (Perez 2019) is largely due to the unreflected masculine identity of the design experts, the institutionalised tradition of the ICT industry for preferring men employees, the gender blind promissory discourses of this industry, and the long-term tradition for using (some) men as representative of humanity. This does not mean that in general, smart technologies do not fit women and their needs, but rather that such fit is accidental. The unreflected application of the so-called I-methodology of design means that too many artefacts and systems primarily are made to be attractive and useful to the Resource Man and the digital housekeeper. The I-methodology refers to a design practice in which designers consider themselves as representative of the users (Oudshoorn et al. 2004). Akrich (1995: 173) describes it as the "reliance on personal experience, whereby the designer replaces his professional hat by that of the layman".

How may more women be included in the design of smart technologies, as experts as well as potential users? Clearly, this is not achieved through quick fixes. The ICT industry does little to attract women and to make them feel welcome once they are there. Rather, the frequent display of 'Brotopia' cultures alienates many women and makes women into outsiders – thus excluding women from the making of smart technologies. The many initiatives to recruit more women as students of computer science and engineering provide evidence of this pessimistic conclusion, since it appears that only very long-term, long-lasting inclusion efforts have any sizeable effect (Lagesen 2011). Still, they may not lead to an improved gender balance in ICT companies (Simonsen and Corneliussen 2019). The industry claims to initiate actions to recruit more women, but the results are disappointing. Londa Schiebinger directs the EU/US Gendered Innovations in Science, Health & Medicine, Engineering, and Environment Project, which aims to improve gender equality in innovations.[7] This represents an effort to produce an alternative discourse about gender, science, and technology, which also asks the designers of smart technologies to actively change their tacit use of gender bias (Zou and Schiebinger 2018; Schiebinger and Ogawa 2018). The impact of this project will at best be long term.

The concern for gender equity is also integrated in the policy for Responsible Research and Innovation (RRI), which the European Union has adopted together with several other countries.[8] Considerable efforts have been put into the development of tools to do RRI, which includes quite a few initiatives to address gender issues.[9] Bührer and Wroblewski (2019) show, using a survey conducted among European researchers, that two items dominate in the gender

7 http://genderedinnovations.stanford.edu/ (accessed May 14, 2020)

8 https://ec.europa.eu/programmes/horizon2020/en/h2020-section/responsible-research-innovation (accessed May 14, 2020)

9 See, e.g., Fit4RRI (https://fit4rri.eu/), RRI Tools (https://www.rri-tools.eu/), and FOSTER (https://www.fosteropenscience.eu/). (Accessed May 14, 2020).

equality activities reported: encouragement of gender-balanced teams and provision of specific support for women within teams. However, few respondents said that they explicitly dealt with gender issues in their research.

The editorial in the *Economist* that we quoted in the introduction identifies a lack of concern for women's needs and preferences as a missed business opportunity. Maybe this will be more widely recognised by the industry and initiate the comprehensive reforms that are called for to ensure a healthier and more including tech industry. Governments could also be mobilised, but as Palmén et al. (2019: 162) argue, based on a study of seven European countries, the "countries surveyed are currently reluctant to impose bureaucratic gender equality obligations on the business sector". Thus, unfortunately, we may have to expect that the problem of harmful gendering of smart technologies will remain with us for quite some time unless both industry and government seize to take serious action. The temptation to use femininity just as a superfluous design and advertising move may be too great (Sørensen and Lagesen 2005).

Acknowledgements

We have received no funding for this research. We are grateful to the editors and an anonymous reviewer for very useful comments.

References

Abbate, J. (2012): *Recoding Gender: Women's Changing Participation in Computing*. Cambridge, MA: The MIT Press.

Ahuja, M. K. (2002): Women in the Information Technology Profession: A Literature Review, Synthesis and Research Agenda, *European Journal of Information Systems* 11, 20–34.

Akrich, M. (1995): User Representations: Practices, Methods and Sociology, in Rip et al. (eds.): *Managing Technology in Society: The Approach of Constructive Technology Assessment*. London: Pinter Publishers, 167–184.

Archibald, J., Emms, J., Grundy, F., Pane, J. and Turner, E. (eds.) (2005): *The Gender Politics of ICT*. London: Middlesex University Press.

Barker, J. L. and Aspray, W. (2006): The State of Research on Girls and IT, in Cohoon, J. M. and Aspray, W. (eds.): *Women and Information Technology: Research on Underrepresentation*. Cambridge, MA: The MIT Press, 3–54.

Bell, D. and Kennedy, B. M. (eds.) (2000): *The Cybercultures Reader*. London: Routledge.

Berg, A.-J. (1994): A Gendered Socio-technical Construction: The Smart House, in Cockburn, C. and Fürst-Dilić, R. (eds.): *Bringing Technology Home: Gender and Technology in a Changing Europe*. Milton Keynes: Open University Press, 165–180.

Blickenstaff, C. J. (2005): Women and Science Careers: Leaky Pipeline or Gender Filter? *Gender and Education* 174, 369–386.

Bührer, S. and Wroblewski, A. (2019): The Practice and Perceptions of RRI – A Gender Perspective, *Evaluation and Program Planning* 77, 101717

Cassell, J. and Jenkins, H. (1998): Chess for Girls? Feminism and Computer Games, in Cassel, J. and Jenkins, H. (eds.): *From Barbie to Mortal Kombat, Gender and Computer Games*. Cambridge, MA: MIT Press, 2–45.

Cassidy, M. F. (2001): Cyberspace Meets Domestic Space: Personal Computers, Women's Work, and the Gendered Territories of the Family Home, *Critical Studies in Media Communication* 18(1), 44–65.

Chang, E. (2018): *Brotopia. Breaking up the Boys' Club of Silicon Valley*. New York: Penguin

Cockburn, C. (1983): *Brothers; Male Dominance and Technological Change*. London: Pluto Press.

Cockburn, C. (1985): *Machinery of Dominance*. London: Pluto Press.

Cockburn, C. and Fürst-Dilic, R. (eds.) (1994): *Bringing Technology Home: Gender and Technology in a Changing Europe*. Maidenhead, Berkshire, UK: Open University Press.

Cockburn, C. and Ormrod, S. (1993): *Gender & Technology in the Making*. Thousand Oaks, CA: Sage.

Cohoon, J. M. and Aspray, W. (2006): A Critical Review of the Research on Women's Participation in Postsecondary Computing Education, in Cohoon, J. M. and Aspray, W. (eds.): *Women and Information Technology: Research on Underrepresentation*. Cambridge, MA: The MIT Press, 137–180.

Cowan, R. S. (1983): *More Work for Mother*. New York: Basic Books.

Cronin, C. and Roger, A. (1999): Theorizing Progress: Women in Science, Engineering, and Technology in Higher Education, *Journal of Research in Science Teaching* 36(6), 636–661

Dubrofsky, R. E. and Magnet, S. A. (eds.) (2015): *Feminist Surveillance Studies*. Durham, NC: Duke University Press.

Ensmenger, N. (2010): *The Computer Boys Take Over: Computers, Programmers, and the Politics of Technical Expertise*. Cambridge, MA and London: The MIT Press.

Everett, A. (2004): On Cyberfeminism and Cyberwomanism: High-tech Mediations of Feminism's Discontents, *Signs: Journal of Women in Culture and Society* 30(1), 1278–1286.

Faulkner, W. (2001): The Technology Question in Feminism: A View from Feminist Technology Studies, *Women's Studies International Forum* 24(1), 79–95.

Faulkner, W. (2009): Doing Gender in Engineering Workplace Cultures: Part II – The in/visibility Paradox, *Engineering Studies* 1(3), 169–189.

Fortunati, L. (2018): Robotization and the Domestic Sphere, *New Media & Society* 20(8), 2673–2690

Fox, M. F., Sonnert, G. and Nikiforova, I. (2009): Successful Programs for Undergraduate Women in Science and Engineering: Adapting versus Adopting the Institutional Environment, *Research in Higher Education* 50, 333–353.

Frieze, C. and Quesenberry, J. L. (eds.) (2019): *Cracking the Digital Ceiling: Women in Computing Around the World*. Cambridge, UK: Cambridge University Press.

Gansmo, H. J., Lagesen, V. A. and Sørensen, K. H. (2003a): Forget the Hacker? A Critical Re-appraisal of Norwegian Studies of Gender and ICT, in Lie, M. (ed.): *He, She and IT Revisited: New Perspectives on Gender in the Information Society*. Oslo: Gyldendal Akademisk, 34–68.

Gansmo, H. J., V. A. Lagesen and Sørensen, K. H. (2003b): Out of the Boy's Room? A Critical Analysis of the Understanding of Gender and ICT in Norway, *Nora* 11(3), 130–139.

Hacker, S. (1989): *Pleasure, Power, and Technology: Some Tales of Gender, Engineering and the Cooperative Workplace*. Boston, MA: Unwin Hyman.

Hacker, S. (1990): *Doing It the Hard Way: Investigations of Gender and Technology*. Edited by Smith, D. E. and Turner, S. M. Boston. MA: Unwin Hyman.

Haraway, D. (1991): A Cyborg Manifesto: Science, Technology and Socialist-feminism in the Late Twentieth Century, in Haraway, D. (ed.): *Simians, Cyborgs and Woman: The Reinvention of Nature*. London: Routledge, 149–181.

Hawthorne, S. and Klein, R. (eds.) (1999): *Cyberfeminism*. North Melbourne: Spiniflex.

Horowitz, R. (ed.) (2001): *Boys and Their Toys: Masculinity, Class and Technology in America*. New York: Routledge.

Husu, L. (2001): On Metaphors on the Position of Women in Academia and Science, *NORA: Nordic Journal of Women's Studies* 93: 172–181.

Jasanoff, S. (2004): Ordering Knowledge, Ordering Society, in Jasanoff, S. (ed.): *States of Knowledge: The Co-Production of Science and Social Order*. London: Routledge, 13–45.

Johnson, N., Garcia, J. and Seppi, K. (2019): Women in CS: Changing the Women or Changing the World? *2019 IEEE Frontiers in Education Conference (FIE)* IEEE, 1–8.

Kafai, Y. B., Heeter, C., Denner, J. and Sun, J. Y. (eds.) (2008): *Beyond Barbie and Mortal Kombat. New Perspectives on Gender and Gaming*. Cambridge, MA: The MIT Press.

Kennedy, J., Nansen, B., Arnold, M., Wilken, R. and Gibbs, M. (2015): Digital Housekeepers and Domestic Expertise in the Networked Home, *Convergence* 21(4), 408–422.

Kleif, T. and Faulkner, W. (2003): "I'm No Athlete [but] I Can Make This Thing Dance!" – Men's Pleasures in Technology, *Science, Technology, & Human Values* 28(2), 296–325.

Lagesen, V. A. (2007): The Strength of Numbers: Strategies to Include Women into Computer Science, *Social Studies of Science* 37(1), 67–92.

Lagesen, V. A. (2008): A Cyberfeminist Utopia? Perceptions of Gender and Computer Science among Malaysian Computer Science Students, *Science, Technology, & Human Values* 33(1), 5–27.

Lagesen, V. A. (2011): Getting Women into Computer Science, in Sørensen, K. H., Faulkner, W. and Rommes, E. (eds.): *Technologies of Inclusion. Gender in the Information Society*. Trondheim: Tapir Academic Press, 147–169.

Lagesen, V. A. (2012): Reassembling Gender: Actor-network Theory (ANT) and the Making of the Technology in Gender, *Social Studies of Science* 42(3), 442–448.

Lagesen, V. A. and Sørensen, K. H. (2009): Walking the Line? The Enactment of the Social/technical Binary in Software Engineering, *Engineering Studies* 1(2), 129–149.

Latour, B. (2005): *Reassembling the Social: An Introduction to Actor-Network-Theory*. Oxford, UK: Oxford University Press.

Lie, M. and Sørensen, K. H. (eds.) (1996): *Making Technology Our Own? Domesticating Technology into Everyday Life*. Oslo: Scandinavian University Press.

Maalsen, S. (2020): Revising the Smart Home as Assemblage, *Housing Studies* 35(9), 1534–1549.

Margolis, J. and Fisher, A. (2002): *Unlocking the Clubhouse: Women in Computing*. Cambridge, MA: The MIT press.

Mellström, U. (1995): *Engineering Lives: Technology, Time and Space in a Male-Centred World*. Linköping, Sweden: Linköping studies in arts and sciences.

Mellström, U. (2003): *Masculinity, Power and Technology: A Malaysian Ethnography*. Aldershot, Hampshire, UK: Ashgate.

Merchant, C. (1980): *The Death of Nature: Women, Ecology and the Scientific Revolution*. San Francisco, CA: Harper & Row.

Millar, J. and Jagger, N. (2001): *Women in ITEC Courses and Careers*. London: Department of Trade and Industry.

Misa, T. J. (ed.) (2010): *Gender Codes. Why Women are Leaving Computing*. Hoboken, NJ: Wiley

Moratti, S. (2020): What's in a Word? On the Use of Metaphors to Describe the Careers of Women Academics, *Gender and Education* 327, 862–872.

Mundy, L. (2017): Why is Silicon Valley so awful to Women? *The Atlantic* 319, 60–73.

Noble, D. F. (1992): *A World Without Women: The Christian Clerical Culture of Western Science*. New York: Alfred A. Knopf.

Nomura, T. (2017): Robots and Gender, *Gender and the Genome* 1(1), 18–25.

Nordli, H. (2003): The Net is Not Enough. Searching for the Female Hacker, Ph.D. thesis, STS report 61/2003. Trondheim, Norway: Centre for Technology and Society, NTNU.

Norris, P. (2001): *Digital Divide. Civic Engagement, Information Poverty, and the Internet Worldwide*. West Nyack, NY: Cambridge University Press.

Oldenziel, R. (1999): *Making Technology Masculine: Men, Women and Modern Machines in America 1870–1945*. Amsterdam: Amsterdam University Press.

Oudshoorn, N., Neven, L. and Stienstra, M. (2016): How Diversity Gets Lost: Age and Gender in Design Practices of Information and Communication Technologies, *Journal of Women & Aging* 28(2), 170–185.

Oudshoorn, N., Rommes, E. and Stienstra, M. (2004): Configuring the User as Everybody: Gender and Design Cultures in Information and Communication Technologies, *Science, Technology, & Human Values* 29(1), 30–63.

Palmén, R., Schmidt, E. K., Striebing, C., Reidl, S., Bührer, S. and Groó, D. (2019): Measuring Gender in R&I – Theories, Methods, and Experience, *Interdisciplinary Science Reviews* 44(2), 154–165.

Perez, C. C. (2019): *Invisible Women. Exposing Data Bias in a World Designed for Men*. London: Chattoo & Windus.

Phelps, E. S. (1972): The Statistical Theory of Racism and Sexism, *The American Economic Review* 62(4), 659–661.

Plant, S. (1997): *Zeroes and Ones: Digital Women and the New Technoculture*. London: Doubleday and Forth Estate.

Rasmussen, B. and Håpnes, T. (2003): Gendering Technology; Young Girls Negotiating ICT and Gender, in Lie, M. (ed.): *He, She and IT Revisited; New Perspectives on Gender in the Information Society*. Oslo: Gyldendal Akademisk, 173–197.

Robertson, J. (2010): Gendering Humanoid Robots: Robo-sexism in Japan, *Body & Society* 16(2), 1–36.

Rommetveit, K, Dunajcsik, M. Tanas, A. Silvast, A. and Gunnarsdóttir, K. (2017): *The CANDID Primer: Including Social Sciences and Humanities Scholarship in the Making and Use of Smart ICT Technologies (edited by K Gunnarsdóttir)*. CANDID (H2020-ICT-35–2016) D5.4. Available at: http://candid.no/progress [25.05.2020]

Schiebinger, L. and Ogawa, M. (2018): Gendered Innovations in Medicine, Machine Learning, and Robotics, *Trends in the Sciences* 23(12), 12_8–12_19.

Silva, B. N., Khan, M. and Han, K. (2018): Towards sustainable smart cities: A review of trends, architectures, components, and open challenges in smart cities, *Sustainable Cities and Society* 38, 697–713.

Simonsen, M. and Corneliussen, H. G. (2020): Can Statistics Tell Stories about Women in ICT? Tracing men and women's participation in fields of ICT through statistics for Norway, Nordic countries and

Europe. Available at: www.vestforsk.no/sites/default/files/2020-02/Simonsen_Corneliussen_2019_Can_Statistics_Tell_Stories_about_Women_in_ICT_1.pdf [25.05.2020]

Singleton, V. (1995): Networking Constructions of Gender and Constructing Gender Networks, in Grint, K. and Gill, R. (eds.): *The Gender/Technology Relation*. London: Taylor & Francis, 146–173.

Søraa, R. A. (2017): Mechanical Genders: How Do Humans Gender Robots? *Gender, Technology and Development* 21(1–2), 99–115.

Sørensen, K. H. (1992): Towards a Feminized Technology? Gendered Values in the Construction of Technology, *Social Studies of Science* 22(1), 5–31.

Sørensen, K. H. and Berg, A.-J. (1987): Genderization of Technology among Norwegian Engineering Students, *Acta Sociologica* 2, 151–171.

Sørensen, K. H., Faulkner, W. and Rommes, E. (2011): *Technologies of Inclusion. Gender in the Information Society*. Trondheim: Tapir Academic Press

Sørensen, K. H. and Lagesen, V. A. (2005): Cyberfeminism Revisited: Is ICT Either Feminine or Masculine? *STS Working Paper* 1/2005. Trondheim: NTNU. http://dx.doi.org/10.13140/RG.2.2.28487.83367

Spender, D. (1995): *Nattering on the Net: Women, Power and Cyberspace*. North Melbourne, Victoria, Australia: Spinifex Press.

Spilker, H. and Sørensen, K. H. (2000): A ROM of One's Own or a Home for Sharing? Designing Inclusion of Women in Multimedia, *New Media & Society* 2(3), 268–285.

Spilker, H. and Sørensen, K. H. (2002): Feminism for Profit? Public and Private Gender Politics in Multimedia, in Sørensen, K. H and Williams, R. (eds.): *Shaping Technology/Guiding Policy: Concepts, Spaces and Tools*. Cheltenham, UK: Edward Elgar, 243–263.

Steinmetz, K. F., Holt, T. J. and Holt, K. M. (2020): Decoding the Binary: Reconsidering the Hacker Subculture through a Gendered Lens, *Deviant Behavior* 41:8, 936–948.

Strengers, Y. (2013): *Smart Energy Technologies in Everyday Life: Smart Utopia?* Houndmills, Basingstoke, UK: Palgrave Maxmillan.

Strengers, Y. (2014): Smart Energy in Everyday Life: Are You Designing for Resource Man? *Interactions* 21(4), 24–31.

Strengers, Y. and Kennedy, J. (2020): *The Smart Wife: Why Siri, Alexa, and Other Smart Home Devices Need a Feminist Reboot*. Cambridge, MA: The MIT Press.

Strengers, Y., Kennedy, J., Arcari, P., Nicholls, L. and Gregg, M. (2019): Protection, Productivity and Pleasure in the Smart Home: Emerging Expectations and Gendered Insights from Australian Early Adopters, *Proceedings of the 2019 CHI Conference on Human Factors in Computing Systems*, 1–13.

Strengers, Y. and Nicholls, L. (2018): Aesthetic Pleasures and Gendered Tech-work in the 21st-Century Smart Home, *Media International Australia* 166(1), 70–80.

Talari, S., Shafie-Khah, M., Siano, P., Loia, V., Tommasetti, A. and Catalão, J. P. (2017): A Review of Smart Cities Based on the Internet of Things Concept, *Energie* 10(4), 421.

Turkle, S. (1984): *The Second Self: Computers and the Human Spirit*. New York: Simon and Schuster.

Turkle, S. (1988): Computational Reticence; Why Women Fear the Intimate Machine, in Kramarae, C. (ed.): *Technology and Women's Voices: Keeping in Touch*. London: Routledge and Kegan Paul, 41–61.

Turkle, S. (2011): *Alone Together. Why We Expect More from Technology and Less from Each Other*. New York: Basic Books.

Uteng, T. P. (2019): Smart Mobilities: A Gendered Perspective, *Kart og Plan* 112(04), 258–281.

van Zoonen, L. (2002): Gendering the Internet. Claims, Controversies and Cultures, *European Journal of Communication* 17(1), 5–23.

Vitores, A. and Gil-Juárez, A. (2016): The Trouble with 'Women in Computing': A Critical Examination of the Deployment of Research on the Gender Gap in Computer Science, *Journal of Gender Studies* 25(6), 666–680.

Wajcman, J. (2004): *Technofeminism*. Cambridge, UK: Polity Press.

Webster, J. (1996): *Shaping Women's Work: Gender, Information Technology and Employment*. London: Longman.

Weizenbaum, J. (1976): *Computer Power and Human Reason: From Judgment to Calculation*. San Francisco, CA: W. H. Freeman.

Wilson, C., Hargreaves, T. and Hauxwell-Baldwin, R. (2015): Smart Homes and their Users: A Systematic Analysis and Key Challenges, *Personal and Ubiquitous Computing* 19(2), 463–476.

Woolgar, S. (1990): Configuring the User: The Case of Usability Trials, *The Sociological Review* 38(1_suppl), 58–99.

Zou, J. and Schiebinger, L. (2018): AI Can Be Sexist and Racist – it's Time to Make it Fair, *Nature* 559, 324–326.

PART 3

Smart technologies and social and economic transformation

12
ARTIFICIAL INTELLIGENCE

Fredrik Heintz

Introduction

Artificial intelligence (AI) is a general-purpose technology that will impact most, if not all, aspects of both our society and our personal everyday life. AI technology has enabled applications such as speech interfaces, vision-based object recognition, and machine translation. AI technology also makes recommendations about music, books, and movies for you, decides whether you will get a bank loan, and controls what posts you see on social media, all of which can have a major impact on your life. It is clear that AI technology will play a central role in most aspects of our professional and private lives as well as society at large. Kevin Kelly predicted that "The business plans of the next 10,000 start-ups are easy to forecast: Take X and add AI" (Kelly 2016). Andrew Ng says that AI is the new electricity, it is a fundamental part of almost everything (Lynch 2017).

The field of artificial intelligence was founded in 1956 with the famous Dartmouth conference. From the start, the goal was to make computers do everything that humans can do. To quote the proposal:

> The study is to proceed on the basis of the conjecture that every aspect of learning or any other feature of intelligence can in principle be so precisely described that a machine can be made to simulate it. An attempt will be made to find how to make machines use language, form abstractions and concepts, solve kinds of problems now reserved for humans, and improve themselves. We think that a significant advance can be made in one or more of these problems if a carefully selected group of scientists work on it together for a summer.
>
> *(McCarthy et al. 1955)*

However, the quest for understanding and mechanising human intelligence started much earlier (McCorduck 2004; Nilsson 2010). For example, Aristotle studied what reasoning is and came up with three types of reasoning – deductive, abductive, and inductive – which covers much of what is being done today in both machine reasoning and machine learning.

From a pragmatic perspective, an AI system can be described as a system that receives data, makes decisions, and acts on this information. Sense-Plan-Act is another way to describe such

 DOI: 10.4324/9780429351921-16

a system. It senses the environment, it makes a plan for how to achieve its goals, and then acts based on the plan (Russell and Norvig 2016). In many cases, these systems can learn from data to improve over time. They are often called agents, as they have a sense of agency which differentiates them from other computer programs. This also gives rise to a cognitive and social view on computation (Shoham 1997).

Machine learning is currently seen as the most interesting part of AI, both because many consider it an essential part of intelligence and because it allows computer programs to improve over time based on experience. This is important since it is hard for people to specify exactly what we want a system to do. Instead, the machine can partly learn what to do and how to do it, as well as improve over time, by collecting data and modifying its behaviour (Brynjolfsson and Mitchell 2017).

An interesting example is AlphaGo which plays the Chinese board game Go better than any other human. In 2015, it beat one of the best humans through a combination of learning from existing games and self-play (Silver et al. 2016). In 2017, an improved version called AlphaGo-Zero beat its previous version in 100 games out of 100 purely through self-play (Silver et al. 2017). The same system can also play chess better than all existing chess playing programs. The same type of techniques has also been used by Google to reduce the energy consumption in their data centres by 40%.[1]

By observing the human expert commentators, and most importantly, their surprise at some of the moves, it is clear that AlphaGo does not purely replicate human strategies, rather it extends existing strategies and creates new ones. This leads to interesting questions about how to validate decisions made by an AI system and how to maintain meaningful human control over a system capable of making better decisions than we are.

The purpose of this chapter is to introduce artificial intelligence (AI) from a non-technical perspective. After reading this chapter you should have a basic understanding of what AI is and where it comes from; what the main concepts and methods are; some of the applications; and some of the future research challenges.

What is AI?

Artificial intelligence is about understanding intelligence sufficiently well to be able to recreate it in machines. Another way to describe AI is as systems taking input, analysing this data, making decisions, and then acting based on these decisions. This approach is often called the Sense-Plan-Act approach (Russell and Norvig 2016). In many cases, these systems learn to improve their performance over time either from data given to them or from data collected during its execution.

A challenge with the definition of AI is that there is no commonly agreed upon definition of human intelligence (Legg and Hutter 2007). A computer can often do things that we assume requires intelligence without any effort, like solving difficult mathematical problems. At the same time, computers are often very poor when it comes to doing what appears to be really simple things like learning a new concept from abstract descriptions, like a zebra is a horse with black-and-white-striped fur, or manipulation tasks, like folding a blanket or tying a pair of shoes. A consequence is the recognition that human intelligence is not necessarily the right baseline to compare against. There is an inherent anthropocentrism in the word artificial, assuming that intelligence can only originate in a human body.

1 https://deepmind.com/blog/article/deepmind-ai-reduces-google-data-centre-cooling-bill-40 [10.03.2021]

Therefore, the focus is often on building systems that behave intelligently, rather than claiming that the systems *are* intelligent. It is also common to focus on particular cognitive functionalities such as planning, natural language processing, or perception. Cognitive science, on the other hand, tries to understand how human cognition works and also tries to replicate the functionality through computer models, which then aim at emulating human cognition with both its strengths and weaknesses (Thagard 2008). The engineering approach to AI, on the other hand, is to develop methods, algorithms, and programs that exhibit intelligent behaviour based on computer science rather than cognitive science or neuroscience, even though nature is often an important source of inspiration. Classical examples of this are artificial neural networks and genetic algorithms. Artificial neural networks were developed in the 1950s based on a simple model of how researchers thought that human neurons worked (Hubel and Wiesel 1962; Rumelhart et al. 1986) and are today implemented as high dimensional matrix operations running on massively parallel GPUs.

AI has many subfields that study different aspects of intelligent behaviour and cognition. Common topics at the main AI conferences include machine learning, knowledge representation and reasoning, heuristic search, planning and scheduling, natural language processing, computer vision, robotics, and multi-agent systems. All of these topics have been studied since the 1950s. Most of them were in fact discussed already at the seminal Dartmouth conference in 1956.

Two of the most important subfields are machine learning and knowledge representation and reasoning. Knowledge representation and reasoning is the scientific study of how to represent knowledge in a computer and how to reason with this knowledge to draw valid conclusions. Machine learning is the scientific study of how a computer can learn things such as finding patterns, recognising objects, and acting to achieve specific goals. Machine learning is mostly based on statistics and correlations (*black box* models) while knowledge representation and reasoning are mostly based on explicitly modelling cause and effect (*white box* models). One of the major consequences is that machine learning is mostly data driven while reasoning is mostly knowledge driven, where explicit knowledge has to be elicited from domain experts. The second major consequence is related to interpretability, explainability, and thus trustworthiness and reliability, where white box models by design provide an explicit representation of the reasoning that facilitates understanding, while black box models are very hard to understand, as the rules are implicit and often distributed in the model.

Currently, most of the attention is focused on machine learning, while knowledge representation and reasoning were the focus in the 1980s and 1990s, often in the form of expert systems. The next big step is likely the combination and integration of reasoning and learning, maybe in a similar manner as we humans do it with two separate but somehow connected systems (Kahneman 2011). System I is the fast, automatic, and opaque system for perception and intuition with very limited introspection, which shares many similarities with data-driven machine learning approaches. System II is the slow, deliberate, and explicit system for analytical thinking and planning with a high degree of introspection, which corresponds roughly to formal, symbolic, reasoning-based approaches. Another significant trend is to study the implications of AI and to make sure that AI is developed in a way that benefits all. The EU is for example putting its weight behind the concept of Trustworthy AI, which requires AI systems to follow the applicable rules and regulations, live up to four ethical principles, and to have a robust and safe implementation (High-Level Expert Group on AI 2019). A consequence of this is that the field of AI is broadening and today includes researchers from a wide variety of scientific disciplines, not only computer scientists.

The rest of this chapter will provide more details about the main topics of reasoning, learning, Trustworthy AI, and human-AI interaction.

Reasoning

Reasoning is mainly about inferring implicit information. An everyday example is to fill in the missing numbers in a game of Sudoku. Another, classical, example is the logical rule of *modus ponens*, which, given two facts – (1) *p* implies *q* and (2) p – allows the conclusion that *q* holds to be drawn. It is not explicitly stated, but it is implicitly true given the standard interpretation of logic. The fact that *q* holds is called a *consequence* and that the conclusion is *entailed* by the two statements. This is an example of classical logic. The study of logic is basically the study of what valid conclusions can be drawn from a set of statements and its roots go back to the ancient Greeks.

The study of formal logics has also provided us with many new insights and scientific facts of formal systems such as G¨odel's incompleteness theorem that states that any formal system that is powerful enough to encode the natural numbers will contain truths that cannot be proven by the system (Arbib 1987). This means that there are known truths that cannot be proven using the system; in other words, it is *incomplete*. Another important result is that first-order logic is *undecidable*, that is, it is not generally possible to prove that a given first-order statement is true or false relative to a set of first-order statements.

Reasoning is often related to knowledge representation. The question is basically: what is knowledge, how can we represent knowledge in a computer, and what can we do with this knowledge? (Brachman and Levesque 2004).

Five important types of reasoners are SAT-solving that encode problems as satisfiability problems (Biere et al. 2009), CSP-solving that encode problems using finite domain constraints (Fru¨hwirth and Abdennadher 2003), model checking that check if a system often described as a timed automaton satisfies temporal logical properties such as safety and liveliness properties (Clarke et al. 2018), automated theorem proving that prove logical statements (Newborn 2000), and planning that find sequences of actions that satisfy a goal given a domain model (Ghallab et al. 2004).

The most common technique to solve reasoning problems is *search*. It is a systematic way of trying different combinations until a solution is found. With a complete search, every possible combination is eventually tried. A concrete example is filling in an empty cell in a partially filled Sudoku. Try 1, check if the row, column, and square conditions are satisfied. If they are, then fill in the next cell, otherwise, try the next number. If there are no more numbers to try, then the partial Sudoku is not correctly filled in and cannot be solved. Through a technique called *backtracking*, a general procedure for solving Sudokus can be constructed which fills in the empty cells one by one until one of them cannot be filled in, in which case the algorithm has to backtrack by "unfilling" the previous cell and restarting the search from there. Another common technique is *mini-max* search for solving two-player games.

Formal reasoning is very useful and powerful in those applications where the domain can be represented formally using logics, finite domain constraints, or similar. Examples include formal verification of microprocessors, communication protocols, planning elevator rides, and configuration management. The main research challenge is the trade-off between expressivity and efficiency – the more expressive the formal language, the longer it takes to compute the answers, and in many cases, it is not even possible to guarantee an answer. Due to this, an important scientific endeavour is to find what is called *maximally tractable subsets of logic*, which allow all valid conclusions to be drawn for an infinite time for some fragment of the original logic (and that the fragment is as large as possible). From an application point of view, the main challenge is usually how to formally model the thing of interest.

Learning

Learning is fundamentally about how to improve given information.

Machine learning can be divided into three areas: supervised learning, unsupervised learning, and reinforcement learning (Bishop 2007). Supervised learning techniques are by far the most used in practice. Based on large collections of input-output examples, called *training data*, these methods try to find an underlying model that generalises the set of training data until it achieves acceptable performance on a different set of *test data*. To verify that the model actually generalises beyond the training and test data, its final performance is evaluated against *validation data*.

The three most common model types are classification, regression, and generative models. A *classifier* is a model that takes an individual, usually described by a vector of features, a *feature vector*, and determines which of a finite set of *classes* it belongs to, or, *classifies* it. Object recognition is a classic example of classification. A *regression* model is a mapping from an input domain to an output domain, which means that it is capable of computing the output for every possible input. Estimating the price of a good based on a feature vector or predicting the weather tomorrow based on the weather the last seven days are two common examples of regression models. Common regression models are neural networks, support vector machines, and Gaussian Processes. The third type, *generative model*, learns a probability distribution over the features, which means that it is possible to generate instances from the model, hence the name generative. The two most common generative models are Variational Auto Encoders (VAE) and Generative Adversarial Networks (GAN) (Goodfellow et al. 2014).

Deep learning is probably the technique that has received the most attention. Normally, deep learning refers to an artificial neural network with many layers (Goodfellow et al. 2016). A *neural network* consists of two or more layers of *neurons* connected by *weights*. The network is trained by changing the weights between the neurons. In a *feed forward* neural network, the neurons in layer n are connected to neurons in layer $n + 1$. In a *recurrent* neural network (RNN), neurons can be connected both to the next layer and to themselves. An important type of recurrent neural network is the *Long Short-Term Memory* (LSTM) which is very good for learning sequences to sequence models, where one sequence is mapped to another sequence such as translating between languages (Hochreiter and Schmidhuber 1997).

Neural networks have been studied since the 1950s (Schmidhuber 2015), starting with the Perceptron. However, at this time, it was not known how to train neural networks with more than two layers. As it was shown in 1969 that these two layer networks, the Perceptrons, could not represent non-linear functions such as XOR (Minsky and Papert 1969), the interest in neural networks dramatically decreased until the mid-1980s when *backpropagation* for propagating the error backwards through several layers was developed (Rumelhart et al. 1986). Backpropagation was actually first discovered by a Finnish master's student (Linnainmaa 1970). In the early 1990s convolutional neural networks (CNN) was developed to effectively represent dynamic filters often applied in computer vision (LeCun et al. 1990). Neural networks have been used commercially, for example for character recognition since the 1990s, but it took another 20 years before its big breakthrough. The reason is that a large number of training examples is needed which requires significant computational resources to train. The major breakthrough came with AlexNet (Krizhevsky et al. 2012) which significantly outperformed all existing object recognition methods on the ImageNet challenge (Deng et al. 2009) after several successive improvements (LeCun et al. 2015).

Deep neural networks have had a profound impact on many areas including computer vision, speech recognition, machine translation, and natural language processing (LeCun et al. 2015; Brynjolfsson and Mitchell 2017).

Reinforcement learning is a technique for learning how to act in an environment through trial and error. The basic idea is that the system learns a mapping from state action pairs to expected cumulative rewards, representing an estimation of how good it is to perform a certain action in a certain state. To update the mapping, the system interacts with the environment by trying different actions and receiving positive or negative rewards. Learning to ride a bike is a good example of reinforcement learning, where you learn how to control your body to not fall while riding. Since the size of the state-action mapping in most applications is too large to be explicitly represented, it is often approximated used a deep neural network, called *deep reinforcement learning*. Reinforcement learning is for example used in robotics but also to play computer games such as Atari (Mnih et al. 2015), Go (Silver et al. 2016), and Star-Craft (Vinyals et al. 2019).

Even though machine learning techniques such as deep learning have been very successful in tasks such as classifying objects, playing games, and translating between languages, there are also many challenges. Some of them are to learn more general representations (Bengio et al. 2013), to avoid overfitting (Srivastava et al. 2014), to learn multiple things at the same time (Caruana 1997), to *explain* why the network made certain recommendations (Gunning et al. 2019), to combine learning and reasoning (Bottou 2014), and to combine learning and logics (Richardson and Domingos 2006). One open question is whether it is sufficient to use neural networks, or if some innate knowledge, structures, or algorithms are also required, to learn everything that we humans can do (Marcus 2018; Dietterich 2017).

Trustworthy AI

To maximise the opportunities and minimise the risks, Europe has decided to focus on human-centred Trustworthy AI based on strong collaboration among key stakeholders. Trustworthiness is a prerequisite for people and societies to develop, deploy, and use AI systems. Without AI systems – together with humans – being demonstrably worthy of trust, unwanted consequences may ensue and its uptake might be hindered, preventing the realisation of the potentially vast social and economic benefits brought by AI systems (High-Level Expert Group on AI 2019).

Trust in the development, deployment, and use of AI systems concerns not only the technology's inherent properties but also the qualities of the socio-technical systems involving AI applications. Analogous to questions of (loss of) trust in aviation, nuclear power, or food safety, it is not simply components of the AI system but the system in its overall context that may or may not engender trust. Striving towards Trustworthy AI hence concerns not only the trustworthiness of the AI system itself, but requires a holistic and systemic approach, encompassing the trustworthiness of all actors, and processes that are part of the system's socio-technical context throughout its entire life cycle (High-Level Expert Group on AI 2019).

According to the High-Level Expert Group on AI, Trustworthy AI has three main aspects, which should be met throughout the system's entire life cycle (High-Level Expert Group on AI 2019):

1 It should be *Lawful*, ensuring respect for all applicable laws and regulations;
2 It should be *Ethical*, ensuring adherence to ethical principles and values; and
3 It should be *Robust*, both from a technical and social perspective, since, even with good intentions, AI systems can cause unintentional harm.

Each of these three components is necessary but not sufficient on its own to achieve Trustworthy AI. Ideally, all three reinforce each other. However, in practice, there may be tensions between these elements, for example when breaking the law might be necessary to save lives or when the scope and content of existing law might be misaligned with ethical norms. According to the ethical guidelines, it is our individual and collective responsibility as a society to work towards ensuring that all three components help to achieve Trustworthy AI. The ethical principles are *respect for human autonomy*, *prevention of harm*, *fairness*, and *explicability*. Based on the three components and the four ethical principles, the High-Level group defined seven key requirements for Trustworthy AI: (1) human agency and oversight, (2) technical robustness and safety, (3) privacy and data governance, (4) transparency, (5) diversity, non-discrimination, and fairness, (6) environmental and societal well-being, and (7) accountability. To assist organisations using AI systems and developers to build Trustworthy AI systems, the High-Level group developed an assessment list for Trustworthy AI (High-Level Expert Group on AI 2020).

To achieve these requirements robustly there are many technical research challenges. Some of them are *fairness*, *explainability*, *transparency*, and *safety*.

Fairness can be defined as "absence of any prejudice or favouritism toward an individual or a group based on their inherent or acquired characteristics" in the context of decision-making (Mehrabi et al. 2019). AI-based complex socio-technical systems may amplify data biases, and also introduce new forms of biases (Osoba and Welser 2017). The reason is that AI systems usually rely on data, which may be biased in ways that are socially significant. One source of bias is that data generation is often a social phenomenon full of human biases. This bias may carry over to the decision-making of the AI system in ways that are unfair to the subjects of the decision-making process. For example, it has been shown that automated methods applied to language necessarily learn human biases inherent in our use of language (Caliskan et al. 2017). Fairness-aware machine learning algorithms seek to provide methods under which the predicted outcome of a classifier operating on data about people is fair or non-discriminatory. Broadly, fairness-aware machine learning algorithms have been categorised as those pre-processing techniques designed to modify the input data so that the outcome of any machine learning algorithm applied to that data will be fair, those algorithm modification techniques that modify an existing algorithm or create a new one that will be fair under any inputs, and those post-processing techniques that take the output of any model and modify that output to be fair. Many associated metrics for measuring fairness in algorithms have been explored (Mehrabi et al. 2019).

Explainability can be defined as "the act or process of making something clear or easy to understand".[2] A more specific definition by DARPA of Explainable AI is as "AI systems that can explain their rationale to a human user, characterize their strengths and weaknesses, and convey an understanding of how they will behave in the future" (Gunning et al. 2019). Explainable AI is essential if users are to understand, appropriately trust, and effectively manage AI systems. Post hoc and transparent design explanations can be used to help AI explain its strengths and weakness and convey how it understands a concept (Guidotti et al. 2018). In post hoc explanation, given an opaque AI model, a so-called *black box*, the aim is to reconstruct its logic either by mimicking the overall behaviour of the opaque model with a transparent classifier (*global explanation*) or by focusing on the construction of a local explanation for a specific instance (*local explanation*). On the other hand, transparent design explanations aim to develop a model that is explainable on its own. Some of the research challenges are developing methods for learning more explainable models, designing effective explanation interfaces, and understanding the psychological requirements for effective explanations (Miller et al. 2017). There are also challenges related to designing metrics to measure the performance of these methods.

2 www.merriam-webster.com/dictionary/explanation

Transparency in AI plays a very important role in the overall striving to develop more Trustworthy AI as applied to markets and in society. It is one of the seven key requirements for Trustworthy AI, and one of the key five principles emphasised in the vast number of ethical guidelines addressing AI on a global level (Jobin et al. 2019). It is particularly trust and issues of accountability that drive the contemporary value of the concept, including the narrower scope of transparency found in Explainable AI (Ribeiro et al. 2016; Miller 2019). AI transparency takes a system's perspective rather than focusing on the individual algorithms or components used (Larsson and Heintz 2020). It is therefore a less ambiguously broad term than algorithmic transparency (Diakopoulos and Koliska 2017). In order to understand transparency in AI as an applied concept, it has to be understood in context, mitigated by literacies, information asymmetries, "model-close" explainability, as well as a set of competing interests. Transparency in AI, consequently, can best be seen as a balancing of interests and a governance challenge demanding multidisciplinary development to be adequately addressed (Larsson 2019; Larsson and Heintz 2020). *Safety*. AI systems should be conceived to be safe for humans, be robust against perturbations, varying contexts, malicious attacks, and have fallback plans. As AI systems become more complex, to achieve safety and robustness we need to re-understand their evaluation to (1) verify a system under acceptable assumptions (verifiability), or (2) precisely assess how often, how much, and when the system may fail (calibration, profiling, and context-dependent evaluation) (Hicks 2018). One very important area is the area of safe reinforcement learning (García and Fernández 2015) where an agent or robot learns to achieve goals in a guaranteed safe manner. In areas where verification is not possible or practicable, AI can be used to systematically validate other AI systems through simulating the interaction of AI systems with realistic, AI-learned models of the environment that an AI system will have to interact within reality (Dahmen et al. 2019).

A promising direction of active research is to develop methods for learning causal models which allows us through reasoning to explain the past and to predict the future (Pearl and Mackenzie 2018). It is also crucial that these causal models can be reasoned about in order to analyse, for example, their fairness properties. For this to be achieved, the major scientific challenge is how to integrate learning and reasoning in a principled manner while retaining the explanatory power of more structured, often logical, approaches together with the adaptability and efficiency of data-driven machine learning approaches. This is also in line with how people think, using two different systems (Kahneman 2011). Many data-driven machine learning approaches could be seen as example instances of System I, which often work but without an explicit understanding of exactly what they base their decisions on. On the other hand, most knowledge-based machine reasoning approaches could be seen as instances of System II, which often produce high-quality explicit and formally guaranteed results, but only if the encoded knowledge is correct and representative of reality. A major research challenge is how to combine these different methods into working AI systems.

Human-AI interaction

An interesting question related to AI is how this influences the role of humans. Humans and computers are fundamentally good at different things, which makes humans and computers complementary (Kamar 2016). Instead of complete automation, where we hand over the control completely to the computer, it is better if humans and computers solve problems together. Even if a computer is good at recognising objects and classifying images, humans are many times still even better and definitely more general. The role of humans then becomes to train and teach AI algorithms right from wrong and monitor that they are actually doing the right thing. The training most likely never will be completely finished, but rather incremental and continuous as new concrete examples of incorrect decisions and situations where the computer does now know what to do are collected. In these cases, we humans have to take over and provide the correct answer. A challenge for us humans then becomes what we think is right, given our different perspectives and backgrounds.

An interesting example is chess. We humans have no chance against the best chess computers, and have not had a chance for over 20 years (Siegel 2016). At the same time, the quality of human chess playing is increasing, as we are practising against chess computers. Some claim that Magnus Carlsen is the best chess player in the world since he is the human who is the best at playing like a computer. This is natural to him as he has been practising against the computer since he was a young child. What is even more interesting is that if you combine humans and chess computers, the joint team, called a centaur, becomes better than both the best humans and the best computers. It is even the case that the team becomes even better if you include several people (Kasparov 2017). This is a concrete example of how the result improves when humans and computers collaborate to solve complex problems. There is no dichotomy between humans and computers, it is not a question of either/or, but rather humans and computers. Simplified, one can say that computers are good at doing, while humans are good at what should be done and why. We are good at asking questions, and computers are good at answering them. Examples are question answering systems that are great at answering questions and planning systems that can generate elaborately detailed plans for how to achieve goals, but the questions and the goals have to be provided to the systems by human users.

An important observation is that it is a different skill to play chess with a computer compared to playing chess on your own. This means that even if you are a really good expert and you are provided with the best possible tool, it is not necessary that the result improves. You might still perform worse compared to a person who is less of an expert in the subject but more of an expert on using the tool effectively. To really leverage the computational power, we need to both educate people in solving problems with AI tools and adapt the way we work to truly leverage the tools. Thus, relevant education, changed ways of working, and new organisational forms are required; see for example Wagner (2020a). A central capability is to transform business problems into computational problems. That is, to formulate problems in such a way that computers or computer tools can assist (Brynjolfsson and Mitchell 2017).

The social and economic pattern that derives from the complementary nature of human and machine intelligence is one that has been prevailing and gaining momentum since the beginning of the industrial revolution: specialisation and division of labour. Specialised machine intelligence allows this process to further accelerate, especially in the field of knowledge work, leading to micro-division of labour (Wagner 2020b).

By interpreting AI as a new type of agent within a company, it has been argued that a number of effects on companies can be supported (Wagner 2020c):

1 AI intensifies the effects of economic rationality on the firm;
2 AI introduces a new type of information asymmetry;
3 AI can perforate the boundaries of the firm;
4 AI can create triangular agency relationships; and
5 AI has the potential to remove traditional limits of integration.

Computational thinking captures this general skill of solving problems in a way that computers can assist (Wing 2006, 2011). For computers to help us, we have to be better at understanding how a computer solves problems. Thus, computational thinking is to a large extent about learning to understand how a computer "thinks" when it solves a problem.

When you solve problems with a computer, it is often about describing to the computer what should be done, rather than doing it yourself. Programs are descriptions of how to solve something that computers understand. Traditionally, humans have to describe every step of the process in great detail. AI actually reduces this by enabling the computer to fill in many of the details. It is also interesting to consider the macro-level. Instead of looking at the individual actor (micro-level), we take the perspective of groups rather than individuals. An interesting example is human-agent collectives which are presented as a new class of socio-technical systems in which humans and smart software (agents) engage in flexible relationships in order to achieve both their individual and collective goals (Jennings et al. 2014). Sometimes the humans take the lead, sometimes the computer does and this relationship can vary dynamically.

AI and computational thinking can actually be seen as two sides of the same coin. AI is about enabling the computer to solve problems we consider requires intelligence, or casually speaking, enable computers to "think". Computational thinking turns this around and asks the question: how can people become better at solving problems by learning from how computers do?

Conclusion

Artificial intelligence is a fascinating topic, which has both a technical side rooted in computer science and mathematics and a humanistic side rooted in cognitive science and sociology. This chapter tries to provide an overview of the subject, mostly from a computer science perspective, but with several connections to social sciences. The dynamic and productive research field will most likely continue to evolve through the stimulation and interaction with many different fields of research. Even though a lot has been accomplished, the most important and fascinating breakthroughs probably lie ahead of us. AI is still a young research discipline that has much to offer to every researcher that wants to explore what it means to be human and what the enigma of intelligence really is.

References

Arbib, M. A. (1987): *G¨odel's Incompleteness Theorem*. New York: Springer US, 162–188.

Bengio, Y., Courville, A. and Vincent, P. (2013): Representation Learning: A Review and New Perspectives, *IEEE Transactions on Pattern Analysis and Machine Intelligence* 35(8), 1798–1828.

Biere, A., Heule, M., Van Maaren, H. and Walsh, T. (2009): *Handbook of Satisfiability*. Amsterdam: IOS Press.

Bishop, C. M. (2007): *Pattern Recognition and Machine Learning*. Heidelberg: Springer.

Bottou, L. (2014): From Machine Learning to Machine Reasoning: An Essay, *Machine Learning* 94, 133–149.

Brachman, R. and Levesque, H. (2004): *Knowledge Representation and Reasoning*. San Francisco: Morgan Kaufmann Publishers Inc.

Brynjolfsson, E. and Mitchell, T. (2017): What Can Machine Learning Do? Workforce Implications, *Science* 358(6370), 1530–1534.

Caliskan, A., Bryson, J. J. and Narayanan, A. (2017): Semantics Derived Automatically from Language Corpora Contain Human-like Biases, *Science* 356(6334), 183–186.

Caruana, R. (1997): Multitask Learning, *Machine Learning* 28(1), 41–75.

Clarke, E., Grumberg, O., Kroening, D., Peled, D. and Veith, H. (2018): *Model Checking*. Cambridge, MA: MIT Press.

Dahmen, T., Trampert, P., Boughorbel, F., Sprenger, J., Klusch, M., Fischer, K., Ku¨bel, C. and Slusallek, P. (2019): Digital Reality: A Model-based Approach to Supervised Learning from Synthetic Data, *AI Perspectives* 1(1). https://doi.org/10.1186/s42467-019-0002-0

Deng, J., Dong, W., Socher, R., Li, L.-J., Li, K. and Fei-Fei, L. (2009): ImageNet: A Large-Scale Hierarchical Image Database, in *IEEE Computer Society Conference on Computer Vision and Pattern Recognition Workshops (CVPR Workshops)*, Miami: FL, 248–255. https://doi.org/10.1109/CVPR.2009.5206848

Diakopoulos, N. and Koliska, M. (2017): Algorithmic Transparency in the News Media, *Digital Journalism* 5(7), 809–828.

Dietterich, T. G. (2017): Steps toward Robust Artificial Intelligence, *AI Magazine* 38(3), 3–24.

Frühwirth, T. and Abdennadher, S. (2003): *Essentials of Constraint Programming*. Berlin, Heidelberg: Springer.

García, J. and Fernández, F. (2015): A Comprehensive Survey on Safe Reinforcement Learning, *Journal of Machine Learning Research* 16(1), 1437–1480.

Ghallab, M., Nau, D. and Traverso, P. (2004): *Automated Planning – Theory and Practice*. San Francisco: Morgan Kaufmann.

Goodfellow, I., Bengio, Y. and Courville, A. (2016): *Deep Learning*. Cambridge, MA: MIT press.

Goodfellow, I. J., Pouget-Abadie, J., Mirza, M., Xu, B., Warde-Farley, D., Ozair, S., Courville, A. and Bengio, Y. (2014): Generative Adversarial Nets, *Advances in Neural Information Processing Systems* 27. Available at: https://papers.nips.cc/paper/2014 [10.03.2021]

Guidotti, R., Monreale, A., Ruggieri, S., Turini, F., Giannotti, F. and Pedreschi, D. (2018): A Survey of Methods for Explaining Black Box Models, *ACM Computing Surveys* 51(5). https://doi.org/10.1145/3236009

Gunning, D., Stefik, M., Choi, J., Miller, T., Stumpf, S. and Yang, G.-Z. (2019): XAI – Explainable Artificial Intelligence, *Science Robotics* 4(37). https://doi.org/10.1126/scirobotics.aay7120

Hicks, D. (2018): The Safety of Autonomous Vehicles: Lessons from Philosophy of Science, *IEEE Technology and Society Magazine* 37, 62–69.

High-Level Expert Group on AI (2019): Ethics Guidelines for Trustworthy AI. Report. Brussels: European Commission.

High-Level Expert Group on AI (2020): Assessment List for Trustworthy AI (ALTAI). Report. Brussels: European Commission.

Hochreiter, S. and Schmidhuber, J. (1997): Long Short-term Memory, *Neural Computation* 9(8), 1735–1780.

Hubel, D. and Wiesel, T. (1962): Receptive Fields, Binocular Interaction, and Functional Architecture in the Cat's Visual Cortex, *Journal of Physiology* 160, 106–154.

Jennings, N., Moreau, L. Nicholson, D., Ramchurn, S., Roberts, S., Rodden, T. and Rogers, A. (2014): Human-agent Collectives, *Communications of the ACM* 57(12), 80–88.

Jobin, A., Ienca, M. and Vayena, E. (2019): The Global Landscape of AI Ethics Guidelines, *Nature Machine Intelligence* 1(9), 389–399.

Kahneman, D. (2011): *Thinking Fast and Slow*. Canada: Doubleday.

Kamar, E. (2016): Directions in Hybrid Intelligence: Complementing AI Systems with Human Intelligence, *Proceedings of the International Joint Conference on AI (IJCAI)*'16, 4070–4073.

Kasparov, G. (2017): *Deep Thinking: Where Machine Intelligence Ends and Human Creativity Begins*. New York: PublicAffairs.

Kelly, K. (2016): *The Inevitable*. New York: Viking Press.

Krizhevsky, A., Sutskever, I. and Hinton, G. E. (2012): Imagenet Classification with Deep Convolutional Neural Networks, *Advances in Neural Information Processing Systems* 25. https://doi.org/10.1145/3065386.

Larsson, S. (2019): The Socio-legal Relevance of Artificial Intelligence, *Droit et Société* 103(3), 573.

Larsson, S. and Heintz, F. (2020): Transparency in Artificial Intelligence, *Internet Policy Review* 9(2), 1–16.

LeCun, Y., Bengio, Y. and Hinton, G. (2015): Deep Learning, *Nature* 521(7553), 436–444.

LeCun, Y., Boser, B., Denker, J., Henderson, D., Howard, R., Hubbard, W. and Jackel, L. (1990): Handwritten Digit Recognition with a Back-propagation Network, *Advances in Neural Information Processing Systems 2*. Available at: https://papers.nips.cc/paper/1989 [10.03.2021]

Legg, S. and Hutter, M. (2007): Universal Intelligence: A Definition of Machine Intelligence, *Minds and Machines* 17(4), 391–444.

Linnainmaa, S. (1970): *The Representation of the Cumulative Rounding Error of an Algorithm as A Taylor Expansion of the Local Rounding Errors*. Master's thesis, Univ. Helsinki.

Lynch, S. (2017): Andrew Ng: Why AI is the New Electricity, *Insights by Stanford Business* 11. Available at: www.gsb.stanford.edu/insights/andrew-ng-why-ai-new-electricity [10.03.2021]

Marcus, G. (2018): Deep Learning: A Critical Appraisal, *ArXiv* abs/1801.00631.

McCarthy, J., Minsky, M., Rochester, N. and Shannon, C. (1955): A Proposal for the Dartmouth Summer Research Project on Artificial Intelligence, *AI Magazine* 27(4), 12. https://doi.org/10.1609/aimag.v27i4.1904

McCorduck, P. (2004): *Machines Who Think*. Natick, MA: A.K. Peters, Ltd.

Mehrabi, N., Morstatter, F., Saxena, N., Lerman, K. and Galstyan, A. (2019): A Survey on Bias and Fairness in Machine Learning. Available at: https://arxiv.org/abs/1908.09635 [10.03.2021]

Miller, T. (2019): Explanation in Artificial Intelligence: Insights from the Social Sciences, *Artificial Intelligence* 267, 1–38.

Miller, T., Howe, P. and Sonenberg, L. (2017): Explainable AI: Beware of Inmates Running the Asylum or: How I Learnt to Stop Worrying and Love the Social and Behavioural Sciences. Available at: https://arxiv.org/abs/1712.00547 [10.03.2021]

Minsky, M. and Papert, S. (1969): *Perceptrons: An Introduction to Computational Geometry*. Cambridge, MA: MIT Press.

Mnih, V., Kavukcuoglu, K., Silver, D., Rusu, A. A., Veness, J., Bellemare, M. G., Graves, A., Riedmiller, M., Fidjeland, A. K., Ostrovski, G., Petersen, S., Beattie, C., Sadik, A., Antonoglou, I., King, H., Kumaran, D., Wierstra, D., Legg, S. and Hassabis, D. (2015): Human-level Control through Deep Reinforcement Learning, *Nature* 518(7540), 529–533.

Newborn, M. (2000): *Automated Theorem Proving – Theory and Practice*. New York: Springer.

Nilsson, N. (2010): *The Quest for Artificial Intelligence*. Cambridge, MA: Cambridge University Press.

Osoba, O. and Welser, W. (2017): *An Intelligence in Our Image – The Risks of Bias and Errors in Artificial Intelligence*. Santa Monica, CA: RAND Corporation.

Pearl, J. and Mackenzie, D. (2018): *The Book of Why*. New York: Basic Books.

Ribeiro, M. T., Singh, S. and Guestrin, C. (2016):"Why Should I Trust You?": Explaining the Predictions of Any Classifier, *Proc. Mining. ACM SIGKDD International Conference on Knowledge Discovery and Data*, 1135–1144. https://doi.org/10.1145/2939672.2939778

Richardson, M. and Domingos, P. (2006): Markov Logic Networks, *Machine Learning* 62(1–2), 107–136.

Rumelhart, D. E., Hinton, G. E. and Wilson, R. J. (1986): Learning Representations by Back-propagating Errors, *Nature* 323, 533–536.

Russell, S. J. and Norvig, P. (2016): *Artificial Intelligence: A Modern Approach*. Malaysia: Pearson Education Limited.

Schmidhuber, J. (2015): Deep Learning in Neural Networks: An Overview, *Neural Networks* 61, 85–117.

Shoham, Y. (1997): An Overview of Agent-oriented Programming, in Bradshaw, J. (ed.): *Software Agents*. Cambridge, MA: AAAI Press, 271–290.

Siegel, R. (2016): 20 Years later, Humans Still No Match for Computers on the Chessboard. *NPR.org*. Available at: www.npr.org/sections/alltechconsidered/2016/10/24/499162905/20-years-later-humans-still-no-match-for-computers-on-the-chessboard?t=1615312678700 [28.01.2021]

Silver, D., Huang, A., Maddison, C. et al. (2016): Mastering the Game of Go with Deep Neural Networks and Tree Search, *Nature* 529, 484–489.

Silver, D., Schrittwieser, J., Simonyan, K., Antonoglou, I., Huang, A., Guez, A., Hubert, T., Baker, L., Lai, M., Bolton, A., Chen, Y., Lillicrap, T., Hui, F., Sifre, L., van den Driessche, G., Graepel, T. and Hassabis, D. (2017): Mastering the Game of Go without Human Knowledge, *Nature* 550, 354–359.

Srivastava, N., Hinton, G., Krizhevsky, A., Sutskever, I. and Salakhutdinov, R. (2014): Dropout: A Simple Way to Prevent Neural Networks from Overfitting, *Journal of Machine Learning Research* 15(56), 1929–1958

Thagard, P. (2008): Cognitive Science, in Zalta, E. N. (ed.): *The Stanford Encyclopedia of Philosophy*. Stanford: Stanford University.

Vinyals, O., Babuschkin, I., Czarnecki, W. et al. (2019): Grandmaster Level in Starcraft II Using Multi-agent Reinforcement Learning, *Nature* 575, 350–354.

Wagner, D. N. (2020a): Augmented Human-centered Management Human Resource Development for Highly Automated Business Environments, *Journal of Human Resource Management* XXIII(1), 13–27.

Wagner, D. N. (2020b): Economic Patterns in a World with Artificial Intelligence, *Evolutionary and Institutional Economics Review* 17(1), 111–131.

Wagner, D. N. (2020c): The Nature of the Artificially Intelligent Firm – An Economic Investigation into Changes that AI Brings to the Firm, *Telecommunications Policy* 44(6), 101954.

Wing, J. (2006): Computational Thinking, *Communications of the ACM* 49(3), 33–35.

Wing, J. (2011): Computational Thinking: What and Why. Available at: www.cs.cmu.edu/link/research-notebook-computational-thinking-what-and-why [28.01.2021].

13

THE SCIENCE SPACE OF ARTIFICIAL INTELLIGENCE KNOWLEDGE PRODUCTION

Global and regional patterns, 1990–2016

Dieter F. Kogler, Adam Whittle, and Bernardo Buarque

Introduction

How will automation affect human well-being? This question arguably dates back as far as the first industrial revolution with the invention of the steam engine, and beyond. It is also a question that has more recently recaptured the attention of social scientists following the advent of Industry 4.0, and in particular the emergence of Artificial Intelligence (AI). Most noticeably from an economics perspective, several recent publications have sought to evaluate how AI will transform labour productivity (Brynjolfsson et al. 2018), economic growth (Aghion et al. 2017), international trade (Goldfarb and Trefler 2018), and employment (Arntz et al. 2016; Agrawal et al. 2019).

To put it succinctly, while AI has the potential to result in economic growth, prosperity, and positive change, it could just as easily produce job displacement and income inequality (Korinek and Stiglitz 2017). That is, despite the excitement over its many promises, the recent debate regarding AI seems to focus on the future of work "in a world in which computer algorithms can perform many of the functions that a human can" (Furman and Seamans 2019: 161). Indeed, comparable to the emergence of any other disruptive general-purpose technology (GPT), the rise of AI will have a profound impact on our daily lives and well-being. As Buarque et al. (2020: 175) state, it will eventually "lead to significant shifts in employment and income distributions across and within society, particularly when the gains are concentrated in AI-producing regions or sectors".

Irrespective of whether your point of view of AI is optimistic or pessimistic, it is unquestionable that policies enacted today will shape how AI impacts society tomorrow (Agrawal et al. 2019). As interests around AI have begun to increase, countries around the world have started developing their own AI programmes with an eye on becoming a market leader in this disruptive technology (Dutton 2018). Equally, scholars in world-class institutions are also actively working to examine, propose, and implement policies that can enhance the benefits offered by AI, while mitigating against any of its negative consequences. Instrumental in this arena was a conference organised by the OECD in 2017, "AI: Intelligence Machines, Smart Policies", with the sole aim of mobilising social, economic, and political responses to the transformation of society brought on by the advent of AI technologies. Therefore, we are not only observing a

DOI: 10.4324/9780429351921-17

mere surge in AI advancement, but also rising political concern about how to respond to it, and moreover how it should be managed. However, before it is possible to design fit-for-purpose policies, we must first understand, in an in-depth manner, the evolution and diffusion of AI systems as well as their many socio-economic consequences that are just now unfolding.

It is strikingly obvious that AI has become a hot topic and a frequently used buzzword. Nevertheless, despite its growing popularity, one dimension which remains unclear is how we can accurately measure the creation and diffusion of AI. In fact, whilst there seems to be a consensus that AI will transform our daily lives, it remains to be seen how these transformations will manifest in space, that is, through economic growth or productivity. As a consequence, AI risks becoming a policy ahead of the theory initiative, based primarily on speculative analysis and anecdotal evidence. Moreover, to the best of our knowledge, the relevant literature continues to even lack a precise definition of AI. Therefore, it is critical for AI's successful implementation that there is an accurate depiction of its creation and development. Further still, to produce reliable inferences about how AI impacts our economy and society, we also need to develop robust data on its spatial and temporal diffusion; otherwise, our understanding of AI will remain speculative at best.

Against this backdrop, the present handbook chapter is a structured attempt to inform AI discourse and provide both a review of the relevant literature as well as a novel methodological axiom to analyse the creation and diffusion of AI technology. Essentially, our objective is to construct a relational database composed of academia publications on AI derived from Web of Science (WoS) data. Thereafter, this information is used to graph the distribution of AI knowledge in both the global and EU scientific communities. In doing so, it is possible to address the issue of when and where AI is created, as well as to identify potential trends in the evolution of this new disruptive technological domain.

As previously mentioned, very few empirical studies have managed to accurately disentangle the relationship between AI and socio-economic outcomes, primarily due to the lack of necessary data required. Furthermore, those notable exceptions that have addressed this issue have primarily looked at the impact of automation on labour outputs by using proxy data for the local exposure of AI methods. For instance, Acemoglu and Restrepo (2017) and Graetz and Michaels (2018) use data from the International Federation of Robotics (IFR) to estimate the regional and industrial exposure to robots and thus determine their impact on the local economy. Yet, their dataset by no means captures all the dimensions of AI; it merely proxies a fast-growing technology for the presence of robots in the industry. Hence, it does not actually allow the authors to correctly infer the multiple aspects or consequences of Artificial Intelligence in an economy.

Another common approach is to measure the likelihood that certain occupations will become automated by advances in the field of AI. Frey and Osborne (2017) famously pioneered this method when they gathered data on the probability that different cognitive tasks would become "computerised" in the future. Thereafter, they combined this information with the O*NET dataset, which describes the dependency of 702 distinct occupations on each of these tasks. Using both sets of data, they were able to estimate the prospect of automation for all 702 occupations. The task-based approach to measuring the *risk of automation* has since become a popular strategy for scholars looking to evaluate the relationship between AI and the labour market (OECD 2017; Acemoglu and Restrepo 2019). For example, both Arntz et al. (2016) and Nedelkoska and Quintini (2018) applied this approach to study the likelihood of automation for the OECD member nations.

However, despite its merits, these task-based approaches all suffer from the same empirical shortcoming in that they do not enable a thorough and detailed investigation into the presence and diffusion of AI across local economies. As a consequence, they are not suitable methods for evaluating the determinants of AI knowledge production and its many implications on society. In fact, these task-based approaches only permit a high-level view of how AI-driven automation "might" affect one specific aspect of the economy: the local labour market.

Given the previously stated, it therefore appears that existing data are not sufficient to fully appraise the spread of AI, and as a consequence we are severely limited in our capacity to understand the drivers of this change and its impact on our society. Notwithstanding these numerous shortcomings, authors have begun exploiting recent advances in the field of text-analytics to circumvent these issues. Namely, previous research used text-analysis to classify patents and other documents into unique "technological" groups and used this information to infer the extent of innovation in a given domain, such as measuring environment-related technologies (Haščič and Migotto 2015). Following this logic, Mann and Puttmann (2018) applied a machine learning algorithm to a dataset consisting of texts from American patents to sort them into automation and non-automation innovations. Once they identified the automation patents, the authors could (geo)locate them in time and space. Further still, they could begin to expose the relationship between the volume of automation patents and local employment outputs.

Similarly, Cockburn et al. (2018) conducted a keyword search on a corpus of publications and American patents to distinguish which "inventions" should be classified as symbolic systems, robotics, and deep learning. Thereafter, the authors compared the diffusion of these methodologies across fields, regions, and time. Moreover, at least for the scope of this analysis, Buarque et al. (2020: 176) employed a list of technological classes and keywords to identify European patents that are associated with AI methods. In doing so, their goal was to "build a comprehensive data set of AI patents, which will enable us to study AI knowledge production and how it is distributed across the different regions and technological sectors of the European economy".

Following Buarque et al. (2020), the present investigation employs a list of identifying keywords from WIPO (2019) to map the creation of scientific knowledge. Therefore, unlike most prior studies the focus here is not on patents, but rather on the scientific literature that concerns Artificial Intelligence. As the initial step, we created a subsample of academic documents from the Web of Science (WoS), our primary source of bibliographic data. WoS indexes approximately 280,000 scientific journals, as well as several conference proceedings and books.[1] As such, it provides valuable information on academic publications, authors, institutions, and citations. Most importantly, however, the WoS also collects data on the keywords for each document – as provided by the authors. Exploring the information available in these keywords, we performed a search algorithm to identify and classify all the AI-relevant documents. More precisely, we looked within the publications for keywords that describe an AI method, like "Neural Networks" or "Genetic Algorithms". We then classified and sampled a document as AI whenever it includes at least one keyword associated with the technology. Next, we used the metadata of these AI publications to graph the development of Artificial Intelligence in space and time. Thus, providing a valuable map of the creation and diffusion of Artificial Intelligence among the global scientific community.

While the present investigation should prove very useful, it does not provide empirical evidence on the determinants of AI creation, nor does it provide estimates on its potential implication for local economies. Instead, the objective is to offer a first glance at the creation

1 The present analysis is based on data retrieved from the following Web of Science bibliographic databases: "1980–2017 – Annual Science Citation Index Expanded and Proceedings-Science Combined".

and diffusion of AI methods in the scientific world and on a variety of spatial scales – global, national, and regional. In turn, we hope to inspire and support more detailed empirical investigations into this emergent and meaningful technology.

Taking advantage of the proposed data and methods, future investigations in the field might shine a light on how to foster the development of AI as well as produce essential estimates on the impact of AI on social inequality and human well-being. Following this approach, further analysis will potentially contribute not only to our understanding of this general-purpose technology, but also inform policymakers seeking to design "smart policies" in the age of Artificial Intelligence. Most policy briefings currently emphasise the economic opportunities brought about by AI systems, and the need to educate displaced workers for the jobs of the future (OECD 2017). The methodological framework developed by Buarque et al. (2020), which we also employ here, will allow one to touch – at least marginally – on both of those issues. First, mapping the evolution of AI would allow for the recognition of the sectors/regions with related core competencies most suited for building a development pathway into this rapidly growing technology field (Hidalgo et al. 2007; Kogler et al. 2017; Whittle 2020). That is, mapping the diffusion of AI across regions and sectors would enable one to recognise opportunities to "invest in and develop AI for its many benefits" (The White House, Executive Office of the President 2016). Second, the AI "knowledge-space" (Kogler et al. 2013) could identify which sectors/regions are more likely to be affected by this expanding technology. In other words, studying the diffusion of AI could help to identify those more vulnerable to job displacement and other negative consequences.

Identification strategy

The first phase of our analysis involves identifying AI documents. The database which supports our analysis is a raw Web of Science (WoS) corpus containing articles from over 46m journals, books, and conference proceedings. Following, we will use all documents in that database that have been published over the period 1990–2016, which is about 38m records. While WoS shares commonalities with other bibliometric databases, including Google Scholar, Scopus, or Microsoft Academic Knowledge Graph, two noticeable differences are pertinent for our investigation. Firstly, WoS has a proclivity to favour journal articles over other outlets. Further, WoS has a bias towards the "Hard Sciences" (i.e. the natural sciences, engineering, and biomedical research, at the expense of social science and arts and humanities) (Mongeon and Paul-Hus 2016). Nevertheless, for the purpose of the present investigation these issues are not that detrimental. In fact, given the substantive nature of AI and the fact that journal articles are the preferred outlet for dissemination across relevant disciplines, this bias may even serve to our advantage and increase our overall coverage.

Turning to the information contained within each document, WoS lists the titles, journal names, year of publication, authors, and their affiliations, among other data. Most importantly, WoS provides a list of keywords for each document, as determined by the authors. For these reasons, WoS can rightfully be considered a strong medium for analysing the creation, integration, and evolution of Artificial Intelligence throughout space and time.

Given its inherently fuzzy nature, there is no easy way to identify the AI documents. To tackle this problem, we adopt a commonly used technique in bibliographic studies and apply a keyword identifier to the WoS database. Namely, we search across the keywords section of each document for AI-specific terms, such as "Machine Learning" or "Supervised Learning". In turn, this approach enables us to classify all documents in our database as either AI or non-AI.

Naturally, the choice of AI identifiers will heavily influence our results. For this analysis, we follow the example set by Buarque et al. (2020) and borrow a list of AI-related keywords

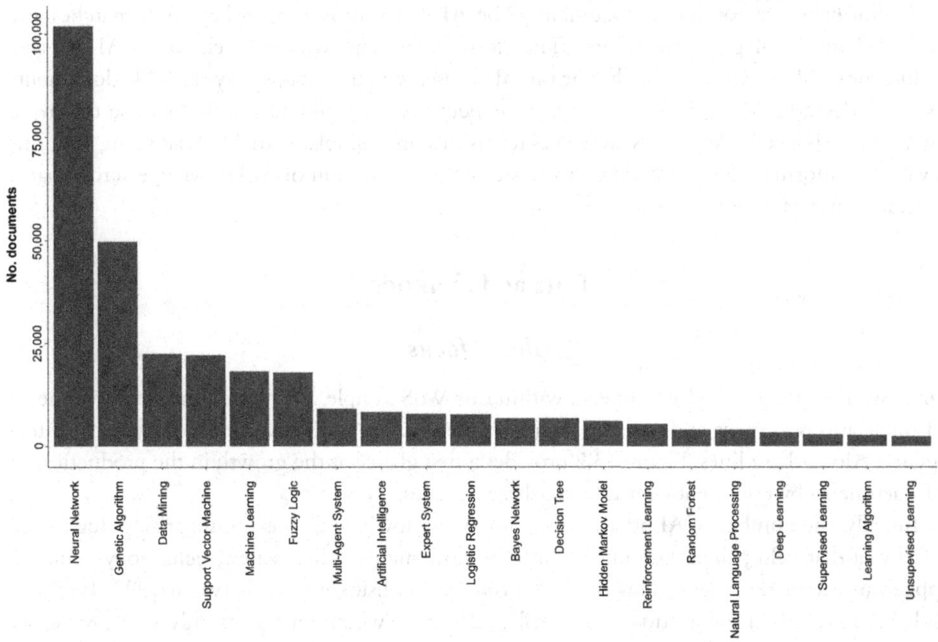

Figure 13.1 Frequency of AI keywords in searched publications

Source: Authors' calculation

produced by the World Intellectual Property Organization (WIPO 2019). We believe WIPO's identification gives a more recent and specific definition of AI than other alternatives. Seeking to identify AI-related patents in Europe, Buarque et al. (2020) employed a list of 43 n-grams[2] that are indicative of modern AI technology. Thus, we adopt the same list of words to recognise AI-specific knowledge in the WoS database, which includes the terms: "Artificial Intelligence", "Data Mining", and "Learning Algorithm" (see Appendix A for a full list of terms).

Using this methodology, we identified 260,351 documents as AI – out of the 38 million possible documents in our WoS database. Thus, although everyone seems to be discussing AI, very few have been able to seriously engage with the technology so far. To explore this issue further, Figure 13.1 plots the 20 most frequent keywords associated with AI. The histogram is positively skewed with "Neural Networks" being the most frequently occurring keyword. This is immediately followed by "Genetic Algorithm", "Data Mining", and "Support Vector Machine".

It should be noted that we have carried out a thorough and careful stemming and cleaning process on our entire list of AI documents. These processes are necessary given the fact that inconsistencies are a common feature with any large-scale dataset. Moreover, one of the most common difficulties is that different elements actually represent the same thing. For example, the same keyword may be reported in a variety of ways (e.g. Neural Network or Neural Networks or "NN"). For the analysis reported here, we have applied a stemming technique to all of the keywords with the aim to minimise duplications.

2 A n-gram consists of a list of "n" items from a sample text. For example, in this analysis, we say that "machine learning" is a 2-gram.

Additionally, we consider a document to be AI if at least one of its keywords matches any of our 43 unique n-gram identifiers.[3] Therefore, documents which we classify as AI can also include non-AI keywords. As such, our final database, which contains only the "AI" documents also includes non-AI keywords. Indeed, we expect every document in our database to have a mixture of AI-specific keywords, as well as terms that are not related to AI. That being said, this is valuable information since it allows us to study the integration of AI knowledge across space, time, and subject matter.

Data and methods

Global focus

Once we identified the AI documents within our WoS sample, we plotted the total volume of AI publications in space and time to obtain a better picture of the evolution of Artificial Intelligence. Along these lines, Figure 13.2 provides a first glance at the growth in the production of AI documents by continent over the past three decades.

Initially, the number of AI publications grew very slowly, which is not surprising due to the novelty of that field paired with uncertainties of how such a radically new technology could be applied in the market place. This is evident once we considered that between 1990–1999 the global number of AI publications was equal to 21,531 – which represents only about 8% of all documents in our sample and 22% of documents produced in the following decade. Nevertheless, even at this primitive stage, it is still possible to identify the Western economies of Europe and the Americas as key regional players, whereby they account for 5,632 (26%) and 5,555 (25%) respectfully. Finally, although not as prominent in the early years of the first period, it is also possible to identify a nascent cluster in Asia (19%).

In the second period (2000–2009) the first "real" surge in AI publications is evident. During this time, a total of 95,813 papers were published, signifying an almost five-fold increase over the first period. From a path-dependence perspective, Europe and America continue to dominate the initial years of the decade, accounting for 28% and 23% respectfully. However, this position is transformed following the emergence, and thereafter the dominance, of Asia. Throughout the decade, the Asian economy accounts for almost half (50%) of the global output, and from 2002 onwards reports a doubling of its publication output relative to its Western counterparts.

Equally important, during this period, AI begins to formalise as a discipline. Frequency analysis based on the journals' keywords (see the following subsection) indicates that topics such as "Neural Networks", "Machine Learning", "Genetic Algorithms", "Pattern Recognition", "Fuzzy Logic", and "Data Mining" began to emerge in documents during this time. More formally, these advances served to establish the foundation on which modern AI is based and thus can be regarded as setting up the pre-conditions for the final stage.

The third and final stage refers to the years 2010–2016, and it is during this time period that the bulk of the data lies. During this stage, we observe a further increase in the divergence between Eastern and Western economies. Focusing on the global crisis of 2008, we see that Asia was initially impacted by this crisis, but quickly recovered. In contrast, whereas Europe and

3 Over the period of analysis, the mean number of keywords listed on journal articles remained constant at 4.8. Therefore, for an article to be identified as an AI document, one of its keywords would need to match our identifier, but the remaining three or four keywords do not. This approach enables us to create a robust picture of how AI is beginning to integrate with other research fields.

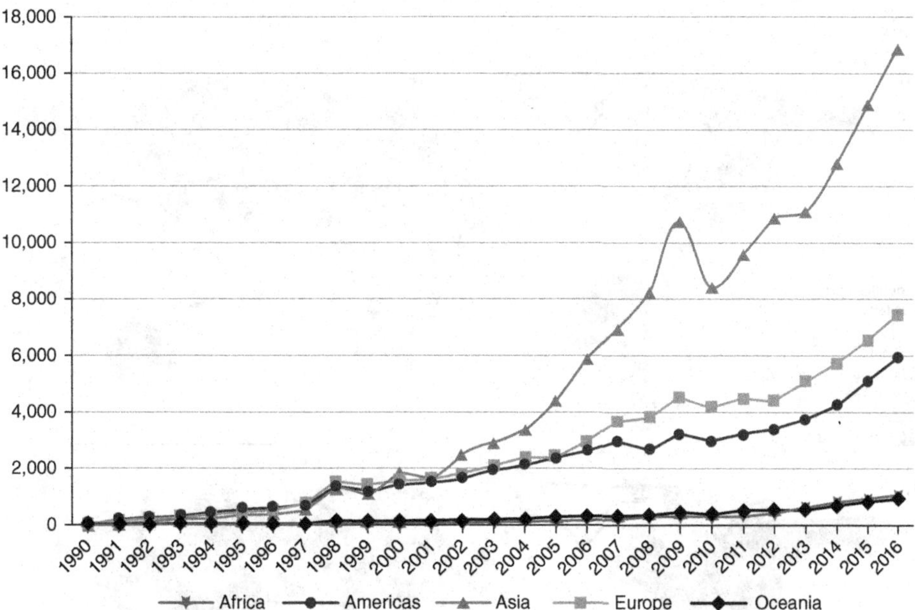

Figure 13.2 Number of AI documents by continent

Source: Authors' calculation

the Americas appeared to be less affected in the years preceding and during the crisis, they have struggled to increase their total number of publications since then.

Country-level analysis

Expanding the initial analysis beyond the continental level, it is possible to use the authors' affiliations on each WoS document to map the spatial distribution of Artificial Intelligence across countries. Doing so enables us to view continents in terms of their countries rather than as a collective. To illustrate, Figure 13.3 (Panels a, b, and c) displays the spatial distribution of AI-specific documents across countries for the three time periods discussed previously. Using this approach, we observe that Europe's dominance in AI is partially the result of five countries (United Kingdom, France, Spain, Germany, and Italy), which make up nearly 62% of the continent's total output. A similar pattern emerges in Asia, where China (47%) clearly dominates while other countries like Taiwan (7%), India (12%), Iran (7%), and Japan (7%) also make a noticeable contribution. The same pattern does not emerge for the Americas where the United States is continually the primary producer of AI with 70% of the continent's contribution.[4]

Comparing between periods, we can further observe the path dependency process mentioned in the previous section. Namely, those countries with a historical advantage in AI production, that is, those leading the development of AI in our first period, continue to dominate in terms of the overall share of documents. Our results, thus, seem to corroborate past work in

4 As with any country level analysis the problem with the United States is that its sheer size distorts the innovative potential of the individual US states. Given the space constraints here, we do not go into detail as to which states in the United States are responsible for its dominance in AI, but future work should engage with these questions.

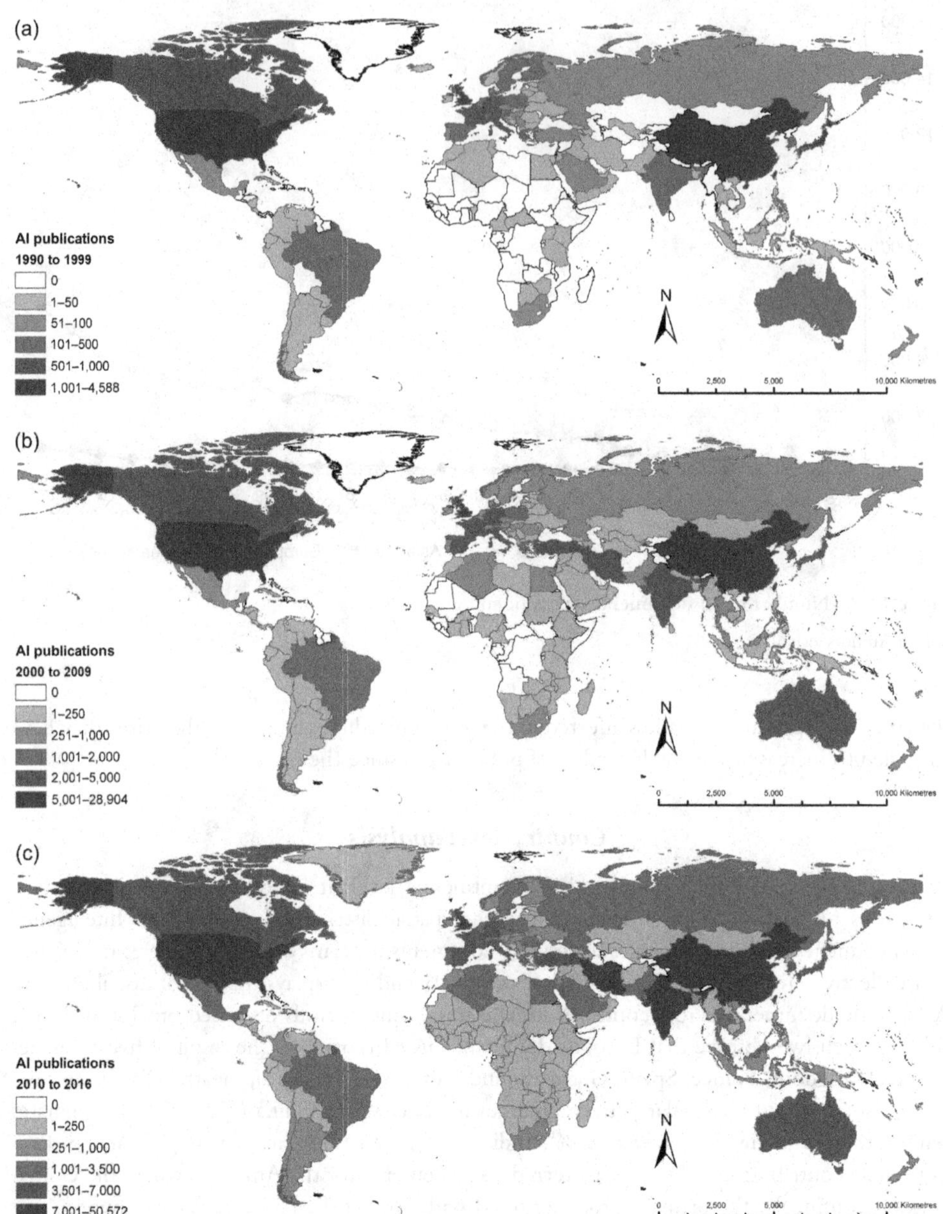

Figure 13.3 Distribution of AI documents by country and time period
(a) 1990–1999
(b) 2000–2009
(c) 2010–2016

Source: The authors; based on Web of Science records derived from the "Science Citation Index Expanded" and the "Conference Proceedings Citation Index". GIS shapefile source is EUROSTAT.

Evolutionary Economic Geography (EEG), which highlights that regional innovation is driven by a path-dependent process (Martin and Sunley 2006; Kogler 2016). Furthermore, it is in line with the findings of Buarque et al. (2020) who showed the same tendency amid the AI-specific patents of Europe; those regions that excel in computing technology in the early stages of the analysis are also those with the largest share of AI patents in the end period.

On the other hand, our maps show the emergence of Asian economies, China in particular, as key players in the scientific development of AI technologies. One might explain the rapid rise of China as a by-product of global geopolitics. After all, as China marches to become the largest economy in the world, it is only natural that it will control the technologies of the future, for example, AI. Nevertheless, it is more likely that the growth in China is a result of the country's ambitious policies regarding the development of AI (Dutton 2018). Since 2017, when the "New Generation of Artificial Intelligence Development Plan" was implemented, China has even invested more in the production of AI for the future than it has in the past. Thus, China's continuous dominance in this technology domain will most likely prevail for quite some time.

Country collaboration network

Measuring the distribution in time of AI documents across countries is surely informative. Nonetheless, to obtain a better picture of the creation and diffusion of AI, we also need to account for international collaborative efforts. The seminal contribution of Wuchty et al. (2007) highlights how journal articles are increasingly found to be the result of collaborative efforts involving teams of researchers. Taking an Evolutionary Economic Geography stand, one possible reason for their findings is that there is an upper echelon limiting the extent an individual scientist, firm, region, or country can create all the knowledge they require internally; this seems to be particularly valid for the creation of general purpose or complex technologies whose production is the result of the recombination of multiple parts (Whittle 2019). Besides, Buarque et al. (2020: 177) have further commented that "AI is best developed when well connected to other research and development activities within the larger regional knowledge production ecosystem". From this perspective, one might expect that collaboration between institutions, regions, and countries is at the heart of developing Artificial Intelligence knowledge. Further, to understand the creation of AI, it is paramount that we study cross-country collaboration networks in our WoS subsample.

Along these lines, Figure 13.4 uses the information on the co-location of authors listed within the same WoS document to generate a global collaboration network for AI. Country nodes are coloured in shades of grey according to the continent they belong to, whereas their size indicates the number of AI publications in that country. To draw the networks, we used a force-directed algorithm to ensure the position of the nodes is proportional to their graph distances. Hence, from this network, we can deduce that countries that are closer together collaborate more frequently than those further apart. Moreover, as one might expect, the major AI producers we identified in the last sections (e.g., China and the United States), also occupy the centre of the collaborative network.[5] That is, these big AI producers also are among the most influential nation-states in the international collaboration network.

Beyond this, we observe that a large proportion of international collaboration occurs between countries within the same continent. Cultural and social arguments from economic

5 One potential reason why Europe might appear as less innovative to America (which is in contrast to Figures 13.1 and 13.2) is because Europe is a collection of many individual countries whereas the Americas are largely dominated by the United States.

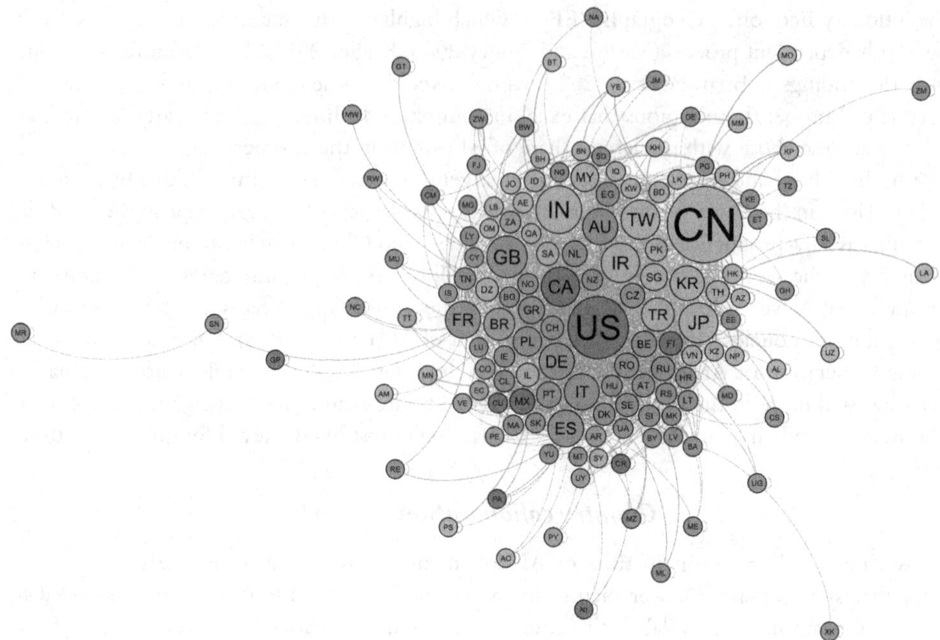

Figure 13.4 Global AI science/knowledge collaborator network

Source: Authors' calculation/illustration

geography may further help to explain these trends. Particularly, we expect that collaborations and interactions are more likely to occur between agents (e.g., individuals, firms, countries) that share the same language, customs, and routines (Boschma 2005). Finally, we must acknowledge those countries scattered on the periphery of the collaboration network who have yet to establish a serious footing in AI and may have only published a few articles. Many of these are either African or those smaller Asian economies. From an evolutionary perspective, it will be interesting to see how these economies develop over the coming years and whether they will enter the global AI collaboration network.

Evolution of keywords

As a methodological axiom, co-occurrence analysis is a valuable research strategy. It has found residence in a wide variety of fields, including economic geography (Kogler et al. 2013), regional development (Hidalgo et al. 2007), and scientometrics (Leydesdorff 2007). Likewise, with the advent of big-data, co-occurrence analysis based on the frequency of (key)words that occur in the same publication has also been identified as a burgeoning research field. Indeed, while earlier research sought to measure similarity across authors and fields using the co-citations networks (McCain 1990), more recent analysis has drawn on advances in text-mining and text-analytics to map knowledge structures using the co-occurrence of words. These developments are particularly helpful when tracing the evolution and intellectual structures of emerging new fields, such as the Internet of Things (Yan et al. 2015) and Infometrics (Sedighi 2016). Further, the methodology can also be used to produce bibliometric data on particular journals (Ravikumar et al. 2015), and to produce systematic literature reviews (Zhu et al. 2019).

Following this line of inquiry, Figure 13.5 illustrates the keyword co-occurrence network for the AI documents in the sample. It splits the analysis into the previously discussed periods in order to examine the changing research frontier of Artificial Intelligence. Here, each node represents a keyword with its size being proportional to the frequency at which it occurs in journal articles. Like before, when drawing these networks, we ensured that keywords that frequently co-occur across our AI data sample are closer together than those that do not. Doing so reveals a core-periphery structure with the most focal concepts at the centre. At last, for a better visualisation, the ten most common nodes are highlighted in red and have been labelled.

Although Artificial Intelligence initially developed slowly (see previous subsection about "global focus"), it still produced a very dense network of approximately 1,600 nodes and 32,000 edges. A plausible explanation is that, due to its infancy, many researchers were actively experimenting and trying to find applications for such a radical technology. In Figure 13.5(a), the three largest nodes are "Neural Network", "Genetic Algorithm", and "Pattern Recognition", which,

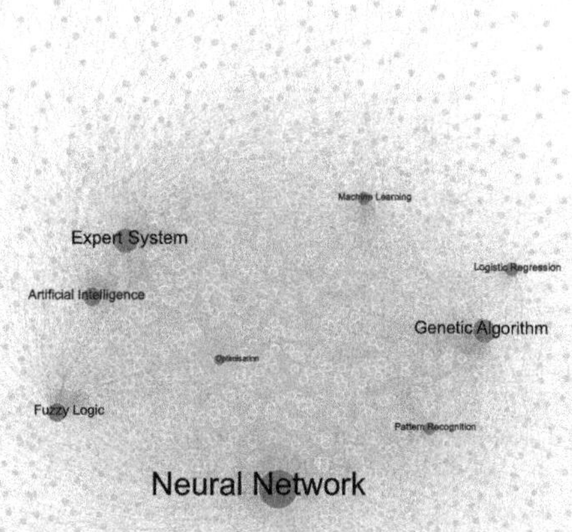

Figure 13.5a The changing research frontier of AI and the evolution of AI keywords: 1990–1999

Source: Authors' calculation/illustration

to this day, are considered frontier concepts in the field. Furthermore, other concepts such as "Optimisation", "Fuzzy Logic", and "Classification" have also begun to emerge – although in a far smaller capacity. Like before, these concepts are also inherent to AI and, in particular, its trial-and-error processes. Lastly, whilst it is often very informative to know which nodes are present, it is equally valuable to know which nodes have yet to appear. To these ends, concepts relating to "Decision Trees", "Data Mining", or "Unsupervised Learning" are still missing at this first stage.

Moving to Figure 13.5(b), what immediately becomes clear is that there are significantly more nodes. Indeed, even after filtering we observe 400 more nodes than previously, which indicates that the network has grown. Likewise, we observe a significant shift regarding how said networks are wired. That is, comparing the two periods, not only the keyword co-occurrence network has increased in size, but it also has become denser and more concentrated. As with the period before, "Neural Network" remains the largest node, which is unsurprising given its focal positioning in the study of Artificial Intelligence (see Figure 13.1). Other noticeable changes include the introduction of keywords like "Data Mining", "Support Vector Machines", and "Reinforcement Learning". Colloquially, these tokens are commonly used to explain Artificial Intelligence, so it is not that surprising they appear here. Beyond this, there was also an obvious concentration around the concepts of "Fuzzy Logic", "Classification", and "Optimisation", which again have a strong resonance with AI.

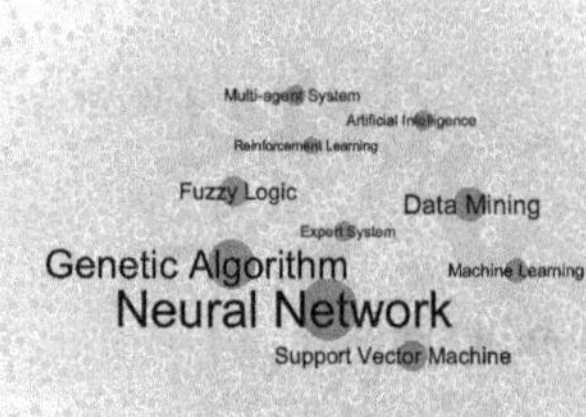

Figure 13.5b The changing research frontier of AI and the evolution of AI keywords: 2000–2009

Source: Authors' calculation/illustration

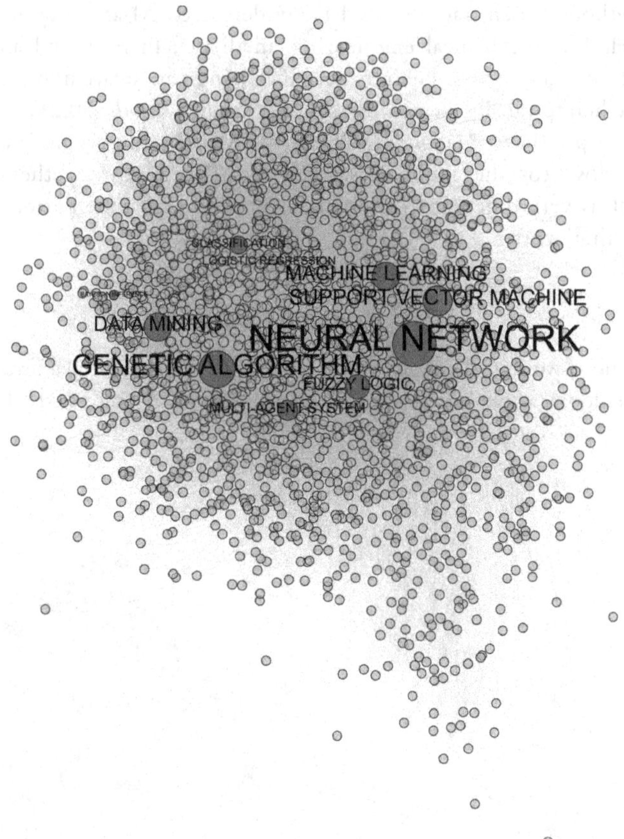

Figure 13.5c The changing research frontier of AI and the evolution of AI keywords: 2010–2016

Source: Authors' calculation/illustration

However, perhaps the most important shift between the two time periods was the concentration around the keyword "Genetic Algorithm". Although inherently different from Neural Network, the two terms co-occur rather frequently on the same documents.[6] Indeed, throughout the period in consideration, a little over 2,000 documents listed both terms as keywords – which represents nearly 10% of all "Genetic Algorithm" articles in the period. In turn, this event highlights the potential for combining different AI methods to achieve faster and better algorithms, particularly for these two widespread methods.

Moving to the last stage, one can see that the network has become increasingly dense. This is largely because while the shift from Figure 13.5(b) to 13.5(c) reports only a small increase in the number of nodes, it sees a massive increase in the overall number of edges, or connections between nodes, which now stands at over 157,000. The surge in edges implies that AI has become increasingly intertwined with other disciplines. Otherwise stated, the

6 Neural networks are a sub-form of deep-learning where the algorithms are inspired by the structure of the human brain. In short, neural networks are trained to identify patterns in data (text, audio, visuals, etc.) and then predict outputs for a new set of similar data. Genetic Algorithms on the other hand reflect the processes of natural selection where the fittest mutations are selected for producing the next generation.

theory and methods which traditionally have underscored AI are now finding residence in new areas, including mechanical engineering, medicine, finance, and automation. These changes are the driving factors behind the self-driving car, smart home technologies, and mechanical medicine. Finally, keywords such as *Neural Network* (50,237), *Genetic Algorithm* (26,242), and *Support Vector Machine* (15,250) remain vitally important. We also see they are now produced closer together in the network; that is, they appear together on the same publications, which is very different than observed in the initial time period where they were distinct and further apart.

Regional focus

In this final section, we shift the focus of our analysis to the subnational level. That is, we will focus on the temporal evolution and spatial distribution of AI across 318 European NUTS2

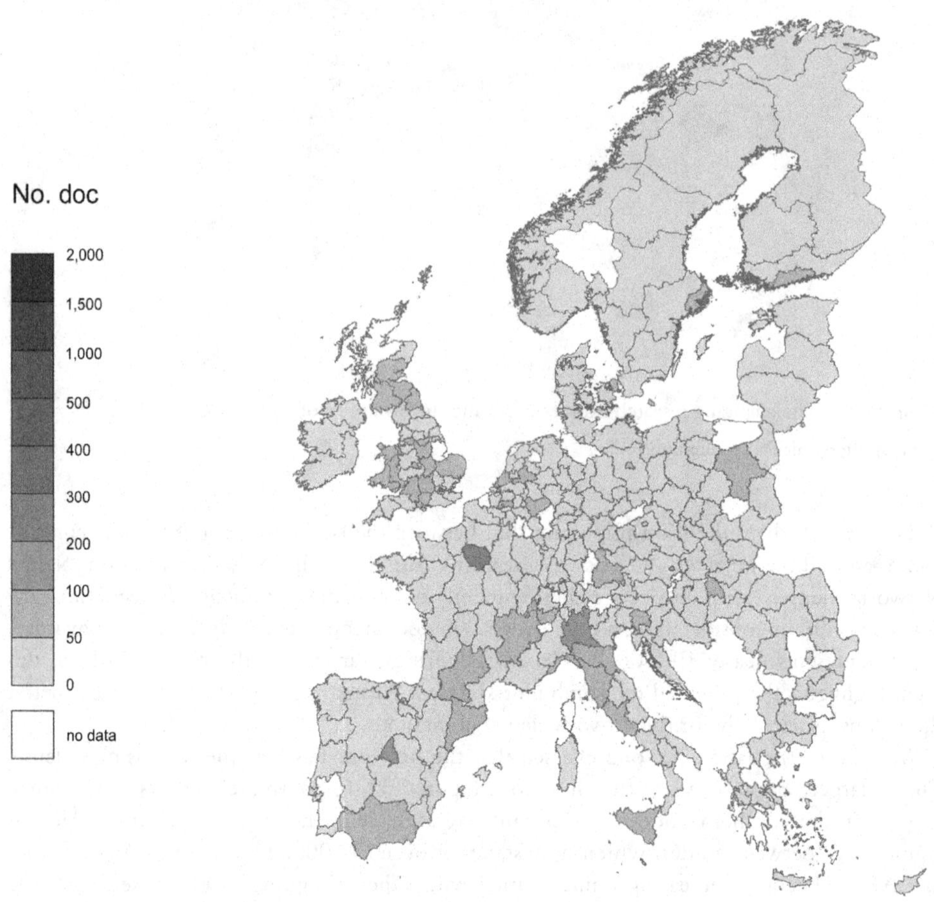

Figure 13.6a The spatial and temporal evolution of Artificial Intelligence science across European regional economies: 1990–1999

Source: Authors' calculation/illustration

regions. In this context, Figure 13.6 compares the total number of AI articles in our dataset over the NUTS2 areas in the initial period 1 (1990–1999) compared to the final period 3 (2010–2016). Supplementary information about the top AI producing regions of Europe is also provided in Table 13.1.

Surprisingly, Figure 13.6(a) shows that most regions appear to have had an early start publishing in AI. During the first period, the median number of AI documents was 68, which in part illustrates the technology's novelty. At the same time, it is possible to identify some early AI "hotspots" – with the Île de France (FR10) appearing as a driving force. By the same token, there are clusters of activity in the South-East of England (UK13), Northern Italy (ITC4), and in Central Spain (ES30).

Moving between Figures 13.6(a) and 13.6(b), several differences are immediately apparent. Early hotspots like Île de France, Madrid, Lombardy, and London retained their status as leaders.

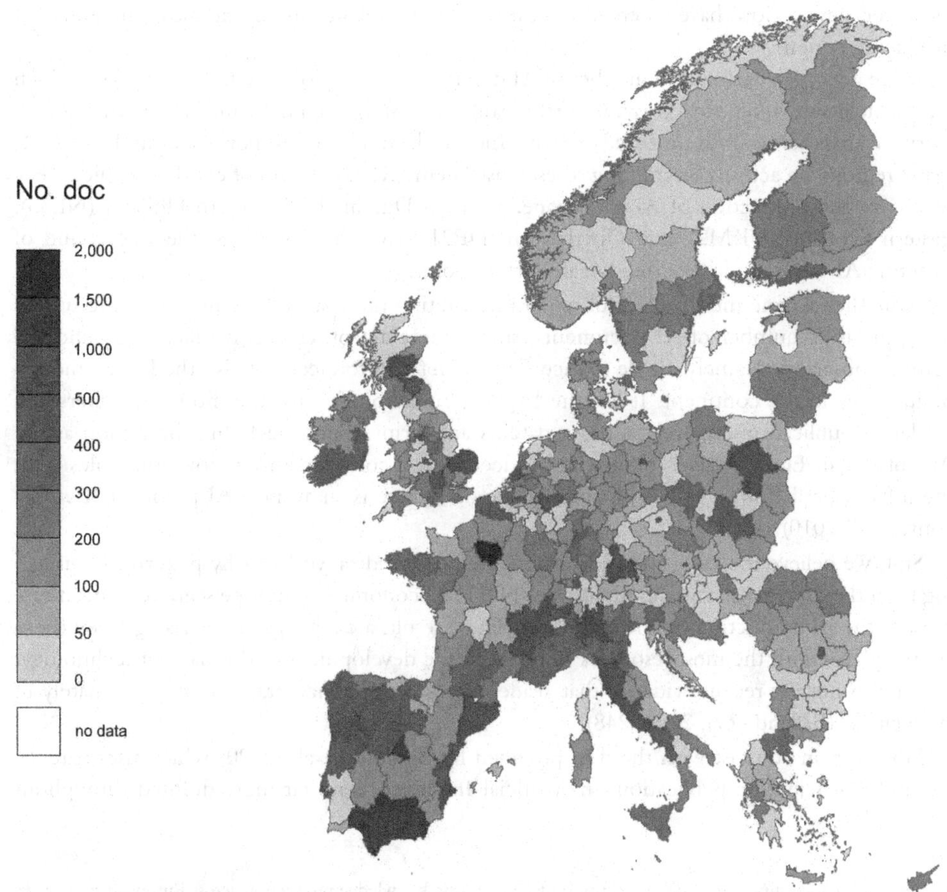

Figure 13.6b The spatial and temporal evolution of Artificial Intelligence science across European regional economies: 2010–2016

Source: Authors' calculation/illustration

Table 13.1 Top AI producing regions

NUTS2 region	*Period 1*	*NUTS2 region*	*Period 3*
Île de France (FR10)	250	**Île de France** (FR10)	1,516
Lombardia (ITC4)	127	**Madrid** (ES30)	1,264
Inner London – West (UKI3)	118	**Andalucía** (ES61)	1,123
Madrid (ES30)	101	**Inner London – West** (UKI3)	1,021
South Yorkshire (UKE3)	96	**Cataluña** (ES51)	863
South Holland (NL33)	92	**Norte** (PT11)	765
West Midlands (UKG3)	91	**Lombardia** (ITC4)	714
Rhône-Alpes (FR71)	87	**Fermo** (ITI4)	684
Southern West Scotland (UKM3)	84	**Rhône-Alpes** (FR71)	662
Attica (EL30)	83	**Mazowieckie** (PL12)	645

Source: Authors' calculation

However, new regions have emerged indicating a restructuring among the most influential AI producing regions.[7]

More generally, the median number of AI documents per region rose from 68 to 344, which represents nearly a five-fold growth. Further still, this rise is potentially indicative of achieving a critical mass across both time and regions. Indeed, beyond the hotspots listed in Table 17.1, many regions are actively attempting to establish themselves as centres of excellence and as key players in the production of AI in Europe. Visually, Dublin (IE02), South Holland (NL33), Eastern Scotland (UKM2), and Oberbayern (DE21) have all moved into the foreground of scientific AI research, despite their weak starting point.

Nonetheless, one must be careful when interpreting the results. Foremost, because we are using the total number of AI documents, surely there are other decisive factors behind the patterns observed. Namely, Île de France – our chief AI producer – is also the largest metropolitan area in the continent. It has the largest population density, the most researchers and academic publications, as well as more patents and firms. Thus, declaring the region as the AI hotspot in Europe seems premature. Indeed, if we consider how innovation scales with the urban population, it might just be that Île de France is an average AI producer (Bettencourt et al. 2010).

Still, we believe the table and graphs presented earlier do reveal worthy patterns. Continuing from the previous paragraph, the largest NUTS2 economies in Europe seem to concentrate most of the AI production on the continent. These results are perhaps unsurprising given those are the places with the most resources to invest in the development of this nascent technology; but, it also mirrors recent evidence that academic research "concentrates disproportionately in large cities" (Balland et al. 2020: 248).

However, in contrast with the data provided by Buarque et al. (2020) which uses patents, we find that scientific publications in Artificial Intelligence are far more diffused throughout

7 The results concerning the spatial distribution of AI scientific knowledge production across European regions, in particular for period 3 (2010–2016), are not necessarily what we would have expected at the onset based on prior research on the distribution of AI technical knowledge production (Buarque et al. 2020). We've conducted some further investigation into the validity of these results; all records for region PT11 (Portugal, North) were reviewed manually to ensure consistency in our geocoding approach, which confirmed the findings.

the continent. Smaller regions, and those often considered marginal in the European market, are producing a lot more scientific knowledge on AI than one might expect from the patent data. In turn, this could reveal inherent differences on how far and fast scientific or practical knowledge travels.

While the previous paragraphs describe the distribution of AI documents in Europe, the present analysis seeks to go beyond this and examine how AI-specific knowledge connects to other sectors of the regional economy. That is, we wish to estimate how AI is embedded in the local knowledge-producing and innovative environment. Since many AI-promoting policies seek to specifically develop with an eye towards multiple commercial products and processes (Dutton 2018), we need to understand how AI interacts with other domains of knowledge.

To construct a measure of AI embeddedness, we follow the approach introduced by Kogler et al. (2013, 2017) and produce a regional knowledge space for each European region. Using data contained in patent documents, these authors discern a measure of technological relatedness based on the co-occurrence of "Cooperative Patent Classification" (CPC) classes. Focusing on the US metropolitan areas, they discovered a link between higher levels of technological relatedness and faster rates of patenting per worker. In a subsequent analysis, Rigby (2015) found that technologies related to a region's pre-existing knowledge were more likely to enter the said region than those that were further apart from the region's expertise. Since then, geographers and regional scientists have applied these general principles to examine the innovative ability of cities, regions, and countries using a variety of indicators (see Whittle and Kogler 2020 for an overview of these methods).

Nonetheless, for the present examination, we are particularly interested in the recent works by Feldman et al. (2015) and Buarque et al. (2020). The first examined the diffusion of rDNA technology and illustrated how cognitive and geographical proximity affects the spread of this revolutionary method. Whereas Buarque et al. (2020) illustrated how bibliographical analysis alongside embeddedness studies can be used to measure the creation and integration of Artificial Intelligence in Europe.

However, despite their value, these contributions have exclusively used patent data to compute the innovative performance of regions. Whilst earlier research has recognised patents as an excellent proxy for innovation, especially on a regional scale (Acs et al. 2002), they also have significant limitations. Particularly, it has been argued that patents are the result of R&D and therefore reflect the innovative output potential of a region. In turn, we wish to study more succinctly the inputs of knowledge creation. To this end, by looking at academic publications, which form the bedrock on which many patents are created, we have a more accurate picture of the creation and diffusion of AI in the regional economies.

We proceed by employing the methods mentioned in the above paragraphs, only this time, we do so to our AI sample of the Web of Science. In other words, using detailed information within each article, we can produce graphs that map the co-occurrence of keywords across all the documents in a NUTS2 region – a local science space. Every WoS entry has at least one keyword, but most have between four and five. These keywords are signifiers and provide a snapshot of the document's underlying knowledge. Thus, by examining the frequency at which individual keywords occur together in our data sample, we can generate matrices of how related these AI keywords are to one another. Namely, as we did for all the AI documents, we can assume that keywords that often co-occur together are more related than those that do not.

Hence, we graphed the AI scientific space for all the 318 European NUTS2 in our sample. To be specific, to make these graphs, we used all documents flagged as AI that contained at least one author residing in the region at the time of publication. Next, we plotted the data from these documents in a network, where each keyword is a node and, each time two keywords show on the same publication, we create an edge between the two. Then, to understand how the production and integration of AI knowledge vary across Europe, we collect several network characteristics data for all the regions' scientific space. In other words, we can measure how dense, centralized, clustered, and how long it takes to traverse each AI regional network. In turn, we expect these statistics will allow us to infer how embedded Artificial Intelligence knowledge is across the European regions. For example, we imagine that where the AI scientific space is denser and longer, the technology is more rooted in the local innovation environment – as it seems to be more connected to other sectors.

Along these lines, Figure 13.7 shows the distribution of two network statistics from the AI scientific spaces for all the regions with at least 100 documents published between 1990–2016. On the top graph, we show the distribution of network density – a measure of the proportion of existing edges out of all possible links in the network. And on the bottom, we display the degree centralization of these scientific spaces, which captures how central the most central node is, compared to all other nodes in the network. To measure centralization, we first calculate the sum in differences in centrality between the most central node and all other nodes, then divide this value by the theoretically largest possible difference in any network of the same size. As one may conclude, the European regions are quite different regarding their "AI embeddedness." Indeed, it seems like AI-specific knowledge is far more centralized in some regions than others. The region's AI scientific space is more concentrated in a few keywords, and it lacks some potential applications and alternative methods. On the other hand, some regions have a denser network – thus demonstrating a more connected AI knowledge space. The result is not surprising in light of research by Buarque et al. (2020) who measured a very skewed distribution for the AI centrality index in patents.

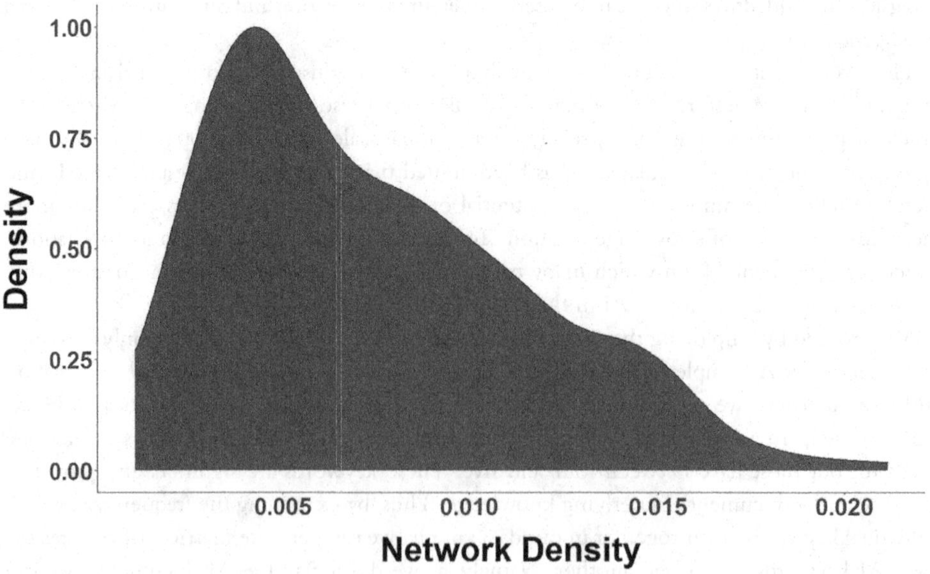

Figure 13.7 AI science space network statistics

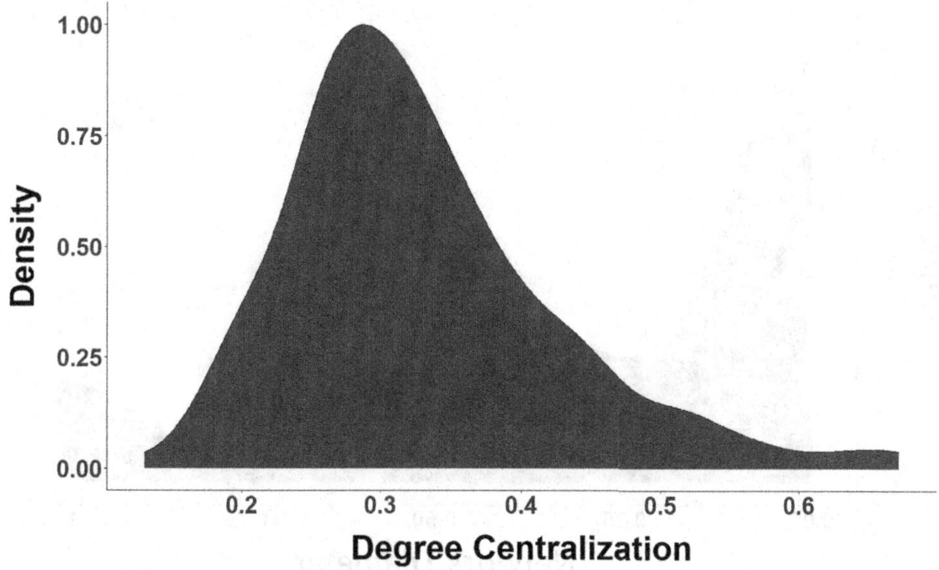

Figure 13.7 Continued

Source: Authors' calculation/illustration

Note: The graphs display the density distribution for two network statistics collected from the science space of 213 European NUTS2 regions. We include in these figures all the places with at least 100 documents published during the 1990-2016 period. The x-axis shows the values for the descriptive statistics, while on the y-axis, we display the number of places presenting that range of values. We scale the y-axis between one and zero, where the statistic with the highest number of regions is equal to one. Network density refers to the proportion of existing edges out of all possible links in the network. We measure degree centralization according to the formula $\frac{\sum_{i=1}^{N} c_x(p_*) - c_x(p_i)}{\max \sum_{i=1}^{N} c_x(p_*) - c_x(p_i)}$
The nominator represents the sum in differences in centrality between the most central node and all other nodes, and the denominator the theoretical maximum value of such differences.

We must remark, however, that it can be challenging to make meaning from these network statistics. First, it can be complex to interpret the results. For example, what conclusions can we make from finding that a region has a degree centralization of 0.5? Without a proper benchmark, these network variables do not tell us much. More importantly, there is ample evidence that network statistics vary significantly even when the networks come from the same model. And that these metrics are highly dependent on the networks' sizes (Anderson et al., 1999; Faust and Skvoretz, 2002; Van Wijk et al., 2010). Along these lines, we understand that "to make comparisons, the measures must be on the same scale and have the same meaning across the various networks; however, when the networks being compared vary in size, often the values of these network statistics can be dominated by size effects" (Smith et al., 2016). Therefore, for better metrics, we follow the seminar work by Anderson et al. (1999) and first compare the regional networks to random graphs (Erdős-Rényi) of the appropriate size. That is to say, for every NUTS2, we generate 1,000 random networks containing the same number of nodes and density. We then collect the same statistics from these random graphs and normalize the metrics from our regional science space against their equivalent random distribution. We thus obtain scale metrics between one and zero, and Figure 13.8 displays the density distribution for two such values – network diameter and average path length. Both these statistics offer us information about how

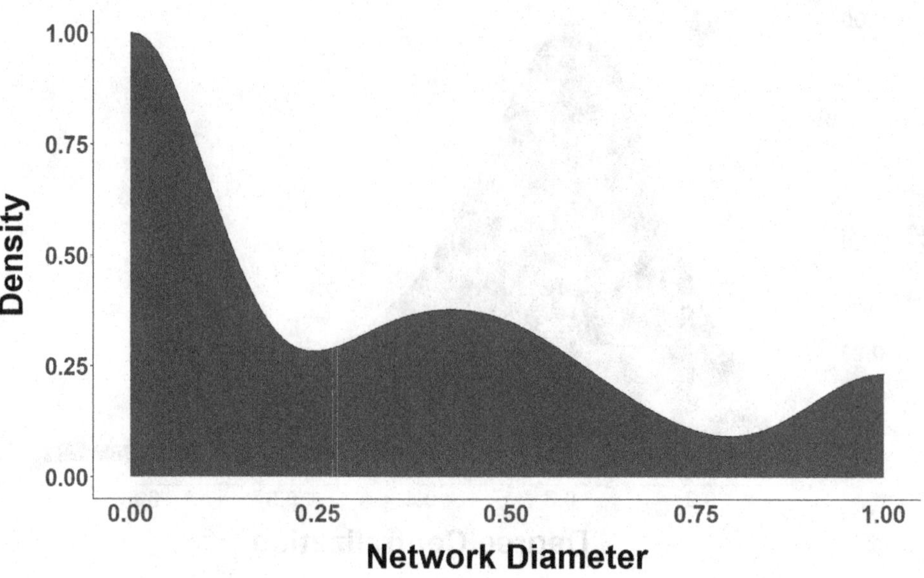

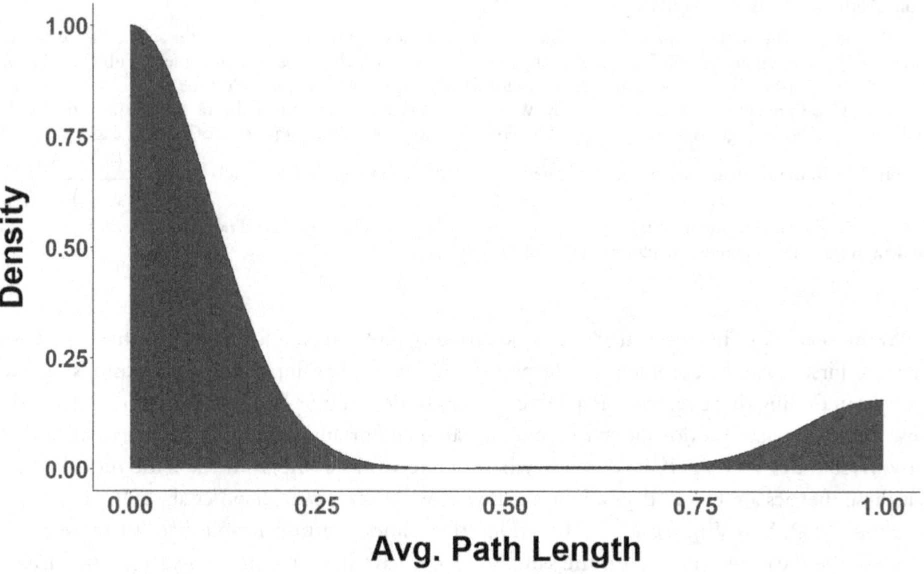

Figure 13.8 AI science space network statistics scaled

Source: Authors' calculation/illustration

Note: Note: The plots display the density distribution for two network statistics from the science space of 214 European NUTS2 regions – those with at least 100 documents during 1990–2016. The x-axis shows the values for the descriptive statistics, where we scale these values against the distribution of 1,000 equivalent statistics from random graphs of similar sizes (Anderson et al., 1999). The y-axis represents the number of places presenting that range of values. We scale the y-axis between one and zero, where the statistic with the highest number of regions is equal to one. Network diameter refers to the shortest distance between the two most distant nodes of the network. And average path length is the average number of steps along the shortest paths of all possible pairs of nodes.

difficult it is to transverse the regional science spaces. And once again, we find the European regions have very different AI networks regarding their connectivity and span.

Given the significant divergence across the European regional networks, one might propose to examine how the level of AI embeddedness relates to the amount of AI knowledge produced by the NUTS2. To these ends, Buarque et al. (2020:186) sought to expose the correlation between the number of AI patents and their relative importance for the region's knowledge space. Accordingly, they show "there is a positive correlation between those regions where AI patents are most prevalent and those for which AI is most embedded." We must remark, nonetheless, that our methodology diverges from Buarque et al. (2020) in one vital detail. They estimated the value of AI patents in the regional knowledge space by artificially removing said patents when building the graphs and thus observing the impact of this exclusion on the networks' structures. In contrast, the present analysis exclusively represents the scientific space for the AI documents. We consider only the co-occurrence of keywords within articles that we identified as AI and do not estimate the region's overall scientific space - i.e., our focus concerns the AI-specific knowledge.

We must equally recognize that we are focusing on differences regarding the networks' global statistics. These provide valuable insights into the overall structure of the regional science spaces and allow us to reflect on how inclusive or rooted is AI knowledge. But there are limitations to using these methods. For example, as mentioned before, the global statistics do not necessarily "yield robust results," and "often fails in catching important local features" like communities and others (Tantardini et al., 2019:3). Of course, the literature offers several other techniques one could use to measure diversity across the AI science spaces, and Tantardini et al. (2019) provides a valuable review on these many alternatives. Although beyond the scope of this initial examination, one could readily employ any prefered "network comparing method" to strengthen our comprehension of how distinct are the European NUTS regarding their AI production. Likewise, they could measure how similar the regional networks are to one another. Hence, future research can profit from our approach coupled with robust statistics to examine what distinguishes the AI key players. What are the driving forces behind the different AI embeddedness? Is there notable structural variance across the regions – e.g., some are more focused on applied systems than others? And even study the role of social and geographical proximity in shaping these regional networks – i.e., do regions nearby exhibit more similar AI science spaces?.

Along these lines, to further substantiate the value of our science space statistics, Figure 13.9 illustrates two regional AI networks we built using data from 2010-2016. On the top, we have Dublin (IE02) and on the bottom Vienna (AT13). For comparative purposes both regions belong to high-income countries, have a similar number of universities, and enjoy a very high standard of living. As depicted earlier, both regions are significant producers of Artificial Intelligence and have roughly the same amount of journal articles, IE02 (422) and AT13 (381). However, despite their commonalities, these regions produce very different network structures with AI occupying a more central position in Dublin.

In terms of the sheer number of AI-specific keywords, Dublin and Vienna are once again very similar. Of the 36 keywords[8] listed by the WIPO (2019), Vienna has published with 23 of them, whereas Dublin published with 26. Although the volume of documents and keywords in a region is indicative of its capacity to produce AI knowledge, in order to understand the region's full potential we must also account for the links between keywords and the interconnections among the different kinds of knowledge produced. Along these lines, comparing the

8 The World Intellectual Property Organization (WIPO 2019) provides 43 n-grams, which we used to identify the AI documents. We grouped the different n-grams into 36 keywords. Namely, we grouped together terms that refer to the same or very similar method – such as, "Supervised Training" and "Supervised Learning".

Figure 13.9a Dublin and Vienna science space of AI: Dublin (IE02)

Source: Authors' calculation/illustration

two networks shown in Figure 13.9a, we first observe that Dublin's scientific AI space has more nodes overall and is denser than Vienna's. That is, visually, we can conclude that Dublin combines more industries and sectors into its AI network; thus leading to a more diverse and applied technology when compared to Vienna.

Furthermore, in both networks under consideration, we highlight "Neural Networks" and "Machine Learning" as the most traversed nodes, that is, the most relevant keywords. For Dublin, you can see that these keywords are closely connected, indicating they frequently occur in the same publications. Moreover, these keywords are also tightly surrounded by other nodes (both AI and non-AI), further demonstrating their recombinatorial potential. Recall from the previous sections how AI has become increasingly intertwining with other sectors of the economy (Frey and Osborne 2017) and technological frontiers (Buarque et al. 2020); it is precisely this recombination that is driving AI policy and regional development (Clifton et al. 2020; Acemoglu and Restrepo 2020).

Conversely, whilst "Neural Network" and "Machine Learning" are also the most connected nodes in Vienna's network, they are not as embedded in the region's scientific space. Thus, it seems that the region has been unable to connect distinct research frontiers in AI, which significantly hampers its ability to harness the capabilities of Artificial Intelligence. Insights from

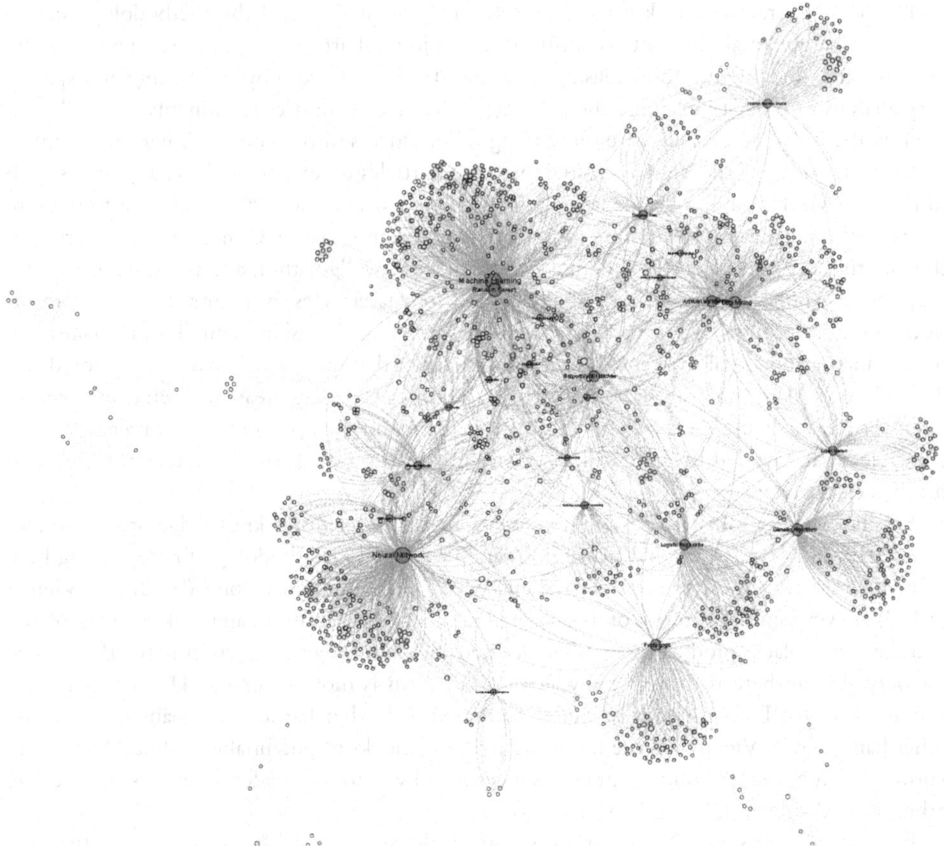

Figure 13.9b Dublin and Vienna science space of AI: Vienna (AT13)
Source: Authors' calculation/illustration

Evolutionary Economic Geography (Kogler 2016) further substantiate this point, illustrating that though Vienna might have the necessary building blocks, it fails to connect them in a meaningful way and as a result their network remains sparsely connected.

Conclusion

Artificial Intelligence is currently one of, if not the most, widely debated science-technology fields in business and policy circles, and the rush to develop and market AI-related technology is palpable. Since its emergence in the early 1990s, governments around the world have been keen to develop strategies in order to harness and capitalise on its obvious societal and economic potential. In this context, the purpose of the present contribution is to provide insights into the spatial and temporal evolution of AI scientific knowledge production over the past three decades. Following this vision, the objective we set out with was to make a series of connected contributions to the relevant literature, both theoretical and empirical, all of which should inspire and support further work on this relevant topic. In particular, the study provides insights into three aspects that should advance this line of inquiry.

Firstly, to overcome a lack of precise definition, we augmented the methodology developed by Buarque et al. (2020) to identify AI-based journal articles and indexed approximately 260,000 such documents. To establish a solid foundation, we began by conducting an exploratory analysis in order to examine the path-dependence of countries (continents), as well as to identify the jurisdictions that were the leading AI producers throughout each period in time.

Secondly, utilising the keywords listed on journal articles we examined the changing research frontier of AI. Through a detailed analysis of their co-occurrence, it was possible to explore how AI is simultaneously becoming more concentrated and diverse. Concentrated by virtue of the fact that the core concepts – neural networks, genetic algorithm, and machine learning – that define AI are appearing more frequently on journal articles over time. Diverse in terms of the number of non-AI keywords that are also appearing alongside them. This indicates AI's recombinatory potential whereby theories and methods that traditionally have underscored and defined core AI research are now also finding residence in new areas, including mechanical engineering, medicine, finance, and automation. As mentioned previously, it is precisely these principles that have led to the creation of the self-driving car, smart home technologies, and mechanical medicine.

Finally, we positioned these AI documents into the scientific knowledge space (Kogler et al. 2013, 2017) of two capital EU regions and developed a methodology for describing how embedded AI is in these regions. The results reveal that although Dublin (IE02) and Vienna (AT13) are very similar in terms of their overall number of publications and AI keywords, by the end these two places produce very different scientific knowledge production networks. A preliminary finding here is that AI knowledge production is more central in Dublin's network, and as such, that Dublin might be better equipped to further harness its capabilities. On the other hand, whilst Vienna has the necessary building blocks to potentially exploit AI scientific knowledge, it has yet to connect these in a meaningful way to its broader network structure (i.e. other non-AI subjects).

In terms of next steps, an obvious direction would be to extend the methodology that was utilised in this study and to include information embedded in the relevant publications regarding authors and their institutions. The addition of this micro-dimension would permit a more thorough and detailed analysis of both the creation and diffusion of AI focusing specifically on those actors involved. In doing so, it would be possible to analyse not just which countries are collaborating, but also the institutions and individuals embedded in these countries. Such an analysis would be of critical importance in identifying those institutions that are at the forefront of AI scientific knowledge production and could be used by policymakers and funding agencies when targeting specific investment opportunities. Similarly, by focusing on authors, it would possible to discern (at the institutional/departmental level) who is collaborating with whom. For example, if a researcher in computer science (i.e. AI research), is engaged in a collaborate process with a colleague in medicine, the expectation then would be that scientific AI knowledge gets applied to a specific problem, and in turn provides inputs to generate a solution. It is these inter-disciplinary collaborations that provide the opportunity to produce recombinant knowledge with the potential to push forward technological change and the research frontier in the science/technology knowledge space (Kogler et al. 2013, 2017; Kedron et al. 2020).

This point also speaks more broadly to a crucial methodological contribution in the present investigation. In particular, we further substantiate the viability of text-matching and text-analysis methodologies for identifying and analysing the creation and diffusion of science and technologies when they are not easily identifiable by traditional means. Haščič and Migotto (2015) provide an additional example that follows this approach where a text-matching algorithm is employed in order to identify "green" technologies. We hope that the present study will inspire other scholars to further explore AI scientific knowledge production as well as other non-standardised technology fields at the intersection of traditional domains.

References

Acemoglu, D. and Restrepo, P. (2017): Robots and Jobs: Evidence from US Labor Markets. Technical report. *NBER Working Paper No. w23287*. Cambridge, MA: National Bureau of Economic Research. Available at: www.nber.org/papers/w23285 [30.01.2021]

Acemoglu, D. and Restrepo, P. (2019): Automation and New Tasks: How Technology Displaces and Reinstates Labor, *Journal of Economic Perspectives* 33(2), 3–30.

Acemoglu, D. and Restrepo, P. (2020): The Wrong Kind of AI? Artificial Intelligence and the Future of Labour Demand, *Cambridge Journal of Regions, Economy and Society* 13(1), 25–35. https://doi.org/10.1093/cjres/rsz022

Acs, Z., Anselin, L. and Varga, A. (2002): Patents and Innovation Counts as Measures of Regional Production of New Knowledge, *Research Policy* 31(7), 1069–1085. https://doi.org/10.1016/S0048-7333(01)00184-6

Aghion, P, Jones, B. and Jones, C. (2017): Artificial Intelligence and Economic Growth. Technical report. *NBER Working Paper No. w23928*. Cambridge, MA: National Bureau of Economic Research. Available at: www.nber.org/papers/w23928 [30.01.2021]

Agrawal, A., Gans, J. S. and Goldfarb, A. (2019): Artificial Intelligence: The Ambiguous Labor Market Impact of Automating Prediction, *Journal of Economic Perspectives* 33(2), 31–50.

Anderson, B. S., Butts, C., and Carley, K. (1999): The Interaction of Size and Density with Graph-level Indicess. *Social Networks*, 21(3), 239–267.

Arntz, M., Gregory, U. and Zierahn, U. (2016): The Risk of Automation for Jobs in OECD Countries: A Comparative Analysis. *OECD Social, Employment and Migration Working Papers, No. 189*, OECD Publishing, Paris. https://doi.org/10.1787/5jlz9h56dvq7-en

Balland, P. A., Jara-Figueroa, C., Petralia, S. G., Steijn, M. P., Rigby, D. L., and Hidalgo, C. A. (2020): Complex Economic Activities Concentrate in Large Cities, *Nature Human Behaviour* 4(3), 248–254.

Barabási, A.-L. and Albert, R. (1999): Emergence of Scaling in Random Networks. *Science* 286(5439), 509–512.

Boschma, R. (2005): Proximity and Innovation: A Critical Assessment, *Regional Studies* 39(1), 61–74. https://doi.org/10.1080/0034340052000320887

Brynjolfsson, E., Rock, D. and Syverson, C. (2018): Artificial Intelligence and the Modern Productivity Paradox: A Clash of Expectations and Statistics, in Agrawal, Ajay, Gans, Joshua and Goldfarb, Avi (eds.): *The Economics of Artificial intelligence: An Agenda*. Chicago: University of Chicago Press, 23–66.

Buarque, B, Davies, D., Hynes, R. and Kogler, D. F. (2020): OK Computer: The Creation and Integration of AI in Europe, *Cambridge Journal of Regions, Economy and Society* 13(1), 175–192. https://doi.org/10.1093/cjres/rsz023

Clifton, J., Glasmeier, A. and Gray, M. (2020): When Machines Think for Us: The Consequences of Work and Place, *Cambridge Journal of Regions, Economy and Society* 13(1), 3–23. https://doi.org/10.1093/cjres/rsaa004

Cockburn, I. M., Henderson, R. and Stern, S. (2018): The Impact of Artificial intelligence on Innovation. Technical report. *NBER Working Paper No. w24449*. Cambridge, MA: National Bureau of Economic Research. Available at: www.nber.org/papers/w24449 [30.01.2021]

Dutton, T. (2018): An Overview of National AI Strategies, in Medium. Available at: https://medium.com/politics-ai/an-overview-of-nationalai-strategies-2a70ec6edfd. [30.01.2021]

Faust, K. and Skvoretz, J. (2002): Comparing Networks across Space and Time, Size and Species. *Sociological Methodology*. 32(1), 267–299.

Feldman, M., Kogler, D. F. and Rigby, D. (2015): rKnowledge: The Spatial Diffusion and Adoption of rDNA Methods, *Regional Studies* 49(5), 798–817. https://doi.org/10.1080/00343404.2014.980799

Frey, C. and Osborne, M. (2017): The Future of Employment: How Susceptible are Jobs to Computerisation? *Technological Forecasting and Social Change* 114, 254–280. https://doi.org/10.1016/j.techfore.2016.08.019

Furman, J. and Seamans, R. (2019): AI and the Economy, *Innovation Policy and the Economy* 19(1), 161–191. https://doi.org/10.1086/699936

Goldfarb, A. and Trefler, D. (2018): AI and International Trade, in Agrawal, A., Gans, J. and Goldfarb, A. (eds.): *The Economics of Artificial intelligence: An Agenda*. Chicago: University of Chicago Press, 463–493.

Graetz, G. and Michaels, G. (2018): Robots at Work, *Review of Economics and Statistics* 100(5), 753–768. https://doi.org/10.1162/rest_a_00754

Haščič, I. and Migotto, M. (2015): Measuring Environmental Innovation Using Patent Data. Technical report. Paris: OECD Environment Working Papers, No. 89, OECD Publishing. https://doi.org/10.1787/19970900

Hidalgo, C., Klinger, B., Barabási, A. and Hausmann, R. (2007): The Product Space Conditions the Development of Nations, *Science* 317(5837), 482–487.

Kedron, P., Kogler, D. F., and Rocchetta, S. (2020): Mind the gap: Advancing evolutionary approaches to regional development with progressive empirical strategies. *Geography Compass* 14(9).DOI: 10.1111/gec3.12501.

Kogler, D. F. (ed.) (2016): *Evolutionary Economic Geography: Theoretical and Empirical Progress*. New York: Routledge.

Kogler, D. F. (2017): Relatedness as Driver of Regional Diversification: A Research Agenda – a Commentary, *Regional Studies* 51(3), 365–369. https://doi.org/10.1080/00343404.2016.1276282

Kogler, D. F., Essletzbichler, J. and Rigby, D. (2017): The Evolution of Specialization in the EU15 Knowledge Space, *Journal of Economic Geography* 17(2), 345–373. https://doi.org/10.1093/jeg/lbw024

Kogler, D. F., Rigby, D. and Tucker, I. (2013): Mapping Knowledge Space and Technological Relatedness in US Cities, *European Planning Studies* 21(9), 1374–1391. https://doi.org/10.1080/09654313.2012.755832

Korinek, A. and Stiglitz, J. (2017): Artificial Intelligence and its Implications for Income Distribution and Unemployment. Technical report. *NBER Working Paper No. w24174*. Cambridge, MA: National Bureau of Economic Research. Available at: www.nber.org/papers/w24174 [31.03.2021]

Leydesdorff, L. (2007): Betweenness Centrality as an Indicator of the Interdisciplinarity of Scientific Journals, *Journal of the American Society for Information Science and Technology* 58(9), 1303–1319. https://doi.org/10.1002/asi.20614

Mann, K. and Püttmann, L. (2017): Benign Effects of Automation: New Evidence from Patent Texts. Technical report. SSRN 2959584. Social Science Research Network. Available at: https://ssrn.com/abstract=2959584 [31.03.2021]

Martin, R. and Sunley, P. (2006): Path Dependence and Regional Economic Evolution, *Journal of Economic Geography* 6(4), 395–437. https://doi.org/10.1093/jeg/lbl012

McCain, K. (1990): Mapping Authors in Intellectual Space: A Technical Overview, *Journal of the American Society for Information Science* 41(6), 433–443.

Mongeon, P. and Paul-Hus, A. (2016): The Journal Coverage of Web of Science and Scopus: A Comparative Analysis, *Scientometrics* 106(1), 213–228. https://doi.org/10.1007/s11192-015-1765-5

Nedelkoska, L. and Quintini, G. (2018): Automation, Skill Use and Training. OECD Social, Employment, and Migration Working Papers No 202, OECD Publishing, Paris. https://doi.org/10.1787/1815199X

OECD (2017): Future of work and skills. OECD Employment Outlook Technical report, OECD. https://doi.org/10.1787/19991266

Ravikumar, S., Agrahari, A. and Singh, S. (2015): Mapping the Intellectual Structure of Scientometrics: A Co-word Analysis of the Journal Scientometrics (2005–2010), *Scientometrics* 102(1), 929–955. https://doi.org/10.1007/s11192-014-1402-8

Rigby, D. (2015): Technological Relatedness and Knowledge Space: Entry and Exit of US Cities from Patent Classes, *Regional Studies* 49(11), 1922–1937. https://doi.org/10.1080/00343404.2013.854878

Sedighi, M. (2016): Application of Word Co-occurrence Analysis Method in Mapping of the Scientific Fields (case study: the field of Informetrics), *Library Review* 65(1), 52–64. https://doi.org/10.1108/LR-07-2015-0075

Smith, A., Calder, C. A., and Browning, C. R. (2016): Empirical Reference Distributions for Networks of Different Size. *Social Networks* 47, 24–37.

Tantardini, M., Ieva, F., Tajoli, L., and Piccardi, C. (2019): Comparing Methods for Comparing Networks. *Scientific Reports* 9(1), 1–19.

Van Wijk, B. C., Stam, C. J., and Daffertshofer, A. (2010): Comparing Brain Networks of Different Size and Connectivity Density Using Graph Theory. *PloS One* 5(10), e13701.

Watts, D. J. and Strogatz, S. H. (1998): Collective Dynamics of 'Small- World' Networks. *Nature* 393(6684), 440–442.

Whittle, A. (2019): Local and Nonlocal Knowledge Typologies: Technological Complexity in the Irish Knowledge Space, *European Planning Studies* 27(4), 661–677. https://doi.org/10.1080/09654313.2019.1567695

Whittle, A. (2020): Operationalizing the Knowledge Space: Theory, Methods and Insights for Smart Specialisation, *Regional Studies, Regional Science* 7(1), 27–34. https://doi.org/10.1080/21681376.2019.1703795

Whittle, A. and Kogler, D. F. (2020): Related to What? Reviewing the Literature on Technological Relatedness: Where We Are Now and Where Can We Go? *Papers in Regional Science*, 97–114. https://doi.org/10.1111/pirs.12481

World Intellectual Property Organization (WIPO) (2019): Background Paper for WIPO Technology Trends 2019: Artificial Intelligence. Technical report. Geneva: World Intellectual Property Organization. Available at: www.wipo.int/publications/en/details.jsp?id=4386 [15.03.2021]

Wuchty, S., Jones, B. and Uzzi, B. (2007): The Increasing Dominance of Teams in Production of Knowledge, *Science* 316(5827), 1036–1039.

Yan, B, Lee, T. and Lee, T. (2015): Mapping the Intellectual Structure of the Internet of Things (IoT) Field (2000–2014): A Co-word Analysis, *Scientometrics* 105(2), 1285–1300. https://doi.org/10.1007/s11192-015-1740-1

Zhu, S., Jin, W. and He, C. (2019): On Evolutionary Economic Geography: A Literature Review Using Bibliometric Analysis, *European Planning Studies* 27(4), 639–660. https://doi.org/10.1080/09654313.2019.1568395

Appendix A

Table 13.2 List of AI keywords

artificial intelligence	computation intelligence	neural network bayes	network
bayesian network	rankboost	semi-supervised connections	decision model
deep learning genetic	data mining	semi-supervised training	inductive program
machine learning	natural language generation	natural language generation	reinforcement learning
unsupervised learning	unsupervised training	semi-supervised learning	algorithm
inductive logic	expert system	random forest decision tree	transfer learning
learning algorithm	learning model	support vector machine	adaboost
gradient tree boosting	chatbot	natural language processing	xgboost
logistic regression	stochastic gradient descent	multilayer perceptron	latent semantic analysis
latent dirichlet allocation	multi-agent system	hidden markov model	fussy logic

Source: WIPO (2019)

14
STRUCTURAL DYNAMICS IN THE ERA OF SMART TECHNOLOGIES

Ariel L. Wirkierman

Introduction

Will our jobs be replaced by robots or adapt and complement intelligent machines? How will the share of (human) labour income be affected? Who is entitled to the income generated by a machine learning algorithm? What is the value of data? Will our economy dematerialise as we increasingly digitalise production and consumption?

The economic analysis of structural dynamics may be applied to explore possible answers to these questions. It does so by describing the structure of the economy in terms of its sectoral composition (e.g. primary sector, manufacturing and services), its production factors (e.g. land, capital and labour) and their income entitlements (e.g. rent, profits and wages). Sectors of the economy are connected by buyer-supplier relations, factors are part of production techniques and their income entitlements have different origins, depending on the economic perspective considered.

But there is an inherent tension in the term structural dynamics. *Structure* conveys the idea of relations between parts which have a rather pervasive, time-invariant nature. Instead, *dynamics* refers to processes of change within a system. This apparent tension stems from the confusion between a method of enquiry and the phenomena which it intends to analyse.

As has been noted by Frisch (1929), it is our method which is *static* or *dynamic*, economic phenomena being either *stationary* or *evolutionary*. While a static analysis focuses on comparison of alternatives, a dynamic one formulates quantitative relationships between rates of change of economic magnitudes through time. When adopting a dynamic method, structural dynamics may be seen as an offspring of growth theory.

And growth theory has frequently been formulated on the basis of a set of *benchmark* phenomena – Kaldor's 'stylised facts' (Kaldor 1961: 178) – which alternative frameworks try to either comply with, accommodate or challenge: (1) steady trend growth of labour productivity (i.e. output-per-worker), (2) increasing degree of mechanisation (i.e. capital-to-labour ratio), (3) steady rate of profits on capital, (4) steady (or at least no clear long-term trend in) capital intensity (i.e. capital-to-output ratio) and (5) high correlation between income share of profits and (net) output share of investment, implying a steady wage share and a real wage increasing vis-à-vis average labour productivity.

These facts express relationships between growth rates (e.g. of output, capital and employment) and their implied effects on aggregate ratios of the economy (e.g. capital intensity, degree of mechanisation and distributive shares). They are 'stylised' because they are considered to

 DOI: 10.4324/9780429351921-18

apply across advanced industrial economies. But as 'facts' they convey quasi-stationary phenomena: growth of all magnitudes proceeds at a uniform rate and distributive shares remain unaltered as accumulation proceeds. They imply a *balanced* growth path.

Unless technical progress is assumed to be 'purely labour-augmenting' – increasing the efficiency of each worker but leaving capital intensity and distributive shares unaltered (if competitive conditions prevail) – the evolution (or revolution) in technological possibilities poses a challenge to dynamic analyses based on balanced growth. Smart technologies are no exception.

But the era of smart technologies is ongoing. To pin down an abstract definition and periodisation with respect to previous technological and industrial revolutions is challenging, especially because "ideas about radical technological change are based on *ex-post rationalisations of historical events*" (Kurz et al. 2018: 551, italics added). However, by illustrating their character, we may link smart technologies with their representation within economic structure.

Smart technologies cover mechanical, digital and biological realms, and crucially rely on digitalisation of information. Their adoption is triggering changes in production by industries and consumption by households. Following UNIDO (2019), advanced manufacturing comprises advanced digital production (ADP, hereinafter) technologies, nanotechnology (e.g. nanoelectronics), biotechnology (e.g. genetic engineering) and new materials (e.g. carbon fibre reinforced plastics). Clear-cut distinctions are difficult to make, as these areas intertwine (e.g. new 'nanomaterials' derived from renewable resources) or are interlinked (e.g. nanoscale processes used for quantum computing). In particular, our focus will be on:

> **Advanced digital production [ADP] technologies**: Technologies that combine hardware (advanced robots and 3D printers), software (big data analytics, cloud computing and artificial intelligence) and connectivity (the Internet of things). Advanced digital production technologies are the latest evolution of digital technologies applied to production, a core technological domain associated with the fourth industrial revolution. They give rise to smart production – also referred as the smart factory, or Industry 4.0.
>
> *(UNIDO 2019: xvi)*

This focus is motivated by the topics we aim to analyse.

Robot deployment, additive manufacturing (i.e. 3D printing) and cloud computing are changing the nature and function(s) of (fixed) capital, whereas the looming scenario of technological unemployment stems from the fear of robotisation. These aspects are considered in the following section. To explore the degree to which human workers and intelligent machines will complement or compete with each other, the section afterwards unbundles the task content of labour as a factor of production.

Branches of artificial intelligence – such as machine learning – and big data analytics imply that machines use massive volumes of *human* data as their input. Thus, the valuation of such digital data acquires great relevance. Indeed, digitalisation is blurring the boundaries between material and immaterial output, with an ensuing change in the sectoral composition of the economy. These issues are explored in the section before the conclusion.

Humans and machines: technological unemployment

The emergence and adoption of successive vintages of fixed capital as input in production has been a key driving force of structural dynamics.[1] This section *conceptually* explores three

1 Fixed capital consists in (non-financial) assets used as inputs in production over several accounting periods (more than one year), such as machines, equipment and industrial plants (UN 2009: 8).

chronological stages of technical change involving fixed capital: mechanisation, computer-based automation and robotisation, coupled with cloud computing.

Mechanisation and employment reabsorption

The race between the 'human' and the 'machine' – leading to the (human) fear of technological unemployment – has been present at least since British textile labourers and weavers resisted the adoption of knitting frames and mechanised looms in the 19th century.

Such a fear had been almost immediately labelled (the 'Luddite fallacy') and intellectually counteracted (Babbage 1832: 330): while introducing a machine threw out workers directly involved in a given production process, the increase in demand due to the reduced price of the commodity under the mechanised technique would reabsorb part (or even the whole) of the displaced labour force.

The logical steps of the thought experiment leading to the (partial or full) reversal of direct employment losses explicitly relied on the institutional mechanism of capitalist *competition*: under free entry, the generalised diffusion of the labour-saving technique across producers would drive down extra profits, reducing the commodity's price, thereby increasing real incomes, expanding demand, output and employment.

Insightful as they might be, thought experiments are not flawless. Already Ricardo (1821, chapter XXXI) illustrated how the introduction of a *more mechanised* technique – a technique with an associated higher fixed capital-labour ratio in value terms (Kurz 1984: 219) – could lead to a shrinking *gross* output in the economy, being "injurious to the labouring class" (Ricardo 1821: 390).

Ricardo's conclusion evinced that when a relatively more mechanised technique is introduced in an industry, employment reabsorption is likely to occur in *different* industries than the one which adopted the new technique. By changing the input proportions in an industry, mechanisation activates output from different sectors. And the higher income resulting from productivity increases associated with mechanisation will be spent in different proportions, according to the distribution of the fruits of technical progress between social classes.

Computer-based automation: Input-Output approaches

Therefore, to assess the comprehensive effects of mechanisation (and automation in general) on employment, an approach based on the interdependence between sectors of the economy is required. This interdependence can be quantified by the *productive* ties between industries and articulated into an Input-Output (I-O, hereinafter) table. An I-O table is the matrix representation of the bilateral flows of commodities in terms of monetary units between industries: "[t]he double-entry bookkeeping of the input-output table thus reveals the fabric of our economy, woven together by the flow of trade which ultimately links each branch and industry to all others" (Leontief 1986: 5).

The analysis of *potential* technological unemployment due to computer-based automation by means of dynamic I-O models pioneered by Leontief and Duchin (1986), and further refined – especially in terms of investment hypotheses – by Kalmbach and Kurz (1990), evince the importance of structural dynamics in projecting societal transformations.

A crucial distinction in this regard is that between *technical progress* and *technological change*. Whilst the former refers to the "emergence of new technical opportunities of production" (Pini 1997: 76), the latter concerns the "progressive adoption and diffusion of these opportunities in the economic system" (Pini 1997: 76). Thus, technical progress is only a necessary – rather than sufficient – condition for technological change: for novel production methods to become widespread, interacting economic and institutional mechanisms involving, for example, profitability, competition, R&D, intellectual property and product standards, need to unfold.

Mindful of this distinction, Leontief and Duchin (1986) devised a set of scenarios differing in the *pace of diffusion* of computer-based automation across industries, quantifying model-implied changes in the sectoral and occupational structure of employment between 1963 and (a projection onto) the year 2000, for the US economy. In the majority of sectors, accelerated diffusion of new technologies would lead to output increases accompanied by employment reductions. Whilst computers would mainly disrupt office work and education, robotisation would affect production workers in manufacturing, and computer numerically controlled (CNC) machine tools would mostly affect metal-working human operations. In terms of sectoral composition, automation would decelerate the transfer of employment from manufacturing to services, given the increased production of new vintages of capital goods, coupled with substantial labour-saving trends in services, due to office automation.

Inspired by Leontief and Duchin (1986), ensuing contributions applied a similar framework to other countries (e.g. McCurdy 1989; Matzner et al. 1990) and/or refined the theoretical structure of the dynamic I-O model (Kattermann and Kurz 1988).

In particular, given that computer-based automation is diffused through new vintages of capital goods, the dynamics of investment demand is crucial. In this sense, Kalmbach and Kurz (1990: 372) introduced a two-step decision process involving an accelerator principle (investment demand is driven by expected sales) and a capacity planning norm, in which the (increasingly automated) 'best practice technique' is gradually diffused. Depicting alternative *diffusion* scenarios for West Germany by means of comparative dynamics (in the sense of Hicks 1983: 109), they suggested that an accelerated diffusion path might affect employment levels to a *lesser* extent than a slower one. Moreover, in line with Leontief and Duchin (1986), the diffusion of micro-electronic-based new technologies increased the economy-wide labour intensity of construction and electronic data processing industries. However, demand compensation effects would be insufficient to revert the overall labour-saving trend in the economy.

While the previously mentioned I-O contributions focus on the changes in the volume and composition of employment due to automation, technological change is also bound to upset relative prices and income distribution.

Under the condition that a uniform rate of profits prevails across industries, the relative price structure emerging from the cost-minimising choice of available techniques guides the assessment of the potential distributive consequences of competing technologies (Cesaratto 1995). This assessment can be done by depicting the factor price frontier (FPF, hereinafter) associated with each alternative I-O technique. In an economy where prices can be *reduced* into a wage and profit component, the FPF specifies the *inverse* functional relation between the real wage rate (w) and the rate of profits (r) – that is the map $w(r)$ indicates the real wage rate that may be obtained at the rate of profits r, for a *given* I-O technique (Kurz 1990).

As a *dual* exercise to Leontief and Duchin (1986), Leontief (1985) compared the FPFs for the US economy between the technique in use at the end of the 1970s and "the economic recipes that could prevail by the year 2000 as a result of the introduction of computer-based automation" (Leontief 1985: 41), evincing that the incentive to switch from the 'old' to the 'new' technology depends on the *actual* configuration of distributive variables.

In fact, the changing shape of the FPF can be used to analyse *historical* forms of technical progress (Schefold 1976). To illustrate this, we may compare two techniques, α and β, each characterised by alternative skill composition of tasks, labour and capital input requirements.

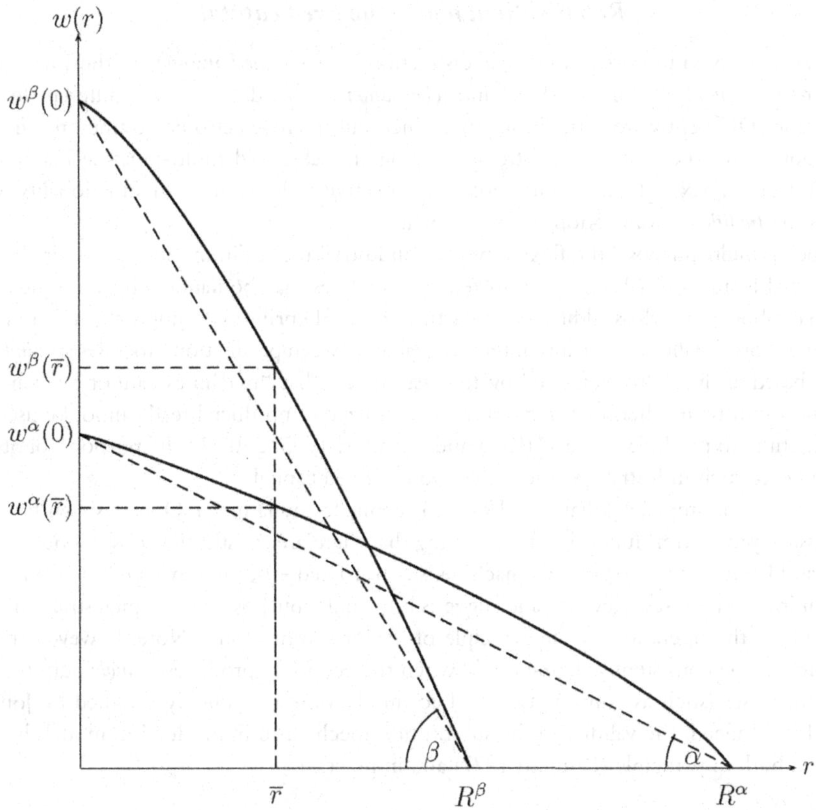

Figure 14.1 Factor Price Frontiers (FPFs) before (α) and after (β) the introduction and diffusion of relatively more *automated* I-O techniques in the economy

We may assume β represents the new technology, whereas α is the incumbent one. As depicted in Figure 14.1, by increasing the fixed capital intensity of production, automation implies a reduction in the maximum rate of profits ($R^{\beta} < R^{\alpha}$) but an increase in the maximum real wage rate ($w^{\beta}(0) > w^{\alpha}(0)$) which – for a given, fixed composition of the standard of value – means that (skill-adjusted) labour requirements are decreasing. Thus, automation will normally be accompanied by a rise in the degree of mechanisation of the economy.

Interestingly, the extent to which $w(0)$ increases depends both on reductions in labour input requirements but, *also*, on how the skill composition of occupations across industries changes with the introduction of new equipment. A sharp increase in the skill content of tasks of automating industries may counteract the fall in labour input requirements, *taming* the expansion of distributive possibilities due to the new technology β.

Moreover, automation renders clear that new technology adoption depends not only on technical conditions, but also on income distribution: the technology providing a higher w for a given r depends on whether the *actual* value of $r - \bar{r}$ in Figure 14.1 – is to the left (technique β is preferred) or to the right (technique α is preferred) of the intersection point between the two FPFs. And the decrease in $w^{(\beta)}(0)$ due to a higher skill content of labour tasks in automated industries may *widen* the range where the switch between preferred techniques occurs.

Robotisation: malleable fixed capital

However, there is an important *conceptual* distinction between *mechanisation* of the 19th century, *computer-based-automation* of the 1980s and *robotisation* of this day: they are different forms of automation. Differently from traditional machines and micro-electronic computers, an industrial robot is an "automatically controlled, reprogrammable and multi-purpose manipulator" (UNIDO 2019: xix). It has an autonomy, connectivity, flexibility and functionality which exceeds the *traditional* conception of fixed capital.

By being multi-purpose, the flexibility of an industrial robot diminishes the required pace of fixed capital formation. Multiple automated product lines may be handled by a unique device. Industrial robots – as well as additive manufacturing (i.e. 3D printing) – imply that fixed capital is becoming more "malleable", that is, interchangeable between production processes and *industries*.

I-O-based analyses are facilitated by the assumption that "machines cannot be transferred from one sector to another, that is, an oven once utilized to produce bread cannot be used during its lifetime to produce biscuits" (Kurz and Salvadori 1995: 250). Multi-purpose robots pose a challenge to such industry-specific conception of fixed capital.

In fact, the attempt by Johansen (1959) to reconcile (dynamic) I-O analysis with smooth Neoclassical production functions, by assuming that there are substitution possibilities between capital and labour *ex ante* – before a machine is constructed – but not *ex post* – once a machine has been installed – loses relevance and suggests how multi-purpose fixed capital strengthens the argument for the operation of the principle of 'factor substitution'.[2] Note, however, that this principle may be consistently defined only when the economy produces a *single* output satisfying all final uses (such as consumption and accumulation), as explicitly assumed by Johansen (1959: 158). Indeed, the validity of the substitution mechanism in production models has been criticised, both in principle (Pasinetti 1977) and in practice.[3]

Robotisation: empirical evidence

Pertinent as these conceptual observations might be, at any rate, the most heavily debated aspect of the growth of industrial robots remains its associated effects on employment and the wage share. However, as documented by UNIDO (2019: 53), trade in capital goods intensive in ADP took off only around 2002, patenting of ADP intensive technologies started to accelerate only around 2007, and global annual installation of industrial robots took off around 2005 (UNCTAD 2017: 47), making the diffusion of ADP technologies a phenomenon that dates back to less than two decades. Thus, empirical evidence at this early stage may not be considered conclusive.

As evinced by Gort and Klepper (1982), there are lags of variable length between stages in the life cycle of innovations. And while the diffusion (and imitation) time interval has been declining systematically over time, we may not have seen yet a stage in which, due to robotisation, "successful innovators [intended as industrial robot *adopters*] within an industry may be increasing employment but require less employment than unsuccessful firms that contract and exit" (Haltiwanger 2018: 69). Crucially, concerns about job displacement effects of robotisation

2 Within a Neoclassical system, the principle of 'factor substitution' states that changes in factor prices (r, w) exactly correspond to changes in relative proportions of capital (K) to labour (L), in equilibrium. Therefore, the degree of mechanisation of the economy (the ratio of the value of capital to labour input) is inversely and monotonically related to the factor price ratio (r/w).

3 As stated by Leontief (1951: 39): "the concept of technical substitution and the law of variable proportions – if applied to aggregative industries – have in the main no other function than to conceal the non-homogeneous character of the conventional industrial classification".

revolve around the time-path of the *traverse*, the *transitional* dynamics between the old and new technologies (Haas 2018).

This alerts on the caution needed when presented with evidence on the debate. In fact, two of the most influential empirical studies so far (Acemoglu and Restrepo 2017; Graetz and Michaels 2018) cover a period between the early 1990s and 2007, *before* the accelerated growth in industrial robot deployment.

Using a panel of 17 countries across 14 industries from 1993 to 2007, Graetz and Michaels (2018) find that industrial robot densification is associated with *increases* in labour productivity, Total Factor Productivity (TFP, hereinafter)[4] and average wage rates, with a *decrease* in output prices, but *no* statistically significant implications for changes in the wage share and overall employment.

On the contrary, for the same period but focusing on US local labour markets more intensively exposed to industrial robot deployment, Acemoglu and Restrepo (2017) find that "one additional robot per thousand workers . . . *reduces aggregate* employment to population ratio by 0.34 percentage points and *aggregate* wages by 0.5 percent" (Acemoglu and Restrepo 2017: 36, italics added).

In between these two contributions, using a long-run distributed lag framework which considers TFP growth (rather than robotisation) and a wider time span (1970–2007), Autor and Salomons (2018) find that automation (proxied by industry-level TFP changes) has been associated with *increases* in *industry* employment (mainly due to TFP growth in supplier sectors) but to an *erosion* of the wage share. Note, though, that by using reduced-form econometric methods, results convey conditional correlations rather than causal effects.

In assessing the employment effects of robotisation, the employment-to-population ratio is often used (e.g. Acemoglu and Restrepo 2017). A key related debate concerns the denominator – rather than the numerator – of such ratio, for the long-period trend of a demographic decline in advanced industrial economies paves the way for the deployment of robots. Declining working-age populations alert on the need for an acceleration of labour productivity growth to sustain current standards of living in advanced countries (Leitner and Stehrer 2019). Thus, a positive correlation between robot deployment and labour productivity growth would provide a rationale for robot densification.

Interestingly, using a panel for 60 countries between 1993 and 2013, Abeliansky and Prettner (2020) find that a faster pace of population growth tends to be related to a reduction in the growth rate of robot deployment. It would be important, though, to be careful when interpreting such correlation. A declining population and the scale of robot deployment might be complementary in advanced countries, but this may not be the case in developing economies, where traditional *mechanisation* is still the prevalent form of automation (UNCTAD 2017: 39).

Cloud computing and capital services: outsourcing fixed capital

The diffusion of ADP technologies leads to rethink the role of capital as a factor of production beyond robotisation. Cloud computing, that is, renting a computing environment and associated storage space hosted in equipment operating elsewhere, crystallises the approach in the latest System of National Accounts (SNA, hereinafter) to measuring 'productive' capital using

4 Within the Neoclassical approach, TFP is assumed to measure "how productively the economy uses all the factors of production" (Aghion and Howitt 2009: 106), its changes being measured as a *residual*; *assuming* factor markets are perfectly competitive, TFP growth rates are obtained by subtracting the growth rate of the degree of mechanisation (the capital-to-labour ratio) from the growth rate of real income (i.e. double-deflated value added).

capital *services* (UN 2009: chapter 20): the sum across capital goods weighted by their rental price, that is, by the price that would have to be paid to hire the asset for a period. By rendering the hiring process explicit, cloud computing avoids imputation difficulties when valuing capital services for owners and users of Information and Communication Technology (ICT, hereinafter) equipment.

If cloud computing becomes generalised, we may expect a redistribution of gross fixed capital formation by *destination* industry: rather than investing in ICT equipment themselves, industries will purchase an *intermediate* service to data processing, hosting, renting and leasing activities.[5] These service industries will, thus, increase their relative importance as activating demand sources of physical ICT infrastructure, having as a counterpart a whole new stream of intermediate consumption transactions with users of cloud computing services.

Interestingly, this (potential) process of fixed capital *outsourcing*, may resemble (or at least be analysed as) the process of outsourcing of labour from manufacturing into business-related service industries during the 1980–1990s (see, e.g. Franke and Kalmbach 2005).

Tasks, jobs and occupations: the content of labour content

The asymmetry between process and product innovations (Pini 1997: 65–66) maintains its relevance when assessing the introduction and diffusion of ADP technologies. The labour-saving, cost-reducing potential of process innovations should be weighed against potential job-creation effects of product innovations, which involve, at least, two channels: (1) employment induced by new product markets (e.g. smart devices) which require a whole range of supporting functions (via I-O linkages), and (2) human-robot complementarity in the workplace requiring *new* occupations (e.g. software developers, data analysts).

But assessing the potential for human-robot complementarity requires to unbundle the *content* of labour as a factor of production. To begin with, there are conceptual differences between the notions of task, job and occupation. Tasks represent granular activities at the workplace. A job may be seen as a set of tasks, whereas an occupation represents a set of jobs whose main tasks have a high degree of similarity. Empirical work rendering operational these distinctions has been made possible by the Occupational Information Network (O*NET) programme, which has developed a *granular* database mapping tasks to occupations for the US economy,[6] as well as by the European classification of Skills, Competences and Occupations (ESCO), a database identifying, describing, codifying and classifying occupations and skills across the European Union (EU).[7]

Task-based, skill-biased theory of automation

It is by no means simple, and in many ways arbitrary, to *classify* tasks. The analytical taxonomy introduced by Autor et al. (2003) distinguishes, first, between 'routine' and 'non-routine' tasks. A 'routine' task is "sufficiently well understood that [it] can be fully specified as a series of instructions to be executed by a machine" (Acemoglu and Autor 2011: 1076). Essentially, a

5 ISIC Rev. 4 categories 66 and 77. See UN (2008) for details.

6 For details, please see: www.onetcenter.org. The ONET database is based on the US-BLS Standard Occupational Classification (SOC), which may be converted into the ILO's International Standard Classification of Occupations (ISCO).

7 For details, please see: http://https://ec.europa.eu/esco/portal. The current version at the time of writing (ESCO v. 1.1) articulated 2,942 occupations and 13,485 skills linked to them across the EU.

routine task is highly *codifiable*. Non-routine tasks, instead, can be sub-classified as 'manual' or 'abstract'. The former require "situational adaptability, visual and language recognition, and in-person interactions" (Acemoglu and Autor 2011: 1077), whereas the latter require "problem-solving, intuition, persuasion, and creativity" (Acemoglu and Autor 2011: 1076).

The rationale for this taxonomy is that of associating a *skill set* to each of these three categories, allocating the *highest* skill set to non-routine abstract tasks and the *lowest* skill set to routine tasks. Thus, if ADP technologies (such as robots) were substitutes to (human) routine tasks but complementary to non-routine abstract tasks, increased relative labour demand for occupations intensive in the latter task type would widen the wage gap between workers with highest and lowest skill sets. Therefore, there would be a 'skill bias' associated with ADP technology adoption (and recent automation in general) explaining widening wage inequalities.

The unequal skill profile of labour demand associated with automation triggers a 'race between education and technology'. In particular, Acemoglu and Restrepo (2018a, 2018b) have applied the idea of a mismatch between skill requirements and their availability to formulate a 'task-based' theory of technological unemployment and declining wage share.

In their framework, services embodying a range of tasks are combined to produce aggregate final output, used both for consumption and accumulation. Rather than purely labour-augmenting innovations, they assume that machines and labour are perfect substitutes to produce (a range of) tasks, so that cost-minimisation implies that labour will be selected for those tasks in which it has a higher relative productivity (with respect to machines), that is, tasks in which humans have a *comparative advantage*. Automation, thus, is represented by an expansion in the set of tasks autonomously performed by (intelligent) machines.

Equilibrium conditions imply that automation will *always* be wage-share-reducing, and its impact on labour demand (and wages) depends on how the *displacement* effect of workers from automated tasks is counteracted by a *productivity* effect, given by the gap between the productivity-to-input-price ratios of machines and labour: only when the 'benefit-to-cost' ratio of machines is notoriously higher than for labour, automation will increase labour demand (see Acemoglu and Restrepo 2018a: 19, for details).[8]

Moreover, by restricting the range of tasks that low-skill workers may perform, while assuming that the share of high-skilled workers in the economy is *lower* than the share of tasks that only they can produce (i.e. high-skilled workers are relatively *scarce*), the labour demand profile triggered by automation is biased towards higher skill sets, increasing the wage gap between worker types.

This renders clear the predicted outcomes of automation 'at the extensive margin' (i.e. through expansion of the share of tasks produced by machines). However, it is also possible for automation to work 'at the intensive margin' (i.e. when machine productivity increases in tasks which had already been automated). In this case, employment should expand and the wage share should remain unaltered.

Interestingly, this distinction between *extensive* and *intensive* margins of automation *echoes* the Marxian extraction of *absolute* and *relative* surplus value from a (human) worker, but applied to a robot. Hence, unless robot 'exploitation' is sufficiently high, extensive automation will reduce

8 The compensation between the 'displacement' and 'productivity' effects operates as follows: if the gap between the productivity of machines ('benefit') and their rental rate ('cost') is *sufficiently high* with respect to the gap between the productivity of labour ('benefit') and the wage rate ('cost') for the same set of *automated* tasks, the substitution of machines for labour – under competitive conditions – reduces production costs *and* prices, increasing per-capita real incomes channelled towards higher demand for products, triggering further demand for labour. This demand for labour exceeds the originally displaced workers substituted by machines performing the automated tasks.

the wage share, displace workers and increase wage gaps between workers with different skill sets. To partially revert this trend, Acemoglu and Restrepo (2018a: 22) assert that technological advances ought to bring about the creation of new tasks in which human labour has a comparative advantage. In this way, with extensive robot deployment displacing low-skill workers and the emergence of new tasks for high-skilled workers (i.e. human-robot complementarity), the economy is predicted to reach a *balanced* growth path (provided the high-skill worker supply adjusts accordingly).

Changes in the labour process: routine intensity of tasks

The creation of new tasks alerts on the importance of studying the (potential) effect of ADP technology adoption beyond employment levels, focusing also on changes in the labour process itself, in the *content* of labour content.

A first challenge, though, is to empirically distinguish between routine and non-routine tasks (of both manual and abstract types). In an attempt to address it, Marcolin et al. (2016) introduce a 'routine-task intensity index' (RII, hereinafter), quantifying the degree to which a task can be routinised and – after aggregating across tasks for each occupation – identify occupation × industry combinations particularly intensive in routine-based tasks.[9]

Table 14.1 suggests that the RII is generally higher for manufacturing industries, and highest in food processing, textiles/apparel and transport equipment, which evinces the importance that cross-country structural differences – in terms of the sectoral composition of the economy – may have in assessing potential effects of automation.

Equally relevant, the RII by occupation (Marcolin et al. 2016: 17, Table 3a) is highest for elementary occupations, plant operators and services/sales workers (ISCO-08 categories 9, 8 and 5, respectively) which, again, evinces the relevance of the occupational structure *supporting* the sectoral composition of the economy.

The RII was highest in occupations with lowest *skill levels*, according to ILO's ISCO-08 classification (ILO 2012: 14, Table 1). This raises a key issue: what is the relationship between (non-)routine tasks and the skill content of occupations?

If we take at face value the skill content of occupations as allocated by the ISCO-08 classification, Table 14.2 suggests that non/low-routine intensive tasks are predominantly done by workers with the highest skill level. Instead, medium/high-routine intensive tasks are predominantly carried out by workers with *medium*-level skills. Interestingly, it is medium – rather than low – skill-level workers currently employed in routine-based tasks.

Thus, a point open to debate is whether ADP technologies will be accompanied by a generalised increase in the *skill threshold* of the workforce *or*, instead, we will see a job *polarisation* process, with the hollowing out of human occupations in the middle range of skills,[10] and at the same time, a proliferation of complementary low-skill and high-skill jobs functional to the deployment of intelligent machines.

9 The four dimensions used to capture the routine-intensity of a task concern "the frequencies with which individuals may, respectively: [(1)] choose the sequence of the tasks involved by the job; [(2)] change the content of work or how this is carried out; [(3)] plan their own work activities; and [(4)] organise their own working time" (Marcolin et al. 2016: 9). Data has been obtained from the OECD Programme for the International Assessment of Adult Competencies (PIAAC) survey.

10 Occupations such as plant and machine operators, assemblers, services and sales workers, involving tasks such as operating mechanical machinery/electronic equipment, as well as information manipulation, ordering and storage.

Table 14.1 Routine Intensity Index (RII) by industry (22 OECD countries; years 2011–2012)

Sector	*ISIC Rev. 4*	*Mean*	*SD*
Agriculture	01–03	2.26	1.14
Mining	05–09	2.29	1.00
Food, Beverages & Tobacco	10–12	2.75	1.28
Textiles, Apparel & Leather	13–15	2.66	1.29
Wood & Paper	16–18, 58	2.31	1.13
Chemicals	19–23	2.37	1.17
Basic & Fabricated Metals	24, 25	2.40	1.16
Machinery n.e.c.	28	2.22	1.07
Electrical Equipment	26, 27	2.35	1.15
Transport Equipment	29, 30	2.61	1.22
Manufacturing n.e.c.	31–33	2.35	1.16
Utilities	35, 36	2.12	0.91
Construction	41–43	2.25	1.04
Trade & Hotels	45–47, 55, 56, 95	2.41	1.12
Transport & Telecom	49–53, 61, 79	2.59	1.20
Finance	64–66	1.99	0.88
Business Services	62, 63, 68, 69–78, 80–82	2.00	0.95
Personal Services	37–39, 59, 60, 84–88, 90–94, 96	2.17	0.95

Notes: Column ISIC Rev. 4 reports the 2-digit classes composing the reported sectors

Source: Adapted from Marcolin et al. (2016: 17, Table 3b)

Table 14.2 Employment by skill and routine intensity (22 OECD countries; years 2011–2012)

Routine intensity / *Skill level*	*NR* *Non-routine*	*LR* *Low routine*	*MR* *Medium routine*	*HR* *High routine*
Low	0.00	0.01	0.07	0.24
Medium	0.09	0.30	0.68	0.73
High	0.91	0.69	0.25	0.03

Notes: Skill levels correspond to one-digit ISCO-08 categories
1 to 3: managers, professionals, technicians and associate professionals (high)
4 to 8: clerical support/services/sales, skilled agriculture, crafts, machine operators (medium)
9: elementary occupations (low)

Source: Adapted from Marcolin et al. (2016: 20, Table 5)

Skill content of a task or skill level of a worker?

So far, though, we have not enquired into how skill levels are allocated to workers. For one thing is to measure the *skill content* of a task within an occupation, and another is to measure the *skill level* of the worker performing the task. The most diffused *proxy* of skill level for the latter is the level of educational attainment, *assuming* that individuals awarded a higher education qualification possess the *skill content* of high-skill tasks. But this can be misleading.

First, because while a higher education qualification might be a *sufficient* condition to perform certain high-skill tasks, it may not be a *necessary* one. For example, tasks requiring "factual, technical and procedural knowledge in a specialized field" (ILO 2012: 13, ISCO-based Skill Level 3) can be performed by university graduates, but may often be proficiently carried out by workers who have completed specialised vocational education and on-the-job training.

Second, because it validates the conception that education *mostly* represents an *investment* with an expected return, to be reflected in a wage premium. But this seems to confuse the true source of extra income (the high-skill content of a task) with the *commodity* that is perceived to command it (a higher education qualification). It is institutional specificities (and historical trajectories) of advanced industrial societies which have led high-skill occupations to be remunerated relatively more than low-skill ones, not an explicit societal consensus about the 'discounted present value' of expected income streams of a human asset who enrolled into university. In fact, when high-skill occupations are confused with the advanced educational attainment of the worker who performs them, there is a risk of interpreting tasks performed *by* university graduates as necessarily *knowledge*-intensive, because of the length of formal instruction of their degree.

The structural transformation in the skill content of tasks is prone to widen wage inequalities if the correlation between skill content and remuneration across occupations continues to be strongly positive, in a context where routine-intensive tasks with medium-skill requirements disappear. But can we be sure that high-skill tasks can be safeguarded from automation? Machine learning techniques have improved robot performance in both non-routine *abstract* tasks, such as natural language processing, image/video/speech recognition, as well as in non-routine *manual* tasks involving robot dexterity (UNIDO 2019: 75).

Physical and digital: the value of data and the changing composition of output

By means of machine learning techniques, robots are beginning to perform non-routine *abstract* tasks, automating decision-making. Beyond robotics, computer-aided design and manufacturing (CAD-CAM) are automating the production not only of physical goods (by controlling machine tools) but also of *immaterial* goods (such as drafting industrial designs).

However, robots need data as input to learn from experience and improve their performance. Thus, while humans use robots as a fixed capital input in the production process, robots use *human* data as *their* input. In fact, "data are a *productive resource* that fuels the learning of machines" (Kurz et al. 2018: 571, italics added).

Therefore, a key question associated with the structural dynamics of prices triggered by the diffusion of ADP technologies is: what is the *value* of the productive resource labelled as 'data'? A single sentence may contain multiple (and often contrasting) views on the underlying source(s) of data value:

> [D]ata can be used multiple times (e.g. in different contexts) without *inherently diminishing their value*. In principle, data can be exploited and re-exploited infinitely at low *marginal cost*; it is *data infrastructure and analytics* that are the *primary costs* related to data re-use.
>
> *(OECD 2019: 240, italics added)*

This sentence implies, at least, three alternative conceptions for data value: (1) value as a substance embodied in commodities, (2) value deriving from scarcity and (3) value as interdependence.

Claiming that the inherent value of data is not consumed through repeated usage conveys the idea of data as a substance transferred to products. The low marginal cost of extremely abundant data exploitation relates to its consideration as a scarce resource. Finally, data value may derive from the *structure of interdependence* between fixed capital ('data infrastructure') and labour ('data analytics') requirements to reproduce it.

Data as a non-produced asset

Given the difficulty to reconcile contrasting theoretical perspectives, one interesting route is to explore current proposals (Ahmad and van de Ven 2018; Mitchell 2018; OECD 2019; Mitchell 2020) and discussions within international statistical organisations (mainly the OECD and the UN Statistics Division) aimed at building a specific digital economic account. Starting from digital data measurement within the System of National Accounts (SNA, hereinfater) may prove enlightening to distil how alternative conceptions of value underpin such proposals.

Currently, digital data is considered as a *non-produced asset* in the SNA; while it appears in the balance sheet of firms, its very production does not increase GDP. Data in itself will only be valued when a *market transaction* occurs, recording the monetary amount under the category of a non-financial, non-produced asset.

This convention has implications for both the production and asset boundaries of the SNA. Crucially, it is argued that "not to treat the data, in and of itself, as produced does not mean that data has no value" (Ahmad and van de Ven 2018: 5). But where does this 'value' derive from?

From a *Classical* perspective, value derives from *production*, whereas from a *Marginalist* (or Neoclassical) viewpoint, it derives from *exchange* under conditions of relative *scarcity*. In fact, the (relatively) undisputed character of GDP as an *observable* indicator of economic value derives from its being *exchanged production*. So how may data be a 'productive resource' without being itself produced? The Neoclassical reconciliation guiding the SNA framework could be that data can have value as it is exchanged, but does not contribute to GDP because it is not produced.

This, however, overlooks whether digital data may be considered a *substance* embodied in commodities. Just as human labour is measured in *hours*, data is measured in *bytes*. It would be difficult, though, to discern value differences between two products based on the *volume* of bytes used for their (re)production. In essence, the underlying value of data reflects its *knowledge* content. But while knowledge is embodied in data, it is not *apparent* just by accessing data. Hence, the value of data is separated from the reproduction costs of its storage medium, a database, for example.

But, then, if it is the underlying knowledge content of data that has value, its *digital* nature cannot be a necessary condition for rendering data valuable. For instance, physical record files in an archive would possess a similar knowledge content, albeit in a different storage medium. Thus, it still remains an open question how to distil the *uniqueness* of *digital* data stored electronically as a source of economic value.

Probably, as convincingly argued by Ahmad and van de Ven (2018: 13, italics added):

> "the decision not to treat data as produced was in large part a function of the fact that to do so would lead to an *implicit recognition that all knowledge was produced*, and as such should be valued as contributing to GDP."

This would significantly expand the *production* boundary of the SNA, generating new income entitlements derived from knowledge creation.

Indeed, even before national accounts acknowledge it, machine learning methods are already disrupting the idea that the entitlement to shares in income are distributed in proportion to factor contributions. When an algorithm learns *by itself* how to exploit new arbitrage opportunities, who should be rewarded with the additional value added or net product generated?

Is it the labourer who designed and codified the algorithm? Is it the owners of the computing equipment on which the algorithm runs, learning from experience? Alternative theories of value would reply differently to these questions (Savona 2019).

Keeping data with its embodied knowledge out of income *generation* avoids having to discuss whether intelligent machines should be granted (human) agency. Being considered a non-produced asset renders *data* similar to *land* as a production factor. In fact, land made available for productive uses generates *rents* rather than *rentals*. The former are part of the primary income *distribution* account of the SNA, and need not be financed out of value added. The latter, instead, belong to the income *generation* account of the SNA, being part of the added value of the economy.

In a nutshell, current practice in the national accounts implies that when digital data is created there is no immediate impact on GDP, effects may be indirectly traced when data is used to produce other products within the production boundary.

In order to trace these indirect effects through a network of money flows, a framework of 'Digital Supply-Use Tables' (SUTs) has been proposed (Mitchell 2018, 2020). In this way, an economy-wide digital economic account would articulate data-related money flows into a cross-tabulated classification of digital industries and products. Such a classification would codify data circulation, and the circulation of mutually dependent flows allows to derive economic value.

Immaterial goods are not services

The emphasis on digital over physical outputs, pervasive in conceptualising ADP technologies, echoes the divide between material and immaterial (or intangible) products, prominent in decades-long discussions on the transition towards a 'service economy' (Walker 1985). Material production was associated with manufacturing, whereas immaterial output with services. Hence, the increasing share of services in (nominal) value added and employment suggested that the economy was gradually 'dematerialising' as it was 'deindustrialising' (in relative terms).

At least two points emerge from these debates. First, the need to conceptually clarify the notions of digital (as immaterial) and physical (as material) production in terms of structural analysis. Second, the need to explain the changing sectoral income shares in the economy.

On the first point, Parrinello (2004) insightfully clarified the difference between goods and services in relation to (im)material production: "commodities include goods and services, goods can be material or immaterial, but services are not immaterial goods" (Parrinello 2004: 389). By analytically dissecting a uniform production-consumption period into a series of independent processes at a sufficiently granular level, he singled out two relationship types between processes, *serial* and *parallel* I-O relations.

Serial I-O relations mean that today's outputs are tomorrow's inputs. This time-lag in production also implies that an inventory of inputs can be maintained and restored. On the other hand, in parallel I-O relations, the output of a provider process 'serves' as input to the user process, during the *same* period. Quantities resulting from serial relations, albeit dated, have no intrinsic time dimension and may be accumulated. They represent *goods*. Instead, those from parallel relations may only be defined during a time period and cannot be accumulated. They represent *services*.[11]

11 Parrinello (2004: 389) also distinguishes a service from a pure perishable good, as the latter – though not storable – is first produced through a *serial* I-O relation, before its consumption.

This distinction might be useful to show that 'service' industries produce both services *and* immaterial goods, whereas 'manufacturing' industries generate a sizeable amount of services in the process of production of physical goods. As an example of the former, the output of a firm in a service industry consisting of a patented industrial design represents an immaterial good, whilst if the same firm provides time for analytical activities to another one – without generating vendible intellectual property as an output – then it is supplying a service. As an example of the latter, consider specialised repair services across manufacturing firms.[12]

Conceptual distinctions between product types become relevant to avoid the commonplace (mis)conception that the increasing weight of service industries will necessarily render the economy more knowledge-intensive. As lucidly put forward by Parrinello (2004: 396):

> "[t]he myth [of a *post-industrial* economy] rests upon a sort of deduction from two spurious premises: (i) services are immaterial *goods* (ii) immaterial goods are fragments of knowledge and information; hence (iii) more services reflect more knowledge and more information."

Just as was argued – in the previous section – that tasks are not necessarily knowledge-intensive due to the length of formal instruction of the university graduates who perform them, service activities are not necessarily knowledge-intensive due to their being confused with immaterial goods.

Thus, digital products involve both immaterial goods and services. Services are functional to the production of both physical and immaterial goods. And the changes in relative shares between manufacturing and services should not be seen as a direct consequence of increasing digitalisation, as evinced by the fact that ADP technology development is, to a great extent, conditioned by human learning within manufacturing industries (UNIDO 2019: 61).

Sectoral income shares: Baumol's cost disease

Debates around the underlying cause(s) and measurement of the changing sectoral income shares of manufacturing and service industries have been a long-standing feature of structural dynamics, especially since Baumol (1967). In his framework, faster productivity growth in manufacturing vis-à-vis services under *competitive* conditions imply: (1) a *relative* unit labour cost and price increase for services, and (2) labour-displacement in manufacturing and labour-absorption in services, *if* demand for the latter is (sufficiently inelastic to be) maintained (despite a higher relative price). As a consequence, while the manufacturing-to-services output ratio may remain constant, the nominal income share of services will increase, as well as its share in employment. Thus, '*balanced growth* in a world of *unbalanced productivity*' requires a progressive slowdown of aggregate labour productivity, labelling the predicted dynamics as Baumol's 'cost-disease'.

Under this view, manufacturing industries represent *progressive* activities, whereas service industries *stagnant* ones. Baumol et al. (1985) empirically confirmed model-implied trends for the US (between 1947 and 1976). They did so by extending the original framework through the introduction of *asymptotically stagnant* industries (i.e. sectors using inputs from progressive and stagnant industries in fixed proportions). In such sectors, the stagnant labour-intensive component gradually assumes a greater share of the unit cost, eventually rendering the activity stagnant.

12 Interestingly, the latest ISIC Rev. 4 classification (UN 2008: 161), has moved "Repair and installation of machinery and equipment", which consists of a *service* output, within the umbrella of manufacturing industries.

This third industry type becomes particularly relevant when considering ADP technology diffusion, as the authors argue precisely that 'data processing (computing services)' represents a prime example of an asymptotically stagnant sector: *software* takes over *hardware* in unit costs and "[s]oftware development remains essentially a handicraft activity, and is, *so far*, a stagnant service" (Baumol et al. 1985: 813, italics added). It remains an open question whether the automation of non-routine abstract tasks, such as component-driven software development through machine learning methods, will overcome the *predicted* asymptotically stagnant character of (at least, some) digital industries.

Interacting productivity, demand and income: mechanisms of structural dynamics

Baumol's cost disease suggests that widespread adoption of ADP technologies would deepen the uneven dynamics of sectoral productivities, slowing down aggregate labour productivity growth, if relative *output* shares remain (approximately) constant, that is, if a balanced growth path prevails.

A challenge to a world in which there is convergence towards a balanced growth path is the approach of structural economic dynamics introduced by Pasinetti (1981, 1993). By means of a dynamic I-O model with uneven sectoral dynamics of per-capita consumption and labour productivities, Pasinetti specifies a *normative* benchmark in which technological unemployment and unbalanced growth are the normal state of the economy.

The mechanism of *structural* dynamics implied by the benchmark configuration of this approach may be framed as follows: uneven sectoral productivity changes modify *relative* production costs, but if average productivity gains accrue to labour, (aggregate) price dynamics is slower than nominal income expansion, increasing real incomes. As real income increases, consumption patterns change – as predicted by Engel curves (Moneta and Chai 2014) – altering the compositional structure of household expenditure. Hence, gross output induced by household expenditure has a changing commodity composition. This generates a potential mismatch between activating sources of demand and activated sources of employment, as sectors for which consumption demand is growing faster (slower) than productivity will expand (contract) employment.

Therefore, if the adoption of ADP technologies accelerates productivity growth in branches of the economy for which the corresponding demand for its final output is stagnant (such as traditional motor vehicles), or if the expansion rate of household demand for digital outputs is short of productivity increases in its supplying sectors, technological unemployment is bound to increase.

But whilst Pasinetti (1981) considers a system where sector-specific per-capita consumption and labour productivity are continuously changing at uneven rates, these are considered to be exogenously given. In particular, no explicit link is specified between final demand expansion and productivity growth. However, at an *aggregate* level, Verdoorn (1949) already documented an empirical, positive relationship between labour productivity growth and output expansion, whereas Kaldor (1966) argued that the relationship is particularly associated with 'secondary' activities, especially with manufacturing. More importantly, Kaldor emphasised that it is labour productivity growth which is a *positive function* of the growth rate of manufacturing output, and argued against the reverse direction of causality, which would mainly operate through relative price adjustments.

In this way, the Kaldor-Verdoorn mechanism allows to (partly) *endogenise* productivity dynamics based on the evolution of demand-induced output. Lorentz and Savona (2008) take this *aggregate* relationship to firm and industry-level dynamics, formulating a simulation model – calibrated with German data – in order to study *tertiarisation* patterns. More in general, the logic of the Kaldor-Verdoorn mechanism implies that shifts in the composition of final expenditures, as well as *autonomous* determinants of *both* technical progress – such as advances in scientific and technological knowledge – *and* the level of aggregate demand – such as public expenditures – shape the evolution of productivity growth.

And not only of productivity growth, but also of productivity *decline*. Because a symmetric application of the Kaldor-Verdoorn mechanism suggests that a slow growth of actual output is conducive to a labour productivity slowdown. This might help explain the 'productivity puzzle' (ONS 2020) facing some advanced industrial economies since the Great Recession of 2008–09.

Conclusion

This chapter has explored how advanced digital production (ADP) technologies disrupt three main axes of economic structure: (1) the changing nature and function of fixed capital in relation to (human) job displacement, (2) the changing content of labour tasks complementing automated production and (3) the evolving distinction between physical and digital output and assets.

From mechanisation debates in the 19th century to the Input-Output (I-O) studies of computer-based automation of the 1980s, job displacement effects due to the diffusion of automated production techniques were not fully compensated by mechanisms of capitalist competition. And while preliminary evidence on the employment and distributive consequences of robotisation since the 1990s is still not conclusive, industrial robots and cloud computing are accelerating a trend towards multi-purpose, malleable and outsourced fixed capital.

Hence, the degree of human-robot complementarity – and the extent of job displacement – depends on the skill set required by tasks which characterise those occupations interacting with new vintages of fixed capital goods. By mapping the skill content of tasks to their relative codifiability, the empirical application of a 'routine-task intensity index' across selected advanced economies suggests that transport equipment, food processing and textiles/apparel are industries with highest *routinisation* potential. Note that the latter two sectors are amongst those with lowest labour share and highest share of female workforce in the economy (UNIDO 2019: 81). Crucially, as routinisation potential predicts the technical feasibility of robotisation, industry-level differences suggest that the impact of robot deployment on the economy's wage share depends on its structural composition (UNCTAD 2017: 41).

But will non-routine, high-skill tasks be safeguarded from automation? The improvement of robot performance in abstract and manual tasks by means of machine learning techniques cast doubts. A key novelty brought about by ADP technologies is a potential reversal of roles in human-machine complementarity. Traditionally, humans have used fixed capital as a productive input. Instead, machine learning allows for robots to use *human data* as their *input*. Hence, the valuation of *data* is a crucial (still open) question for the structural dynamics of prices. The current practice in the System of National Accounts (SNA) considers data as a *non-produced* asset, that is, data is not part of value added generation in the economy. Keeping data with its embodied knowledge out of income generation avoids having to discuss whether intelligent machines should be granted (human) agency. Being considered a non-produced asset renders data, in some respects, similar to land as a production factor.

Despite the fact that data is not considered an output in itself, digital products based on data have been pervasive to conceptualise ADP technologies. In fact, the ongoing changing sectoral composition of the economy requires to go beyond the dichotomy between manufacturing and service industries. This is because digital products involve both immaterial goods (such as an industrial design or a software package) and services (such as cloud computing), whilst manufacturing industries remain at the core of human learning conducive to novel ADP technologies, resulting in new digital products (UNIDO 2019: 61).

In hindsight, to understand the unfolding dynamics of economic structure, it is worthwhile to glimpse at its historical development. We have traditionally described the structure of the economy in terms of its sectoral composition (e.g. primary sector, manufacturing, services), its production factors (e.g. land, capital and labour) and their income entitlements (e.g. rent, profits and wages). The assumption that one factor is more intensively employed in each sector has been instrumental to identify the *privileged* income entitlement in each stage of structural transformation.

In this way, the primary sector, reliant upon *biological* processes on land, has privileged rent. Manufacturing took over, articulated around the *transformative* power of machines, privileging profits. Finally, services, anchored in active *human* labour, have privileged wages (with compounding hierarchies and widening gaps across occupations).

But technological change has increasingly blurred the neat mapping between factors of production and sectors of the economy, as well as the clear-cut entitlement to factor payments. With the mechanisation of agriculture, profits became prominent in the primary sector. With the servicification of manufacturing, the physical transformation of goods has been bundled with labour-intensive tasks.

So how will ADP technologies *and* digital outputs alter these mappings? Rent payments to grant the mining of digital (identity and footprint) human data may represent a new cycle in the loop, making *data rentiers* a prominent social group owning 'lands' of data. Moreover, if the income streams attributable to a machine learning algorithm operating on an industrial robot accrue to owners of robots as profits and rents to owners of the embodied intellectual property, we may head into an era of 'automated inequality'.

At a deeper level, what is called into question is what might be the role of *human activity* in value generation and its share in income. In an *extreme* scenario, data resulting from human *consumption* may become a *productive input* into robotised processes, which require a tiny fraction of the workforce to run. And while consumers may embrace a 'rentier' future of digital existence, in which they are remunerated for the data they generate, doing without the *indispensable* role of labour in production is not without consequences (Zalai 1989).

Several research avenues remain open. For example, the role of digitalisation in deepening financialisation deserves to be explored, as financial services have 'leveraged on' digital media beyond any other sector of the economy (Mitchell 2018: 28).

Moreover, global robot production is currently highly concentrated,[13] and so is the *capitalisation* of digital assets (Tambe et al. 2020). This dichotomy between (1) the highly *centralised* production of industrial robots and accumulation of data assets and (2) the highly *decentralised* consumption of smart devices and digital outputs, alerts on the need to carefully analyse the market structure implications of current trends.

Finally, the international dimension. ADP technology adoption has been sharply uneven across countries (Ghodsi et al. 2020). Deepening asymmetries in *functional* specialisation of labour might hinder wage upgrading through Global Value Chain participation, whereas a *robotised reshoring* of internationally fragmented production might not boost employment in

13 To the extent that "China, Germany, Japan and Republic of Korea [. . .] accounted for about 83 per cent of the global production of industrial robots in 2015" (UNCTAD 2017: 46).

advanced economies, while lowering income in developing ones. Thus, wider implications of ADP technologies for *global* structural change are still awaiting to be drawn.

References

Abeliansky, A. and Prettner, K. (2020): Automation and Demographic Change. *Global Labor Organization (GLO) Discussion Paper* 518.

Acemoglu, D. and Autor, D. (2011): Skills, Tasks and Technologies: Implications for Employment and Earnings, in Ashenfelter, O. and Card, D. (eds.): *Handbook of Labor Economics*. Vol. 4b. Amsterdam: Elsevier, 1043–1171.

Acemoglu, D. and Restrepo, P. (2017): The Race between Machine and Man: Implications of Technology for Growth, Factor Shares and Employment. *NBER Working Paper 22252*.

Acemoglu, D. and Restrepo, P. (2018a): Artificial Intelligence, Automation and Work. *NBER Working Paper 24196*.

Acemoglu, D. and Restrepo, P. (2018b): Modeling Automation. *NBER Working Paper 24321*.

Aghion, P. and Howitt, P. W. (2009): *The Economics of Growth*. Cambridge, MA: MIT Press.

Ahmad, N. and van de Ven, P. (2018): Recording and Measuring Data in the System of National Accounts. Working Party on National Accounts, Meeting of the Informal Advisory Group on measuring GDP in a digitalised economy, 9 November 2018, OECD Conference Centre, SDD/CSSP/WPNA(2018)5.

Autor, D. and Salomons, A. (2018): Is Automation Labor-displacing? Productivity Growth, Employment, and the Labor Share. *Brookins Papers on Economic Activity (BPEA) Conference Draft, Spring*.

Autor, D. H., Levy, F. and Murnane, R. J. (2003): The Skill Content of Recent Technological Change: An Empirical Exploration, *The Quarterly Journal of Economics* 118(4), 1279–1333.

Babbage, C. (1832): *On the Economy of Machinery and Manufactures*. 2nd enlarged ed. London: Charles Knight.

Baumol, W. J. (1967): Macroeconomics of Unbalanced Growth: The Anatomy of Urban Crisis, *The American Economic Review* 57(3), 415–426.

Baumol, W. J., Blackman, S. A. B. and Wolff, E. N. (1985): Unbalanced Growth Revisited: Asymptotic Stagnancy and New Evidence, *The American Economic Review* 75(4), 806–817.

Cesaratto, S. (1995): Long-period Method and Analysis of Technological Change: Is There Any Inconsistency? *Review of Political Economy* 7(3), 249–278.

Franke, R. and Kalmbach, P. (2005): Structural Change in the Manufacturing Sector and its Impact on Business-related Services: An Input-Output Study for Germany, *Structural Change and Economic Dynamics* 16(4), 467–488.

Frisch, R. (1992[1929]): Statics and Dynamics in Economic Theory, *Structural Change and Economic Dynamics* 3(2), 391–401. Translated from the original Norwegian article which appeared in Nationalokonomisk Tidsskrift, Vol. 67, 1929.

Ghodsi, M., Reiter, O., Stehrer, R. and St¨olinger, R. (2020): Robotisation, Employment and Industrial Growth Intertwined Across Global Value Chains. *The Vienna Institute for International Economic Studies (wiiw), Working Paper*, 177.

Gort, M. and Klepper, S. (1982): Time Paths in the Diffusion of Product Innovations, *The Economic Journal* 92(367), 630–653.

Graetz, G. and Michaels, G. (2018): Robots at Work, *The Review of Economics and Statistics* 100(5), 753–768.

Haas, D. (2018): Diffusion of a New Intermediate Product in a Simple 'Classical-Schumpeterian' Model, *Metroeconomica* 69(2), 326–346.

Haltiwanger, J. (2018): Comments and Discussion: Is Automation Labor-displacing? Productivity Growth, Employment, and the Labor Share, *Brookins Papers on Economic Activity (BPEA) Conference Draft, Spring*, 64–71.

Hicks, J. R. (1983): *Classics and Moderns. Collected Essays on Economic Theory, Vol III*. Oxford: Basil Blackwell.

ILO (2012): *International Standard Classification of Occupations (ISCO-08), Volume 1 – Structure, Group Definitions and correspondence Tables*. Geneva: International Labour Office.

Johansen, L. (1959): Substitution versus Fixed Production Coefficients in the Theory of Economic Growth: A Synthesis, *Econometrica* 27(2), 157–176.

Kaldor, N. (1961): Capital Accumulation and Economic Growth, in Lutz, F. A. and Hague, D. C. (eds.): *The Theory of Capital*. London: The Macmillan Press, 177–222.

Kaldor, N. (1966): *Causes of the Slow Rate of Economic Growth of the United Kingdom*. Cambridge: Cambridge University Press.

Kalmbach, P. and Kurz, H. D. (1990): Micro-electronics and Employment: A Dynamic Input-output Study of the West German Economy, *Structural Change and Economic Dynamics* 1(2), 371–386.

Kattermann, D. and Kurz, H. D. (1988): Technological Change and Employment, in Flaschel, P. and Kruger, M. (eds.): *Recent Approaches to Economic Dynamics*. Frankfurt and New York: Peter Lang, 397–411.

Kurz, H. D. (1984): Ricardo and Lowe on Machinery, *Eastern Economic Journal* 10(2), 211–229.

Kurz, H. D. (1990): Factor Price Frontier, in Eatwell, J., Milgate, M. and Newman, P. (eds.): *Capital Theory. The New Palgrave*. London: Palgrave Macmillan, 155–160.

Kurz, H. D. and Salvadori, N. (1995): *Theory of Production – A Long-period Analysis*. Cambridge, UK: Cambridge University Press.

Kurz, H. D., Schu¨tz, M., Strohmaier, R. and Zilian, S. (2018): Riding a New Wave of Innovations – A Long-term View at the Current Process of Creative Destruction, *Wirtschaft und Gesellschaft* 44(4), 545–583.

Leitner, S. and Stehrer, R. (2019): The Automatisation Challenge Meets the Demographic Challenge: In Need of Higher Productivity Growth. *The Vienna Institute for International Economic Studies (wiiw), Working Paper*, 171.

Leontief, W. (1951): *The Structure of American Economy, 1919–1939 – An Empirical Application of Equilibrium Analysis*. New York: Oxford University Press.

Leontief, W. (1985): The Choice of Technology, *Scientific American* 252(6), 37–45.

Leontief, W. (1986): *Input-Output Economics*. 2nd ed. New York, Oxford: Oxford University Press.

Leontief, W. and Duchin, F. (1986): *The Future Impact of Automation on Workers*. New York: Oxford University Press.

Lorentz, A. and Savona, M. (2008): Evolutionary Micro-dynamics and Changes in the Economic Structure, *Journal of Evolutionary Economics* 18, 389–412.

Marcolin, L., Miroudot, S. and Squicciarini, M. (2016): The Routine Content of Occupations: New Cross-Country Measures Based On PIAAC. *OECD Trade Policy Papers*, No. 188.

Matzner, E., Schettkat, R. and Wagner, M. (1990): Labour Market Effects of New Technology, *Futures* 22(7), 687–709.

McCurdy, T. H. (1989): Some Potential Job Displacements Associated with Computer-based Automation in Canada, *Technological Forecasting and Social Change* 35(4), 299–317.

Mitchell, J. (2018): A Proposed framework for Digital Supply-Use Tables. *Working Party on National Accounts, Meeting of the Informal Advisory Group on Measuring GDP in a Digitalised Economy, 9 November 2018, OECD Conference Centre, SDD/CSSP/WPNA(2018)3.*

Mitchell, J. (2020): Guidelines for Supply-Use tables for the Digital Economy. *Working Party on National Accounts, Meeting of the Informal Advisory Group on Measuring GDP in a Digitalised Economy, 1–2 July 2019, OECD Conference Centre, SDD/CSSP/WPNA(2019)1/REV1.*

Moneta, A. and Chai, A. (2014): The Evolution of Engel Curves and Its Implications for Structural Change Theory, *Cambridge Journal of Economics* 38(4), 895–923.

OECD (2019): *Measuring the Digital Transformation: A Roadmap for the Future*. Paris: OECD Publishing.

ONS (2020): Productivity Measurement – How to Understand the Data around the UK's Biggest Economic Issue. UK Office for National Statistics. Available at: www.ons.gov.uk/employmentandlabourmarket/peopleinwork/labourproductivity/ [21.03.2021]

Parrinello, S. (2004): The Service Economy Revisited, *Structural Change and Economic Dynamics* 15(4), 381–400.

Pasinetti, L. L. (1977): On 'Non-substitution' in Production Models, *Cambridge Journal of Economics* 1(4), 389–394.

Pasinetti, L. L. (1981): *Structural Change and Economic Growth: A Theoretical Essay on the Dynamics of the Wealth of Nations*. Cambridge: Cambridge University Press.

Pasinetti, L. L. (1993): *Structural Economic Dynamics: A Theory of the Economic Consequences of Human Learning*. Cambridge: Cambridge University Press.

Pini, P. (1997): Technical Change and Labour Displacement – Some Comments on Recent Models of Technological Unemployment, in Antonelli, G. and De Liso, N. (eds.): *Economics of Structural and Technological Change*. London: Routledge, 61–83.

Ricardo, D. (1821): On The Principles of Political Economy and Taxation, Third Edition, in Sraffa, P. and Dobb, M. H. (eds.): *(1951): The Works and Correspondence of David Ricardo, Volume 1*. Cambridge: Cambridge University Press.

Savona, M. (2019): The Value of Data: Towards a Framework to Redistribute It. *SPRU Working Paper Series (SWPS) 2019–21 (October).*

Schefold, B. (1976): Different Forms of Technical Progress, *The Economic Journal* 86(344), 806–819.

Tambe, P., Hitt, L., Rock, D. and Brynjolfsson, E. (2020): Digital Capital and Superstar Firms. *NBER Working Paper*, 28285.

UN (2008): *International Standard Industrial Classification of All Economic Activities – Revision 4.* Department of Economic and Social Affairs, Statistics Division, Statistical Papers, Series M No. 4/Rev.4.

UN (2009): *System of National Accounts 2008.* ST/ESA/STAT/SER.F/2/Rev.5, United Nations, New York.

UNCTAD (2017): *Trade and Development Report 2017: Beyond Austerity: Towards a Global New Deal.* United Nations Conference on Trade and Development, UNCTAD/TDR/2017. New York and Geneva.

UNIDO (2019): *Industrial Development Report 2020. Industrializing in the Digital Age.* United Nations Industrial Development Organization, UNIDO ID/449, Vienna.

Verdoorn, P. J. (1949): Fattori che Regolano lo Sviluppo della Produttivita' del Lavoro, *L'Industria* 1, 3–10.

Walker, R. A. (1985): Is There a Service Economy? The Changing Capitalist Division of Labor, *Science & Society* 49(1), 42–83.

Zalai, E. (1989): Eigenvalues and Labour Values: Contributions to the Quantitative Analysis of Value, *Economic Systems Research* 1(3), 403–410.

15
THE DIFFUSION OF INDUSTRIAL ROBOTS

Bernhard Dachs, Xiaolan Fu, and Angela Jäger

Introduction

At the time of the publication of this book, robots will turn 100 years old. The term "robot" was first used in 1920 by the Czech writer Karel Čapek to describe artificial people created to work. Over the last century, industrial robots have made an impressive career, from the theatre stage and the pages of science-fiction books to the shop floors of the world. The number of robots seems to be growing faster than other categories within information and communication technologies, in particular, faster than personal computers (DeStefano et al. 2017). The International Federation of Robotics (IFR) reports that 303,847 robots were sold globally in 2016, up from 53,409 units in 1993.

This rapid diffusion can be explained by the decreasing price/performance ratio of robots, which is a result of performance increases in computing power, sensors, cameras, communication technologies, and energy storage (Pratt 2015). New technologies such as collaborative robots and improved human-machine interaction have enlarged the potential application areas. Major advances in the future can also be expected from the combination of "machine learning" (artificial intelligence) and robotics (IFR 2019). However, productivity gains will also arise when firms learn to better integrate robots into their production processes. Another organisational innovation where the IFR sees potential for the future is "Robots as a Service".

This chapter will provide an overview of the diffusion of robots in the economy and its economic effects. It is divided into eight sections: the next section discusses the characteristics of industrial robots. In the following sections, we will provide information on the diffusion of robots at country and sectoral level. Both sections will draw on data from the International Federation of Robotics (IFR). These data will be complemented with results from EUROSTAT and from the European Investment Bank in the successive section. The section afterwards will analyse the characteristics of firms using robots, including their size and the characteristics of their production processes. Besides country and sectoral differences, there is also a considerable heterogeneity in the use of robots among different types of firms. Then, we will wrap up studies that investigated the economic effects of robots' use in different countries. The final section closes with a summary and some conclusions.

DOI: 10.4324/9780429351921-19

The evolution of industrial robots

Before we look at the diffusion of robots in the economy, a few remarks are provided on the technology. The International Organization for Standardization (ISO 2012) defines a robot as an "actuated mechanism programmable in two or more axes with a degree of autonomy, moving within its environment, to perform intended tasks". ISO also distinguishes between robots for industrial applications ("industrial robots") and service robots, which "perform useful tasks for humans or equipment excluding industrial automation applications".

Robots are not a new technology. According to Mansfield (1989), the industrial robot was an American invention, and the first commercial robot was put into operation in 1961 by General Motors. The first industrial robots in Europe and in Japan were installed in 1967 (Scalera and Gasparetto 2019). In the following years, robots increasingly found applications in repetitive tasks such as welding, painting, bending, or polishing. Major sectors making use of robots in the mid-1980s were the automotive and the electrical industries (Mansfield 1989) – very similar to the specialisation patterns found today.

The evolution of industrial robots encompasses various generations (Hägele et al. 2016; Scalera and Gasparetto 2019). While the first generation was little more than a programmable mechanical arm with no ability to move and no connection to the external environment, later generations became increasingly adaptive and autonomous due to increasing computing power and the use of various sensors. The second generation (between 1968 and 1977) was already programmable, used microprocessors and electric motors instead of hydraulic systems, had some limited capabilities to recognise their environment, and was able to carry out more difficult tasks. These abilities were further improved in the third generation (1978–1999), which also included limited adaptive capabilities and self-programming. Industrial robots for sorting, picking, and other complex tasks arrived in new application areas such as electronics, the food industry, or machinery.

The current generation of industrial robots is characterised by high computing capabilities, advanced sensors, and intensive data exchange with their environment, including with other robots. Pratt (2015) identifies eight drivers determining the capabilities of today's industrial robots: (1) exponential growth in computing performance and (2) in global computing power; (3) improvements in electromechanical design tools and numerically controlled manufacturing tools; (4) improvements in electrical energy storage and (5) in electronics power efficiency; (6) exponential expansion of the availability and performance of local wireless digital communications; (7) exponential growth in the scale and performance of the internet; (8) and exponential growth of worldwide data storage. He predicts a "Cambrian Explosion" of robotics technology in the future, due to the combination of exponential growth in computing power ("Cloud Computing") and applications of artificial intelligence ("Deep Learning"), which will lead to a virtuous cycle of explosive growth.

The worldwide diffusion of robots

Prospects for robots seem bright, and the diffusion of robots so far may only be the prelude to things to come. Since the year 1993, annual shipment of industrial robots worldwide has grown from 53,409 units to 303,847 in 2016, which means more than a five-fold increase during a 23-year timespan. In particular, the shipment of industrial robots gradually increases, with some fluctuations, since 1993 and reaches its first and second peaks in 2000 and 2005 respectively (Figure 15.1).

The annual shipment of industrial robots remains stable from 2005 to 2008 and is confronted with a sharp downturn to a level of 60,018 units in 2009 in response to the global financial crisis. Nevertheless, worldwide production of industrial robots recovers immediately after 2009 and starts to grow again with ever stronger momentum. From Figure 15.1, it seems that the global diffusion of robots starts to take off after the global financial crisis. The global labour force, on the other hand, increases more steadily from roughly 2.4 billion in 1993 to approximately 3.4 billion in 2016.

With regard to the geographical distribution of shipments of industrial robots, all of the four regions, Asia, America, Europe, and Africa, witness significantly higher shipments in 2016 compared to those in 1993 (Figure 15.2). The sudden and dramatic downturn in the year 2009, when the impact of the financial crisis spreads to every corner of world economy, hits robot diffusion in all four regions. Throughout the whole period, annual industrial robot shipments in Asia outnumber those in the other three regions, and this gap further enlarges after the recovery from the global financial crisis. In 2016, Asia ends up with 200,042 units, almost twice the total shipments in the rest of the world. The shipment of industrial robots in Europe slightly exceeds those in America, with a similar annual change in both regions.

Admittedly, Africa has the lowest volume of shipments, with no more than 5,000 units annually, even in 2016. However, the growth rate in this region exceeds that of the rest of the world by far, especially in the first decade of the 21st century. Before 1999, there are no shipments of industrial robots in Africa, while the stock almost triples in 2001 compared to the previous year. Since then, industrial robot shipments show a high growth rate overall until 2016.

The uneven geographical distribution of industrial robots is even more pronounced at the country level. Among the total shipments of industrial robots during the period of 1993–2016 in Asia, Japan takes up 48.6%, nearly equalling all the other countries and districts in Asia

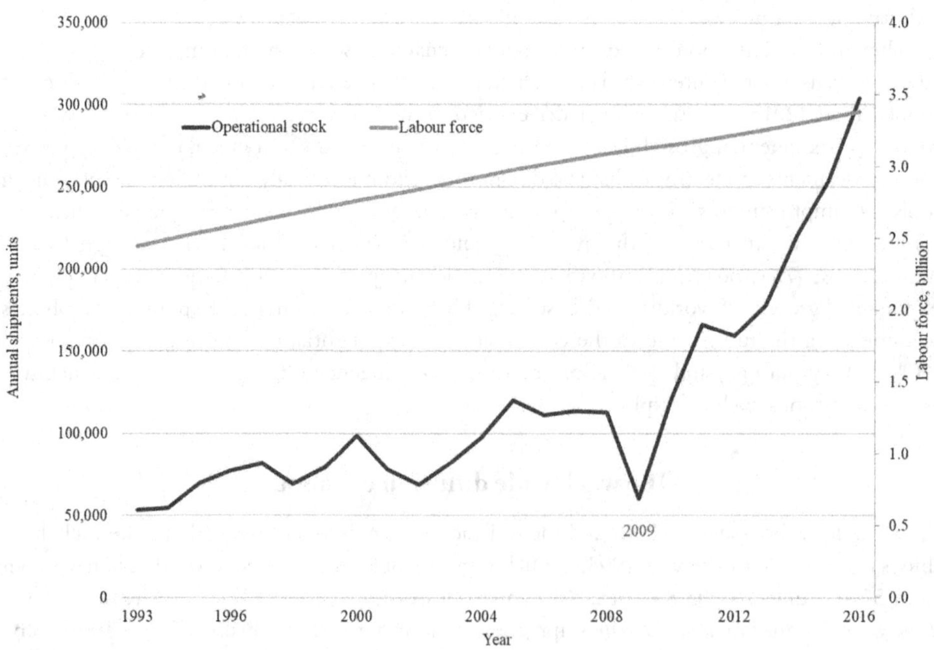

Figure 15.1 Worldwide shipments of industrial robots and labour force

Source: IFR, World Bank

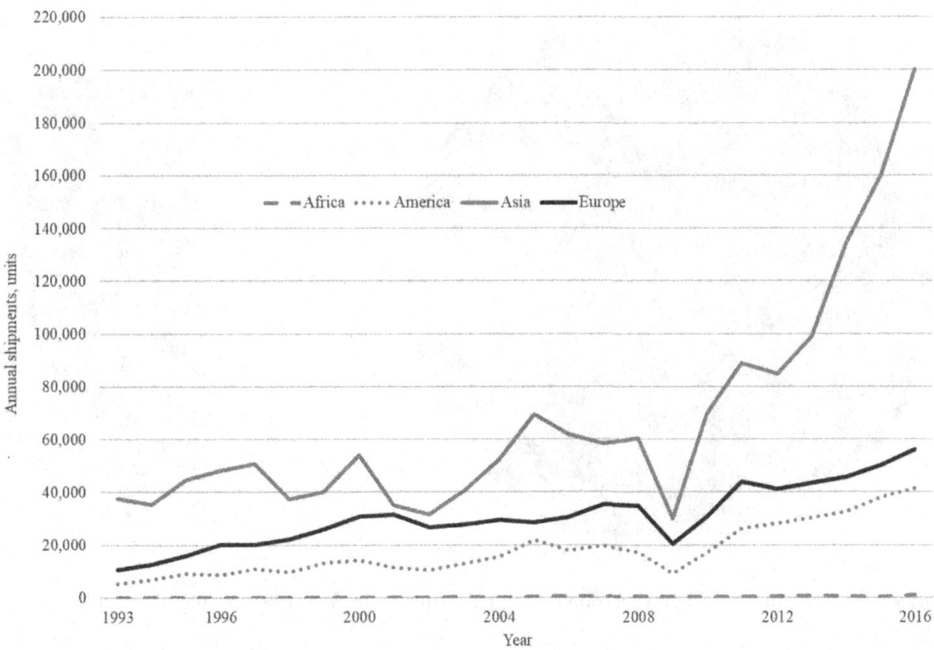

Figure 15.2 Worldwide distribution of industrial robots

Source: IFR

(excluding Hongkong, Macau, and Taiwan); China comes in second place with 23%, followed by the Republic of Korea (Korea in the remainder of the text) with 18.3%. These three industrial economies together share 89.9% of Asian shipments of industrial robots, while the next seven countries in total account for 6.4%. Thailand, Singapore, and India each account for ~1–2% of the total shipments; for the other top ten countries, however, no single country takes up a share higher than 1% of total shipments to Asia. Indonesia has the fewest shipments of industrial robots among the Asian top ten countries, receiving only 0.4% of the total production (Figure 15.3).

America displays a greater concentration of industrial robot shipments. The US alone is responsible for 87.7% of the total shipments across the continent, followed by Mexico with a share of only 4.8%. The remaining shipments go to Canada (3.3%) and other North and South American countries including Brazil (3.3%), Argentina (0.3%), Chile (0.02%), Columbia (0.01%), Puerto Rico (0.01%), and the rest of South America (0.7%).

By contrast, shipments of industrial robots are relatively evenly distributed in Europe with five countries taking up around 90% of total European shipments during the years 1993–2016 (Figure 15.4). Particularly, Germany stands out among European economies by receiving 42.2% of the total shipments, more than twice as much as Italy (15.7%) which ranks in second place in the region. The share of shipments to France and Spain are similar, 8.4% and 7.2% respectively, followed by the United Kingdom as the fifth largest market with 4.4% of European total shipments. The remaining industrial robot shipments in Europe go to Sweden (2.7%), Czech Republic (2%), Belgium (2.0%), Netherlands (1.9%), and Austria (1.8%).

In a longitudinal perspective, the ranking of the ten largest markets worldwide remains relatively stable, with a few variations, as emerging economies and developing countries play an increasing role in technological development and industrialisation. In 1993, all of the top

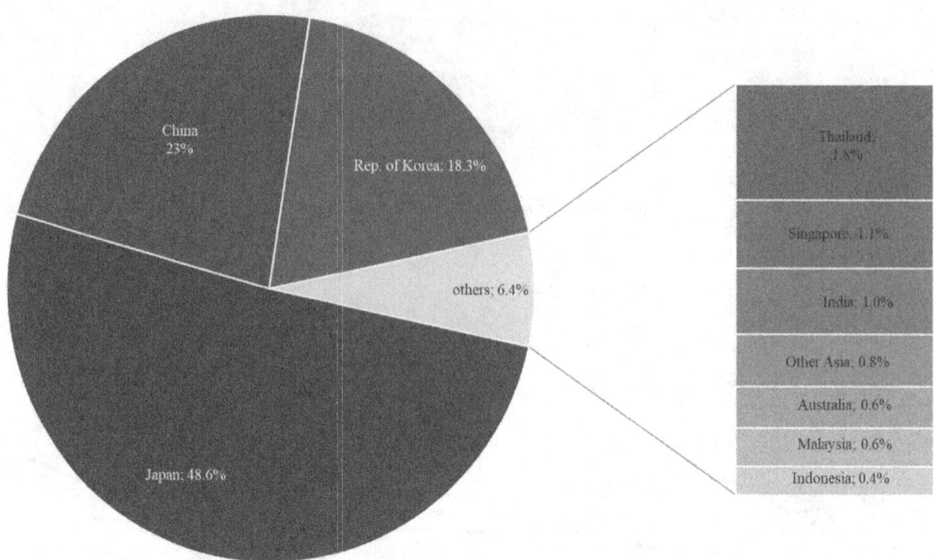

Figure 15.3 Shipments of industrial robots – top ten countries in Asia, 1993–2016

Source: IFR

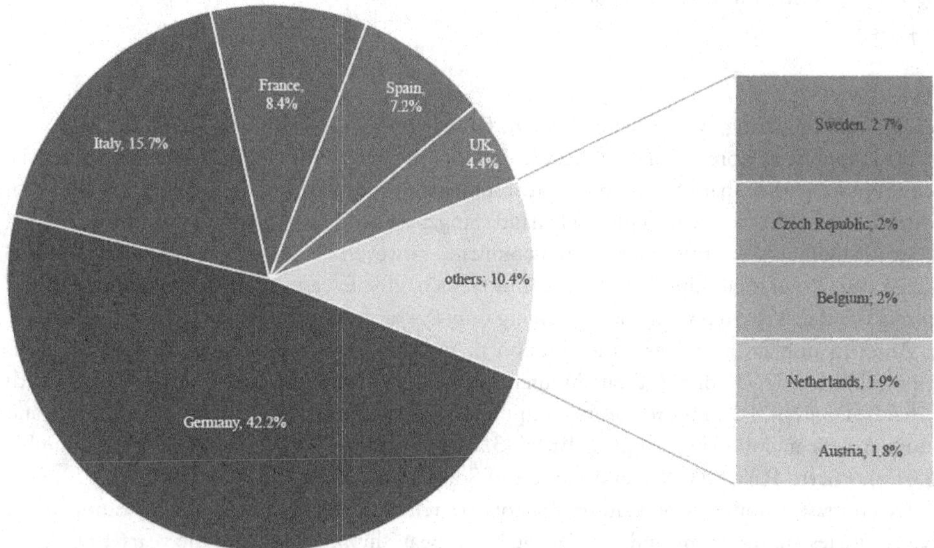

Figure 15.4 Shipments of industrial robots – top ten countries in Europe, 1993–2016

Source: IFR

ten markets were developed countries. China entered as the tenth market for industrial robots in 2001 and Thailand and Mexico joined the ranking in 2009 and 2016, respectively. It is worth noting that China rose to be the fifth largest market after the global financial crisis and leapt to replace Japan as the world's largest market in 2016. With 96,500 units, Chinese firms install more than twice the number of industrial robots as Korea, Japan, the US, and Germany.

Table 15.1 Worldwide shipments of industrial robots (annually, '000 of units)

Rank	*1993*		*2001*		*2009*		*2016*	
1	Japan	33.5	Japan	28.4	Japan	12.8	China	96.5
2	US	5.2	Germany	12.7	Germany	8.5	Korea	41.4
3	Germany	4.3	US	10.8	US	8.4	Japan	38.6
4	Korea	2.7	Italy	6.4	Korea	7.8	US	31.4
5	Italy	2.5	Korea	4.1	China	5.5	Germany	20.1
6	France	1.0	Spain	3.6	Italy	2.9	Italy	6.5
7	UK	0.6	France	3.5	France	1.5	Mexico	5.9
8	Spain	0.5	UK	1.9	Spain	1.3	France	4.2
9	Singapore	0.4	Sweden	0.9	Thailand	0.8	Spain	3.9
10	Belgium	0.3	China	0.7	Belgium	0.7	Thailand	2.6

Source: IFR

Countries such as Belgium and Sweden, initially in the list of top markets for industrial robots, gradually fade away from the list (Table 15.1).

It is no coincidence that the largest users of robots in Europe and the world are also the largest car manufacturing countries. In regard to robot shipments, the top ten countries very closely resemble the top ten car producers worldwide, with the exception of India, which is not among the top ten in the diffusion of robots. Thus, the rise of China, Korea, Mexico, and Thailand in the ranking of Table 15.1 is closely linked to the evolution of the automotive industry in these countries. The close relationship between robots and the automotive industry will be explored in the following section.

Let us draw a focused comparison between the four leading economies where the adoption of industrial robots is concentrated: China, Japan, the US, and Europe. In 1993, Japan is the country adopting the most industrial robots (33,502 units), more than twice as many as the EU (10,509 units) and the US (5,246 units) combined. In contrast, there is no industrial robot adoption in China until 1999 (Figure 15.5).

The diffusion of industrial robots is far from being a linear process. From 1993 to 2008, installations of industrial robots increase in the EU, the US, and in China at different speeds. Figure 15.6 shows that positive growth dominates in these three economies, despite occasional declines before and after the millennium. Moreover, investment in robots is strongly affected by the business cycle. The economic downturns of 2001 and 2009 are clearly visible in the shipments of industrial robots, and this points to the importance of the business cycle for investment decisions in this technology. The fluctuation of shipments in China during the crisis is even more dramatic, with the adoption of industrial robots dropping 11% after an 84% surge in the previous year, and soon soars 178% in the following years. This pro-cyclical pattern of robot shipments indicates that firms consider robots as a long-term investment and do not expect short-term benefits like cost cutting; otherwise, investment in robots to reduce production costs would be an appropriate response to declining demand during a crisis. The slump in demand for robots can also be related to labour hoarding, the strategy of firms to keep their workforce stable and tolerate decreasing productivity as demand for their products falls into an economic crisis (Leitner and Stehrer 2012). Thus, investment in robots is not the first priority for firms in a crisis.

IFR does not provide data on the main producers of industrial robots at the country level. We also found no other public source for this information. However, from the world's largest firms in the robotics industry, we may infer that most robots are manufactured in Japan, Europe, and China.

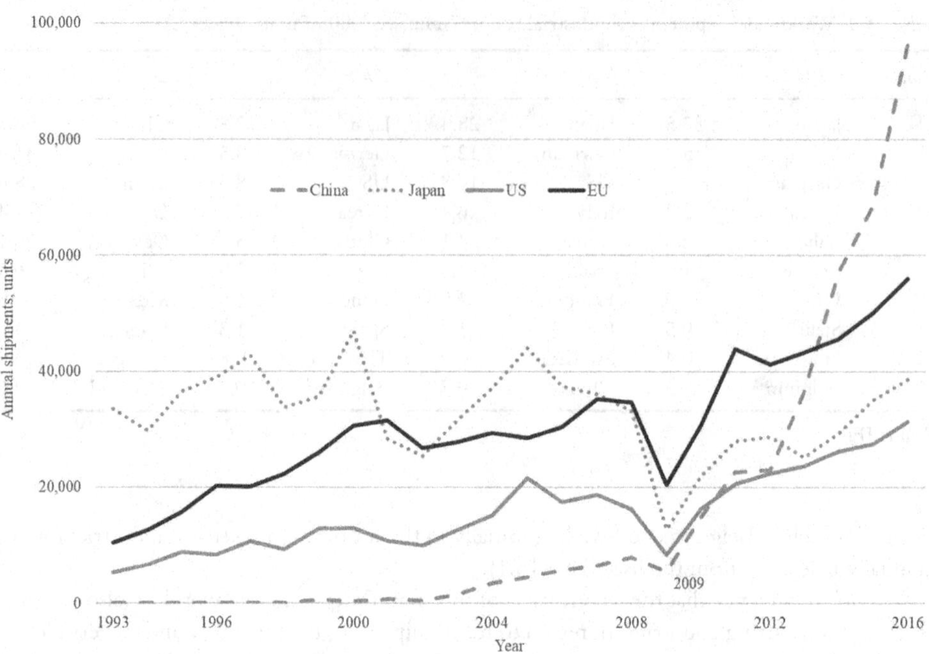

Figure 15.5 Distribution of industrial robots in four regions

Source: IFR

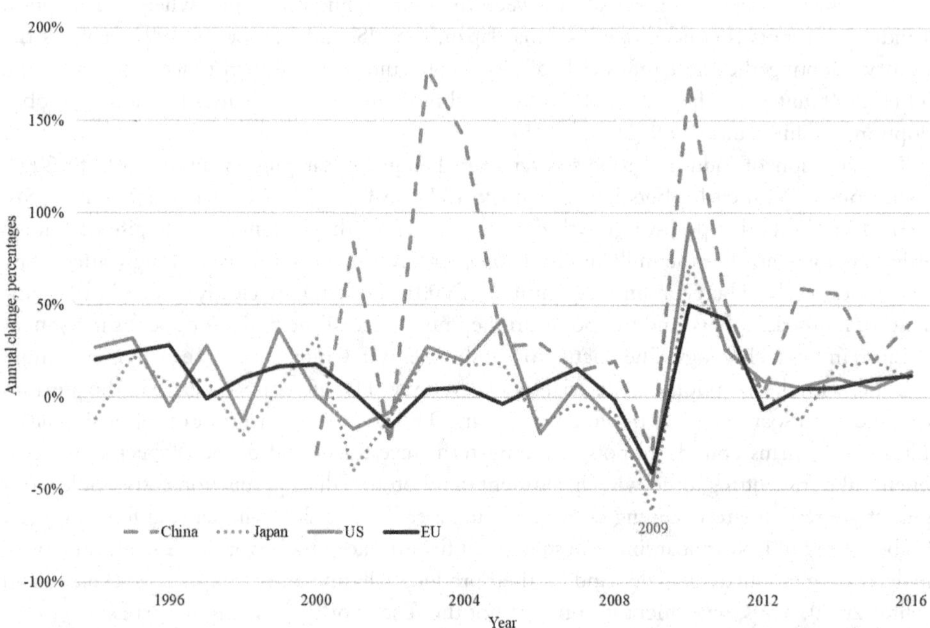

Figure 15.6 Distribution of industrial robots in four regions, by change rate

Source: IFR

Sectoral patterns of the diffusion of robots

Apart from regional diversity, global industrial robot distribution also varies considerably between industrial sectors and sub-sectors. In all countries, industrial robots are predominantly installed by manufacturing firms, and only rarely distributed to other sectors (i.e. the education/research sector, agriculture, forestry and fishing industry, construction industry, electricity, gas and water supply industry, and mining and quarrying industry).

Within the manufacturing sector, the automotive industry displays the highest adoption of industrial robots with 44,400 shipments in 2004, followed by the electrical and electronics industry with 15,600 units, which is only slightly more than one-third of the former. The shipment of industrial robots to the automotive industry dropped to 38,600 in 2010, two years after the global financial crisis in 2008. It is worth noting that the decline in robot shipments during the global financial crisis was due to a drop in demand in the automotive industry, as this is the only industry that has experienced a downturn in the adoption of industrial robots between 2004 and 2010. Despite the shock, the annual use of industrial robots in the automotive industry soon recovers, rising to 103,300 units by 2016 – more than twice as many as in the years 2004 to 2010.

The electrical and electronics industry is the second largest user industry of industrial robots. However, the gap to the automotive industry is large: in 2004, only 15,600 units were applied in this industry. The number doubled in 2010 compared to 2004 and tripled in 2016. Since 2010, the gap between the electrical/electronics industry and the automotive industry has been narrowing (Figure 15.7). This indicates that, over time, more application areas for robots outside the automotive industry have emerged. Examples for such new application areas of industrial

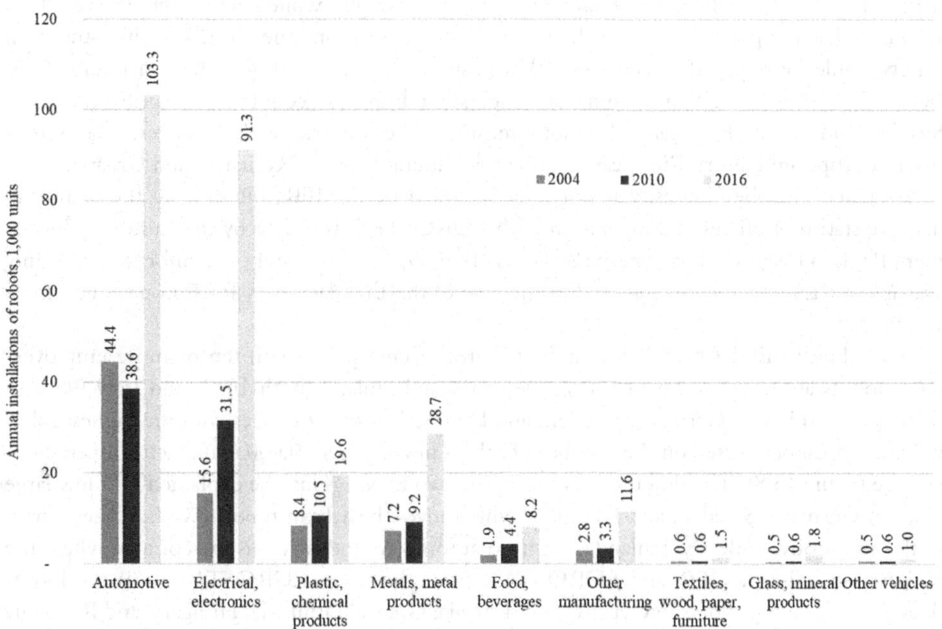

Figure 15.7 Industrial robots installed in manufacturing industries worldwide

Source: IFR

robots outside the automotive industry are the manufacturing of metal products, of plastic products, the chemical industry, and the manufacturers of food and beverages.

In most industries, industrial robot adoption accelerates over time, just like in the automotive and in the electrical and electronic industry. Medium and high technology-intensive industries use more robots than low-tech branches, disregarding the difference in the size of the sub-industries. For example, 7,200 units of industrial robots were installed in the metal products industry in 2004, and only 2,000 more units joined the stock in 2010. However, six years later, the annual adoption of industrial robots has more than tripled, reaching a total of 28,700 units. This is the same for plastic products and for the chemical industry, the fourth highest adoption branch in manufacturing sectors, where the use of industrial robots nearly doubles in the second period. Similar trends can be observed in industries where robots are only applied to a limited extent, such as the industries of wood and furniture, paper, textiles, other vehicles, glass, ceramics, stone, mineral products, and all other manufacturing branches. For these categories, there are hardly any scenarios for the use of industrial robots. For instance, even in the year 2016, only 300 units are adopted in the textiles industry. The only exception is the food and beverage sector, where growth in the adoption of industrial robots doubles in the second period (Figure 15.7).

Robot' intensity across countries

In a discussion about the diffusion of robots, not only should absolute numbers be considered, but also how intensively firms in different countries use robots. Intensity is usually measured by the number of installed robots per 10,000 workers or the ratio of firms using robots to all firms in a country.

The ranking of the countries according to robot intensity clearly differs from that in Table 15.1. Singapore is the country with the highest robot density, followed by Korea and Germany. Here, some small countries have a much higher weight, while neither the US nor China are in the leading positions. Given the sheer size of these economies, a massive investment in robots would be required to achieve such a position. Europe (dark grey bars in Figure 15.8) lags behind Japan, the US, and China, although some European countries, especially Germany, Sweden, and Denmark, reveal high robot intensities. The countries with the lowest robot intensity in Europe are Eastern European countries including Estonia, Romania, and Croatia.

Evidence on robot intensity is not only provided by the IFR, but also by the European national statistical offices in cooperation with EUROSTAT, as well as by the European Investment Bank in its annual Investment Report (EIB 2019). Unfortunately, this information is only available for European countries, and in the case of the EIB data, only for European countries and the US.

According to EUROSTAT, the highest shares of enterprises with ten or more employees using robots are found in Spain (11%), Denmark and Finland (both 10%), and Italy (9%). In contrast, the IFR sees Germany, Sweden, and Denmark as the countries with the highest robot intensity in Europe, based on the number of robots installed per 10,000 manufacturing employees (see Figure 15.8). The difference between the two rankings may be explained by a few large firms in Germany, Sweden, and Denmark which invest heavily in robots. We find these firms in the automotive industry. Finland – with no automotive industry – is the country where the rankings according to IFR and EUROSTAT differ the most. EUROSTAT finds the lowest shares of firms using robots in Cyprus (1%), Estonia, Greece, Lithuania, Hungary, and Romania (Figure 15.9), which is consistent with IFR data.

Another interesting result from EUROSTAT data is that there is no difference in the diffusion of robots between Western European member states and the EU-28 average (two right-hand

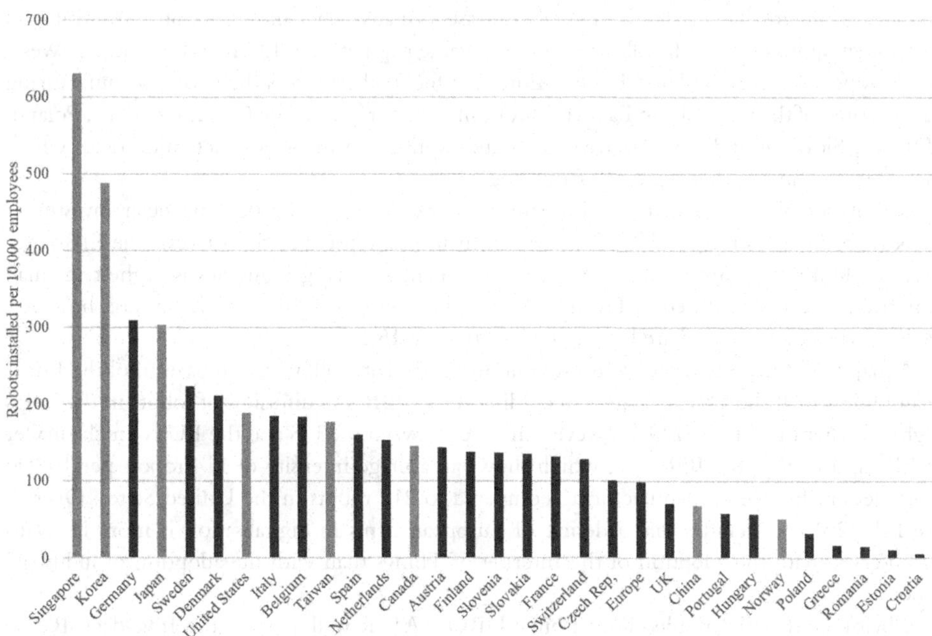

Figure 15.8 Number of robots per 10,000 employees in manufacturing, 2016

Source: IFR

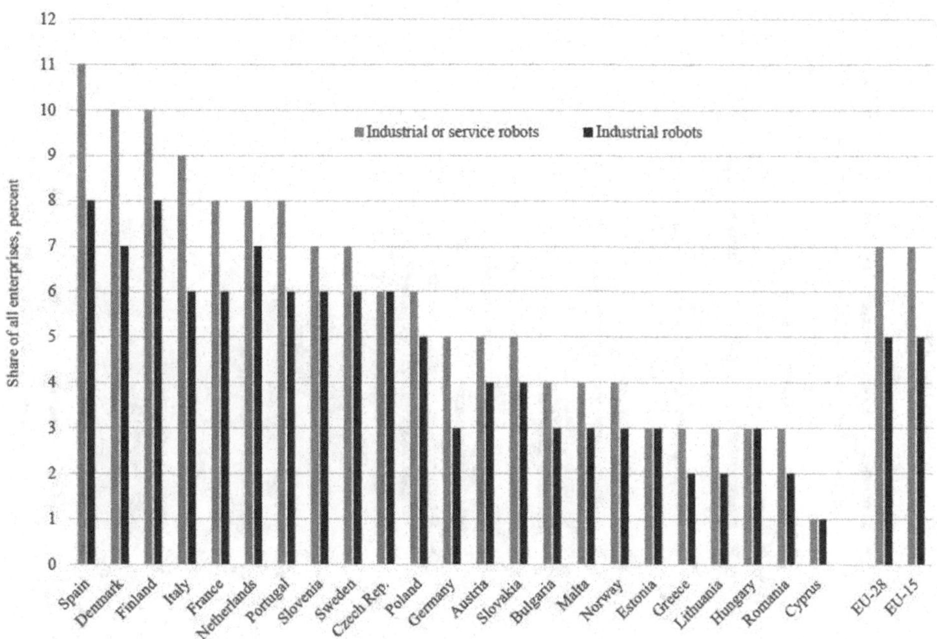

Figure 15.9 Share of enterprises in different European countries that use industrial or service robots, 2018

Source: EUROSTAT, ICT usage in enterprises

columns in Figure 15.9). This is in stark contrast to many other indicators of innovation and technology diffusion, which still show a considerable gap between EU member states in Western Europe and in Central and Eastern Europe. One explanation is the strong manufacturing base in some of the Central and Eastern European member states. The Czech Republic, Poland, Hungary, Slovakia, and Slovenia have large automotive and metal products industries, which manifests itself in a higher share of firms using robots.

Robots are also covered in the Investment Survey 2019/20 by the European Investment Bank (EIB 2019). A total of 12,672 European firms participated in this survey. The EIB notes that the diffusion of advanced robots in the EU manufacturing industries is highest in small countries: Slovenia, followed by Finland, Austria, Denmark, and Sweden. In contrast, however, Malta, Cyprus, and Ireland are lagging behind (Figure 15.10).

A major advantage of the EIB Investment Survey is that it allows a comparison of the European Union with the United States. According to the EIB, the diffusion of robots in the US is slightly higher than in the EU. However, the gap between the US and the EU is much smaller in EIB data than in the IFR data, which shows an average intensity of 114 robots per 10,000 employees in European manufacturing compared to 217 robots in the United States. Overall, the EIB (2019) concludes that a deficit of European firms in digitalisation is more likely to be observed with the adoption of the Internet of Things than with the adoption of industrial robots.

Obviously, the EIB results differ from EUROSTAT as well as from the IFR data. Readers will also find that the numbers reported by the EIB are considerably higher than those reported by EUROSTAT, although both surveys cover the year 2018. The most likely reasons for these discrepancies are different survey techniques and different questions about robotics'

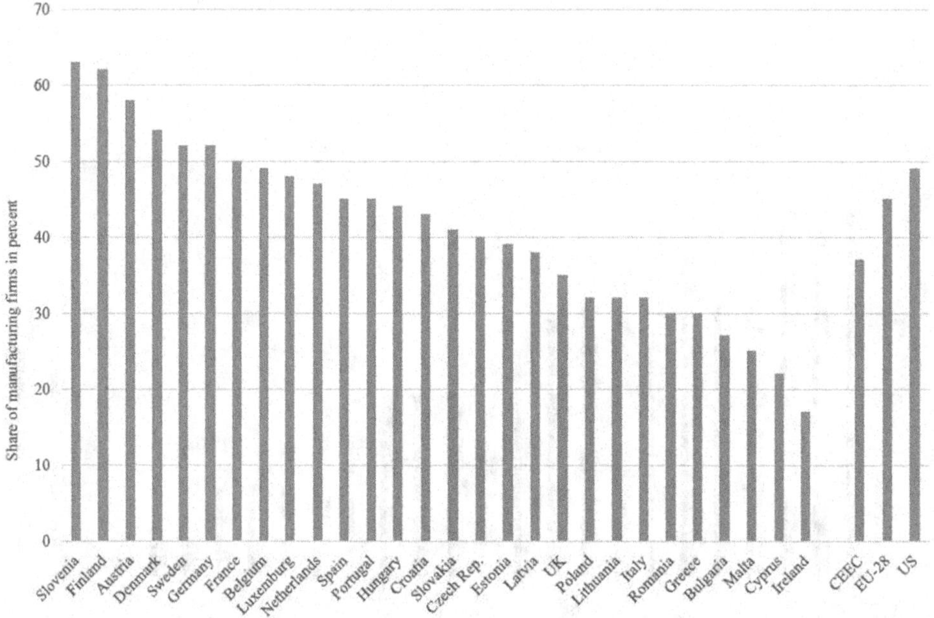

Figure 15.10 Share of manufacturing firms that use advanced robots, 2018

Note: CEEC are Central and Eastern European Countries

Source: European Investment Bank, Investment Survey 2019/20

use. EUROSTAT's survey asks whether the firm uses industrial or service robots, including some potential applications. Moreover, as the diffusion of robots increases with the firm size, the share of small firms in the EIB survey may be lower compared to the EUROSTAT survey. The IFR is based on actual shipments, which is a different concept than the robots installed in firms. Shipments may go to intermediaries or be forwarded from the headquarters of multinational firms to subsidiaries in other countries.

To compare IFR, EUROSTAT, EIB data, we normalise the country values to the EU or European average, so that countries with a higher diffusion than the EU average have a value larger than one and vice versa (Figure 15.11). The country rankings are quite inconsistent between the three surveys. Only five countries (Finland, Netherlands, France, Sweden, and Denmark) are at or above the EU average in all three surveys. These countries can be regarded as the leaders in the application of robots in Europe. Germany is difficult to include, as the divergence between IFR and the other two datasets is largest here. Germany scores well on the IFR and the EIB data but is below EU average in the EUROSTAT data. A careful examination of the data would be necessary to understand the reasons for these differences. However, it seems conceivable that German firms are among the most sophisticated users of robots worldwide. Poland, Romania, Hungary, Greece, and Estonia rank below the EU or European average in all three surveys, so it seems fair to say that these countries are lagging behind in the diffusion of robots.

With a Spearman correlation index of 0.63, the EIB and the EUROSTAT survey are the most comparable, followed by EIB-IFR (correlation of 0.62) and EUROSTAT-IFR (correlation of 0.6). However, some small EU countries are not included in the IFR data, which makes the comparison difficult. Nevertheless, due to their methodology, the data on the diffusion of robots at country level reveal large differences. From a methodological point of view, a survey at

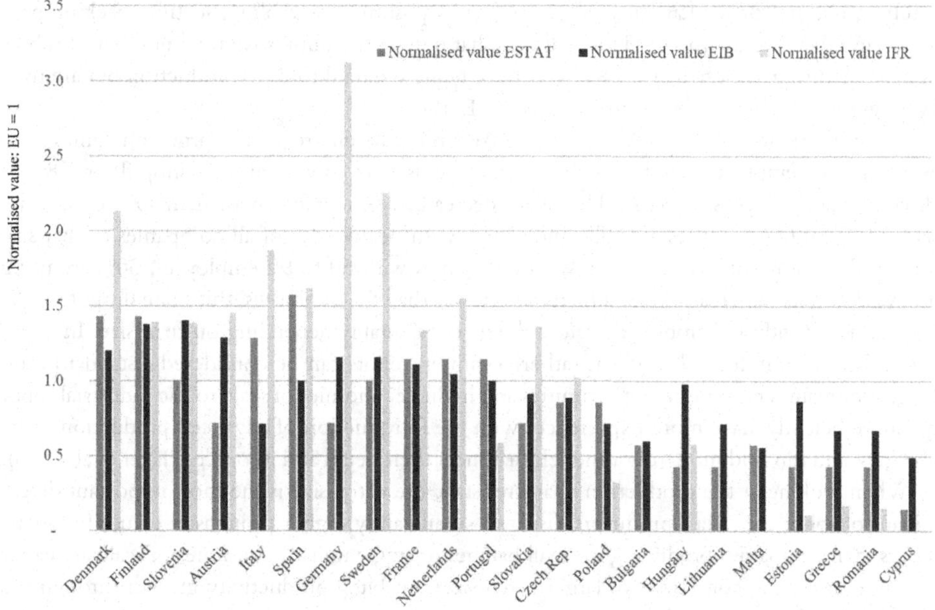

Figure 15.11 Normalised values for IFR, EUROSTAT, and EIB data on robot diffusion, 2016 and 2017

Source: IFR, EUROSTAT, EIB, own calculations

the firm level, such as the EIB or EUROSTAT surveys, is preferable to shipment data because it is easier to match with other data at the firm level. This also puts some question marks against cross-country studies using IFR data because they cannot take into account the concentration of robots in some very robot-intensive firms and industries.

Robots in manufacturing: some firm-level evidence

In manufacturing, robots are applied in many different processes and production contexts. The IFR data provides insights from the perspective of robot producers and indicates who buys these robots. Thus, IFR data offer an interesting insight into the robot market but provide only little information about the users of industrial robots, that is, those companies that purchase the robots. When discussing the diffusion of robots, it is also of great interest to place the firms that use robots into the context of their non-using counterparts. Therefore, we add an additional perspective by analysing the use of industrial robots in manufacturing at firm level. This approach allows us to clarify structural barriers and delivers an insight from the perspective of robot users.

The *European Manufacturing Survey* (EMS) provides this perspective. EMS investigates the use of techno-organisational innovations in manufacturing at the level of individual manufacturing sites (Jäger and Maloca 2016). In each country, the survey comprises a large random sample of manufacturing firms with at least 20 employees covering the entire manufacturing sector. EMS is organised by a consortium of research institutes and universities from across Europe. The EMS 2015 data includes information from Germany, Austria, Spain, Croatia, the Netherlands, Slovenia, Switzerland, and the Republic of Serbia, containing a sample of over 2,750 manufacturing companies. Thus, EMS enables us to draw a reliable picture of industrial companies that use robots.

On average, one-third of all manufacturing firms with at least 20 employees in these eight European countries use industrial robots either for manufacturing processes (e.g. welding, painting, cutting) or for handling processes (e.g. depositing, assembling, sorting, packing processes) in 2015 (Figure 15.12). This indicates that despite the impressive total number of robots in manufacturing, industrial robots are far from being standard tools in production but are used by a significant sub-group of manufacturers in Europe.

One important sub-group is large firms. According to Figure 15.12, large companies with 500 and more employees are by far the most active users of robots on their shop floor (78% of all companies in that size class). This share decreases considerably with firm size: companies with 100 to 499 employees are still quite active with nearly 50% of all companies of that size using industrial robots. Among industrial companies with 50 to 99 employees, 30% are using industrial robots, whereas, in small firms with fewer than 50 employees, this share drops to 23%.

The use of industrial robots therefore differs considerably depending on firm size. In larger firms, the use of industrial robots is rather common; robots can be considered a standard tool. Larger companies have more opportunities and higher economies of scale to use industrial robot systems efficiently, have more experiences with the introduction of advanced production technologies, and can find more robot models that meet their needs and resources (Jäger et al. 2016).

When looking at the production characteristics, the batch size is the most important determinant of robot use. Mass producers display a significantly higher propensity to use industrial robots than companies producing in small batches or single units. Economies of scale are easier to realise under the conditions of larger batch sizes, enabling productivity growth through the automation of repetitive tasks.

Slight differences can be found regarding the product complexity and the degree of customisation. Figure 15.13 shows that 36% of all companies that produce medium complex products

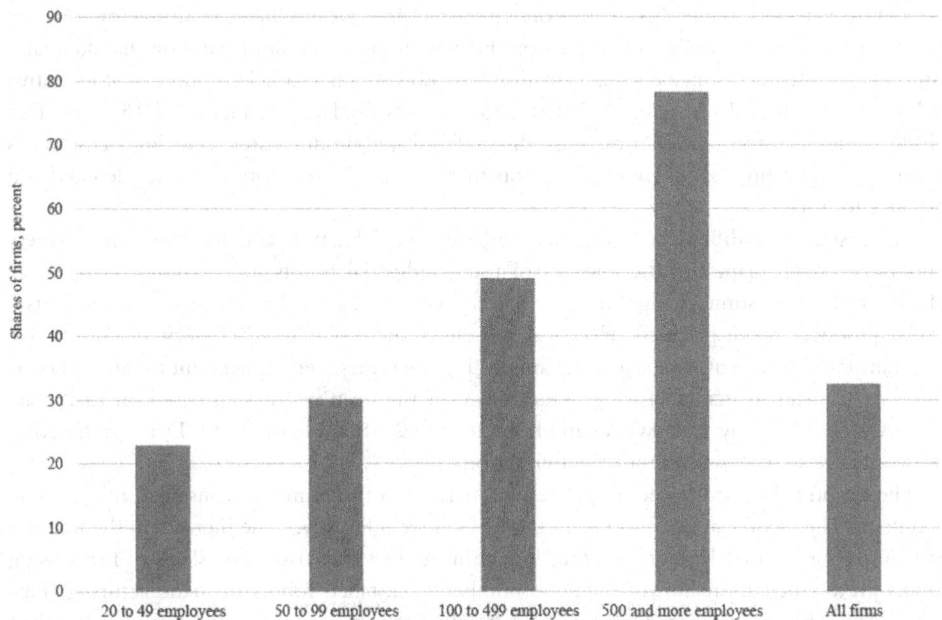

Figure 15.12 Share of European manufacturing firms that use robots in different firm sizes, 2015
Source: European Manufacturing Survey 2015, eight countries, compiled by Fraunhofer ISI, weighted data

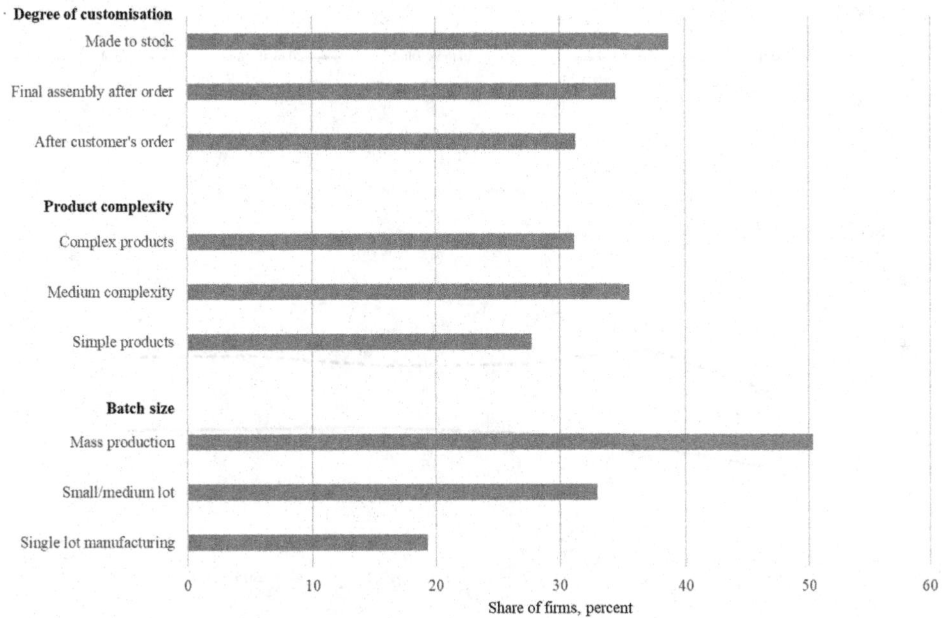

Figure 15.13 Share of European manufacturing firms that use robots by production characteristics, 2015
Source: European Manufacturing Survey 2015, eight countries, compiled by Fraunhofer ISI, weighted data

use industrial robots, while this share is only 30% and 33% for producers of simple or complex products, respectively. Medium complex products with high volumes have more handling and assembly tasks that are suitable for automation but also contain a sufficient number of repetitive tasks to be automated (Jäger et al. 2015). Moreover, as displayed in Figure 15.13, firms that produce after a customer's order or finish the prefabricated product after receiving a customer's order are significantly less likely to use robots than firms that anticipate customer demand and produce to stock.

These structural differences, together with sectoral affiliation and location, are of great importance in determining the chances of using industrial robots in a manufacturing firm. However, there are some changes in these determinants over time. Producers of discrete parts – plastic products, metal products, electrical and electronic products, machinery, or automotive and transport equipment – are a good example. Together, these sectors account for around two-thirds of all firms in manufacturing. The shifts over time can be assessed based on EMS data from Austria, Germany, and Switzerland from 2003 to 2018 in Figure 15.14. Data for the other five countries are not available for the full period.

The chapter already showed an impressive increase in the number of installed robots in all countries. This trend can be observed in the EMS as well. However, the increase in the number of firms using industrial robots is rather low compared to IFR data. The share of firms using robots grew much more slowly than the number of installed robots in manufacturing (Figure 15.14). This indicates that the boom in installed robots is a capital deepening rather than broadening issue.

Figure 15.14 also shows that the increase of firms using robots can mainly be related to large firms; the share of robot users in this group rose from 60% to over 80% between 2003 and 2018. The expectation that robots will become more common in manufacturing over time and that

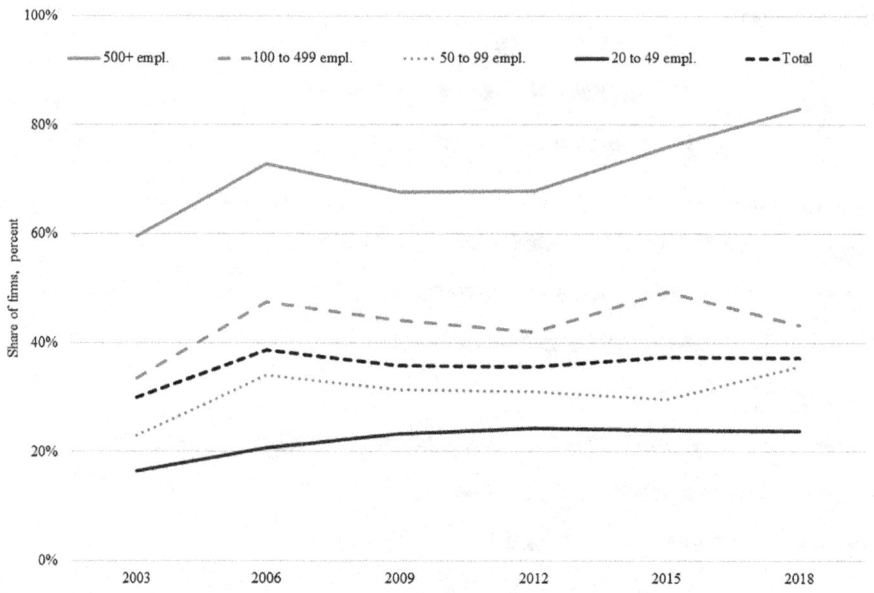

Figure 15.14 Share of robot users among Austrian, German, and Swiss manufacturers of discrete parts, differentiated by firm size groups, 2003–2018

Source: European Manufacturing Survey AT, CH, DE – 2003–2018, compiled by Fraunhofer ISI

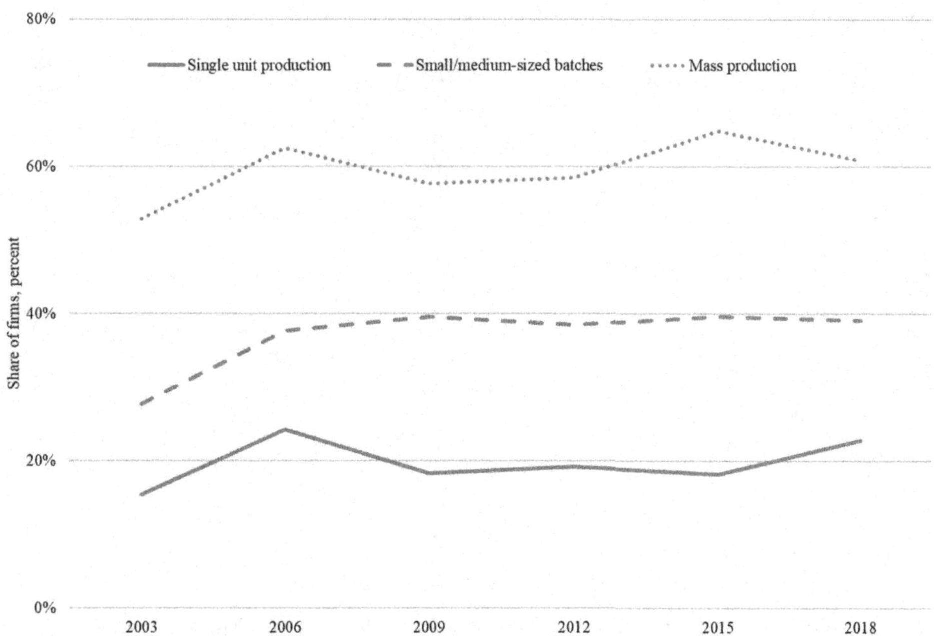

Figure 15.15 Share of robot users among Austrian, German, and Swiss manufacturers of selected core sub-sectors, differentiated by batch size, 2003–2018

Source: European Manufacturing Survey AT, CH, DE – 2003–2018, compiled by Fraunhofer ISI

the gap will decrease between larger and smaller firms in the diffusion of robots has not been fulfilled. On the contrary, the advantage of larger firms has become even greater. The diffusion among small firms remains weak.

Another difference in the diffusion of industrial robots which persists over time is between firms producing single pieces, small batches, and mass production (Figure 15.15). Mass producers use industrial robots much more frequently than manufacturers with smaller batches, and this gap does not narrow over time. Between 2003 and 2009, the use of industrial robots increased among manufacturers producing small or medium-sized batches, but the diffusion stagnated after 2009, indicating that the promise of flexibility given by Industry 4.0 is still not fulfilled. The development of industrial robots that are easier to program and more flexible to integrate into not (fully) automated systems has been successful to some extent (Kinkel and Weißfloch 2009). Nevertheless, the use of industrial robots by companies producing single units remains fairly constant at around 20%.

A similar trend can be observed among firms that produce in varying degrees of customisation (Figure 15.16). A larger share of companies that produce to stock still use industrial robots in their production, as opposed to companies that start production only after receiving the customers' order. However, it appears that manufacturers that only perform final production steps after receiving customer orders and use prefabricated parts from stock have been able to benefit more from robots in the last 15 years. The distribution among those producers increased disproportionately by 10 percentage points.

In contrast, the convergence in the use of industrial robots between manufacturers of products of varying degrees of complexity, as shown in Figure 15.17, indicates a clear technological shift during the last 15 years. At the beginning of the century, industrial robots were mainly

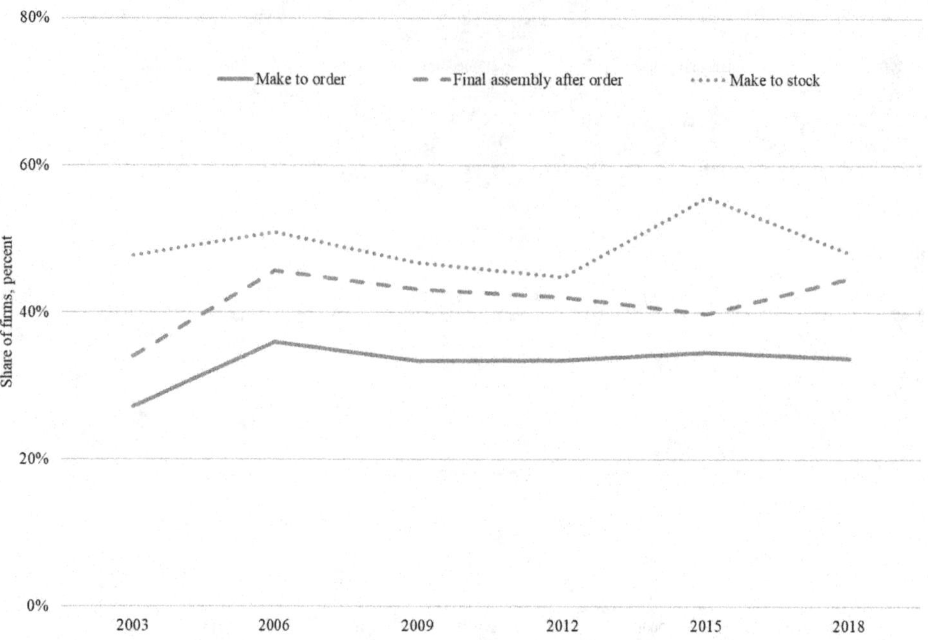

Figure 15.16 Share of robot users among Austrian, German, and Swiss manufacturers of selected core sub-sectors, differentiated by degree of customisation, 2003–2018

Source: European Manufacturing Survey AT, CH, DE – 2003–2018, compiled by Fraunhofer ISI

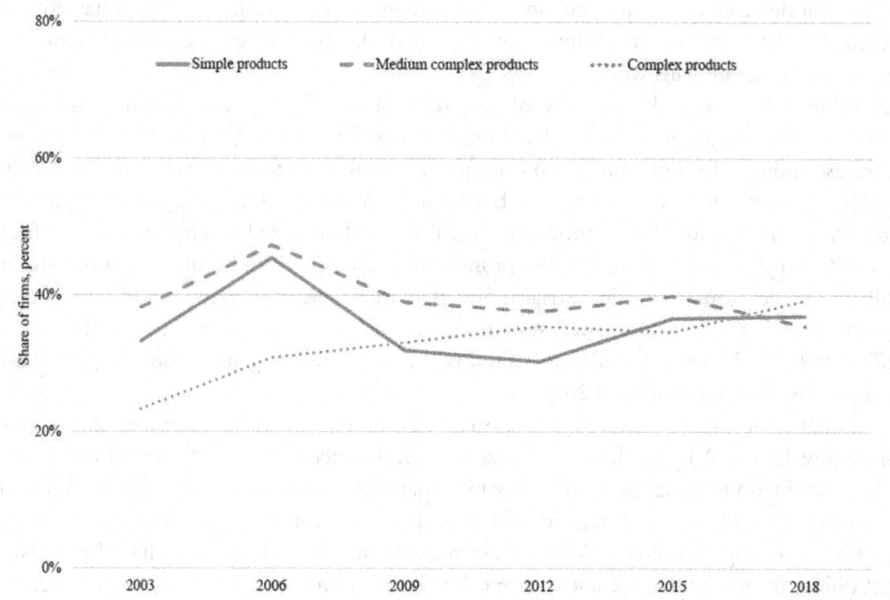

Figure 15.17 Share of robot users among Austrian, German, and Swiss manufacturers of selected core sub-sectors, differentiated by product complexity, 2003–2018

Source: European Manufacturing Survey AT, CH, DE – 2003–2018, compiled by Fraunhofer ISI

used by producers of simple products, but not by producers of complex products. Today, the differences between manufacturers of products with different degrees of complexity have disappeared. In 2003, the share of robot users among producers of complex products was much lower. Since then, robots have been used quite consistently by companies producing simple products, complex products, or products of medium complexity. All in all, the potential of automation today is less determined by the complexity of the manufactured or assembled product.

These results show that industrial robots became more flexible and functional during the last 15 years. Today, robots are used both by manufacturers of complex products as well as by producers of simple products and are equally suitable for customer-specific assembly processes based on prefabricated pieces as well as for make-to-stock production. These seemingly minor changes should not be underestimated, as they were only possible due to significant technical developments in the manufacturing of industrial robots on the one hand and in pricing and thus dimensioning of industrial robots on the other.

However, robots are still far from being a general-purpose technology. In light of the results of this section, industrial robots still look like a tool for automated mass production in the paradigm of the Third Industrial Revolution, rather than a tool of Industry 4.0 with its promise to combine customisation with the cost advantages of mass production (Lichtblau et al. 2015). The vision of a new generation of universally applicable industrial robots, which act as a "third hand" of the workers and can be easily programmed for very different tasks by non-professionals, has not yet been realised (Kinkel and Weißfloch 2009). Robots continue to be specialised in individual tasks, and their main application area is still production processes in which a large number of similar products are manufactured, benefitting from economies of scale. The applications of robots for production in smaller batches or single unit production processes is still considerably less common.

Moreover, in the last decade, smaller companies have not introduced robots on a larger scale, despite the consistent findings and assumptions of a positive impact on productivity even when all costs are taken into account (Horvat et al. 2019). The gap between large and small companies in the use of industrial robots has not (yet) been diminished by the high edge innovations of mobile, collaborating, autonomous robots nor by a decade of digitisation of production processes. The barriers for SMEs to adopt robots remain high: investment costs, uncertainty about the potential economic effects, and a lack of skilled personnel (Kroll et al. 2016).

Economic effects of robots: some results from recent studies

Robots have existed for more than 50 years, but their capabilities have only grown exponentially over the last ten years due to rapid improvements in their components. Thus, we ask whether the impact of robots is already visible in the economy and what effects can be observed through the use of robots. Information and communication technologies (ICT) have made important contributions to aggregate productivity growth in the last decades (Brynjolfsson and Hitt 1996; Oliner et al. 2008; van Ark et al. 2008), and we may observe similar effects as a result of the exponential growth of robots.

The paper by Acemoglu and Restrepo (2020) has perhaps received most attention so far. They estimate the impact of industrial robots on wages and employment for local US labour markets between 1990 and 2007 using IFR data and find a robust negative effect of robot intensity on employment and wages. One additional robot per 1,000 employees reduces employment by 0.2 percentage points. Acemoglu and Restrepo point out that this effect is distinct from the impacts of foreign trade, offshoring, the decline in routine jobs, or the effects of other types of ICT. Productivity effects of robots are an implicit part of the model, since a high productivity growth from robots would compensate for the negative employment effect. However, the productivity gains from robots are too small to compensate for such an effect. Using the same methodology and data, Giuntella and Wang (2019) find a large negative effect for China as well.

Dauth et al. (2018) apply the analysis to data from Germany. In contrast to Acemoglu and Restrepo, they do not find a negative effect on employment. The authors explain this result with the tendency of German workers to accept wage cuts in order to stabilise jobs in the light of the threat from automation. In a later version of the paper (Dauth et al. 2019), they observe a substantial shift in employment away from manufacturing towards business services. Hence, productivity improvements in their core business might be outweighed by the expansion of firms to lower productivity activities such as administrative or other service activities and a higher demand for these activities.

Graetz and Michaels (2018) investigate the impact of robots on employment at the industry level for the period 1993 to 2007 for 17 countries with IFR data. They find a positive effect of the increased use of robots on productivity, wages, and subsequently on GDP growth, but find no evidence of a negative effect of robots on aggregate employment. The exceptions are low-skilled workers and a weaker negative effect for middle-skilled employment.

Fu et al. (2020) offer some consistent evidence based on an industry-level and cross-country panel dataset of 74 economies between 2004 and 2016. The paper concludes that the adoption of industrial robots is related to significant gains in labour productivity and total employment in developed economies, whereas increased robot adoption decreases the share of labour and GDP in developing economies. Increased robot adoption is linked to significantly higher income inequality in both developed and developing economies, although there is no evidence of technological unemployment over the same period.

The World Bank (2019) concludes that most studies on the employment effects of robots overestimate the adverse effects, and that robots are hardly responsible for job losses during the past years. The World Bank mentions that the US and the UK have lost industrial jobs due to robots, while the growth of manufacturing in East Asia has more than compensated for this loss. Flexible skills and less specialisation in individual tasks or narrow industries are key to making robots more employment-oriented, but much of the aggregate effect of robots remains uncertain. Another recent paper (Berg et al. 2018) is more pessimistic. The authors investigate the effects of robots in a dynamic general equilibrium model and find that robots put pressure on real wages, substantially decreasing the labour share, and consequently increase overall inequality.

A main drawback of the studies mentioned is their level of aggregation. Restrictions in the data force the authors to investigate the effects of robots at the sectoral or economy-wide level and ignore the heterogeneity in the diffusion and the effects of robots, as well as selection effects in robot use. One exception is Koch et al. (2019) who use firm-level data on robot adoption from Spain. They show that firms with robots expand employment and turnover, while non-adopters shrink and destroy employment by shifting business to adopters. Moreover, Koch et al. (2019) find ex-ante selection effects – firms that adopt robots are already larger and more productive than non-adopters. Südekum and Woessner (2019) reveal that the top 20% of firms in manufacturing benefit disproportionally from robots, while the other firms show no effect. This is in line with an earlier paper by the OECD that demonstrated that the productivity gap between frontier firms and laggards has widened, possibly due to the use of ICT (Andrews et al. 2016). It is also in line with insights from firm-level evidence from Shi et al. (2020) which show that innovation in digital production technologies (robotics included) tends to increase wages on average, and this wage-boosting effect tends to be particularly significant in the high-tech manufacturing sectors.

The considerable variability of the results suggests that the final verdict on the economic effects of robots has not been decided yet. New datasets will enlarge our understanding of these effects in the future. Moreover, there are still considerable blind spots in the literature. Micro studies on the effects of robots, for example, are largely absent in the literature. In addition, the economic policy literature on Industrie 4.0 suggests that flexibility and customisation are the main effects of these technologies – a claim that has not yet been empirically investigated. A better understanding of these effects will also shed some light on the overall productivity effects of robots – one can assume that more flexibility and a higher degree of customisation will eventually turn into higher productivity. It may therefore be a fruitful way to break down overall productivity into its sub-effects at the firm level. Additionally, robots are just one of several technologies known under the common heading of "Industrie 4.0". These technologies may reveal substantial cross-fertilisation when they are employed together – for example, modern logistics and seamless data exchange can multiply the productivity of robots by enabling firms to synchronise their production with suppliers and customers. Such cumulative productivity effects are not yet considered in the literature.

Conclusions

Over the last 20 years, robots have made an impressive career from the pages of science-fiction books to the shop floors of the world. Industrial robots have become more flexible, adaptive, and powerful, and have moved into new application areas. New developments such as mobile, collaborating, and autonomous robots, as well as the convergence of robotics and artificial intelligence will provide further stimuli for their diffusion.

From 1993 to 2016, the worldwide diffusion of industrial robots increased significantly, especially after the global financial crisis of 2008/09. Despite this growth, industrial robots are unevenly distributed across countries, sectors, and firms. Asian countries install more industrial robots than America, Europe, and Africa combined. In 2016, China was the world's largest market for industrial robots, followed by Korea and Japan. The manufacturing sector is by far the major application domain for robots, in particular the automotive and the electrical/electronic industries. This uneven distribution suggests that robots have not yet emerged into a general-purpose technology that can be applied in all sectors of the economy. This is also evident from data at the firm-level: the main application of robots is still the mass production of goods with a low degree of customisation. Firm size is still an important determinant of robot use, and robots are only slowly diffusing among small firms.

Evidence of an uneven diffusion of robots can also be found in comparisons of the intensity of robot use across countries. Country rankings based on the share of firms using robots lead to results that differ from rankings based on the absolute number of robots per 10,000 employees in manufacturing, since robots are concentrated in large firms and in a few manufacturing sectors. This makes the number of installed robots a somewhat misleading indicator for comparisons between countries and challenges the results of some of the empirical analyses on robots, productivity, and employment.

Firm-level data on the diffusion of robots as well as other new technologies are also better suited to analyse the economic effects of robots. Results to date on the employment and productivity effects of robots are mixed; some studies find such effects, while others fail to establish a link between employment, productivity, and the use of robots. This is a gap in the literature that needs further research.

References

Acemoglu, D. and Restrepo, P. (2020): Robots and Jobs: Evidence from US Labor Markets, *Journal of Political Economy* 128(6), 2188–2244.

Andrews, D., Criscuolo, C. and Gal, P. N. (2016): *The Best versus the Rest.* OECD Productivity Working Papers, 2016–05, Paris: OECD Publishing. https://doi.org/10.1787/63629cc9-en

Berg, A., Buffie, E. F. and Zanna, L.-F. (2018): Should We Fear the Robot Revolution? (The Correct Answer is Yes), *Journal of Monetary Economics* 97, 117–148. https://doi.org/10.1016/j.jmoneco.2018.05.014

Brynjolfsson, E. and Hitt, L. (1996): Paradox Lost? Firm-Level Evidence on the Returns to Information Systems Spending, *Management Science* 42(4), 541–558. https://doi.org/10.1287/mnsc.42.4.541

Dauth, W., Findeisen, S., Südekum, J. and Woessner, N. (2018): German Robots – The Impact of Industrial Robots on Workers. *CEPR Discussion Papers 12306*, London.

Dauth, W., Findeisen, S., Südekum, J. and Woessner, N. (2019): The Adjustment of Labor Markets to Robots. Unpublished Manuscript. Available at https://sfndsn.github.io/downloads/AdjustmentLaborRobots.pdf [30.01.2021].

DeStefano, T., De Backer, K. and Moussiegt, L. (2017): Determinants of Digital Technology Use by Companies. *OECD Science, Technology and Innovation Working Papers No 40.* Paris: OECD Publishing.

EIB (2019): European Investment Bank Investment Report 2019/2020: Accelerating Europe's Transformation, in *European Investment Bank* (ed.). Luxembourg.

Fu, X. Q., Bao, Q., Xie, H. J. and Fu, X. L. (2020): Diffusion of Industrial Robots and Inclusive Growth: Labour Market Evidence from Cross-Country Data, *Journal of Business Research* 122, 670–684. Available at: www.sciencedirect.com/science/article/abs/pii/S0148296320303544 [30.01.2021]

Giuntella, O. and Wang, T. (2019): Is an Army of Robots Marching on Chinese Jobs? *IZA Discussion Paper No. 12281*, Bonn.

Graetz, G. and Michaels, G. (2018): Robots at Work, *Review of Economics and Statistics* 100(5), 753–768.

Hägele, M., Nilsson, K., Pires, J. N. and Bischoff, R. (2016): Industrial Robotics, in Siciliano, B. and Khatib, O. (eds.): *Springer Handbook of Robotics.* 2nd ed. Berlin, Heidelberg: Springer International Publishing, 1385–1422.

Horvat, D., Kroll, H. and Jäger, A. (2019): Researching the Effects of Automation and Digitalization on Manufacturing Companies' Productivity in the Early Stage of Industry 4.0., *Procedia Manufacturing* 39, 886–893.

IFR (2019): *IFR Press Conference.* Shanghai: International Federation of Robotics.

ISO (2012): ISO 8373, Robots and Robotic Devices – Vocabulary. International Organization for Standardization. Geneva. Available at: www.iso.org/standard/55890.html [30.01.2021]

Jäger, A. and Maloca, S. (2016): *Dokumentation der Umfrage Modernisierung der Produktion 2018.* Karlsruhe: Fraunhofer ISI.

Jäger, A., Moll, C. and Lerch, C. (2016): *Analysis of the Impact of Robotic Systems on Employment in the European Union. 2012 data update.* Update of the Report for the European Commission, DG Communications Networks, Content & Technology, Brussels.

Jäger, A., Moll, C., Som, O., Zenker, C., Kinkel, S. and Lichtner, R. (2015): *Analysis of the Impact of Robotic Systems on Employment in the European Union.* Report for the European Commission, DG Communications Networks, Content & Technology, Brussels.

Kinkel, S. and Weißfloch, U. (2009): Estimation of the Future User Potential of Innovative Robot Technologies in SMEs – Promising Prospects, in IFR (ed.): *World Robotics 2009 Industrial Robots: Statistics, Market Analysis, Forecasts, Case Studies and Profitability of Robot Investment.* Frankfurt am Main: VDMA, 376–381.

Koch, M., Manuylov, I. and Smolka, M. (2019): Robots and Firms. *CESifo Working Papers 7608.* Munich.

Kroll, H., Copani, G., van de Velde, E., Simons, M., Horvat, D., Jäger, A., Wastyn, A., PourAbdollahian, G. and Naumanen, M. (2016): *An Analysis of Drivers, Barriers and Readiness Factors of EU Companies for Adopting Advanced Manufacturing Products and Technologies*, Report on behalf of the European Commission, DG Growth, Brussels.

Leitner, S. and Stehrer, R. (2012): Labour Hoarding during the Crisis: Evidence for selected New Member States from the Financial Crisis Survey. *wiiw Working Paper 84.* Vienna.

Lichtblau, K., Stich, V., Bertenrath, R., Blum, M., Bleider, M., Millack, A., Schmitt, K., Schmitz, E. and Schröter, M. (2015): *Industrie 4.0-Readiness.* Aachen: IMPULS-Stiftung des VDMA.

Mansfield, E. (1989): The Diffusion of Industrial Robots in Japan and the United States, *Research Policy* 18(4), 183–192. https://doi.org/10.1016/0048-7333(89)90014-0

Oliner, S. D., Sichel, D. E. and Stiroh, K. J. (2008): Explaining a Productive Decade, *Journal of Policy Modeling* 30(4), 633–673. https://doi.org/10.1016/j.jpolmod.2008.04.007
Pratt, G. A. (2015): Is a Cambrian Explosion Coming for Robotics? *Journal of Economic Perspectives* 29(3), 51–60. https://doi.org/10.1257/jep.29.3.51
Scalera, L. and Gasparetto, A. (2019): A Brief History of Industrial Robotics in the 20th Century, *Advances in Historical Studies* 8(1), 24–35.
Shi, L., Li, S. and Fu, X. (2020): The Fourth Industrial Revolution, Technology Innovation and Firm Wages: Empirical Evidence from OECD Economies, *Revue d'économie industrielle* 169(1), 89–127.
Südekum, J. and Woessner, N. (2019): *Robots and the Rise of European Superstar Firms. European Economy – Discussion Papers 2015–118*, Directorate General Economic and Financial Affairs (DG ECFIN), Brussels.
van Ark, B., O'Mahoney, M. and Timmer, M. P. (2008): The Productivity Gap between Europe and the United States: Trends and Causes, *Journal of Economic Perspectives* 22(1), 25–44. https://doi.org/10.1257/jep.22.1.25
World Bank (2019): *The Changing Nature of Work. World Development Report 2019.* International Bank for Reconstruction and Development. Washington, DC: The World Bank.

16

THE TRIPLE BOTTOM LINE OF SMART MANUFACTURING TECHNOLOGIES

An economic, environmental, and social perspective

Thorsten Wuest, David Romero, Muztoba Ahmad Khan and Sameer Mittal

Introduction

Smart technologies and *digital transformation* are impacting many realms of modern society, including but not limited to the economy and its different industrial sectors. Manufacturing is no exception, given its stature as one of the oldest and most critical industrial sectors for a majority of global economies. *Smart manufacturing* and the *Industry 4.0 paradigm* build upon the opportunities presented by the emerging digital, or smart, technologies, including but not limited to artificial intelligence (AI), the Industrial Internet of Things (IIoT), and Additive Manufacturing (AM) (Thoben et al. 2017).

The manufacturing sector has seen a rapid (digital) transformation and adoption of smart technologies across the board. Both companies and (local and federal) governments globally have invested heavily in new *smart technologies*, *processes*, and *tools*. However, many are still struggling with this *digital transformation*. Many ambitious projects aimed at the adoption and implementation of these new smart technologies are finding themselves stuck in a so-called prototype or pilot purgatory, never deploying on a large scale on the shop floor or in the supply chain. There are many reasons for this struggle, including the sheer complexity of the task at hand, undefined responsibilities, the novelty of these new smart technologies, and unclear business cases and value propositions. Small and medium-sized manufacturers are especially struggling with this transition, given their limited resources and other limiting factors (Mittal et al. 2018a).

When we look at smart technologies within the manufacturing domain, the majority are digital in nature and heavy on information and communication technologies (ICTs). This requires a different *(digital) skill set* compared to traditional manufacturing operations. While *digital transformation* continues to shape the industry, manufacturing operators have to be onboarded and have their voices heard when it comes to new technologies adoption and or the question of alignment with current and future (re-)designed processes. Overall, a top-down approach of forcing smart technologies only, without re-thinking current processes and workflows, has proved to be very problematic (Romero et al. 2019a; Sinha et al. 2020).

DOI: 10.4324/9780429351921-20

A question that has so far not been completely addressed is the impact of these new smart manufacturing technologies on the triple bottom line (TBL[1]) of the sustainability of businesses and supply chains. Interestingly, the origin of "smart manufacturing" can be traced back to smart and sustainable manufacturing – with a strong emphasis on "energy-efficient manufacturing" and facilitation of more "efficient manufacturing processes". However, when we look at sustainability more holistically – focusing on the triple bottom line – there has not been much work dedicated to exploring this phenomenon. Just recently, some attention has been put on the TBL perspective with regards to the smart manufacturing paradigm in professional and scientific literature, for example, in Forbes business magazine (Clemons 2019) and selected academic journals (Abubakr et al. 2020). However, these recent works focus not solely on smart technologies but more on the overall smart manufacturing paradigm.

This chapter aims to close this gap and opens the discussion by seeking to answer the aforementioned question: *What is the impact of new smart manufacturing technologies on the triple bottom line of the sustainability of businesses and supply chains?*

The chapter is structured as follows. We first provide an overview based on an explorative literature review of the key terminology, technological developments, and on-going paradigm evolution within the emerging smart manufacturing systems. In the following section, we discuss the impact of Industry 4.0 and smart manufacturing paradigms from a TBL perspective before honing in on the smart technologies that are commonly associated with smart manufacturing systems. After this, we review recent scientific and grey literature and identify 10 key smart manufacturing technology clusters and discuss each with an emphasis on use cases and their economic, social, and environmental impacts. The next section takes a step back and highlights the barriers and challenges faced by organisations when engaging in their smart manufacturing journey and the adoption of these key technologies. The penultimate section takes a bold look at future technological developments and provides an outlook as to how this diffusion of smart technologies will shape the manufacturing sector and our society overall. The final section concludes the chapter and addresses the limitations of this work.

Literature review

Smart manufacturing and Industry 4.0 paradigms

The manufacturing sector has always been at the forefront when it comes to technological innovations. Over the last centuries, there have been four distinctive events, commonly referred to as industrial revolutions. During the *First Industrial Revolution*, the manufacturing sector became industrialised, driven by the mechanisation of labour, exemplified by the introduction of the power loom. As a result, a myriad of machines, tools, and inventions blossomed and spurred tremendous economic growth. Then, the *Second Industrial Revolution* introduced mass production and the moving assembly line. Most notably, Henry Ford's Model T production facilities are credited with initiating and exemplifying the Second Industrial Revolution. The *Third Industrial Revolution* came into the picture with the advent of numerical control and computer systems and their introduction to the shop floor. Finally, around 2011, the term "Industry 4.0"

1 The *triple bottom line (TBL) framework*, in the manufacturing context, accounts for three forms of sustainability outcomes when producing a product: (a) environmental – e.g., emissions, hazardous waste, and (natural) resources usage; (b) economic – the flow of capital and infrastructure investment; and (c) social – e.g., safe working conditions and fair wages (Elkington 2018).

was born, indicating the start of the *Fourth Industrial Revolution*, centred around the introduction of Cyber-Physical Systems (CPS) in the manufacturing domain.

The term *Industry 4.0* originates in politics, and therefore different terms have been coined globally to represent it. For instance, in the US it is often synonymously used with *smart manufacturing*, whereas in South Korea, it is referred to as *smart factory* (Thoben et al. 2017). However, unlike the three earlier industrial revolutions, "Industry 4.0" came popular prior to its full manifestation in the industrial landscape. Industry 4.0 is not a technology; rather, it denotes an umbrella of technologies that will lead to connected systems and data-driven decision-making environments (Mittal et al. 2018a). Although "data" has always been present, with the advances in and convergence of information and operational technologies (IT/OT) and the consistently decreasing associated cost of installing both, Industry 4.0 has become quite popular in the last few years.

Implementing *Industry 4.0* or *smart manufacturing technologies* can have many advantages; these have motivated a number of organisations and the governments of many countries to act and engage in activities aimed at implementing them into their processes (Tao et al. 2018):

1 *Real-time awareness*. With the help of Industry 4.0 technologies, such as the *Industrial Internet of Things (IIoT)*, all systems and subsystems, and their elements, within an organisation or across a supply chain can be connected. As a result, if there is an issue anywhere, the entire system can be notified and thus is enabled to make an informed decision or take action.
2 *Involvement of the customer during product design*. The involvement of the customer during the product design phase, also known as "co-design", was previously difficult, as the customer was required to visit the plant more frequently. However, with the advent of Industry 4.0 technologies such as *virtual and augmented reality (VR/AR)* and *digital twins*, this is now possible to a much greater degree.
3 *Predictive maintenance*. The availability of *big data* thanks to *smart sensors and networks*, and the seamless transmission, storage, and processing of data in *the cloud*, or *edge*, enabled the emergence of new *data analytics techniques* supporting new, more foresight-oriented maintenance strategies.
4 *Employees and smart machines as co-workers*. The use of *data* also ensures that employees, who are now working in parallel with "smart" machines, and all stakeholders can learn from each other. As a result, the degree of employee productivity and employee satisfaction will increase, a paradigm known as "Operator 4.0" (Romero et al. 2016).
5 *Sustainable production*. Industry 4.0 technologies enable data-driven decision-making environments and, therefore, can make the best use of any production resource while reducing waste through the use of different optimisation techniques. As a result, the application of Industry 4.0 technologies also leads to more sustainable production systems such as Digital Lean Manufacturing Systems (Romero et al. 2019b).

Economic, environmental, and social impact of Industry 4.0

"Sustainability" and "Industry 4.0" are trending in the advanced manufacturing technologies literature. The technological state of the art includes AM applications (Ford and Despeisse 2016) and emerging production processes and systems based on flexible automation, collaborative and cognitive robotics, cloud computing, big data analytics, and CPS (Kiel et al. 2017; Bonilla et al. 2018; Jena et al. 2020). On the one hand, manufacturing businesses are expecting TBL-related benefits from Industry 4.0 technologies in terms of:

1 *Economic gains*, enabled by savings through more accurate and precise production planning and control systems, as well as shorter lead times thanks to the development of agile manufacturing capabilities on the shop floor.
2 *Environmental gains*, enabled by increased energy efficiency in manufacturing operations; real-time monitoring of production assets' energy consumption helps to avoid spikes, reduce base loads, evaluate peak hours, and identify irregularities and thus optimise operational scheduling. Further, a decrease of manufacturing scrap waste is enabled by employing advanced total quality management systems in order to improve process efficiency and product quality while minimising and eliminating quality defects and process errors.
3 *Social gains*, enabled by wearable and collaborative technologies increase the safety, health, and welfare of workers on the shop floor (Ejsmont et al. 2020).

On the other hand, selected Industry 4.0 technologies adoption implies and raises several sustainability concerns about

1 *Economic factors*, due to their cost-intensive nature and difficulties in estimating their full financial benefits (ROI) and economic effectiveness.
2 *Environmental factors*, which are related to their energy consumption and electro-waste at their end-of-lifecycle.
3 *Social factors*, regarding human–robot interaction issues, unemployment threats, and privacy issues (Ejsmont et al. 2020).

Hence, both benefits and concerns should be clearly pictured in a TBL perspective and within a Sustainable Industry 4.0 Framework (Kamble et al. 2018).

Smart technologies in Industry 4.0

The past three industrial revolutions were supported by disruptive production technologies and power sources of the time that significantly altered manufacturing processes and systems. For example, the First Industrial Revolution was promoted by water- and steam-powered mechanical manufacturing facilities; the Second Industrial Revolution was supported by electrically powered mass-production lines; and the Third Industrial Revolution was led by the widespread usage of electronics and ICTs to further automate manufacturing systems (Liao et al. 2017). Today's emerging "smart" manufacturing technologies are mainly fuelled by data. These data-fuelled smart technologies, such as the Industrial Internet of Things (IIoT), CPS, and big data analytics (BDA), are paving the way for the Fourth Industrial Revolution that sets the frame for the smart manufacturing paradigm.

Smart manufacturing is driven by data and data analytics. The successful operation of any *smart manufacturing system* depends on the generation, transmission, storage, processing, and application of data-driven decisions in operations management, which are greatly facilitated by various *smart manufacturing technologies* (Klingenberg et al. 2019). For instance, smart sensor systems, embedded systems, and RFID technologies are critical for data generation. For data transmission, some of the important technologies are IIoT, smart sensor networks, and cellular networks, such as 5G. In order to store and process the transmitted data, we can utilise various technologies such as cloud storage and -computing, blockchain, big data analytics, machine learning, artificial intelligence, virtual and augmented reality, and cybersecurity. Finally, relevant technologies that can facilitate data application may include robotics, connected factories, smart products, and, ultimately, *smart manufacturing*.

Relevant smart manufacturing technology clusters in literature

Several researchers have previously identified a wide variety of technologies related to smart manufacturing systems. Mittal et al. (2019) presented a comprehensive list of characteristics, technologies, and enabling factors frequently associated with smart manufacturing. Klingenberg et al. (2019) identified 111 Industry 4.0 technologies through a systematic literature review, while Frank et al. (2019) compiled a list of six technology clusters associated with smart manufacturing. Ghobakhloo (2019) provides a schematic presentation of the key information and digital technology trends that enable smart manufacturing. However, the smart technologies mentioned in the available scientific literature do not always match.

A refined list of smart manufacturing technology clusters[2] built on previous work is presented in Table 16.1, which intentionally excludes some smart technologies illustrated in previous publications. Examples of the omitted technologies are (a) *CAx technologies* (Dalenogare et al. 2018; Mittal et al. 2019; Ghobakhloo 2019); (b) *vertical and horizontal systems integration* (BCG 2018; Dalenogare et al. 2018; Klingenberg et al. 2019); and (c) *energy management* (Mittal et al. 2019; Frank et al. 2019). While the reasons for exclusion vary, they are based on either the reviewing methodology of the original sources (e.g., energy management) or disagreement with their classification as a technology rather than as an enabling factor (e.g., vertical and horizontal systems integration). In the table, we mapped the alignment with previous work on the topic to ground our resulting list in the state of the art. We decided to add 5G networks (Table 16.1 – No. 10) despite this not being mentioned in previous work, as we believe this will have a significant impact on future developments in the smart manufacturing domain.

In the following section, we discuss each of the *10 key smart manufacturing technology clusters* presented in Table 16.1 in more detail.

Smart manufacturing technology clusters: a triple bottom line (TBL) discussion

The *10 key smart technology clusters* associated with smart manufacturing and Industry 4.0 are diverse in nature and include manufacturing processing, data analytics, and information and communication technologies. We present each in the following subsections with an emphasis on their more significant use cases and economic, social, and environmental impacts to adequately reflect the TBL perspective.

Artificial intelligence (AI), machine learning (ML), and advanced simulation

Artificial intelligence (AI) and other analytical and data-driven approaches are at the core of the smart manufacturing and Industry 4.0 paradigms (Kusiak 2017). Therefore, we consider AI, *machine learning (ML)*, and advanced *simulation* technologies as the foundational pillars and thus the basic technologies in the smart manufacturing domain. Several other key technologies, listed later, aim to either feed data into analytical models or to build on actionable insights derived from data.

2 The refined *Smart Manufacturing Technology Clusters* were created using established ontologies in relation to smart manufacturing vocabulary (e.g., glossaries) and the consideration of the semantic distance between the terms being classified (Harispe et al. 2015).

Table 16.1 Relevant smart manufacturing technology clusters in the literature

No	*Smart manufacturing technology cluster*	*Technology*	*Mittal et al. 2019*	*Frank et al. 2019*	*Ghobakhloo et al. Ghobakhloo 2019*	*Klingenberg et al. 2019*	*Bai et al. 2019*	*BCG 2018*	*Dalenogare et al. 2018*
1	Artificial intelligence, machine learning, & advanced simulation	Artificial intelligence	x	x	x	x	x		
		Big data analytics	x	x	x	x	x	x	x
		Machine learning	x		x	x			
		Machine vision	x						
		Advanced simulation	x	x	x	x	x	x	x
2	Cloud, fog, & edge computing	Cloud computing	x	x	x	x	x	x	x
		Edge computing							
3	Additive manufacturing	Additive/Hybrid manufacturing	x	x	x	x	x	x	x
4	Industrial Internet of Things & Cyber-Physical Systems	Industrial Internet of Things (IIoT)	x	x	x	x	x	x	
		Cyber-Physical Systems (CPS)	x	x	x	x			x
5	AR, VR, & DTs	Augmented, virtual, & mixed reality	x	x	x	x	x	x	
		Digital twins	x		x				
6	Automation & robotics	Automated guided vehicles & autonomous mobile robots	x	x	x		x	x	
		Collaborative robots					x	x	
		Drones				x	x		
7	Cybersecurity	Cybersecurity	x		x	x	x	x	x
8	Blockchain	Blockchain			x	x	x		
9	Smart sensor systems	Industrial sensors, actuators, & PLCs		x	x	x	x		
		Smart products, machines, & wearables	x	x	x	x			
10	5G networks	5G							

Key use cases. AI and data analytics are broadly applied across the various levels within a manufacturing environment. Use cases range from (a) *predictive maintenance*, where AI predicts the moment in the future when a machine tool will fail, through (b) continuous improvement of the *production planning and scheduling* with advanced simulation and optimisation techniques, to (c) *in-situ real-time monitoring and prediction of process efficiency and product quality* using machine learning. In the future, even personalised automated design will be AI-powered, and on a digital supply network level, the supply and demand risks will be predicted and managed by AI.

Economic impact. The economic impact of AI and other data-driven tools is significant. Recent studies suggest that AI can help make products lighter and stronger, as well as reduce demand forecast errors by 30% and demand planners' workload by 50% (Columbus 2020). Overall, AI is expected to have a transformational impact on the manufacturing sector as a whole, and organisations failing to adopt this technology are predicted to fall behind their competitors (Bughin and Seong 2018).

Environmental impact. The environmental impacts of AI come down to the energy that computer systems use to train AI/ML models and supply the necessary training data for model creation. While running a classification model is comparably energy efficient, training a new, complex AI/ML prediction model can be five times the life-time emissions of an average car (284t CO_2) (Lu 2019).

Social impact. The social impact of AI changes the workflow and how operators approach their different manufacturing tasks. Before the introduction of AIs on the shop floor, operators spent a significant amount of time on analysing and interpreting data. In the future, with AI-based analytics monitoring of manufacturing processes in real time, the operators will be alerted to any future issues. However, often the AI models are black-boxes that do not provide insight into how the predictions are derived – which can cause stress and anxiety for the human operators. Furthermore, human operators might struggle with accepting tasks from an artificial system. Despite the widespread fear that AI is eliminating many jobs in manufacturing, newer studies do not support this (Scott and Shaw 2020).

Cloud, fog, and edge computing

Cloud computing and *industrial internet platforms* have transformed the way we operate and approach storage, communication, and access to data, as well as its complex analysis (Menon et al. 2019). The ability to scale data storage and processing capabilities without having to add, manage, and maintain servers in-house is already a game-changer. However, the ability to rely on dedicated cybersecurity experts as well as the flexibility to adjust to peak computing demands makes cloud platforms a core smart manufacturing technology. While cloud computing is at the core, fog, edge, and hybrid variants and extensions are expanding the data management abilities for different shop floor applications that are driven by demanding requirements (e.g., low latency) or security measures (e.g., defines manufacturers). Hence, *fog* and *edge computing* enhance the cloud computing paradigm by adding data processing and analytics (physically) closer to the data origin (Sinha et al. 2020). While often mistakenly used interchangeably, *fog computing* describes the processing of data in the local network, and *edge computing* focuses on processing on or connected to the sensor system itself. Moreover, *hybrid computing* describes a combination of different layers, for instance, where some parts of the network are hosted locally, and other data is stored in a cloud environment. The reasons for a hybrid approach are manifold and may include cybersecurity concerns, economic considerations, and/or technical (computing) requirements.

Key use cases. Cloud, fog, edge, and hybrid computing in smart manufacturing range from process planning applications, to process data storage, to data exchange, to advanced analytics.

Quality monitoring and *predictive maintenance* are two applications that heavily rely on industrial internet platforms and extending data analytics across different nodes of the shop floor and supply chains, which are mostly made possible by leveraging cloud infrastructures.

Economic impact. The economic impact of cloud infrastructures and industrial internet platforms on the manufacturing sector are manifold. The main appeals of cloud computing for manufacturers include zero upfront cost, rapid scalability, reduced redundancy requirements (increasing resiliency), and dedicated cybersecurity. While these are all considered positive, and also lower the barriers for small and medium enterprises (SMEs) to engage in smart manufacturing applications, some economic aspects are less appealing. There are only a limited number of industrial internet platform providers, and as such, there might be some form of lock-in possible in the future (Menon et al. 2020).

Environmental impact. From an environmental perspective, utilising cloud services has a significant impact on energy consumption. On the one hand, centralised hosting can leverage economies of scale and scope and focus on renewable energy usage. A 2020 study by Microsoft (Microsoft 2020) found that cloud hosting is 20–93% more energy-efficient than local hosting. This is promising. On the other hand, however, the ease of use and ability to scale entices users to store more data and use the systems' processing capabilities to a greater degree. However, Masanet et al. (2020) found that while demand for cloud services surged, the energy demand has grown at a significantly lower rate.

Social impact. The social impact of cloud technologies is largely considered positive. Cloud computing enables remote access and, as such, remote work and more flexible workplace models that benefit underdeveloped areas and underrepresented groups – for example, working parents in rural areas. On the other hand, decentralised storage and access from anywhere do not align with higher costs of living in some areas/countries and might lead to more competition in the labour market and lower wages. However, at this point, there is no evidence that this might materialise at scale. The other aspect at the intersection of social and economic impact is (digital) skills development and retention of critical knowledge within the organisation. When manufacturers outsource storage, processing, and analytics to external platforms, there is a risk of losing critical skill sets within their own workforce. Given that data analytics and data-driven decision-making are core skills that will define the future of competitive success, this might lead to significant problems sooner rather than later.

Additive Manufacturing (AM)

Traditional subtractive manufacturing processes such as milling, turning, and drilling remove material from a blank to create a product. Deformative processes such as forging or casting change the shape without adding or removing material. *Additive Manufacturing (AM)* deposits material layer by layer and creates a 3D-object with the help of a CAD model and is therefore also referred to as 3D printing. Subtractive and deformative manufacturing processes utilise digital technologies, such as Computer Numerical Control (CNC), but are not dependent on them. AM, on the other hand, is digital in nature and cannot be operated without digital input. Based on the strength requirement of a final product, various AM methods may be used, such as powder bed fusion, vat photopolymerisation, binder jetting, and direct metal deposition.

Key use cases. AM methods can be seen across various industries, such as automotive, aerospace, biomedical devices, and even construction, due to their unique capability to enable the manufacturing of complex geometries and rapid prototypes (Ngo et al. 2018).

Economic impacts. AM is still in its early adoption phase, and there are various barriers hampering its widespread adoption and implementation, such as a comparably slow speed of printing, high cost of printers, materials, and maintenance, CAD software complexities, copyright infringement, and lack of expert designers (Shukla et al. 2018). As a result, only selected industries have started to use AM at scale. However, the rapid pace of innovation in the AM domain will likely help to overcome many of these barriers. In the future, more and more industries will find AM technologies to be an affordable option that provides easy access to customisation, personalisation, and rapid prototyping capabilities.

Environment impact. AM technologies, in general, are significantly more environmentally friendly compared to subtractive manufacturing technologies, as in an AM process there is no wasted (raw) material generated. Even the powders or resins used as raw materials during AM processes in most cases are recyclable. In consensus with earlier statements, an innovative strategy combining both additive and subtractive manufacturing processes, known as hybrid manufacturing, has suggested that such a strategy will demand low levels of energy and create less environmental impact compared to the impact created by using subtractive manufacturing processes alone (Thao Le et al. 2017). This environmental impact will be further reduced when AM is used by itself.

Social impact. When AM technologies become more common and are adopted on a mass scale, there is potential to revolutionise many apparel industries. For instance, the footwear industry providing customised footwear will share the design of the footwear with the customer, while the customers themselves can 3D-print their own shoes. Similar changes will be observed around customised clothes and automobile spare parts as well, thus affecting the entire business model and supply chain. For example, with lead time reduced significantly, products will not take too long to arrive.

Industrial Internet of Things (IIoT) and Cyber-Physical Systems (CPS)

The *Industrial Internet of Things (IIoT)* is about connecting "things" used in an industrial environment such as the manufacturing shop floor or a warehouse. These things may be in the form of people, machines, materials, or products. Furthermore, there is now an extension that includes immaterial "things" such as service, sometimes referred to as the Industrial Internet of Things and Services (IIoTS). When these "things" are used in the consumer goods space, we refer to the widely known Internet of Things (IoT). When various systems are interconnected with the help of IoT devices and can-do computations and make complex decisions, we refer to them as *Cyber-Physical Systems (CPS).* IoT devices lay the foundation for CPS and therefore are considered within the same technology cluster. By remaining connected, "things" are able to develop real-time awareness of each other and thus enable informed decision-making.

Key use cases: IIoT and CPS regularly shape our daily lives. With the help of IoT devices, the real-time location of an order can be monitored by the manufacturer without contacting the supplier. These connected smart devices also let us know what the weather will be in a region, and if needed, the route of the vehicle transporting the order from supplier to manufacturer may be changed to avoid delays due to inclement weather.

Economic impact. IIoT and CPS require less investment compared to other smart manufacturing technologies. The typical requirements are sensors, network, and cloud, which are available at affordable prices. The advantage of IIoT and CPS is that due to the availability of a wide variety of sensors, everything can be connected. As a result, organisations can become more efficient and utilise their production and logistics resources optimally. The literature also shows how traditional lathe and mill machines, omnipresent in most manufacturing workshops, can be upgraded and made smart without significant investments (Mittal et al. 2018b).

Environmental impact. IIoT and CPS can make the lives of shop floor managers easier and contribute to reducing machine tool energy consumption. However, cybersecurity becomes a potential threat. If a hacker gains access to a smart machine, they might be able to manipulate it remotely and cause significant harm to the production, machine, or even operators. Similarly, environmental hazards are created by batteries and other electronic waste, which will be produced in greater quantities than before, and disposal will become an issue that needs to be addressed. The signals emerging from the operation of these smart devices can affect populations of birds and insect pollinators, as their eggs may be damaged. Similarly, this overall development will also further increase screen time for humans, thus affecting their health, including their eyes.

Social impacts. IIoT and CPS will have a tremendous impact on smart factories and the future workplace (Autor 2015; Romero et al. 2020). On a smart shop floor enabled by IoT devices, for instance, a worker will be able to (remotely) operate production resources (viz., smart machines, robots, and computer systems) present at the manufacturing site by opening an app on a mobile phone. As a result, if a person forgot to shut off a machine, he/she would be able to do that from any remote location.

Augmented reality (AR), virtual reality (VR), and digital twins (DTs)

Augmented reality (AR) is a technology that overlays computer-generated virtual objects into the physical environment, whereas *virtual reality (VR)* replaces the physical environment with a computer-simulated one that uses sight and sound to produce an immersive experience for its user. Furthermore, *digital twins (DTs)* are virtual replicas of physical devices used to run advanced simulations before actual devices are built and deployed, and are able to mirror in real time the static and dynamic characteristics of its physical twin as a result of seamless data exchange with its virtual replica.

Key use cases. Among the main VR and AR industrial applications, virtual engineering, virtual training environments, and digital assistance systems stand out. Two of the biggest industrial use cases of AR technology are in the maintenance, repair, and overhaul of complex equipment and in the assembly of complex products. In both cases, AR technology enriches the real world with virtual objects and digital information that is overlaid in real time in the operator's field of view for error-proofing the execution of different manual operations. Some VR technology industrial use cases include, but not limited to, virtual engineering and virtual training environments. The former supports the modelling, simulation, and visualisation of products and systems behaviour under real-world operating conditions. The latter facilitates the transfer and assessment of procedural knowledge and technical skills for performing assigned tasks. Both virtual environments provide a combination of interactive virtual reality and advanced simulations of realistic scenarios for optimised decision-making and action-taking. Furthermore, some DT use cases may include virtual simulations supporting the development of new or re-design of existing products and product variants; quality assurance by real-time monitoring of every part of a product, asset, or production process; and system planning and control by modelling and simulating different scenarios for a manufacturing system in order to optimally configure, update, and provide feedback for its production plan and schedule. In any of these three cases, DTs use advanced simulations and other prediction models to proactively identify and correct performance issues.

Economic impacts. According to PwC (2020), VR and AR technologies could bring net economic benefits of $1.5 trillion by 2030 thanks to their productivity uplifting capabilities.

On the one hand, AR technology can improve worker productivity, efficiency, and accuracy on the shop floor by means of digital assistance systems; on the other hand, VR environments can lessen the costs of product prototyping and testing as well as of expensive training scenarios. At the same time, they reduce the time to market for new product developments and offer a speed-up of the learning curves of the workforce. In relation to the economic impacts of DTs, GMInsights (2019) claims that their market size is expected to grow to $20 billion during the period from 2019 to 2025, driven by their potential to enhance operational efficiency, predictive maintenance, and dynamic simulations and to reduce overall design and engineering costs.

Environmental impact. VR and DT technologies have the potential to contribute towards more environmentally friendly design, engineering, and production operations. This is mainly based on making use of virtual environments and advanced simulations for prototyping and testing new product designs in a resource-efficient way and commissioning new production plans and schedules that have been optimised for the efficient consumption of commodities, resources, and additives, energy demand, waste accumulation, and emissions generations. Furthermore, AR technology can be seen as a path towards a paperless shop floor when it comes to the delivery of work instructions.

Social impacts. When it comes to occupational safety and health (Romero et al. 2018), VR technology has created safer (virtual) training environments in cases where new procedures for dangerous tasks should be learned and learning rules and regulations may not be enough to guarantee the safety of a worker in training. Moreover, on the job, AR technology can provide the worker with detailed, step-by-step breakdowns of the tasks at hand in his/her peripheral vision through a digital overlay in order to encourage follow-up of the safe work procedure and to be alert to potential dangers that could be digitally highlighted. Lastly, human DTs tracking health-related metrics of workers could allow the real-time monitoring of the workforce's physical and cognitive workload during its work-shift and set alerts and warnings to manage proper levels of occupational effort and stress.

Automation and robotics

While *automation* refers to the process of using physical machines, software, and other technologies to perform tasks with no (or minimal) human intervention, *robotics* refers to the process of designing, creating, and using robots to perform a certain task or set of tasks on their own or collaboratively with humans or other robots. In both cases, these processes make use of control systems to reduce or eliminate the need for human work in the execution of a certain task. Their required degree of intervention varies based on the different levels of automation (from completely manual to fully automatic).

Key use cases. There is an endless number of use cases available to exemplify automation and robotics in a manufacturing setting, from autonomous robots aimed at taking over dangerous tasks that humans cannot or should not do, to collaborative robots (cobots) designed to work side by side with humans to increase their productivity. Both types of robotic system focus on allowing human workers to move into higher-value work. Some industrial use cases of robotic systems include welding, painting, assembly and disassembly, palletising, packaging and labelling, materials handling, pick and place, and load/unload as well as transport in the case of automated guided vehicles (AGVs), autonomous mobile robots (AMRs), and drones.

Economic impact. Automation and robotics systems continue to improve with regard to their functionalities while at the same time, their costs decline. Their impact on shop floor productivity increases when they are applied to tasks that they perform more efficiently and to a higher and more consistent level of quality compared to human operators. This allows humans to transition from low-value tasks to middle- or higher-value tasks such as problem-solving, finding creative solutions, and developing new ideas (Acemoglu 2002; Autor et al. 2003; Autor 2015; Goos et al. 2014).

Environmental impact. Present and future automation and robotics systems are designed to become more energy-efficient and will continue to help minimise the need for larger, less-efficient machines and eliminate waste with fewer human errors.

Social impact. There may be a negative effect on some labour segments due to automation and robotic systems replacing low-skilled workers and automating the tasks that they previously performed. At the same time, these same systems create new, high-paying and desirable jobs. For example, autonomous, collaborative, or mobile robots are now performing menial tasks such as raw materials sorting, transporting, and stocking, while higher-skilled roles are focusing on quality-related tasks and designing the robotic systems and their applications themselves.

Cybersecurity

While advancements in AI, smart sensor systems, and cloud technologies are driving innovation in the manufacturing sector, the integration of IIoT, cloud databases, industrial robots, and wireless networks have made smart manufacturing systems more vulnerable to cyber-attacks (Wu et al. 2018). A compromised device, machine, robot, or computer system within a smart manufacturing system can lead to significant economic losses, environmental damage, and unsafe working conditions resulting in bodily injuries or even deaths (Leander et al. 2020). Therefore, *cybersecurity* has become a primary technology of smart manufacturing systems. *Cybersecurity technologies* help to prevent or mitigate the effect of cyber-attacks. Emerging cybersecurity technologies in smart manufacturing include AI for intrusion detection and handling, end-to-end encryption to prevent data being read or secretly modified, machine authentication for authorising automated human-to-machine (H2M) or machine-to-machine (M2M) communication, rule-based access control for restricting system access to only authorised users, and blockchain for security and traceability of sensitive manufacturing information (Wu et al. 2018; Leander et al. 2020).

Key use cases. Cybersecurity technologies are used to monitor smart manufacturing systems and to detect unusual activity in control systems to prevent loss of critical data such as source files or intellectual property, bodily injuries or other safety issues on the shop floor, spyware, ransomware, and negative environmental impacts (Huelsman et al. 2016; Tsoutsos et al. 2020). For example, companies implement automated audit policies to monitor and signal when a system is operating out of tolerance so that the associated risk can be avoided or mitigated (Huelsman et al. 2016).

Economic impacts. According to a report from the Center for Strategic and International Studies (CSIS) and McAfee, cyber-attacks cost the world almost $600 billion annually (Lewis 2018). The manufacturing sector, as a critical part of economic growth, is among one of the most frequently hacked industries, second only to healthcare (Wu et al. 2018), and is increasingly attracting more cyber-attacks (Huelsman et al. 2016). For example, in 2014 cyber-attackers infiltrated the network of a German steel mill and eventually took command of their industrial control system. As a result, the control system could not shut down the blast furnace, which led to significant property damage and economic loss to the company (Lee et al. 2014).

Environmental impact. The consequence of a cyber-attack can extend beyond economic loss and safety hazards to significant environmental damage. In absence of proper cybersecurity measures, hackers might be able to manipulate control systems and actuators of a smart manufacturing system to release toxic emissions into the environment. For example, in 2000, a hacker caused a spill of 264,000 gallons of untreated sewage to flood local parks and waterways of Maroochy Shire in Australia (Sayfayn and Madnick 2017).

Social impact. The social impact of cybersecurity in smart manufacturing systems can be considered from the perspective of safety-related workplace issues (Romero et al. 2018; Romero et al. 2020). For example, the safety of assembly workers, machine operators, or maintenance staff who work close to autonomous robots and machines may be compromised when there is a lack of appropriate cybersecurity measures. From another viewpoint, different cybersecurity measures are impacting society positively. For example, manufacturing companies have started to provide workforce training programs related to cybersecurity, ultimately improving skills and awareness of operators and other employees.

Blockchain

Blockchain can be perceived as a chronologically ordered set of time-stamped blocks, where each block contains transaction data and is managed by a cluster of nodes in a peer-to-peer network instead of by a central authority (Nofer et al. 2017). Since chained blocks are time-stamped and the data within the blocks are stored on a peer-to-peer network, transaction records become traceable, verifiable, and nearly impossible to alter or delete illicitly. Based on these characteristics, *blockchain technology* provides a transparent, secure, and reliable mechanism to store, integrate, and communicate transactional data between stakeholders and authorised organisations without relying on a third-party trusted authority. *Blockchain technology* also enables self-executing contracts, known as smart contracts, that automatically execute when a set of previously agreed-upon terms and conditions are met.

Key use cases. As manufacturers move towards the usage of *robotics processes automation (RPA)*, the potential impact of blockchain technology is becoming more prevalent. *Supply chain management (SCM)* plays a critical role in ensuring the effectiveness of a smart manufacturing system. The combination of blockchain and IIoT enables a transparent and immutable record of information, inventory, and monetary transactions that results in faster and cost-effective product delivery, better product traceability, streamlined value exchange, and enhanced coordination among the supply chain partners while maintaining respective confidentiality (Bai et al. 2019; Gaur and Gaiha 2020). Similarly, blockchain can be used for tracking manufacturing processes to quickly identify the fault source when a quality-related issue arises (Westerkamp et al. 2020). Furthermore, blockchain-based smart contracts between collaborative partners or among various devices within connected factories facilitate process automation in smart manufacturing systems (Assaqty et al. 20209.

Economic impact. According to Gartner, the global business value-added of blockchain technology is projected to grow more than $176 billion by 2025 and $3.1 trillion by 2030 (Lovelock et al. 2017). Blockchain technology allows manufacturers to build confident relationships with supply chain partners; as a result, they can eliminate the cost of trust, also known as "trust-tax", and reduce verification costs (Ko et al. 2018; Zhang et al. 2019). Moreover, blockchain-based smart contracts may help manufacturers to automate value-adding processes, thereby reducing transaction costs, minimising paperwork, and accelerating turnaround time substantially (CIOL Bureau 2019).

Environmental impact. When integrated with IIoT, blockchain can trace product lifecycle and provide accurate and tamperproof data that can facilitate optimal end-of-life decisions and make long-term circular economy planning more effective (Kouhizadeh 2019). Accurate product usage information when retrieved from a blockchain can lead to increased adoption of remanufactured products by end customers (Tozanlı et al. 2020). However, blockchain-based solutions in general consume more energy than centralised architectures (Sedlmeir et al. 2020).

Social impacts. Blockchain technology allows supply chain partners, including customers, to detect counterfeit items and unethical suppliers. It also assures that products and components are produced under safety standards, the fair use of labour, human rights, and work practices (Saberi et al. 2019).

Smart sensors systems

Smart sensors systems are a collection of smart materials and products, smart wearables, industrial sensors, actuators, and PLCs (programmable logic controller). Smart materials and products have built-in decision-making capabilities on their own, and unlike IoT devices connected to a network, these do not necessarily require constant connectivity to make decisions. Industrial sensors, actuators, and PLCs are deployed to collect data, control manufacturing processes, and increasingly perform analytics and decision-making on the edge. With the help of smart sensor systems, small batches, down to batch-size-one, may be produced without compromising on the use of resources, and smart sensor systems therefore become a vital technology for future smart manufacturing systems.

Key use cases. Smart sensor systems are not limited to in-situ systems embedded in the smart manufacturing system but rather encompass a wide variety of applications, including wearables such as fit-bits, heart rate monitors, etc. These devices can help operators to monitor occupational health and safety. Another example of a smart sensor systems' use case is smart materials used in building structures that change their configuration based on the season and the weather outside (Hu 2009).

Economic impacts. Smart sensor systems will need more investment compared to IIoT and CPS. However, their deployment will optimise the use of resources and lead to the desired results. Therefore, the requirement of quality inspections, repair, and maintenance will be eliminated, as smart sensor systems will be able to rectify the defects themselves.

Environmental impact. Smart sensor systems will help to optimise the input resources required and thus reducing the use of energy and reduction in the waste of raw materials. However, at times smart sensor systems consist of sensors and actuators made from rare earth minerals. These emit harmful radiation during their lifecycle, and there is no environmentally friendly method to decompose them.

Social impact. Smart wearable health monitoring devices can support workers independently. Therefore, these devices can be of great help to elderly workers or workers that operate in stressful or unsafe environments. Smart sensor systems can contribute to identifying and avoiding repetitive and monotonous tasks performed by humans. This will help to reduce labour injuries, for example, while lifting heavy objects in domains such as manufacturing and mining. However, it may also reduce opportunities for unskilled and semi-skilled labour. Another potentially harmful aspect is reduction in social interaction between humans.

5G networks

5G is the fifth-generation technology standard for broadband cellular networks and is emerging as a key enabling technology for future smart manufacturing systems. On the one hand, it supports complex autonomous, collaborative, or mobile automation and robotic systems where ultra- Reliable and Low Latency Communications (uRLLC) will enable fully automated systems and safe collaboration between humans and robots on the shop floor. On the other hand, it enables massive Machine Type Communications (mMTC) that focus on leveraging the full potential of the Industrial Internet of Things (IIoT), where vast numbers of connected devices can communicate with each other (Burow et al. 2019). Moreover, according to Nokia (2020), 84% of manufacturing decision-makers at the global level are considering deployment of their own local, private 4G/5G networks.

Key use cases. 5G private cellular networks will be a fundamental part of the rise of smart factories, allowing the secure connection of all manufacturing equipment on the shop floor and enabling the deployment of autonomous mobile robots, remote machines and robot operation, manufacturing processes monitoring, predictive maintenance, and other IIoT applications.

Economic impact. 5G connectivity will be a catalyst for economic growth in the Fourth Industrial Revolution with an estimated $13.2 trillion of global economic value reached by 2035 (IHS 2019). Through utilising 5G network communication and data exchange capabilities, production and logistics operations at the factory floor and supply chain levels will be more easily coordinated and therefore will enhance manufacturers' and suppliers' productivity gains in terms of agility, flexibility, efficiency, safety, and security (Tech4i2 2019; WEF 2020).

Environmental impact. 5G networks will enable an energy-efficient Industrial Internet of Things by reducing the energy consumption of its connected devices. Moreover, such connected IIoT devices will track in real-time each production resource performance in terms of raw materials, energy, and water utilisation towards higher production efficiency levels (Tech4i2, 2019; WEF 2020)

Social impact. Smart workplaces (Autor 2015; Romero et al. 2020) powered by 5G capabilities will support workers to maximise their productivity and performance by optimising the efficiency and effectiveness of business processes, assets, and services in the emerging Internet of Things, Services, and People (IoTSP) (Tech4i2 2019; WEF 2020).

Barriers and challenges to the adoption of smart manufacturing technologies

The appeal for manufacturers to adopt smart manufacturing technologies is gradually changing from "nice-to-have" to a necessity to remain competitive. However, there are several challenges and barriers which prevent large-scale adoption across the board. First and foremost, smart manufacturing technologies are not a silver bullet that will automatically solve all shortcomings of a company's processes. Smart manufacturing technologies are tools that have to be strategically selected based on the value they add and the problems they address or solve. Moreover, just implementing smart technologies without critically assessing and reimagining one's own processes and priorities is doomed to fail (Romero et al. 2019a; Sinha et al. 2020).

Before illustrating specific barriers and challenges of individual technology clusters (see Table 16.2), we will discuss their general and typical adoption obstacles.

1 *Technology awareness.* Described as the lack of awareness and knowledge of smart manufacturing technologies, this is a major barrier for many manufacturers, especially SMEs. This directly impacts the ability to envision the value-adding use of the technology within their operations. For example, business managers and production planners are not aware of the importance, cost-benefits, and use cases of technologies such as blockchain and cybersecurity. Recently, discussion on smart manufacturing technologies and their productive use on the shop floor has somewhat been democratised – thus expanding the reach from predominantly R&D savvy organisations and research institutes to a broader audience.

Table 16.2 Potential barriers to different smart manufacturing technology clusters adoption

Smart manufacturing technology cluster	*Technology adoption barriers*						
	Awareness		*Piloting and scaling*	*Investment*		*Technology acceptance*	
	Awareness	*Knowledge*	*Difficult to scale*	*Financial resources*	*IT resources*	*Leadership support*	*Operator acceptance*
AI, ML, & Advanced Simulation		x	x	x	x		x
Cloud, Fog, & Edge Computing	x	x				x	
Additive Manufacturing		x		x			
IIoT & CPS	x	x	x		x		
AR, VR, & DTs		x	x	x	x		
Automation & Robotics		x		x	x		x
Cybersecurity		x	x	x	x		
Blockchain	x	x	x	x	x	x	
Smart Sensor Systems	x	x			x		x
5G Networks	x	x		x	x		

To further overcome this barrier, use cases, and case studies from different industries, smart manufacturing technologies, and company sizes are necessary.

2 *Technology piloting and scaling*. The so-called prototype or pilot purgatory can be considered a major barrier to the adoption of smart manufacturing technologies. Many manufacturing companies have started testbeds and prototype projects to evaluate the adoption and value of smart manufacturing technologies. However, often these testbeds and prototypes, while successful demonstrating the principal value of smart manufacturing technologies, fail to scale and transition to regular operation. This can be attributed to several factors, including missing top-level support, lack of acceptance, unclear cost-centre/ownership, or lack of strategic and tactical planning.
3 *Technology investment*. Many of the smart manufacturing technologies require significant capital investment and other resources that represent a key barrier for many manufacturers, again SMEs in particular. Together with the at-times unclear economic benefits associated with adoption and the difficulty of calculating the ROI, this prevents many companies from investing.
4 *Technology acceptance*. This involves the leadership and workforce operators as perceiving the value of, and their attitudes towards, adoption of smart manufacturing technologies.

A different kind of sustainability challenge associated with adoption of smart manufacturing technologies is that the continuous increase in productivity and automation in developed, industrial economies through the productive use of smart manufacturing technologies may have adverse effects on developing economies. Smart manufacturing technologies reduce the amount of manual labour required in many cases and thus the incentive to outsource activities to developing economies with access to lower-cost labour.

In the following, we present a closer look at the 10 key smart manufacturing technology clusters. We structured the barriers and challenges around three general ones but have split some up to reflect more detail. For example, required resources investment is split into financial and IT resources in Table 16.2.

Future developments

Industry 4.0 and *digital transformation* are still in their infancy, especially considering the adoption of smart manufacturing technologies among SMEs. While there has been rapid progress and advances made across the board, the expectation is that this will accelerate even further. During the recent COVID-19 pandemic, we have already seen the potential of digital, smart manufacturing technologies in helping companies to become more resilient and agile (Wuest et al. 2020). The expectation today is that adopting smart manufacturing technologies such as AI is no longer optional but a matter of survival for manufacturers in the competitive marketplace (Taisch et al. 2020).

Necessity during the pandemic has led to an increased desire of employees to work from home. While for most office functions this "new work" is more of a cultural and organisational matter, for shop floor operators there are significant hurdles and challenges to overcome to even consider remote work. Smart manufacturing technologies such as AR, predictive maintenance, and connected systems (IIoT) are one way to facilitate machine operators, maintenance personnel, or production planners to work from home.

An additional aspect that needs to be considered, in the midst of rapid change brought forth by digital transformation with regards to the TBL perspective, is the ability to "replace" (or simplify) complex, multi-tier supply chains with new manufacturing technologies such as AM. AM can facilitate customised or even personalised production locally without needing to use a large number of sources from around the world. Additive or hybrid manufacturing furthermore has the ability to disrupt traditional industries, such as foundries (casting), that today are largely operated overseas (not in the US or EU) today. Given the emphasis on local production and more control over critical supply chains, smart manufacturing technologies not only enable the production of complex parts locally but also do so in a more environmentally friendly and socially responsible way. Local production follows the often more rigid policies and laws, and close proximity to consumers adds another layer of oversight. The abilities of smart manufacturing technologies to drive energy efficiency and automation allows for the running of production at competitive economics despite the higher wage levels. Furthermore, reduced logistics has a positive impact on the carbon footprint, and adding desirable career options to local communities has the potential to improve the social fabric further. Expectations that the introduction of AI, robotics, and automation will lead to large-scale reduction in the workforce are overstated; a shift in the tasks for human operators is more likely to happen. This shift is considered desirable, as mainly strenuous, dangerous, and repetitive tasks will be replaced with tasks more aligned with human requirements, such as creativity, problem-solving, and human ingenuity (Wuest 2020).

Another driver of future development in the manufacturing sector is the servitisation of products and industries. The ability to sensorise smart products and to collect large amounts of data during their operation enables advanced non-ownership business models that are considered as drivers of the circular economy and sharing economy. A positive side effect is that products designed for such non-ownership business models are not designed following the "planned obsolescence" paradigm but are rather built for lasting and efficient operations (Khan et al. 2018).

Overall, the manufacturing sector will continue to transform over the next decade, with smart manufacturing technologies as a driving force. With regard to the triple bottom line, smart manufacturing technologies have the potential to elevate the industrial sector to new heights, with more desirable career options and safer, more digital tasks. Furthermore, the option to locate manufacturing operations closer to consumer markets closes the loop of trade imbalance and adds tax income to the local economy for more sustainable investments. At the same time, the application of smart manufacturing technologies allows us to optimise energy usage and operations to improve the environmental impact of manufacturing operations. Ability to automate, reduction of waste (of energy, materials, etc.), and transition to non-ownership business models provide a positive economic outlook for manufacturing that is largely based on smart manufacturing technologies.

Lastly, while the future seems bright due to the positive potential offered by smart manufacturing technologies for supporting and achieving the triple bottom line, we have to keep in mind that the three dimensions interact, overlap, and at times cause conflict. Hence, organisations face the challenge of acting holistically to pursue environmental, economic, and social bottom lines in all their operations since each dimension represents a necessary, but not sufficient, condition for achieving true sustainability.

Conclusions and limitations

Smart manufacturing systems are built on the foundation of emerging Industry 4.0 and smart manufacturing technologies. However, none of these technologies can solve the sustainable production challenge by themselves without integrating with others in a systems approach. Selected smart manufacturing technologies that are strategically aligned can create significant opportunities to develop a more sustainable industrial environment aimed at achieving the United Nations' Sustainable Development Goal for 2030 of "Responsible Consumption and Production". Nevertheless, even though the emergence of smart manufacturing technologies creates the opportunity to leverage greater production efficiency and other sustainability benefits, the full potential and impact of the smart manufacturing and Industry 4.0 paradigms on the TBL are still unknown. This chapter aims to offer a more comprehensive understanding of how smart technologies can contribute to the triple bottom line of manufacturing and facilitate a continuous discussion among academics and industrial leaders.

Even though this chapter is not based on a formal systematic literature review, but rather contains an explorative review, it builds on the authors' previous work (Mittal et al. 2019), where we conducted a systematic review and an exhaustive characterisation of smart technologies. In this follow-up research work, we provide an update and a critical look at the results reflected in our previous work from a triple bottom line perspective.

Acknowledgements

This work was supported by the J. Wayne & Kathy Richards Faculty Fellowship in Engineering at West Virginia University.

References

Abubakr, M., Abbas, A. T., Tomaz, I., Soliman, M. S., Luqman, M. and Hegab, H. (2020): Sustainable and Smart Manufacturing: An Integrated Approach, *Sustainability* 12, 2280.

Acemoglu, D. (2002): Directed Technical Change, *The Review of Economic Studies* 69, 781–809.

Assaqty, M. I. S., Gao, Y., Hu, X., Ning, Z., Leung, V. C., Wen, Q. and Chen, Y. (2020): Private-Blockchain-Based Industrial IoT for Material and Product Tracking in Smart Manufacturing, *IEEE Network* 34(5), 91–97.

Autor, D. H. (2015): Why Are There Still So Many Jobs? The History and Future of Workplace Automation, *Journal of Economic Perspectives* 29, 3–30.

Autor, D. H., Levy, F. and Murnane, R. (2003): The Skill Content of Recent Technical Change: An Empirical Exploration, *Quarterly Journal of Economics* 118, 1279–1334.

Bai, C., Dallasega, P., Orzes, G. and Sarkis, J. (2020): Industry 4.0 Technologies Assessment: A Sustainability Perspective, *International Journal of Production Economics* 229, 107776.

Bai, L., Hu, M., Liu, M. and Wang, J. (2019): BPIIoT: A Light-Weighted Blockchain-based Platform for Industrial IoT, *IEEE Access* 7, 58381–58393.

BCG – Boston Consulting Group (2018): Embracing Industry 4.0 and Rediscovering Growth: Nine Technologies Transforming Industrial Production. Available at: www.bcg.com/en-us/capabilities/operations/embracing-industry-4.0-rediscovering-growth/ [20.11.2020]

Bonilla, S. H., Silva, H. R. O., da Silva, M. T., Gonçalves, R. F. and Sacomano, J. B. (2018): Industry 4.0 and Sustainability Implications: A Scenario-based Analysis of the Impacts and Challenges, *Sustainability* 10, 3740.

Bughin, J. and Seong, J. (2018): How Competition Is Driving AI's Rapid Adoption, *Harvard Business Review*, 17 October 2018.

Burow, K., Franke, M. and Thoben, K.-D. (2019): 5G-Ready in the Industrial IoT-Environment: Requirements and Needs for IoT Applications from an Industrial Perspective, in *Production Management for the Factory of the Future*, IFIP, AICT 566, Part I, Springer, 408–413

CIOL Bureau (2019): Three Blockchain Use Cases that will Accelerate Industry 4.0 Journey for Manufacturers. Available at: http://ciol.com/three-blockchain- use-cases-will-accelerate-industry-4–0-journey-manufacturers/ [20.11.2020]

Clemons, J. (2019): Smart Manufacturing and Sustainability: Adding Value to the Triple Bottom Line. *Forbes*, 13 February 2019.

Columbus, L. (2020): 10 Ways AI Is Improving Manufacturing In 2020. *Forbes*, 18 May 2020.

Dalenogare, L. S., Benitez, G. B., Ayala, N. F. and Frank, A. G. (2018): The Expected Contribution of Industry 4.0 Technologies for Industrial Performance, *International Journal of Production Economics* 204, 383–394.

Ejsmont, K., Gladysz, B. and Kluczek, A. (2020): Impact of Industry 4.0 on Sustainability – Bibliometric Literature Review, *Sustainability* 12, 5650.

Elkington, J. (2018): 25 Years Ago I Coined the Phrase "Triple Bottom Line". Here's Why It's Time to Rethink It, in *Harvard Business Review*, 25 June 2018.

Ford, S. and Despeisse, M. (2016): Additive Manufacturing and Sustainability: An Exploratory Study of the Advantages and Challenges, *Journal of Cleaner Production* 137(29), 1572–1587.

Frank, A. G., Dalenogare, L. S. and Ayala, N. F. (2019): Industry 4.0 Technologies: Implementation Patterns in Manufacturing Companies, *International Journal of Production Economics* 210, 15–26.

Gaur, V. and Gaiha, A. (2020): Building a Transparent Supply Chain: Blockchain can Enhance Trust, Efficiency, and Speed, *Harvard Business Review* 98(3), 94–103.

Ghobakhloo, M. (2019): Determinants of Information and Digital Technology Implementation for Smart Manufacturing, *International Journal of Production Research* 58(8), 2384–2405.

GMInsights (2019): Digital Twin Market Size by Application. Available at: www.gminsights.com/industry-analysis/digital-twin-market [20.11.2020]

Goos, M., Manning, A. and Salomons, A. (2014): Explaining Job Polarization: Routine-Biased Technological Change and Offshoring, *American Economic Review* 104, 2509–2526.

Harispe, S., Ranwez, S., Janaqi, S. et al. (2015): Semantic Similarity from Natural Language and Ontology Analysis, *Synthesis Lectures on Human Language Technologies* 8, 1–254.

Hu, J. (2009): The Use of Smart Materials in Cold Weather Apparel, in Williams, J. T. (ed.): *Textiles for Cold Weather Apparel*. Oxford et al.: Woodhead Publishing, 84–112.

Huelsman, T., Peasley, S., Powers, E. and Robinson, R. (2016): Cyber Risk in Advanced Manufacturing. Deloitte: Basel, Switzerland. Available at: https://www2.deloitte.com/us/en/pages/manufacturing/articles/cyber-risk-in-advanced-manufacturing.html [20.11.2020]

IHS (2019): The 5G Economy: How 5G will Contribute to the Global Economy. Available at: www.qualcomm.com/media/documents/files/ihs-5g-economic-impact-study-2019.pdf [30.11.2020]

Jena, M. C., Mishra, S. K. and Moharana, H. S. (2020): Application of Industry 4.0 to Enhance Sustainable Manufacturing, *Environmental Progress & Sustainable Energy* 39, 13360.

Kamble, S. S., Gunasekaran, A. and Gawankar, S. A. (2018): Sustainable Industry 4.0 Framework: A Systematic Literature Review Identifying the Current Trends and Future Perspectives, *Process Safety and Environmental Protection* 117, 408–425.

Khan, M., Mittal, S., West, S. and Wuest, T. (2018): Review on Upgradability – A Product Lifetime Extension Strategy in the Context of Product-Service Systems, *Journal of Cleaner Production* 204, 1154–1168. https://doi.org/10.1016/j.jclepro.2018.08.329.

Kiel, D., Müller, J. M., Arnold, C. and Voigt, K.-I. (2017): Sustainable Industrial Value Creation: Benefits and Challenges of Industry 4.0, *International Journal of Innovation Management* 21, 1740015.

Klingenberg, C. O., Borges, M. A. V. and Antunes Jr, J. A. V. (2019): Industry 4.0 as a Data-driven Paradigm: A Systematic Literature Review on Technologies, *Journal of Manufacturing Technology Management*, online first.

Ko, T., Lee, J. and Ryu, D. (2018): Blockchain Technology and Manufacturing Industry: Real-time Transparency and Cost Savings, *Sustainability* 10(11), 4274.

Kouhizadeh, M., Sarkis, J. and Zhu, Q. (2019): At the Nexus of Blockchain Technology, the Circular Economy, and Product Deletion, *Applied Sciences* 9(8), 1712.

Kusiak, A. (2017): Smart Manufacturing must Embrace Big Data, *Nature* 544(7648), 23–25.

Leander, B., Čaušević, A., Hansson, H. and Lindström, T. (2020): Access Control for Smart Manufacturing Systems, in *European Conference on Software Architecture*. Cham: Springer, 463–476.

Lee, R. M., Assante, M. J. and Conway, T. (2014): German Steel Mill Cyber Attack, *Industrial Control Systems* 30, 62.

Lewis, J. (2018): Economic Impact of Cybercrime, No Slowing Down. McAfee. Available at: https://csis-website-prod.s3.amazonaws.com/s3fs-public/publication/economic-impact-cybercrime.pdf [30.11.2020]

Liao, Y., Deschamps, F., Loures, E. D. F. R. and Ramos, L. F. P. (2017): Past, Present and Future of Industry 4.0 – A Systematic Literature Review and Research Agenda Proposal, *International Journal of Production Research* 55(12), 3609–3629.

Lovelock, J. D., Reynolds, M., Granetto, B. and Kandaswamy, R. (2017): Forecast: Blockchain Business Value, Worldwide, 2017–2030. *Gartner*, Stamford.

Lu, D. (2019): Creating an AI can be Five Times Worse for the Planet than a Car, *New Scientist*, 6 June 2019.

Masanet, E., Shehabi, A., Lei, N., Smith, S. and Koomey, J. (2020): Recalibrating Global Data Center Energy-use Estimates, *Science* 367(6481), 984–986.

Menon, K., Kärkkäinen, H. and Wuest, T. (2020): Industrial Internet Platform Provider and End-user Perceptions of Platform Openness Impacts, *Industry and Innovation* 27(4), 363–389.

Menon, K., Kärkkäinen, H., Wuest, T. and Gupta, J. (2019): Industrial Internet Platforms: A PLM Perspective, *Part B: Journal of Engineering Manufacture* 233(5), 1390–1401.

Microsoft (2020): The Carbon Benefits of Cloud Computing (Whitepaper). Available at: http://download.microsoft.com/download/7/3/9/739BC4AD-A855-436E- 961D-9C95EB51DAF9/Microsoft_Cloud_Carbon_Study_2018.pdf [30.11.2020]

Mittal, S., Khan, M. A., Romero, D. and Wuest, T. (2018a): A Critical Review of Smart Manufacturing & Industry 4.0 Maturity Models: Implications for Small and Medium-sized Enterprises (SMEs), *Journal of Manufacturing Systems* 49, 194–214.

Mittal, S., Khan, M. A., Romero, D. and Wuest, T. (2019): Smart Manufacturing: Characteristics, Technologies and Enabling Factors. Proceedings of the Institution of Mechanical Engineers, Part B, *Journal of Engineering Manufacture* 233(5), 1342–1361.

Mittal, S., Romero, D. and Wuest, T. (2018b): Towards a Smart Manufacturing Toolkit for SMEs, in *IFIP International Conference on Product Lifecycle Management*. Cham: Springer, 476–487.

Ngo, T. D., Kashani, A., Imbalzano, G., Nguyen, K. T. and Hui, D. (2018): Additive Manufacturing (3D-printing): A Review of Materials, Methods, Applications and Challenges, *Composites Part B: Engineering* 143, 172–196.

Nofer, M., Gomber, P., Hinz, O. and Schiereck, D. (2017): Blockchain, *Business & Information Systems Engineering* 59(3), 183–187.

Nokia (2020): Nokia and ABI Research Identify Key Trends in Manufacturing Investment to Enable Industry 4.0. Available at: www.nokia.com/about-us/ news/releases/2020/05/06/nokia-and-abi-research-identify-key-trends-in-manufacturing-investment-to-enable-industry-40/ [30.11.2020]

PwC (2020): Why We Believe VR/AR will Boost Global GDP by $1.5 trillion. Available at: www.pwc.co.uk/services/economics/insights/vr-ar-to-boost-global-gdp.html [20.11.2020]

Romero, D., Flores, M. Herrera, M. and Resendez, H. (2019a): Five Management Pillars for Digital Transformation Integrating the Lean Thinking Philosophy. *25th International ICE-Conference on Engineering, Technology and Innovation*, IEEEXplorer, 1–8.

Romero, D., Gaiardelli, P., Thürer, M., Powell, D. and Wuest, T. (2019b): Cyber-Physical Waste Identification and Elimination Strategies in the Digital Lean Manufacturing World, in Ameri, F., Stecke, K., von Cieminski, G. and Kiritsis, D. (eds.): *Advances in Production Management Systems. Production Management for the Factory of the Future. APMS 2019. IFIP Advances in Information and Communication Technology*, vol 566. Cham: Springer. https://doi.org/10.1007/978-3-030-30000-5_5

Romero, D., Mattsson, S., Fast-Berglund, Å., Wuest, T., Gorecky, D. and Stahre, J. (2018): Digitalizing Occupational Health, Safety and Productivity for the Operator 4.0, in Moon, I., Lee, G., Park, J., Kiritsis, D. and von Cieminski, G. (eds.): *Advances in Production Management Systems. Smart Manufacturing for Industry 4.0. APMS 2018. IFIP Advances in Information and Communication Technology*, vol 536. Cham: Springer. https://doi.org/10.1007/978-3-319-99707-0_59

Romero, D., Stahre, J. and Taisch, M. (2020): The Operator 4.0: Towards Socially Sustainable Factories of the Future, *Computers & Industrial Engineering* 139, 106128.

Romero, D., Stahre, J., Wuest, T., Noran, O., Bernus, P., Fast-Berglund, Å. and Gorecky, D. (2016): Towards an Operator 4.0 Typology: A Human-Centric Perspective on the Fourth Industrial Revolution Technologies. *International Conference on Computers & Industrial Engineering (CIE46)*, 1–11.

Saberi, S., Kouhizadeh, M., Sarkis, J. and Shen, L. (2019): Blockchain Technology and its Relationships to Sustainable Supply Chain Management, *International Journal of Production Research* 57(7), 2117–2135.

Sayfayn, N. and Madnick, S. (2017): Cybersafety Analysis of the Maroochy Shire Sewage Spill. Interdisciplinary Consortium for Improving Critical Infrastructure Cybersecurity, *MIT Management Sloan School, Working Paper CISL*, 9.

Scott, K. and Shaw, G. (2020): Automation May Take Jobs – but AI Will Create Them. *Wired*, April 3, 2020. Available at: www.wired.com/story/automation-may-take-jobs-but-ai-will-create-them/ [20.11.2020]

Sedlmeir, J., Buhl, H. U., Fridgen, G. and Keller, R. (2020): The Energy Consumption of Blockchain Technology: Beyond Myth, *Business & Information Systems Engineering* 62(2), 1–10.

Shukla, M., Todorov, I. and Kapletia, D. (2018): Application of Additive Manufacturing for Mass Customisation: Understanding the Interaction of Critical Barriers, *Production Planning & Control* 29(10), 814–825.

Sinha, A., Bernardes, E., Calderon, R. and Wuest, T. (2020): *Digital Supply Networks: Transform Your Supply Chain and Gain Competitive Advantage with New Technology and Processes*. New York: McGraw-Hill.

Taisch, M., Casidsid, M., May, G., Morin, T., Padelli, V., Pinzone, M. and Wuest, T. (2020): *The 2020 World Manufacturing Report: Manufacturing in the Age of Artificial Intelligence*. Milan, Italy: World Manufacturing Foundation.

Tao, F. Qi, Q., Liu, A. and Kusiak A. (2018): Data-driven Smart Manufacturing, *Journal of Manufacturing Systems* 48, 157–169.

Tech4i2 (2019): 5G Socio-Economic Impact in Switzerland. Available at: https://asut.ch/asut/media/id/1465/type/document/Study_Tech4i2_5G_socio-economic_ impact_switzerland_February_2019.pdf [30.11.2020]

Thao Le, V., Paris, H. and Mandil, G. (2017): Environmental Impact Assessment of an Innovative Strategy based on an Additive and Subtractive Manufacturing Combination, *Journal of Cleaner Production* 164, 508–523.

Thoben, K.-D., Wiesner, S. and Wuest, T. (2017): "Industrie 4.0" and Smart Manufacturing – A Review of Research Issues and Application Examples, *International Journal of Automation Technology* 11(1), 4–19.

Tozanlı, Ö., Kongar, E. and Gupta, S. M. (2020): Trade-in-to-Upgrade as a Marketing Strategy in Disassembly-to-Order Systems at the Edge of Blockchain Technology, *International Journal of Production Research* 58(23), 1–18.

Tsoutsos, N. G., Gupta, N. and Karri, R. (2020): Cybersecurity Road Map for Digital Manufacturing, *Computer* 53(9), 80–84.

WEF – World Economic Forum (2020): The Impact of 5G: Creating New Value across Industries and Society. Available at: http://www3.weforum.org/docs/WEF_The_Impact_of_5G_Report.pdf [30.11.2020]

Westerkamp, M., Victor, F. and Küpper, A. (2020): Tracing Manufacturing Processes using Blockchain-based Token Compositions, *Digital Communications & Networks* 6(2), 167–176.

Wu, D., Ren, A., Zhang, W., Fan, F., Liu, P., Fu, X. and Terpenny, J. (2018): Cybersecurity for Digital Manufacturing, *Journal of Manufacturing Systems* 48, 3–12.

Wuest, T. (2020): To Program Robots is Human, to Problem-Solve is Divine, *Smart Manufacturing Magazine* 5(6), 48.

Wuest, T., Kusiak, A., Dai, T. and Tayur, S. (2020): Impact of COVID-19: The Case for AI-Inspired Digital Transformation, in *INFORMS ORMS Today* 47(3), https://doi.org/10.1287/orms.2020.03.16.

Zhang, Y., Xu, X., Liu, A., Lu, Q., Xu, L. and Tao, F. (2019): Blockchain-based Trust Mechanism for IoT-based Smart Manufacturing System, *IEEE Transactions on Computational Social Systems* 6(6), 1386–1394.

17

FROM SMART TECHNOLOGIES TO VALUE CREATION

Understanding smart service systems through text mining[1]

Chiehyeon Lim and Paul P. Maglio

Introduction

Service systems in the transportation, retail, healthcare, entertainment, hospitality, and other sectors are configurations of people, information, organisations, and technologies that operate together for mutual values (Maglio et al. 2009). Service systems have become "smarter" over time as advanced technologies have been used increasingly in the systems (Larson 2016; Watanabe and Mochimaru 2017). Smart service systems can be found in homes (Alam et al. 2012) and in the energy (Strasser et al. 2015), healthcare (Raghupathi and Raghupathi 2014), and transportation sectors (Pelletier et al. 2011), among many others. As the concepts of service system and smartness intertwine, academia, industry, and government pay great attention to the concept of smart service systems (e.g., Maglio et al. 2015; Larson 2016; NSF 2016; Lee and Lim 2021). We believe this concept is meaningful in the development and use of smart technologies, such as the Internet of Things (IoT), artificial intelligence, and blockchain, as it represents the ultimate application and integration of technology for value creation.

Yet despite its obvious importance, the concept of "smart service system" is not clear. What is it? What are smart service system research areas, and how are they related? What are the key technology factors of smart service systems? How does the technology create economic and social values? What are the main application areas, and what does a smart service ecosystem look like? How can we define "smart service system"? The functions and operations of smart service systems depend on sensing (Sim et al. 2011), big data (Maglio and Lim 2016), computation (Lee et al. 2012), and automation (Jacobsen and Mikkelsen 2014), and should consider the customer (Wünderlich et al. 2013) and business aspects (San Román et al. 2011). A search for "{TOPIC: (smart service system)}" in the Web of Science generates more than 5,000 results across engineering, computer science, information systems, control, transportation, healthcare, and other fields. Yet despite widespread application and the importance of research on smart service systems, in-depth understanding of such systems remains lacking in the literature.

1 This chapter is attributed to "Service Science. Copyright ©2018 The Author(s). https://doi.org/10.1287/serv.2018.0208, used under a Creative Commons Attribution License: https://creativecommons.org/licenses/by/4.0/".

 DOI: 10.4324/9780429351921-21

A unified understanding of smart service systems across different fields may facilitate development and innovation. Furthermore, such understanding would promote the use, integration, and improvement of smart technologies from a broad and application-oriented perspective to create economic and social values. Specifically, a generic definition or representation of a smart service system will promote mutual understanding among people with different backgrounds, thereby facilitating collaborative analysis and the development of such systems across multiple disciplines. Similarly, a comprehensive categorisation of applications related to smart service systems can contribute to our understanding of system variety and lead to the creation of synergy between different applications. However, such integrative work is not easy to achieve because of the variety and volume of the studies and applications related to smart service systems.

In this chapter, we develop an understanding of smart service systems by mining text related to these sorts of systems. The interdisciplinary text corpus we analysed includes scientific literature and news articles. The former discuss research topics and technology factors of smart service systems, whereas the latter describe the application areas and business aspects. To capture the essence of the data, our analytics method uniquely incorporates metrics to statistically measure the importance of word-features of the data and unsupervised machine learning algorithms, such as spectral clustering (Von Luxburg 2007) and topic modelling (Blei et al. 2003). Our analysis of 5,378 scientific articles and 1,234 news articles identified statistically significant keywords, research topics, technology factors (e.g., sensing and context-aware computing), and application areas and provided a definition of smart service systems. Furthermore, we developed a conceptual framework of smart service systems and a hierarchical structure of smart service system applications by integrating our findings and those of existing studies.

Our work is unique in that it uses a *data-driven* approach to incorporate a broad range of studies on smart service systems from technology and engineering to business and social science, thereby filling the gap between existing studies on smart technologies and on economic and social value creation. In terms of methodological contribution, it is a first study to accomplish data-driven understanding of a service research topic. Our research methodology can be applied to understanding and describing other topics as well.

This chapter is organised as follows. First, we review studies related to smart service systems and provide the conceptual foundations for this work. Afterwards, we describe the research methodology, including the data collection and analysis methods. Then, we describe the findings. In the penultimate section, we discuss the theoretical implications of our work. Finally, we conclude this chapter.

Literature review

As service systems become increasingly smarter through technological advances in many industries, researchers have begun to investigate common characteristics of smart service systems, and so far, several definitions and descriptions of smart service systems exist (see Table 17.1). For example, a smart service system is a service system capable of learning, dynamic adaptation, and decision making (Medina-Borja 2015) that requires an intelligent object (Allmendinger and Lombreglia 2005; Wünderlich et al. 2015) and involves intensive data and information interactions among people and organisations (Maglio and Lim 2016; Lim et al. 2018a). Smart service systems incorporate technologies for sensing, communication, and control, among others (NSF 2016), to consider the needs and context of stakeholders effectively and efficiently (Lim et al. 2016). Other keywords describing smart service systems in the literature include cognition, sustainability (Spohrer and Demirkan 2015), self-reconfiguration, connection (Carrubbo et al. 2015), interaction (Barile and Polese 2010), people, real time (Gavrilova and Kokoulina 2015),

Table 17.1 Existing definitions or descriptions of smart service system

Source	*Definition or description*
Wünderlich et al. (2013)	Services delivered to or through intelligent products that feature awareness and connectivity are called "smart services" (Allmendinger and Lombreglia 2005). . . . The implementation of smart services is expected to result in substantial efficiency gains on both the provider's and the user's side from benefits such as cost reductions, increased flexibility, increased access, and time savings. . . . Smart interactive services comprise not only an embedded technology within the product that communicates object-to-object but also personal interactions between the user and the service provider employee as part of the smart service delivery process.
Carrubbo et al. (2015)	Smart service systems can be understood as service systems that are specifically designed for the prudent management of their assets and goals while being capable of self-reconfiguration to ensure that they continue to have the capacity to satisfy all the relevant participants over time. They are principally (but not only) based upon ICT as enabler of reconfiguration and intelligent behaviour in time with the aim of creating a basis for systematic service innovation in complex environments. Smart service systems are based upon interactions, ties and experiences among the actors. Of course, among these actors, customers play a key role, since they demand a personalized product/service, high-speed reactions, and high levels of service quality; despite customer relevance, indirectly affecting every participating actor, smart service systems have to deal to every other actor's behaviour, who's expectations, needs and actions directly affect system's development and future configurations. The smarter approach applied to healthcare is called "smarter healthcare". As IBM highlights, a smarter healthcare system is obtained through better connections for faster, more detailed analysis of data.
Gavrilova and Kokoulina (2015)	The term "smart" implies two main properties. First, it highlights anthropomorphic features of the smart service. For example, technology research company Gartner, Inc. claims that smart technologies are "technologies that do what we thought only people could do. Do what we thought machines couldn't do" (Austin 2009). Second, term "smart" is usually related to artificial intelligence (i.e. intelligence of machine) "because it is impractical to deploy humans to gather and analyse the real-time field data required, smart services depend on "machine intelligence" (Allmendinger and Lombreglia 2005). . . . Smart service systems often have the following characteristics of the intelligent system: Self-configuration (or at least easy-triggered reconfiguration), Proactive behaviour (capability for prognosis or preventive actions, as opposed to the reactive behaviour), Interconnectedness and continuous interactivity with internal and external system elements. . . . Smart service attributes include dynamic properties (without modelling of the changing environment; past-based modelling; stochastic modelling), intelligence (knowledge-based; data-based; content-based), Knowledge awareness (context-oriented; explicit knowledge; business intelligence), IT platform (mobile; SaaS; hybrid cloud; corporate servers), and elements (IT; people; hybrid).

(*Continued*)

Table 17.1 (Continued)

Source	*Definition or description*
Spohrer and Demirkan (2015)	Smart service systems are ones that continuously improve (e.g., productivity, quality, compliance, sustainability, etc.) and co-evolve with all sectors (e.g., government, healthcare, education, finance, retail and hospitality, communication, energy, utilities, transportation, etc.). . . . Because of analytics and cognitive systems, smart service systems adapt to a constantly changing environment to benefit customers and providers. Using big data analytics, service providers try to compete for customers by (1) improving existing offerings to customers, (2) innovating new types of offerings, (3) evolving their portfolio of offerings and making better recommendations to customers, (4) changing their relationships to suppliers and others in the ecosystem in ways their customers perceive as more sustainable, fair, or responsible.
NSF (2016)	A "smart" service system is a system that amplifies or augments human capabilities to identify, learn, adapt, monitor and make decisions. The system utilizes data received, transmitted, or processed in a timely manner, thus improving its response to future situations. These capabilities are the result of the incorporation of technologies for sensing, actuation, coordination, communication, control, etc.
Larson (2016)	Until service science achieves the respect that comes with maturation and productive interaction among researchers from its sub-disciplines, our ability to foster substantial innovation in service systems will remain limited. That would be unfortunate, because such innovation could make these systems "smarter" – more sophisticated, nuanced, tuned to human needs, and augmenting human capabilities with supporting technology. Smarter systems, in turn, could help drive sustained economic growth by increasing our national productivity. They could create a steady stream of new jobs in numerous sub-disciplines. And they could help us address serious social problems in arenas like healthcare, education, urban infrastructure, and national and global security. The definition of "service" can be further applied to artefacts such as "smart refrigerators" and "smart homes" that provide services that increase human quality of life.
Moldovan et al. (2017)	An elastic system (i.e., which may be equivalent smart service system in this paper) should leverage and combine developments from multiple fields of computer science, to achieve its goals. This multi-disciplinary approach provides to elastic systems the necessary capabilities to adapt and change with respect to the concerns and requirements of people, processes, and things over both physical and cyber worlds. . . . An elastic system is composed of heterogeneous units (i.e., people, processes, and things) working together. . . . Elastic systems should be built from replaceable self-contained units of functionality, each unit exposing its functionality through a well-defined interface. . . . Elastic systems must have a strong focus on change, from design time, when elasticity capabilities are defined, to run-time and operation, when desired changes occur by enforcing the capabilities of different units. . . . Elastic systems must also consider stakeholders' business requirements for achieving desired business goals.

user-centric (Geum et al. 2016), dynamic experience (Ostrom et al. 2015), autonomy (Maglio 2017), and elasticity (Moldovan et al. 2017). Different types of smart service systems, such as those found in homes (Alam et al. 2012), healthcare (Raghupathi and Raghupathi 2014), buildings (Agarwal et al. 2010), and transportation (Pelletier et al. 2011), have been realised in various forms in smart cities (Lim et al. 2018c).

In addition, IoT (Atzori et al. 2010), big data (Lim et al. 2018b), cloud computing (Iyoob et al. 2013), wearable devices (Patel et al. 2015), cyber-physical systems (Lee 2008), Industry 4.0 (Lee and Lim 2021), blockchain (Kshetri 2018), IoT-enabled servitisation (Lim et al. 2015), and service-oriented data use (Lim et al. 2018a) are all related to smart service systems. These terms emphasise the technological aspects of specific applications, whereas the term "smart service system" places the technology and people in the context of value creation. The smart service system concept may be comparable to existing technology-based service concepts in which "there is an interaction between customers and technology-embedded objects such as a computer, the Internet, or a machine at the moment of truth" (Noh et al. 2016), such as technology-mediated service (Schumann et al. 2012) and information-intensive service (Lim and Kim 2014) concepts. Technology-mediated service (Schumann et al. 2012) or mobile service (Heo et al. 2017) emphasises efficient delivery of information to people, whereas information-intensive service focuses mainly on creating information that helps people achieve their goals (e.g., exercise and transportation) and on collecting the right data to create useful information (Lim and Kim 2015; Lim et al. 2018b). Such traditional technology-based service concepts focus on specific resources or activities of service systems, whereas the smart service system concept in the current study focuses on the entire system of resources and activities to consider how capabilities operate together to increase mutual value. In summary, the smart service system concept depends on both systems thinking (Frost and Lyons 2017) and value creation (Maglio and Lim 2016).

The studies cited suggest that there are various aspects of smart service systems. However, each study represents only the specific perspectives of the respective authors. No study provides an overview of the research and application areas of smart service systems, despite the merit of understanding the wide range of applications and research on these systems as a whole. As described previously, our objective is to develop a unified understanding of smart service systems based on a data-driven approach to reduce subjectivity and increase inclusiveness. Thus, we used a text mining method to aggregate knowledge and information from thousands of scientific and news articles on technologies and value creation.

Research methodology

Developing a unified conceptualisation of smart service systems is not easy, given the variety and volume of related studies and applications. Thus, we relied on a text mining method to develop our understanding of smart service systems. Text mining, or text data mining, is a process to discover previously unknown knowledge from textual data (Bird et al. 2009). Text mining methods have been used for many purposes on a wide range of document types, such as technology trend analysis with patent data (Yoon and Park 2004), customer understanding with customer opinion (Jansen et al. 2009), feedback (Ordenes et al. 2014), review (Mankad et al. 2016), or even related patent data (Lim et al. 2017), and understanding specific research fields with scientific documents (Jo et al. 2007). Recent studies apply machine learning algorithms to text data to discover previously unknown knowledge from the data or to automate the application of existing knowledge to the data. For example, a clustering algorithm can be applied to identify unknown categories of documents (e.g., Aggarwal and Zhai 2012), and a classifier can be applied to automate the classification of spam mail and to identify significant words (Yu and Xu 2008). A notable algorithm for text mining is the topic modelling algorithm, which discovers hidden topics of a set of documents, using for example, latent Dirichlet allocation (LDA) (Blei et al. 2003) and non-negative matrix factorisation (NMF) (Lin 2007). Visualisation is key to interpretation of text data analysis results because most of the text mining cases involve numerous features and require semantic interpretation (Lim et al. 2017). For example, a visualisation of the network between different keywords is useful for understanding the key links between the keywords and categorising the keywords efficiently (e.g., Park and Yoon 2015).

Text mining is an appropriate method for achieving our research objective because we aim to explore aspects and areas of smart service systems in a comprehensive manner, and such work is difficult for people to do alone. In addition, despite insights from experts, the subjective identification and categorisation of key aspects and areas of smart service systems can be difficult to evaluate. A data-driven approach such as text mining with metrics and machine learning algorithms can be an excellent alternative (Blei et al. 2003; Mankad et al. 2016; Antons and Breidbach 2017; Zhuge and Wilks 2017). Moreover, human analysis built upon analytical findings from massive amounts of documents often generates rich insights and implications (Ordenes et al. 2014; Lim et al. 2017).

In this context, we collected and analysed a comprehensive set of 5,378 scientific articles and 1,234 news articles. The interested reader is referred to the Methodology section and Appendices in Lim and Maglio (2018)[2] for a detailed description of our data collection and analysis.

Collection of scientific and news data

Text data obtained from the scientific literature and news articles contain useful information about smart service systems. In this study, the sources of scientific data included titles, abstracts, and keywords of journal articles, review papers, proceedings papers, and book chapters about smart service systems.[3] The news data included introductions, compliments, and critiques about smart service systems. The literature data represent the research topics and technology factors of smart service systems, whereas the news data show the application areas and business aspects. Hence, these two types of data complement each other in strengthening our understanding of the academic and practical aspects of smart service systems, from which theoretical and managerial insights can be derived.

Table 17.2 gives an overview of the data collection in this study. The five queries used for literature data collection were {TOPIC: (smart) AND TOPIC: (service) AND TOPIC: (system)}, {TOPIC: ("smart service")}, {TOPIC: ("smart services")}, {TOPIC: ("smart system")}, and {TOPIC: ("smart systems")}. A total of 6,488 items were collected, including 2,688 articles, 149 reviews, 3,851 conference papers, and 147 book chapters. As of May 7, 2016, these data were the "full" population (i.e., not a sample) of the smart service system literature identified from the Web of Science Core Collection databases of the Science Citation Index Expanded (SCIE) (1945–), Social Sciences Citation Index (SSCI) (1987–), Conference Proceedings Citation Index – Science and Social Science & Humanities (1990–), Book Citation Index – Science

2 This chapter is attributed to "Service Science. Copyright ©2018 The Author(s). https://doi.org/10.1287/serv.2018.0208, used under a Creative Commons Attribution License: https://creativecommons.org/licenses/by/4.0/".

3 Note that collecting full text data from academic articles is difficult because of data protection and cost. Although we may have access to data, many publishers do not provide data in the same HTML format, requiring additional processing. Above all, we did not have access to all data because of cost, which is a fundamental reason why we selected the Web of Science service to download available data.

Table 17.2 Overview of data collection

Data type	*Data source*	*Collection method*	*Content of data*	*Collected items*
Literature data	Web of Science Core Collection	Download all data found with the five queries through the Web of Science data provision service	The title, abstract, and keywords of article	6,488 items
News data	Websites of news service providers	Search for 126 keywords through the Google News service and manually copy the title and content of news from the websites of news service providers	The title, content, and keywords (if specified) of article	1,256 items

and Social Science & Humanities (2005–), and Emerging Sources Citation Index (2015–). The Web of Science Core Collection includes quality papers, and all the papers show characteristics of smart service systems. Unlike existing systematic reviews of selected papers, we collected all available data from the databases first and then excluded little relevant data and non-important features in a semi-automatic and statistical manner.

For news data collection, we searched for relevant two-word phrases that contain "smart" that we identified in the scientific data, particularly those that indicate "application areas" of smart service systems (e.g., "smart grid", "smart power", "smart home", "smart transportation", and "smart parking"), resulting in a list of 352 phrases. We then searched news articles from Google News (language = "English", region = "USA") using this list. Here, we relied on the Google News platform service for data collection to ensure the freshness, diversity, textual richness, originality, and consistency of the content (Google 2017a, 2017b). We finally obtained a set of 126 keywords that was useful in analysing the application areas in terms of number, technological intensity, and overlap of search results (i.e., we found that 226 phrases were not used much in news articles, do not represent technology-based service, or overlap with others). Data were collected from the websites of news service providers, using 126 keywords, by copying and pasting the title and content. Articles that could not be used without permission were excluded. We collected the 10 most relevant items for each keyword, according to Google, only if the article actually discussed an application related to smart service systems. Data were not collected if the article merely introduced goods or showed financial information.

Analysis of the scientific and news data

Figure 17.1 shows the analysis process of the literature data. Before conducting data collection and analysis, we performed multiple pilot studies using small sets of literature and news data to develop effective data analysis methods. This approach enabled us to develop a method for finding a "core vector space" of the smart service system literature. One challenge in text mining is to identify meaningful data and word-features in the context in question; for example, the initial set of 6,488 scientific articles included numerous minimally relevant data and non-important word-features. A core vector space is one without non-relevant data and non-important word-features, namely, a homogenous vector space in terms of smart "service" systems for "people". We used bibliographical and statistical analyses to find a core vector space of the initial data.

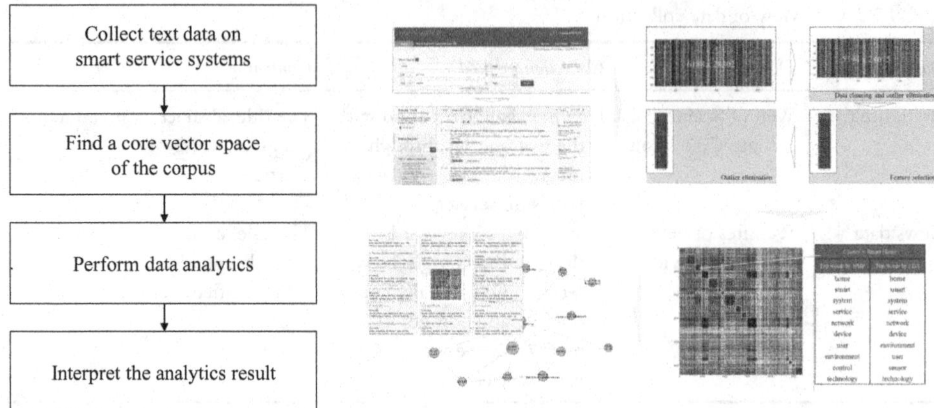

Figure 17.1 Overview of data analysis

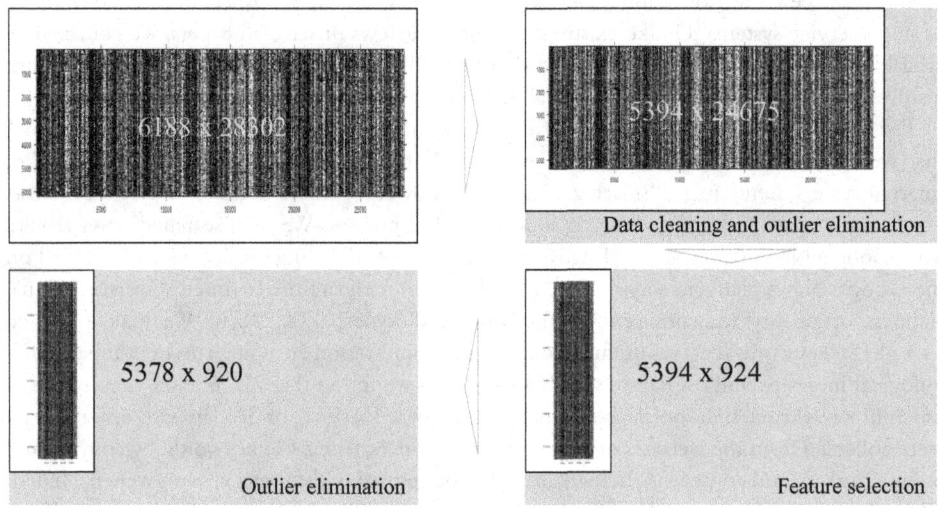

Figure 17.2 An illustration of how to find a core vector space

Figure 17.2 illustrates the process for finding a core vector space in the scientific article data, which consists of three steps. The black cell indicates that the corresponding word (column) appears in the datum (row). First, we deleted data where the abstract or title information was missing or was a duplicate. Then, we cleaned and pre-processed the data. Some popular techniques exist to pre-process textual data and to transform the data into a numerical form, such as stop word elimination, lemmatisation, and stemming to exclude non-meaningful words and standardise the form of words in the documents; vector space representation of the data, which usually refers to the creation of a document-term matrix; the use of term frequency–inverse document frequency (TF–IDF) rather than simple frequency to reflect the feature value of an item meaningfully; and the measurement of the similarity between data using metrics such as cosine similarity, the Jaccard coefficient, and the Pearson correlation coefficient (Bird et al. 2009). As a result of the first step, a total of 5,394 items were chosen from the original 6,188 items.

Second, we selected only the word-features that may represent "smart service systems" based on TF–IDF values, which measure the importance of a word by considering it in both the article and the entire dataset. For example, "external", "require", and "provide" may not be important in an article entitled "Robotic Automated External Defibrillator Ambulance for Emergency Medica Service in Smart Cities" (Samani and Zhu 2016); we found the top four words of this article were "city", "ambulance", "emergency", and "robotic", according to their TF–IDF values across the 5,394 articles. We used the normalised TF–IDF calculation from the Python Scikit-Learn Library (Pedregosa et al. 2011) after testing its performance in pilot studies. From the union of top four words in the data, we selected the words that appeared at least twice across the dataset.

Third, we eliminated outliers by calculating the cosine similarities between data items after recalculating TF–IDF values with the 924 word-features that represent "smart service systems". The pairwise similarity comparison matrix (i.e., 5,394 x 5,394 matrix) was derived and the mean of vectors was calculated, with smaller mean value indicating less similarity. We then checked the least similar data (articles) by manually reading the content of each article, deleting those that were not relevant. Sixteen outliers were eliminated in total, resulting in a core vector space with 5,378 articles and 920 word-features.

After finding the core vector space, we examined the data through descriptive analyses, such as representative word analysis (e.g., ranking the 920 words and visualising word clouds), using several relevance measures (see Appendix A in Lim and Maglio 2018) and the association rule mining (Agrawal and Srikant 1994). We then performed unsupervised machine learning, network analysis, and factor analysis. First, we performed spectral clustering (Von Luxburg 2007) to identify key research topics of smart service systems. Spectral clustering is based on graph partitioning and uses the Laplacian matrix derived from a similarity matrix of data. The graph-partitioning problem in spectral clustering is an NP-hard problem, requiring use of a heuristic algorithm, meaning that the clustering result and the score of the mean silhouette coefficient (Rousseeuw 1987) change with each run. We checked the average of 10 iterations; Appendix A in Lim and Maglio (2018) shows that the average values based on cosine and Euclidean distance are high when the number of clusters is 14 and 16. We compared both cases by checking the representative words determined by the NMF algorithm (Lin 2007). We also reviewed data of each cluster for manual evaluation. Finally, we arrived at an optimal number of 16 clusters.

In interpreting the 16 clusters, we first identified the top 100 word-features with clear differences among the clusters. These 100 features include the words of a specific application area of smart service system (e.g., "city" and "home") and the general words used for a smart service system (e.g., "cloud", "scheme", and "sensor"). This simple analysis implies that the clusters may represent topics such as "smart city", "smart home", "cloud computing", and "sensing". We then used topic modelling algorithms, namely, NMF and LDA, to identify sets of top words that represent each cluster. We also reviewed the contents of the data of each cluster during interpretation, as well as the data sources (e.g., journals and proceedings) of each cluster to assess the homogeneity of the cluster; a cluster may be homogeneous if the data discuss a similar topic and the data sources are very much correlated.

We also used a visualisation method adapted from Longabaugh (2012) to interpret the result of spectral clustering (see Figure 17.3). The left side of Figure 17.3 shows a binary adjacency matrix. Each cluster is highlighted by a square border. The density of the area represents homogeneity of the cluster, whereas the size indicates the number of items in the cluster. Thus, the cluster may indicate a broad topic if the density is low and the size is large. Opposite results are found for specific topics. The homogeneity and size were considered when each cluster was interpreted and named based on its top representative words determined by the NMF and LDA

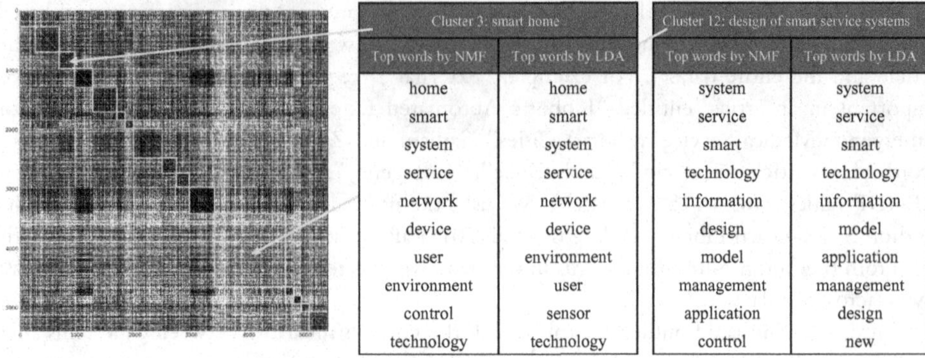

Cluster 3: smart home	
Top words by NMF	Top words by LDA
home	home
smart	smart
system	system
service	service
network	network
device	device
user	environment
environment	user
control	sensor
technology	technology

Cluster 12: design of smart service systems	
Top words by NMF	Top words by LDA
system	system
service	service
smart	smart
technology	technology
information	information
design	model
model	application
management	management
application	design
control	new

Figure 17.3 Interpretation of clustering result

algorithms and five metrics for assessing their relative importance.[4] The right side of Figure 17.3 illustrates the interpretation and naming. We also performed a network analysis and an exploratory factor analysis on the clustering result for further interpretation, which will be introduced in the following section in detail.

The analysis of news article data was conducted in a similar manner. In this case, the core vector space contained 1,234 data (rows) and 256 features (columns). Unsupervised machine learning was applied to the vector space, similar to the scientific data. After examining and interpreting the clusters in detail, 55 areas of smart service system application were identified.

Findings

This section describes the findings of our data analysis. First, we describe the basic attributes of smart service systems identified by the statistical identification of keywords of smart service systems and their relationships. Then, we describe 16 research topics related to smart service systems, and their relationships, that were identified by clustering the scientific literature data. Afterwards, we propose four factors that constitute a smart service system along with a definition of "smart service system" based on the factors identified by analysis of seven clusters representing generic aspects of smart service systems. We then further describe 13 application areas consisting of 55 sub-areas related to smart service systems and their ecosystems based on clustering of news data.

Basic attributes of smart service systems

We rank the 920 word-features from the scientific literature on smart service systems using scaled standardised values of five relevance evaluation metrics (see note 4). Figure 17.4 shows the words – word clouds – we identified for smart service systems based on overall score; the right side shows the word cloud with the full list of 920 word-features, and the left shows the top 100 words. The figure suggests that our work captures meaningful keywords. Table 17.3

4 These are: (1) mean of the TF–IDF scores of a word-feature across data; (2) mean of the cosine similarities of a word-feature to other features; (3) cosine similarity between a word-feature and the centroid of features; (4) mean of the dot product scores of a word-feature to other features; (5) latent factor weight score of a word-feature in NMF for the single topic of entire dataset. For more details, see Appendix A of Lim and Maglio (2018).

Word cloud with the full list of top 100 words

Word cloud with the full list of 920 words

Figure 17.4 Word clouds of the smart service system literature measured based on the five metrics

Table 17.3 A number of top rules in the result of association rule mining sorted by the lift value

Rank	*LHS*	=>	*RHS*	*Support*	*Confidence*	*Lift*
1	{smart, user}	=>	{service}	0.284	0.936	1.069
...	...	...	...	...	...	...
9	{device, smart}	=>	{service}	0.234	0.916	1.046
...	...	...	...	...	...	...
13	{communication, smart}	=>	{service}	0.228	0.911	1.040
...	...	...	...	...	...	...
19	{network, smart}	=>	{service}	0.319	0.909	1.038
...	...	...	...	...	...	...
22	{information, smart}	=>	{service}	0.282	0.908	1.037
...	...	...	...	...	...	...
29	{control, system}	=>	{smart}	0.203	0.958	1.030
...	...	...	...	...	...	...
45	{data, system}	=>	{service}	0.270	0.889	1.015
...	...	...	...	...	...	...
50	{new}	=>	{service}	0.247	0.887	1.013
...	...	...	...	...	...	...
53	{application, smart}	=>	{service}	0.299	0.887	1.012
...	...	...	...	...	...	...

shows the result of association rule mining on the 920 word-features. The support, confidence, and lift values represent the degree of association between the words in the left-hand side (LHS) and right-hand side (RHS). The rows of Table 17.3 are ordered by the lift value, showing a dependency between the two sides. Table 17.4 shows the top 10 research areas in Web

Table 17.4 Top 10 research areas and journals related to smart service systems

Web of Science research area	*Frequency*	*Source*	*Frequency*
Engineering, electrical, & electronic	2117	*IEEE Transactions on Smart Grid*	74
Telecommunications	1316	*Journal of Medical Systems*	52
Computer science, information systems	1257	*International Journal of Distributed Sensor Networks*	52
Computer science, theory, & methods	1148	*IEEE Transactions on Consumer Electronics*	51
Computer science, artificial intelligence	736	*Wireless Personal Communications*	48
Computer science, hardware, & architecture	602	*IEEE Communications Magazine*	35
Computer science, software engineering	444	*Renewable & Sustainable Energy Reviews*	34
Computer science, interdisciplinary applications	377	*2012 IEEE Power and Energy Society General Meeting*	24
Energy & fuels	345	*Personal and Ubiquitous Computing*	23
Automation & control systems	317	*Multimedia Tools and Applications*	22

of Science and the top 10 sources (i.e., journals, proceedings, or books) of the 5,378 scientific articles.

These tables and figures show some of the basic attributes of smart service systems. For example, Table 17.3 indicates that a smart service system requires technologies for networking, data and information processing, control, communications, devices, and applications to provide specific functions to system users. Table 17.4 shows relevant research fields (e.g., telecommunications and automation) and specific application areas (e.g., energy and health). Figure 17.4 shows these attributes in more detail; the top words to describe smart service systems include "network", "data", "user", "application", "technology", "information", "device", "grid", "energy", "power", "environment", "communication", "mobile", "management", "sensor", and "control" "ubiquitous", "real-time", and "interaction".

Sixteen research topics related to smart service systems

Figure 17.5 shows 16 research topics related to smart service systems (i.e., 16 clusters of literature data). Some topics represent generic aspects of smart service systems, whereas others address specific application areas. The seven generic and nine application clusters are described in the next two paragraphs. The studies we refer to for each cluster were selected from the top 10 representative items of the cluster, as determined by mean cosine similarity of an item to other items, and reviewed by the authors.

The seven generic topics (clusters) are as follows: Topic 8 (Cluster 8) is "smart service systems in general", which addresses context-awareness (Gu et al. 2005), delivery to users (Dimakis et al. 2009), smart devices and environment (Cho and Yoe 2010; Crotty et al. 2008), and other general characteristics of smart service systems (e.g., ubiquitous computing and pervasive environment). Topic 12 is "design of smart service systems", which addresses components (Gessner et al. 2009), engineering (Lopes and Pineda 2013), design strategy (Kreuzer and Aschbacher 2011), customer perspective (Wünderlich et al. 2015), and other knowledge for the design of smart service systems (e.g., design model, approach, and process). Topic 16 is "sensing", which addresses the user-centric sensor design (Chen et al. 2004), data gathering (Neves et al. 2010), aggregation (Sim et al. 2011), and other issues on sensing and data monitoring in smart service

(1) Smart card

Keywords:
data, card, transit, security, public, user, bus, transport, passenger, secure, access, …

(2) Wireless networks/communications

Keywords:
network, wireless, communication, traffic, radio, mobile, qos, packet, security multimedia, …

(3) Smart home

Keywords:
home, network, device, user, environment, sensor, control, activity, monitoring, interaction, …

(4) Smart energy management

Keywords:
energy, grid, power, building, demand, control, electricity, storage, renewable, market, cost, …

(5) Mobile devices

Keywords:
mobile, device, user, application, phone, location, communication, wireless, content, platform, …

(6) Cloud computing/environment

Keywords:
cloud, computing, application, user, mobile, security, data, resource, infrastructure, platform, …

(7) Smart antenna

Keywords:
antenna, capacity, wireless, mobile, performance, adaptive, communication, cdma, access, signal, …

(8) Smart service systems in general

Keywords:
contextaware, user, environment, computing, space, ubiquitous, pervasive, interaction, data, …

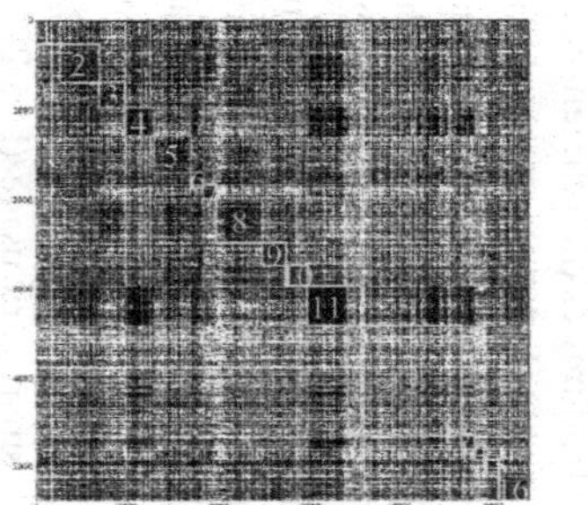

(9) Smart health

Keywords:
health, patient, healthcare, care, medical, data, sensor, monitoring, home, elderly, wearable, …

(10) Internet/Web of Things

Keywords:
web, thing, internet, iot, device, data, application, object, architecture, semantic, communication, …

(11) Smart grid

Keywords:
grid, power, communication, network, distribution, control, energy, generation, electricity, meter, …

(12) Design of smart service systems

Keywords:
technology, information, design, model, application, new, product, development, user, business, …

(13) Use of electric vehicles

Keywords:
ev, vehicle, electric, grid, power, control, energy, station, load, regulation, demand, battery, …

(14) Security

Keywords:
authentication, user, attack, card, rfid, password, key, secure, protocol, anonymity, identity, …

(15) Smart City

Keywords:
city, urban, development, data, public, information, application, infrastructure, citizen, future, iot, …

(16) Sensing

Keywords:
sensor, network, monitoring, wireless, data, bridge, realtime, sensing, device, health, power, …

Figure 17.5 16 research topics of smart service system

systems. Topic 10 is "Internet/Web of Things", which addresses the IoT (Singh et al. 2014), Web of Things (Mainetti et al. 2015), and other issues on the connectivity of objects in smart service systems. Topic 2 is "wireless networks/communications", which addresses traffic control (Malavasi et al. 2003), optimal allocation (Levorato and Mitra 2011), multi-access (Blum et al. 2011), and other issues on wireless networking and communications in smart service systems. Topic 5 is "mobile devices", which addresses mobile phone use (Yada and Naik 2013), information distribution (Noor 2009), location-based functions (Fei et al. 2015), and other issues on mobile devices and networks for smart service systems. Topic 6 is "cloud computing/environment", which addresses cloud computing availability (Lee et al. 2012), security (Getov 2012), authentication mechanism (Kim and Moon 2014), and other issues on infrastructure or platform that enable cloud functions in smart service systems.

The nine application topics (clusters) are as follows: Topic 3 is "smart home", which addresses the elements (e.g., Yu et al. 2011), design (Anbarasi and Ishwarya 2013), trends (Alam et al. 2012), and ecology (Crowley and Coutaz 2015) of home monitoring, control, and automation using specific sensors, devices, and environment. Topic 9 is "smart health", which addresses the management of patient health (Mukherjee et al. 2014), collection and usage of personal health records (Chung and Park 2016), required devices (Goyal et al. 2012) in health care and management systems, and other issues in technology-based health monitoring and care. Topic 4 is "smart energy management", which addresses the efficient use of energy (Colak et al. 2012), distribution of renewable energy (Byun et al. 2011), regional energy management (Cai and Li 2014), system design (Strasser et al. 2015), and other issues pertaining to efficient control of energy demand and consumption. Topic 11 is "smart grid", which addresses aspects of power and communication (Lo and Ansari 2012), security (Amin 2012), services (Chen 2011), trend (El-Hawary 2014) in smart grids, and other issues on energy control through distributed infrastructure network. Topic 13 is "use of electric vehicles", which addresses fleet management (Hu et al. 2016), interactions through smart grids (Mwasilu et al. 2014), business models (San Román et al. 2011), efficient charging (Sbordone et al. 2015), sharing (Lee and Park 2013) of electric vehicles, and other issues on modern vehicle use systems. Topic 14 is "security", which addresses authentication (Lee 2013a), smart card usage (Lee 2013b) in security systems, applications in the health industry (Mishra et al. 2014), and other issues on smart schemes against specific attacks. Topic 15 is "smart city", which addresses the concept (Su et al. 2011), architecture (Anthopoulos and Fitsilis 2014), worldwide applications (e.g., Lim et al. 2018c) of smart cities, and other issues on (public) data-driven digital urban systems and the infrastructure or environment for these systems. Topic 1 is "smart card", which addresses applications and benefits (Bagchi and White 2005; Pelletier et al. 2011) of smart cards. Topic 7 is "smart antenna", which addresses the design (Kawitkar and Shevgaonkar 2003), evaluation (Wong et al. 1998), use (Herscovici and Christodoulou 2001), and other aspects of smart antennas.

Figure 17.6 displays a network of the relationships among the 16 research topics. We first identified the centroids within the clusters for the network analysis and computed the cosine similarities between the centroids. Given that all clusters are highly related, the similarity scores are generally high and all nodes (clusters) are connected; thus, we identified the top 3 most relevant clusters from each cluster and connected only these to observe the most significant relationships between the research topics; size of node represents the network degree (i.e., relationship strength). The nodes in the centre represent generic aspects of smart service systems (e.g., sensing and wireless communications), whereas the nodes in the boundaries pertain to specific application fields (e.g., smart home and energy). The former nodes have strong relationship values (i.e., connected with many other nodes), whereas the latter ones are weaker and connected through the former. The relationship between a specific pair of nodes has

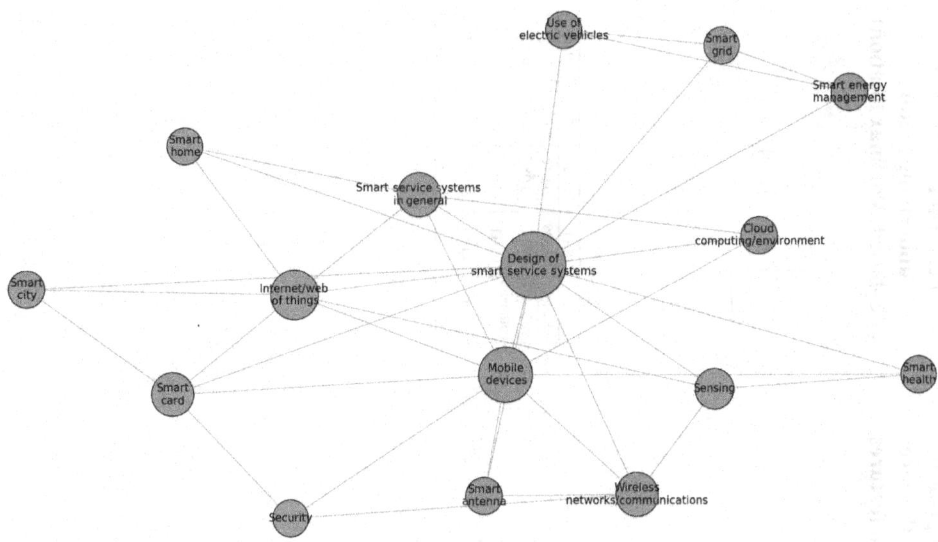

Figure 17.6 Relationship among the 16 topics

implications. For example, "smart grid", "smart energy management", and "use of electric vehicles" are highly relevant.

Factors and definition of smart service systems

Having identified some attributes and research topics of smart service systems, we now ask what technology factors constitute the "core" structure or architecture of a smart service system. What are the factors that constitute the smartness in service systems? Figure 17.7 illustrates our process of identifying the key factors of smart service system. We identified a set of 58 word-features, which are the union of the top 20 word-features in the seven generic clusters using the geometric mean of the standardised five metric values (i.e., overall score). We then excluded "home", "energy", "traffic", "study", and "smart" from the full set of 58 word-features because they may be considered too general or too application-oriented. The remaining 53 generic word-features, such as "user", "thing", "internet", "context-aware", "control", "sensing", "wireless", "location", "interaction", "access", "communication", "computing", "data", and "architecture", may represent the core structure of smart service systems. We took the data of 53 generic word-features from the vector space of 920 features to perform an exploratory factor analysis (Thompson 2004). PCA-based factor identification with varimax rotation enabled us to identify underlying factors. The elbow analysis on the eigenvalue graph in the right part of Figure 17.7 indicated four was an appropriate number of factors. According to the factor loading of the 53 word-features, the four factors are "sensing", "connected network", "context-aware computing", and "wireless communications".

Thus, we propose a data-driven definition of smart service system based on the four factors and the list of 53 generic words: *A smart service system is a service system that controls things for the users based on the technology resources for sensing, connected network, context-aware computing, and wireless communications*. Examples of resources include specific environment, infrastructure, devices, and applications (software). Examples of things to be controlled include specific objects, processes, and users. The definition and examples were derived from a data-driven approach and

53 generic word-features from the union of the top 20 word-features of the seven generic clusters that connect the other nine application clusters

"management", "user", "resource", "thing", "process", "monitoring", "scheme", "new", "infrastructure", "performance", "object", "contextaware", "service", "information", "context", "design", "system", "pervasive", "distributed", "control", "approach", "sensing", "wireless", "framework", "internet", "location", "interaction", "access", "communication", "structure", "computing", "use", "environment", "ubiquitous", "web", "cloud", "sensor", "iot", "space", "semantic", "quality", "application", "network", "platform", "phone", "development", "device", "data", "technology", "model", "structural", "mobile", "architecture"

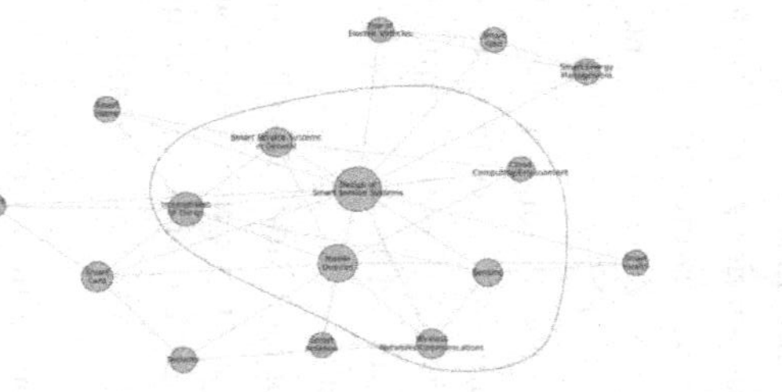

Take the data of 53 generic word-features from the core vector space with 920 features

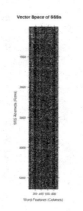

Vector Space of SSSs

Perform EFA to find underlying factors (PC-based, Varimax lotation)

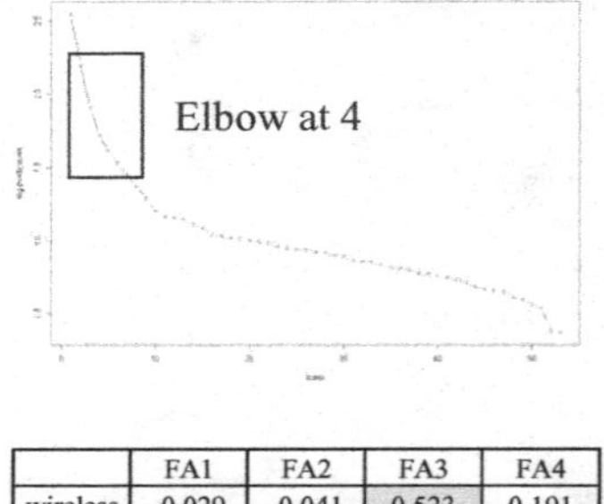

	FA1	FA2	FA3	FA4
wireless	-0.029	-0.041	0.523	0.191
network	0.007	-0.087	0.429	0.109
communi cation	0.011	-0.069	0.331	-0.005
mobile	-0.04	0.154	0.319	-0.217
...	...	...	...	...

Figure 17.7 Factor identification of smart service system

consist of important words identified from the literature data in a statistical way; our findings indicate that representative examples include smart home, energy management, health, and city systems. More examples are introduced in the next section.

Application areas and an ecosystem of smart service systems

This section describes the results of our analysis on the news data. As described previously, 55 clusters were identified from the news data. Cluster interpretation was performed using the same approach applied to the scientific data. We categorised the 55 clusters into 13 areas by analysing the pair-wise cosine similarities between clusters and by reviewing the original 126 keywords used for the collection of news data. We also relied on our reading of each item (i.e., knowledge from the data collection process) in the categorisation. Appendix B in Lim and Maglio (2018) shows the 13 application areas and 55 sub-areas related to smart service systems.

Figure 17.8[5] visualises the relationship among the 55 sub-areas. Network analysis was performed applying the approach used in the case of scientific literature data, except that node size here represents the number of items (i.e., size) of the cluster rather than network degree (strength of relationship). The figure represents an *ecosystem* of smart service systems. Some key implications are as follows. First, smart devices are critical resources that facilitate delivery of various smart service systems to users. Second, smart homes incorporate various technologies positioned at the centre of different smart service systems to make living smarter. Third, a smart service system is related to other systems across different contexts of the users (e.g., smart health in smart home and smart transportation in smart city) and resources (e.g., smart device and environment). Thus, achieving synergy between different smart service systems will effectively streamline the development and operations of the systems.

Discussion

Connection, collection, computation, and communications for co-creation

This section aims to clarify the concept of smart service systems by integrating the four technology factors identified from the scientific literature and findings from existing studies on service systems and smart service systems. Customers create their own value based on offerings from firms (Normann and Ramirez 1993), thus enabling them to apply their competences in the context of firm resources (Prahalad and Ramaswamy 2000). This means that value may be co-created between customers and firms (Prahalad and Ramaswamy 2002) or generally among multiple actors (Lim et al. 2012). As such, service is "the application of competences (knowledge and skills) by one entity for the benefit of another" (Vargo and Lusch 2004), and service systems are "value creation configurations of resources, including people, information, and technology" (Vargo et al. 2008). This fundamental concept of a service system shows that, in smart service systems, resources are integrated for four technology factors (i.e., sensing – collection, connected network, context-aware computing, and wireless communications) to enhance the socio-economic factor (co-creation) for system participants, such as customers and providers.

5 See Figure 8 in Lim and Maglio (2018) for the version in colour.

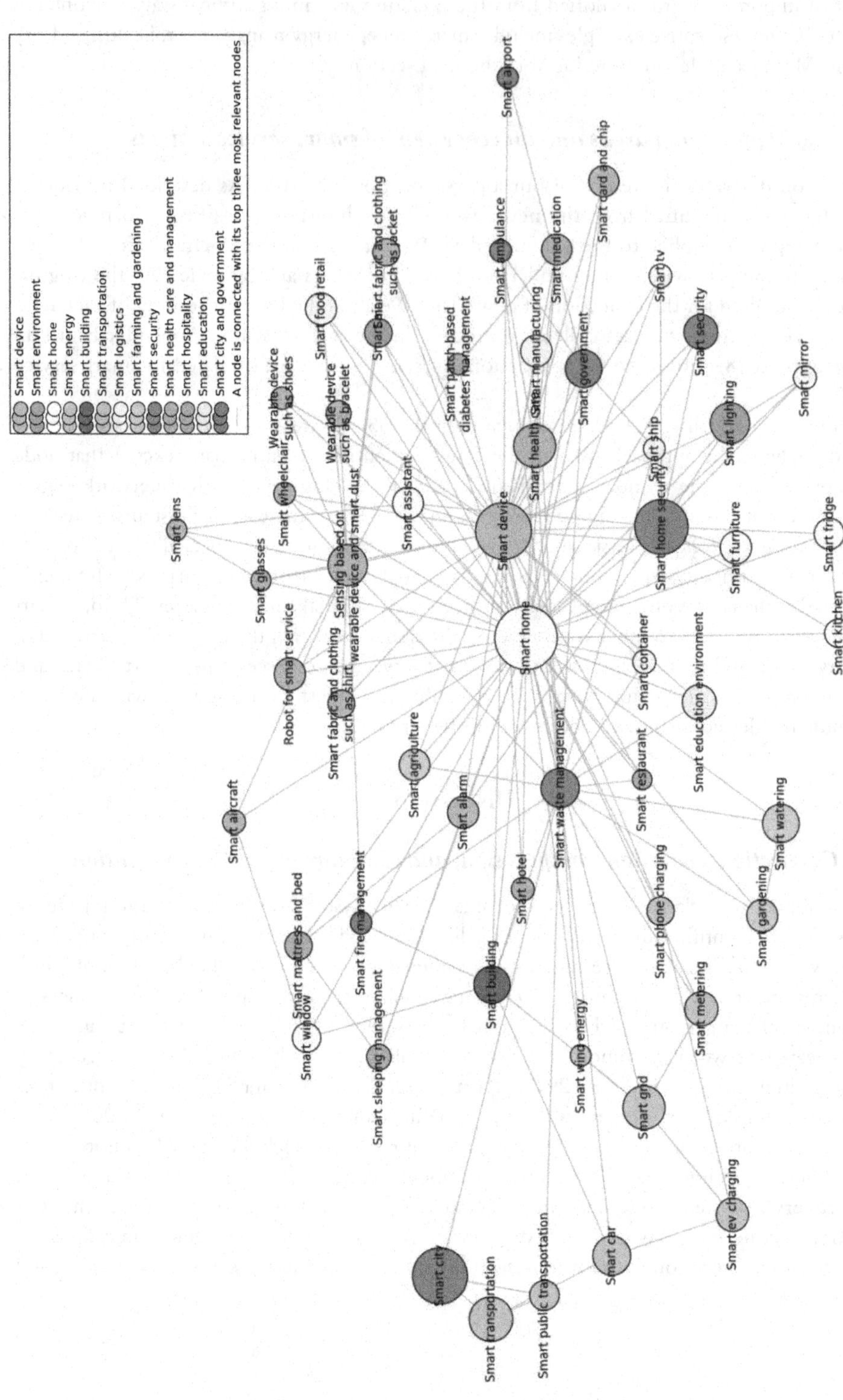

Figure 17.8 Ecosystem of smart service systems

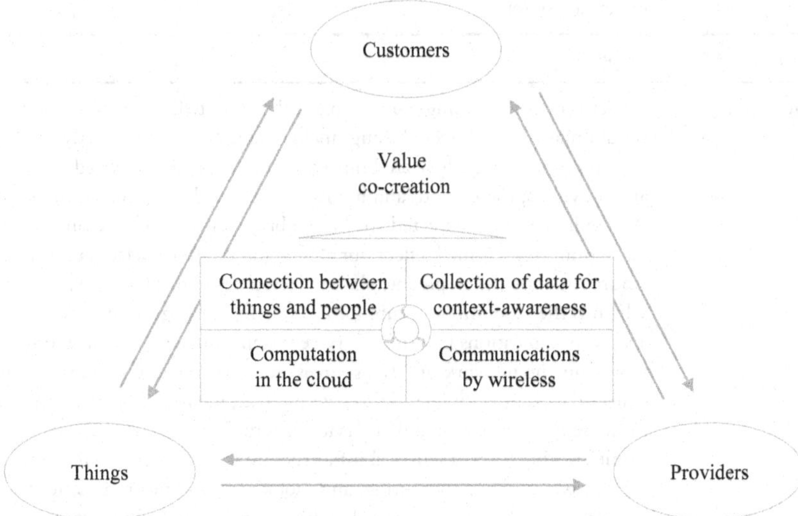

Figure 17.9 Conceptual framework for smart service systems

Thus, we propose that smart service systems may be understood along five dimensions: (1) connection, (2) collection, (3) computation, (4) communications, and (5) co-creation. Figure 17.9 illustrates smart service system mechanisms (i.e., how a smart service system works) based on these "5Cs", and Table 17.5 provides a detailed description of the 5Cs. The lines in Figure 17.9 represents data and information interactions. Smart service systems combine technology resources (e.g., specific device, environment, infrastructure, and software) for the (1) connection of things and people, (2) collection of data for context awareness, (3) computation in the cloud, and (4) communications to automate or facilitate for (5) the value co-creation activities between customers and providers. The circle at the centre represents continuous system control and development, as data and information interactions within a service system are iterative and stakeholders can develop their relationships and improve value co-creation continuously through a cycle of monitoring and learning. This feature shows the importance of service system thinking for using various technologies. The direction of the evolution of smart service systems is clear: continuous development of the value co-creation loop by integrating technologies for connection, collection, computation, and communications.

The first four dimensions represent technological resources of smart service systems, whereas the fifth represents the application objective. The first four dimensions contribute to increasing opportunities for active value co-creation. As we become more connected, encounters for value co-creation increase; as we collect and compute more quality data (quality in terms of variety and volume), the informational or intellectual resources for value co-creation increase; and as we communicate more efficiently and effectively, the frequency and intensity of value co-creation increase. In short, the 5Cs describe theoretically what constitutes "smartness" in modern service systems. The real advantage of considering the full-service system in terms of the 5Cs is that it helps us focus on value creation among stakeholders, which can be facilitated by means of technology. Smart systems co-create value by connecting relevant things and people's concerns, by collecting and computing data, and by communicating with things and people to address concerns. This mechanism applies in describing and developing any type of smart service system (see Table 17.5). Our perspective is consistent with existing definitions of smart

Table 17.5 5Cs of the smart service system

5Cs	*Description*
Connection	Connection between things and people is the first attribute a smart service system should manage. Connected things include tangible goods directly used by customers, as well as dedicated infrastructures generally required by customers and providers; these goods and infrastructures can be connected to other things. We are living in a connected world; the buzzwords "IoT", "Connected Car", or "Connected Home" reflect our ability and desire to better control things around us. The development of a connected network of people and things, which is the base infrastructure for the system, is the groundwork for collection and communications in a smart service system. In fact, a connected network represents the network of "data sources" for smart service systems. Where to collect data is directly relevant to data use (i.e., purpose of service system) and to the scope and potential of a service system. IoT matters because IoT is really about creating a cyber-physical infrastructure for connection. Technologies for data analytics, cloud computing, and mobile communications, among others, can effectively work together only with a connection infrastructure.
Collection	Collection of data from connected people and things is the second attribute of a smart service system. Data include condition traces of engineering systems, event logs of business processes, health and behavioural records of people, and bio-signals of animals. Given our capability for continuous monitoring and learning from data, data are the core resources for context awareness. The term "smart" mainly pertains to information actions rather than to physical or interpersonal actions; hence, this term is inevitably related to the use of data. A major distinction between traditional and recent data collection is the data source (i.e., engineering systems versus human systems). In other words, current sensing methods include physical plus social sensing. In this chapter, physical sensing refers to a process conducted using physical sensors, whereas social sensing includes any type of sensing enabled or conducted by people without using physical sensors. Examples of social sensing include data collection from social network services, surveys, interviews, queries, and documents. Physical and social sensing from things and people within a service system produce data that indicate behaviours and operations of people, operations and condition management of organisations and things, and interactions within a service system. Data use contributes in making these effective and efficient.
Computation	Computation is the third key attribute of smart service systems. Computational processes involve the use of specific algorithms and expert knowledge for decision making. Computation is the prerequisite for data and information communication in a connected network because these processes transform raw data into standardised data or information that enable machine-understandable data or human-understandable information. The key functions of smart service systems, such as context awareness, predictive and proactive operations, adaptation, real-time and interactive decision making, self-diagnosis, and self-control, can be created only through computation on specific data. This often requires several pre-tasks for data analytics, such as analysis planning, data cleaning, anonymisation, aggregation, integration, and storage. Two of the key requirements of computation of smart service systems are cloud computing availability and security because of the distributed nature of connections in a smart service system.

5Cs	Description
Communications	Communications by wireless between people and things is the fourth attribute of a smart service system. The contexts of communications include both machine-to-machine actuation and machine-to-human guidance; thus, the issues of this attribute encompass not only the issues of communications of machine-understandable data but also human-understandable information, such as visualisation methods and other information delivery methods through auditory, olfactory, palate, and tactile stimulation in physical, virtual, and augmented reality. Although the same goods, infrastructures, and stakeholders can be involved in multiple service systems, interactions are relatively unique in each service system. Although technologies for connection, collection, and computing are fulfilled in a specific service system, the key to transforming such a system into a smarter service system or to creating a new smart service system lies in improving the unique interactions within the system in question. As such, the communications technology that facilitates interactions is crucial in any smart service system; the communications technology circulates the blood of the system.
Co-creation	Co-creation of value between customers and the provider of a service system is the fifth attribute of a smart service system. Value creation is the core purpose and central process in economic exchange. Any type of socio-technical service system involves value co-creation that brings different stakeholders together to jointly produce a mutually valued outcome. In this respect, the development and use of technologies ultimately aim for enhanced value creation or for creation of new value. Examples of value co-creation stakeholders include customers of IT goods, manufacturers, government agencies of infrastructure, and application developers. The first four attributes represent the technological resources for smart service systems, whereas the co-creation attribute represents the application objective of various resources. In fact, the first four attributes of smart service systems contribute to increasing the opportunities for active value co-creation. As we become more connected, encounters for value co-creation increase; as we collect and compute more quality data (quality in terms of variety and volume), the informational or intellectual resources for value co-creation increase; and as we communicate more efficiently and effectively, the frequency and intensity of value co-creation increases.

service system (for example, see the definitions of Spohrer 2013 and Barile and Polese 2010). Compared with existing studies on smart service systems, our clarification is significant because it confirms, aggregates, and simplifies insightful but subjective expert perspectives *based on data.*

Table 17.6 shows that the 5Cs are useful in describing the main characteristics of the 11 types of smart service systems categorised from the news data analysis. Describing smart service systems using the 5Cs provides a basis for interconnecting different fields with emphasis on applications. Recent concepts, such as IoT, big data management, AI, cloud computing, blockchain, and wearable devices, are related to smart service systems, and each corresponds to one or more system attributes. For example, wearable devices, such as smartphones, wristbands, and watches, which serve mainly as data collection and information delivery channels, are related to collection and communications. AI is linked to computation, whereas IoT and blockchain relate more to connection. Moreover, each of the research fields related to smart service systems, such as electronic engineering on connection and collection, computer science and industrial

Table 17.6 Main characteristics of the 11 types of smart service systems

Smart service system	*What the service system automates or facilitates (according to the 5Cs)*
Smart home	Value co-creation activities of residents and related stakeholders through in-home or home-around connectivity, collection of living-related data, computation for context awareness, and wireless communications within or through a technology-equipped house
Smart energy	Value co-creation activities of energy users, producers, and other stakeholders through connectivity, collection of energy operations data, computation to optimise energy usage, and communications between machines, facilities, etc.
Smart building	Value co-creation activities of building occupants, managers, and other stakeholders through connectivity, collection of work-related and building operations data, computation for comfort and performance optimisation, and communications within or through a technology-equipped building
Smart transportation	Value co-creation activities of drivers, riders, and other stakeholders through connectivity between vehicles, roads, and other infrastructures; collection of vehicle operations and health data; computation for safety and efficiency, and communications between vehicles, people, etc.
Smart logistics	Value co-creation activities of manufacturers, distributors, and other stakeholders through connectivity between facilities, vehicles, and goods; collection of production and logistics data; computation for optimal operations management; and communications between facilities, vehicles, people, etc.
Smart farming	Value co-creation activities of farmers, agriculture companies, and other stakeholders through connectivity between living properties and farming equipment, collection of condition and environment data, computation for optimal health management, and communications within or through a technology-equipped farm
Smart security	Value co-creation activities of property owners and protectors by the connectivity, collection of property condition and environment data, real-time computation for surveillance, and real-time communications between the stakeholders
Smart health	Value co-creation activities of patients, healthy people, healthcare providers, and other stakeholders through connectivity between people, devices, and healthcare environment; collection of health-related data; computation for diagnosis and prognosis; and communications within or through technology-equipped people, living, and care environment
Smart hospitality	Value co-creation activities of guests and service providers through connectivity between people and service environment, collection of stay-related data, computation for context awareness, communications within or through technology-equipped hospitality environment
Smart education	Value co-creation activities of students, teachers, and other stakeholders through connectivity between people, devices, and education environment; collection of study-related data; computation for maximal learning and satisfaction; communications within or through technology-equipped education device and environment
Smart city and government	Value co-creation activities of citizens, public infrastructures, government agencies, and other stakeholders through connectivity among people and organisations, collection of data for public purposes, computation for optimal administration and living conditions of citizens, and communications between stakeholders

engineering on computation and communications, and marketing and business on co-creation, may focus on one or some of the system attributes and seek synergy with different fields related to other attributes.

As with any large-scale initiative for change, the transition to smarter service systems is difficult. Nevertheless, a strategy toward smarter service systems can be developed based on the 5Cs. First, the native value co-created through the service system in question must be identified because it represents the mission or target of technology development and use. Second, connectivity within the service system must be improved in terms of scope, size, depth, or trust because the connected network represents the capability and potential of the service system. Third, the right kind of data must be collected through physical and social sensing. Moreover, the quality of the data must be managed because data content is directly connected to the scope and potential of the service system. Fourth, data must be integrated and analysed through computation using appropriate data analytics processes, because the computing method determines the smartness and accuracy levels of the service system; the creation of information from data is directly related to the value and attractiveness of any service system. Fifth, communications between machines and humans must be improved to facilitate easy and spontaneous data collection as well as to increase the efficiency and acceptability of information delivery to customers.

We do not propose the 5Cs as the exhaustive list of dimensions of smart service systems. Researchers may identify different dimensions from different studies. Human actions in service systems can be categorised as informational, physical, and interpersonal (Apte and Mason 1995). The automation of such actions, which has evolved from automated teller machines of banking services to warehouse robots of shipping services, has made these services smarter (Maglio and Lim 2016). We think the current status of most smart service systems is mainly at the level of the automation of informational actions. The emergence of autonomous service systems (Maglio 2017) with the automation of physical actions and interactions, such as self-driving cars and fully automated buildings, may force us to extend our 5Cs of smart service systems to include *control*, a 6th "C" (Maglio and Lim 2018). And automation of interpersonal (emotional) actions in the distant future may force us to extend further with *care*, a 7th "C".

Hierarchical structure of smart service system applications

The 13 application areas in the previous section can be distinguished according to the type of application. Smart device and environment are resource-type areas, which are required in any kind of smart service system. Smart home, energy, building transportation, logistics, farming and gardening, security, health care and management, hospitality, and education are business system-type areas. Smart city and government systems are a public administration-type area. As shown in Appendix B in Lim and Maglio (2018), common keywords of the 55 subareas include "device", "product", "app", "data", and "information". This list implies that the essence of smart service system applications, which have been discussed in various news articles, is to use a device or product with a smartphone application to collect data from people and objects and to deliver information to them.

Based on the categorisation of the 13 areas and the conceptual framework in the previous subsection, Figure 17.10 shows a hierarchical structure of smart service system applications. Things, companies, and customers are connected by using smart devices and environment, which are key resources that collect data and facilitate the delivery of various smart business systems to customers. The stakeholders then co-create value through smart business systems. Smart cities and governments comprise 10 types of smart business systems, namely, smart home, health, transportation, energy, building, logistics, farming, security, hospitality, and education service systems (Lim et

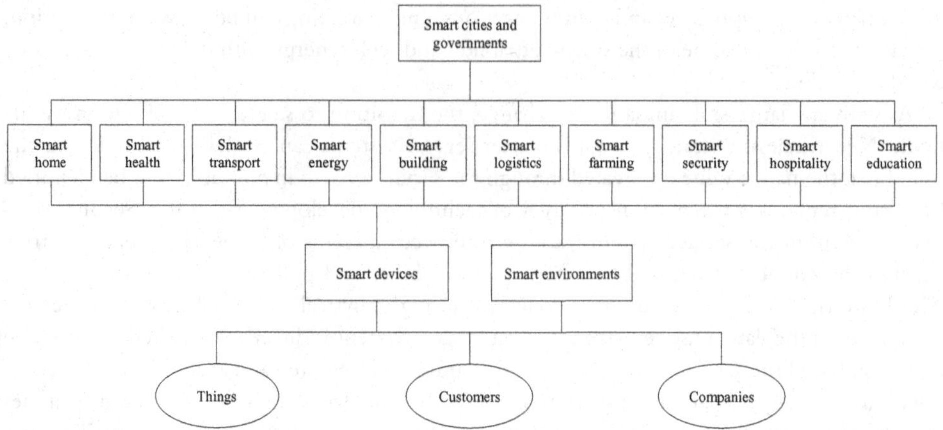

Figure 17.10 Hierarchical structure of smart service system applications

al. 2021). Figures 17.9 and 17.10 are useful for describing a smart service system; for example, we can define a smart home as a service system that automates or facilitates value co-creation activities (e.g., lighting, cooking, temperature control, garage opening, and exercising) between residents and related stakeholders through connectivity enabled by in-home or home-around devices and environment, collection of living-related data, computation for context awareness, and wireless communications achieved within or through a technology-equipped house.

Concluding remarks

Our work advances understanding of smart service systems by mapping dispersed knowledge from scientific literature and news data to achieve a more systemised and integrated conceptualisation. From a research methodology perspective, our study provides a successful example of the collection, analysis, and interpretation of large text databases for understanding a service research topic. This study is unique because we aimed to minimise subjectivity in processing and aggregating text about smart service systems. In doing so, our work identifies key research and application areas of smart service systems and recognises the multi-dimensional smartness (i.e., the 5Cs) of modern service systems. Such findings are consistent with the information we can read from existing studies or news. For example, the journal names and research areas in Table 17.4 are highly relevant to our findings. This fact implies that our findings from a semi-automated analysis of big text data make sense, as it naturally and clearly reflects the existing structure of smart service system research and application. As such, our contribution is to aggregate and confirm the key concepts and areas of broad studies and applications of smart service systems based on data.

Nonetheless, we see several limitations to our work that can be addressed in future research. First, our findings depend on data, which means the results may change if the source and time of data collection change. Likewise, different search keywords could be used for news data collection. We collected literature and news data from May to June 2016; data in 2021 will undoubtedly show further research topics (in particular virtual reality and blockchain), application areas, and factors of smart service system. Nevertheless, we think that our work provides a comprehensive picture of smart service systems in the early formation phase of this important research area.

Second, our work did not provide detailed information of each research topic, application area, or factor of smart service systems. A systematic review of existing literature for a specific topic may be valuable to analyse smart service systems in more detail. The analysis of abstracts, titles, and keywords was appropriate to achieve our research objective (i.e., examining wide studies on smart service system and developing a high-level concept of the entire literature), but full text analysis may be recommendable if the research objective is focused on a specific topic.

Third, although this study analysed the broad aspects of smart service systems, an analysis of user perspectives on smart service systems is missing. Surveys or interviews with users may be done in the future to further develop our understanding of smart service systems. Patent data can be analysed to do an in-depth analysis of technological aspects of smart service systems. Other types of data on smart service systems, such as company profiles, can be used to investigate other aspects of smart service systems.

Finally, the findings obtained from text mining should be incorporated in real research and development projects related to smart service systems. We have conducted such projects with the industry and the government, such as those about smart cars and transportation, health, and building systems (e.g., Lim and Kim 2014; Lim et al. 2015; Kim et al. 2018; Lim et al. 2018a, 2018b; Lim et al. 2019; Lee et al. 2021a; 2021b). An integration of the current study and such projects will further facilitate the development of smart service systems.

We believe our work is important because an intensive discussion has been going on among researchers about the definition of "smartness" in modern socio-economic systems. We hope our work will bring some clarification and elaboration to this issue. In addition, our data-driven research method can be used in future studies to understand other emerging technology paradigms.

References

Agarwal, Y., Balaji, B., Gupta, R., Lyles, J., Wei, M. and Weng, T. (2010): Occupancy-driven Energy Management for Smart Building Automation, *Proceedings of the 2nd ACM Workshop on Embedded Sensing Systems for Energy-Efficiency in Building* (ACM), 1–6.

Aggarwal, C. C. and Zhai, C. (2012): A Survey of Text Clustering Algorithms, in Aggarwal, C. C. and Zhai, C. (eds.): *Mining Text Data.* Boston: Springer, 77–128.

Agrawal, R. and Srikant, R. (1994): Fast Algorithms for Mining Association Rules, *Proceedings of 20th International Conference on Very Large Databases* (VLDB) 1215, 487–499.

Alam, M. R., Reaz, M. B. I. and Ali, M. A. M. (2012): A Review of Smart Homes: Past, Present, and Future, *IEEE Transactions on Systems, Man, and Cybernetics, Part C* 42(6), 1190–1203.

Allmendinger, G. and Lombreglia, R. (2005): Four Strategies for the Age of Smart Services, *Harvard Business Review* 83(10), 131.

Amin, S. M. (2012): Smart Grid Security, Privacy, and Resilient Architectures: Opportunities and Challenges, *Proceedings of 2012 IEEE Power and Energy Society General Meeting* (IEEE), 1–2.

Anbarasi, A. and Ishwarya, M. (2013): Design and Implementation of Smart Home Using Sensor Network, *Proceedings of International Conference on Optical Imaging Sensor and Security* (IEEE), 1–6.

Anthopoulos, L. and Fitsilis, P. (2014): Exploring Architectural and Organizational Features in Smart Cities, *Proceedings of 16th International Conference on Advanced Communication Technology* (IEEE), 190–195.

Antons, D. and Breidbach, C. (2017): Big Data, Big Insights? Advancing Service Innovation and Design with Machine Learning, *Journal of Service Research*, Online First, 1–23.

Apte, U. M. and Mason, R. O. (1995): Global Disaggregation of Information-Intensive Services, *Management Science* 41(7), 1250–1262.

Atzori, L., Iera, A. and Morabito, G. (2010): The Internet of Things: A Survey, *Computer Networks* 54(15), 2787–2805.

Austin, T. (2009): *The Disruptive Era of Smart Machines Is Upon Us.* Gartner, Inc.

Bagchi, M. and White, P. R. (2005): The Potential of Public Transport Smart Card Data, *Transport Policy* 12(5), 464–474.

Barile, S. and Polese, F. (2010): Smart Service Systems and Viable Service Systems: Applying Systems Theory to Service Science, *Service Science* 2(1), 21–40.

Bird, S., Klein, E. and Loper, E. (2009): *Natural Language Processing with Python*. Sebastopol, CA: O'Reilly Media, Inc.

Blei, D. M., Ng, A. Y. and Jordan, M. I. (2003): Latent Dirichlet Allocation, *Journal of Machine Learning Research* 3, 993–1022.

Blum, N., Yamada, J., Fukayama, A., Magedanz, T. and Uchida, N. (2011): A Smart Information Sharing Architecture in a Multi-access Network, Multi-service Environment, *Proceedings of 2011 IEEE International Conference on Communications* (IEEE), 1–6.

Byun, J., Hong, I., Kang, B. and Park, S. (2011): A Smart Energy Distribution and Management System for Renewable Energy Distribution and Context-aware Services Based on User Patterns and Load Forecasting, *IEEE Transactions on Consumer Electronics* 57(2), 436–444.

Cai, X. and Li, Z. (2014): Regional Smart Grid of Island in China with Multifold Renewable Energy, *Proceedings of 2014 International Power Electronics Conference* (IEEE), 1842–1848.

Carrubbo, L., Bruni, R., Cavacece, Y. and Moretta Tartaglione, A. (2015): *Service System Platforms to Improve Value Co-creation: Insights for Translational Medicine, System Theory and Service Science: Integrating Three Perspectives in a New Service Agenda*. Napoli: Giannini Editore.

Chen, T. M. (2011): Services in Smart Grids, in Prasad, A. R., Buford, J. F. and Gurbani, V. K. (eds.): *Advances in Next Generation Services and Service Architectures*. Aalborg: River Publishers, 65–79.

Chen, Y. F., Wu, W. J., Chen, C. K., Wen, C. M., Jin, M. H., Gau, C. Y., Chang, C. C. and Lee, C. K. (2004): Design and Implementation of Smart Sensor Nodes for Wireless Disaster Monitoring Systems, *Proceedings of SPIE 5391* (International Society for Optics and Photonics), 798–805.

Cho, Y. and Yoe, H. (2010): A Smart Service Model Using Smart Devices, in Kim, T., Lee, Y., Kang, B. H. and Ślęzak, D. (eds.): *Future Generation Information Technology. FGIT 2010. Lecture Notes in Computer Science*. Berlin, Heidelberg: Springer, 661–666.

Chung, K. and Park, R. C. (2016): PHR Open Platform Based Smart Health Service Using Distributed Object Group Framework, *Cluster Computing* 19(1), 505–517.

Colak, I., Wilkening, H., Fulli, G., Vasiljevska, J., Issi, F. and Kaplan, O. (2012): Analysing the Efficient Use of Energy in a Small Smart Grid System, *Proceedings of 2012 International Conference on Renewable Energy Research and Applications* (IEEE), 1–4.

Crotty, M., Taylor, N., Williams, H., Frank, K., Roussaki, I. and Roddy, M. (2008): A Pervasive Environment Based on Personal Self-improving Smart Spaces, in Gerhäuser, H., Hupp, J., Efstratiou, C. and Heppner, J. (eds.): *Constructing Ambient Intelligence*. Berlin Heidelberg: Springer, 58–62.

Crowley, J. L. and Coutaz, J. (2015): An Ecological View of Smart Home Technologies, *Proceedings of European Conference on Ambient Intelligence* (Springer International Publishing), 1–16.

Dimakis, N., Gkekas, G., Karame, G., Karachristos, T., Tsolakidis, S., Soldatos, J. and Polymenakos, L. (2009): Facilitating Human-centric Service Delivery Using a Pluggable Service Development Framework, *International Journal of Ad Hoc and Ubiquitous Computing* 4(3–4), 223–236.

El-Hawary, M. E. (2014): The Smart Grid: State-of-the-art and Future Trends, *Electric Power Components and Systems* 42(3–4), 239–250.

Fei, N., Zhuang, Y., Gu, J., Cao, J. and Yang, L. (2015): Privacy-preserving Relative Location Based Services for Mobile Users, *China Communications* 12(5), 152–161.

Frost, R. and Lyons, K. (2017): Service Systems Analysis Methods and Components: A Systematic Literature Review, *Service Science* 9(3), 219–234.

Gavrilova, T. and Kokoulina, L. (2015): Smart Services Classification Framework, *FedCSIS Position Papers*, 203–207.

Gessner, T, Baum, M., Gessner, W. and Lugert, G. (2009): Smart Integrated Systems: From Components to Products, in Zhang, G. Q. and Roosmalen, A. (eds.): *More than Moore*. London, New York: Springer, 33–62.

Getov, V. (2012): Security as a Service in Smart Clouds – Opportunities and Concerns, *Proceedings of 2012 IEEE 36th Annual Computer Software and Applications Conference* (IEEE), 373–379

Geum, Y., Jeon, H. and Lee, H. (2016): Developing New Smart Services Using Integrated Morphological Analysis: Integration of the Market-Pull and Technology-Push Approach, *Service Business* 10(3), 531–555.

Google (2017a): Ranking. Available at: https://support.google.com/news/publisher/answer/68292?hl=en [30.04.2021]

Google (2017b): Google Algorithm Change History. Available at: https://moz.com/google-algorithm-change [30.04.2021]

Goyal, D., Bhaskar, J. and Singh, P. (2012): Designing the Low Cost Patient Monitoring Device (LCPMD) & Ubiquitous Based Remote Health Monitoring and Health Management System using Tablet PC, *Proceedings of 2012 2nd IEEE International Conference on Parallel Distributed and Grid Computing* (IEEE), 7–11.

Gu, T., Pung, H. K. and Zhang, D. Q. (2005): A Service-Oriented Middleware for Building Context-aware Services, *Journal of Network and Computer Applications* 28(1), 1–18.

Heo, J., Lim, C. and Kim, K. (2017): Scales for Measuring Mobile Service Quality: A Literature Review and Identification of Key Dimensions, *International Journal of Services and Operations Management* 27(4), 524–548.

Herscovici, N. and Christodoulou, C. (2001): Potentials of Smart Antennas in CDMA Systems and Uplink Improvements, *IEEE Antennas and Propagation Magazine* 43(5), 172–177.

Hu, J., Morais, H., Sousa, T. and Lind, M. (2016): Electric Vehicle Fleet Management in Smart Grids: A Review of Services, Optimization and Control Aspects, *Renewable and Sustainable Energy Reviews* 56, 1207–1226.

Iyoob, I., Zarifoglu, E. and Dieker, A. B. (2013): Cloud Computing Operations Research, *Service Science* 5(2), 88–101.

Jacobsen, R. H. and Mikkelsen, S. A. (2014): Infrastructure for Intelligent Automation Services in the Smart Grid, *Wireless Personal Communications* 76(2), 125–147.

Jansen, B. J., Zhang, M., Sobel, K. and Chowdury, A. (2009): Twitter Power: Tweets as Electronic Word of Mouth, *Journal of the Association for Information Science and Technology* 60(11), 2169–2188.

Jo, Y., Lagoze, C. and Giles, C. L. (2007): Detecting Research Topics via the Correlation between Graphs and Texts, *Proceedings of the 13th ACM SIGKDD International conference on Knowledge Discovery and Data Mining* (ACM), 370–379.

Kawitkar, R. S. and Shevgaonkar, R. K. (2003): Design of Smart Antenna Testbed Prototype, *Proceedings of 6th International Symposium on Antennas, Propagation and EM Theory* (IEEE), 299–302.

Kim, K., Kim, K., Lim, C. and Heo, J. (2018): Development of a Lifelogs-Based Daily Wellness Score to Advance a Smart Wellness Service, *Service Science* 10(4), 408–422.

Kim, J. M. and Moon, J. K. (2014): Secure Authentication System for Hybrid Cloud Service in Mobile Communication Environments, *International Journal of Distributed Sensor Networks*, 1–7.

Kreuzer, E. and Aschbacher, H. (2011): Strategy-Based Service Business Development for Small and Medium Sized Enterprises, in Snene, M., Ralyté, J. and Morin, J.-H. (eds.): *Proceedings of International Conference on Exploring Services Science*. Berlin, Heidelberg: Springer, 173–188.

Kshetri, N. (2018): Blockchain's Roles in Meeting Key Supply Chain Management Objectives, *International Journal of Information Management* 39(2), 80–89.

Larson, R. C. (2016): Commentary – Smart Service Systems: Bridging the Silos, *Service Science* 8(4), 359–367.

Lee, C. and Lim, C. (2021): From Technological Development to Social Advance: A Review of Industry 4.0 through Machine Learning, *Technological Forecasting & Social Change* 167, 120653.

Lee, E. A. (2008): Cyber Physical Systems: Design Challenges, *Proceedings of 2008 11th IEEE International Symposium on Object and Component-Oriented Real-Time Distributed Computing* (IEEE), 363–369.

Lee, J. and Park, G. L. (2013): Planning of Relocation Staff Operations in Electric Vehicle Sharing Systems, in Pan, J.-S., Chen, S.-M. and Nguyen, N. T. (eds.): *Proceedings of Asian Conference on Intelligent Information and Database Systems*. Berlin, Heidelberg: Springer, 256–265.

Lee, S., Park, H. and Shin, Y. (2012) Cloud Computing Availability: Multi-Clouds for Big Data Service, in Lee, G., Howard, D. and Ślęzak, D. (eds.): *Proceedings of International Conference on Hybrid Information Technology*. Berlin, Heidelberg: Springer, 799–806.

Lee, Y. C. (2013a): Smart-card-loss-attack and Improvement of Hsiang et al.'s Authentication Scheme, *Journal of Applied Research and Technology* 11(4), 597–603.

Lee, Y. C. (2013b): Weakness and Improvement of the Smart Card Based Remote User Authentication Scheme with Anonymity, *Journal of Information Science and Engineering* 29(6), 1121–1134.

Lee, C., Kim, S., Jeong, S., Kim, J., Kim, Y., Lim, C. and Jung, M. (2021a): MIND Dataset for Diet Planning and Dietary Healthcare with Machine Learning: Dataset Creation using Combinatorial Optimization and Controllable Generation with Domain Experts, Proceedings of the 35th Conference on Neural Information Processing Systems (NeurIPS) Datasets and Benchmarks Track.

Lee, C., Kim, S., Lim, C., Kim, J., Kim, Y. and Jung, M. (2021b): Diet Planning with Machine Learning: Teacher-forced REINFORCE for Composition Compliance with Nutrition Enhancement, Proceedings of the 27th ACM SIGKDD International Conference on Knowledge Discovery & Data Mining, 3150–3160.

Levorato, M. and Mitra, U. (2011): Optimal Allocation of Heterogeneous Smart Grid Traffic to Heterogeneous Networks, *Proceedings of IEEE International Conference on Smart Grid Communications* (IEEE), 132–137.

Lim, C. and Kim, K. (2014): Information Service Blueprint: A Service Blueprinting Framework for Information-intensive Services, *Service Science* 6(4), 296–312.

Lim, C. and Kim, K. (2015): IT-enabled Information-intensive Services, *IT Professional* 17(2), 26–32.

Lim, C., Kim, K., Hong, Y. and Park, K. (2012): PSS Board: A Structured Tool for Product-service System Process Visualization, *Journal of Cleaner Production* 37, 42–53.

Lim, C., Kim, K., Kim, M., Kim, K. and Maglio, P. P. (2018b): From Data to Value: A Nine-factor Framework for Data-based Value Creation in Information-intensive Services, *International Journal of Information Management* 39(2), 121–135.

Lim, C., Kim, K. and Maglio, P. P. (2018c): Smart Cities with Big Data: Reference Models, Challenges, and Consideration, *Cities* 82, 86–99.

Lim, C., Kim, M., Heo, J. and Kim, K. (2015): Design of Informatics-based Services in Manufacturing Industries: Case Studies Using Large Vehicle-related Databases, *Journal of Intelligent Manufacturing*, Online First, 1–12.

Lim, C., Kim, M., Kim, K. and Kim, K. (2018a): Using Data to Advance Service: Managerial Issues and Theoretical Implications from Action Research, *Journal of Service Theory and Practice* 28(1), 99–128.

Lim, C. and Maglio, P. P. (2018): Data-driven Understanding of Smart Service Systems through Text Mining, *Service Science* 10(2), 154–180.

Lim, C., Maglio, P. P., Kim, K., Kim, M. and Kim, K. (2016): Toward Smarter Service Systems through Service-oriented Data Analytics, *Proceedings of 2016 IEEE International Conference on Industrial Informatics*, 1–6.

Lim, J., Choi, S., Lim, C. and Kim, K. (2017): SAO-based Semantic Mining of Patents for Semi-Automatic Construction of a Customer Job Map, *Sustainability* 9, 1386, 1–17.

Lim, C., Kim, M., Kim, K., Kim, K. and Maglio, P. (2019): Customer Process Management: A Framework for Using Customer-related Data to Create Customer Value, *Journal of Service Management* 30(1), 105–131.

Lim, C., Kim, J. and Cho, G. (2021): Understanding the Linkages of Smart-city Technologies and Applications: Key Lessons from a Text Mining Approach and a Call for Future Research, *Technological Forecasting& Social Change* 170, 120893.

Lin, C. J. (2007): Projected Gradient Methods FOR Nonnegative Matrix Factorization, *Neural Computation* 19(10), 2756–2779.

Lo, C. H. and Ansari, N. (2012): The Progressive Smart Grid System from Both Power and Communications Aspects, *IEEE Communications Surveys & Tutorials* 14(3), 799–821.

Longabaugh, B. (2012): Visualizing Adjacency Matrices in Python. Available at: http://sociograph.blogspot.com/2012/11/visualizing-adjacency-matrices-in-python.html [31.03.2021]

Lopes, A. J. and Pineda, R. (2013): Service Systems Engineering Applications, *Procedia Computer Science* 16, 678–687.

Maglio, P. P. (2017): New Directions in Service Science: Value Cocreation in the Age of Autonomous Service Systems (Editorial), *Service Science* 9(1), 1–2.

Maglio, P. P., Kwan, S. J. and Spohrer, J. (2015): Toward a Research Agenda for Human-Centered Service System Innovation, *Service Science* 7(1), 1–10.

Maglio, P. P. and Lim, C. (2016): Innovation and Big Data in Smart Service Systems, *Journal of Innovation Management* 4(1), 11–21.

Maglio, P. P. and Lim, C. (2018): On the Impact of Autonomous Technologies on Human-centered Service Systems, in Vargo, S. L. and Lusch, R. F. (eds.): *Handbook of Service Dominant Logic*, London: SAGE, 689–700.

Maglio, P. P., Vargo, S. L., Caswell, N. and Spohrer, J. (2009): The Service System is the Basic Abstraction of Service Science, *Information Systems and e-business Management* 7, 395–406.

Mainetti, L., Mighali, V. and Patrono, L. (2015): A Software Architecture Enabling the Web of Things, *IEEE Internet of Things Journal* 2(6), 445–454.

Malavasi, F., Breveglieri, M., Vignali, L., Leaves, P. and Huschke, J. (2003): Traffic Control Algorithms for a Multi Access Network Scenario Comprising GPRS and UMTS, *Proceedings of the 57th IEEE Semiannual Vehicular Technology Conference* (IEEE), 145–149.

Mankad, S., Han, H. S., Goh, J. and Gavirneni, S. (2016): Understanding Online Hotel Reviews Through Automated Text Analysis, *Service Science* 8(2), 124–138.

Medina-Borja, A. (2015): Smart Things as Service Providers: A Call for Convergence of Disciplines to Build a Research Agenda for the Service Systems of the Future, *Service Science* 7(1), ii–v.

Mishra, D., Srinivas, J. and Mukhopadhyay, S. (2014): A Secure and Efficient Chaotic Map-based Authenticated Key Agreement Scheme for Telecare Medicine Information Systems, *Journal of Medical Systems* 38(10), 1–10.

Moldovan, D., Copil, G. and Dustdar, S. (2017): Elastic systems: Towards Cyber-physical Ecosystems of People, Processes, and Things, *Computer Standards & Interfaces*, Online First, 1–7.

Mukherjee, S., Dolui, K. and Datta, S. K. (2014): Patient Health Management System Using e-health Monitoring Architecture, *Proceedings of 2014 IEEE International Advance Computing Conference (IEEE)*, 400–405.

Mwasilu, F., Justo, J. J., Kim, E. K., Do, T. D. and Jung, J. W. (2014): Electric Vehicles and Smart Grid Interaction: A Review on Vehicle to Grid and Renewable Energy Sources Integration, *Renewable and Sustainable Energy Reviews* 34, 501–516.

Neves, P. A., Esteves, A., Cunha, R. and Rodrigues, J. J. (2010): User-centric Data Gathering Multichannel System for IPv6-enabled Wireless Sensor Networks, *International Journal of Sensor Networks* 9(1), 13–23.

Noh, H., Song, Y., Park, A. S., Yoon, B. and Lee, S. (2016): Development of New Technology-based Services, *The Service Industries Journal* 36(5/6), 200–222.

Noor, A. (2009): Distributed Java Mobile Information System, *Communications of the IBIMA* 10, 127–132.

Normann, R. and Ramirez, R. (1993): Designing Interactive Strategy, *Harvard Business Review* 71(4), 65–77.

NSF (2016): Partnerships for Innovation: Building Innovation Capacity (PFI:BIC). Available at: www.nsf.gov/funding/pgm_summ.jsp?pims_id=504708 [30.04.2021]

Ordenes, F. V., Theodoulidis, B., Burton, J., Gruber, T. and Zaki, M. (2014): Analyzing Customer Experience Feedback Using Text Mining: A Linguistics-based Approach, *Journal of Service Research* 17(3), 278–295.

Ostrom, A. L., Parasuraman, A., Bowen, D. E., Patricio, L., Voss, C. A. and Lemon, K. (2015): Service Research Priorities in a Rapidly Changing Context, *Journal of Service Research* 18(2), 127–159.

Park, H. and Yoon, J. (2015): A Chance Discovery-based Approach for New Product-service System (PSS) Concepts, *Service Business* 9(1), 115–135.

Patel, M. S., Asch, D. A. and Volpp, K. G. (2015): Wearable Devices as Facilitators, Not Drivers, of Health Behavior Change, *JAMA* 313(5), 459–460.

Pedregosa, F., Varoquaux, G., Gramfort, A., Michel, V., Thirion, B., Grisel, O., Blondel, M., Prettenhofer, P., Weiss, R., Dubourg, V., Vanderplas, J., Passos, A., Cournapeau, D., Brucher, M., Perrot, M. and Duchesnay, E. (2011): Scikit-learn: Machine Learning in Python, *Journal of Machine Learning Research* 12, 2825–2830.

Pelletier, M. P., Trepanier, M. and Morency, C. (2011): Smart Card Data Use in Public Transit: A Literature Review, *Transportation Research Part C: Emerging Technologies* 19(4), 557–568.

Prahalad, C. K. and Ramaswamy, V. (2000): Co-opting Customer Competence, *Harvard Business Review* 78(1), 79–90.

Prahalad, C. K. and Ramaswamy, V. (2002): The Co-creation Connection, *Strategy and Business* 27, 50–61.

Raghupathi, W. and Raghupathi, V. (2014): Big Data Analytics in Healthcare: Promise and Potential, *Health Information Science and Systems* 2(3), 2–10.

Rousseeuw, P. J. (1987): Silhouettes: A Graphical Aid to the Interpretation and Validation of Cluster Analysis, *Journal of Computational and Applied Mathematics* 20, 53–65.

Samani, H. and Zhu, R. (2016): Robotic Automated External Defibrillator Ambulance for Emergency Medical Service in Smart Cities, *IEEE Access* 4, 268–283.

San Román, T. G., Momber, I., Abbad, M. R. and Miralles, A. S. (2011): Regulatory Framework and Business Models for Charging Plug-in electric Vehicles: Infrastructure, Agents, and Commercial Relationships, *Energy policy* 39(10), 6360–6375.

Sbordone, D., Bertini, I., Di Pietra, B., Falvo, M. C., Genovese, A. and Martirano, L. (2015): EV Fast Charging Stations and Energy Storage Technologies: A Real Implementation in the Smart Micro Grid Paradigm, *Electric Power Systems Research* 120, 96–108.

Schumann, J. H., Wunderlich, N. V. and Wangenheim, F. (2012): Technology Mediation in Service Delivery: A New Typology and an Agenda for Managers and Academics, *Technovation* 32(2), 133–143.

Sim, S. H., Carbonell-Marquez, J. F., Spencer, B. F. and Jo, H. (2011): Decentralized Random Decrement Technique for Efficient Data Aggregation and System Identification in Wireless Smart Sensor Networks, *Probabilistic Engineering Mechanics* 26(1), 81–91.

Singh, D., Tripathi, G. and Jara, A. J. (2014): A Survey of Internet-of-Things: Future Vision, Architecture, Challenges and Services, *Proceedings of 2014 IEEE world forum on Internet of Things* (IEEE), 287–292.

Spohrer, J. C. (2013): NSF Virtual Forum: Platform Technologies and Smart Service Systems. Available at: http://service-science.info/archives/3217 [30.04.2021]

Spohrer, J. C. and Demirkan, H. (2015): Introduction to the Smart Service Systems: Analytics, Cognition, and Innovation Minitrack, in *System Sciences (HICSS), 2015 48th Hawaii International Conference on IEEE*, 1442–1442.

Strasser, T., Andren, F., Kathan, J., Cecati, C., Buccella, C., Siano, P., Leitão, P., Zhabelova, G., Vyatkin, V., Vrba, P. and Mařík, V. (2015): A Review of Architectures and Concepts for Intelligence in Future Electric Energy Systems, *IEEE Transactions on Industrial Electronics* 62(4), 2424–2438.

Su, K., Li, J. and Fu, H. (2011): Smart City and the Applications, *Proceedings of 2011 International Conference on Electronics, Communications and Control* (IEEE), 1028–1031.

Thompson, B. (2004): *Exploratory and Confirmatory Factor Analysis: Understanding Concepts and Applications.* American Psychological Association. https://doi.org/10.1037/10694-000

Vargo, S. L. and Lusch, R. F. (2004): Evolving to a New Dominant Logic for Marketing, *Journal of Marketing* 68(1), 1–17.

Vargo, S. L., Maglio, P. P. and Akaka, M. A. (2008): On Value and Value Co-creation: A Service Systems and Service Logic Perspective, *European Management Journal* 26(3), 145–152.

Von Luxburg, U. (2007): A Tutorial on Spectral Clustering, *Statistics and Computing* 17(4), 395–416.

Watanabe, K. and Mochimaru, M. (2017): Expanding Impacts of Technology-assisted Service Systems through Generalization: Case Study of the Japanese Service Engineering Research Project, *Service Science* 9(3), 250–262.

Wong, K. K., Letaief, K. B. and Murch, R. D. (1998): Investigating the Performance of Smart Antenna Systems at the Mobile and Base Stations in the Down and Uplinks, *Proceedings of 1998 Vehicular Technology Conference* (IEEE), 880–884.

Wünderlich, N. V., Heinonen, K., Ostrom, A. L., Patricio, L., Sousa, R., Voss, C. and Lemmink, J. G. (2015): "Futurizing" Smart Service: Implications for Service Researchers and Managers, *Journal of Services Marketing* 29(6/7), 442–447.

Wünderlich, N. V., Wangenheim, F. V. and Bitner, M. J. (2013): High Tech and High Touch: A Framework for Understanding User Attitudes and Behaviors Related to Smart Interactive Services, *Journal of Service Research* 16(1), 3–20.

Yada, K. and Naik, V. S. (2013): Empowering Feature Phones to Build Smart Mobile Networked Systems, *Journal of the Indian Institute of Science* 93(3), 521–540.

Yoon, B. and Park, Y. (2004): A text-mining-based patent network: Analytical tool for high-technology trend, *The Journal of High Technology Management Research* 15(1), 37–50.

Yu, B. and Xu, Z. B. (2008): A Comparative Study for Content-Based Dynamic Spam Classification Using Four Machine Learning Algorithms, *Knowledge-Based Systems* 21(4), 355–362.

Yu, Y. C., Shing-Chern, D. and Tsai, D. R. (2011): Smart Door Portal, *Proceedings of IEEE International Conference on Consumer Electronics* (IEEE), 761–762.

Zhuge, H. and Wilks, Y. (2017): Summarization of Things – Call for Papers, *IEEE Intelligent Systems*, Available at: www.computer.org/intelligent-systems/2016/11/07/summarization-of-things-call-for-papers/ [30.04.2021]

18

SMART CITIES, A SPATIAL PERSPECTIVE

On the "how" of smart urban transformation

Elke Pahl-Weber and Nadja Berseck

Introduction

A smart city represents for many people the promise of high-tech cities, with autonomous cars rolling the streets, drones delivering food, and sensors recording information about air quality, temperature, traffic, and so on. However, this representation does not convey the complexity of this field, where many issues – such as economy, environmental and social sustainability, and democratic participation – are intertwined. Moreover, it does not include any spatial perspective, which has long *not* been the focus of debates on how to develop smart cities. A city is a complex entity with a broad range of very different and dynamically changing spaces. There is no smart city development without a spatial perspective.

Adopting a spatial perspective leads to the question of what the planning process to develop a smart city looks like. Is urban planning a way to establish a smart city? Or are the smart city and its spaces developing despite all planning approaches, just by the doing and making of large ICT companies? Or do we follow the motto of Jane Jacobs, who says, "Cities have the capability of providing something for everybody, only because, and only when, they are created by everybody" (Jacobs 1961)? Best practice examples of smart city development can be found all over the world, and smart city researchers have made major efforts to investigate their design and implementation process (e.g., Angelidou 2017; Datta 2015; Mora et al. 2019). However, even with 30 years of literature on this topic, we still do not know how to explain the process of the success of urban environments in becoming smart. This transition can be seen as one of the most complex and wicked problems of our time, requiring a smart city perspective beyond the technological view.

In this chapter, we focus on the process of "how" to develop smart cities, and not so much on the question of "what" the action fields of a smart city are, mostly understood as action fields of IT and IOT. We follow the holistic smart city literature stream (Nielsen et al. 2019) and agree that for a city to be smart, the combination, connection, and integration of all urban aspects (spatial, technical, institutional, legal, economic, environmental, and social) are essential. Therefore, we argue that cities have to be analysed in a context of complex systems. Moreover, the role of space in the city has to be defined, assuming that the spatial perspective is not limited to including the physical or technological parts of the city, as debates on urban development seem to promote. We conclude that there is a need for a new paradigm in urban planning, which has

 DOI: 10.4324/9780429351921-22

been discussed for more than 20 years: Patsy Healey (1997) published her book on collaborative planning in 1997, Batty (2013) addressed the science of cities, and Richard Sennett (2018) discussed the unstable foundations of urban development. Sennett stated that neither the theories of Weber, Marx, and Engels nor the approaches of the "Great Generation" with Haussmann and Cerda could solve the challenges of how masses behave. They "experimented with the city, encountering as any open experiment does, dead ends and defeats as well as successes" (Sennett 2018: 62). Peter Townsend strives for a new approach to new civics for a smart century:

> We have seen that putting the needs of citizens first isn't only a more just way to build cities. It is also a way to craft better technology and do so faster and more frugally. And giving people a role in the process will ultimately lead to greater success in tackling theory urban problems and greater acceptance in the solutions smart cities will offer.
>
> *(Townsend 2013: 282)*

In other words, the literature addresses the people, the masses – let's call it the human factor. And this human factor is essential for the development of the urban space. Smart cities are going to change urban life, the ways of communication, the modes of production, the use of technology; this also refers to the human factor. As urban space is continuously changing, there is a need for an answer – not just about what smart city development will bring, but also how the process of transition will work in urban spaces. This process requires a multi-stakeholder, human-centred, and place-specific approach to understanding wicked urban issues, while also managing a transition to a more integrated way of treating cities.

One approach to deal with the problem of smart city transition is urban design thinking. It takes human-centred design principles often used in product design and related industries and applies it to the shaping of urban space. Through the study of multiple cases, we aim to show how urban design thinking can be used to address challenges in the smart city paradigm in a collaborative setting of diverse stakeholders, generating novel ideas and innovative urban development approaches that always have a spatial impetus. The method has been developed in TU Berlin by a team that consists mainly of urban planners and economic experts. It has been developed in a series of integrated projects in lectures as well as a series of funded research projects. The authors have always been part of these projects and stand for the development of the method.

In the following, we describe the smart city in the context of complex systems, the role of space in smart cities, and conditions for smart city planning. Then, we explain the method of urban design thinking and show how its application in the context of participatory planning and smart city challenges.

Smart cities as complex systems

Urban development is increasingly attempting to be fostered by applying a diverse range of more or less modern and technological concepts, such as free parking space detection, energy and water synergy parks, smart housing concepts, shared mobility solutions, and digital administration. In fact, a smart city is a massive market, often promoting one-size-fits-all solutions rather than catering to local specificities (Kitchin 2019). According to a Frost and Sullivan report (Frost and Sullivan 2019), this market is estimated at 1.56 trillion USD by 2025. The market is based on assets like algorithms and data that are often owned by large ICT companies, creating dependencies on the part of municipalities, administrations, governments, and society (Dembski et al. 2020: 2–3). It is often uncritically assumed that, when emerging technologies are integrated into the operations of city functions, they inevitably succeed in becoming sustainable or smart (Colding et al. 2019). Indeed, the smart city discourse often reduces questions such as "what kind of city would you like to live in?" to technological aspects (e.g., a city with a smooth traffic flow), instead of addressing more profound questions of equity, justice, collaboration, fairness, or governance). As Kitchin (2019) puts it, one gains the impression that the starting point of most smart city initiatives is technology. The question of what core issue the technological intervention might address is secondary. Put differently, "the means is post-justified by ends, rather than the ends shaping the means" (Kitchin 2019: 7).

Often smart city advocates regard cities as a manageable system (or system of systems) that can be controlled through technical levers in a linear and hierarchical way (Kitchin 2019). However, smart cities should not be described as automatic machines that can be steered like a train or bus. This is a lesson we learned from system theory more than half a century ago (Yang and Yamagata 2020: 3). Cities were first formally considered as "systems", defined as

> distinct collections of interacting entities, usually in equilibrium, but with explicit functions that could enable their control, often in analogy to processes of their planning and management.
>
> *(Berry 1964 in Batty 2013: 14)*

This view treated cities as distinct from their wider, largely benign, environment. Their functioning was mainly dependent on re-establishing their equilibrium through several negative feedbacks (Batty 2012; Chadwick 1971). However, cities are not well behaved in the sense that they return to a previous equilibrium in response to an external shock. Nor are they centrally ordered, and their issues cannot simply by solved by technical solutions; they are products of countless individual and collective decisions, from the bottom up and the top down, that do not conform to any grand plan. Cities are containers of innovation. Indeed, they are the product of inventions; they create surprises and catastrophes. They are full of culture, politics, and competing interests. Thus, it has been realised in the last 70 years that the notion of systems was never correct (Batty 2012, 2013). As the science of cities developed, the complexity sciences became an integral part because they allow us to understand the many angles of a cities that echo their diverse and plural character. They are sufficiently open to embracing many different approaches and to incorporating the idea that none of them is predominant (Miller and Page 2009). Smart cities are complex systems par excellence (Bibri 2018: 310). They consist of interrelated subsystems essential for their functioning and performance, such as land use planning, healthcare, public transport, housing, education, public services, government, water, and many others (Chourabi et al. 2012; Pribyl and Svitek 2015). The described shift in thinking is best imagined as the transition from thinking of "cities as machines" to "cities as organisms" (Batty 2013: 14).

Within these organisms, problems are wicked and tangled (Colding et al. 2019). Housing market bubbles, ethnic segregation, automation of human learning, global pandemics, traffic congestion, and aging infrastructures – these kinds of problems are associated with multiple and diverse stakeholders, high levels of interdependence, varying dynamics, competing objectives and values, and social and political complexity (Batty et al. 2012; Chourabi et al. 2012). They lack a simplistic solution and straightforward planning responses and often come about as "management surprises" (Colding et al. 2019: 512). The analysis of such problems and their identification is crucial to the sustainability of smart cities. This is opposed to the promise within the smart city discourse that, given enough data and enough processing power, we could directly compute solutions to any problem (Rae and Singleton 2015).

Understanding smart cities means analysing the structure of coupled systems and how this structure develops. We suggest that, for example, it is not goal-oriented to address mobility topics independently from housing issues, economic growth, environmental, and further concerns. Cities that manage to create a complex system wherein the effectiveness of one system (e.g., the way the mobility solutions operate) increases the returns from another complementary system (e.g., the way the housing system is organised) are more successful because those structures are more valuable than the sum of separate combinations (Berseck and zu Knyphausen-Aufseß 2018). Vice versa, this means that a failure in one system can lead to cascading failures because of the strong coupling of parts (e.g., energy system and economic system), which may have catastrophic consequences (e.g., environmental crisis) on the functioning of the system as a whole (Buldyrev et al. 2010). Understanding the structure of complex systems and dealing with wicked problems are challenging. We all "know" that the way to solve difficult problems is to break them into their component parts and solve each part in isolation. This approach is ingrained in education and scientific knowledge. However, our ability to think in systems is constraining (Mc Phearson 2017). Psychology research in complex problem-solving suggests that the problem-solver has to (1) build a parsimonious and viable representation of the most important and urgent elements and relations (i.e., prioritise), and (2) then search for a solution based on the representation of the problem. Doing this allows decisions suitable to the situation to be taken instead of following a very sharp-shaped planning process without the possibility of reacting to changing circumstances (Dörner 2020).

To sum up, the smart city approach is not only about the application of smart technologies that automatically pave the way for more democratic ways of planning or result in a range of environmental benefits (Colding et al. 2019). We advocate that, for a city to be "smart", we need to understand it as a complex system full of wicked problems and competing interests rather than a steerable, rational machine (Kitchin 2019). However, tackling complexity and uncertainty in urban systems is challenging. It requires a holistic, and not only technology-led, approach with an integrated intervention logic and new levels of collaboration among urban actors equipped with complex problem-solving competencies.

The space as the place of urban development, smart in technics, actors, and planning approach

The spatial perspective and planning processes need to take a closer look at what the urban space is and how it can be defined. Urban space is more than concrete, glass, steel, and green; it is a melting pot of physical, manmade, and naturally developed material and the way people shape, behave, and destroy this material. Consequently, urban space cannot be considered as something developed by experts and then adapted by citizens. Addressing the development of urban space as a challenge to integrated intervention for the improvement of inhabitants' lives is just and only possible if we have an urban planning toolkit that approaches it in an integrated way, including smart elements, and, if we know what will really improve the quality of life for inhabitants. Thus, their demands are a crucial topic for all urban development, including smart city development.

The story of urban space development begins much earlier than the discipline of urban planning, which is around 100 years old. It can even be looked upon as the struggle between physical and human elements. What space might be is subject to a long-lasting discourse. The topic of space is looked upon as a physical and metaphysical thing, and there is an idea of space as a phenomenon; it is as physical as medial, political as geographical, esthetical as social. The discourse had already started 350 years ago with René Descartes and his work on the principles of material things (Descartes 1644 in Dünne 2006). In combination with dynamic power, between physical and human elements, urban space is also a mirror of social constructs. Henri Lefebvre already explained that the production of space is an overlay of several social spaces that always produce a changing and dynamic picture (Lefebvre and Nicholson-Smith 1991: 87).

This dynamic power of space is a crucial insight for the development of a smart city. Technologies do not drive cities and change them, not exclusively; cities also change through social impetus. Smart city technologies might and are supposed to play a role in these dynamics; however, what it produces is not yet defined. We recognise that a lot of smart technologies in the field of digital communication are already used by people; they are familiar with it, and this development has already changed urban space. However, urban space is changing in a dynamic way, not because smart technologies are applied, but because the space as a social product is changing and has its own dynamics. Consequently, urban space will not be changed in a smart city development without recognising that space is a social product. Social sciences has detected the sociology of spaces for analytical approaches; the spatial turn in social sciences has been discussed for more than a decade. Part of the sociology of spaces is the hypothesis that a capability of synthesis must be attributed to people, in order to have space no longer looked upon as isles (Löw 2001: 268), fragmented places, or archipelagos.

Space as social product is well known in the disciplines of urban planning and planning history as well as in planning theories. Related to this, there are a lot of methods and approaches for analysing urban space. But in urban planning, which is recognised as a technical science, this complex insight is not always present. Planning with the human scale is a very well-known impetus of urban planning, but who defines this human scale? In recent years, a book by Jan Gehl has brought light to this complex element of urban planning (Gehl 2013). The spatial perspective for the development of smart cities can learn from Henri Lefebvre's theories concerning the demands of people; the question is how urban planning as a technical discipline can integrate this into processes of urban development.

The space in urban context is more than the neighbourhood. Smart city developments often address the cities as a whole or even a newly developed part of a city. The interrelations between regions, cities, neighbourhoods, and urban units are part of the technical as well as the social process of the construction of the urban context. This process of constitution is related to a broad range of actors, including citizens. An analysis of the field of actors for the different levels of urban development is a topic of discourse that started more than 20 years ago (Healey 1997). Healey strengthened the role of collaborative planning for political agendas, "collaborative efforts in defining and developing policy agendas and strategic approaches to collective concerns about shared spaces for members of political communities serve to build up social, intellectual and political capital which becomes a new institutional resource"(Healey 1997: 311).

This argument builds on both a normative concern with more people-sensitive modes of governance and a practical concern with the management of local environmental change in situations of multiple, and often conflicting, stakeholders as typical examples of "shared-power worlds" (Healey 1997: 205). Stakeholders' interests in urban development processes are often not clear. When a planning process starts, it is crucial to understand each stakeholders' concern, how they are met, how they conflict, and how they could be brought together. Arguments for including citizens as the most important stakeholder group (their lives shall be improved) is the topic of numerous approaches to citizen participation. Yet, more than participation is needed in a planning process, which is set up by experts, for a real collaborative process to start. We have learned from Richard Sennett (2018) that citizens have the power to be partners in this collaborative process. In his book, *Building and Dwelling: Ethics for the City*, he coins the term "the competent urbanite" (Sennett 2018: 171) and explains through examples that walking in the city and using the city's environment happens on the basis of an "embodied knowledge" (Sennett 2018: 174).

> A stream of consciousness implies awareness of the context – where you are, who is with you, what they or you are doing when you have a particular thought, feeling or sensation. This awareness of context is what embodies a thought: it is matter of sensing the physical circumstances within we are thinking; the thought becomes full of sensate associations. Only when these circumstances change does consciousness begin to stream; it does not flow, as in Descartes, independently, of its own accord.
>
> *(Sennett 2018: 175)*

Knowledge about the place is part of the everyday life of the people; they often have it but cannot express or even have it in mind explicitly and yet have it embodied as part of the consciousness of the "competent urbanites". It comes to light when people begin to talk about their lives; they are not just part of an interview. They also develop a story themselves to recognise implicitly knowing things. Practising dialogue is an ethical practice of communication. This communication encompasses not just mean talking but also practising and changing physical circumstances. This is the point at which the "how" of urban development processes is asked to develop a process able to refer to these insights: a process that makes these competencies available for the planning process. Through this, the spatial perspective is included in the urban development process. This inclusion is not new; the constitution of space as a social product has its tradition (Pahl-Weber 2020: 43), but the process of urban planning has, up to now, not referred to it in an adequate way.

Urban design thinking for planning smart cities

From the previous two sections, we derive the following four conditions for urban planning that enable a smart urban transformation:

1 **Integrated.** Conventional planning tools and methods fail to deal with the complexity of smart cities (Dembski et al. 2020: 4). We need to improve our understanding of the complex problems occurring in today's urban systems and have to analyse what is happening in cities in system terms instead of just planning support for a single policy area or actor. It means that we need tools that help "to connect the dots", i.e., the integration of systems and actors to enable systemwide effects to be tracked, understood, and built into the very responses that characterise the operations and functions of smart cities (Berseck and zu Knyphausen-Aufseß 2018). This requires planning processes that are sensitive to distinct properties of complex systems, such as feedbacks, nonlinearity, unpredictability, and adaptation (Condorelli 2016). Here emerging technologies support integrated planning by connecting data analytics and projecting future scenarios based on urban modelling and simulation of flows (Yang and Yamagata 2020).

2 **Citizen-centred.** We need approaches that allow for participatory and collaborative processes that prioritise people in urban planning. Smart city planning must be citizen-centred rather than driven by the application of new technologies and the generation of business cases thereof, because citizens are the primary beneficiary of smart city initiatives. We need to understand their heterogenous needs, desires, habits, and practices and include local communities to tackle wicked urban issues (Dembski et al. 2020). Whereas in the past, participatory planning was related to a more passive involvement of users, citizens,

or civil society (von Wirth et al. 2019), we advocate for more inclusive and empowering co-creation approaches to design immediate living environments. Such an approach needs urbanite competencies, like Sennett calls them – whether citizens, actors of the scientific world, or experts in certain fields – to deal with urban circumstances and to come up with processes and solutions containing their embodied know-how about the place where they live and work. In every community, there are "experts" who can present valuable perspectives and insights about the area's history, culture, functionalities, or any other aspect that's considered meaningful for the people.

3 **Place-specific.** The special knowledge of urbanites about their neighbourhoods – the local communities' assets, inspiration, and potential – is especially important when planning to integrate new solutions into city systems, in order to make sure that these solutions are actually benefiting, and not hindering, community cohesion and place-making. Planning should not solely follow generic, aspatial, and ahistorical "hypes of the day" (e.g., a new Guggenheim museum) but rather allow for solutions tailored to the identity and indigenous qualities of place (Berseck and zu Knyphausen-Aufseß 2018). Successful place-oriented planning does not begin with a list of elements that appear to have worked well somewhere else. It starts with a process of in-depth research to find underutilised space that can be improved and with identifying the talents of the community (Palermo and Ponzini 2014). The two criteria of citizen-centredness and place-orientation are interrelated; the initiatives of citizens have an effect on the space, and the space has an effect on their initiatives (Jeutner 2020: 29). More specifically, getting people to participate in designing the environment gives the space a significance where people interact. When they interact with space, they form meaning surrounding the space.

4 **Consensus-building.** We need tools that help us to reach consensus among stakeholders. This is due to various reasons, such as different stakeholder interests and priorities; their knowledge, understanding, and interpretation of the problems and solutions; competition for resources; sectoral objectives and limitations; or political gains. In the smart city context, there is still a lot of learning to be done: (1) for city administrations to develop smart city strategies, set up procuring processes, and deploy smart city solutions; (2) for companies to gain a better understanding of how cities are managed and function; (3) for communities living with smart city initiatives; and (4) for researchers seeking to understand the multi-levered processes that are unfolding (Kitchin 2019). Hence, techniques to build consensus among stakeholders belonging to different disciplines and sectors emerge as one of the critical needs in building smart cities in multi-stakeholder development contexts. We assume that learning within this context will progress most effectively through co-creation. It is powerful in urban settings because it supports dialogue while working together towards a shared goal and constructive feedback between all stakeholders.

According to these conditions, smart city planning requires a creative, design-led approach starting from a human perspective, developing solutions that are not only technically feasible and economically viable but also, and foremost, desirable for a specific group in a specific space (Plattner et al. 2015). Design thinking is an approach that allows for dealing with ill-defined, wicked problems within their own context (conditions 1 and 3) because it reframes these types of problems in human-centric ways (condition 2) and helps people to generate and explore ideas as a group and eventually build consensus (condition 4) (e.g., Greco and Cresta 2015; Mootee 2013). Usually design thinking is a method known more in the area of industrial innovation and less in the urban planning practice. Indeed, the application of design thinking to urban projects (i.e., urban design thinking) is not always a perfect match. First, in private

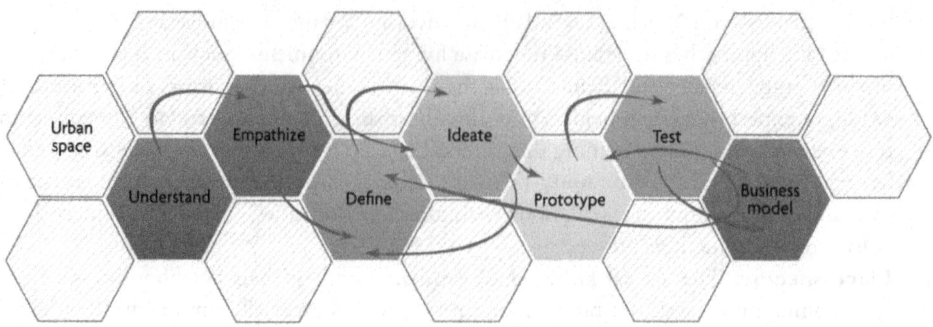

Figure 18.1 Urban design thinking process (own illustration)

companies, the design process focuses on intended users; however, this is challenging to implement in the urban context. While private companies are free to target users as they wish, urban designs have more stakeholders, particularly when tax money is at play. Therefore, we abandon the classic user concept and focus on all actors, whose needs shape urban developments. These actors can be users in the sense of customers, but also citizens, neighbourhood organisers, local politicians, administrators, local companies, and so on (Pahl-Weber et al. 2020: 14). Second, we believe that, in the urban context, the two lenses of design thinking and system thinking need to be combined. The design thinking mindset helps to break down problems in complex systems through a strong hypothesis-driven approach and by reducing them to their individual parts (user needs).[1] Still, according to our first planning condition, we need to have an integrated approach. We cannot understand the behaviour of a system by just studying its parts; we need to study the whole (Mc Phearson 2017).

System thinking performs a comprehensive analysis of the system and the relationships among its components and considers the end-to-end impacts of new concepts or strategies, including unintended consequences (Pourdehnad et al. 2011; Tjendra 2018). Especially in the smart city context, it complements design thinking because it helps us to analyse complex urban settings and develop system-in-systems steered towards the needs of the end-users and other stakeholders.

Given these differences, we are still convinced that urban design thinking can contribute to improving urban societies and building upon common values (Liedtka 2015). It can be used for co-creating smart solutions with stakeholders, co-designing implementation strategies, and building stakeholders' collaborative skills to tackle problems in an interdisciplinary setting. The urban design thinking process is iterative, including different design steps that move from empathising with the end users and stakeholders to defining a specific perspective on the problem, to idea generation, prototyping, and testing to further evaluation and implementation by developing a business or operating model (see Figure 18.1). In addition, the method often asks "what if?" questions to imagine future scenarios freely rather than complying with the way things are done, emphasising the creative and intuitive ways of solving problems (Nielsen et al. 2019).

1 The hypothesis-driven approach to problem-solving is also part of what Dörner describes in his complex situation experiments (e.g., community and humanitarian planning scenarios). He noted that successful participants went on to test proposed hypotheses about the effects of their actions, while unsuccessful ones considered the first proposal they came up with as "truth" (Dörner 1997: 24).

Three urban design thinking case studies – examples for a citizen-centred, spatial perspective in smart city developments

In this section, we present three case studies to demonstrate how urban design thinking can be applied as an approach for planning smart cities. The cases have been selected from the research activity of the Department for Urban Renewal and Sustainable Development at TU Berlin between the years 2015 and 2020.[2] The method of urban design thinking was first tested in interdisciplinary, research-based learning projects with students (urban planning, architecture, engineering, and business students), supervisors, and project partners working together on a semester challenge. Later, its application was extended to an international research project with the city of Mannheim. All three projects used the method to create innovative, liveable environments that are realisable as well as economically and environmentally sustainable. More specifically, we dealt with the issues of smart energy, smart mobility, smart work, and smart participation. Through this study of these particular cases, we do not aim to present specific smart city solutions but rather want to increase general understanding of urban design thinking as a mindset, process, and method in smart city planning processes. Moreover, we show its potential to transform our perspective from technology-centric to a holistic, human-centric, and multi-stakeholder smart city.

Energetic renewal in a residential area

Project description. The first case took place in 2015. It was an interdisciplinary master's student project on the Lichterfelde Süd campus, a residential area in the south of Berlin, owned by the housing co-operative Märkische Scholle. Unlike conventional real estate companies, the members of Märkische Scholle are owners and tenants at the same time, which is common for this type of organisation, a Genossenschaft. Therefore, investments are done according to the needs of the residents. Already in 2013, the process of smart energy modernisation had started at Lichterfelde Süd campus. Old buildings from the 1930s were renovated because of problems such as pipe breaks, high heating costs, or growing mould. Solar thermal systems, underground heat storage, or ambient assisted living technologies were installed. Moreover, 43 new residential units were built. The refurbishment process was planned to take up to 12 years, with expected costs of 74 million Euros. With the re-design of the campus, Märkische Scholle intended to (1) become energetically self-sufficient and (2) to create an attractive neighbourhood, particularly for young families, as almost half of the residents were older than 65. In this context, four design teams received the task of re-designing the campus to achieve intergenerational, energy-efficient housing in an open and vibrant environment for every generation. In five months, they went through the urban design thinking process in three iterations to develop solutions that cater to the unique needs of the community's old and new residents.

Learnings

Discovering place-specific user needs. With wicked problems, we do not always know all of their parameters, plus they will be different with every campus, district, or city. Therefore, the goal of urban design thinking as a process is to investigate all that is known and unknown

2 The three examples can be understood as examples of a western democratic system of planning. The application of the method has also been tested in projects at a global scale. The TU Berlin team has recently dealt with challenges in the cities of New Delhi, Coimbatore, Berlin, and Pune in the context of four Urban Labs in India and Berlin (IGSI 2020). Our approach also seemed to be working in other cultural contexts.

about a situation in order to evoke non-obvious user needs. This is where we find potential (and often unexpected) pathways for reaching our urban development goals (Stimmel 2015: 52). The problem with finding the non-obvious is that users often lack the self-knowledge to give an accurate opinion or to express a wish for something that they cannot imagine. In Lichterfelde Süd, the teams applied a range of design methods to discover the unanticipated insights.

1 They engaged with residents in in-depth, qualitative interviews and got invited to flats to observe residents interacting with the newly installed technology. Instead of asking what they wanted or wished for, the design teams identified the residents' behaviours, feelings, and thoughts about their current living situations and the refurbishment process. Specifically, they learned about residents' lack of awareness regarding the details of the refurbishment process, their insecurities dealing with the newly installed smart technology, their anger about losing the garden and basement storage space, and the lack of community feeling.
2 They immersed how it feels to live in Lichterfelde Süd by attending residents' meetings and videotaping their campus walks.
3 They analysed spatial structures with regard to energy production and consumption, history of the neighbourhood, and green as well as public space. This was important, because it helped the teams to empathise with users within the built environment. Combining these analyses of spatial usage with the needs of the tenants for storage space and community building was key to unveiling valuable insights. The analyses showed how interrelated the subsystems of energy, public space usage, and community are, and how changing heat storage to the underground impacted the other systems – another example of cities as a complex system.

Re-defining the problem. Thus, already at an early stage of the project, it became clear that the technical aspects of the modernisation were less of a problem for residents than the social and spatial ones. As their initial hypotheses about the residents' needs seemed to be wrong, the teams were forced to define and re-define their initial understanding of the situation, which is typical for smart city projects with a fairly unclear or complex problem space. Instead of focusing on technological aspects of the refurbishment process, they focused on the question how the design of public space can enable residents in Lichterfelde Süd to be socially connected and can create more storage possibilities. In the following, the problem statement acted as a filter that retained only the ideas that meet this need. Moreover, it aligned the varying views of the team members, project partners, and further stakeholders. This problem definition process was not linear. Based on residents' feedback and additional insights, the teams realised that tenants might lack an overarching community feeling; however, there existed diverse sub-communities (e.g., gymnastics group). Thus, existing assumption and preconceived ideas were continuously challenged. This iterative, hypothesis-driven problem definition process allowed the teams to define their problem in a truly human-centric way, moving away from a technology-led problem definition.

Co-creating solutions on eye level. Urban design thinking forces an environment where ideas come to light in a multidisciplinary setting. In order to increase empathy among stakeholders and designers, we decided on a collaborative approach where we generated ideas and designed concepts in collaboration with residents, co-op employees, and refurbishment engineers. This is where the system thinking principles came into consideration. We believe that successful design is not imposed on or provided to the community from a source external to the system. The best way to ensure that the solutions serve the community's purpose is to include the stakeholders in their formulation (Pourdehnad et al. 2011). At the co-creation workshop in Lichterfelde Süd, it was important to create confidence that participants can contribute to workshop with their specific experience and knowledge. Second, it was essential for idea generation to step out of the shadows of the status quo, allow even amazingly far-out ideas, and not fixate on obstacles and limitations. This open mindset allowed participants to move away from purely technology-driven ideas to solutions creating outdoor space for neighbourhood life that, at the same time, serve as storage space. The outcome of this collaborative, multidisciplinary approach showed that it is possible to break down silos and realise novel opportunities in a future-oriented way. It also built collaborative skills among the residents in becoming a "competent urbanite" (Sennett 2018: 171–204). When groups begin to understand the interconnected complexity of the problem of place, they are closer to a consensus.

Car-free mobility in a mixed-used residential development

Project description. The second case is also an interdisciplinary master's project about the planning of a smart city district in Berlin Spandau, an area surrounded by waterways on the north-western outskirts of Berlin. The "Gartenfeld project" has an ambitious car-free mobility strategy, which is particularly challenging because the area is an island site on a former industrial area, isolated from its external environment. This strategy is being developed by the private project developer UTB Projektmanagement GmbH, which worked as a partner in this student project. It needs a completely new public transport system that links the new district to the rest of the city and a new approach for mobility and logistics solutions within the district. In this context, our project partner challenged the design teams to develop solutions that reduce motorised individual mobility by dealing with (1) public transport, (2) logistics, (3) cycling, and (4) shared mobility. The teams were supported by a stakeholder network consisting of representatives of Siemens, Volkswagen, and Berlin's public transport service BVG.

Learnings

Discovering needs of future users. As Gartenfeld is still in the planning stage, there are no residents, potential road users, or others whose needs can be discovered. Therefore, we used analogies to find neighbourhoods with similar socio-economic, geographical, and infrastructural properties and worked with future personas to help the design teams to develop new insights. More specifically, teams interviewed residents at the Alt Stralau peninsula, a medium- to high-income neighbourhood with a rate of car ownership above Berlin's average. The design teams wanted to understand what keeps their residents from switching to other modes of transport. Moreover, they went to shared mobility meccas in Kreuzberg with young, hip residents, a target group that might want to move to the city's green outskirts with their young families. They mapped their mobility journeys to discover difficulties with managing car-free family logistics; fear of bike theft, which discourages use of bicycles at all; or car drivers who avoided public transport because of safety concerns. Then, the teams combined their findings with future scenarios of Gartenfeld to create future personas, i.e., fictional individuals with specific mobility needs living in Gartenfeld. This tool allowed us to design for a specific group of stakeholders. Users and objects in urban areas are somehow connected and designing for one group inherently affects another. However, instead of tackling the whole mobility issue at once, the future personas allowed us to progress in incremental steps. We first reduced complexity in order to then eventually develop a comprehensive solution. In addition, this approach forced the teams to understand the heterogenous needs and requirements of citizens surrounding Gartenfeld's mobility challenge, even though the users did not exist yet. We avoided falling into the trap of only emphasising already existing technological solutions.

Prototyping, testing, and iterating. Design teams developed four initial mobility prototypes: (1) an on-demand, autonomous public shuttle bus, (2) an autonomous logistics robot, (3) LED bike lanes, and (4) an intermodal mobility app (including ridesharing). Rather than diving into an endless journey of research and analysis to come up with a single, perfect solution, prototyping allowed them to test the value of their concepts in the above-mentioned neighbourhoods – learning by doing and experiencing. This prototyping approach makes it easier to understand the complexities and linkages of urban systems in the early stage of the project. While testing, they also empathised with other user groups (e.g., elderly), who they did not engage with in the first round. In the urban development context, this is particularly important because we do not want to use empathising/testing as a mechanism to only appeal to the majority or those immediately affected by a problem. This process was iterated several times. Due to the complexity of district-building, the iteration did not take place at the level of the project as a whole; testees solely gave input on concepts and ideas. In other words, we could not test whether future residents will like the new car-free Gartenfeld because building parts of the future district itself to test was too expensive and certainly not possible.

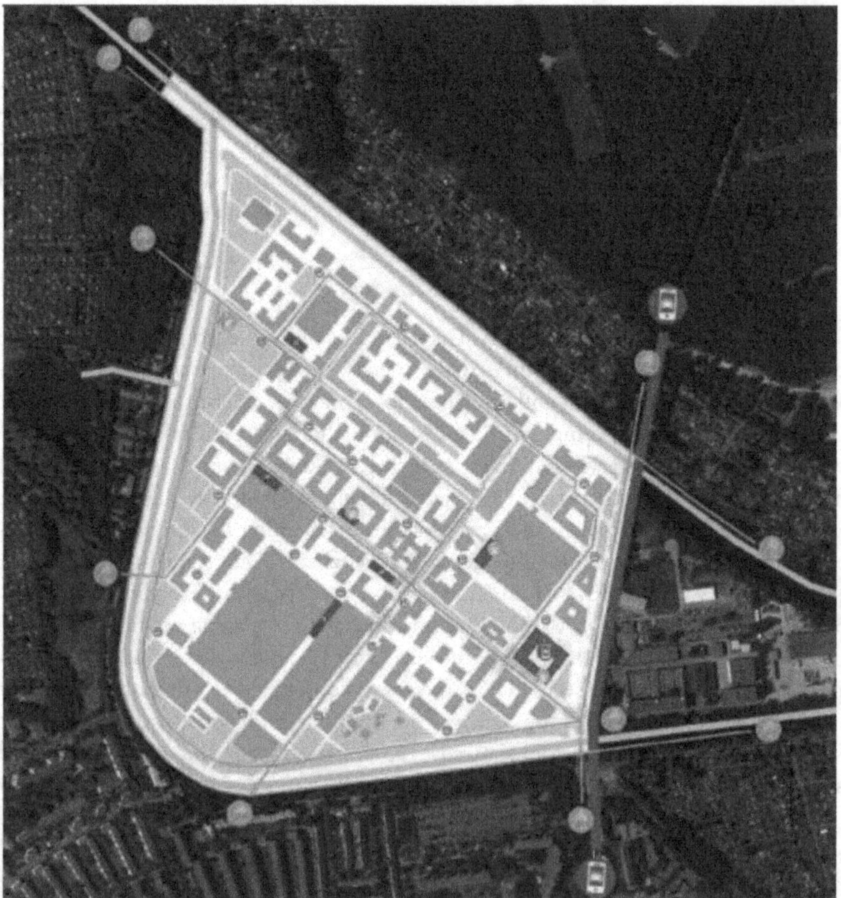

Figure 18.2 Map of Gartenfeld with mobility solutions (own illustration)

Integrating mobility solutions. Even though there were many perspectives involved in the urban design thinking process, the stakeholders gave input from their individual experiences and never saw how this could fit into a whole system. It was the role of the design teams to piece everything together because offering a real alternative to owning a car in Gartenfeld required an integrated mobility concept. To systemise the flow of people, activities, and resources between the individual solutions in order to ensure synchronicity, integration, and consistency in Gartenfeld's mobility concept, teams analysed the specific regulatory and stakeholder framework, set up basic business models for their solutions, and analysed interdependencies in order to specify integrated service packages (e.g., mobility budget as operating costs of housing rent) and operating models that add value to the residents and also the real estate developer.

Urban co-production research project for sustainable transformation

Project description. The third case describes "Perspectives welcome – Migrants4Cities", a research project of TU Berlin, the City of Mannheim, and inter3-institute. It is funded by the German Ministry of Education and Research in the field of sustainable transformation of urban areas (Federal Ministry of Education and Research 2020). The project started in 2016 and lasted for three years. It is proceeded by a follow-up project that started in winter 2019 and goes until 2021. Here, the beginnings of both of projects are referred to. In this project, urban design thinking is defined as a "collaborative process active for the design of the city of Mannheim"

Figure 18.3 Impressions of Mannheim's Urban Labs (own illustrations)

(Jeutner and Thomaier 2019: 14).[3] The university city of Mannheim is the economic and cultural centre of the Rhine-Neckar metropolitan region with 320,000 inhabitants from 173 nations, over 182,000 employees, and around 28,000 students (Migrants4Cities, online). The internationality of urban society is seen as a cultural heritage. For Mannheim, the Migrants4Cities project was the next step to welcome this diversity of perspectives for the sustainable design of the city. The city worked together with migrants to find new ideas for a climate-friendly, socially balanced, economically prosperous, and culturally diverse Mannheim. During the first project phase from 2016 to 2019, 25 migrants worked together with city representatives in culturally, ethnically, and professionally mixed teams of five to seven people. During nine Urban Labs (full-day workshop sessions), they developed solutions in the area of mobility, living, work, and participation, for example a KulturTram for intercultural encounters in everyday life or a work box for flexible working in parks. The teams put their topics on the agenda, discussed and developed ideas together, tested their work results, and presented them to Mannheim's urban society. The follow-up project on the implementation of selected solutions and knowledge transfer started in winter 2019. Both the city and the migrants benefit from this project: The city welcomes the perspectives of its immigrants. The migrants, on the other hand, can actively participate in the design of their city and present their expertise to Mannheim's urban society (Migrants4Cities 2020a). They discussed their processes and results in the UN supported event "Urban Thinkers Campus" in Mannheim in autumn 2019 (Urban Thinkers 2019).

Learnings

Multi-stakeholder collaboration. From our perspective, it was important to put all relevant stakeholders of different sectors and opposing views in the same workshops, to truly understand "the rules of the game". These complex problems need a lot of collaboration from people who usually work individually. This meant to involve the citizen perspective of migrants, who felt valued for being asked about their views and initiated exchange with their neighbours to bring new insights to the next Urban Lab. Second, representatives of the city administration, businesses, and other concerned institutions (e.g., start-up accelerator) were invited as owners of the problems discussed. Here it was important that they actually had the role and position to implement the changes explored. This setup had to be consciously managed and facilitated well by the TU Berlin research team. In our experience, it offered the unique opportunity to improve collaboration in the quadruple helix of citizens, city administration, businesses, and research through co-creation and co-production (see also Kitchin 2019: 12). It also encouraged the transcendence of organisational and procedural silos, established hierarchies, or bureaucratic categories. Some participants indicated a host of preconditions to such a process. It initially took them out of their comfort zones;, adopting urban design thinking means adopting messy failures and the emotions that come with what can sometimes be a fickle creative process. It required them to work cohesively, negotiating on the constraints on and interests of the different perspectives. Nevertheless, our workshop outcomes were similar to Nielsen, Baer, Gohari et al.'s (2019) results, in that urban design thinking clearly had potential for reconciling conflicting feelings and interests, albeit time-consuming and challenging to satisfy all stakeholders.

Information discrepancies and capacity limitations. This particular team setup, consisting of city professionals and citizens as non-experts, accompanied by scientists from the

3 Both authors of this chapter have been part of Elke Pahl-Weber's project team (head of the project at TU Berlin); Nadja Berseck was coach for the team on the Mobile Working Box.

university who structured and coached the process, meant that discrepancies in the level of knowledge were significant. The professional expertise of participants is important for the urban design thinking process. In order for everybody (especially the non-professionals) to be able to follow the discussions, information had to be shared at an early stage and the different abilities to understand and influence the proposed actions had to be recognised. During the course of the project, the scientists realised the importance of accessibility to high-quality information for participants. It showed the necessity of researching enhanced knowledge as a deliberate step before/in between the Urban Labs. However, with full-time jobs, team members usually did not have the capacities to fulfil research or other related tasks on top of their daily responsibilities. Therefore, support structures provided by the project were essential. Moreover, it was difficult to keep their commitment levels up over the course of three years. It showed us the resource limits of collaborative planning[4] and the need for good project structure to avoid putting participants under pressure to perform their tasks additional to their work liabilities.

Implementing and scaling solutions. Our experiences with implementing solutions of the urban design thinking process have been limited up to now. Generally, the method has been criticised for a lack of continuity after prototyping and business modelling, failing to place the solution with the right system to manage change. Often business as usual continues, with new solutions managed by using legacy processes in silos by teams who may not have been trained in urban design approaches (Tjendra 2018). In Mannheim, we realised that to set up this system, the city administration played a vital role. It acted as a coordinator by finding project partners for implementation (e.g., Chamber of Handicrafts, city-owned companies, university) at an early stage of the project and providing funding for the development of high-resolution prototypes and/or further implementation steps. This required a more entrepreneurial, risk-taking role for the administration, applying trial-and-error problem-solving and experimentation. Additionally, in the second project phase starting in 2020, the solutions are developed on a small scale, also to reduce political risks. The idea is to move neighbourhood-embraced best practices upward to greater scales through city-level collaboration. Solutions can be identified and adopted in Mannheim and tuned to fit the unique needs of other urban contexts. Thus, specific neighbourhoods can become hubs for innovation and prove the demand for new solutions (Nielsen, Baer, Gohari, et al. 2019).

Summarising our learnings of the three case studies, Table 18.1 shows a comparison of aspects of urban design thinking with the traditional planning approach.

Shortcomings of urban design thinking

The study of these three cases allowed us to gather insight into how urban design thinking can be successfully applied as an approach for planning smart cities. At the same time, it revealed some shortcomings: As already mentioned in the Mannheim section, it is still a question of how solutions developed with the urban design thinking approach can be taken from concept to reality. This could be called a blind spot in the method, but even so, there are starting points to deal with it in current research projects (Migrants4Cities 2020b). While urban design thinking is good for designing concepts and creating value-based visions for problem-solving, other, more specific approaches are needed to detail and direct specific implementation steps (Nielsen, Baer, Gohari, et al. 2019).

4 During the COVID-19 pandemic, we conducted our first, less resource-intense, virtual labs, which proved to be as good as face-to-face meetings for some phases of the urban design thinking process.

Table 18.1 Comparison urban design thinking and traditional urban planning

	Urban design thinking	*Traditional planning approach*
Result	Business and/or operating model	Plan
Community participation	Community-led by definition	Varying degrees of participation from case to case
Needs assessment	Observing, in-depth interviewing, and immersion, cheap installations to get feedback	Surveys, community workshops, public hearings, etc.
Spatial assessment	Focus on how a place is experienced (e.g., activity mapping, qualitative site analysis, storytelling)	String bias towards quantitative data analysis (e.g., zoning research, census data, traffic/ people moving counts)
Problem-solving approach	Allowing mistakes and thorough testing	Flawless planning and thorough analysis
Feedback	Small experiments, proactive	Presentations, isolated from public
Process structure	Iterative	Mostly linear
Prototypes	Various shapes, interactive, tested in urban space	Models (mainly for experts)
Team setup	Interdisciplinary	Interdisciplinary (to a limited extent)
Facilitated process	Yes, via coaches	No, only evaluation by higher instances
Time planning	Time boxing	Individual planning

Second, urban design thinking supports a small-scale experimentation approach. However, the gradual implementation of successive projects, each one focusing on specific topics, may lead to a sea of unmanageable systems, thus preventing solutions from becoming scalable and potentially limiting the realisation of the smart city concept to its full extent (Cavalcante et al. 2016). To integrate these different projects (and possibly applications and systems), city agencies will need the ability to coordinate their planning, management, and operations to have a holistic view of the city that allows planners to exploit several information sources and better allocate resources (transport, living, water, etc.) (Berseck and zu Knyphausen-Aufseß 2018; Cavalcante et al. 2016). Moreover, they need a common vision and strategy for answering fundamental questions of identity like "Who are we?" and "What is our direction?" to align the agendas of various stakeholders and unveil the wicked problems that are tough to solve.

Third, during the course of an urban design thinking project, there is always the risk of missing a key stakeholder group. Ideally, all stakeholders are involved in the design process from the beginning; however, often there is a design team that includes external perspectives to guide them and none of the stakeholder groups are involved in the entirety of the design process. This means that the design team needs to elicit the other stakeholders' buy-in. We caution that it is much more likely that the ideas generated will be implemented and maintained if the stakeholders involved are those who came up with the solutions in the first place (Pourdehnad et al. 2011).

Why smart cities need a citizen-centred, spatial perspective

The purpose of this chapter has been to set out some of the key shortcomings and risks associated with smart city initiatives and to suggest that we need a better understanding of the process of "how" to develop smart cities rather than "what". Smart cities are complex systems full of wicked problems. Smart cities refer to dynamic processes of change that include the physical, natural, and human parts of the urban entity. Understanding and steering this process is challenging. It requires a holistic and not only technology-led planning approach with (1) an integrated intervention logic, (2) citizen-centeredness, (3) place-specificness, and (4) consensus-building mechanism among various stakeholders. This chapter posits that urban design thinking is an attractive approach for dealing with the problem of smart city transition. By analysing three case studies, we show the potential of the method to identify major challenges that really matter for urban citizens, to work with stakeholders, and to co-create solutions which can be rapidly tested and iterated. More specifically, we conclude the following:

Smart city – sustainable city and the need of tackling complexity

The understanding of urban complexities and what the spatial perspective might be, on the one hand, and the smart city case studies dedicated to the UN millennium goals that strive for sustainable development, on the other hand, show clearly that the integration of digital technologies in the operation of city functions and traditional infrastructure does not automatically contribute to sustainable urban development. The human factor and its role in developing the urban space cannot be denied as part of the planning processes. The instrument and methodology we introduce to follow this view is urban design thinking, which we propose as one tool among others to follow sustainable development goals.

In addition, urban design thinking is a way to deal with a complex system, like the urban context always is, and this is what we intended to express with our case studies. It is an approach that combines the applied and the theoretical and might be understood as a way to show how social capital for smart urban development could be created by paying attention to the "how" of urban development processes.

> An informed public debate provides a good basis for an active civil society; but it is not sufficient to tackle the world´s current challenges in all their complexity. When it comes to shifting the currently dysfunctional world direction into a more functional one, no single actor – neither civil society nor politics nor business – will be able to deliver the entire solution. Instead, each actor must contribute a different but essential piece of knowledge.
>
> *(Von Weizsäcker and Wijkman 2017: 183)*

This statement, in a recent report to the Club of Rome, expresses the need for a new approach to working on complex challenges. It might even stand for an approach for not moving forward with overly specialised strategies. Smart city development is probably not a specialised strategy, but not including the spatial perspective can result in following the path described in the quote.

Smart city – efficient city?

Smart city proponents often claim that ICTs will make our lives easier to live and our cities more sustainable. As highlighted herein, we caution against such simplistic proclamation, especially in smart city development that aims for increased levels of efficiency by reducing alternative options in the operation and functioning of the city. Efficiency is no objective in general. The specific connection between efficiency and sustainability were set by the Club of Rome more than ten years ago (Von Weizsacker et al. 2009), defining a path to come to sustainable prosperity through technological progress, including social values. Present debates refer to data as part of technological progress. The improvement of lives by data needs a social movement to come up with a city that might be smart and improve people's lives (Townsend 2013: 320). Big data is a challenge for the smart city in terms of hosting data or having access in order to analyse it. There might be a connection to the approach of Richard Sennett and his term the "competent urbanite"; the use of data in these circumstances can bring basic know-how of developments and facts in the urban context forward and make it transparent, besides the need of communication.

Smart city – the human factor and the need for communication

Technology has extraordinary potential to make cities safer, healthier, or more affordable, but the diversity of cities makes even fundamental challenges like waste, water, and crime very difficult to address. No two cities hold the same qualities; each city has its own history, population, culture, politics, legacy infrastructures and systems, political and administrative geography, modes of governance, sense of place, hinterlands, interconnections and interdependencies with other places, and so on (Kitchin 2019). Smart city designs that consider technology, as well as the lives of the human beings living in the environments, hold the greatest chance of creating sustainable and vital communities by helping to reframe problems and deliver solutions that bring all stakeholders to the table (Stimmel 2015). The spatial perspective includes the needs of human beings in the given local situation. We hypothesise that people will probably use new technologies and behave in a more sustainable manner if their needs are met. Their needs cannot be explored by advertising, or by marketing surveys, but must instead involve a dialogue and communication process. This process of exploring the people's needs and discussing them gives us the chance to develop new insights.

Coming from these insights to appropriate solutions might be the next step. And this step is not carried out by urban administration, planners, or entrepreneurs alone; they are being developed by a broad range of urban stakeholders. Within the field of smart cities, carried out as a research and practice field by the German Ministry of Interior, some of the cities show that, coming from communicated solution to realisation in the urban space, the field of actors becomes much broader than urban politics and its administration. Working together on a democratic level between elected urban councils and civil initiative paves the way for a process of communication in urban action, like Habermas explained in his work on the theory of communicative action (Habermas 1981).

Contribution of urban design thinking to the smart urban transformation

Finally, there is still a lot of work that needs to be done in order for cities around the world to create smart environments for the people and by the people. We suggest that urban design thinking should be used at a larger scale to develop more flexible approaches to urban planning, regulation, policy, and space making in smart city futures. This way, researchers, citizens, industry, and municipalities can build knowledge and frameworks that contribute to innovation and apply this way of thinking to identifying the key areas and core values for co-creation in smart cities. However, when applying urban design thinking at such a scale, one needs to understand that it is more than just a workshop, more than just a small slice of the whole. We need to acknowledge the complexity behind this human-centred approach. It requires know-how to frame and conduct qualitative and quantitative research, the ability to make sense of data and deduct insights, as well as ideation, visualisation, and detailed design skills. Nevertheless, urban design thinking is not a miracle potion, either. We recommend taking a closer look and identifying the right strategy for the right context. Especially in the smart city context, there exist solutions on the market, and novel practices might not be required. Moreover, urban design thinking is not one-size-fits-all approach. Its strength lies in the fact that is can be adopted, that it acknowledges the specifics of urban space, shaped by technical, functional, material, and human elements. We conclude that the transformation of smart cities could greatly benefit from – and perhaps even base its success on – urban design thinking's mindset and tools. It might in fact distinguish the smart city from the less smart one.

References

Angelidou, M. (2017): The Role of Smart City Characteristics in the Plans of Fifteen Cities, *Journal of Urban Technology* 24(4), 3–28.

Batty, M. (2012): Building a Science of Cities, *Cities* 29, S9–S16. https://doi.org/10.1016/j.cities.2011.11.008

Batty, M. (2013): *The New Science of Cities*. Cambridge, MA: MIT Press.

Batty, M., Axhausen, K. W., Giannotti, F., Pozdnoukhov, A., Bazzani, A., Wachowicz, M. and. Portugali, Y. (2012): Smart Cities of the Future, *The European Physical Journal Special Topics* 214(1), 481–518.

Berry, B. J. (1964): Cities as Systems within Systems of Cities, *Papers in Regional Science* 13(1), 147–163.

Berseck, N. and zu Knyphausen-Aufseß, D. (2018): Resource Orchestration as a Source of Competitive Advantage for Cities, *Managementforschung* 28(1), 87–115.

Bibri, S. E. (2018): *Smart Sustainable Cities of the Future: The Untapped Potential of Big Data Analytics and Context – Aware Computing for Advancing Sustainability*. Cham: Springer International Publishing.

Buldyrev, S. V., Parshani, R., Paul, G., Stanley, H. E. and Havlin, S. (2010): Catastrophic Cascade of Failures in Interdependent Networks, *Nature* 464(7291), 1025–1028.

Cavalcante, E., Cacho, N., Lopes, F., Batista, T. and Oquendo, F. (2016): *Thinking Smart Cities as Systems-of-systems: A Perspective Study*. Paper presented at the Proceedings of the 2nd International Workshop on Smart. https://doi.org/10.1145/3009912.3009918

Chadwick, G. F. (1971): *A Systems View of Planning*. Oxford, UK: Pergamon Press.

Chourabi, H., Nam, T., Walker, S., Gil-Garcia, J. R., Mellouli, S., Nahon, K. and Scholl, H. J. (2012): *Understanding Smart Cities: An Integrative Framework*. Paper presented at the 2012 45th Hawaii international conference on system sciences. https://doi.org/10.1109/HICSS.2012.615

Colding, J., Barthel, S. and Sörqvist, P. (2019): Wicked Problems of Smart Cities, *Smart Cities 2*(4), 512–521.

Condorelli, R. (2016): Complex Systems Theory: Some Considerations for Sociology, *Open Journal of Applied Sciences* 6(7), 422–448.

Datta, A. (2015): New Urban Utopias of Postcolonial India: "Entrepreneurial Urbanization" in Dholera Smart City, Gujarat, *Dialogues in Human Geography* 5(1), 3–22.

Dembski, F., Wössner, U., Letzgus, M., Ruddat, M. and Yamu, C. (2020): Urban Digital Twins for Smart Cities and Citizens: The Case Study of Herrenberg, Germany, *Sustainability 12(6)*, 2307.

Dörner, D. (1997): *The Logic of Failure: Recognizing and Avoiding Error in Complex Situations*. New York City: Basic Books.

Dörner, D. (2020): Interview: Die Logik des Misslingens. Available at: www.geo.de/magazine/geo-wissen/5515-rtkl-interview-die-logik-des-misslingens?utm_campaign=&utm_source=email&utm_medium=mweb_sharing [10.01.2021]

Dünne, J., Günzel, S., Doetsch, H. and Lüdeke, R. (2006): *Raumtheorie. Grundlagentexte aus Philosophie und Kulturwissenschaften*. Orig. Ausg. Frankfurt aM: Suhrkamp (Suhrkamp Taschenbuch Wissenschaft, 1800).

Federal Ministry of Education and Research (2020): Funding Measures – Sustainable Transformation of Urban Areas. Available at: www.fona.de/en/measures/funding-measures/sustainable-transformation-of-urban-areas.php [10.01.2021]

Frost and Sullivan (2019): Frost & Sullivan Value Proposition – Smart Cities. Available at: https://ww2.frost.com/wp-content/uploads/2019/01/SmartCities.pdf [10.01.2021]

Gehl, J. (2013): *Cities for People.* Washington, DC: Island press.
Greco, I. and Cresta, A. (2015): *A Smart Planning for Smart City: The Concept of Smart City as an Opportunity to Re-think the Planning Models of the Contemporary City.* Paper presented at the International Conference on Computational Science and Its Applications. https://doi.org/10.1007/978-3-319-21407-8_40
Habermas, J. (1981): *Theorie des kommunikativen Handelns.* Vol. 2. Frankfurt a. M.: Suhrkamp.
Healey, P. (1997): *Collaborative Planning-Shaping Places in Fragmented Societies. Planning-Environment-Cities.* Hampshire: Mac Millan Press.
IGSI. (2020): Integrated Urban Development and Co-Production for Indian Cities. Available at: www.igsi.info/# [10.01.2021]
Jacobs, J. (1961): *The Death and Life of American Cities.* New York: Random House.
Jeutner, M. (2020): Von Smart cities lernen? Ko-Kreation in sädtischen Innovationsprozessen, in C. Hübel, E. Pahl-Weber and Schön, S. (eds.): *Willkommene Perspektiven, Nachhaltige Stadtentwicklung durch Urban Design Thinking.* Berlin: Universitätsverlag der TU Berlin, 15–38.
Jeutner, M. and Thomaier, S. (eds.). (2019): *Urban Design Thinking als Methode für städtische Ko-Produktion.* Mannheim: Stadt Mannheim.
Kitchin, R. (2019): Reframing, Reimagining and remaking Smart Cities, in Coletta, C., Evans, L., Heaphy, L. and Kitchin, R. (eds.): *Creating Smart Cities.* London, New York: Routledge, 219–230.
Lefebvre, H. and Nicholson-Smith, D. (1991): *The Production of Space.* Vol. 142. Oxford, UK: Blackwell Publishing.
Liedtka, J. (2015): Perspective: Linking Design Thinking with Innovation Outcomes through Cognitive Bias Reduction, *Journal of Product Innovation Management* 32(6), 925–938.
Löw, Martina (2001): Raumsoziologie, Frankfurt a.M.: Suhrkamp.
Martina, L. (2001): *Raumsoziologie.* Frankfurt a.M.: Suhrkamp.
Mc Phearson, T. (2017): Wicked Problems, Social-ecological Systems, and the Utility of Systems Thinking. Available at: www.smartcitiesdive.com/ex/sustainablecitiescollective/wicked-problems-social-ecological-systems-and-utility-systems-thinking/113741/ [10.01.2021]
Migrants4Cities (2020a): Über das Projekt. Available at: www.migrants4cities.de/de/uber-das-projekt-2/ [10.01.2021]
Migrants4Cities (2020b): Willkommene Perspektiven – Über das Projekt. Available at: www.migrants4cities.de/de/uber-das-projekt-2/ [10.01.2021]
Miller, J. H. and Page, S. E. (2009): *Complex Adaptive Systems: An Introduction to Computational Models of social life*: Princeton: Princeton University Press.
Mootee, I. (2013): *Design Thinking for Strategic Innovation: What They Can't Teach You At Business or Design School.* Hoboken, New Jersey: John Wiley & Sons.
Mora, L., Deakin, M. and Reid, A. (2019): Strategic Principles for Smart City Development: A Multiple Case Study Analysis of European Best Practices, *Technological Forecasting and Social Change* 142, 70–97.
Nielsen, B. F., Baer, D., Gohari, S. and Junker, E. (2019): *The Potential of Design Thinking for Tackling the "Wicked Problems" of the Smart City.* Paper presented at the Proceedings of the 24th International Conference on Urban Planning, Regional Development and Information Society.
Nielsen, B. F., Baer, D. and Lindkvist, C. (2019): Identifying and Supporting Exploratory and Exploitative Models of Innovation in Municipal Urban Planning; Key Challenges from Seven Norwegian Energy Ambitious Neighborhood Pilots, *Technological Forecasting and Social Change* 142, 142–153.
Pahl-Weber, E. (2020): Urban Design Thinking: Ein ko-kreativer Weg zu bedarfszentrierter Stadtplanung? in Hübel, C., Pahl-Weber, E. and Schön, S. (eds.): *Willkommene Perspektiven, Nachhaltige Stadtentwicklung durch Urban Design Thinking.* Berlin: Universitätsverlag der TU Berlin, 39–54.
Pahl-Weber, E., Zu Knyphausen-Aufseß, D. and Kratzer, J. S., F. (2020): Introduction – Neighborhoods as Real Laboratories – the Last Mile for Social, Technical and Echological Concepts, in Pahl-Weber, E., zu Knyphausen-Aufseß, D., Kratzer, J. and Straube, F. (eds.): *Kiez-Logistik für die letzte Meile. Das Reallabor Distribute.* Berlin: Universitätsverlag der TU Berlin, 8–46.
Palermo, P. C. and Ponzini, D. (2014): *Place-making and Urban Development: New Challenges for Contemporary Planning and Design.* New York: Routledge.
Plattner, H., Meinel, C. and Leifer, L. (2015): *Design Thinking Research: Making Design Thinking Foundational.* Cham, Switzerland: Springer.
Pourdehnad, J., Wexler, E. R. and Wilson, D. V. (2011): Integrating Systems Thinking and Design Thinking, *The Systems Thinker* 22(9), 2–6.
Pribyl, O. and Svitek, M. (2015): *System-oriented Approach to Smart Cities.* Paper presented at the 2015 Photonics North. https://doi.org/10.1109/ISC2.2015.7428760

Rae, A. and Singleton, A. (2015): Putting Big Data in its Place: A Regional Studies and Regional Science Perspective, *Regional Studies, Regional Science* 2(1), 1–5. https://doi.org/10.1080/21681376.2014.990678

Sennett, R. (2018): *Building and Dwelling: Ethics for the City*. New York: Farrar, Straus and Giroux.

Stimmel, C. L. (2015): *Building Smart Cities: Analytics, ICT, and Design Thinking*. New York: CRC Press.

Tjendra, J. (2018): Systems Thinking Is the New Design Thinking, *Business Innovation Design*. Available at: https://businessinnovation.design/blog/2018/4/25/systems-thinking-is-the-new-design-thinking [10.01.2021]

Townsend, A. M. (2013): *Smart Cities: Big Data, Civic Hackers, and the Quest for a New Utopia*. New York, London: WW Norton & Company.

Urban Thinkers (2019): Urban Design Thinkers Campus 2019. Available at: www.mannheim.de/en/shaping-the-city/urban-thinkers-campus-2019 [10.01.2021]

Von Weizsacker, E. U., Hargroves, C., Smith, M. H., Desha, C. and Stasinopoulos, P. (2009): *Factor Five: Transforming the Global Economy Through 80% Improvements in Resource Productivity*. London: Routledge.

Von Weizsäcker, E. U. and Wijkman, A. (2017): *Come on! Capitalism, Short-termism, Population and the Destruction of the Planet*. Berlin: Springer.

von Wirth, T., Rach, S., Verhagen, M., Buchel, S., de Graaff, S. and Loorbach, D. (2019): Participatory Design of People-centered Cities. Mapping of Scientific Research and Relevant Theories, Scientists and Actors. Available at: www.bosch-stiftung.de/sites/default/files/publications/pdf/2020-04/Participatory%20Design%20of%20People-centered%20Cities_2020_0.pdf [10.01.2021]

Yang, P. P. and Yamagata, Y. (2020): *Urban Systems Design: Shaping Smart Cities by Integrating Urban Design and Systems Science*. Amsterdam, Oxford, UK, Cambridge, MA: Elsevier.

19

PRODUCING THE 'USER' IN SMART TECHNOLOGIES

A framework for examining user representations in smart grids and smart metering infrastructure

Antti Silvast, Robin Williams, Sampsa Hyysalo, Kjetil Rommetveit and Charles Raab

Introduction

This chapter explores the role of smart technologies with the increasing use of information and communication technologies (ICTs) to manage energy production and distribution. It focuses on smart grids (electricity infrastructures that involve such ICTs) and smart-energy meters. The latter are among the prominent components of smart grids, installed in domestic buildings, businesses, and other sites of energy usage; they enable nearly real-time measurements and regular two-way communication between energy suppliers and consumers. We draw upon the sociology of user representations (Oudshoorn and Pinch 2003; Oudshoorn et al. 2004; Stewart and Williams 2005; Hyysalo and Johnson 2015; Hyysalo et al. 2016) to explore how the designs of smart technologies 'embed' specific assumptions about their users: for example, regarding who they are expected to be, what uses these technologies would have, and what the situations of use will be like. The chapter examines these representations of users alongside broader infrastructural visions – such as energy decarbonisation or cost minimisation. These are real-time representations of possible future technological capabilities. Our analytical framework highlights the 'expectation work' – through activities such as pilot projects, simulation models, qualitative scenarios, and cost–benefit calculations – that experts use to mobilise these expectations (Borup et al. 2006; Konrad 2008; Wilkie and Michael 2009). Through these kinds of vision, representations of users in smart technologies may become embedded in whole industrial fields. Such user-related expectations have strong potential impacts on coordinating activities and aligning the interests of actors, for example in energy market regulation, standardisation, and investment.

Over the past decade, smart grids have been the focus of major energy investment and infrastructure modernisation policy programmes (European Parliament and Council 2012; UK National Infrastructure Commission 2016; Pahkala et al. 2018; European Commission 2019). Energy suppliers and distributors are expected to enjoy major benefits from smart grids and meters through improved energy peak-load shifting, emissions reduction, smaller meter-reading

DOI: 10.4324/9780429351921-23

costs, and reduced operational expenses (BEIS 2016). Proposals for 'smarter' energy systems sit alongside a great number of other 'smart' development projects and initiatives – from 'smart' cities to 'smart' health – conceived as ways of resolving an increasing range of issues, including the impacts of climate change, citizen participation, the ageing population, and ensuring economic growth. These ambitions are evident at the European Union (EU) level and in industrial and developing states, and among various industrial sectors, policy-makers, researchers, and expert circles (Rommetveit et al. 2017; van Dijk et al. 2018). The concept of 'smart' has been applied in such a wide range of ways that it cannot be given a single meaning. Attention to specific smart practices and users has given way to a generic message which has become a kind of virtue signalling adopted by developers and promoters of technology (Williams et al. 2005: 150–152). The concept brings together certain key themes, such as the integration of digital and non-digital infrastructures, interactivity, and social and technical problem-solving. In the EU CANDID (Checking Assumptions and Promoting Responsibility in Smart Development) project, we sought to capture these through a conceptual typology of 'smart' (Table 19.1).

Smart-grid and smart-metering projects fit into more than one part of this typology, including new data-driven agency in smart projects and modernising incumbent power grids. Nevertheless, one of these meanings – new forms of consumerism – shapes several of the recent expectations concerning smart energy. The needs and contributions of the energy 'end user', 'final user', 'consumer', 'customer' (or other related terms) figure prominently in smart-energy programmes. In EU policy documents, deploying smart grids and metering promises new possibilities for self-managing energy consumption, improved energy efficiency among final

Table 19.1 Selected common uses and characteristics of 'smart'

Use of 'smart'	*Examples*
Inventory of certain characteristics	Smart as digital, interactive, or user-centred, and pertaining to solutions in markets, such as phones, tablets, energy systems, home management, and transportation.
Intersecting innovations and artefacts	The Internet of Things, RFIDs in networks, and radical expansion of sensors in anything from household appliances to traffic controls, big data and algorithmic decision-making systems.
Continuation of the modernising project	Smart city, where smart co-exists with the digitalisation of city infrastructures and a focus on governance, services, smart regulation and law.
A professional achievement	The making, distribution, promotion, and use of smart solutions by lawyers, engineers, software engineers, and users.
Data-driven agency	Change in legislation and regulation or by the engineering of rights into smart systems and services. Also threatening of privacy, identity, autonomy, and legal rights such as non-discrimination, due process, and the presumption of innocence.
Shifting social and scientific relationships	Unblackboxed domestic energy consumption through increased transparency, accountability, and rendering technology visible, citizen science, do-it-yourself (DIY), peer-to-peer (P2P), co-production, and crowd-sourcing approaches.
New forms of consumerism	Advancements toward the smart society that raise awareness of consumption by aiming to alter consumer behaviour with personalisation, a privatisation of politics, and an appeal to aesthetics

Source: Rommetveit et al. 2017

consumers, and transition to more consumer-centric energy systems (European Parliament and Council 2012; European Commission 2011, 2014, 2019). This became pronounced again in the EU's Clean Energy for All Europeans package in 2019: highlighting how smart-energy systems support active consumers who 'take their own decisions on how to produce, store, sell, or share their own energy', with smart-energy meters that allow these consumers to be 'informed about their energy consumption and costs in real time' (European Commission 2019: 12). Additionally, the United Kingdom's (UK) smart metering programme, the focal point of our present inquiry, has emphasised consumer protection and consumer engagement with the new smart meters (DECC 2012). More recently, it has focused on how households and small businesses might be given 'incentives' to accept smart meters voluntarily, such as accurate automatic billing, cost savings, and real-time energy consumption feedback (BEIS 2016; UK House of Commons 2016; Helm 2017).

Meanwhile, in some countries, people have reacted to and opposed smart-energy meters because of concerns over the protection of their personal data, to avoid information overload, or – particularly in the United States (US) – health (Hess and Coley 2014). The Netherlands, the US, the UK, Norway, and Canada have seen either anti-smart meter campaigns, lawsuits against smart metering, critical reports by consumer organisations, or motions passed by governments (van der Horst et al. 2014). Events like these relate to the supposed 'users' of smart-grid technology in various ways. These range from rejecting 'use' because of intrusion on the right to privacy in the Netherlands (von Schomberg 2011; Hoenkamp et al. 2011) to the emergence of networks of counter-experts and democratic contestation, paving the way to health-oriented precautionary politics in California (Hess and Coley 2014).

In this chapter, we examine these various representations of smart-grid users. We pay close attention to the nature of representations of users in smart-energy development projects, their likely evolution, and hence the mechanisms by which these projections of future use may eventually affect people's lives. The chapter develops Science and Technology Studies (STS) perspectives to address such developments. More specifically, it examines how STS research on user representations may help to understand the anticipated use and users of smarter energy technologies. By studying these uses and users, we also produce a more subtle understanding of how exactly smart-energy technologies might contribute to sustainability in energy systems. In this endeavour, we draw upon and hope to contribute to the considerable body of STS research on user–technology relationships (Oudshoorn and Pinch 2003; Hyysalo et al. 2016) and the technological, social, and policy assumptions and implications of smart grids and meters (Wolsink 2012; Darby 2012; Strengers 2013; Ballo 2015; Bulkeley et al. 2016; Heiskanen and Matschoss 2016; Throndsen 2017; Lovell et al. 2017; Sareen and Rommetveit 2019; Sareen 2020). The chapter addresses two overarching research questions:

1 What kind of user-related expectation work is carried out in smart-grid projects?
2 What do these representations of users and user-related expectations reveal about the evolving social impacts of infrastructural technologies such as smart grids?

We elaborate our approach drawing on findings from the CANDID research project. This project explored, *inter alia*, how European technology developers and experts understand the final use and social impacts of smart grids and smart-energy meters. Inspired by an interest in user representation in smart-grid projects, our analysis of this data focuses on various ways in which smart-grid development experts structure their view of users. While data collection was constrained by the short (12 month) duration of our study, we use cross-country comparison as a 'proxy' for studying the evolution of user representations in smart grids over

Table 19.2 Selected national differences in smart grids and smart meters in UK and Finland

Characteristic	*UK*	*Finland*
Completion of smart meter rollout	2024	2013
Responsibility for smart meter rollout	Electricity suppliers	Electricity distributors
Data and communications infrastructure	Data communications company (under license)	Distribution and transmission systems operators
Remote communication resolution	30 minutes	1 hour
In-home displays of smart meters	Obligatory with detailed technical specifications	Generic technical specifications
Core concern	Smart meter users	Smart grids in future energy markets and society

Sources: Pöyry 2017; Pahkala et al. 2018

time. The UK's smart meter programme will be considered as being in an early adoption stage, while experiences from other European countries – Finland in particular – help us to gather insights into the later adoption stages of smart energy and the evolution of its impacts. The premise of this method is that stronger insights arise from tracing the evolution of technological promises – regarding how they have changed over time, or between various sites of technology such as production and consumption (Williams et al. 2005; Silvast and Virtanen 2019; Hyysalo et al. 2019). We want to emphasise that this study is not a formal cross-national comparison but uses the two countries as exemplars in the stages of technology adoption. An illustrative list of the main national differences in smart-energy rollout between Finland and the UK is in Table 19.2.

This analysis contributes theoretically to the sociology of user representations in complex and long-term infrastructural projects and recapitulates earlier findings on user representation. These showed that there is a range of competing user representations regarding an emerging technology; that the diversity of users and situations of use is reduced to manageable development visions through simplifying user characterisations; and that these representations play roles in the expectation work involved in the development of emerging technologies. Our results also offer the possibility to develop these conceptual tools. In particular, smart-grid user representations draw attention to important new issues in the representation of users in multi-layered infrastructures. There exist vast differences in the likely form and intensity of user engagement with smart grids and meters; differences in whether the representations refer to current smart technologies and uses or to developments that are expected to emerge and become established over the next ten years; and in whether the user is envisaged as the user of smart meters or of an entire layer of new energy services and applications.

The structure of the chapter is as follows. The first section explains our conceptual position on smart-energy technologies, starting from the sociology of user–technology relationships, user representations, and expectations and reviewing initial social science and humanities scholarship on smart energy. It then introduces the CANDID empirical study, presenting it by focusing on key aspects of how 'users' become represented in expectation work – to be explained in the successive section – related to smart grids, smart metering, new energy-saving and load-shifting services, and promises of new kinds of energy consumption and self-production – or 'prosumption' – at the household level. Finally, the conclusion draws out more general implications for understanding user representations and suggests possibilities for further work in research and development of smart grids from the perspective developed in this chapter.

Literature review

Who are the 'users'? User–technology relationships and the sociology of user representations

The high hopes and fears related to smart metering and smart-grid projects may obstruct a more realistic understanding of how these technologies will be applied and their consequences for various stakeholders. By unpicking ideas of 'smartness' and its supposed beneficiaries, we seek to consider more carefully how 'users' – whoever they are – and their 'needs' – whatever they are – may be met by smart-grid projects. STS scholarship has focused on user–technology relationships for several decades (Bijker et al. 1987; Akrich 1992; Oudshoorn and Pinch 2003; Oudshoorn et al. 2004; Stewart and Williams 2005; Hyysalo and Johnson 2015; Hyysalo et al. 2016) and provides a very promising vantage point for exploring this. STS work on 'users' opens up several questions about who or what the 'users' of technology are expected to be; how representations of 'users' feature in developing expectations that mobilise resources and shape the trajectory of future developments; how designers represent and relate to these 'users' and how they understand their characteristics; and how 'users' may use technologies in a different manner than designers envisaged and, in so doing, may re-shape the design.

Considering the meaning of the term 'user', a key contribution of this STS research has been showing that in empirical sites of technology production the 'user' is considered not merely as an identifiable person, but also as a conceptual entity relational to the developers of technology (Akrich 1995; Oudshoorn et al. 2004). This relationality means that researchers and developers tend to represent eventual users during technology design by extrapolating from business models, market studies, consumer panels, co-design workshops, or even just common sense, in the hope that these will 'bridge' towards these users (Hyysalo and Johnson 2015). These user representations link multiple modalities of emerging technologies: visions, requirement specifications, models and prototypes, marketing materials and manuals, pilot assemblies, and eventually uses by concrete people in concrete settings (Hyysalo 2004).

The sociology of user representations (Akrich 1995) examines the processes that lead to particular actor positions becoming embedded or 'scripted' into the characteristics of technology (Woolgar 1990). These characteristics concern who the users are, what they are supposed to do with that particular technology, and in what situations. This sociology examines how configurations of users come into being and how they may become accepted, altered, or rejected by eventual adopters (Akrich 1992; Williams et al. 2005). Initial concepts characterising these relationships – such as the concept of 'script' in technology or Woolgar's (1990) suggestion that designers seek to 'configure the user' – targeted relatively discrete and clearly delineated objects, such as solar batteries, computer hardware, or even door closers (Akrich 1992; Woolgar 1990). These mostly featured 'closed' – i.e., limited and singular – scripts for specific actions rather than 'open' scripts and 'circumscriptions' that would cater for an unspecified but wide array of possible user actions (Johnson/Latour 1988; Akrich and Latour 1992).

When considering more complex technologies, the notion of a closed script was not sophisticated enough. In such technologies, designers' 'configuration of the user' was rather like a preconfiguration (Williams et al. 2005; Hyysalo 2010): an expression of how designers might prefer the users to act with the technology. Yet, such idealisations were confronted by and have merged with various other preconfigurations in users' settings – from procedures to habits, norms, conventions, and beyond – leading to various 'reconfigurations'. Moreover, when it came to the influence of 'smart' interactive information technologies on users, the configuration process did not play out as a simple one-time contestation but tended to evolve across cycles of development and use, and also to evolve in conjunction with the development of adjoining systems and services (Hyysalo 2004; Williams et al. 2005; Hyysalo 2010; Johnson 2013).

This newer understanding of developer–user relations has been paired with more sophisticated understandings of user representations. One important source of understandings arises from the efforts of designers and developers to develop more effective representations to inform their design and development work (Oudshoorn et al. 2004; Hyysalo 2004). There is considerable variation as to how specific and detailed the representations are: some are based on clearly delineated user demographics and specific use cases and are typically found in specific application contexts or as clearly targeted parts of larger systems (Johnson 2013; Hyysalo and Johnson 2015). However, in mass-produced consumer goods and services and large systems, the diversity among users increases beyond what can be meaningfully responded to by means of segmentation, need analysis, or product differentiation. A common developer response has been to no longer insert specific users but simply actions that users would perform with the technology, en masse (Oudshoorn et al. 2004; Johnson 2013).

Furthermore, STS research on information infrastructure (Edwards et al. 2007; Pollock and Williams 2010) has shown that the diversity of potential future users and uses often overwhelms its developers. The risk is that they will roll out what are already (or will soon become) legacy systems. To this end, those involved in building such information systems have adopted more experimental development strategies that continuously monitor, configure, and manage their 'communities' of 'users' (Stewart and Williams 2005; Pollock et al. 2016; Mozaffar 2016).

However, these user representations are not held only by design teams, and not all end up as features in products. Some user representations circulate among particular companies but may never result in design features (Hyysalo 2010), and yet others concern whole technology fields (Stewart 2003; Williams et al. 2005; Konrad 2008). Konrad (2008), for instance, observes how representations of users and the ensuing 'scenarios of use' for Interactive TV remained remarkably stable for over a decade around the turn of the millennium, despite considerable technological advances and the surrounding development of Internet technology. Meanwhile, the evolution in e-commerce scenarios of use was so rapid that the producer companies did not even have time to test their adequacy in real-life applications in their race to adapt and shape the visions that drove the overall development of this business field. In such field-shaping efforts, wide arrays of user representations can be purposively cast to have a mobilising effect on a range of actors in industry and policy (Williams et al. 2005; Konrad 2008; Wilkie and Michael 2009).

Williams et al. (2005) and Stewart and Hyysalo (2008) observed information technology development further, outlining various kinds of layering in these contexts, including horizontal and vertical layering of technologies as well as users, and creating correspondingly layered user representations. First, ICT technologies featured intermediary service providers who assembled component technologies into product and service offerings; intermediate users such as work organisations and bespoke service providers that configured systems for final users; and proxy users who informed the development. Second, the technologies featured several layers: an infrastructural hardware layer of routers, servers, electricity networks, and beyond; a software infrastructure consisting of, for example, Internet protocols, low-level computer software, and operating systems; and various application software that 'end users' would mostly be using rather than the infrastructures working 'invisibly' at another layer. All these layers would have completely or somewhat different user bases and consequently different representations of their use.

Similarly, unpacking smart grids reveals a large cast of characters. This includes energy suppliers, electricity distribution operators, 'next-generation intermediaries', automation analysis software providers, price-comparison websites, energy regulatory agencies, marketing and promotional actors, and several others as well coupled to the promises of smart grids. These promises are shaped by broader infrastructural visions, such as decarbonisation and cost minimisation. Smart meters are also meant to operate along with other 'smart' home technologies, interacting with everyday objects and the Internet of Things, thereby paving the way for standardisation, new markets, and business models.

Such visions and field-wide arrays of user representations are part of expectation work that is important in coordinating actors and aligning interests, convincing investors, achieving standardisation, and developing regulation. This topic has been addressed within STS work on the sociology of expectations (Konrad 2008; Wilkie and Michael 2009; Borup et al. 2006). Expectations refer to 'real-time representations of future technological situations and capabilities' and 'wishful enactments of a desired future' (Borup et al. 2006: 286). By guiding activities, expectations are fundamentally generative or 'performative': they not only describe desired future technological and economic situations, but may also partially bring them about, at least for a time (Silvast 2017a). Expectation work refers to the activity – ranging from roadmaps and qualitative scenarios to cost–benefit analyses, pilot projects, and simulation models – that produces these representations of futures in practice.

These insights into the representation of the 'user' and expectation work are also directly relevant in thinking about large energy systems and infrastructures (Silvast 2017b; 2018). In fact, the literature on user representation has been actively employed in making sense of the evolving smart metering and smart-grid systems.

Social science and humanities research on smartness and smart grids

In earlier social sciences and humanities research, an important context of smart grids and smart meters has been seen as the wider drive to make many aspects of our societies 'smart'. This 'smartness' is defined by a certain kind of modernism, meaning increased rationalisation and individuality. In these accounts, smart grids, their 'users', and uses often relate to, for example, projects of smart urbanism, resilience and self-healing of energy systems, integration of smart grids with smart home technologies, integration of digital and non-digital infrastructure, and data sharing among various actors. Studies indicate that smart energy constitutes an entire 'smart' ontology that constructs new data-driven relationships between energy producers and users (Strengers 2013). Smart grids are expected to enable more intelligent management of electricity networks, especially to cope with intermittent renewable energy sources such as wind and solar (Verbong et al. 2013). Smart grids and smart meters are key in wider transitions to 'smart' urbanism (Luque 2014; Powells et al. 2016). At the level of households, smart meters also promise to be central in enabling information flows in the 'smart' home (Hargreaves and Wilson 2013; Wilson et al. 2015).

Ballo (2015: 12) applies the concept of socio-technical imaginaries in order to 'explore how actors produce future visions or imaginaries that describe desirable or and feasible futures' in smart-grid projects. The Norwegian smart-grid initiatives that she studied mainly draw upon two separate 'visions' that she labels technological (e.g., adding electricity grids with intelligence, utilising new data) and economical (e.g., combining electricity grids with a market logic, creating new possibilities of action for energy consumers) that are also nationally bounded. Similarly, a study in Washington State found distinct meanings of the smart meter concept among utilities, academic research and development laboratories, regulators, and technology firms. On different occasions, the smart meter was conceptualised as a tool for economic efficiency, as a democratisation project, or as part of a vision of a new machine-governed society, while the necessity of interoperability among the components of the future's smart grid was also flagged (Frickel et al. 2016). Other works have used the concept of imaginaries to explore how smart grids also produce emergent visions of very different smart-grid 'users' (Throndsen 2017). One key expectation has been that individuals' behavioural changes will be triggered by smart technologies such as smart meters, driven by energy and cost savings (Strengers 2013).

Similar 'preconfiguration' of certain kinds of user in smart-energy industry standards and energy policies has also received attention (Darby 2012; Pullinger et al. 2014; Schick and Gad 2015). Throndsen (2017), reviewing literature on smart-grid research papers, argues that preconfigurations of the residential smart-grid user fall within three wide categories: economic configuration (making 'users' more active in an economically rational way); technical configuration (automation of energy consumption and bypassing active forms of use); and social science configuration (comparing visions of imagined users to 'real' users, whoever they may be). To this list, we can add the legal configuration of users as holders of specific rights, such as consumer rights (Rommetveit et al. 2017; van Dijk et al. 2018). A study of Danish smart-grid experimentation projects also discovered three main 'scripts' that inform the interaction that these projects expect between household consumers and future smart grids: an economic incentives script, an automation script, and an information and visualisation script (Hansen and Borup 2018). At the same time, many social science scholars of smart grids and smart meters have reacted to these types of preconfiguration of the user and their models of rationality by highlighting the diverse and often unpredictable ways in which consumers engage with smart energy, most typically in their homes (Heiskanen and Matschoss 2016; Hargreaves and Wilson 2017; Kahma and Matschoss 2017; Winther and Bell 2018), but also partially as part of wider settings such as residential areas (Bulkeley et al. 2016).

These are valuable insights, especially for understanding why supposed 'users' may wish to become 'non-participants' in the smart grid, for example in smart-grid demonstration sites (Throndsen 2017) or by not being involved in smarter energy services (Kahma and Matschoss 2017). Yet, these examples and literature call for a more dynamic way of characterising the relationship between producers and users in smart-grid projects, and more extensive engagement with the sociology of user representations and its importance for the development of complex new infrastructural projects that we outlined in the previous subsection. To this end, we build on empirical work on the representation of users and uses in smart grids and smart metering to develop a theoretical contribution to this debate.

Methodology

We advance a new, more dynamic position on user–technology relationships by focusing on the diversity of representations of 'user' in smart-grid projects, and by highlighting the kind of attention given to 'final users' and use among expert participants and project developers. Our analysis draws empirically from 21 qualitative accounts given by a diverse selection of smart-technology experts from different parts of Europe. A pool of 71 subjects was selected to encompass a range of respondents, academics and non-academics as well as engineers and social science and humanities researchers. Our research drew on social scientific interview and survey research techniques but built upon these to go beyond the conventional single-site qualitative study (Silvast and Virtanen 2019; Hyysalo et al. 2019). We wanted to involve social science and humanities scholars as well as engineers in our consultation and also include those advising people on smart meters, information campaigners, and the electricity industries, among many others. The project received 21 responses, giving a response rate of 30 percent. Sixteen respondents were academic researchers on smart technology. The whole dataset has a mix of engineering experts (12) and those representing social science and humanities (9). Outside universities, the experts studied (5) worked for associations (e.g., energy conservation, smart-energy campaigns, citizens' advice) and a consultancy. Men were overrepresented (13 men, as compared with 8 women). Most of the experts were based in the UK (15), which was our main focus, but the materials also contain responses from Finland (2), Austria (2), Italy (1), and Australia (1).

Results

Layered infrastructure and layered 'use': smart meter users or users of smart electricity services?

In current policy programmes, marketing campaigns, and some development projects, smart meter users have become very central loci of intervention and discussion. Policy pronouncements and promotional materials on smart meter rollouts frequently emphasise that smart grid and smart metering will allow end-users to monitor their energy use and thereby potentially save money. This is expected to apply in (or is at least promoted in relation to) ordinary households. The informants in our study typically referred to these behavioural visions as their starting point. As a representative of a publicity campaign for smart meters expressed it:

> Smartness has two components: enabling responsive consumers and near real-time information. It allows you to change your behaviour. It allows you to do things that you couldn't do before by responding to data, such as refurbish your home's heating systems you could not do in the same way by using the 'dumb' meter.

Two academic researchers who responded also exemplified this premise of putting rational individuals at the centre of 'smartness'. One of them thought that 'people say they want information that they can use, so if they know that they're spending a relatively large amount of money on say taking really long showers, then they know they can do something about it'. Another concluded similarly: 'You could say I'm going to have an intelligent battery that just charges when it's cheap and then I'm going to spend the electricity later. Then you can think of very smart things'.

A response from Finland differs from these expectations in some respects, though it also shares the key premises on 'smartness' benefiting the 'final user'. A researcher of smart-grid innovations pointed out that the financial benefits of active energy management are mostly for major commercial energy users, such as large retail chains that can shift the time when they use their refrigerators, for example. There is a similar benefit for large companies, for whom relevant arrangements in getting compensated for their flexibility in energy use predate the smart grids by several years. However, current smart-grid devices, smart meters in particular, have in fact shown relatively limited capabilities in changing energy consumption only at that layer of the energy infrastructure. This has been evidenced in countries such as Finland and Sweden that have already rolled out smart meters. There, patterns of electricity consumption have remained stable and predictable with little evidence of the anticipated new forms of consumer response. This corresponds with a pertinent finding by sociological studies. Namely, consumers engage with new smart technologies in diverse and often unpredictable ways, with their consumption deeply ingrained in everyday habits most of the time; thus, the experts' faith in 'rational individuals' conveys an idealisation (Heiskanen and Matschoss 2016; Hargreaves and Wilson 2017; Kahma and Matschoss 2017; Winther and Bell 2018). That being said, adding 'smartness' to power grids is not merely about new forms of consumption as it could also bring about new smarter energy services for these energy users.

Energy suppliers and distribution operators can access a considerable amount of new data from smart meters and apply that to various ends. As our informants envisaged, these include improvements to supply–demand balancing, infrastructure reinforcements, network optimisation, planning, and identification of power outages. In the related sector of smart water meters, infrastructure managers anticipate significant benefits arising from more detailed knowledge about water use, improvements in efficiency via reduced consumption, reduced costs, and capturing the variation of water use during the week and days as part of infrastructure planning (March et al. 2017). However, infrastructure suppliers could also use the increased 'smartness' to offer new added-value services to their customers. One relatively simple and common example is time-of-use tariffs of electricity that vary according to the hour; this is still consistent with the visions of more rational consumers whose interest is in saving money. More complex demand-side flexibility services might put buffers between energy production and energy use by creating services that automate the switching of tariffs or even steer household appliances in ways largely invisible to the consumers. These expectations of users of smart-energy services, rather than just users of smart meters, also have very particular implications for the design of smart-grid infrastructure, as we show later in this chapter.

In addition to conventional energy suppliers and distribution operators, there could be an opportunity for a host of service providers to use the smart grid to enter the energy supply markets. Developers of ICT systems, or so-called next-generation intermediaries, have already emerged in Finland (Hyysalo et al. 2017) and Norway (Sareen and Rommetveit 2019; Sareen 2020) with fully digital automatic meter-reading across the country. There, a number of new automation and analysis programmes now offer to draw on the data from the smart meters and translate it for the use of consumers in a variety of ways (Hyysalo et al. 2017). These include the formation of 'virtual power plants' that reduce demand by pooling stocks of households (typically some hundreds or thousands) among whom (invisible) load shifting in use of electric heating and electric appliances can be achieved.

Differing time spans of benefits: users of current smart-grid configurations or users of future intelligent networks?

The difference between present-day users of smart meters and users of emerging energy services leads to a point on different temporalities of benefits of 'smartness'. It would seem that concerns over how users could save money are mainly represented as important in the early stages of smart-meter rollout. As the smart-grid infrastructure evolves, however, new differences could emerge between the users of the current smart grids and the more flexible users of future intelligent energy networks.

Rather than merely seeing them as optimising their energy use, the subjects related the representation of smart-grid users with broader infrastructural visions, technological developments, and associated issues. The subjects anticipated that peak consumer demands in electricity will become an increasing problem. For example, with the adoption of electric vehicles, the load caused by drivers arriving home at the same time may strain the electricity grid. Similar problems, but on the generation side, were seen to arise from the transition to renewable energy sources (e.g., wind and solar) where output is intermittent depending on weather and season and often difficult to foresee, creating issues for supply–demand balancing as well as for maintaining stable electricity grids. Altogether, a whole chain of policy problems and visions could shape the importance of smart grids for society over the coming decades. The head of an energy institute in mainland Europe explained this policy dynamic:

> We should ask though why decarbonisation is so important. Decarbonisation is following the top priority of addressing climate change. Here we have an important

> driver for decarbonisation itself. The other key issues in the decarbonisation target of the European Commission and at world level include the environmental aspect of pollution, so several new countries are deciding one after the other a switch off date for gasoline cars. The main driver here is not climate change but environmental issues, in order to prevent the consequences of having polluted cities.

These policy priorities put a particular strain on the infrastructure of the electricity grid to withstand unpredictability in supply and demand. The informants thought that smartness offers a key pathway for minimising the risk of outages and inadequate service. Smart grids thus are not merely meters added with ICTs, but as one researcher working in Austria put it, they 'are intelligent grids that are made to be stable – the aspiration of electrical engineering is avoidance of faults and of excessive production or consumption'. The users of these future intelligent networks will obviously be very different from the users of current smart-grid configurations, particularly those restricted to just having smart meters. Evidently, a number of new roles are available for the future smart-grid 'users' across the energy supply chain.

Differing perceptions of what is a smart grid: everyone is in a grid or specific and differentially impacted groups of users?

At this point, there is a blurring in the smart-grid concept between suppliers and customers, as indeed, potentially everyone is a 'user' of 'smartness' in the future's more intelligent energy networks. In these scenarios, smart-energy services are not necessarily meant just for affluent populations, for example those owning electric vehicles, ground heat pumps, or 'smart' homes (Schick and Gad 2015). Smart grids can, but do not necessarily have to, follow upon the earlier logics of liberalisation, where energy markets were segmented so that those willing to pay can be offered higher-quality infrastructure services (Graham and Marvin 2001). Smart-grid infrastructures do not inevitably lead either to reinforcing differences or to equalisation. Indeed, our respondents had no single view on these issues. To some respondents, it seems, variations of service were inherent in all liberalised provisions. Other informants recognised the potential 'sorting' of customers as an issue and reflected on it extensively, suggesting new forms of knowledge and interventions that could help to understand specific and differentially impacted groups of users.

While our respondents often spoke about different industrial 'users' of smart energy, in the majority of cases, when our informants mentioned the 'user', they actually meant final users such as households, individuals, or, in some cases, energy-using communities. Many of our experts also wanted to pursue specific work to identify their needs and build a 'bridge' toward achieving them. Their concept of the 'user' clearly changed in explaining this issue: away from the abstract diagrammatic idea of the user in the two-way web of smart-energy systems, to associating users very closely with 'real' flesh-and-blood people, drawing upon their socio-demographic and attitudinal characteristics such as type of household, age, region, income, consumer interest, trust and risk perception, and even their vulnerability. An economist who had run public workshops on this topic in the UK summed up:

> I think a factor might be consumers' interest towards reducing their bills, however it was when we did public workshops for example on that, people don't see the trade off in terms of higher electricity bills versus flexibility that they need to create.

Another recent focus group on smart meters discovered that consumers are positively predisposed to demand-side management and smart technologies in general, but, at least in a context in which they lack first-hand experience of smart-energy services, are reluctant to accept automated systems if they are controlled by energy suppliers (Mesarić et al. 2017). Such socio-economic variables could explain not only smart metering use and acceptance, but also the protests against them. At least according to one informant, such protests are shaped by 'the feeling in the local community' concerning infrastructure projects rather than opposition to the dangers of technology as such.

There have been long-term attempts within industries to comprehend the needs of energy consumers: energy supply companies engage in market research that aims to segment customers, and to identify their interest in different products and their preference for billing models that are also linked with their acceptance of 'smartness'. Indeed, in temporal terms, many of our subjects suggested that such conventional knowledge still shapes smart-energy innovations: namely, that users are brought into the smart-energy development at a late stage, where they can tweak functionalities but not thoroughly change designs (Heiskanen and Matschoss 2016). This tendency can be reinforced by the fragmentation of roles and responsibilities in liberalised energy supply and distribution (Silvast 2017b). Thus, meter manufacturers have little direct contact with end-users. As a representative of meter manufacturers told us:

> Making smart energy meters often has little direct relation to energy end users in their homes. First, the manufacturers do not sell smart meters directly to the public. The end users, for their part, will be very rarely looking directly at their energy meter even if it is a 'smart' meter. The meter may be placed under the stairs or other difficult locations to see. There are other companies making In-Home Displays (IHDs) and apps on the phone for these smart meters. They do engage in market research. There have been a few companies that both manufactured meters and the IHD for it but in the UK's next Smart Metering Equipment Technical Specification (SMETS) 2.0, no companies are present that do both.

The 'active and rational consumer' is at once at the centre of many smart-grid programmes; yet, as this quote suggests, the research on smart-energy services is only done by select organisations. How much do we know about these 'active users', and what other metaphors might inform smart development projects?

Varying certainty of user and technology representations in different parts of smart grids

The anticipated 'users' of smart technologies fall within a wide range, including many intermediaries between 'producers' and 'users'. In this fashion, the umbrella term 'user' masks a range of emerging interests, from households to microbusinesses, new energy service providers, consumption aggregators, price-comparison web sites, and even the energy systems operators themselves. However, the kinds of 'user' that benefited most directly, according to our informants, were professionals, including electricity suppliers and network operators, meter manufacturers, ICT companies, and other intermediaries. Benefits to consumers for 'self-managing' energy were anticipated, but also sometimes markedly difficult to evidence with certainty.

According to our respondents, this matter could be resolved by long-term research that seeks to understand better the end-users and uses of energy more generally. Our research informants suggested that the possibilities that people will anticipate in smart-energy technologies are shaped by the types of housing stock in different regions, the access to infrastructure (especially Internet connections, although interestingly not required by smart meters in the UK), lifestyle-based choices, household economics, and even the very fact that it is not just individuals, but often every person in a household, who affect energy choices. One technical researcher thought that more knowledge of energy consumption in the household could even expose new conflicts in the family. In a recent study, anthropologists showed similarly how new home displays of smart meters in Norway and the UK brought complex social dynamics into being, 'regarded as an ally by those members most concerned with saving on household electricity consumption by providing objective evidence of costs linked to specific practices' (Winther and Bell 2018: 13).

Vulnerable customers who have difficulty in paying their energy bills pose further questions to the smart-energy infrastructure along these lines. Low-income people, especially those experiencing fuel poverty, have much to gain by reducing their bills, but our informants recognised that they will need very specific support to make use of the smart-meter information. As a researcher of smart-energy innovation pointed out, low-income people 'often struggle with so many other problems that this is not their main priority'. Building a bridge between preconceptions about 'users' and actual situations of use requires dedicated work in cases like this. Many ways to iterate between technologists and 'users' were suggested in our materials: for example, sending field agents to low-income families to discuss their budget management or creating special 'energy offices' that translate the ideal of money saving into the contexts of management of energy consumption at home.

With examples such as this, the aspiration of understanding the 'user' has taken almost a full turn from the infrastructural visions and promises that started our analysis. It seems that while smart projects are typically conceptualised around an abstract, singular representation of a 'typical' user (the user as everybody) as they proceed, they also tend to uncover the variability of users and reveal how well technology designers have understood the different conditions of use. Yet, even when developers address these issues, they are still 'representing' the user in a certain way that remains open to further discussion. For example, we would claim that even in this case, the developers and experts assumed the ideal of active, rational consumers; that is why they believed that financial counselling of residential consumers will help address problems with fuel poverty. Smart-grid developers could pay explicit attention to recognising these various descriptions of the 'user' and their underpinning assumptions, and also ask whether they correspond with the kinds of energy service that the smart grid can offer in the future, as we discuss next.

Metaphors of system evolution: active, rational consumer or routinised user of optimisation services?

For a number of years, social scientists have argued that the 'rational consumer' or 'resource man' envisaged by many smart development programmes conveys an idealised image (Strengers 2013; Winther and Bell 2018). We now know that relatively few people may have the interest or the capacity to analyse their energy meter readings all the time or very frequently in their everyday lives. We also know that householders experience infrastructures as part of their everyday life, relating to various social practices, ingrained habits, household technologies, and socio-cultural contexts of daily life (Shove and Warde 2001; Shove et al. 2012). Considering these contexts, persuading users to become more active participants in 'smartness' tends to be time-consuming, and some informants found appeal in innovations that partially sidestep these issues and automate energy use: for example, by letting an automated agent switch residential energy tariffs.

A developer of such automation said that it is an essential element in smartness, notwithstanding the few exceptional individuals who are willing to give constant attention to their energy usage: 'I think you have to have automation where you have that sort of variability. . . . With the increased complexity of variable use, variable tariffs, you have to have the corresponding automation'. As this example shows, automation offers a more plausible way for users to control electricity use. Rather than imagining a domestic consumer who checks prices all the time, it could provide a space for consumers to deliberate about the kinds of price/service trade-off they would like, for example delaying their use of household devices, or consenting to an energy tariff switch that was, as such, decided by autonomous software. Stepping away from the restricted focus on micro-decisions may also allow more space for providing information on other aspects that interest people in energy issues: such as the source of electricity or volumes of energy trade across national borders. Automation, then, does not simply mean that the end users become disinterested in energy. Indeed, studies have shown that some smart-grid 'users', especially innovative 'lead user-consumers', have considerable capability to stimulate industry-wide innovation and articulate the social and societal responsibility relevant to energy markets (Heiskanen and Matschoss 2016).

In any case, whether designers focus on active and rational consumers or try to support routinised users of optimisation services has important repercussions for the evolution of energy systems. Many of the current ambitions, such as more or less real-time measurements of energy consumption, are not possible with today's smart-energy meters, which measure energy by a 15-min or 30-min interval. If the ambitions are to be reached, then, this means future systems should be designed in a way that allows for information to be gathered in much finer detail than before: for example, via new devices called submeters, which allow for nearly real-time readings. This would enable suppliers to gather data about consumers, but also involve consumers to generate more nuanced demand-side responses to the signals that they receive from their meter. As a technologist explained, however, enabling consumers' demand-side responses need not only 'fast and reactive' real-time measurements, but also energy providers gaining access to the final appliances at home so that they can partly control them. Such solutions have been tested but are not available on the mass markets. These considerations show that an emphasis on the directly active and rational consumer has to be paired with one equipped with smart systems that measure and control energy use down to the detail of final appliances, potentially requiring relatively sporadic active engagement.

This development could, however, open up privacy and data protection issues. In the example of the Netherlands, lawyers deemed that smart-meter readings more frequent than every 15 minutes infringe the European Convention on Human Rights (Cuijpers and Koops 2013). Additionally, the access of network operators, suppliers, and third parties to smart-metering data triggered public opposition (Hoenkamp et al. 2011). A researcher of smart systems in the UK agreed about such concerns by noting that 'there is a lot of personal information that one can get from a smart meter so they're absolutely right to be worried about the privacy of it'. As another researcher of smart technologies noted, protests belong to democratic contestation and can bring these kinds of issue to the fore: their key value is 'getting something into the agenda or calling attention to something'.

In contrast, an emphasis on the routinised users of optimisation services has different implications for the evolution of the energy system. This focus includes consideration of energy sources – especially intermittent renewable energy – but also the electricity distribution system where the 'smartness' is envisaged to provide major changes. In this case, smart grids and smart meters will themselves act as infrastructures for demand-response and optimisation services in the future. This means that, for operators, service providers, 'third party intermediaries', utilities, and developers, the focus should be on the interoperation of smart grids and other infrastructures and linking the abundance of new data from smart meters to advances in the expanding systems made by ICT developers, who our respondents saw as significant drivers of smart-energy developments.

Altogether, the expert considerations in country contexts that do not as yet have two-way metering systems in place resemble the past discourse in settings such as Finland, where remote real-time control of both heat and electricity is already possible. Experts in these countries – indicated at least in Finland as well as Norway – see the future quite differently by now. The major thrust appears to be that home energy consumption is and will remain highly routinised and that smart automation systems will not primarily operate by way of decision systems that offer price signals and expect people to respond. Such information guidance is likely to feature as one part of the service. However, we could think of examples where it will remain reserved for informing consumers about the rare extremely high-peak energy prices (tens or hundreds of times the average). Otherwise, they will proceed by two to five pre-set programmes by which the energy consumption of the household is organised (such as 'no constraints', 'weekday pattern', 'max load shifting', 'away'), between which consumers switch via control panels and mobile phone apps (Hyysalo et al. 2018).

Cui bono? What can early assessment assert about the harms and benefits of smart-grid projects?

In current debates concerning the spread of smart meters across societies, a key concern of social scientists has been linked with questions posed by the sociology of user representations and the sociology of expectations. These issues are related to the manner in which the 'user' is envisaged in smart development projects and to the kind of industry-wide expectations of 'uses' that give shape to these representations. Recently, one main concern in this respect has been that smart metering should be 'socially' integrated into everyday life, to avoid causing potential resistance when it is not 'socially' integrated enough and leading to a need to inquire more closely into how users and use are conceived (Sovacool et al. 2017). As a way of counter-argument, public campaigns on smart meters have drawn upon independent surveys (Smart Energy GB 2018) to measure consumer acceptance: for example, to indicate that almost all the population, such as in the UK, are aware of smart meters, the majority of whom would recommend smart meters to their peers, with many of them expecting to get a smart meter soon (Electronics Weekly 2017). In these ways, the meters are 'socially' integrated into everyday life insofar as, according to surveys, they would be accepted by the surveyed population.

The two arguments are obviously different. However, they share one key assumption about evaluation: what happens is an early assessment of the harms and benefits of smart meters as they are being rolled out, whether this is done by social science analysis of policy programmes or large-scale consumer surveys. This premise is indeed very helpful especially for understanding the user–technology relationships in currently existing smart meters and smart grids. If smart grids and smart meters will, rather, become an infrastructure for a host of future intelligent optimisation and other services, then it might not yet be fully known who will 'benefit' and who will 'lose' from them.

Whereas certain problems may recede into the background (for example, those predicated on creating rational, interactive consumers), others may intensify (such as concerns with privacy and human rights). Therefore, as in any other large-scale system and infrastructure, room should be preserved for the user representation to evolve along with technological advances and experiences from early assessments. Early in the European smart meter rollout, a report to the European Consumer Organisation, BEUC, argued that in order to avoid technological lock-ins, 'beyond information to consumers, which is necessary but not sufficient, we suggest that consumers should be allowed to experiment with different configurations of the smart meters while still in the process of invention' (Klopfert and Wallenborn 2012: 2). This insight into learning and the co-evolution of technological innovations, users, and uses remains very relevant today.

Researchers and policy-makers pursuing smart grids might develop concepts and analyses of how such user representations evolve alongside the smart-grid infrastructure and its multiple potential uses, and how these have played out in countries where smart grids are already further developed. In so doing, they might also keep a close eye on how that affects the risks that opponents of smart grids and meters claim to be inherent in these technologies.

Discussion

By studying smart-grid development projects and drawing upon insights from the social sciences and humanities as well as the experience of experts working in these fields and ICTs, we can now bring forward a number of views that take these issues further.

This chapter focused on how smart grid development projects represent the 'user' of smart-grid and smart-metering technologies. We showed that there are varieties of 'user' of smart grids in such projects, only some of them conventional 'final users' such as households. Energy suppliers, electricity distribution operators, automation analysis software providers, price comparison websites, smart homes, and others are linked to the smart-grid promises, themselves shaped by wider visions about decarbonisation and cost minimisation. Yet, in the majority of cases, the experts involved in the CANDID research still envisaged the user – often a household – as actually the 'final user', or as a person in the fuller sociological meaning. In different ways, this user was expected to become more active than the current largely routinised energy consumers are (Shove and Warde 2001) and motivated to save money and energy by using smart technologies, or even becoming a more active participant in managing energy provision altogether.

At this point, smart systems encounter a difficulty about which social scientists know much, including those in our study and many energy experts in industries and governments. The problem corresponds to the 'users' and uses of most large-scale infrastructures, as exemplified by the Internet (Edwards et al. 2007). In a system that envisages connecting anyone who uses energy – in practice, millions of consumers – it is very hard to know what each 'user' needs and to tailor the offered system to these (Oudshoorn et al. 2004; Rommetveit et al. 2017). With these issues in view, industries and manufacturers have needed to 'produce' different 'proxy' users, thus helping them to design, produce, and market their products. In contrast to what is sometimes assumed, these analysts are not merely suggesting that all these 'proxy' users be replaced with input from 'actual users' or 'real users'. As Hyysalo and Johnson (2015) argue, such a move could restrict choices in the human-centred design of technology. This chapter has developed a conceptual framework, informed by STS-inspired studies of design and users, and has applied it to study the developers of smart-grid infrastructures.

Working from these perspectives, we observed a number of the features identified in earlier work on the sociology of user representations to be at play with smart-grid development as well. These include discovering a range of competing user representations in an emerging technology; observing that developers reduce the complexity of these representations through simplifying user characteristics; and noting that user representations are not mere abstract exercises but play important roles in the expectation work of mobilising resources and support as well as for developing emerging technologies such as smart grids and smart meters.

Beyond this, smart-grid user representations draw attention to a set of new issues in how users are represented in emerging and multi-layered infrastructures:

1 The likely intensity of users' engagement with smart-grid technologies varies considerably: at the one end of the spectrum, users draw on smart technologies to carry out limited optimisation of their energy usage; at the other end are 'prosumers' who actively optimise their own production, storage, and home energy use and transit electricity themselves.
2 There is a significant temporal difference: the represented user could be expected to engage with currently existing smart-metering solutions, or with intelligent optimisation software and services that are envisaged to be in place once the smart grid is mature sometime in the 2020s.
3 These differences find their corollary in whether the user is seen just in terms of the smart meter (and perhaps supplemented by visualisation tools), or more comprehensively through the adoption of further layers of services and applications.

Many of these representations would entail a dramatic increase in the presumed rationality of users. Others are compatible with the concept of the routine consumer who chooses a contractual regime that determines how much optimisation and demand-response takes place and how often s/he can switch between pre-set options. This latter representation also includes consumers who are motivated in part by saving money – they are just doing it in very different terms from the envisaged active users. However, we add an important caveat to these findings. These 'preconfigurations' of 'the user' capture important aspects of active decision-making and habits as part of the user–technology relationship. They might be less sensitive to the experiences of those users and public that become resistant to the capacities of the enrolled technologies. Examining these issues further may require a shift in terminology, from 'user' or 'consumer' to 'citizen', 'right holder', or 'public', in future studies.

All of this has significant implications for understanding how adequate our conceptualisations of the smart-grid user are, and also how well we are able to assess the eventual impacts of these technologies. If one asserts that the smart grid consists primarily of added remote metering, the root image would veer towards the active and rational consumer who is busy optimising energy usage and planning. On the other hand, if one projects that smart meters are only the infrastructural basis for arrays of optimisation and demand-response services to come and are likely to feature considerable contract and service variety, the images of represented users change. The range of users grows much wider and use becomes potentially more voluntary, subject to service contracts purchased from new markets. The eventual impacts on people are likely to evolve, reflecting pertinent regulatory and technological decisions made over time. Importantly, this provides room to learn from the manifested impacts so that adverse effects can be remedied, with the strong points of smart-grid technologies being capitalised on in both regulatory regimes and digitalised service development.

Conclusions

Understanding what the different representations of 'users' are provides important input for designing smart-energy systems and discussing how 'responsible' and 'responsive' these innovations are likely to become for different users and toward other societal actors. Our analysis highlights the inadequacy of an implementation strategy that merely introduces into existing networks technical capacities for remote sensing and two-way communication. We have argued that such a limited conception of the functioning of smart-grid infrastructures is not likely to be the end-state of their evolution and appears not to be the desired state. This situation calls for a systematic exploration of optimisation and demand-response systems that would bring wider and different arrays of benefits to different consumer segments.

To steer the development of smart-grid infrastructures and associated evolving user representations, infrastructural evolution should also be periodically monitored from the point of view of the various consumers. Such anticipatory assessment might usefully include monitoring for developments that veer towards partisan or unduly narrow models of infrastructural operation, service provision, or ownership, and give balanced attention to disbenefits and harms as well as benefits sought earlier on. Policy-makers, innovators, and researchers pursuing smartness could explore evidence from the varying experimental development strategies in different contexts and countries that, for example, seek to grow systems incrementally by solving current demands and attending to future needs. Such long-term periodic assessment and steering strategy has been practiced within Constructive Technology Assessment (Rip et al. 1995) and in the management of radical innovations (Raven et al. 2009). If coupled with systematic consultations in different country settings, they could fit particularly well with the governance of smart grids being developed in somewhat different configurations in many settings for a foreseeably long time.

By deploying the approach developed in this chapter, smart-grid developers and experts could give more explicit attention to recognising the descriptions of 'users' in smart grid projects. They could ask what level of engagement is expected of these 'users', and how feasible these expectations of 'use' are in comparison to the possibilities and limits of existing, or future, energy services and applications. Conversely, examination of user representations can highlight the need for further technology and service development if some of the envisaged user profiles and user actions appear unrealistic for presently available technologies. Smart-grid user representations clearly indicate the need to move beyond smart meters and focus development efforts for further layers of services available for differently oriented consumers. These descriptions of the 'user' should be assessed not just once, but also periodically updated as development projects evolve and the related smart technologies mature. Cross-country comparisons, which we have started to conduct in this chapter, would provide critical input to these assessments, helping to discover previously overlooked representations of 'users' by drawing on contexts where the smart grid is configured very differently. Much research and development work remains to be done in understanding the emerging usage, services, regulatory choices, and implications of smart-energy technologies from these perspectives.

Acknowledgement

This research was funded by the Checking Assumptions and Promoting Responsibility in Smart Development (CANDID) Projects EU research programme. European Commission Horizon 2020 (2017, grant number 732561, www.candid.no). Antti Silvast wishes to acknowledge the funding from the NTNU Energy Transition Initiative at the Norwegian University of Science and Technology (NTNU). Kjetil Rommetveit also wishes to recognise the funding provided by the Norwegian Research Council for the ERA-Net project PARENT, project number 259777. Sampsa Hyysalo acknowledges funding from Strategic Research Council of Finland grant number 133143251, 'Smart Energy Transition'.

An earlier version of this chapter was published in *Sustainability* in 2018, 'Who "Uses" Smart Grids? The Evolving Nature of User Representations in Layered Infrastructures' (Silvast et al. 2018).

References

Akrich, M. (1992): The De-Scription of Technical Objects, in Bijker, W. and Law, J. (eds.): *Shaping Technology/Building Society: Studies in Sociotechnical Change*. Cambridge, MA: MIT Press, 205–224.

Akrich, M. (1995): User Representations: Practices, Methods and Sociology, in Rip, A., Misa, T. J. and Schot, J. (eds.): *Managing Technology in Society: The Approach of Constructive Technology Assessment*. London: Pinter Publisher, 167–184.

Akrich, M. and Latour, B. (1992): A Summary of a Convenient Vocabulary for the Semiotics of Human and Nonhuman Assemblies, in Bijker, W. E. and Law, J. (eds.): *Shaping Technology/Building Society: Studies in Sociotechnical Change*. Cambridge, MA: MIT Press, 259–264.

Ballo, I. F. (2015): Imagining Energy Futures: Sociotechnical Imaginaries of the Future Smart Grid in Norway, *Energy Research & Social Science* 9, 9–20.

BEIS (UK Department of Business, Energy & Industrial Strategy) (2016): Smart Meter Rollout Cost-Benefit Analysis Part I. Available at: https://assets.publishing.service.gov.uk/government/uploads/system/uploads/attachment_data/file/567167/OFFSEN_2016_smart_meters_cost-benefit-update_Part_I_FINAL_VERSION.PDF [28.08.2020].

Bijker, W., Hughes, T. P. and Pinch, T. (eds.) (1987): *The Social Construction of Technological Systems: New Directions in the Sociology and History of Technology*. Cambridge, MA: MIT Press.

Borup, M., Brown, N., Konrad, K. and van Lente, H. (2006): The Sociology of Expectations in Science and Technology, *Technology Analysis and Strategic Management* 18, 285–298.

Bulkeley, H., Powells, G. and Bell, S. (2016): Smart Grid and the Constitution of Solar Electricity Conduct, *Environment and Planning A* 48, 7–23.

Cuijpers, C. and Koops, B.-J. (2013): Smart Metering and Privacy in Europe: Lessons from the Dutch Case, in Gutwirth, S., Leenes, R., de Hert, P. and Poullet, Y. (eds.): *European Data Protection: Coming of Age*. Berlin: Springer, 269–293.

Darby, S. J. (2012): Metering: EU Policy and Implications for Fuel Poor Households, *Energy Policy* 49, 98–106.

DECC (UK Department of Energy and Climate Change) (2012): Smart Metering Implementation Programme. Available at: https://assets.publishing.service.gov.uk/government/uploads/system/uploads/attachment_data/file/68976/Smart_metering_programme_update_-_April_2012.pdf [28.10.2020]

Edwards, P. N., Jackson, S. J., Bowker, G. and Knobel, C. B. (2007): Understanding infrastructure: Dynamics, tensions, and design. Report of a Workshop on 'History & Theory of Infrastructure: Lessons for New Scientific Cyberinfrastructures'. Available at: https://deepblue.lib.umich.edu/bitstream/handle/2027.42/49353/UnderstandingInfrastructure2007.pdf?sequence=3&isAllowed=y [01.11.2020]

Electronics Weekly (2017): Smart Energy GB Hits Back at Smart Meter Research. Available at: www.electronicsweekly.com/news/smart-energy-gb-hits-back-smart-meter-research-2017-09/ [01.11.2020]

European Commission (2011): Smart Grid: From Innovation to Deployment. Available at: https://eur-lex.europa.eu/LexUriServ/LexUriServ.do?uri=COM:2011:0202:FIN:EN:PDF [28.10.2020]

European Commission (2014): Benchmarking Smart Metering Deployment in the EU-27 with a focus on electricity. Available at: https://eur-lex.europa.eu/legal-content/EN/TXT/?uri=COM%3A2014%3A356%3AFIN [28.10.2020]

European Commission (2019): *Clean Energy for all Europeans*. Luxembourg: Publications of the Office of the European Union.

European Parliament and Council (2012): Directive 2012/27/EU on Energy Efficiency, Amending Directives 2009/125/EC and 2010/30/EU and repealing Directives 2004/8/EC and 2006/32/EC. Available at: https://eur-lex.europa.eu/legal-content/EN/TXT/?uri=CELEX%3A02012L0027-20200101 [28.10.2020]

Frickel, S., Wuhr, D., Horne, C. and Kallman, M. E. (2016): Field of Visions: Interorganizational Challenges to the Smart Energy Transition in Washington State. *Brookings Law Review* 82, 693–724.

Graham, S. and Marvin, S. (2001): *Splintering Urbanism: Networked Infrastructures, Technological Mobilities and the Urban Condition*. London: Routledge.

Hansen, M. and Borup, M. (2018): Smart Grids and Households: How are Household Consumers Represented in Experimental Projects? *Technology Analysis and Strategic Management* 30, 255–267.

Hargreaves, T. and Wilson, C. (2013): Who Uses Smart Home Technologies? Representations of Users by the Smart Home Industry. Available at: www.eceee.org/library/conference_proceedings/eceee_Summer_Studies/2013/6-appliances-product-policy-and-ict/who-uses-smart-home-technologies-representations-of-users-by-the-smart-home-industry/ [01.11.2020]

Hargreaves, T. and Wilson, C. (2017): Domestication of Smart Home Technologies, in Hargreaves, T. and Wilson, C. (eds.): *Smart Homes and Their Users*. Berlin: Springer, 75–90.

Heiskanen, E. and Matschoss, K. (2016): Consumers as Innovators in the Electricity Sector? Consumer Perceptions on Smart Grid Services, *International Journal of Consumer Studies* 40, 665–674.

Helm, D. (2017): Cost of Energy Review: Independent Report for the UK Government. Available at: www.gov.uk/government/publications/cost-of-energy-independent-review [28.10.2020]

Hess, D. J. and Coley, J. S. (2014): Wireless Smart Meters and Public Acceptance: The Environment, Limited Choices, and Precautionary Politics, *Public Understanding of Science* 23(6), 688–702.

Hoenkamp, R., Huitema, G. B. and de Moor-van Vugt, A. J. (2011): The Neglected Consumer: The Case of the Smart Meter Rollout in the Netherlands, *Renewable Energy Law and Policy Review* 4, 269–282.

Hyysalo, S. (2004): Uses of Innovation: Wristcare in the Practices of Engineers and Elderly. Ph.D. Thesis, University of Helsinki, Helsinki, Finland.

Hyysalo, S. (2010): *Health Technology Development and Use: From Practice-Bound Imagination to Evolving Impacts*. London: Routledge.

Hyysalo, S., Jensen, T. E. and Oudshoorn, N. (eds.) (2016): *The New Production of Users: Changing Innovation Collectives and Involvement Strategies*. London: Routledge.

Hyysalo, S. and Johnson, M. (2015): The User as Relational Entity: Options that Deeper Insight into User Representations Opens for Human-Centered Design, *Information Technology & People* 28, 72–89.

Hyysalo, S., Marttila, T., Temmes, A., Lovio, R., Kivimaa, P., Auvinen, K., Pyhälammi, A., Lukkarinen, A. and Peljo, J. (2017): Uusia Näkymiä Energiamurroksen Suomeen – Murrosareenan tuottamia kunnianhimoisia energia- & Ilmastotoimia Vuosille 2018–2030. Available at: www.smartenergytransition.fi/tiedostot/murrosareena-loppuraportti.pdf [01.11.2020]

Hyysalo, S., Pollock, N. and Williams, R. A. (2019): Method Matters in the Social Study of Technology: Investigating the Biographies of Artifacts and Practices, *Science & Technology Studies* 32(3), 2–25.

Johnson J./Latour B. (1988): Mixing Humans and Nonhumans Together: The Sociology of a Door-closer, *Social Problems* 35, 298–310.

Johnson, M. (2013): How Social Media Changes User-Centered Design – Cumulative and Strategic User Involvement with Respect to Developer – User Social Distance. Ph.D. Thesis, University of Helsinki, Helsinki, Finland.

Kahma, N. and Matschoss, K. (2017): The Rejection of Innovations? Rethinking Technology Diffusion and the Non-Use of Smart Energy Services in Finland, *Energy Research & Social Science* 34, 27–36.

Klopfert, F. and Wallenborn, G. (2012): Empowering consumers through smart metering. Report for BEUC, the Bureau Europeen Des Unions Des Consommateurs. Available at: www.beuc.eu/publications/2012-00369-01-e.pdf [01.11.2020]

Konrad, K. (2008): Dynamics of Type-based Scenarios of Use: Opening Processes in Early Phases of Interactive Television and Electronic Marketplaces, *Science and Technology Studies* 21, 3–26.

Lovell, H., Pullinger, M. and Webb, J. (2017): How do Meters Mediate? Energy Meters, Boundary Objects and Household Transitions in Australia and the United Kingdom, *Energy Research & Social Science* 34, 252–259.

Luque, A. (2014): The Smart Grid and the Interface between Energy, ICT and the City, in Dixon, T., Eames, M., Hunt, M. and Lannon, S. (eds.): *Urban Retrofitting for Sustainability: Mapping the Transition to 2050*. London: Routledge, 159–173.

March, H., Morote, Á. F., Rico, A. M. and Saurí, D. (2017): Household Smart Water Metering in Spain: Insights from the Experience of Remote Meter Reading in Alicante, *Sustainability* 9, 582–600.

Mesarić, P., Đukec, D. and Krajcar, S. (2017): Exploring the Potential of Energy Consumers in Smart Grid Using Focus Group Methodology, *Sustainability* 9,1463–1480.

Mozaffar, H. (2016): User Communities as Multi-Functional Spaces: Innovation, Collective Voice, Demand Articulation, Peer Informing and Professional Identity (and more), in Hyysalo, S., Jensen, T. E. and Oudshoorn, N. (eds.): *The New Production of Users: Changing Innovation Collectives and Involvement Strategies*. London: Routledge, 219–248.

Oudshoorn, N. and Pinch, T. (2003): *How Users Matter: The Co-construction of Users and Technology*. Cambridge, MA: MIT Press.

Oudshoorn, N., Rommes, E. and Stienstra, M. (2004): Configuring the User as Everybody: Gender and Design Cultures in Information and Communication Technologies, in *Sci. Technol. Hum. Values* 29, 30–63.

Pahkala, T., Uimonen, H. And Väre, V. (2018): Flexible and Customer-centred Electricity System- Final Report of the Smart Grid Working Group. Helsinki: Ministry of Education and Employment in Finland. Available at: https://julkaisut.valtioneuvosto.fi/handle/10024/161147 [01.11.2020]

Pollock, N. and Williams, R. (2010): E-infrastructures: How Do We Know and Understand Them? Strategic Ethnography and the Biography of Artefacts, *Computer Supported Cooperative Work* 19, 521–556.

Pollock, N., Williams, R. and D'Adderio, L. (2016): Generification as a Strategy: How Software Producers Configure Products, Manage User Communities and Segment Markets, in Hyysalo, S., Jensen, T. E. and Oudshoorn, N. (eds.): *The New Production of Users: Changing Innovation Collectives and Involvement Strategies*. London: Routledge, 178–208.

Powells, G., Bulkeley, H. and McLean, A. (2016): Geographies of Smart Urban Power. In Marvin, S., Luque-Ayala, A. and McFarlane, C. (eds.): *Smart Urbanism: Utopian Vision or False Dawn?* London: Routledge, 141–160.

Pöyry (2017): Seuraavan sukupolven älykkäiden sähkömittareiden vähimmäistoiminnallisuusvaatimukset. Available at: https://tem.fi/documents/1410877/3481825/AMR+2.0+loppuraportti+15.12.2017/6a2df7e6-a963-40c0-b4d8-d2533fbca488 [28.10.2020]

Pullinger, M., Lovell, H. and Webb, J. (2014): Influencing Household Energy Practices: A Critical Review of UK Smart Metering Standards and Commercial Feedback Devices, *Technology Analysis and Strategic Management* 26, 1144–1162.

Raven, R. P., Jolivet, E., Mourik, R. M. and Feenstra, Y. C. (2009): ESTEEM: Managing Societal Acceptance in New Energy Projects: A Toolbox Method for Project Managers, *Technological Forecasting and Social Change* 76, 963–977.

Rip, A., Misa, T. J. and Schot, J. (1995): *Managing Technology in Society. The Approach of Constructive Technology Assessment*. London: Pinter Publisher.

Rommetveit, K., Dunajcsik, M., Tanas, A., Silvast, A. and Gunnarsdóttir, K. (2017): The CANDID Primer: Including Social Sciences and Humanities scholarship in the making and use of smart ICT technologies. CANDID (H2020-ICT-35–2016). Available at: http://candid.no/progress [28.10.2020]

Sareen, S. (2020): Social and Technical Differentiation in Smart Meter Rollout: Embedded Scalar Biases in Automating Norwegian and Portuguese Energy Infrastructure, *Humanities and Social Sciences Communications* 7(1), 1–8.

Sareen, S. and Rommetveit, K. (2019): Smart Gridlock? Challenging Hegemonic Framings of Mitigation Solutions and Scalability, *Environmental Research Letters* 14(7), 075004.

Schick, L. and Gad, C. (2015): Flexible and Inflexible Energy Engagements – A Study of the Danish Smart Grid Strategy, *Energy Research & Social Science* 9, 51–59.

Shove, E., Pantzar, M. and Watson, M. (2012): *The Dynamics of Social Practice: Everyday Life and How it Changes*. London: Sage Publishing.

Shove, E. and Warde, A. (2001): Inconspicuous Consumption: The Sociology of Consumption, Lifestyles and the Environment, in Dunlap, R. E., Buttel, F. H., Dickens, P. and Gijswijt, A. (eds.): *Sociological Theory and the Environment: Classical Foundations, Contemporary Insights*. Lanham: Rowman & Littlefield Publishers, 230–251.

Silvast, A. (2017a): Energy, Economics, and Performativity: Reviewing Theoretical Advances in Social Studies of Markets and Energy, *Energy Research & Social Science* 34, 4–12.

Silvast, A. (2017b): *Making Electricity Resilient: Risk and Security in a Liberalized Infrastructure*. London: Routledge.

Silvast, A. (2018): Co-constituting Supply and Demand: Managing Electricity in Two Neighbouring Control Rooms, in Shove, E. and Trentmann, F. (eds.): *Infrastructures in Practice: The Evolution of Demand in Networked Societies*. London: Routledge, 171–183.

Silvast, A. and Virtanen, M. J. (2019): An Assemblage of Framings and Tamings: Multi-sited Analysis of Infrastructures as a Methodology, *Journal of Cultural Economy* 12(6), 461–477.

Silvast, A., Williams, R., Hyysalo, S., Rommetveit, K. and Raab, C. (2018): Who 'Uses' Smart Grids? The Evolving Nature of User Representations in Layered Infrastructures, *Sustainability* 10(10), 3738.

Smart Energy GB (2018): Smart Energy Outlook. Available at: www.smartenergygb.org/en/resources/press-centre/press-releases-folder/smart-energy-outlook-march18 [01.11.2020]

Sovacool, B. K., Kivimaa, P., Hielscher, S. and Jenkins, K. (2017): Vulnerability and Resistance in the United Kingdom's Smart Meter Transition, *Energy Policy* 109, 767–781.

Stewart, J. (2003): The Social Consumption of Information and Communication Technologies (ICTs): Insights from Research on the Appropriation and Consumption of New ICTs in the Domestic Environment, *Cognition Technology & Work* 5, 4–14.

Stewart, J. and Hyysalo, S. (2008): Intermediaries, Users and Social Learning in Technological Innovation, *International Journal of Innovation Management* 12, 295–325.

Stewart, J. and Williams, R. (2005): The Wrong Trousers? Beyond the Design Fallacy: Social Learning and the User, in Rohracher, H. (ed.): *User Involvement in Innovation Processes: Strategies and Limitations from a Socio-Technical Perspective*. Munich: Profil, 195–221.

Strengers, Y. (2013): *Smart Energy Technologies in Everyday Life: Smart Utopia?* Berlin: Springer.

Throndsen, W. (2017): What Do Experts Talk about When They talk about Users? Expectations and Imagined Users in the Smart Grid, *Energy Efficiency* 10, 283–297.

UK House of Commons. (2016): Evidence Check: Smart Metering of Electricity and Gas. Science and Technology Committee. Available at: https://publications.parliament.uk/pa/cm201617/cmselect/cmsctech/161/161.pdf [28.10.2020]

UK National Infrastructure Commission. (2016): Smart Power. Available at: https://assets.publishing.service.gov.uk/government/uploads/system/uploads/attachment_data/file/505218/IC_Energy_Report_web.pdf [28.10.202]

van der Horst, D., Staddon, S. and Webb, J. (2014): Smart Energy, and Society? *Technology Analysis and Strategic Management* 26, 1111–1117.

van Dijk, N., Tanas, A., Rommetveit, K. and Raab, C. (2018): Right Engineering? The Redesign of Privacy and Personal Data Protection, *International Review of Law, Computers & Technology* 32(2–3), 230–256.

Verbong, G., Beemsterboer, S. and Sengers, F. (2013): Smart Grid or Smart Users? Involving Users in Developing a Low Carbon Electricity Economy, *Energy Policy* 52, 117–125.

von Schomberg, R. (2011): Towards Responsible Research and Innovation in the Information and Communication Technologies and Security Technologies Fields, in Von Schomberg, R. (ed.): *Towards Responsible Research and Innovation in the Information and Communication Technologies and Security Technologies Fields* Brussels: European Commission Services, 7–16.

Wilkie, A. and Michael, M. (2009): Expectation and Mobilisation: Enacting Future Users, *Science, Technology, & Human Values* 34, 502–522.

Williams, R., Stewart, J. and Slack, R. (2005): *Social Learning in Technological Innovation: Experimenting with Information and Communication Technologies*. Cheltenham, UK: Edward Elgar Publishing.

Wilson, C., Hargreaves, T. and Hauxwell-Baldwin, R. (2015): Smart Homes and their Users: A Systematic Analysis and Key Challenges, *Personal and Ubiquitous Computing* 19, 463–476.

Winther, T. and Bell, S. (2018): Domesticating in Home Displays in Selected British and Norwegian Households, *Science & Technological Studies* 31, 19–38.

Wolsink, M. (2012): The Research Agenda on Social Acceptance of Distributed Generation in Smart Grid: Renewable as Common Pool Resources, *Renewable and Sustainable Energy Reviews* 16, 822–835.

Woolgar, S. (1990): Configuring the User: The Case of Usability Trials, *Sociological Review* 38, 58–99.

PART 4

Smart technologies, governance and institutions

20

DIGITAL TRANSFORMATION AND THE SOVEREIGNTY OF NATION STATES

Richard Sturn

Introduction

The crypto currency Bitcoin and the Facebook project Libra represent two different directions of technologically supported institutional innovation in the transition from public to private governance. Both developments are part of the digital transformation and both challenge traditional concepts and boundaries of public and private sphere, implying the potential of fatally undermining the ever challenged ideal of state sovereignty – a legal–institutional artefact which has been a crucially important element in the development of constitutionally framed democracies and market societies in the west.

On the one hand, challenges are brought about by the multi-faceted governance ecosystems becoming visible in governance-related research on Blockchain and smart contracts based on contract theory, mechanism design, and the economics of institutions. They may change the polycentric governance architecture of contemporary economies in unexpected ways. On the other hand, the growth dynamics and monopoly power of digital monopolies reach a new quality through a combination of different (partly well-known and partly new) factors changing the governance architecture and the general background conditions of market exchange. To be sure, both developments may interact with each other. However, as a first step, it is useful to analyse their basic logic as distinct developments – and the nature of the implied challenges.

This chapter is organised as follows. First, a few main tendencies pertinent to the issue of sovereignty in the digital age are briefly sketched. This is followed by four sections, which explain the underlying institutional and political challenges with some theoretical backgrounds (referring to economics as well as to discussions in political theory and constitutional law) in greater detail: the institutional architecture of modernity and the causes of its instability – with special reference to the digital transformation – are discussed. Also, arguments suggesting that Blockchain is not simply a transaction cost-reducing "exchange technology", but may change the institutional architecture, are sketched. The implications of "incomplete contracts" and the "background conditions" of exchange and contract regarding sovereignty are also discussed. The conclusions put forward for discussion are highlighted in terms of three scenarios of vanishing state sovereignty. Seen together, those scenarios elucidate the logic of trade-offs, but also systemic interactions between "surveillance capitalism" and surveillance state.

 DOI: 10.4324/9780429351921-25

Challenging state sovereignty: general overview of main tendencies

The institutional and organisational horizon of digital technologies extends far beyond crypto currencies and the market power of digital monopolies in the traditional sense. Developments such as Bitcoin and other applications of Blockchain, on the one hand, give a new turn to the perspective that essential parts of the institutions framing the conditions under which market transactions are carried out emerge in the private sector, created by an invisible hand in the course of the evolution of digitally based institutional ecosystems. While the evolution of the contract-mediated market economies always tended to bring about profound unintended side-effects on future background conditions of exchange (not least because market outcomes are systematically related to wealth distribution), the evolution of digitally based ecosystems implies new potentials of crowding out public institutions and circumventing their rule-setting and regulatory functions. Marcella Atzori (2015) sums up these perspectives as follows:

> This process might rapidly change even the tenets that underpin existing political systems and governance models, calling into question the traditional role of State and centralized institutions. Indeed, many Blockchain advocates claim that the civil society could organize itself and protect its own interests more effectively, by replacing the traditional functions of State with Blockchain-based services and decentralized, open source platforms.

Indeed, the impact on sovereignty of the invisible hand driving such ecosystems requires a thorough analysis of how Blockchain is functioning as a substitute for known institutional background conditions of exchange, including apparent paradoxes. The challenge of developing adequate degrees of centralisation (given the scope of coordination problems) is exacerbated by problems of accountability and distribution. On the other hand, the monopolistic rent potentials accompanying the digital transformation (dealt with by Luigi Zingales's 2017 political theory of the firm) tend to undermine state sovereignty in a way which is easier to grasp. Pertinent aspects of the digital transformation exacerbate the tension between private governance and public characteristics of technologies and their development as social forces of production. Ubiquitous network externalities and non-rivalry properties of data are only two aspects of those public characteristics. Information-intensive goods have high fixed and low marginal costs; in combination with artificial intelligence, data volumes show increasing economies of scale. There is a tendency towards winner-takes-all industries and information complementarities, as aptly illustrated by Zingales (2017: 121):

> The value of the data derived from Facebook and Instagram combined is likely to be higher than the sum of the value of the data derived from Facebook and Instagram separately, since the data can be combined and compared. Thus, Facebook is likely to be the higher-value user of Instagram data, even ignoring any potential market power effect. If you add market power effects, the momentum toward concentration might be irresistible.

This gives additional leverage to pre-existing trends towards globally operating corporations with annual revenues higher than those of many national governments. However, the political dimensions of this trend are not only associated with the sheer size, with concomitant lack of competition, or with political influence activities in the traditional sense. Moreover, they differ from the challenges of the military-industrial complex famously invoked in President

Eisenhower's farewell address. First, those well-known challenges change their nature up to a point where traditional modes of antitrust policy become obsolete or self-defeating (see e.g. Basu's 2019 argument on vertically serrated industries). Second, the specific core of the political dimension of digital monopolies is related to their increasingly profound and far-reaching public and medial functions. In their role as platform and network operators, they act as providers of services with public characteristics and often are, in fact, creators and enforcers of norms and standards. For instance, they have a considerable regulatory function with regard to weighing (sometimes conflicting) basic values: which postings are permissible within the framework of the right to freedom of expression and which are to be eliminated? While the fact that commercial companies determine and provide quality weighting/impact factors for scientific publications may be a secondary facet,[1] those roles in standard-setting and evaluation processes, however, are exemplary cases of (semi-)public goods provided by private sector mechanisms.

Better prospects for private provision of public goods may appear as unambiguously good news. However, private economy accounting mechanisms are not designed for public/political accountability. They are based on narrower, specifically targeted controlling and feedback mechanisms which are undoubtedly superior for organisations dealing with standard private goods. However, they become increasingly problematic with increasing importance of public characteristics, including the functionality of the respective goods in supporting the liberal order of a sovereign state. Private modes of provision are moreover associated with the fact that what is actually provided are mixed goods or private substitutes for public goods. Their use characteristics will tend to be biased in favour of specific interests according to willingness-to-pay. This, in accordance with the logic of private provision, may be associated with selective (non-open) access to certain uses of these goods.

Third, a combination of current trends may induce a politico-economic process which gives a dramatic turn to the long-standing problem of political influence activities. Zingales (2017) refers to the digitally induced version of this turn as the "Medici vicious circle" of the mutual reinforcement of political and economic power.[2] This implies that choices affecting everybody (thus collective by nature) are made not by way of explicit political choices, but by privileged norm makers, whereas all others are norm takers. In analogy to the term shadow economy, one could speak of shadow politics in this respect (cf. Sturn 2021). With these tendencies, democratically legitimised rule development, regulatory policy, and political mechanisms of accountability might be rendered powerless or crowded out: more far-reaching "collective choices" are imposed in the sphere of shadow politics by privileged norm-makers on the one hand and by the invisible hand of spontaneously developing institutional ecosystems on the other, both circumventing the constitutional and political mechanisms of public actors whose sovereignty gradually erodes.

1 It is nevertheless an instructive case study. The operative concretisation and adaptation take place in negotiation processes in which profit-oriented private actors (in this case scientific publishers and companies such as Clarivate Analytics) play a decisive role, whereas it is not always clear how effectively the scientific communities are able to organise the science-relevant public.

2 Zingales (2017: 119) aptly refers to political implications of incomplete contracts: "If rents are not perfectly allocated in advance by contracts and rules, there is ample space for economic actors to exert pressure on the regulatory, judiciary, and political system to grab a larger share of these rents. . . . If the ability to influence the political power increases with economic power, so does the need to do so, because the greater the market power a firm has, the greater the fear of expropriation by the political power. Hence, the risk of what I will call the 'Medici vicious circle'". It is obvious that the same logic applies to more traditional kinds of rents, e.g. certain types of the resource curse (rent-grabbing agents with questionable legitimacy have the need and the incentive to capture the political system) or scenarios of crony privatisation conducive to a regime of oligarchs.

State sovereignty is an integrative part of the institutional architecture under modern conditions of polycentric (public, private, and intermediate sector) governance, which developed along with a specifically pronounced separation of public and private spheres/sectors. In the context of the modern state, the distinctively modern ideal of sovereignty plays a crucial role: according to the Austro-German public lawyer Georg Jellinek (1921: ch. 14, 1911: vi.34), the author of one of the most insightful accounts of the function of state sovereignty within the complex architecture of modern institutions, this legal concept was crucial for state agency, legitimacy, and authority, including the self-limitation of states under the rule of law in a world of strategic uncertainty. Sovereignty means non-domination by external and internal powers. However, it does not imply but rather rules out absolute power. It is a legal construct with some analogies to moral autonomy, which does not mean that you can do whatever you want. In an illuminating comparison with classical antiquity, Jellinek points out that sovereignty is categorically different from the notion of autarchy, which according to Aristotle is the defining characteristic of a self-sufficient polity. For different reasons, the notion of sovereignty could play a role in neither the relatively small Greek city states nor the Roman Empire with its monopolistic position. As explained by Jellinek by way of those and further instructive contrasts (such as the Hanse, a federation of non-sovereign commercial cities in Northern Germany and the Baltic), sovereignty is not a necessary condition for provision of certain public functions such as the support of trading networks. For instance, certain kinds of security may be provided by non-sovereign entities and even as a by-product of feudal domination. In our time, the quest for global governance triggered by digital transformation and climate change adds a new dimension to sovereignty and polycentric governance, over and above the institutional connotations of the geopolitical notion of a multi-polar world.

Those observations are important in order to avoid the following fallacy: collapse of sovereignty does not imply collapse of all public functions. However, sovereignty was crucial for developing and sustaining a form of governance neutralising threats and influencing activities from within and outside in core areas of modern state functions - to an extent such that a truly public sector could gradually supersede privilege-ridden forms of governance and create the degree of openness characteristic for modernity.

The analysis provided in the present entry supports the conclusion that neither the concepts of sovereignty (or a modernised analogue) nor the associated distinction of public and private sectors becomes functionally obsolete under digital transformation: quite to the contrary, a public sector with correspondingly developed forms of political accountability is more than ever necessary. In absence of this, the digitally supported increase of the extent of market mechanisms (the expansion of the principle of free contract) may become a one-sided appropriation mechanism for oligarchs, rather than unfolding its potential as a socially productive force. Without an effective public sector, the potential of digitalisation with regard to the storage, aggregation, availability, processing, and development of knowledge and information not only is used sub-optimally, but also is being perverted. Instead of the differentiation between a functionally specific public sector based on collective choice and political accountability and private entrepreneurial economy based on commercial accounting, precarious forms of semi-public governance, shadow politics, and compromised privacy may emerge.

While sovereignty of public agency will not become functionally obsolete, the potentials, as well as the challenges, connoted by digital transformation will drastically modify the setting in which sovereignty and the public sector can fulfil their function. First, the social-territorial incongruence of public tasks and statehood (that has emerged already in the face of globalisation and the climate problem) is virulent anew in the context of digitalisation. What do digital business models mean for taxation and social policy? Can political accountability be de-territorialised in a problem-oriented way? Is it possible to organise political accountability along functionally specific overlapping federal structures, or cosmopolitically? How are the current tendencies of re-nationalisation to be assessed?

Second, the potentials of digitalisation in terms of making information available should be seen under perspectives including both good news (improvement of allocation and feedback mechanisms) and bad news (emergence of new power asymmetries). Consider the good news first. Due to new possibilities of aggregation, processing, and analysis of different information/data, digitalisation has a considerable potential with regard to two forms of decision-supporting accountability and feedback which co-evolved with modern market societies. Both forms of accountability and feedback are likely to become even more important in a period of accelerated dynamics of innovation accompanying digital transformation, including multi-level social learning processes: (1) the discursive-political legitimacy of public choices in combination with electoral mechanisms in the public sphere against the background of a cosmos of pluralistic, latently conflicting values and interests; (2) private sector accounting/controlling against the background of one-dimensional (typically monetary) indicators of success.[3] Even though this is good news, there is a non-negligible risk: if the pertinent potential of digital technologies is not embedded in the constitutional-political architecture supporting those feedback processes in suitable ways, keeping public concerns and the domain of private entrepreneurship apart, there is a danger (to be discussed later) that it will develop in a way that ultimately leads to an accountability deficit while at the same time compromising the private sphere through a kind of "surveillance capitalism", possibly combined with surveillance statism.

As mentioned earlier, one specific problem is centralisation without accountability through private monopolies along the lines outlined by Zingales (2017). Another specific problem arises through decentralisation à la Blockchain, associated with a spontaneous evolution of platforms behind the back of the actors. In both cases, optimists rightly stress the efficiency gains and in particular the huge potentials for everybody by facilitating many kinds of decentralised transactions by "small" market participants. (Indeed, this also applies to platforms run by monopolists.) Both tendencies were unambiguously beneficial if (in a co-evolutionary process, as it were) mechanisms emerged that secured the public character of the provision of exchange-supporting services. Unfortunately, in the history of capitalist market economies, the public character of exchange-supporting institutions has hardly ever been the result of a smooth, spontaneous co-evolutionary process: it had to be deliberately pursued by political agents and defended against special interests. It is hard to see why the evolution of governance "ecosystems" without formal hierarchies (let alone of digital monopolies) should be immunised to the dangers of rent-seeking and shadow politics by some miraculous invisible hand guided by public interest. While divergence in wealth accumulation and the development of de facto privileges can easily be modelled as the outcome of spontaneous processes, adequate distributive regulation and social policies are among the core functions of a genuinely public sector – an insight epitomised by theories such as Lorenz von Stein's account of the Sozialstaat, which according to Stein is a necessary complement to the Rechtsstaat (rule of law) as foundational prerequisite of modern market societies.

Moreover, spontaneous counter-reactions to distributive polarisation and Medici vicious circles of economic and political power may degenerate into the kind of pathological countermovements already described by Karl Polanyi (1944). Instead of breaking a vicious circle, phenomena such as populism just add another (possibly dangerous) loop to the Medici vicious

3 See Onora O'Neill (2005, 2017) and Colin Crouch (2004) on problems of one-dimensional surrogate indicators in multidimensional targets.

circle and (while operating in the shadow of accountability by polemically targeting unaccountable elites) fully destroy the public sphere and effective mechanisms of accountability. The latter is a predictable consequence of populist currents becoming hegemonic in the political system (as a reaction to these deficits), whose political business model is the transformation of the public sphere into an arena where friend-and-foe polarisation rules the roost. Those political forces will not deal with the systematic reasons for the accountability deficits to which they partly owe their boom, but will seek salvation in illiberal variants of capitalism with a tight association of political and economic power (for some background see, among others, Zingales 2017; Streeck 2013; Schmitt 1933; Müller 2013). Developments of this kind have come a long way in some places in recent decades.

Given the political and artificial character of sovereign bodies capable of consistently pursuing the public interest, we might face an accountability paradox: while digitalisation supports the development of refined and expanded information systems as a prerequisite for refined micro-accountability, by undermining public–private architecture it could exacerbate existing accountability deficits and/or create new ones. All this gives rise to the conjecture that, in the face of digital challenges, the vulnerable civilisational constructs of state sovereignty and the specifically modern public–private distinction could experience its finest hour – or demise.

So, what is the historical starting point for analysis of governance in the age of digital transformation? Accountability deficits and (related to them) the mutual interpenetration of private and public spheres focused on by Zingales (2017) are already politically virulent, including various forms of populism as a problematic counter-movement to allegedly unaccountable elites. In view of a prospective digital accountability paradox, things could become even more difficult: political business models capitalising on such difficulties already have gained momentum. However, the current visibility of the problems of such developments also may induce us to take more seriously concomitant challenges to foundational aspects of modern governance such as sovereignty.

The institutional architecture of modernity

Given what has been argued so far, two questions ought to be kept in mind when dealing with optimistic views of current trends. Putting digital potentials and challenges into the perspective of the institutional architecture of modernity, we need to ask: (1) is there any evidence for the obsolescence of sovereignty and the public–private dichotomy (associated with the digital transformation), for instance because Blockchain supports decentralised coordination processes delivering everything that the public sector used to deliver, but in a hierarchy-free manner based on voluntary transactions? (2) To what extent are the dangers disappearing that previously made sovereign public agency unstable? Put another way, will public agency become a self-propelling force in the course of the digital revolution, because the preconditions for permanent privileges, asymmetries, and special interest rent-seeking are vanishing in the digital world?

An optimistic literature implicitly answers both questions in the affirmative. However, I already have indicated some reasons for negative answers. Here are a few additional backgrounds and thoughts motivating a sceptical view. Ad (1): Joseph Schumpeter (1942: 197–198) concisely points to the sharpness with which the public sphere in modernity became demarcated from the private sphere and developed into a separate, functionally complementary social sector. The modern public–private dichotomy is semantically comparable to dichotomies from classical antiquity (Polis vs. Oikos, privatus vs. publicus). However, it is fundamentally different not only in terms of creating the preconditions for sector-specific mechanisms, institutions, and actor types. The sectoral differentiation between public and private is constitutive for sovereign modern states in market societies. The development of two sociologically different sectors is accompanied by the evolution of institutionally differentiated frameworks for complementary, problem-adequate accountability mechanisms. The modern public sector, moreover (guided by ideas such as Adam Smith's, 1776, system of natural freedom), promotes the development of the market as a productive force in an economy dynamised by a cumulative process of division of labour (which, according to Smith, is limited by the extent of the market). Providing or supporting exchange infrastructures and dismantling of trade barriers and privileges is a crucial element in this process.

Thus, the private sector mechanism of free exchange became a dynamic social productive force when the state began to influence the background conditions of exchange in a systematic way "creating markets", not least by providing constitutional rules and public infrastructures and dismantling the "aggregate of privileges" that Hegel identified as a characteristic of pre-modern governance and background conditions. The regulative ideas of a public sphere are associated with openness, free access, non-discrimination, ideas of "equal freedom", and the rejection of privileges. Complementary to this are the spheres of private living and the private economic sphere for the management of private goods and for entrepreneurial experimentation. According to this kind of strong complementarity, the private sphere can only become truly private when the public sphere becomes truly public.

Ad (2): In view of the indicated risks of the digital transformation, it is of central importance to emphasise the historically contingent, unstable character of state sovereignty and of the separation of private and public spheres. A public sector capable of appropriate action can neither be taken for granted as a natural outcome of evolution nor viewed as an exogenous, God-given instance. Moreover, on the whole the mechanisms of the public sphere are unlikely to become a smoothly operating, unified machinery unambiguously implementing rational policies. In a sense, Buterin et al. (2018: 36) have a point when characterising the interactions between state, society, and economy as "awkward dance of capitalist atomisation coupled with checks and balances among various rigid levels of collective organisation". Indeed, this observation captures the darker side of the ways in which the power and stability problems outlined in the introduction have been dealt with to date. However, the causes for the "awkward dance" and the functional background of the checks and balances should be kept in mind as background when considering alternatives. This includes coping with Medici vicious circles, viz. the socio-economic forces systematically amplifying and petrifying power asymmetries endangering the kind of openness stressed by North et al. (2009). In the presence of such forces, the functional scope of elegant mechanism-based solutions will be limited.

In other words: while it is questionable that the functions attributed to the public sector can be dealt with by mechanisms fundamentally different from the "awkward dance", in view of the manifold problems of traditional state–society–economy interactions it is certainly worthwhile (as suggested by Buterin et al. 2018) to think about the potential of decentralised, hierarchy- and state-free orders (e.g. on the basis of Blockchains) in which all formerly public services are provided privately or by "voluntary associations" or commons. However, such imaginations are question-begging if they are not based on a clear view of the problems which represent the raison d'être of the institutional architectures of modernity and which probably do not vanish into thin air in the face of digital technologies.[4]

Envisaging phenomena such as the Medici vicious circle, the interpenetration of private and public spheres should be analysed in a more general systemic context, including two aspects:

4 It is noteworthy that authors such as the Ethereum developer Vitalik Buterin take a substantial part in the discussion about broader political–economic perspectives.

first, stressing the characteristic sharpness of the modern distinction between public and private must not lead to ignoring the numerous and inevitable interfaces and frictions between the two spheres. These include the possibility of precarious forms of mutual interpenetration. A classical area of such interfaces is covered by public finance. It arose from the need to develop a doctrine of the public economy, including specific interfaces such as taxation and public procurement. Taxation and public credit represent characteristically friction-laden interfaces, complemented by the various forms of involvement of private actors in public tasks. Properly mediating those frictions is a precondition for state stability, agency, and sovereignty. Second, apart from those interfaces, the "public sector" mechanisms are associated with problems and pitfalls of their own. Political processes must not be idealised as tension-free mechanics of the public sphere. The concept of the state as an agency of collective rational design, as in utilitarianism, can at best serve as a point of reference, but is way too simplistic even as a regulative idea. Paradigms of politics as a space of discursively mediated responsibility, and/or of the development of dialogical reason or Aristotelian idealisations of the polis may be valuable as inspiration. However, they require a critical perspective in view of the tasks to be delivered by a modern state in pluralist societies. Indeed, the seemingly cumbersome institutional architectures of modern public sectors can only be understood out of the difficulties in reconciling tension zones created by divergent interests and cultural heterogeneity. The state is not an exogenously given and smoothly functioning actor – as the conception of government in mainstream economics[5] would have it, for example. Rather, according to Ernst-Wolfgang Böckenförde (1976: 60), the modern constitutional state finds itself in the dilemma that it cannot itself provide the preconditions on which it is based. In a nutshell: a functioning public sector is indispensable, but it cannot be provided by technocratic fiat. It is a higher-order public good based on complex, artificial constructs, including sovereignty and the separation of public and private. The institutionalisation of the latter has the function of counteracting pathogenic forms of mutual penetration of these spheres, such as the colonisation of the public sphere by certain classes or interest groups.

This architecture requires built-in stabilisers, including not only constructs such as the separation of powers, but also a minimum of convergence of mental models and common basic norms, for otherwise the foundational understanding of the public sphere dissolves. Without such an understanding, the accountability mechanisms themselves sooner or later are being captured by special interests or distorted by group-specific idiosyncrasies. There is a danger of successive degeneration towards either the Medici vicious circle or repressive variants of socialism that seek collectivistic closure, which fail in the long term due to a lack of openness in the development of social productive forces. In short, the modern institutional form of this dichotomy is an *artefact rich in prerequisites*, based on checks and balances and unstable due to the inevitable existence of the interfaces mentioned.

While the institutional architecture of the modern liberal order is an *artefact rich in prerequisites*, a crucial aspect of its complexity is related to its development as a mix of spontaneous process and politically "planned" market-making on several levels. In keeping with Karl Polanyi (1944), this process can be described as "disembedding". However, this term only refers to the most striking aspect of a multi-level transformation of background conditions of market exchange, partly politically implemented, partly spontaneously evolving. A decisive aspect of

5 Robert Sugden (2004: 3) characterises this position as follows: "Most modern economic theory describes a world presided over by a *government* (not, significantly, by *governments*), and sees this world through the *government's* eyes. The *government* is *supposed* to *have* the *responsibility*, the *will* and the *power* to *restructure society* in *whatever way maximizes social welfare*; *like* the *US Cavalry* in a *good Western*, the *government* stands ready to rush whenever the market 'fails', and the economist's job is to advise it on when and how to do so." See also Smith (1790: VI.ii.2).

this transformation is the parallel movement of the state and the market: the traditional social embeddings are pushed back, as it were, from those two sides (unleashing market competition; superseding traditional norms by state regulation). Thinkers such as Lorenz von Stein[6] have worked out the functionally necessary basic principles and specifics of modern state governance, which is geared to the gradual, discontinuous, and difficult overcoming of privilege-ridden governance: what Stein calls the Rechtsstaat (rule of law) is a necessary, albeit insufficient, prerequisite for this modernisation. According to Stein, it must be accompanied by a Sozialstaat (welfare state) embedded in the public economy. An essential feature of such constructs, which evolved towards the "democratic capitalisms" politically framed by sovereign states in the 20th century whose major successes came in the period after 1945 (Streeck 2013), is the development of background conditions striking a balance between unleashing and socially stabilising the market as a social force of production.

The liberal order organising this balance includes a set of politically produced background conditions of exchange. It is an irreducibly public good of higher order: at this level, private substitutes are categorically impossible. Moreover, this irreducibly public good is closely associated with the public provision of certain core first-order public goods, whose selective or biased provision is incompatible with the liberal order. In other words, public provision is systemically relevant. Hence, ruling out private substitutes at the level of those core first-order public goods supporting the liberal order is a crucial test of state sovereignty. Private substitutes counteract the openness that is constitutive of liberal orders (cf. North et al. 2009) and undermine sovereignty. The development of this order depends on filter and balance mechanisms such as separation of powers, which may appear as "awkward dance" but prevent phenomena such as the Medici vicious circle. Sovereignty is thus dependent on the public provision of certain first-order public goods. Private substitutes are possible (e.g. some kind of security as a by-product of feudal allegiance), but their discriminatory character will tend to undermine the higher-order public good.

Tendencies towards the erosion of the public sphere are omnipresent even if we put digitalisation aside. In this respect, the development of mental models in the past decades must be taken into consideration. There are signs of a diminishing problem sensorium with regard to precarious interpenetration of the public and private spheres. Mechanisms of public accountability tend to be downgraded to dispensable decorative accessory without any real meaning, or to an annoying cost factor and an obstacle to efficiency, thus paving the way for shadow politics of all kinds. Problematic aspects of the role of private actors in the provision of public services or the development of private or semi-private substitutes for public goods are not well understood. If disruptive changes in this complex architecture are to be expected as a result of digitalisation, this will happen in a phase in which the public sector is in jeopardy anyway. As Atzori (2015: 31) rightly points out:

> When assessing risks and benefits of Blockchain applications, we cannot overlook the fact that to overthrow the State and to absorb its functions is a profitable business. While the Blockchain was originally created to eliminate the need of a third party in transactions, the paradox is that stakeholders now involved in Blockchain governance play the classical role of *tertius gaudens* . . ., a "rejoicing third" that attains economic benefits by replacing the State in some or all its functions; even worse, these agents may also intentionally pursue a strategy of *divide et impera* (divide and rule) between

6 A good overview can be found in the chapter on Lorenz von Stein in Böckenförde (1976).

> civil society and State, aimed to undermine the traditional democratic order, modify the existing balance of power and achieve a dominant position in society. If it is true that "the neo-liberal ascendancy and its corporate agenda are producing its own version of democracy" (Marden 2003: xiv), it is not unreasonable to assume that this will take on the features of an algorithm-based decentralized society. In such scenario, to advocate the idea of State means to reaffirm the primacy of politics over economics and to recognize the need for a coordination point in society, in which the tensions between individual interests and common good find a constructive, political compromise.

Whatever "primacy of politics" means in the context of the richer concept of sovereignty: in the following, the analytical focus is the search for valid clues to the indispensability of political accountability and the "coordination points in society" that do justice to the reach of pure public goods in the digital transformation. But first we need to look at another aspect. Atzori emphasises the profit opportunities of privatisation. One could now object to this: so what? In a market economy, profit opportunities indicate previously untapped, socially valuable potentials – analogous to the market for corporate control, which, according to the textbook, leads to the untapped potentials of companies being leveraged when they are taken over by the highest bidder. The quest for the primacy of politics and sovereignty depends on answers to questions such as: what are the core public tasks that are neglected, undermined, or even counteracted under certain digital transformation scenarios? The answer is much deeper than the level of neglected rail infrastructure or the like. In the digital context, it is linked to two aspects: (1) the public characteristics of digital technologies and their tension with private accumulation and (2) the role of the state as part of the background conditions of contracts, to which we now turn.

Widening the scope of private mechanisms

> *What you are, you are only through contracts.*
>
> *Richard Wagner, Das Rheingold, Prelude, 2nd Scene*

> *We may even go so far as to abstract from entrepreneurs and simply consider the productive services as being . . . exchanged directly for one another.*
>
> *Léon Walras (1969: 225)*

Exchange and contract are indispensable in modern societies for the mediation of various interdependencies. They do this on the basis of a diverse and complex structure of background conditions, a large part of which have been created politically and are condensed into differently situated and differentiated layers of formal and informal norms, rules (such as labour law or collective bargaining), standards, or taxes, but also authority relationships in companies or co-evolving behavioural dispositions such as reciprocity (cf. Bowles 1998). Moreover, the background conditions of market exchange include the respective alternatives in the bargaining situation that precedes the contract. The attractiveness of alternatives influences the bargaining power. Wealth distribution thus determines structural background conditions: those who have no viable alternatives are more likely to accept an "immoral offer". Modern statehood as an effective institutionalisation of the public sphere is an essential element of background conditions. Without it, the market would not have been able to develop as a dynamic social productive force. All in all, the state-supported part of background conditions provides "solutions" to a series of different strategic interaction problems. Some of those problems could not be solved at all, some only with great difficulty, by socially dis-embedded individuals who egoistically maximise utility in absence of institutional frameworks. (In societies with rudimentary institutional background conditions, exchange and market were marginal phenomena.)

Digital technologies such as Blockchain can now be considered as technological possibilities to support contract-like interaction without state-supported background conditions. The simplest concept for capturing the part of background conditions represented by formal and informal institutions is that of transaction costs. Transaction cost reduction is an obviously interesting potential of technologies such as Blockchain. However, technological transaction cost reduction is not a specific feature of Blockchain. It is not new that technologies change patterns of transaction costs, thus affecting the extent to which the price system is economically viable to coordinate economic activities. For example, the introduction of electronic toll systems facilitates pricing the use of urban traffic areas (even with efficient price discrimination), which would previously have been impossible in view of the high transaction costs – inducing an increase in the range of advantageous applications of price-based mechanisms. Davidson et al. (2018: 3), however, argue that Blockchain is not merely an exchange technology in the sense of Coase's paradigm of reducing transaction costs, but an "institutional technology". It extends the portfolio of modern institutions (company, state, commons, and relational contracting) by a new type of institutional coordination.

The observation that the functioning of Blockchain cannot be reduced to the reduction of transaction costs is correct and important. Additional functions come into view when we systematically use the theoretical concept of incomplete contracts.[7] Davidson et al. (2018: 3) also refer to this concept, in two almost obvious directions: (1) Blockchains are part of a digital evolution towards more complete contracts; and (2) Blockchains lead to a kind of disintermediation, i.e. previously required institutions which complement incomplete contracts in the mediation of interactions become superfluous.

> Cryptographically secured Blockchains are said to be "trustless" because they do not require third-party verification (i.e. trust), but instead use high-powered crypto-economic incentive protocols to verify the authenticity of a transaction in the database (i.e. to reach consensus). This is how Blockchains can disintermediate a transaction (a consequence of which is lowered transaction costs), resulting in new forms of organization and governance. . . .
>
> The implication is that Blockchains may not compete head-to-head with firms, but rather may carve out those parts of firms that can be rendered as complete contracts. . . . For instance, Blockchain-enabled smart contract-facilitated transactions should in principle experience fewer efficiency problems due to information asymmetries – adverse selection (prior to a transaction) and moral hazard (following a transaction). Smart contracts could also be effective ways to load significant numbers of low-probability state contingencies into contracts. These could function like open-source libraries able to be inserted into machine-readable contracts, reducing the complexity cost of writing large state-contingent contracts, and so lowering transaction costs. Both ex ante contractual discovery and ex post contractual renegotiation costs

7 These are contracts in which not all relevant eventualities are (or can be) contractually determined ex ante because this would be too expensive or also because relevant aspects, in which the contracting parties are very interested (e.g. work-effort), are difficult to observe and not verifiable at all (e.g. before a court). Employment contracts are typical examples of this, with regard also to the long-term nature of incomplete contracts (relational contracting).

(i.e. bargaining and haggling costs) are an expected consequence of incomplete contracts. Such contracts have dynamic benefits, enabling adaptation, but in the shadow of these expected but uncertain costs all parties will contract less than is optimal. Blockchains potentially enable the known parts of these relationships to be carved out efficiently from the unknown parts, and executed automatically based upon state conditionals, increasing the range to which economic coordination can extend into the future.

Both the "dis-intermediation" and the enhancement of complete contracts (which as "known parts of these relationships" are separated from the "unknown parts" with respect to which contracts remain incomplete) are associated with a further change in the institutional architecture. The socio-economic functions of Blockchain thus cannot be sufficiently understood with the concept of transaction cost reduction. It not only increases the number but also widens the scope of transaction types for which voluntary exchange conditioned by entry and exit can be considered as an unambiguously advantageous, non-hierarchical mode of self-regulating interaction: nobody is forced to participate in Ethereum. Those who do so, do so to their advantage. The "principle of free contract" underlying those attractive properties belongs to the DNA of economists (Basu 2007): voluntary contracts of responsible actors should not be obstructed, as long as they do not have a detrimental effect on third parties. Voluntary exchange generates a surplus that is distributed among the exchange partners in such a way that each of them is better, or at least not worse, off than with the next best alternative ("outside option"). If they failed to benefit or were made worse off, they would not agree to the exchange. Related principles have a long tradition, including "caveat emptor" or "volenti non fit iniuria" (cf. Sturn 2009, 2017). In the context of the present problems, the following two problems need to be discussed now:

1 Why does the extension of the range and the depth of the market pose problems? Why are transaction cost-reducing advances in contract technology ambivalent?
2 What are possible problems of changes in the institutional architecture that go beyond transaction cost reduction?

The answer to the first question seems to be simple in view of the explanations given in earlier. The function of institutions is not only to support the emergence of "desirable" (Pareto-improving) interactions and cooperation but also to reduce "undesirable" transactions (with negative third-party effects). The latter include the entrepreneurial marketing of private substitutes for public goods, when these private substitutes compete with core public goods and displace them, as the entire complex of background conditions may be moved outside the field of political collective decisions and corresponding forms of accountability.

The issue of distribution is at the core of the answer to the second question. Much beyond the fact that the incentive mechanisms underlying Blockchains are operating in the world with significant socio-economic asymmetries, this issue requires a more detailed analysis of the background conditions in light of the theory of incomplete contracts. This reflects the fact that background conditions *typically combine two functions: (1) facilitating efficiency-enhancing contracts; (2) mediating distribution problems*. However, pertinent considerations are often guided by transaction cost economics, which captures the functions and effects of institutions only under the efficiency aspect (cf. Williamson 2000). In the following section, it will be shown that the concept of incomplete contracts is of crucial importance for the assessment of the opportunities and risks of digital transformation, as it allows for capturing distribution aspects.

Utopias, dystopias, reclaiming state sovereignty: the role of incomplete contracts

As explained by A. Smith and Schumpeter, modern capitalism amounts to a process of the division of labour associated with specialisation, waves of innovation, unplanned redistribution, and concomitant uncertainties. These environmental conditions are conducive to persistent contract incompleteness. Governance is thus inevitably polycentric, as coordination tasks as well as the fuel for rent-seeking and the sources of factual power are co-evolving with economic change: new patterns of (quasi-)political governance are emerging in the public and the private sector, giving rise to polycentricity as a social fact. The latter is ambivalent, reflecting the duality of forces driving polycentric governance innovations: on the *functional side*, they are necessitated by incomplete contracts and other coordination problems associated with novelty in the economic domain. However, specificities of emerging polycentric governance solutions will reflect the *distributive interests* of those who have a more active (perhaps entrepreneurial) role in promoting them – possibly interacting with existing tendencies towards power concentration and distributive polarisation. In this setting, sovereign public agency has a crucial function: promoting *ordered polycentricity* by procedurally qualified and legitimatised rule-making, filtering distributive biases and precluding the cumulation of economic and political power.

Digital technologies such as Blockchain are often referred to as game changers. However, the nature of envisaged change depends on whether they lead us to a world of complete contracts – or whether the dynamism of novelty along with asymmetries in knowledge and power and the increasing complexity of contracts implied by widening extent of markets keeps setting the stage for persistent contract incompleteness, despite all progress in contract governance. Here are the two contrasting visions.

1 *Re-inventing sovereign public agency*. We basically remain in a world where governance is somehow attuned to incomplete contracts. The dynamism of division of labour and novelty keeps nourishing the forces of polycentric governance, which are additionally boosted by technologies such as Blockchain. We thus remain in the setting where the modern concept of sovereignty (sketched earlier) became pivotal, including procedures of legitimate rule-setting and enforcement in a broad sense. This sovereignty presupposes a public sphere: its raison d'être is developing adequately centralised coordination points and innovation-friendly distributive regulation precluding phenomena such as Medici vicious circles and oligarchic clusters of power.
2 *Radical game change*. Contract incompleteness is eliminated/marginalised due to technological perfection of contract governance. The architecture of governance becomes attuned to complete contracts. Under complete contracts, the private sector is (as construed by influential theoretical idealisations of the market) nothing but a power-free machinery for mutual advantage. However, this scenario gives rise to sharply contrasting visions, depending on whether issues of distribution do or do not matter – in the sense that distribution affects economic outcomes and becomes politically virulent at local or wider levels. Assuming irrelevance of distribution, the expansion of complete contractability (including contract-mediated arrangements for public goods) prepares the ground for market anarchism, or a static minimal state providing enforcement of property rights and contracts. In contrast, ambivalent perspectives arise if the issue of distribution is still factually virulent – perhaps because the blessings of Blockchain unfold in an already unequal society (and not in some harmonious Golden Age), or because the polycentric dynamism of division of labour and novelty brings about new inequalities and/or is distorted by rent-seeking.

> If distribution matters, the complete contract model calls for conceptions of statehood/public authority contradicting the logic of sovereignty valid under polycentric governance. It is absolute rather than sovereign and may be complemented by different political visions.

To analyse the ambivalence of this radical game change scenario, consider public sector functions in a market economy in general and somewhat vague terms. *Developing the background conditions of exchange and contract in such a way that the socially advantageous potential of contractually mediated interactions (the principle of free contract as a social productive force) comes to be employed as well as possible.* While there will be broad agreement on this, transaction cost economics neglects the analytical potential of its factorisation in two subfunctions. *Subfunction 1*: as many people as possible in as many situations as possible benefit from the advantages of free exchange. *Subfunction 2*: the cases in which exchange occurs under background conditions giving rise to "coercive offers" and one-sided rent-extraction are constrained by innovation-friendly distributive regulation.

While systematic one-sided rent extraction is part of the regime of accumulation in nascent capitalism, maturing capitalism as well as the foundational principles of liberal democracy are associated with such distributive regulation. Transitory Schumpeterian innovation rents become the systemically dominating form of private rent-appropriation by entrepreneurs. This is done by a range of institutions combining subfunctions 1 and 2, with sovereign public agency at their core. Problematic forms of rent appropriation are mitigated by distributive regulation legitimised by due procedures, tending to neutralise rent-seeking as fuel for governance distortions. Enforcing such regulation requires sovereignty but is also an expression of sovereignty.

This background is pivotal for institutions that support the sustainable functioning of systemically relevant markets, notably the labour and capital market. Starting from incomplete contracts, the functionality of various norms, institutions, governance mechanisms, and forms of social embedding can be explained by their role in supporting the principle of free contract and making it sustainable by preventing contract-mediated exploitation. As mentioned, not all of those contract governance institutions are part of the state sector. Private sector and intermediary institutions are also involved in contract governance combining subfunctions 1 and 2. In this sense, firms can be understood as political institutions (Sturn 1994; Bowles 2004; Bowles and Gintis 2000: 1425). While the "private sector" is not a peaceful and harmonious sphere of solved political problems and harbours endogenous forces which may have irritating political potential, it also may be the locus of political solutions, including mechanisms of collective bargaining etc.

Polycentric governance thus includes mechanisms that endogenously promote contract enforcement when third-party enforcement is too costly or impossible (due to a lack of an "external" enforcement agency, or due to the relevance of non-verifiable information which is not usable by an enforcement agency such as a court), typically also combining subfunctions 1 and 2. The diversity and complexity of basic institutions, governance mechanisms, norms of behaviour, trust, and the limited yet important role of moral motivations ("fairness") in modern market economies are all linked to "endogenous" enforcement. In other words, incomplete contracts are a key to the functions and ambivalences of complex background conditions supporting exchange in modern economies – and ultimately the awkward dance alluded to earlier.

Suppose now that the background conditions (formal institutions, legal rules enforced by the judicial apparatus, trust etc.) are substituted by privately operated technologies, including complete smart contracts supported by Blockchain mechanisms. Notice that pertinent claims by Blockchain enthusiasts typically apply to subfunction 1, while they are often unaware of subfunction 2.

Subfunction 1 is *carved out* from an institutional complex in which subfunctions 1 *and* 2 were merged. This kind of carving out may have more dramatic implication than the carving out discussed in the previous section: it may crowd out sovereign public agency, giving rise to either utopian or dystopian perspectives. Suppose that the process of such technological carving out is completed. Then we are in the complete contract scenario: subfunction 1 is fulfilled by Blockchains combined with smart contracts and new contractual forms. This could promote tendencies towards authoritarian, absolutist, or even totalitarian forms of statehood, unless we either live in a world where no power/distribution problems are disturbing the perennial harmony of voluntary contracts, or else those problems are perfectly solved by morally elevated public agency.

To see this, suppose a world where the support of digital technologies is allowing for perfectly complete contracts, including perfectly efficiency-enhancing background conditions based on Blockchain etc. Further suppose that the natural course of things in that world (due to technological and other factors nourishing tendencies of divergence) leads to distributive polarisation and/or systemically problematic forms of contract-mediated exploitation/rent appropriation. As digital technologies such as Blockchain enable complete contracts for all economically viable activities across the board (and remaining incompleteness can be dealt with using economic mechanisms), a public institution such as the state would "only" be confronted with subfunction 2: the challenge of distribution – this, however, in its entirety, in full sharpness, and in an environment that includes the technological possibilities of digitalisation, while at the same time the status of classic modes of peaceful political problem-solving is downgraded because they are no longer needed for non-distributive functions. Background conditions for market-based exchange would essentially consist of Blockchain-based ecosystems, given the distributive patterns that determine the individuals' alternative options and their bargaining power. The state would have a reduced and at the same time notoriously difficult, politically contested role to play: fixing distribution problems. Thus, ubiquitous complete contracts are not only questionable as utopias because of their lack of realism but also (should Blockchain and the "new contractual forms due to better monitoring" discussed by Varian 2014, carry the day) promote dystopic tendencies even beyond Zuboff's (2015) diagnoses and projections of "surveillance capitalism": this world of complete contracts may be worse off as far as the background conditions of exchange are concerned – especially concerning the prospects of enlightened political reform of background conditions. It would be a world in which state sovereignty loses its self-limiting aspects (lucidly explained in quoted works by Georg Jellinek) and its association to the public interest combined with effective agency. Improving the background conditions would rely on redistribution according to the logic of the Second Theorem of Welfare Economics – in a world where Blockchain tends to amplify the opportunities for shifting economic activities away from spheres accessible as tax bases. Given that some redistribution is systemically essential, this could lead to an awkward dance even worse than the one that Buterin et al. (2018) wish to get rid of: the state employing all the power of surveillance technologies, alleviating the constraints on redistribution imposed by generalised digital tax evasion.

To grasp some features of worlds where this may happen, canonical idealisations implying complete contracts are useful as theoretical background. Consider static models of private property market economies à la Walras. In such a world, well-defined private property rights (i.e. background conditions set once and for all) are almost synonymous with the full utilisation of the welfare-enhancing coordination potential of voluntary exchange by rational market participants. Politics is an exogenous agency "solving" distribution problems, while the working of the market itself never entails "political" problems, i.e. problems in which "political forms of mediation" (including: argumentation/deliberation, negotiation, voting, fighting; cf. Elster 2000) play a role. This model is the basis of the separation between economics and politics that Lerner (1972) summarised in the statement that economics is the queen of the social sciences – on the basis of solved political problems – i.e. on the basis of the determination of the distribution of property rights, which serves as a starting point for mutually beneficial exchange. This also corresponds to a common demarcation of the subject areas of both subjects among political scientists (Bowles 2004: 171–172 cites Lasswell and Kaplan, for example) and economists (J.St. Mill, Pareto, Robbins). Although unrealistic, this model illustrates some crucial conditions of a peaceful, depoliticised market world that is determined by the principle of free contract *without leaving room for rents and distribution struggles* – or other problems that ultimately require political mediation.

In combination with the conjecture that ubiquitously complete contracts are an unrealistic vision, their dystopic connotations lead to the conclusion that modern democracies should stick to the guiding idea of well-ordered polycentric governance by sovereign, accountable public agency. The use of digital technologies should not be seen primarily in the light of the welcome approximation to the ideal of complete contracts familiar from model theory. It rather should be seen in the context of the functionalities of and complementarities to historically evolved background conditions and their reform, including its role for a problem-oriented modernisation of public sector institutions and their capabilities: indeed, sovereign public agency needs to be re-invented in the global digital transformation (see Sturn 2021).

To be sure, the re-invention of accountable sovereignty will not be easy going. While the functionality of mechanisms of public choice and accountability may be augmented by digital technologies, some known problems cropping up in a setting of incomplete contracts may become more severe. This includes the vicious circles of data capitalism à la Zingales (2017) – referring to the rule-making potential of economically most powerful agents – as well as non-Walrasian behaviours such as price-setting, including personalised take-it-or-leave-it offers or exploiting the benefits of first Mover Advantages/Stackelberg Leadership. Asymmetric digital capability of data-based "personalization and costumization" (Varian 2014) may play a role, which may be used for the accurate determination of rent-absorbing offers and may interact with a cumulative causation mechanism that makes rent-appropriators even more powerful through network externalities.

State-mediated polycentric governance was key to the development of background conditions contingently justifying trust in the time-honoured principle of free contract. Despite all difficulties, modern democratic sovereign states under the rule of law developed background conditions enhancing the power of this principle. Modern legal systems ensured that more contracts could be concluded and that they were less incomplete. This increased the scope of contract mediation viz. the extent of the market. In addition, various constitutional and legal rules (e.g. labour law and consumer protection) and welfare state institutionalisations strengthened the position of disadvantaged market participants in the face of the dangers of background conditions triggering "coercive offers". Non-state institutions and informal norms relieved/supplemented the state, which in turn promoted ordered polycentrism, by and large aligning the working of non-state governance and informal institutions with subfunction 1 and subfunction 2 as defined earlier. Finally, the scope of what can be contracted was deliberately limited in some cases. This applied, for example, to contracts with harmful third-party effects, but also to contracts (e.g. slavery contracts) in which the principle of free contract becomes questionable/incredible despite formal free consent.

In order to elucidate the challenges ahead, the following section concludes by sketching sets of conditions preparing the ground for three different visions of how sovereign public agency may disappear, taking digital-based new contractual forms and fictional worlds of complete contracts as analytical reference point.

Varieties of post-sovereignty: techno-libertarianism, techno-liberalism, and techno-oligarchy

To summarise: digital technologies are game-changers. However, they are not likely to put the basic constellation of modern dynamism on its head, which required a specifically sovereign public agency coping with the challenges of polycentric governance and incomplete contracts. Re-inventing sovereign public agency hence should thus be the guiding idea for policy. In absence of public agency, the coordination and distribution problems in the transformation towards a digital economy and low-carb mode of production cannot be coped with. These transformations would be distorted by Medici vicious circles of economic and political power, inducing socially undesirable lock-ins and coordination failures. If arguments à la Zingales (2017) are sound, the latter will be much worse than problems of coordination and power in the epoch of Railroadisation and the Gilded Age (cf. White 2011: 110–111).

While state sovereignty will undergo major transformations, post-sovereignty is misleading as a guiding idea or leitmotif of institutional innovation. However, it may nonetheless become reality. Variation of assumptions regarding contract (in)completeness and polycentric governance is thus useful in view of diagnosis and scrutiny of current trends and visions, including varieties of post-sovereignty. Suppose that contemporary societies fail to re-invent sovereign public agency under conditions of the digital transformation. Taking on board the insights on new contractual forms and their implications for (in)complete contracts, we consider three different perspectives faced by market societies: techno-libertarianism, techno-liberalism, and techno-oligarchy. While techno-libertarianism and techno-liberalism are resting on the assumption of pervasive complete contracts, techno-oligarchy is framed by the following (fairly realistic) combination:

1 Significant digital technology-supported increase of the extent of markets.
2 Higher numbers of both complete and incomplete contracts.
3 Mainly private development of background conditions such as Blockchains and economic mechanisms relevant to incomplete contracts.

The combination 1–3 is the most likely scenario accompanying vanishing sovereign public agency. Despite Blockchains and new contractual forms, a comprehensively complete contractual world is unrealistic in view of capitalist innovation dynamics whose risky, uncertain, and unknown effects preclude scenarios of contractual perfectibility. Understanding the forces driving this scenario (such as some kind of techno-oligarchy emerging from a Medici vicious circle à la Zingales) is a top priority. Coping with tendencies of degenerate oligarchic polycentrism (possibly aggravated by populist counter-movements operating "in the shadow of accountability"; cf. Sturn 2021) is the main political challenge – enhancing conditions for re-invigorating public sector sovereignty and strengthening the capacity of public agencies.

Nonetheless, it is worthwhile to discuss the two other market-based scenarios (techno-libertarianism and techno-liberalism), as they exhibit radically different directions of superseding sovereignty, both playing a significant role in contemporary discussions. Indeed, views like free-marketeering techno-libertarianism and techno-liberalism are major trends among digital avantgardes. Considered as political agenda under realistic assumptions, they may be accompanied by unintended consequences: while techno-liberal recipes may trigger post-sovereign forms of state governance, techno-libertarianism may boost oligarchic polycentrism as unintended consequence.

Techno-libertarianism/anarchism. Suppose a scenario of complete contracts, supported by Blockchain-based decentralised enforcement mechanisms. Digital market anarchists believe that nothing else is needed for supporting all desirable contract-mediated transactions, while libertarian minarchists attribute a residual role to the state regarding the protection of private property rights. While one might discuss the case for minarchist libertarianism à la Nozick (1974) vs. market anarchism under digital premises, we will not deal with this issue here, as two specific differences common to both are centre stage: (techno-)libertarianism *and* anarchism imply that contract incompleteness/contested exchange does not (no longer) compromise the advantages of voluntary contracts. In particular, *in sharp contrast to techno-liberalism*, techno-libertarianism and anarchism both imply that in contract-mediated exchange no problems occur which would require or justify distributive regulation.

In getting a libertarian position off the ground, two levels of arguments play a crucial role, including normative theory on the one side and positive theory and empirics of market processes on the other. Regarding normative theory, market libertarianism traditionally connotes a specific set of assumptions regarding the background conditions of exchange: while libertarians may admit the possible existence of "coercive offers" in contractual settings, objectionable coercion is diagnosed only in case of offers based on morally objectionable behaviour by one of the parties (notably "force and fraud", e.g. threats made credible by a gun put under your nose, accompanying the take-it-or-leave-it offer "money or life"). However, background conditions which are the basis of diagnoses of "coercive exchange" are broader and more multi-faceted than environments where "money or life offers" are common – including in particular distributive asymmetries (Zimmermann 1981; Scanlon 1988; Peter 2002; Sturn 2009 and 2017). For libertarians, distributive asymmetries (the existence of disadvantaged parties lacking reasonable alternatives to accepting some offers), however, never pose problems justifying political reform of background conditions, including distributive rules.

Such a view gains in prima facie plausibility when we assume a specific positive theory and empirics of market processes: a libertarian view is easier to sustain if market remunerations are determined in a competitive way, mainly reflecting relative scarcities. According to this theory, systemically entrenched divergences in income, wealth, and power (as e.g. discussed by Zingales 2017 or Piketty 2014) do not occur. Incidental income inequalities are triggered by individual preferences and hence are not morally objectionable, exemplified by Nozick's (1974) famous Wilt Chamberlain case (who earns a lot of money due to the willingness-to-pay of his fans who like watching him play basketball). On the grounds of such an optimistic view of competitive markets and of the incidence of exchange-mediated coercion, libertarians welcome withering away of sovereign state agency, once property rights and contract enforcement are managed by technologies like Blockchain.

With regard to background conditions and their dynamics, the problem of market libertarian interpretations of the principle of free contract can be described as follows. The more the principle of free contract is taken at face value without reflection of the background conditions related to distribution, the less problems of lacking consent come into view. However, even when starting with a scenario of pervasive complete contracts, in the context of the potentials of digital markets, such problems should be considered in view of asymmetries triggered by the following developments:

1 The range of market-based mediation increases, i.e. "new markets" also emerge in the area of what was previously not considered tradable.

2 Digital technologies (Blockchain, digitally supported monitoring) increase the potential for complete contracts.
3 The potential for asymmetries in the "private" development of background conditions is considerable, whether through asymmetries between platform developers, managers, and users, or through unintended consequences in the spontaneous further development of the "ecosystem" of platforms.

If 1–3 and/or other factors render distribution sufficiently unequal such that privileged agents can constantly make the dispossessed successful take-it-or-leave-it offers, (a) complete contracts are insufficient for the political neutrality of the exchange process and (b) a dystopia threatens: complete contracts fix an iron cage of asymmetrical bondage in which the libertarian principle of free contract becomes an ideology falsely attributing all this to the action of anonymous economic forces of nature or fancy preferences like those in the Wilt Chamberlain case.

Techno-liberalism. Techno-liberalism (see e.g. Posner and Weyl 2018) can be considered as the search for a way out of those problems. Consider again the extreme scenario of ubiquitous complete contracts, assuming that digital technologies extend the reach of complete contracts to cover all economically relevant interdependencies and goods. Even pure public goods are provided by decentralised mechanisms à la Buterin et al. (2018).

Notice that the textbook ideal of a depoliticised market sphere (cf. v. Weizsäcker 1999; Lerner 1972) thus becomes reality. In contrast to the techno-libertarian scenario, however, techno-liberals believe that distributive problems are still virulent and in part exacerbated by the just-mentioned tendencies of the digital world: for them, the market sphere is fully depoliticised on the basis of complete contracts if and only if suitable mechanisms of distributive regulation are implemented, preventing the just-sketched drawbacks of libertarianism. The distribution of individual "endowments" is considered a political issue.

As indicated earlier, in a world of complete contracts, political agency confronted with this issue faces specific difficulties. Digitally depoliticised background conditions of a perfect market world leave the state (apart from maintaining its own authority against internal and external threats) only one essential area of responsibility: pure distribution policy. Two implications follow. (1) The entire problem of power and distribution is shifted to the political system. Private governance or intermediate institutions no longer are part of the mediation of distribution problems: that is, negotiation processes in firms, collective bargaining between social partners, but also certain moralised motives such as fairness or reciprocity (which sometimes support contract efficiency while at the same time addressing distributive problems) simply have no function in a complete contract world. (2) Government action is strictly limited to correcting the distribution. The form of its action would somehow resemble the pure redistribution of the textbook according to the Second Theorem of Welfare Economics. Moreover, in such a world redistribution is an all-purpose weapon against contractual asymmetries and exploitation in all their forms. However, there would be no room for more far-reaching social or educational policies, or for other background conditions whose function is to deal with incomplete contracts. Put another way, in the case of complete contracts, the distribution-related outside options in exchange situations become the defining moment of the background conditions: the more the ideal type of complete contracts permeates a society, the more important the structure of alternative options becomes as a regulator for the distributional outcomes of exchange, and also for the asymmetries in the co-evolution of the economic system and the other social spheres. Notice that this logic explicitly shapes the arguments in favour of an unconditional basic income, as elaborated in a rigorous economic–philosophical analysis by Van Parijs (1995). An unconditional basic income would systematically push back coercive offers.

Let us now look more closely at political processes that deal with redistribution in such a scenario. What about the functioning of a political sphere that is a priori specialised at pure distribution politics as a zero-sum game? It seems obvious that organising discursive political accountability with regard to such a zero-sum game will be demanding. Distributional discourses entail conflicting interests and often polarising ideas of justice. Even if it is not true that this constellation can ultimately only be mastered by fighting (Friedman 1953: 5) or that fighting is the only elixir of life of "the political" (Schmitt 1933), acceptable modes of distribution (through argumentation, bargaining, and electoral mechanisms) are more likely to emerge if it is worked out to what extent such modes are the basis for dealing with cooperation problems in a way that is socially advantageous in the sense of a positive-sum game. Rawls (1971) articulates this impressively in his conception of the economic circumstances of justice under conditions of society as a cooperative venture. The specific principles of justice derived by Rawls are not in the focus of our interest. What matters here is rather his approach of emphasising the socio-economic contextualisation of market–societal distribution problems: as stressed by Rawls, those problems are not pure cake-distribution zero-sum games, but games in which the conditions of the cooperative production of the cake must be taken into account. As Rawls's exercise shows, it turns out that (somehow paradoxically) acceptable distributive principles are easier to derive when distributive problems are intertwined with the solution of cooperation problems. Accordingly, "successful" redistribution is more difficult if the state is reduced to an isolated redistribution function. This, however, puts the state in a dilemma – especially when initial endowments are very unevenly distributed, or when endogenous polarisation-promoting mechanisms obtain (cf. Zuboff 2015; Piketty 2014; Zingales 2017). It is then impossible to embed distributive regulation in a generally advantageous reform of the background conditions and thus communicate it as a long-term win–win situation (as is the case with certain social and educational policies) – unless perhaps with the dubious argument of revolution prevention.

Institutionally, pure distribution policy requires a lean, yet specifically powerful public agency, specialised in the implementation of a tax-and-transfer scheme such as a tax-financed basic income. This agency must be sufficiently informed and powerful in particular to prevent the erosion of tax bases by tax evasion. The latter are most likely an even greater problem than in the past, because Blockchain-like technologies and digital business models may favour the shadow economy and tax evasion. In this scenario, all politics would entail a problematic starting point: redistribution will be associated with an "excess burden" due to induced adjustment reactions, unless the state has a perfectly informed and effective redistribution bureaucracy. The model of "markets as constraints" of politics as elaborated in Hammond (1987) would become more and more real. The public sector must react with surveillance state means – or wither away.

Under such conditions, an unconditional basic income à la Van Parijs (1995) would be a redistribution machinery that most closely corresponds to egalitarian-liberal values (cf. Posner and Weyl 2018). However, its effective implementation (like other tax and transfer policies under these conditions) would require an effective surveillance state, unless an "ethos of solidarity" (which Van Parijs 1995: 230 wishes to promote "resolutely") or generalised reciprocity is sufficiently strong to minimise tax evasion and avoidance. Otherwise, this model would be associated with an amalgam of surveillance capitalism and surveillance statism, with respect to which an additional question crops up: in what socio-geographical context could such an arrangement best be organised? An answer would go beyond the scope of this entry. It seems obvious, however, that the relatively small European nation states, but also cosmopolitan models, would find it difficult to meet the challenges of such a scenario.

Techno-oligarchy. As stated by way of introduction of this section, the most realistic post-sovereignty scenario assumes that although digital support for more complete contracts is effective, incomplete contracts remain an important phenomenon, while unpolitical "private" governance is virulent as ideological fiction permanently nourishing talks about the obsolescence of state regulation – entailing an overall tendency undermining sovereignty. Policy tends to be pushed (with similar implications as in the techno-liberalism scenario) towards the distribution function, yet in a much less systematic manner than in techno-liberalism à la Posner and Weyl (2018). The classic mediation mechanisms of politics are eroding – combined with the erosion of political accountability mechanisms. Potentials of deliberative democracy (Elster 2008; Habermas 1997) remain untapped or are not developed further on an appropriate scale (Habermas 2013). On the other hand, the political problems of incomplete contracts (cf. Zingales 2017) remain virulent. Compared with the previous extreme scenarios, the more realistic scenario is more ambivalent with regard to the levels of interplay of neo-feudalist power and surveillance state. In this scenario, complex background conditions of incomplete contracts would form a tense environment of crisis-threatened economy and crisis-threatened statehood.

Even before digitalisation, neither the market nor the state could generate the preconditions of their stability by their own means, but instead had to rely on other sub-systems of society for stabilisation. Ultimately, re-inventing sovereignty will be contested by amalgams of interests and ideologies. Making the development of the background conditions a public matter under relevant technological and ecological premises will thus be a major challenge. However, sovereign public agency promoting ordered polycentrism by accountable rule-setting and regulation will be necessary for

- Preventing coordination gaps endangering the market system.
- Enhancing the innovation-promoting social productive power of free markets.
- Precluding markets from operating as one-sided appropriation mechanisms.

Two aspects of such a milieu should be kept in mind: (1) the state is complemented in its distribution and regulatory function by intermediary institutions and (2) the interplay between distribution problems, regulation, and the provision of public goods is addressed in a problem-responsive way. Conflicting situations can thus be transformed into win–win scenarios, as illustrated by the prisoner dilemma in public goods games. In this process of strengthening the public sector, two theoretical insights play a significant role. First, consistently complete contracts are not only empirically implausible but also have an ideological potential and are questionable as a regulative idea. Second, state sovereignty, including public institutions with collective decision-making and political accountability, is a historically contingent, difficult, and unstable construct. However, the development of social productive forces/public goods will not occur without it, not least in the light of the digital and low-carb technology transformation. Like the great transformation of the Industrial Revolution, the transformation processes of the 21st century require the institutionalisation of the public sphere adequate to the far-reaching challenges.

References

Atzori, M. (2015): Blockchain Technology and Decentralized Governance: Is the State Still Necessary? Available at SSRN: https://ssrn.com/abstract=2709713 or http://dx.doi.org/10.2139/ssrn.2709713

Basu, K. (2007): Coercion, Contract and the Limits of the Market, *Social Choice and Welfare* 29, 559–579.

Basu, K. (2019): New Technology and Increasing Returns: The End of the Antitrust Century. *IZA Policy Paper No. 146.*

Böckenförde, E.-W. (1976): *Staat, Gesellschaft, Freiheit.* Frankfurt a. M.: Suhrkamp.

Bowles, S. (1998): Endogenous Preferences: The Cultural Consequences of Markets and Other Institutions, *Journal of Economic Literature* 36, 75–111.

Bowles, S. (2004): *Microeconomics*. Princeton: Princeton University Press.
Bowles, S. and Gintis, H. (2000): Walrasian Economics in Retrospect, *The Quarterly Journal of Economics* 115(4), 1411–1439.
Buterin, V., Hitzig, Z. and Weyl, G. (2018): *Liberal Radicalism: Formal Rules for a Society Neutral Among Communities*. Mimeo. Available at: https://www.researchgate.net/publication/327742727_Liberal_Radicalism_Formal_Rules_for_a_Society_Neutral_among_Communities/citation/download [18.10.2021]
Crouch, C. (2004): *Post-democracy*. Oxford: Polity Press.
Davidson, S., de Filippi, P. and Potts, J. (2018): Blockchains and the Economic Institutions of Capitalism, *Journal of Institutional Economics* 14(4), 639–658.
Elster, J. (2000): Arguing and Bargaining in Two Constituent Assemblies, *University of Pennsylvania Journal of Constitutional Law* 2, 345–421.
Elster, J. (ed.) (2008): *Deliberative Democracy*. Cambridge: Cambridge University Press.
Friedman, M. (1953): *Essays in Positive Economics*. Chicago: University of Chicago Press.
Habermas, J. (1997): Popular Sovereignty as Procedure, in Bohman, J. and Rehg, W. (eds.): *Deliberative Democracy. Essays on Reason and Politics*. Cambridge, MA: MIT Press, 35–91.
Habermas, J. (2013): Demokratie oder Kapitalismus? Vom Elend der nationalstaatlichen Fragmentierung in einer kapitalistisch integrierten Weltgesellschaft, *Blätter für deutsche und internationale Politik* 13, 59–70.
Hammond, P. (1987): Markets as Constraints: Multilateral Incentive Compatibility in Continuum Economies, *Review of Economic Studies* 54, 399–412.
Jellinek, G. (1911): *Ausgewählte Schriften und Reden*. Berlin: Häring.
Jellinek, G. (1921): *Allgemeine Staatslehre*. 3rd ed. Berlin: Springer.
Lerner, A. P. (1972): The Economics and Politics of Consumer Sovereignty, *The American Economic Review* 62(2), 258–266.
Marden, P. (2003): *The Decline of Politics*. Aldershot, Hants, England: Ashgate.
Müller, J.-W. (2013): *Contesting Democracy*. Princeton: Princeton University Press.
North, D. C., Wallis, J. and Weingast, B. (2009): *Violence and Social Orders. A Conceptual Framework for Interpreting Recorded Human History*. Cambridge: Cambridge University Press.
Nozick, R. (1974): *Anarchy, State and Utopia*. Oxford: Basil Blackwell.
O'Neill, O. (2005): Gerechtigkeit, Vertrauen und Zurechenbarkeit, in Neumaier, O., Sedmak, C. and Zichy, M. (eds.): *Gerechtigkeit*. Frankfurt: Ontos, 33–55.
O'Neill, O. (2017): Accountable Institutions, Trustworthy Cultures, *Hague Journal on the Rule of Law* 9, 401–412.
Peter, F. (2002): Wahlfreiheit versus Einwilligung – Legitimation in Markt und Staat, *Jahrbuch für normative und institutionelle Grundlagen der Ökonomik vol. 1*, 153–172.
Piketty, T. (2014): *Capital in the 21st Century*. Cambridge, MA: Harvard University Press.
Polanyi, K. (1944): *The Great Transformation*. New York: Rinehart.
Posner, E. and Weyl, G. (2018): *Radical Markets*. Princeton: PUP.
Rawls, J. (1971): *A Theory of Justice*. Cambridge, MA: Harvard University Press.
Scanlon, T. M. (1988): The Significance of Choice, in Sen, A. and McMurrin, S. M. (eds.): *The Tanner Lectures on Human Values 8*. Salt Lake City: University of Utah Press, 149–216.
Schmitt, C. (1933): *Der Begriff des Politischen*. Hamburg: Hanseatische Verlagsanstalt.
Schumpeter, J. A. (1942): *Capitalism, Socialism and Democracy*. New York: Harper.
Smith, A. (1776): *An Inquiry into the Nature and Causes of the Wealth of Nations*. London: W. Strahan and T. Cadell.
Smith, A. (1790): *Theory of Moral Sentiments*. 6th ed. London: Millar.
Streeck, W. (2013): *Gekaufte Zeit. Die vertagte Krise des demokratischen Kapitalismus*. Berlin: Suhrkamp.
Sturn, R. (1994): The Firm as a Political Institution: Economic Democracy and the Tradability of Labor, in Biesecker, A. and Grenzdörffer, K. (eds.): *Soziales Handeln in ökonomischen Handlungsbereichen*, Bremen: Donat, 107–131.
Sturn, R. (2009): Volenti non fit iniuria? *Analyse und Kritik* 31, 81–99.
Sturn, R. (2017): Agency, Exchange, and Power in Scholastic Thought, *The European Journal of the History of Economic Thought* 24, 640–669.
Sturn, R. (2021): Markt- und Staatsversagen in großen Transformationen, *Jahrbuch normative und institutionelle Grundfragen der Ökonomik* 19.
Sugden, R. (2004): *The Economics of Rights, Cooperation and Welfare*. London: Palgrave Macmillan.
Van Parijs, P. (1995): *Real Freedom for all*. Cambridge: Cambridge University Press.
Varian, H. R. (2014): Beyond Big Data, *Business Economics* 49(1), 27–31.

Walras, L. (1969): Elements of Pure Economics. New York: W. Kelly.
Weizsäcker, C. C. (1999): *Logik der Globalisierung*. Göttingen: Vandenhoek & Rupprecht.
White, R. (2011): *Railroaded*. New York: Norton.
Williamson, O. (2000): The New Institutional Economics: Taking Stock and Looking Ahead, *Journal of Economic Literature* 38, 595–613.
Zimmermann, D. (1981): Coercive Wage Offers, *Philosophy and Public Affairs* 10, 121–145.
Zingales, L. (2017): Towards a Political Theory of the Firm, *Journal of Economic Perspectives* 31(3), 113–130.
Zuboff, S. (2015): Big Other: Surveillance Capitalism and the Prospects of an Information Civilization, *Journal of Information Technology* 30(1). https://doi.org/10.1057/jit.2015.5.

21

ANTITRUST LAW AND DIGITAL MARKETS

Viktoria H.S.E. Robertson

Introduction: antitrust scrutiny of digital markets

Antitrust law is a set of rules that wants to ensure competitive markets that bring about benefits for customers and consumers in terms of choice, innovation, price, and quality. This area of the law is also referred to as competition law. In the European Union (EU), competition law was introduced with the 1957 Treaty of Rome,[1] and the legal provisions on competition law have, in substance, remained unchanged ever since.[2] In the 21st century, digital markets are transforming the economy, and this upheaval in the markets also has profound repercussions on European competition law as such. Digital markets comprise all those business activities which make use of digital technology, either to support existing business ventures or to grow new ones (Competition and Markets Authority 2019a: 5). This definition applies to an increasing amount of economic activity (Akman 2019: 5).

In digital, data-driven markets, many of the basic assumptions underlying European competition law need to be questioned, including the analytical tools and the substantive theories of harm that competition law is premised upon. Over the past years, a plethora of reports have dealt with the question of whether and how to adapt competition law to new digital market realities. These reports were published by competition authorities, antitrust legislators, and independent experts, allowing for a good mix of opinions and suggestions. Table 21.1, albeit non-exhaustive, lists some of the most prominent reports in the field in chronological order.

The findings from these reports are now slowly being put into action: in December 2019, Margrethe Vestager, Vice-President of the European Commission, announced that the Commission would reassess its market definition guidance in the face of the changes which digital markets have brought about (Vestager 2019). That same month, the Japan Fair Trade Commission published guidelines on the abuse of a superior bargaining position in the digital realm and thereby took an important first step towards implementing the advice on competition law in digital markets

1 Treaty establishing the European Economic Community (25 March 1957), art 85 (anti-competitive agreements) and 86 (abuse of a dominant position).

2 Compare the provisions from 1957 with the almost identically worded provisions in today's Articles 101 (anti-competitive agreements) and 102 (abuse of a dominant position) of the Consolidated Version of the Treaty on the Functioning of the European Union [2016] OJ C202/47 (TFEU).

DOI: 10.4324/9780429351921-26

Table 21.1 Reports on competition law in digital markets (as of 1 January 2021)

Title of the report	*Lead author(s) or institution*	*Commissioning institution*	*Release date*
Competition Law and Data	French Autorité de la concurrence, German Bundeskartellamt		May 2016
Report of Study Group on Data and Competition Policy	Competition Policy Research Center	Japan Fair Trade Commission	June 2017
Big Data and Innovation: Key Themes for Competition Policy in Canada	Competition Bureau Canada		February 2018
Modernising the Law on Abuse of Market Power	Heike Schweitzer, Justus Haucap, Wolfgang Kerber, Robert Welker	German Federal Ministry for Economic Affairs and Energy	August 2018
Unlocking Digital Competition: Report of the Digital Competition Expert Panel	Jason Furman et al.	UK Government	March 2019
Competition Policy for the Digital Era	Jacques Crémer, Yves-Alexandre de Montjoye, Heike Schweitzer	European Commission	April 2019
Digital Platforms Inquiry – Final Report	Australian Competition and Consumer Commission		July 2019
A New Competition Framework for the Digital Economy – Report by the Commission 'Competition Law 4.0'	Martin Schallbruch, Heike Schweitzer, Achim Wambach	German Federal Ministry for Economic Affairs and Energy	September 2019
Digital Era Competition: A BRICS View	Ioannis Lianos, Alexey Ivanov	BRICS Competition Law and Policy Centre	September 2019
Stigler Committee on Digital Platforms – Final Report	Nolan McCarty, Guy Rolnik, Fiona Scott Morton, Lior Strahilevitz	The University of Chicago Booth School of Business	September 2019
Algorithms and Competition	French Autorité de la concurrence, German Bundeskartellamt		November 2019
Online Platforms and Digital Advertising	UK Competition and Markets Authority		July 2020
Investigation of Competition in Digital Markets	Subcommittee on Antitrust, Commercial and Administrative Law	US House of Representatives	October 2020

through soft law (Japan Fair Trade Commission 2019). It also emerged that the United Kingdom (UK) is considering the creation of a technology regulator in early 2021 that should be tasked with watching over big tech companies, based on recommendations in the UK's Furman

Report (Murgia and Beioley 2019; Competition and Markets Authority 2019b). In June 2020, the European Commission went public with its plans to introduce a new competition tool to tackle structural risks for competition that have been observed in digital markets (European Commission 2020b, 2020c). In January 2021, the German legislature adopted an amendment to the German Competition Act, not only in order to implement the ECN+ Directive 1/2019[3] but also so as to incorporate findings from the German reports in the area of digital markets.[4] In the area of digital markets, the new legislation contains several important changes, which are discussed below. Finally, in December 2020, the European Commission published its proposal for a Digital Markets Act containing new do's and don'ts for digital gatekeepers as well as a scaled down version of a market investigation mechanism for digital markets (European Commission 2020e).

The present chapter focuses on the main implications of the digital economy for five crucial areas of competition law: market definition, market power assessments, anti-competitive agreements, abusive conduct, and merger control. For each of these areas, it will identify the particular issues that competition law needs to address in digital markets, discuss some of the European experience in dealing with those issues and highlight how each particular area of competition law may (need to) evolve in the near future. This includes a discussion of the proposals made in three of the digital competition reports: the Furman Report, the EU Report, and the Competition Law 4.0 Report.

The Furman Report of March 2019 was commissioned by the UK Government and contains recommendations by a digital competition expert panel headed by Jason Furman and composed of experts in economics, law, and computer science.

The EU Report of April 2019 was written by three Special Advisors to Commissioner Vestager, Jacques Crémer, an economist; Heike Schweitzer, a competition law professor; and Yves-Alexandre de Montjoye, a data scientist.

Finally, the Competition Law 4.0 Report of September 2019 was commissioned by the German Federal Ministry for Economic Affairs and Energy in preparation for its Council Presidency in the second half of 2020. The Report was prepared with a view to updating EU competition law and was compiled by a team headed by Martin Schallbruch, a computer scientist; Heike Schweitzer – the same competition law professor that co-authored the EU Report; and Achim Wambach, an economist and then head of the German Monopoly Commission.

The relevant market in the digital sphere

In competition law, the delineation of the relevant market represents one of the basic analytical tools which is employed when analysing a case, whether in the area of anti-competitive agreements, abuse of a dominant position, or merger control. While market definition has its roots in economics, once this concept is received into competition law it becomes a distinct legal concept that builds upon the economic one but takes on its own legal conception (Robertson 2019b). Under EU competition law, the relevant market is relied upon to determine market share thresholds and market power, to provide the economic context of a particular case, and to inform the competition theory of harm. The European Commission has laid down its approach to market definition in its Market Definition Notice of 1997 (European Commission 1997), which is based

3 Directive (EU) 2019/1 of the European Parliament and of the Council to empower the competition authorities of the Member States to be more effective enforcers and to ensure the proper functioning of the internal market [2019] OJ L 11/3 (ECN+ Directive).

4 Gesetz zur Änderung des Gesetzes gegen Wettbewerbsbeschränkungen für ein fokussiertes, proaktives und digitales Wettbewerbsrecht 4.0 und anderer Bestimmungen (GWB-Digitalisierungsgesetz), Federal Law Gazette I 2021/1, 2.

on the case law of the Court of Justice of the European Union (CJEU) while at the same time being infused with a more economic approach to market definition (see Robertson 2020a: 50–52).

The issues

Market definition is based on substitutability tests which help to identify a relevant antitrust product market, and these tests were established against the background of static markets. The dynamic characteristics of digital markets make the delineation of the relevant market particularly demanding (OECD 2014: paras. 141–147). Digital markets bear a number of characteristics that are markedly different from more traditional static markets. These characteristics need to be reflected in the delineation of relevant product markets for competition law purposes. They include the fast-moving nature of digital markets, the existence of zero price markets or market sides, the 'winner takes all' nature of some digital markets, the propensity of digital platform markets to tip based on network effects and resulting in user lock-in, and competition for the market as a particular feature of competition in digital markets (Baye 2008: 640; EU Report 2019: 42–47). Big data and big analytics are cornerstones of the data-driven digital market environment, and it is paramount to get a good understanding of them. In addition, platforms (also called multi-sided markets) have become a popular business model that needs to be understood, together with the direct and indirect network effects such a platform relies upon.[5] Closely related to platforms, digital ecosystems are being created that can lead to a lock-in of consumers and a reduction of competition. This goes hand in hand with an increasingly conglomerate company structure (Bourreau and de Streel 2019). Secondary markets that are specific to digital ecosystems may also need to be considered, especially as they relate to data (Competition Law 4.0 Report 2019: 29; EU Report 2019: 88–90). Digital market environments may significantly impact competitive relationships (Ezrachi and Stucke 2016: 149, 157; Competition Law 4.0 Report 2019: 28), something that needs to be reflected in market definition. The digitalisation of the economy has led to markets in which identical or similar offerings can be bought 'offline' (e.g., printed books in a book shop) or 'online' (e.g., printed books through an online retailer, or e-books). Here, the question arises whether online and offline markets may be converging or whether they constitute separate relevant markets (Robertson 2017: 146–149). Where one market side receives services in exchange for data rather than against payment of a monetary price, the traditional economic tools for market definition cannot apply in a straightforward manner (Vestager 2019; Robertson 2020a: 249 ff).

As set out earlier, digital platforms are frequent in the digital sphere. These are markets in which a platform occupies an intermediary position between several market sides ('multi-sided markets'). Take for instance Facebook's social media platform, which is an interface for a number of market sides: users interacting with friends and family, developers wanting to attract custom, content providers showcasing their offerings to an audience, and advertisers showing ads to their target audience. Frequently, one of the market sides (usually: the user side) receives services such as the social media functionality against exchange of their data and attention, leading to the question whether attention markets (Newman 2016, 2020) or data markets (Jones Harbour and Koslov 2010) could be at issue. Another option is to view each market side individually as the relevant antitrust market. Economists, however, have suggested that a platform's

5 In multi-sided platforms, both direct and indirect network effects can be observed. On a social media platform, for instance, the platform will become increasingly valuable to users as more and more users (friends, family, colleagues) join the network (direct network effect). On the other hand, an increase in the number of users will attract more advertising customers to the social media platform (indirect network effect).

market sides cannot be properly assessed in isolation (e.g., Caillaud and Jullien 2003; Rochet and Tirole 2003; Armstrong 2006). Such a narrow approach would miss the big picture of how a digital platform works, losing from view important insights that should be harnessed for the antitrust analysis (Thépot 2013). On the other hand, a relevant market that encompasses the entire platform may be too broad for traditional antitrust analysis – and may include relevant markets that are not substitutable with each other. Such an approach would require significant changes in how competition law relies on market definition.

The EU's experience

Platform markets are an important feature of digital markets, yet so far neither the European Commission nor the CJEU have properly incorporated multi-sided market theory into their market definition framework. In a number of cases, they alluded to the two-sided nature of a market, for instance in *Google/DoubleClick* (2008) or in *MasterCard* (2014).[6] Even in more recent cases such as *Google Shopping* (2017),[7] however, the Commission merely referred to a market's multi-sidedness rather than incorporating this knowledge into the market definition. This points to the lack of a coherent market definition framework as far as platforms are concerned.

Zero price markets (or market sides) that form part of a digital platform are now regularly accepted as commercial activities that can and should be scrutinised under competition law. In its *Google Shopping* decision, the European Commission relied on a zero-price market, namely general online search, as the relevant market. The Commission argued that this was justified because users paid with their data when using Google's search engine. In addition, the free user side was part and parcel of Google's platform business model, and price was not the most important competitive parameter in general online search.[8]

Despite the importance of data in digital markets, the European Commission has not yet delineated a data market where companies were not in the business of selling data. In the merger case of *Facebook/WhatsApp* (2014), for instance, the Commission refrained from delineating a market for data because the two parties only used their data in-house.[9] Nevertheless, it remains an option to delineate potential markets under such circumstances, which would allow for an antitrust approach that takes into account the importance of data in digital markets.[10] In the recent *Google/Fitbit* merger review,[11] the Commission had announced that it would examine the effects of the combination of Fitbit's and Google's 'databases and capabilities in the digital healthcare sector' (European Commission 2020d). While this merger was cleared subject to Google's behavioural commitments, the European Commission would have had the opportunity to not only earnestly look at an actual or potential health data market in its analysis, but also to consider the privacy implications of ever-growing digital conglomerates (Bourreau et al. 2020).

In order to grasp market-related innovation that has not yet reached the stage of a product that could be delineated as a relevant antitrust product market, the European Commission has resorted to the identification of 'innovation spaces' in cases such as *Dow/DuPont* (2017) and

6 *Google/DoubleClick* (Case COMP/M.4731) Commission Decision of 11 March 2008 [2008] OJ C184/10, para 290; Case C-382/12 P *MasterCard v Commission* EU:C:2014:2201, para 237.

7 *Google Search (Shopping)* (Case AT.39740) Commission Decision of 27 June 2017 [2018] OJ C9/11, para 159.

8 *Google Search (Shopping)* (Case AT.39740) Commission Decision of 27 June 2017 [2018] OJ C9/11, paras 158–160.

9 *Facebook/WhatsApp* (Case COMP/M.7217) Commission Decision of 3 October 2014, paras 70, 72.

10 Case C-418/01 *IMS Health v NDC* EU:C:2004:257, para 44.

11 *Google/Fitbit* (Case COMP/M.9660) Commission Decision of 17 December 2020.

Bayer/Monsanto (2018).[12] It understands innovation spaces as spaces in which innovation competition occurs before a relevant antitrust market emerges. Such an approach may also be feasible where digital innovation is concerned.

How market definition needs to adapt to digital markets

The EU Report notes that digital markets may not allow us to arrive at clear market boundaries under the present market definition framework, concluding that 'in digital markets, less emphasis should be put on the market definition part of the analysis' (2019: 45–48, direct quote at 46). It also suggests that in digital markets, market definition may differ significantly depending on whether an ex ante (Article 102 of the Treaty on the Functioning of the European Union or TFEU) or an ex post (merger control) view is taken (p. 47). In view of the digital ecosystems which digital platforms are developing, one may need to delineate secondary markets that are specific to a particular ecosystem in cases where users are locked into such an ecosystem (p 48, Punkt). Under the current EU competition law regime, this would entail very narrowly defined secondary markets and, consequently, strict antitrust standards.

The Furman Report acknowledges the importance of market definition, for instance in order to arrive at concentration levels that competition economics frequently rely upon in merger control. While it highlights that market definition in digital markets is fraught with difficulties (2019: 24, 30, 89), it does not provide a solution to this issue.

The Commission Competition Law 4.0 recommends revising the Commission's Market Definition Notice of 1997 in order to reflect the intricacies of digital markets (2019: 31). Bearing in mind the characteristics of digital markets, it suggests that the European Commission may want to issue separate guidelines on how to delineate markets in the digital environment. This first recommendation has already borne fruit: on 9 December 2019, Margrethe Vestager, Executive Vice-President of the Commission and Commissioner for Competition, announced that the European Commission was beginning to review its Market Definition Notice against the background of digital markets. She stressed that well-established methods for defining antitrust markets may no longer apply in the digital environment, especially in multi-sided platforms and zero price markets (Vestager 2019). The Commission's review of its 1997 Notice is now underway, with both an evaluation of the proposed roadmap and a public consultation concluded over the summer of 2020 (European Commission 2020a).

In 2017, the German legislator adopted an amendment to its competition code, which now states that the fact that a product is provided without monetary remuneration does not invalidate the assumption that an antitrust market exists.[13] In EU law, data and attention are also increasingly seen as a consumer's counter-performance for the receipt of digital services.[14] Together with the *Google Shopping* decision discussed earlier, these developments show that zero price markets are no longer an obstacle to antitrust enforcement.

The challenges that market definition is faced with in digital environments lead to the question of whether this analytical tool can continue to fulfil its traditionally assigned role of

12 *Dow/DuPont* (Case M.7932) Commission Decision of 27 March 2017 [2017] OJ C353/9, para 350; *Bayer/Monsanto* (Case M.8084) Commission Decision of 21 March 2018 [2018] OJ C456/10, para 80, fn 23.

13 German Restriction of Competition Act (Gesetz gegen Wettbewerbsbeschränkungen or GWB), Federal Law Gazette I 2114/2005, as amended, § 18(2a).

14 Directive (EU) 2018/1972 establishing the European Electronic Communications Code [2018] OJ L321/36, Recital 16; Directive (EU) 2019/770 of 20 May 2019 on certain aspects concerning contracts for the supply of digital content and digital services [2019] OJ L136/1, Article 3(1).

assessing market power in these markets from a primarily quantitative point of view (Competition Law 4.0 Report 2019: 30). It is very well possible that market definition will need to focus on its second major role in digital markets: that of characterising the market so as to provide the necessary background to understanding the markets at issue and developing a coherent theory of harm in those markets (Robertson 2020a: 318).

Market power assessments in digital markets

The CJEU holds a dominant position to be

> a position of economic strength enjoyed by an undertaking which enables it to prevent effective competition being maintained on the relevant market by affording it the power to behave to an appreciable extent independently of its competitors, its customers and ultimately of the consumers.[15]

In terms of market shares, the CJEU presumes a dominant position from market shares of 50%.[16]

The existence, creation, or strengthening of a company's dominant position on a relevant market often represents the crux of a competition law case. This is so because Article 102 TFEU[17] forbids the abuse of a dominant position, which in turn presupposes the existence of such a dominant position. The EU Merger Regulation (EUMR)[18] views the creation or strengthening of a dominant position as one of the main impediments of effective competition that can lead to the prohibition of a merger (Article 2 para 3 EUMR). Market power also plays a major role under Article 101 TFEU, which prohibits anti-competitive agreements: block exemption can be obtained for a range of agreements if certain market share thresholds are not surpassed, as low market shares are considered to point to a lack of market power, alleviating competition concerns.[19]

The issues

In digital markets, market shares as the traditional measure for market power (or for the absence of market power) may no longer apply with full force. Market shares and concentration ratios based on market shares are not meaningful measures in a market environment that is characterised by its dynamic nature and in which the very tool that market share calculation relies upon – market definition – is increasingly complex.

Market power in digital markets shares an intricate relationship with user data, and the importance of big data for digital markets in all its dimensions – volume, variety, velocity, and veracity – cannot be overstated. Where a digital platform has the capability to both harvest up-to-date data on a large scale and analyse it with appropriate tools, it may be found to be

15 Case 85/76 *Hoffmann-La Roche v Commission* EU:C:1979:36, para 38.

16 Case C-62/86 *AKZO Chemie v Commission* EU:C:1991:286, para 60. For legal purposes, 'market shares for each supplier can be calculated on the basis of their sales of the relevant products in the relevant area' (European Commission 1997, para 53).

17 Consolidated Version of the Treaty on the Functioning of the European Union [2016] OJ C202/47 (TFEU).

18 Council Regulation (EC) 139/2004 on the control of concentrations between undertakings [2004] OJ L24/1 (EU Merger Regulation).

19 E.g., see Commission Regulation (EU) No 330/2010 on the application of Article 101(3) of the Treaty on the Functioning of the European Union to categories of vertical agreements and concerted practices [2010] OJ L102/1, art 3 (30% market share threshold for both buyer and supplier).

in a dominant position for competition law purposes (European Data Protection Supervisor 2014: para 60). There is also a self-reinforcing data advantage which stems from network effects in digital platforms (Ezrachi and Robertson 2019: 8) as well as from the 'winner takes all' type of competition found in digital markets that are prone to tipping (OECD 2013: 5; Bundeskartellamt 2016: 45).

Where digital platforms possess large troves of up-to-date user data and users are locked into a particular platform, 'we might be witnessing a rise in power over consumers even when, seemingly, market power relating to a specific antitrust market is not there (yet)' (Ezrachi and Robertson 2019: 17). This power over consumers by a handful of 'data-opolies' (Stucke 2018) often stems from the practice of third-party tracking, whereby a tracker scoops up a user's digital footprint in order to build a comprehensive user profile (Binns et al. 2018). The power of platforms over their market sides also extends to customers that may sometimes also be the platform's competitor, for instance in the case of online marketplaces on which the platform provider itself simultaneously acts as a retailer and therefore competes with other retailers (its customers) selling through its marketplace.

The EU's experience

The European Commission and the General Court have recognised that market shares may not adequately reflect the existence of market power in the digital market environment. In *Cisco Systems* (2013), the General Court agreed with the Commission that the consumer communications sector at issue was

> a recent and fast-growing sector which is characterised by short innovation cycles in which large market shares may turn out to be ephemeral. In such a dynamic context, high market shares are not necessarily indicative of market power and, therefore, of lasting damage to competition which Regulation No 139/2004 [i.e., the EUMR] seeks to prevent.[20]

Shortly thereafter, in *Facebook/WhatsApp* (2014), the Commission relied on an almost identical wording, also in the context of consumer communication services.[21] In *Apple/Shazam* (2018), the Commission again highlighted that 'market shares may not be a perfect proxy for measuring market power in recent and fast-growing sectors characterised by frequent market entry and short innovation cycles'.[22] However, in the case at hand, it characterised the market as a mature one that was not subject to this logic.

All three of the cases just discussed were merger decisions in the digital environment, in which an ex ante assessment of the proposed merger was carried out. To date, this logic does not seem to have been applied in ex post assessments. In *Google Shopping* (2017), for instance, the Commission emphasised that very large market shares were 'save in exceptional circumstances, evidence of the existence of a dominant position'.[23] With reference to *Cisco Systems*, the Commission held that under Article 102 TFEU, the fact that a market is highly dynamic does not preclude reliance on market shares, particularly where a market has not shown instability during

20 Case T-79/12 *Cisco Systems v Commission*, EU:T:2013:635, para 69.

21 *Facebook/WhatsApp* (Case COMP/M.7217) Commission Decision of 3 October 2014, para 99.

22 *Apple/Shazam* (Case COMP/M.8788) Commission Decision of 6 September 2018, para 162.

23 *Google Search (Shopping)* (Case AT.39740) Commission Decision of 27 June 2017 [2018] OJ C9/11, para 266.

times of fast economic growth.[24] Overall, the Commission took the following factors into account when assessing Google's dominant position on the market for general online search: its market shares, barriers to entry and expansion, the lack of multi-homing users, brand effects, and the lack of countervailing buyer power.[25]

Adapting market power assessments to the digital age

Market power in digital markets is most often seen through the lens of digital platforms. While a first difficulty lies in adequately delineating the relevant market(s) that a digital platform is operating on, the next hurdle consists in assessing the platform's market power on any given market. Frequently, digital platforms do not conform to traditional notions of market power, and the fact that digital platforms can create self-sufficient ecosystems comes into play when trying to evaluate their market power. Where digital platforms act as intermediaries between various market sides, they may enjoy something akin to intermediation power if they can control suppliers' access to certain distribution channels or to certain customer groups (Schweitzer/Haucap/Kerber/Welker 2018: 66 ff). In a similar vein, other reports have referred to digital platforms as unavoidable trading partners (EU Report 2019: 49) and have problematised the bottleneck power that digital platforms have over single-homing users (Stigler Report 2019: 32, 105). The Furman Report introduces the concept of a strategic market status, which it understands to consist in the ability 'to exercise market power over a gateway or bottleneck in a digital market, where [those companies] control others' market access' (2019: 41, direct quote at 55).

What these analyses have in common is a concern about whether digital platforms and their specific market power may remain below the radar of current methods for assessing market power because the kind of market power they can exert over users, customers, and competitors is different from traditional markets. Therefore, the Competition Law 4.0 Report believes that the concept of market power needs to be further clarified in the digital sphere, also calling for separate Commission guidance on assessing market power in digital platforms and further research into cross-market foreclosure strategies in the digital economy (2019: 32 f).

The German legislature reacted to the difficulties of assessing market power in digital platforms at an early stage: the 2017 amendment to the German Competition Act introduced a provision stating that in the assessment of market power in multi-sided markets and platforms, one needs to take into account (1) direct and indirect network effects, (2) the parallel use of services from different providers and the switching costs for users, (3) the undertaking's economies of scale arising in connection with network effects, (4) the undertaking's access to data relevant for competition, and (5) innovation-driven competitive pressure (§ 18 para 3a GWB, i.e., the German Act against Restraints of Competition). When it prohibited Facebook's data practices in February 2019, the German Bundeskartellamt relied on these new provisions in order to assert the social network's market power. It found that Facebook had a market share in excess of 95% among daily active users of private social networks and that strong direct network effects were experienced by users, with advertisers being exposed to strong indirect network effects. Based on the exit of competing social networks, the authority assumed that it was witnessing a market in the process of tipping.[26]

In January 2021, the German legislature adopted its long-awaited amendment to bring Germany's competition code up to date with the digitalisation of markets. The law now contains

24 *Google Search (Shopping)* (Case AT.39740) Commission Decision of 27 June 2017 [2018] OJ C9/11, para 267.
25 *Google Search (Shopping)* (Case AT.39740) Commission Decision of 27 June 2017 [2018] OJ C9/11, paras 271–318.
26 Bundeskartellamt, *Facebook* (B6–22/16, 6 February 2019), paras 422–521.

a new § 18 para 3b GWB: when assessing the market position of an undertaking that acts as an intermediary on a multi-sided market, the importance of those intermediary services for access to supply and sales markets needs to be taken into account. In addition, § 18 para 3 nr 3 GWB now refers to a company's access to data relevant for competition as a factor to consider when assessing market dominance, irrespective of whether or not the undertaking is active on a multi-sided market.[27]

In addition, the European Commission has proposed new rules for gatekeepers, i.e., providers of certain platform services that are an important gateway for business users to reach end users and that enjoy an entrenched and durable position in their operations (European Commission 2020e: Art 3). With these proposed rules, the Commission strays away from traditional notions of market dominance under the competition laws and instead tries to capture major big tech platforms through alternative criteria.

Anti-competitive behaviour in digital markets

Under Article 101 TFEU, companies are not allowed to enter into anti-competitive agreements with each other. In para 1, the provision lists a number of agreements that are considered to be anti-competitive, such as price fixing, limiting or controlling markets or technical development, market sharing, discriminating among trading parties, and tying or bundling. As this list is merely exemplary, it can be adapted to the specificities of digital markets. So far, the application of Article 101 TFEU in digital markets has not required extraordinary efforts, as anti-competitive agreements appear to take similar forms in the online and offline world. It is presumably for this very reason that the reports introduced at the outset of the present chapter almost exclusively focus on abuses of a dominant position or on merger control, rather than on anti-competitive agreements. Therefore, the following sections will equally be very concise.

The issues

Algorithms may allow a company to engage in dynamic pricing, adapting prices to the current demand and supply situation or to competitor behaviour (Autorité de la concurrence and Bundeskartellamt 2019). Similarly, pricing algorithms may rely on user data to discriminate among consumers (Ezrachi and Stucke 2016: 89, 107). The antitrust liability for such behaviour very much depends on whether algorithms are relied upon in order to collude with other companies and their algorithms, in which case Article 101 TFEU can come into play. Where algorithms are used in order to apply dissimilar conditions to equivalent transactions with other trading parties, the identically worded Article 101(1)(d) TFEU or Article 102(c) TFEU may apply. In *MEO* (2018), the CJEU held that discriminatory pricing is only considered abusive where it 'tends to distort competition'.[28] As Article 101(1)(d) TFEU and Article 102(c) TFEU explicitly refer to 'other trading parties' (Graef 2018: 558), their applicability to price discrimination involving final consumers needs to be further tested.

Online marketplaces have become important retail outlets in the digital sphere. Both suppliers' restrictions on selling through such platforms and these marketplaces' own restrictions on their suppliers have attracted antitrust scrutiny. Some have observed a certain friction between both approaches, criticising that competition authorities were promoting marketplace platforms

27 For undertakings active on multi-sided markets, this was already required under § 18 para 3a GWB since the 2017 amendment; this provision remains in force.

28 Case C-525/16 *MEO v Autoridade da Concorrência* ECLI:EU:C:2018:270, para 26.

through the prohibition of platform bans in distribution contracts, while at the same time challenging the way that marketplace platforms conduct their business (Ibáñez Colomo 2018). However, this does not necessarily represent a contradiction. While retailers should remain free to sell through various kinds of sales channels, marketplace platforms must abide by the antitrust rules.

Digital platforms such as hotel booking platforms frequently require their customers, i.e., hotels, to agree to most-favoured-nation or MFN clauses. In these contractual clauses, the hotels promise not to offer their services for a lower price through any other platform, sales channel, or their own website (see Ezrachi 2015; EU Report 2019: 55–57). MFN clauses can be assessed under Article 101 TFEU and under Article 102 TFEU. While they may occur in both digital platforms and offline scenarios, the frequent use of MFNs by digital platforms has brought them under antitrust scrutiny.

The advent of blockchain technology represents another interesting development for competition law. As blockchain technology relies on a decentralisation of databases, which replaces trust among parties with technology and provides transparency in the distribution of information, there is scope for collusion through blockchains, including cartels or information sharing (Ristaniemi and Majcher 2018). This would be caught out under Article 101(1) TFEU like any other cartel. Under Article 102 TFEU, blockchain technology may make it more difficult to determine what constitutes a dominant market position, or to attribute liability for the abuse of such a position to a specific company (Schrepel 2019). Scenarios may be plausible in which access to private blockchains is refused, thus triggering an intervention under Article 102 TFEU akin to refusals to supply in the pre-blockchain world. Blockchains can also implement other forms of anti-competitive behaviour, such as tying or price discrimination (Schrepel 2019). Importantly, anti-competitive practices may be executed through blockchain technology – and can then be caught out by competition law. For antitrust enforcement, a number of positives are associated with blockchain technology (Tulpule 2017): it may create a level playing field in terms of access to information, reduce transaction costs in merger cases, or allow competition authorities to monitor commitments made by parties (Ristaniemi and Majcher 2018).

The EU's experience

Concerning online marketplaces, the CJEU dealt with the application of competition law to distribution agreements in the online sphere in a series of preliminary rulings. It highlighted that a supplier could not place an absolute ban on online sales on its distributor; this would generally not be acceptable under Article 101(1) TFEU.[29] More recently, however, the Court's preliminary ruling in *Coty v Parfümerie Akzente* (2017) found that manufacturers could foresee certain restrictions regarding online sales channels, such as marketplace platforms, if they were dealing with luxury goods, and if the preservation of the luxury character of those goods required such restrictions.[30]

In 2018, the European Commission revealed that it was investigating Amazon's dual role as the provider of a sales platform for merchants and, in parallel, as these merchants' direct competitor in a wide array of product markets. In particular, it voiced concerns whether Amazon may be using data on its merchants, to which it has access as the sales platform provider, in order to improve its own position as their competitor (Schechner and Pop 2018). While the

29 Case C-439/09 *Pierre Fabre Dermo-Cosmétique* EU:C:2011:649, para 2.

30 Case C-230/16 *Coty Germany v Parfümerie Akzente* EU:C:2017:941, paras 37–58.

Commission's press release of November 2020 suggests that it is currently investigating this case under Article 102 TFEU (European Commission 2020f),[31] the case could equally shed light on how the dynamics of a digital platform come into play in a case under Article 101 TFEU. The same applies to the Commission's investigations into Amazon's activities related to the Buy Box and access to its Amazon Prime customers for retailers, as the restrictions being investigated relate to the contractual terms between Amazon and independent retailers.[32]

In the summer of 2020, the European Commission opened an investigation into Apple and its alleged refusal to grant third parties access to the tap and go functionality on iPhones, as well as to its refusal to grant access to its mobile payment solution, Apple Pay. This is being investigated as a potential anti-competitive agreement, with an alternative of constituting an abuse of a dominant position.[33]

The rise of MFN clauses in some digital markets, most notably on hotel booking portals, has led to a number of antitrust investigations in the *Booking.com* cases (2015)[34] and in the German *HRS* case.[35] MFN clauses have also been at issue in relation to e-books: in the *Amazon E-books* case (2017), Amazon offered commitments in order to counter the European Commission's concerns as to the anti-competitive nature of its MFN clauses. Amazon had required e-book suppliers '(i) to notify Amazon of more favourable or alternative terms and conditions they offer elsewhere, and/or (ii) to make available to Amazon terms and conditions which directly or indirectly depend on the terms and conditions offered to another E-book Retailer'.[36] In light of Amazon's position on the markets for the retail distribution of English and German e-books to consumers, the Commission considered this as constituting an abuse of a dominant position (Article 102 TFEU). However, this case could also have been decided under Article 101 TFEU. In its commitments, Amazon agreed not to enforce any parity clauses already contained in agreements and not to conclude any e-book agreements containing such clauses.

The future of anti-competitive agreements in digital markets

As stated earlier, the application of Article 101 TFEU to cases arising in digital markets has been very much in keeping with well-established case law. Two topics are likely to receive more attention in times to come. The first is the question of how collusion through algorithms may be caught under Article 101 TFEU, given that this provision requires some form of concurrence of wills or concerted practices above the threshold of tacit collusion (see Autorité de la concurrence and Bundeskartellamt 2019). Second, the question of online distribution will certainly play an important role in the ongoing revision of the Vertical Agreements Block Exemption Regulation, a legal instrument adopted by the European Commission which sets out under which conditions distribution agreements are not considered to be anti-competitive.[37] When

31 *Amazon Marketplace* (Case AT.40462) Commission Decision pending.

32 *Amazon Buy Box* (Case AT.40703) Commission Decision pending.

33 *Apple Pay* (Case AT.40452) Commission Decision pending.

34 Konkurrensverket, Case 596/2013 *Booking.com* (15 April 2015); Autorité de la concurrence, Case 15-D-06 *Booking.com* (21 April 2015); Autorità Garante della Concorrenza e del Mercato, Case I779 *Booking.com* (21 April 2015); Bundeskartellamt, Case B 9–121/13 *Booking.com* (22 December 2015); Higher Regional Court Düsseldorf, Case VI – Kart 2/16 (V) *Booking.com* (4 June 2019).

35 Higher Regional Court Düsseldorf, Case VI – Kart 1/14 (V) *HRS* (9 January 2015).

36 *E-Book MFNs and related matters (Amazon)* (Case AT.40153) Commission Decision of 4 May 2017 [2017] OJ C264/7.

37 Commission Regulation (EU) No 330/2010 on the application of Article 101(3) of the Treaty on the Functioning of the European Union to categories of vertical agreements and concerted practices [2010] OJ L102/1.

describing the purpose for the evaluation of the Vertical Agreements Block Exemption Regulation, the European Commission highlighted that it thought a revision might be necessary due to 'the increased importance of online sales and the emergence of new market players such as online platforms' (European Commission 2018b: 1).

Abuse of a dominant position in the digital sphere

Under Article 102 TFEU, companies with market power (i.e., with a dominant position on the relevant market) may not engage in anti-competitive unilateral behaviour. As instances of such anti-competitive behaviour, the provision lists the imposition of excessive prices and unfair trading conditions, the limiting of markets or technical development, discriminating among trading parties, and tying or bundling. However, the list contained in Article 102 TFEU is not exhaustive. For the application of competition law to digital markets, this means that unilateral anti-competitive conduct that occurs in the digital sphere can either be caught by already established categories of abuse, or new types of abuse can be developed against the specific background of digital markets.

The issues

A distinctive feature of big tech companies that operate digital platforms is that they operate in a great number of different markets (Bourreau and de Streel 2019). By creating entire digital ecosystems, they are able to leverage their market power from one market into adjacent or perhaps even rather distant markets (EU Report 2019: 47–48, 65 ff). In addition, the user data that digital platforms accumulate in one relevant market is of a multi-purpose nature and can prove useful in entirely different markets. While Article 102 TFEU also applies where a dominant undertaking leverages its market power in markets in which it is not (yet) dominant, these dynamics of competition raise the question of whether competition law is fit for purpose in the face of digital ecosystems. Similarly, the frequent acquisitions of (potential) competitors by big tech companies also needs to be scrutinised with a view to maintaining competition (see the following section).

Based on the data-centric nature of digital markets, the question arises whether privacy-related abuses are sanctionable under the current competition rules. Privacy-related abuses may, for instance, relate to a reduction of quality of the platform's services, to the excessive gathering of user data that digital platforms require in return for digital services (Ezrachi and Robertson 2019), or to low data protection standards that user data is awarded. The European Commission considers that privacy-related issues can be of relevance 'in the competition assessment to the extent that consumers see [privacy] as a significant factor of quality' (European Commission 2016). In addition, however, the 'normative backdrop' (Costa-Cabral and Lynskey 2017: 14) of the competition provisions in the EU is outspokenly privacy-friendly, including the fundamental right to privacy and data protection enshrined in the Fundamental Rights Charter[38] and other legislative instruments, such as the General Data Protection Regulation (GDPR)[39] and the proposed ePrivacy Regulation.[40] This could lead to EU competition law being infused with

38 Charter of Fundamental Rights of the European Union [2016] OJ C 202/389, art 8.

39 Regulation (EU) 2016/679 of 27 April 2016 on the protection of natural persons with regard to the processing of personal data and on the free movement of such data (General Data Protection Regulation, GDPR) [2016] OJ L119/1.

40 European Commission, Proposal for a Regulation concerning the respect for private life and the protection of personal data in electronic communications (ePrivacy Regulation) COM(2017) 10 final.

a more privacy-driven outlook. At the same time, a breach of data protection laws should not automatically be understood as an infringement of the competition rules, as the two sets of rules protect two different legal interests (Robertson 2020b: 188 f).

The EU's experience

In the past, the European Commission has primarily relied upon already established categories of abuse of dominance in the digital sphere. The *Google* cases showcase what kind of behaviour the Commission considers to constitute an abuse of a dominant position by a dominant digital platform, both when translating established abuses of dominance into the digital sphere and when establishing new types of abuses. In *Google Shopping* (2017),[41] the European Commission found that Google was using its market dominance in general online search in order to systematically place its own comparison-shopping service at or near the top of the search results (self-preferencing; leveraging). In addition, Google demoted competing comparison-shopping services in the generic results of its search engine. Together, these two practices stifled competition by giving Google's own comparison-shopping service an advantage and foreclosing competing shopping comparison sites – which were among the complainants that had brought the case before the Commission – from the market. In the appeal case before the General Court,[42] the Commission's reliance on self-preferencing as the theory of harm will be tested. Some have argued that by favouring its own business, Google is merely competing on the merits; under that view, self-preferencing as a theory of harm is not compatible with Article 102 TFEU (Vesterdorf 2015: 6–8). Others have criticised that the Commission did not clearly spell out which legal test it applied to Google's self-preferencing, instead relying on cases that referred to different theories of harm such as refusal to supply, tying, and margin squeeze (Ibáñez Colomo 2019). Yet, others have argued that Google's leveraging strategy was anti-competitive based on a manipulation of information by the dominant platform (Colangelo and Maggiolino 2019: 12). Now it is for the General Court to weigh in. In the meantime, the proposed Digital Markets Act, if and when adopted, would also prohibit gatekeepers from engaging in self-preferencing, among others (European Commission 2020e: Art 6(1)(d)).

In its biggest fining decision to date, *Google Android* (2018), the European Commission penalised Google's anti-competitive behaviour with a €4.34 billion fine. The Commission looked, among others, at the licensing terms of Google's Android mobile operating system. It found that Google engaged in anti-competitive tying by requiring manufacturers of smartphones to pre-install its search and browser apps if they wanted to license Google's popular Play Store. In addition, Google made illegal payments to some manufacturers and mobile network operators for exclusively pre-installing its search app. Furthermore, it engaged in another instance of anti-competitive tying by requiring manufacturers to install the Google-approved version of Android if they wanted to pre-install Google apps. Thereby, it also obstructed the development and distribution of competing Android versions, so-called forks (European Commission 2018a).[43] The case is currently on appeal before the General Court.[44] Its theory of harm relies on classic tying and other single branding measures, all of which are well established under EU competition law. The *Google Android* case simply applies those to digital markets.

41 *Google Search (Shopping)* (Case AT.39740) Commission Decision of 27 June 2017 [2018] OJ C9/11, see esp. para 334.

42 Case T-612/17 *Google and Alphabet v Commission.*

43 *Google Android* (Case AT.40099) Commission Decision of 18 July 2018.

44 Case T-604/18 *Google and Alphabet v Commission.*

In the third *Google* case, *Google AdSense* (2019), the European Commission believes that Google's behaviour as regards display search advertisements falls foul of EU competition law. In agreements with large clients, Google ensured that these clients did not obtain search ads from any of Google's competitors (exclusivity). It did this by requiring these clients to grant premium placement to a certain number of Google search ads, and by obliging its clients to obtain its approval for changing the display of competing search ads. This effectively foreclosed actual and potential competitors from this lucrative market (European Commission 2019).[45] Again, this case applies well-established theories of harm relating to single branding/exclusivity to a digital market environment. When the General Court rules on *Google AdSense*,[46] it is therefore not expected that the case will generate novel antitrust theories of harm specific to the digital environment.

As discussed earlier, the Commission is currently also investigating two ongoing *Amazon* cases under Article 102 TFEU as well as Apple's behaviour relating to Apple Pay as either an anti-competitive agreement or an abuse of dominance (see the previous section).

A new type of abuse is currently being 'tested' in Germany, albeit under national competition law rather than under EU competition law:[47] in February 2019, the Bundeskartellamt issued a prohibition decision against Facebook based on that social media platform's collection of user data.[48] In the eyes of the German competition watchdog, Facebook's collection and use of user data from outside the social platform went against the values enshrined in the GDPR.[49] The Bundeskartellamt understood this to also constitute a violation of the German provision against abuse of dominance (Robertson 2019a). It thereby conflated the dimensions of data protection and competition law. Facebook appealed that decision before the Higher Regional Court Düsseldorf and applied for the suspensory effect of its appeal. When deciding on the latter, the Higher Regional Court Düsseldorf voiced strong legal concerns about the Bundeskartellamt's legal analysis in its *Facebook* decision, and granted the suspensory effect.[50] The case then came before the German Bundesgerichtshof, as the Bundeskartellamt appealed the granting of suspensory effect. In its much-anticipated ruling of June 2020, the Bundesgerichtshof sided with the Bundeskartellamt and ruled that there were no serious doubts about either Facebook's dominant position or its abuse of the latter.[51] However, the Bundesgerichtshof based its decision on a somewhat different reasoning. First of all, it stated that contrary to the Bundeskartellamt's analysis, the question of whether Facebook's data conditions had breached the privacy rules of the GDPR was irrelevant for the competition law assessment. Instead, the Bundesgerichtshof insisted that another question should be at the centre of attention, namely whether Facebook's terms of use did not allow private Facebook users a choice on whether they wanted to use a highly personalised version of the social network (covering their online activities off Facebook) or whether they preferred a personalised version that was limited to the data they divulged on Facebook only. This lack of choice, the Bundesgerichtshof argued, represented an exploitative abuse of users as well as an exclusionary abuse vis-à-vis competitors. The cases before the Higher

45 *Google AdSense* (Case AT.40411) Commission Decision of 20 March 2019.

46 Case T-334/19 *Google and Alphabet v Commission.*

47 For a criticism of this fact, see Wils (2019).

48 Bundeskartellamt, *Facebook* (B6–22/16, 6 February 2019). On this case, see already previously at the previous section.

49 Regulation (EU) 2016/679 of 27 April 2016 on the protection of natural persons with regard to the processing of personal data and on the free movement of such data (General Data Protection Regulation, GDPR) [2016] OJ L119/1.

50 Higher Regional Court Düsseldorf, Case VI – Kart 1/19 (V) *Facebook* (26 August 2019).

51 Bundesgerichtshof, Case KVR 69/19 *Facebook* (23 June 2020).

Regional Court Düsseldorf and the Bundesgerichtshof only concerned the suspensory effect of Facebook's appeal, while the case still needs to be decided in substance. However, it would be surprising if the Higher Regional Court Düsseldorf were to disregard the Bundesge-richtshof's clearly worded judgement of June 2020 when deciding the case.

How to approach abuses of dominance in the digital sphere

In all reports on competition law in digital markets, the issue of how to deal with abuse of dominance in these markets takes centre stage. While the discussion of the European Union's experience with abuse of dominance in digital markets has shown that Article 102 TFEU is able to adapt to the digital market environment in many instances, the specificities of platforms and their gatekeeper or intermediary function have not yet been fully captured. The Furman Report therefore suggests introducing a digital markets unit which would develop a code of conduct for companies with a strategic market status (2019: 59). This code of conduct should especially address how companies with a strategic market status ought to deal with smaller companies and consumers and should be directed by principles that have specific theories of harm in mind (p. 60). For instance, smaller companies that depend on the digital platform with a strategic market status should be given non-discriminatory access, their ranking and reviews should be decided on a fair, consistent and transparent basis, and they should not be forced to single-home on a particular platform (p. 61). This approach is reminiscent of the concept of relative market power that exists in several EU Member States' competition laws, but not at EU level.

The EU Report identifies two abusive practices by digital platforms that require particular attention, namely leveraging and self-preferencing. It stresses that under Article 102 TFEU, self-preferencing is only abusive if it leads to anti-competitive effects (2019: 65–67). It also notes that in digital markets, competitive parameters such as quality and innovation are more important than price-based effects (41). Finally, the EU Report urges that competition law should be applied in digital markets with full force, with a preference for erring on the side of over-enforcement (51).

Similar to the EU Report, the Competition Law 4.0 Report warns that in the light of increasing concentration in digital markets, and due to digital platforms' gatekeeper function, the cost of false negatives would be particularly high (Competition Law 4.0 Report 2019: 49–51). Therefore, the Report recommends introducing a set of legal rules that contains a code of conduct for digital platforms of a certain size. The code of conduct should adapt the provision of Article 102 TFEU to the digital market environment, including the prohibition of unjustified self-preferencing and rules on data portability (53–55). Beyond these suggestions, the Report also recommends setting up a digital markets board that should be tasked with carrying out a comprehensive European digital policy, not limited to competition law (82). The Stigler Report similarly recommends the setting up of a digital authority with rule-making competence in the areas of consumer protection, privacy policies, transparency, data portability, and data and algorithmic access for external auditing and research (Stigler Report 2019: 273).

From the reports, it can be deduced that due to the competitive structure of digital platform markets, the area of abuse of dominance is of particular importance when attempting to adequately apply competition law to digital markets. At the same time, the many calls to introduce new rules or rule-making bodies suggest that this is an area that needs to be further developed, in both practice and research. The reports also make clear that competition law should not be the sole focus of these rule-making bodies, suggesting that a broader picture would be more beneficial to shaping digital markets in a pro-consumer fashion. Both the EU Report and the Competition Law 4.0 Report underline that antitrust enforcers should prefer false positives over false negatives in the digital environment (on error costs, see Easterbrook 1984), as non-intervention or late intervention may not be able to be remedied at a later stage due to the nature of digital platform markets.

As already set out earlier, the European Commission has started its review of the EU competition law framework by tackling digital market definition. It has also become active in the area of unilateral conduct, however: in December 2020, the Commission published a new legislative proposal entitled the Digital Markets Act, which foresees a number of obligations for so-called gatekeepers (Articles 5 and 6 of the Digital Markets Act Proposal). Several of the obligations imposed on gatekeepers seem to have been based on case law that the Commission has gathered experience with in digital markets. The proposal also includes a market investigation mechanism limited to three specific purposes: the designation of what constitutes a gatekeeper from a qualitative rather than a quantitative perspective, the tackling of systemic non-compliance with the Digital Markets Act, and a possible revision of the legal framework in order to address digital markets in which gatekeepers lead to competition issues (on these three purposes, see Articles 15, 16, and 17 of the Digital Markets Act Proposal). In proposing this tool, the Commission is only putting into action a very narrow set of the recommendations found in the EU Report, and is straying quite far from the initial consultation on the new competition tool (European Commission 2020b). It is now for the European Parliament and the Council to take the Digital Markets Act Proposal further in the legislative process.

Germany has also recently changed its competition law provisions. In light of the new issues arising with reference to digital platforms as intermediaries and their role as gatekeepers to entire digital ecosystems, the German legislature adopted an amendment to the German Competition Act in January 2021 that introduces a new § 19a GWB.[52] This provision has already entered into force and prohibits certain anti-competitive conduct by undertakings that have a paramount significance for competition across markets. It gives the German Bundeskartellamt an additional tool in order to rein in digital platforms that are spreading across multiple markets: § 19a GWB does not apply automatically, only if the Bundeskartellamt issues an order to this effect. As a first step, the Bundeskartellamt needs to declare that a company has a paramount significance for competition across markets, based on a number of factors such as its dominant position, its financial strength, its vertical integration, its access to relevant data, and the importance of its activities for third parties (§ 19a para 1 GWB). These criteria show that § 19a GWB intends to catch digital platforms that have a gatekeeper function or that possess intermediation power. Such an order needs to be restricted to five years. Only when a company has been declared to have such a paramount significance for competition across markets can the Bundeskartellamt go a step further and prohibit various kinds of behaviour, including self-preferencing, pre-installation requirements, holding back competitors, use of data in an anti-competitive way, or hindering interoperability of products (§ 19a para 2 GWB). A declaratory order on a company's paramount significance for competition across markets can be combined with such a prohibition (§ 19a para 2 GWB).

A further novelty in the recent German amendment, which entered into force on 19 January 2021, relates to conduct that is prohibited for companies with relative or superior market power (§ 20 GWB), a provision that has now been extended to benefit all companies (previously, it was restricted to small and medium-sized enterprises). The provision now also explicitly mentions economic dependency on intermediaries in multi-sided markets. According to

52 Gesetz zur Änderung des Gesetzes gegen Wettbewerbsbeschränkungen für ein fokussiertes, proaktives und digitales Wettbewerbsrecht 4.0 und anderer Bestimmungen (GWB-Digitalisierungsgesetz), Federal Law Gazette I 2021/1, 2.

§ 20 para 1a GWB, economic dependency may arise from the fact that a company is dependent on access to data controlled by another undertaking for its own activities. Contrary to 'regular' market power, which assesses a company's market position in relation to the market as a whole, relative market power assesses whether a company might have a dominant position vis-à-vis certain customers that depend on the business relationship with the dominant player for their own business. As the EU currently does not contain any provision on relative market power, this proposal is only of interest to jurisdictions that contain such a provision, such as Austria.

This GWB reform hopes to tackle the type of market power that is arising in digital markets and that opens up new possibilities for unilateral anti-competitive behaviour. It also hopes to keep markets competitive and access to essential platforms open (Federal Ministry for Economic Affairs and Energy 2020). Reactions to the new proposal have been mixed, with some scholars praising the new approach (Neuerer 2020; Chazan and Espinoza 2020) and others, most notably the German Monopoly Commission, cautioning that new types of abuse of dominance should not be introduced hastily (Monopoly Commission 2020).

Moving forward, the interplay between the Digital Markets Act – if and when adopted – and national provisions on competition in digital markets will certainly represent an important focal point. In Austria, for instance, the Ministry for Digital and Economic Affairs has started a review of the Austrian competition law framework in order to take digital markets and data into account. When launching this review, the Ministry exchanged views with the German Ministry, the German Monopoly Commission, and the European Commission (Federal Ministry for Digital and Economic Affairs 2020) – a promising start for a pan-European vision of digital competition law.

Digital mergers

The European Union's Merger Regulation 139/2004 (EUMR) sets out the European merger regime. As highlighted previously, in the eyes of the EUMR the creation or strengthening of a dominant position can require the prohibition of a merger or some significant changes to the transaction before it is cleared (Article 2 para 3 EUMR). In addition, the EUMR's test of a significant impediment of effective competition also allows for the consideration of other theories of harm that do not need to be solely based on the merging companies' market power.

The issues

When it comes to mergers in digital markets, two particular issues have come to the forefront of the debate: data concentration and killer acquisitions. Data concentration relates to the combination of data in the hands of a small number of 'data-opolies' (Stucke 2018). Based on the importance of data in digital markets, the question needs to be asked as to whether unchallenged data concentration could allow these data-opolies to engage in an array of anti-competitive behaviour, such as exploitative and exclusionary abuses. As a preventative measure, merger control could seek to prevent such data concentration.

Killer acquisitions refer to the practice whereby a company buys up promising start-ups with the intention of 'killing', i.e., discontinuing, the start-ups' product. This type of behaviour is well known from the pharmaceutical sector (e.g., Cunningham et al. 2021). Where the buyer is a digital platform, however, it will frequently not have the intention to discontinue the start-up's product, but rather to continue its (potential) competitor's promising product for itself (EU Report 2019: 117). Where digital platforms buy up (potentially) competing start-ups, they may do so with the intention of expanding their ecosystem so as to stay ahead in this competition for the market (Stigler Report 2019: 88). The Furman Report notes that the five biggest digital platforms have made more than 400 acquisitions over the last decade (p. 11 f) – none of them having been prohibited by a competition authority, and very few of them scrutinised at all. In terms of competition law, and in light of the issue of digital ecosystems discussed earlier, these acquisitions raise the question of whether they may hinder both competition and innovation in digital markets (EU Report 2019: 118–123; Competition Law 4.0 Report 2019: 65). Based on the low turnover of the targets in these acquisitions, these concentrations are usually not caught by the turnover-based thresholds of the Merger Regulation (Article 1 EUMR), meaning that it is often not possible to scrutinise these transactions from an antitrust perspective at the EU level.

The EU's experience

In the decisional practice of the European Commission, one finds an increasing number of merger cases in which data played a role. Nevertheless, none of these mergers has so far been prohibited. In *Google/DoubleClick* (2008), the European Commission was convinced that any negative impact that the merger may have on user privacy would be held at bay by fundamental rights and the EU's data protection rules.[53] While the Commission acknowledged the data concentration that the merger would lead to, it emphasised that the data Google would have access to was replicable by other online advertising services, thus minimising this concern.[54] The Commission did not consider whether Google's power over users may be such as to prevent competitors from effectively collecting this data. When the same merger was investigated in the US, the Federal Trade Commission also found that it would not negatively affect consumer privacy.[55] In her dissenting statement, FTC Commissioner Pamela Jones Harbour made clear that in her view, Google would become 'a "super-intermediator" with access to unparalleled data sources' after the merger, and questioned whether competitors could harvest data of a comparable scope (Jones Harbour 2007: 8).

The *Facebook/WhatsApp* merger (2014) also required the European Commission to delve into the data issue. It found that Facebook was only one among a number of competitors collecting user data, so the merger would not necessarily reinforce Facebook's market power in targeted advertising.[56] Again, the Commission relegated any privacy concerns to the realm of EU data protection law and did not take them into account as a relevant antitrust concern.[57] During the merger investigation, Facebook claimed that it would not be able to automatically and reliably match Facebook and WhatsApp user accounts. Later, however, it became clear that this was a misleading claim, leading to a €110 million fine on Facebook for providing incorrect or misleading information to the Commission during the merger investigation.[58]

In *Microsoft/LinkedIn* (2016), the Commission held that data concentration could raise competition concerns in two kinds of cases: where market power was derived from data, or where

53 *Google/DoubleClick* (Case COMP/M.4731) Commission Decision of 11 March 2008 [2008] OJ C184/10, para 368.

54 *Google/DoubleClick* (Case COMP/M.4731) Commission Decision of 11 March 2008 [2008] OJ C184/10, para 269.

55 FTC, Statement of the Federal Trade Commission Concerning Google/DoubleClick (File No 071–0170, 20 December 2007) 2–3.

56 *Facebook/WhatsApp* (Case COMP/M.7217) Commission Decision of 3 October 2014 [2014] OJ C417/4, paras 168–191.

57 *Facebook/WhatsApp* (Case COMP/M.7217) Commission Decision of 3 October 2014 [2014] OJ C417/4, para 164.

58 The Commission argued that this misleading information had no impact on the outcome of the merger review, as it had considered this possibility despite Facebook's assertions that it was not technically possible; *Facebook/WhatsApp* (Case COMP/M.8228) Commission Decision of 18 May 2017 [2017] OJ C286/6.

competition to obtain data was eliminated due to the merger. In the case at issue, the Commission was not concerned about the data concentration that would occur through the acquisition. It emphasised that the merging of datasets would be governed by EU data protection law, that 'the combination of their respective datasets does not appear to result in raising the barriers to entry/expansion for other players in this space' due to the replicable nature of the data, and that the merging companies were only small players in this regard.[59]

In 2018, Apple's acquisition of the music app Shazam gave the European Commission another opportunity to assess the importance of user data in digital mergers. The European Data Protection Board had hoped for a careful review of the data concentration that the merger brought about. It held that Apple and Shazam would together hold 'significant informational power' (EDPB 2018) based on the amalgamation of data capabilities through the merger. The Commission acknowledged that user data did indeed play an important role in the business of Apple and Shazam. However, it concluded that Apple's post-merger data advantage would not have anti-competitive effects in this particular case (Zingales 2018: 4).[60] In the recent *Google/Fitbit* merger, the Commission was presented with yet another opportunity to adapt its decisional practice to the specificities of data concentrations in digital markets. However, it approved that merger, to the discontent of many commentators (Bourreau et al. 2020).[61]

While data concentration through mergers has been on the European Commission's radar for a while, it appears that there does not yet exist a convincing theory of harm that would incite the Commission to consider data amalgamation as a significant impediment of effective competition under the Merger Regulation (Article 2 para 2, recitals 25–26 EUMR) (see Robertson 2020c). Similarly, for innovative start-ups with a low turnover threshold, EU competition law does not provide any tools for intervention, apart from the possible referral of cases from national competition authorities to the Commission (Article 22 EUMR).

Data concentration and killer acquisitions: the road ahead

In 2017, two EU Member States – Austria and Germany – for the first time introduced transaction value-based thresholds into their merger regime.[62] Where a merger does not meet the general turnover thresholds, for instance because the target does not yet generate any significant turnover, but the value of the transaction reaches certain thresholds (€200m in Austria; €400m in Germany), it needs to notify the national competition authority. Although intended for the case of digital platforms, it appears that many other types of cases have since been notified under those provisions.

A first consultation by the European Commission on the introduction of transaction value-based thresholds for merger control yielded mixed responses, leading it not to press forward on this issue (European Commission 2017: 4–7). In light of this consultation, both the EU Report (2019: 113–116) and the Competition Law 4.0 Report (2019: 66, 68) suggest that the time is not yet ripe to include such a transaction value-based threshold into European merger control.

59 *Microsoft/LinkedIn* (Case COMP/M.8124) Commission Decision of 6 December 2016, paras 176–181 (direct quote at para 180).

60 *Apple/Shazam* (Case COMP/M.8788) Commission Decision of 6 September 2018, paras 327 f.

61 *Google/Fitbit* (Case COMP/M.9660) Commission Decision of 17 December 2020.

62 German Restrictions of Competition Act (Gesetz gegen Wettbewerbsbeschränkungen), Federal Law Gazette I 2114/2005, as amended, § 35 para 1a; Austrian Competition Act (Kartellgesetz 2005), Federal Law Gazette I 61/2005, as amended, § 9 para 4.

Instead, both Reports urge the European Commission to review its substantive theories of harm in connection with acquisitions of innovative start-ups by powerful digital platforms, also in light of tipping effects that can be observed in platform markets. In particular, such a review should focus on data, innovation, and conglomerate effects (EU Report 2019: 112, 116 f; Competition Law 4.0 Report 2019: 71).

The Furman Report suggests setting up a digital markets unit that should be notified of any intended acquisitions by companies with a strategic market status (2019: 95). While the UK merger regime currently relies on voluntary notification, this change would allow this new type of regulator to review any kind of acquisition that important digital platforms envisage, irrespective of the turnover. The Stigler Report suggests the introduction of transaction value-based thresholds for merger review and of a rebuttable presumption that mergers between dominant platforms and important (potential) competitors are unlawful, thereby shifting the burden of proof (2019: 16, 98). The Digital Markets Act, if adopted in its currently proposed form, would impose a notification requirement on gatekeepers, but without any consequences for the transaction (Article 12 of the Digital Markets Act Proposal).

The recent German Competition Act amendment introduced a new provision that allows the Bundeskartellamt to require a specific company to notify the authority of certain acquisitions, provided that the acquiring company's turnover exceeds €500 million and the target's turnover exceeds €2 million (§ 39a GWB). While not motivated by digital markets (Podszun 2020), this new provision may well be used to target certain acquisitions by big digital players that could lead to anti-competitive outcomes.

Outlook

The specific characteristics of digital markets represent a significant challenge for the delineation of the relevant antitrust product market, the assessment of market power, the understanding of anti-competitive agreements and of abusive practices, and the issue of merger control. As was highlighted throughout this chapter, these are challenges that national competition authorities, the European Commission, national courts, and the CJEU are already facing in their daily work of enforcing EU and national competition law. While there have been voices arguing that EU competition law is already flexible enough to adjust to digital markets without changing the law (Jaeger 2017), this rather pragmatic approach needs to be understood in light of how cumbersome it is to change the EU Treaties. In some aspects, the flexibility of EU competition law may well reach its limits, and alternative ways of fine-tuning competition law to the exigencies of digital markets will have to be found.

In Europe and beyond, a new approach to assessing market power and market behaviour in digital markets is in the process of being carved out. The plethora of reports on making competition law fit for purpose in digital markets, the legislative changes introduced and debated, most notably in Austria and Germany, and the ongoing review of legislative and soft law instruments connected with the digital economy are all important indicators of this developing approach. While the national developments in this respect are encouraging, this does not mean that the EU should let national legislators take the lead on shaping competition law in our digital times. At the level of the European Union, it appears that the European Commission is best placed to introduce necessary adjustments for the antitrust analysis of digital markets through its tried-and-tested soft law approach, as it is currently undertaking in the area of market definition, and through its role of preparing legislative proposals, a path it has taken when proposing the Digital Markets Act.

References

Akman, P. (2019): Competition Policy in a Globalized, Digitalized Economy. *White Paper for the World Economic Forum.*

Armstrong, M. (2006): Competition in Two-Sided Markets, *RAND Journal of Economics* 37, 668–691.

Australian Competition and Consumer Commission (2019): Digital Platforms Inquiry – Final Report. Available at: https://www.accc.gov.au/system/files/Digital platforms inquiry – final report.pdf [10.03.2021]

Autorité de la concurrence and Bundeskartellamt (2016): Competition Law and Data. Available at: www.bundeskartellamt.de/SharedDocs/Publikation/DE/Berichte/Big%20Data%20Papier.pdf?__blob=publicationFile&v=2 [10.03.2021]

Autorité de la concurrence and Bundeskartellamt (2019): Algorithms and Competition. Available at: www.autoritedelaconcurrence.fr/sites/default/files/algorithms-and-competition.pdf [10.03.2021]

Baye, M. R. (2008): Market Definition and Unilateral Competitive Effects in Online Retail Markets, *Journal of Competition Law & Economics* 4, 639–653.

Binns, Reuben et al. (2018): *Measuring Third Party Tracker Power Across Web and Mobile.* arXiv:1802.02507.

Bourreau, M. et al. (2020): Google/Fitbit Will Monetise Health Data and Harm Consumers, in *VoxEU* 30. September. Available at: https://voxeu.org/article/googlefi tbit-will-monetise-health-data-and-harm-consumers [10.03.2021]

Bourreau, M. and de Streel, A. (2019): Digital Conglomerates and EU Competition Policy. Available at: www.crid.be/pdf/public/8377.pdf [10.03.2021]

Bundeskartellamt (2016): *Market Power of Platforms and Networks.* B6–113/15.

Caillaud, B. and Jullien, B. (2003): Chicken & Egg: Competition Among Intermediation Service Providers, *RAND Journal of Economics* 34, 309–328.

Chazan, G. and Espinoza, J. (2020): Tech Companies Face Clampdown in Germany over Competition Fears, *The Financial Times.* 4 February 2020. Available at: www.ft.com/content/39559796-4698-11ea-aee2-9ddbdc86190d [10.03.2021]

Colangelo, M. and Maggiolino, M. (2019): Manipulation of Information as an Antitrust Infringement, *Columbia Journal of European Law* (forthcoming). https://ssrn.com/abstract=3262991

Competition and Markets Authority (2019a): Digital Markets Strategy. Available at: https://assets.publishing.service.gov.uk/government/uploads/system/uploads/attachment_data/file/814709/cma_digital_strategy_2019.pdf [10.03.2021]

Competition and Markets Authority (2019b): CMA Lifts the Lid on Digital Giants, *Press Release* 18. December. Available at: www.competitionbureau.gc.ca/eic/site/cb-bc.nsf/vwapj/CB-Report-Big-Data-Eng.pdf/$file/CB-Report-BigData-Eng.pdf [10.03.2021]

Competition Bureau Canada (2018): Big Data and Innovation: Key Themes for Competition Policy in Canada. Available at: www.competitionbureau.gc.ca/eic/site/cb-bc.nsf/vwapj/CB-Report-BigData-Eng.pdf/$file/CB-Report-BigData-Eng.pdf [10.03.2021]

Costa-Cabral, F. and Lynskey, O. (2017): Family Ties: The Intersection between Data Protection and Competition in EU Law, *Common Market Law Review* 54, 11–50.

Crémer, J., Schweitzer, H. and de Montjoye, Y.-A. (2019): Competition Policy for the Digital Era [EU Report]. *Report for the European Commission.* Available at: https://ec.europa.eu/competition/publica tions/reports/kd0419345enn.pdf [10.03.2021]

Cunningham, C., Ederer, F. and Ma, S. (forthcoming 2021): Killer Acquisitions, *Journal of Political Economy.* https://doi.org/10.1086/712506

Easterbrook, F. H. (1984): The Limits of Antitrust, *Texas Law Review* 63, 1–40.

European Commission (1997): Notice on the Definition of Relevant Market for the Purposes of Community Competition Law [1997] OJ C372/5.

European Commission (2016): Mergers: Commission Approves Acquisition of LinkedIn by Microsoft, Subject to Conditions, *Press Release* Nr IP/16/4284. 6 December 2016. Available at: https://ec.europa.eu/commission/presscorner/detail/en/IP_16_4284 [10.03.2021]

European Commission (2017): Summary of Replies to the Public Consultation on Evaluation of Procedural and Jurisdictional Aspects of EU Merger Control. July. Available at: https://ec.europa.eu/compe tition/consultations/2016_merger_control/summary_of_repli.es_en.pdf [10.03.2021]

European Commission (2018a): Antitrust: Commission Fines Google €4.34 Billion for Illegal Practices Regarding Android Mobile Devices to Strengthen Dominance of Google's Search Engine. *Press Release*

Nr IP/18/4581. 18 July 2018. Available at: https://ec.europa.eu/commission/presscorner/detail/en/IP_18_4581 [10.03.2021]

European Commission (2018b): Evaluation Roadmap – Evaluation of the Vertical Block Exemption Regulation. Ref. Ares(2018)5722104.8.11.2018.

European Commission (2019): Antitrust: Commission Fines Google €1.49 Billion for Abusive Practices in Online Advertising, *Press Release* Nr IP/19/1770. 20 March 2019. Available at: https://ec.europa.eu/commission/presscorner/detail/en/IP_19_1770 [10.03.2021]

European Commission (2020a): Evaluation of the Commission Notice on the Definition of Relevant Market for the Purposes of Community Competition Law. Available at: https://ec.europa.eu/competition/consultations/2020_market_definition_notice/index_en.html [10.03.2021]

European Commission (2020b): Single Market – New Complementary Tool to Strengthen Competition Enforcement. Available at: https://ec.europa.eu/info/law/better-regulation/have-your-say/initiatives/12416-New-competition-tool [10.03.2021]

European Commission (2020c): Antitrust: Commission Consults Stakeholders on a Possible New Competition Tool, *Press Release* Nr IP/20/977. 2 June 2020. Available at: https://ec.europa.eu/commission/presscorner/detail/en/ip_20_977 [10.03.2021]

European Commission (2020d): Mergers: Commission Opens In-Depth Investigation into the Proposed Acquisition of Fitbit by Google, *Press Release* Nr IP/20/1446. 4 August 2020. Available at: https://ec.europa.eu/commission/presscorner/detail/en/ip_20_1446 [10.03.2021]

European Commission (2020e): Proposal for a Regulation of the European Parliament and of the Council on contestable and fair markets in the digital sector (Digital Markets Act). COM(2020) 842 final. 15 December 2020. Available at: https://eur-lex.europa.eu/legal-content/en/TXT/?qid=1608116887159&uri=COM%3A2020%3A842%3AFIN [10.03.2021]

European Commission (2020f): Antitrust: Commission Sends Statement of Objections to Amazon for the Use of Non-public Independent Seller Data and Opens Second Investigation into Its E-commerce Business Practices, *Press Release* Nr IP/20/2077. 10 November 2020. Available at: https://ec.europa.eu/commission/presscorner/detail/en/ip_20_2077 [10.03.2021]

European Data Protection Board (2018): Statement on the Data Protection Impacts of Economic Concentration. 27 August 2018. Available at: https://edpb.europa.eu/sites/edpb/files/files/file1/edpb_statement_economic_concentration_en.pdf [10.03.2021]

European Data Protection Supervisor (2014): Privacy and Competitiveness in the Age of Big Data. 26March 2014. Available at: https://edps.europa.eu/sites/default/files/publication/14-03-26_competition_law_big_data_en.pdf [20.10.2021]

Ezrachi, A. (2015): The Competitive Effects of Parity Clauses on Online Commerce, *European Competition Journal* 11, 488–519.

Ezrachi, A. and Robertson, V. H.S.E. (2019): Competition, Market Power and Third-Party Tracking, *World Competition* 42, 5–19.

Ezrachi, A. and Stucke, M. E. (2016): *Virtual Competition: The Promise and Perils of the Algorithm-Driven Economy*. Cambridge, MA: Harvard University Press.

Federal Ministry for Digital and Economic Affairs, Austria (2020): Kampf um Daten – Wettbewerbsrecht in der modernen Welt, *Press Release*. 10 January 2020. Available at: www.bmdw.gv.at/Presse/AktuellePressemeldungen/BMDW-Kampf-um-Daten.html [10.03.2021]

Federal Ministry for Economic Affairs and Energy, Germany (2020): Entwurf eines Zehnten Gesetzes zur Änderung des Gesetzes gegen Wettbewerbsbeschränkungen für ein fokussiertes, proaktives und digitales Wettbewerbsrecht 4.0 (GWB-Digitalisierungsgesetz). 24 January 2020. Available at: www.bmwi.de/Redaktion/DE/Downloads/G/gwb-digitalisierungsgesetz-referentenentwurf.pdf [10.03.2021]

Furman, J. et al. (2019): Unlocking Digital Competition – Report of the Digital Competition Expert Panel [Furman Report]. Available at: https://assets.publishing.service.gov.uk/government/uploads/system/uploads/attachment_data/file/785547/unlocking_digital_competition_furman_review_web.pdf [10.03.2021]

Graef, I. (2018): Algorithms and Fairness: What Role for Competition Law in Targeting Price Discrimination Towards End Consumers? *Columbia Journal of European Law* 24, 541–559.

Ibáñez Colomo, P. (2018): Amazon's Antitrust Paradox (the Real One): The Strange Case of the Bundeskartellamt', *Chillin' Competition Blog*. Available at: https://chillingcompetition.com/2018/11/30/amazons-antitrust-paradox-the-real-one-the-strange-case-of-the-bundeskartellamt-by-pablo [10.03.2021]

Ibáñez Colomo, P. (2019): Self-Preferencing: Yet Another Epithet in Need of Limiting Principles, *Chillin' Competition Blog*. Available at: https://chillingcompetition.com/2019/04/24/self-preferencing-yet-another-epithet-in-need-of-limiting-principles/ [10.03.2021]

Jaeger, M. (2017): Perspective of the Judiciary. *12th GCLC Annual Conference*. Brussels, 27 January 2017.

Japan Fair Trade Commission (2017): Report of Study Group on Data and Competition Policy. Available at: www.jftc.go.jp/en/pressreleases/yearly-2017/June/170606_files/170606-4.pdf [10.03.2021]

Japan Fair Trade Commission (2019): Guidelines Concerning Abuse of a Superior Bargaining Position in Transactions between Digital Platform Operators and Consumers that Provide Personal Information. 17 December 2019. Available at: www.jftc.go.jp/en/pressreleases/yearly-2019/December/191217DPconsumerGL.pdf [10.03.2021]

Jones Harbour, P. (2007): Dissenting Statement – In the Matter of Google/DoubleClick. File No 071–0170. 20 December 2007.

Jones Harbour, P. and Koslov, T. I. (2010): Section 2 in a Web 2.0 World: An Expanded Vision of Relevant Product Markets, *Antitrust Law Journal* 76, 769–797.

Lianos, I. and Ivanov, A. (2019): Digital Era Competition: A BRICS View. Available at: http://bricscompetition.org/upload/iblock/6a1/brics%20book%20full.pdf [10.03.2021]

Monopoly Commission (2020): Policy Brief Nr 4. Available at: https://monopolkommission.de/images/Policy_Brief/MK_Policy_Brief_4.pdf [10.03.2021]

Murgia, M. and Beioley, K. (2019): UK to Create Regulator to Police Big Tech Companies, *The Financial Times*. 18 December 2019. Available at: www.ft.com/content/67c2129a-2199-11ea-92da-f0c92e957a96 [10.03.2021]

Neuerer, D. (2020): Kartellrechtler lobt Altmaier-Pläne für digitales Wettbewerbsrecht, *Handelsblatt*. 27 January 2020. Available at: www.handelsblatt.com/politik/deutschland/digitalisierung-kartellrechtler-lobt-altmaier-plaene-fuer-digitales-wettbewerbsrecht/25474782.html?ticket=ST-521937-bLZrRorP9uLa346EcQOC-ap1 [10.03.2021]

Newman, J. M. (2016): Antitrust in Zero-Price Markets: Applications, *Washington University Law Review* 94, 49–111.

Newman, J. M. (2020): Attention Markets and the Law. *University of Miami Legal Studies Research Paper*.

OECD (2013): *The Digital Economy*. DAF/COMP(2012)22. Available at: www.oecd.org/daf/competition/The-Digital-Economy-2012.pdf [10.03.2021]

OECD (2014): Data-Driven Innovation for Growth and Well-Being. Available at: www.oecd.org/sti/inno/data-driven-innovation-interim-synthesis.pdf [10.03.2021]

Podszun, R. (2020): Offiziell: GWB10-REFE, *D'Kart Blog*. 24 January 2020. Available at: www.d-kart.de/blog/2020/01/24/offiziell-gwb10-refe/ [10.03.2021]

Ristaniemi, M. and Majcher, K. (2018): Blockchains in Competition Law – Friend or Foe? *Kluwer Competition Law Blog*. 21 July 2018. Available at: http://competitionlawblog.kluwercompetitionlaw.com/2018/07/21/blockchains-competition-law-friend-foe/ [10.03.2021]

Robertson, V. H.S.E. (2017): Delineating Digital Markets under EU Competition Law: Challenging or Futile? *Competition Law Review* 12, 131–151.

Robertson, V. H.S.E. (2019a): The Theory of Harm in the Bundeskartellamt's *Facebook* Decision, *CPI EU News*, 1–4.

Robertson, V. H.S.E. (2019b): The Relevant Market in Competition Law: A Legal Concept, *Journal of Antitrust Enforcement* 7, 158–176.

Robertson, V. H.S.E. (2020a): *Competition Law's Innovation Factor: The Relevant Market in Dynamic Contexts in the EU and US*. Oxford: Hart Publishing.

Robertson, V. H.S.E. (2020b): Excessive Data Collection: Privacy Considerations and Abuse of Dominance in the Era of Big Data, *Common Market Law Review* 57, 161–189.

Robertson, V. H.S.E. (2020c): Marktmacht, Wettbewerb und Digitalisierung, in Sturn, R. and Klüh, U. (eds.): *Blockchained? Digitalisierung und Wirtschaftspolitik*. Weimar: Metropolis, 93–117.

Rochet, J.-C. and Tirole, J. (2003): Platform Competition in Two-Sided Markets, *Journal of the European Economic Association* 1, 990–1029.

Schallbruch, M., Schweitzer, H. and Wambach, A. (2019): A New Competition Framework for the Digital Economy. *Report by the Commission 'Competition Law 4.0'*.

Schechner, S. and Pop, V. (2018): EU Starts Preliminary Probe into Amazon's Treatment of Merchants, *Wall Street Journal*. 19 September 2018.

Schrepel, T. (2019): Is Blockchain the Death of Antitrust Law? The Blockchain Antitrust Paradox, *Georgetown Law Technology Review* 3, 281–338.

Schweitzer, H., Haucap, J., Kerber, W. and Welker, R. (2018): Modernising the Law on Abuse of Market Power. *Report for the German Federal Ministry for Economic Affairs and Energy.*

Stigler Committee on Digital Platforms (2019): Final Report [Stigler Report]. Available at: https://research.chicagobooth.edu/-/media/research/stigler/pdfs/digital-platforms–committee-report–stigler-center.pdf?la=en&hash=2D23583FF8BCC560B7FEF7A81E1F95C1DDC5225E [10.03.2021]

Stucke, M. E. (2018): Should We Be Concerned About Data-opolies? *Georgetown Law & Technology Review* 2, 275–324.

Thépot, F. (2013): Market Power in Online Search and Social Networking: A Matter of Two-Sided Markets, *World Competition* 36, 195–221.

Tulpule, A. M. (2017): Enforcement and Compliance in a Blockchain(ed) World. *CPI Antitrust Chronicle.* Available at: https://ssrn.com/abstract=2906465 [10.03.2021]

Vestager, M. (2019): Defining Markets in a New Age, *Speech.* 9 December 2019. Available at: https://ec.europa.eu/commission/commissioners/2019-2024/vestager/announcements/defining-markets-new-age_en [10.03.2021]

Vesterdorf, B. (2015): Theories of Self-Preferencing and Duty to Deal – Two Sides of the Same Coin, *Competition Law & Policy Debate* 1(1), 4–9.

Wils, W. P. J. (2019): The Obligation for the Competition Authorities of the EU Member States to Apply EU Antitrust Law and the *Facebook* Decision of the Bundeskartellamt. *Concurrences* N° 3–2019, 58–66.

Zingales, N. (2018): *Apple/Shazam*: Data Is Power, but Not a Problem Here, *CPI EU News* 1–7.

22

PLATFORM REGULATION

Coordination of markets and curation of sociality on the internet

Ulrich Dolata

Introduction: private-sector conquering of the internet

The mid-1990s – a time when the commercial utilisation of the internet was already well under way (Amazon was founded in 1994, Yahoo in 1995, and Google in 1997) – were characterised by an influential narrative which advocated that the internet could (or should) be free, decentralised, self-regulated, and managed largely without political or state intervention. It is in this spirit that, on the sidelines of the 1996 World Economic Forum in Davos, John Perry Barlow (1996), one of the founders of the Electronic Frontier Foundation, formulated his "Declaration of the Independence of Cyberspace." The Declaration, marked by remarkable pathos and speaking of an indeterminate "we," called for a decidedly self-regulated web combined with a rejection of all attempts at state control:

> We are creating a world that all may enter without privilege or prejudice accorded by race, economic power, military force, or station of birth. We are creating a world where anyone, anywhere may express his or her beliefs, no matter how singular, without fear of being coerced into silence or conformity. . . . Governments of the Industrial World, you weary giants of flesh and steel, I come from Cyberspace, the new home of Mind. On behalf of the future, I ask you of the past to leave us alone. You are not welcome among us. You have no sovereignty where we gather.

A year and a half earlier, in August 1994, Esther Dyson et al. (1994) presented a "Magna Carta for the Knowledge Age," in which libertarian notions of freedom – "America, after all, remains a land of individual freedom, and this freedom clearly extends to cyberspace" – and the open designability of the web were combined more strongly with neoliberal ideas of the market and a suggested deterministic impact of technological progress on processes of economic demonopolisation and decentralisation:

> In Cyberspace itself, market after market is being transformed by technological progress from a "natural monopoly" to one in which competition is the rule. . . . The advent of new technology and new products creates the potential for dynamic competition.

 DOI: 10.4324/9780429351921-27

This mixture of liberal and emancipatory visions of the web, neoliberal views of the market, and a strong technological determinism – comprising what then became known as the so-called Californian ideology – proved to be an extremely powerful narrative in the following decades. It was successful not least because it was able to bring together the world views of two quite different groups of actors: it fitted both the "freewheeling spirit of the hippies" and the "entrepreneurial zeal of the yuppies" (Barbrook and Cameron 1996: 45). Later, these visions were complemented by the prospect or promise, likewise derived directly from new technically based interaction possibilities, of a sovereignty of action and design capability of Web 2.0 users (O'Reilly 2005; Schrape 2019).

Essential elements of the Californian ideology and its successors were, however, based on storytelling that did not, even then, stand up to critical evaluation. For example, the rejection of political interventions and regulation activities camouflaged the substantial role of the state in the entire process of the creation and development of networked computer systems and the internet. The intensive research funding and coordination by the United States government over several decades and until the recent past has to this day decisively shaped research and innovation as well as academic-industrial knowledge transfer. In the beginning, this research funding came primarily from the Department of Defense and its Defense Advanced Research Projects Agency (DARPA) and was later expanded to include specific technology and industrial policy support programs, for example for start-up companies (Abbate 1999; Mazzucato 2013). The widespread rejection of political intervention was aimed less at any governmental research (funding) policies, from which the internet companies, in particular, have all along benefited, than at regulatory interventions by the state in the free play of (market) forces.

> In place of counterproductive regulations, visionary engineers are inventing the tools needed to create a "free market" within cyberspace, such as encryption, digital money, and verification processes.
>
> *(Barbrook/Cameron 1996: 53)*

Yet even back then, the unspecifically presented "we," and with it the promise of a web that would be open to and potentially designable by everyone, was hardly more than ideology. At the end of the 1990s, Lawrence Lessig (1999) coined his famous adage *code is law*, emphasising that the web is by no means a space void of regulation. He argued that, while not so much regulated by the law, the web is all the more composed of complex information technology architectures, codes, and software applications, whose structuring effects on user behaviour, via social instructions inscribed in technology, can be more rigid even than any political law (Feick and Werle 2010). The "we" of the actors considered capable of substantially participating in the design of the web thus shrank to a small elite of those with the technical skills and resources to develop, implement, and control the corresponding technical specifications.

By the 2010s at the latest, the vision of a decentralised internet economy with free markets and full competition was no longer tenable. In the shadow of the long-time popular notion of self-organisation devoid of any state intervention, the commercial exploration and private-regulatory structuring of the internet, largely carried out by companies from Silicon Valley, had gained momentum and taken shape almost entirely unhindered by social intervention and state-regulatory frameworks (Misterek 2017). Massive concentration processes, the emergence of winner-take-all markets, and the establishment of new natural quasi-monopolies, which characterise the web today both economically and socially, are the widely visible consequences of this large-scale land grab.

Above all, the structuring and regulating influence acquired by the leading US technology groups Amazon, Apple, Google, Facebook, and Microsoft now extends far beyond economic market power and deep into the social fabric. With their platforms, these groups develop and operate the essential technical infrastructures and services of the web, on which not only private users but also many companies and public institutions rely today. As quasi-sovereign actors, they control the central access points to the internet; monitor user activities; and curate and edit content, information flows, and discussions on a large scale. As structure-building economic actors, they aspire toward the complete collection, processing, and valorisation of the data traces that users leave behind on the web. To this end, they have embarked on the large-scale undertaking of measuring and commodifying all social activities and relationships, an endeavour that would have been unthinkable in pre-internet days. Moreover, they no longer act merely as leading and trendsetting market participants but also maintain and regulate their own markets and work relationships, whose participants sometimes reach far beyond their corporate context (Dolata 2018a, 2019).

The technical, economic, and social regulatory sovereignty that has been acquired above all by the large internet corporations (and also, albeit on a smaller scale, by a number of newer and more specialised internet companies such as Uber, Airbnb, Spotify, or Netflix) corresponds with a considerably weaker influence of state or civil society actors on internet structuring and design. The majority of economic activities as well as a great deal of private exchange and the net-based public sphere all today take place in privately organised and designed spaces, and thus within technical and socioeconomic regulatory frameworks set by the companies providing those services. Of course, the internet companies are clearly not outside society with all this: they regularly have to face political interventions, consider the interests of other economic actors, and contend with civil society protest or idiosyncratic user behaviour. However, this does little to change the fact that they have become the decisive proactive and trendsetting actors in the design and regulation of the internet.

This brings me to the main subject of this chapter: the question of how and through what mechanisms the internet companies are fulfilling their role as the structure-forming, rule-setting, and action-coordinating core actors of today's web – in terms of both social and technical levels of structuring and regulation that characterise their platforms. This applies in particular to *two major regulatory areas*, as outlined in the following points:

- the independent *organisation and regulation of markets* for products, services, and labour in which these companies, as platform operators, are able to coordinate economic processes and determine the conditions of competition, as well as the *organisation of macroeconomic interrelationships*, as indicated in their plans to introduce their own digital currencies;
- the extensive *structuring and curation of content, communication and public spheres*, by means of which the platform operators lay the institutional foundations for private expression as well as for public information and discursive possibilities, thereby assuming far-reaching social ordering and regulatory functions on the web.

In today's internet, both of these areas – the organisation of markets and the curation of social relationships – are concentrated on a few privately operated platforms which account for the vast majority of social and economic exchange. Each of these do not simply emerge from the interplay of a multitude of social actors but are above all the result of an intentional structure-building driven by the platform operators. I refer to this as *platform regulation*, which is essentially organised and orchestrated by the platform operators and has so far been characterised by an extreme power asymmetry.

The main part of the text begins in the following section with an exploration of the field and revolves around private-sector internet platforms as the central socio-technical infrastructures of today's consumption- and communication-oriented web. I first discuss relevant platform concepts and then develop my own typology and working definition of the platform, including an outline of its socioeconomic foundations.

The successive section then looks at the two regulatory areas mentioned previously – the coordination of markets and the curation of sociality – which constitute the actually new and disruptive aspects of internet platforms. Based on these two areas of regulation, I ascertain the central importance which platforms, as the essential socio-technical institutions of today's internet, have acquired not only for the organisation of economic processes but also and above all for the shaping and regulation of social conditions and processes. This core part of the chapter aims to condense the empirically traceable forms of structuring and organising, coordination and regulation into distinct patterns and mechanisms of a socio-technically constituted regulation *by* platforms.

Although the aforementioned companies have become core actors in the platform-based regulation of the web, they do not, of course, operate outside societal contexts, social debates, and political intervention. Against the backdrop of the increasingly critical public discussions on the power of internet companies and their platforms, the final section analyses the question of possibilities for intervention in the creative sovereignty of platform operators and discusses approaches to the political containment and regulation *of* platforms.

Conceptualisations, variants and reaches of commercial internet platforms

Conceptualisations: five ways of reading the platform

There are numerous, mostly privately operated services on the internet, performing everything from searches, networking, messaging, and advertising to trade, mediation, and media functions. Since the 2000s, having rapidly taken shape and expanded in reach, these services have become the central infrastructures and hubs of information procurement, communication, publicity, and consumption on the net. In order to characterise or refer to these services, the second half of the 2010s then saw the introduction of the concept of the "platform" – one of those umbrella terms that are initially as inclusive as they are indeterminate and can be concretised and contextualised in very different ways. In the following, I will outline and comment on five readings of the platform relevant to the matter under discussion.

The first reading understands platforms as computer-supported, software-based, programmable, and algorithmically structuring *technological architectures* that currently form the central technical infrastructures of the internet and to which countless specific applications can be added (Gillespie 2010, 2014). Through their technical specifications, they not only shape the possibilities for individual users to express themselves but also structure the options for action of providers of content, cultural or political, for example. Using specific software interfaces, they extend far beyond individual platforms (such as Facebook or Google) and deep into the web, thus enabling the centralised collection and analysis of countless decentralised data sets (Gerlitz and Helmond 2013; Helmond 2015). The many social inscriptions in these technical infrastructures are sometimes mentioned (e.g., in Kitchin 2014: 21-26). However, this reading of the platform does not focus on which agents are socially constructing and implementing these infrastructures or on how they do so.

In the economic literature, platforms are primarily understood as *two- or multi-sided markets* in which the platform operators act as intermediaries or matchmakers, bringing together at least two different market actors – sellers and buyers, users and advertisers (Rochet and Tirole 2003; Evans and Schmalensee 2016, 2005; Haucap and Stühmeier 2016). Typical for many of these markets on the internet are network effects with their concentration-promoting results. The more a digital platform is used and the more active members it has, the more interesting it becomes not only for additional users but also for other actors. The number of regularly active users on one side of the market also increases the platform's commercial attractiveness for advertisers, retailers, or other providers on the other side of the market. The basic principle of multi-sided markets has been known for a long time and has been constitutive for decades of many branches of the economy, including the enterprise of bookselling, music, magazines, radio and television, travel, and ride-hailing agencies. These offers are now, of course, being fundamentally restructured on a new technical basis. The idea of the so-called gift economy (Currah 2007; Elder-Vass 2016) – in other words, the free use of services such as those offered by Google or Facebook, which are financed via the other side of the market, for example through advertising – also has its predecessors: private radio and television have long been operating according to this principle (Evans and Schmalensee 2016: 34, 197–206).

From an organisational perspective, commercial internet platforms are sometimes seen as a *new ideal type of company*, "in which the 'firm' is a set of calls on resources that are then assembled into a performance" (Davis 2016: 513). In the second half of the 2010s, the blueprint for such web page enterprises was often provided by the ride-hailing service Uber – a company that, to this day, has not come close to proving the economic viability of its business model – in particular through its highly technically mediated way of organising and coordinating resources and work processes: "Hiring, scheduling, performance measuring, and evaluation are now largely in the hands of algorithms" (Davis 2016: 511; also Rahman and Thelen 2019; Thelen 2018). These new forms of organising resources and work can be described as the continued development and perfection of neoliberal markets and deregulated employment, using new technical means. These trends have been observed for quite some time; we think only of the proliferation of "temp work." However, most often, the literature discussing these developments remains unclear about the socioeconomic reach of these trends toward web page enterprises. In most cases, reference is made to supposedly paradigmatic individual cases ("Uberisation"), the generalisability of which yet has to be proven empirically.

In a perspective that focuses on fundamental changes in the economy as a whole, platforms are understood as a constitutive expression and core element of *substantial changes in the structure of the capitalist economy* and are labelled with far-reaching terms such as "platform capitalism" (Srnicek 2017; Langley and Leyshon 2016), "digital platform economy" (Kenney and Zysman 2016; Zysman and Kenney 2016), or "digital capitalism" (Staab 2019). In addition to emphasising the platform economy's intensive concentration processes and asymmetric power structures, this literature underscores the role of its participating companies as pioneers in the collection, evaluation, and monopolisation of large data stocks, which are becoming increasingly important for the economy as a whole; as organisers of digital economic circulation processes; as coordinators of working environments, user activities, and the contributions of external producers; and as drivers in expanding the possibilities of value creation to include commodifiable content and communications. Admittedly, all these important building blocks have not yet consolidated into a profound political economy of the platform. Above all, and left unanswered, is the question of the extent to which these mechanisms, undeniably observable on the commercial internet, can be transferred to the economy as a whole and generalised into a new model of capitalism or of a digital economy that encompasses the classical economic sectors as well.

The final reading to be outlined in this section broadens the view to the *social, political, and cultural significance of platforms* (Van Dijck et al. 2018; Van Dijck 2013). It argues that platforms and the social rules and norms inscribed in them have deeply penetrated social contexts and, with their structuring achievements, are changing the overall institutional settings through which modern societies have been organised. According to this reading, this process happens via three mechanisms. Platforms are used to mine and process data on a large scale as raw material, to sort content and user behaviour, and to turn activities, ideas, emotions, and objects into tradable commodities. A platform society, therefore, is understood to be a society in which both economic and social processes are increasingly shaped by globally operating platform companies, which gives rise to a parallel world, organised primarily by the private sector, that complements and increasingly undermines established democratic institutions and processes (see also: Nieborg and Poell 2018; Zuboff 2019).

Concretisation: typology, definition, and socioeconomic reach of the platform

The terrain covered by these readings from various angles is admittedly quite rugged. From an empirical point of view, the numerous platforms on the internet differ significantly from one another, calling for a typifying view. The following characteristics of platforms can be distinguished from one another based on their range of services:

- search platforms that are provided by Google as a monopoly or that are oriented toward Google;
- networking and messaging platforms, such as Facebook (with WhatsApp and Instagram), Twitter, or Snapchat;
- media platforms, such as YouTube, Netflix, Apple, or Spotify;
- Trading platforms, such as Amazon, Alibaba, eBay, or Zalando;
- booking or service platforms, for example, in the area of ride-hailing services (Uber, Lyft), travel and accommodation booking (Airbnb, Expedia, Booking.com), or dating services (Match, Parship);
- cloud platforms, such as Amazon Web Services or Google Cloud Platform, to which individual users and business customers as well as government institutions outsource their data and the processing thereof;
- crowdsourcing and crowdfunding platforms, such as Amazon Mechanical Turk, TaskRabbit (a part of the IKEA Group), Kickstarter, or Indiegogo, which serve as hubs for the competition-based awarding of work orders or in order to finance projects.

Overall, these platforms can be seen to comprise *digital, data-based, and algorithmically structuring socio-technical infrastructures* that facilitate the exchange of information, the structuring of communication, the organisation of work and markets, the provision of a broad spectrum of services, and the distribution of digital and non-digital products (Kenney and Zysman 2016; Srnicek 2017: 43–48). As technical infrastructures, they are based on new possibilities for collecting and processing large amounts of data; the comprehensive digital networkability not only of media, information, and communication but also of material things and production structures; and the sorting and coordination of these processes through learning algorithms (Gillespie 2014, 2016). As socioeconomic units, platforms are not crowd- or sharing-based (Sundararajan 2016) – even if their success (or failure) depends heavily on the number of users and on their personal contributions, communications, ratings, and preferences – but are installed, organised, and controlled top-down by profit-oriented companies.

Table 22.1 Internet companies – key economic data 2020

Company (*Fiscal year end*)	*Revenue* *in billion $US*	*Net income* *in billion $US*	*Core business* *in percent of revenue*	*Employees* *in thousand*
Amazon (12/2020)	386.06	+21.33	Retail sales and subscriptions (88%); cloud (12%)	1,289,000
Apple (9/2020)	274.52	+57.41	Devices (80%); services (20%)	147,000
Google (12/2020)	182.53	+40.27	Advertisement (80%); cloud (7,2%)	135,301
Microsoft (6/2020)	143.00	+44.30	Software and services (66%); cloud (34%)	166,475
Facebook (12/2020)	85.97	+29.15	Advertisement (98%)	60,654
Netflix (12/2020)	25.00	+2.76	Film streaming; subscription	12,135
Uber (12/2020)	11.14	−6.77	Ride-hailing service; booking fees	26,900
Spotify (12/2020)	7.88	−0.58	Music streaming / podcasts; subscription and advertisement	6,554
Airbnb (12/2020)	3,38	−4.58	Accommodation bookings; fees	5,597
Twitter (12/2020)	3.72	+1.13	Microblogging; advertisement	4,600
Snap (12/2020)	2.51	−0.95	Instant messaging; advertisement	2,734

Sources: Annual reports of the companies; press review. Author's compilation

Beyond this lowest common denominator, the field becomes quite heterogeneous. Indeed, the various internet platforms differ significantly from one another not only in terms of classic economic indicators, such as their turnover, profit, or employment (Table 22.1), but also in terms of their economic or social reach and significance (Dolata 2018a, 2019; Van Dijck et al. 2018: 12–22).

The leading internet groups Google, Amazon, Facebook, and Apple offer a broad spectrum of coordinated and networked services and businesses, which they have developed into extensive *socio-technical ecosystems* that extend far beyond their traditional field of activity. Google has long ceased to be just a search engine. It owns YouTube, by far the largest video channel on the net; Google Play, the largest app store next to Apple, offering media content of all kinds; Gmail, the leading email service; Google Maps, the most widely used map service; and Android, the leading operating system for mobile devices. Finally, Google is one of the largest providers of cloud services next to Amazon and Microsoft. Facebook, for its part, together with its subsidiaries WhatsApp and Instagram, is the undisputed leader in social networking and messaging. Over the past decade, Apple and Amazon have also distinguished themselves as full-service providers of a broad range of services and media content, some of which they now produce themselves. The private-sector regulation of the internet is essentially carried out via these broadly based platforms that reach deep into the web and whose services are systematically accessed not only by individual users but also by numerous companies, media producers, government institutions, or other platform companies (Barwise and Watkins 2018).

In contrast, the countless smaller internet companies offer more specific services on their platforms. As a rule, these are *singular and specialised consumer or service offerings* that are either purely consumer-oriented, such as ride-hailing services, travel bookings, room referrals, video-on-demand services, and shopping portals, or, like Twitter or Snapchat, communication-oriented. They offer a limited range of services and can generally be assigned to traditional economic sectors, some of which are radically realigned by the activities of the new players. Uber, for example, has brought new momentum to the markets for ride-hailing services, and Airbnb has brought a new dynamic to the network-based brokerage of accommodations. Over the past decade, Netflix has developed from a classic video rental service to the world's leading film streaming service, with its own film productions. However, many of these platforms are dependent on the infrastructure of the big internet companies. For example, Netflix and Spotify run entirely on the servers of Amazon Web Services and Google Cloud, respectively; and Airbnb and many others integrate the Google Maps' geographical navigation service into their offerings.

From an economic perspective, two things stand out. First, the repertoire of commercially viable *business models* has remained quite limited over the years. The focus is still, as it was by and large in the early 2000s, when platforms were being discussed under the label of "e-commerce" (Zerdick et al. 2001: 167–173), on advertising, trade, subscription models, brokerage fees, the commercial exploitation of databases, and the sale of digital devices. This applies to not only smaller platform companies such as Airbnb, Uber, Spotify, and Netflix but also the leading internet groups (Table 22.1).

It is also remarkable that the *economic and employment effects* which the spread of these platforms has entailed have so far remained rather modest. An empirical study by the Bureau of Economic Analysis at the US Department of Commerce estimated that the total number of people employed in the digital economy, which includes the entire information and communications technology industry, contributed only 3.9% to total employment in the United States in 2016. The share of commercial internet platforms in total employment was less than 1%, in other words significantly even lower (Barefoot et al. 2018). Moreover, a study by the International Monetary Fund to measure the macroeconomic effects of the digital economy comes to the conclusion, for the United States, that online platforms and services contributed only 1.5% to the US gross domestic product (GDP) in 2015 (International Monetary Fund 2018). Hence, the transformation of the economy toward a platform capitalism or a digital platform economy seems to be still a long way off.

However, the extremely low macroeconomic significance of this sub-sector of the (digital) economy, as reflected in the above-mentioned figures, does not adequately reflect both the considerable influence which the leading internet groups wield on the readjustment of economic structures and processes and the extraordinary social and socio-political clout that they have attained. The rapid spread of commercial internet platforms over the past two decades has not only triggered massive upheavals and induced substantial restructuring processes in a number of economic sectors (e.g., retail, advertising markets, media, and various service sectors) but also allowed a number of internet companies to establish themselves as rule-setting coordinators of corporately owned and internationally oriented markets. In addition, large parts of the social exchange on the net, from private communication and personal self-presentation to the most diverse kinds of public spheres, are now bundled, evaluated, and curated by a few commercially operated platforms.

The private platforms' roles as organisers of markets and curators of social contexts are, along with the commodification of user behaviour (Zuboff 2019), the essential characteristics that make them a disruptive force and enable them to act as central regulatory bodies in today's internet. These will be examined in more detail later.

Regulation by platforms: organisation of markets and curation of sociality

Organisation of markets and macroeconomic contexts

To begin, it has to be emphasised that platform-operating internet companies expectedly act as *market participants* and try to capture and dominate new market segments with their strategies for expansion. In doing so, they are in intense competition with one another as well as with traditional companies in the areas they seek to tackle. Smaller internet companies, such as Uber, Airbnb, Spotify, or Netflix, not only have to deal with other new competitors in the markets for drive-hailing services or the brokering of accommodation or of media content, but also have to assert themselves against the established providers and, in some cases, against the leading internet groups. Yet even the latter are by no means operating in non-competitive spheres. While they do dominate important and often highly concentrated markets in one way or another, they do not, as a rule, act as monopolists. This applies to internet advertising and app stores as well as to cloud services, integrated media offerings, and retail, which are characterised by duopolistic or oligopolistic structures and patterns of competition. In addition, the internet groups regularly compete for dominance in new technological trends, such as image and voice recognition, machine learning and virtual reality (Dolata 2018a; Parker et al. 2016: 210–227). Thus, clearly visible tendencies toward concentration in internet-based markets are accompanied by fierce competition and strategies for securing and expanding domains.

However, the internet companies have long since been much more than dominant economic actors who compete with other market players. In addition, they are *operating, coordinating, and controlling their own markets* as well. In these privately owned and online-mediated markets, the internet companies assume the rule-setting role of market coordinators: they do not act merely as intermediaries who simply make market transactions of third parties technically possible, but rather structure, regulate, and monitor the activities of all market participants.

This affects some of the major platforms of the leading internet groups. Indeed, Amazon maintains the largest trading platform for third-party providers on the internet, Amazon Marketplace, which by now generates higher sales than the corporation's own online retail business. Google operates YouTube, a central media platform on the web, and organises the framework conditions and monetisation opportunities for YouTuber and Influencer as well as professional media producers through its YouTube Partner Program. Apple, Google, and Amazon also have large app stores where software developers compete for commercial attention, based on guidelines and commission models set by the market coordinators (Barwise and Watkins 2018; Khan 2018; Dolata and Schrape 2014). While the leading internet groups can largely autonomously implement extensive social rules and algorithmic structurings in their corporate-owned markets, such independent rule-setting is more difficult to achieve for the new online-mediated markets for drive-hailing and accommodation services, mainly represented by Uber and Airbnb. Although these companies also act as rule-setting, coordinating, and sanctioning intermediaries who systematically challenge existing (state) regulations, they are under enormous pressure in terms of public legitimation and political regulation (Thelen 2018). This is, among other reasons, because these companies, although operating in international markets, essentially offer services with strong local or regional connections. After all, taxis are hailed, and accommodations are rented locally.

The company-owned markets outlined here differ from numerous other internet markets, in which the companies, as more or less dominant and trend-setting market participants, offer their own commissioned or licensed products or services, such as music or film streaming, cloud services, and online retail. Amazon, for example, assumes both roles: as an online retailer with commissioned offers, the Group is a player in a market it dominates, while with Amazon Marketplace it also acts as the regulator and coordinator of its own market, which it constitutes and controls. Whereas smaller companies such as Uber or Airbnb are largely coextensive with the markets they organise, for the leading internet groups company-owned markets in the sense described previously represent only an important part in their overall activities.

These company-owned markets are organised and regulated by means of extensive socio-technical regulations – market and competition rules; coordination, control, and exploitation mechanisms – which are laid down in general terms and conditions, partner programs, or developer guidelines as well as in technical programs and instructions. In corporate decisions, the platform owners define the inclusion and exclusion criteria for market participants; formulate the market rules, distribution, and remuneration structures; develop product information, rating, ranking, and performance control systems; guarantee secure forms of payment; and seamlessly mine the data of all participants (Kirchner and Beyer 2016). Unlike in other markets, however, the resulting framework of action for market participants and platform users is not primarily defined by the social enforcement of these social rules but rather by the platform's technical infrastructures and programs, in which the social foundations of the market – its structural, regulatory, and procedural characteristics – are inscribed as technical specifications. The implementation of the market rules, as well as the concrete coordination and handling of all market processes, is largely automated and algorithmically controlled (Gillespie 2014; Kitchin 2014: 15–26, 80–87; Beer 2017).

These privately regulated markets are characterised by strong power asymmetries between the involved actors which manifest at various levels. First, the platform operators have considerable *infrastructural power*. They design and control the technical foundations on the basis of which market processes unfold, and they act as gatekeepers who decide on inclusion and exclusion as well as on the conditions to which market participants are subject (Barzilai-Nahon 2008). Second, the privately organised markets are also characterised by a significant *informational power* held by the platform operators: the latter collect, control, and evaluate all the data of all market participants and thus obtain a complete and exclusive overview of everything that happens on the markets they organise. The (supposed) transparency of the information, rating, and ranking systems goes hand in hand with the systematic opacity of their algorithmic foundations – the conception, modification, and continued processing thereof – which remain a black box for users, providers, consumers, and even state regulatory bodies (Pasquale 2015).

Third, these information asymmetries contribute to the already *market-dominating power* of the platform operators, some of which are also leading players in the same market segment as market participants. Google is both a media group with its own commercial offers and the operator of the media channel YouTube. Apple, Google, or Amazon can view countless third-party software developments via the app stores they control and, if required, draw benefit from them for their own business. Amazon has an overview of all offers from all participants on its marketplace and thus can gain competitive advantages for its own trading business, as Khan (2018: 119) explains:

> Amazon is exploiting the fact that some of its customers are also its rivals. The source of this power is: (1) its dominance as a platform, which effectively necessitates that independent merchants use its site; (2) its vertical integration – namely, the fact that it both sells goods as a retailer and hosts sales by others as a marketplace; and (3) its ability to amass swaths of data, by virtue of being an internet company. Notably, it is this last factor – its control over data – that heightens the anticompetitive potential of the first two.

Fourth, and above all, however, the platform operators have *regulatory and action-structuring power* and assume quasi-sovereign tasks of market structuring and regulation. The more relevant a platform becomes for the visibility and processing of a business offer, the stronger the pressure on market participants to be present on the platform and to adapt their own offerings to the platform's structural characteristics and rules. This affects services such as travel and hotel bookings, which are now hardly ever made through the websites of direct providers but rather on platforms such as Airbnb, Booking.com, or Expedia. It also affects large parts of cultural and media production, such as the offers of traditional media companies, which are significantly decreasing in popularity outside major internet platforms. As a result, culture and media producers not only lose autonomy of action and control over their distribution and communication channels but, as demonstrated by Nielsen and Ganter (2018: 1615), have to adapt the production, distribution, and exploitation of their content quite extensively to the structuring framework and rules of the platforms:

> Today, they have far less control over the distribution of news than they had in the past. They may reach wider audiences than they can through their own websites and apps, but they do it by publishing to platforms defined by coding technologies, business models, and cultural conventions over which they have little influence and are increasingly dependent.

As a result, *privately regulated and socio-technically constituted market regimes* have taken shape on the internet that clearly stand out from other markets. They are neither primarily state-organised, regulated, or guaranteed, nor do they constitute themselves through the self-organised and deliberative interaction of various non-state actors (Aspers 2011: 148–168; Ahrne et al. 2015). Instead, they are installed, operated, and controlled by individual companies. The platform operators act neither as competing market participants nor as neutral intermediaries, but rather as rule-setting and regulatory actors who endow themselves with far-reaching authority and powers of intervention and who thus assume essential functions that are prerequisites for the acceptance, functionality, and reliability of the market. Further, the technical infrastructures provided by the platform operators are not neutral architectures through which connections are merely established. Instead, through the rules inscribed in them, they form these markets' institutional foundation, the basis that guides actions and structures processes and to which providers, consumers, and users must orient themselves if they wish to play a part.

Plans to establish platform-specific private currencies go a significant step further. With this, the privatisation of market regimes described previously could be extended to include the much more far-reaching prospect of *private-sector regulation of macroeconomic interrelationships*. Eventually, sovereign tasks, previously performed primarily by democratically legitimised and politically independent institutions, could be, at least partially, delegated to private companies or consortia. This could concern, for example, the regulation of money supply, interest rate policy, and the safeguarding of price level stability or banking supervision, which have so far been the domain of central banks.

Such plans are most advanced at Facebook. In mid-2019, with the Libra project, the social media company presented not only an initial concept for a digital currency but also an appropriate regulatory and institutional framework (Schmeling 2019; Taskinsoy 2019; Mai 2019). The core organisation slated to spearhead this project was the Libra Association, a consortium of internet companies, payment providers, and other organisations, designed as a private-sector counterpart and parallel structure to the central banks. This body was intended to not only be responsible for the design and enforcement of Libra rules and the technical infrastructure of the digital currency but also for managing the Libra reserve, create Libra money and control the money supply, monitor payment channels, and admit new Libra traders (Libra 2019).

Although these plans have since been scaled back following massive political pressure, their basic direction is clearly recognisable. Their general direction of impact was the bid to relativise the importance of central banks and governments in a central area of macroeconomic management and to supplement or replace these with private-sector forms of macroeconomic regulation. In this sense, the original plan comprised the takeover of quasi-sovereign *economic* regulatory tasks by the private sector, in ways that align with the cornerstones of the libertarian ideology outlined at the beginning and which, as we will see next, will be substantially expanded by the assumption of quasi-sovereign *social* structuring and curating tasks.

Curation of social relationships and processes

In addition to organising and regulating markets, these platforms – in particular the widely built-out and networked ecosystems of the leading internet companies – have taken over essential social ordering and regulatory functions on the internet, which are summarised here as *curation of social relationships and social behaviour* (Figure 22.1). Through their numerous services and offerings, these platforms filter information and communication processes, shape individual behaviour and organisational action, and structure social relationships and public spheres – and do so in a far more comprehensive manner than even large media corporations have ever been able to do (Couldry and Hepp 2017 34–56; Lobigs and Neuberger 2018). While media corporations remain embedded in society and in its institutional structure as powerful opinion-forming actors with a limited reach, the large platforms, with their own rule-setting, structuring,

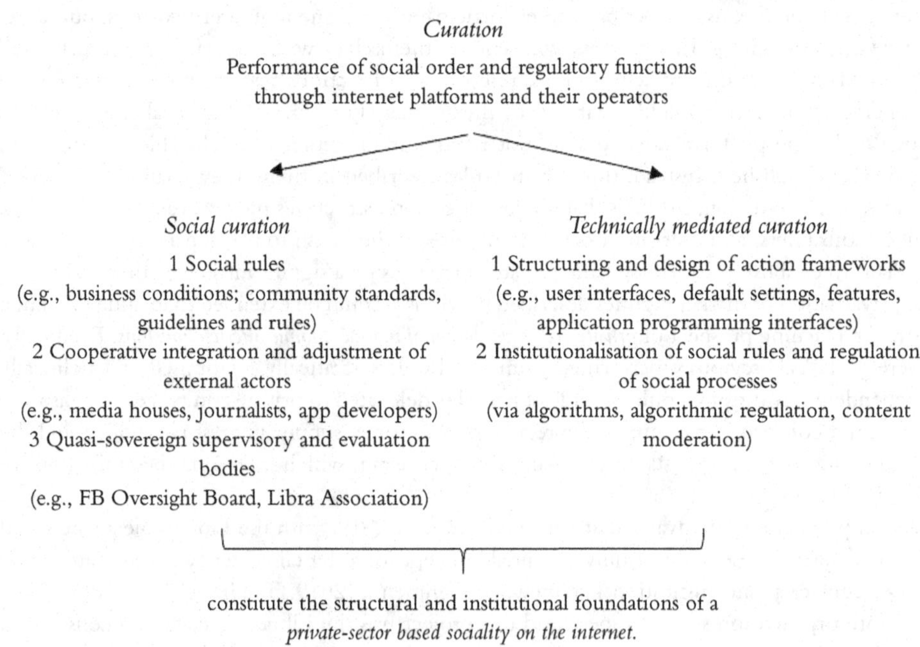

Curation
Performance of social order and regulatory functions through internet platforms and their operators
Social curation
1 Social rules
(e.g., business conditions; community standards, guidelines and rules)
2 Cooperative integration and adjustment of external actors
(e.g., media houses, journalists, app developers)
3 Quasi-sovereign supervisory and evaluation bodies
(e.g., FB Oversight Board, Libra Association)
Technically mediated curation
1 Structuring and design of action frameworks
(e.g., user interfaces, default settings, features, application programming interfaces)
2 Institutionalisation of social rules and regulation of social processes
(via algorithms, algorithmic regulation, content moderation)
constitute the structural and institutional foundations of a
private-sector based sociality on the internet.

Figure 22.1 Social and technically mediated curation

Source: Author's compilation

selection, monitoring, and sanctioning activities, constitute no less than the institutional foundations of a *private-sector sociality on the internet*, which have, over the past two decades, evolved largely decoupled from democratic institutions and state influence.

The basis of curation is formed by binding and sanctionable *social rules*. They are expressed in the general terms and conditions of the companies and, above all, in community standards (Facebook), guidelines, and rules (YouTube; Twitter), in which the platform operators formulate in detail what they consider to be politically unacceptable, a glorification of violence or terrorism, offensive, obscene, erotic, or pornographic. Throughout the ongoing development of their guidelines, which provide the legal and normative framework for all social activities on the platforms, the internet companies do, of course, integrate or consider public opinions and political interventions. However, this does not mean that they have lost sovereignty over rule-making and enforcement on their platforms, on which they alone decide in the last instance.

These guidelines, which form the basis of social curation, are largely translated into technical instructions, structurings, sortings, and rankings, which I refer to as *technically mediated curation*. Research in the sociology of technology has long shown that technology always incorporates social rules, norms, instructions, and control mechanisms which influence the activities and behaviour of their users in a way that sometimes is more rigid than that of social institutions (Dolata 2013: 32–40). In the 1990s, Christiane Floyd (1992) characterised software development as a construction of reality, and the aforementioned Lawrence Lessig (1999), also with regard to software, formulated the metaphor *code is law*, which equates, by virtue of its action-regulating power, all the instructions and procedures inscribed in software with the law and other social systems of rules. Two decades earlier, Langdon Winner (1980: 127f.) had already characterised technical arrangements as structure-forming and rule-setting patterns of social order:

> The things we call "technologies" are ways of building order in our world. . . . In that sense technological innovations are similar to legislative acts or political foundings that establish a framework for public order.

First, in the platform context, this classical view of the structure-forming and institutional effects of technology manifests as a technically mediated *structuring and design of social action frameworks* that both enable and channel the activities of a diverse range of users. This includes the given user interfaces and default settings of the platforms, which have an action-structuring effect by enabling certain activities and excluding or impeding others. The numerous features embedded in the platforms (such as Facebook's Reactions or Twitter's Trending button) can also be summarised as action-orienting and opinion-forming structural elements inscribed in technology. In addition, Application Programming Interfaces (APIs) are used to integrate the web presences of countless third parties into the platforms' scope of action and to establish extensive links between the platforms and external websites, other platforms and apps. Facebook is a good example. An overwhelming number of external websites – e.g., of media organisations, political parties, social movements, public institutions, or companies, to name but a few – are linked to the social media platform via corresponding technical programs and features, thereby providing the internet group with high-quality additional data. This has led to a systematic and large-scale embedding of external technical architectures and thus to a substantial expansion of the reach and social significance of the leading platforms on the internet and is described in the literature as "platformisation." On the one hand, the structuring influence of individual platforms now extends well beyond their original domain and deep into the social web and shapes the scope of action of countless other actors. On the other hand, the integration of third parties enables platform operators to systematically tap into the off-platform data stocks and use them for their own data collection and analysis (Van Dijck 2020; Nieborg and Helmond 2019; Helmond 2015; Gerlitz and Helmond 2013).

Second, these structure-building effects of technology are supplemented by approaches to a technically mediated *institutionalisation of social rules and regulation of social processes*, which is implemented primarily through the use of algorithms and referred to in the literature as algorithmic governance, algorithmic regulation, or algorithmic content moderation (Gillespie 2014, 2016; Kitchin 2014; Just and Latzer 2017; Beer 2017; Yeung 2018; Katzenbach and Ulbricht 2019; Gorwa et al. 2020). Algorithms translate the social rules and norms that are valid on the platforms into technical instructions; monitor and sanction participants' activities; decide what is important and what is not, according to social relevance criteria inscribed in them; select, aggregate, and rank information, news, videos, or photos on this basis; structure private information and communication processes as well as public discourses; and constitute public spheres and communities that would not exist without them. With all this, algorithms essentially become the nucleus of a technically mediated framing, control, and curation of social action on platforms.

The regulatory depth of intervention of algorithms is further augmented by the fact that they can be changed quickly and radically. Corresponding readjustments are regularly made by platform operators (e.g., in the PageRank algorithm of Google searches, the YouTube algorithm, or in the News Feed algorithm of Facebook) and go on to reconfigure the social reality presented on the platforms, in some cases significantly. Changes to the newsfeed algorithm, for example, not only directly affect what users see in personal posts and news but also have a massive impact on the perception and web traffic of public media institutions or private media houses, whose performance is now highly dependent on their presence on these platforms (Nielsen and Ganter 2018; Van Dijck et al. 2018: 49–72). Algorithms that form the basis of all search and information, communication and interaction on these platforms are highly political programs that construct distinct, selective, and increasingly personalised social reality offers based on social criteria that remain completely opaque to individuals, organisations, and political bodies.

Of course, generally speaking, social structures and rules inscribed in technology, with their institutional and regulatory peculiarities, never determine action. Instead, similar to laws, regulations, social norms, or values, they are open to interpretation and are repeatedly adapted, modified, or even suspended, not only by their developers and operators but also as a result of political interventions, social disputes, or idiosyncratic user behaviour. This also applies, more specifically, to algorithms:

> Algorithms are not just what their designers make of them, or what they make of the information they process. They are also what we make of them day in and day out – but with this caveat: because the logic, maintenance, and redesign of these algorithms remain in the hands of the information providers, they are in a distinctly privileged position to rewrite our understanding of them.
>
> *(Gillespie 2014: 187)*

The caveat inserted by Gillespie is important and marks an essential and generalisable difference between technology as an institution and social institutions. While the social institutions of democratic societies generally take shape in and through public discourse and political negotiations and require democratic legitimation, institutional inscriptions in technology are usually the domain of their (private sector) producers and can hardly be publicly negotiated or shaped ex ante.

The two central levels of social and technically mediated curation described here are enriched by two further forms of social curation. On the one hand, the algorithmic structuring and sorting of media content and audiences since the mid-2010s has been supplemented by initiatives of platform operators aimed at a stronger *cooperative integration and platform-oriented alignment of media houses and journalists* (Bell 2018). These include projects such as the Google News Initiative (https://newsinitiative.withgoogle.com/) or the Facebook Journalism Project (www.facebook.com/journalismproject), which are designed to tie media groups and institutions, editorial offices, and media-related organisations more tightly to their platforms and to align them more closely with their operational and exploitation logics, via meetings and training courses organised by the internet companies, through the development of programs to expand digital news services, and through the allocation of grants.

Another major step was the *establishment of a corporate-owned oversight body* at Facebook, responsible for monitoring, moderating, and evaluating content on the platform. The Oversight Board, active since 2020, staffed with external experts and financed by the company, not only seeks to monitor and further develop the implementation of the social rules laid down in the Community Standards but also has the authority to judge disputed content and, if necessary, have it removed from the platform (Harris 2020). In addition to the Libra Association, the group thus has a second body with a quasi-sovereign function, set up as a kind of constitutional court and supervisory committee, albeit without the democratic legitimacy of such bodies or the ability to exert influence on fundamental corporate decisions. While the Libra project has been proactively driven forward by Facebook, the setting up of the Oversight Board is constructed as a domain-securing reaction to increasingly critical political discussions about a stronger regulation of internet platforms. In essence, however, both projects aim to establish extensive quasi-sovereign structures within the platform and parallel to the democratically legitimised societal institutions.

As a result of the combination of these factors, especially the leading internet groups are now far more than infrastructure providers that provide connectivity; media groups that have a broad portfolio of their own media offerings; or advertising, retail, hardware, and service companies that continue to generate the majority of their revenues and profits with their traditional businesses. The few large platforms that today both enable and shape large parts of private and public life on the internet can be understood as *differentiated societal systems with a distinct institutional foundation*, which the companies as platform operators structure and control to a considerable extent and by means of their own rules, regulations, and committees – right up to the assumption of quasi-sovereign tasks by the companies that, hitherto reserved for state authorities, so far largely skirt democratic legitimation and control.

Outlook: regulation of platforms? Possibilities and limits of political intervention

The economic but above all social structuring and regulatory power that the leading internet companies have attained with their platforms is camouflaged rather than disclosed by non-hierarchical notions of an internet governance that focus on "low formalisation, heterogeneous organisational forms, large numbers of actors and massively distributed authority and decision-making power" (Van Eeten and Mueller 2012: 730). This is contrasted by what I have discussed and what I refer to as *regulation by platforms*: the intentional structuring and regulating not only of economic markets but also, and in a much more comprehensive way, of larger societal relations and processes, carried out by internet companies as platform operators and aligned with their economic exploitation interests.

Of course, this does not mean that these regulatory activities could determine the actions of other actors, nor that the internet companies with their platforms could act independently and disregard collective user behaviour, public discourse and opinions, political interventions, or the interests of other economic actors. Power may be distributed very asymmetrically, as in the case here, but it is never absolute or something that some have and others do not. Instead, power is always an expression of complex, often contested, and often volatile societal relations that benefit some more so than others (Dolata and Schrape 2018). These others and their rooms for manoeuvre are at the centre of these concluding remarks. With a focus on the large and most influential platforms, two levels of social and political intervention will be distinguished and four possibilities of political intervention will be explored.

I refer to the first level as *civil society intervention*. Internet companies have to react in rapid succession not only to changes in the very dynamic technological and economic environments in which they operate but also to social or political pressure, which has increased significantly since the 2010s. For one, their large platforms are existentially dependent on the contributions, activity, and acceptance of their users, some of whom adopt the platforms' offerings in rather idiosyncratic ways, repurposing them or even rejecting them, and who must hence be treated with corresponding sensitivity by the platform operators. Second, the leading internet corporations have also been under the intense observation of a more and more attentive media and political public. Investigative journalists, net-political blogs, and the classic media now deal extensively with the various facets of their social and economic might. Among these are: non-transparent business practices and dominant market positions, controversial social guidelines and opaque algorithms, repeated violations of privacy and user surveillance, data scandals (such as those surrounding Cambridge Analytica), the dissemination of fake news, or the use of platforms to influence elections (such as the US presidential election in 2016).

In recent years, the media, in particular, but also other civil society actors have thus contributed to a much more critical assessment of platforms, in both public discourse and the political realm. This cannot simply be ignored by the platform operators, especially when these assessments evolve into serious demands for greater public control and state regulation of the platforms. The internet companies, above all Facebook and Google, have responded to this with a series of transparency initiatives and attempts to integrate civil society actors more closely in the institutional and regulatory structures of their platforms (for example, by setting up the Oversight Board at Facebook) (Gorwa 2019).

The effects that can be achieved by civil society interventions should not be underestimated: in cases where they are brought forward with the appropriate force and met with great social acceptance, they can trigger rapid and, in some cases, substantial adaptation reactions among the internet companies – albeit without calling into question their structuring and regulatory sovereignty. The companies can react to civil society pressure in a voluntary way, according to standards which they themselves set and at a time they consider to be opportune. This remains non-binding and has nothing to do with a *regulation of platforms*, which, in contrast, is essentially based on the enforcement of democratically developed and legally binding public rules with which platform operators must comply.

In parallel to the increase in interventions involving civil society, the second half of the 2010s has also seen – comprising the second level of external influence – an increase in government efforts to achieve *political regulation and control* of the major platforms. In Europe, since the mid-2010s, such activities have been concentrated in two main areas of action:

1 Attempts to *limit economic market power*, brought forward above all by the European Commission. The latter has pursued a series of infringements of EU antitrust law by internet companies and has repeatedly imposed heavy fines, especially on Google and on Facebook, among others, for an abuse of their dominant position in online advertising, with search

engines, or through the mobile operating system Android (Viscusi/Harrington/Sappington 2018: 404–419; Haucap/Stühmeier 2016; European Commission 2019).

2 Efforts for legal and regulatory *intervention in the social regulatory sovereignty of platforms* – for example, in the form of the European General Data Protection Regulation (GDPR); the "right to be forgotten" on the internet, introduced by the European Court of Justice in a landmark decision; or the German Network Enforcement Act (NetzDG), which obliges the providers of leading social networks such as Facebook, YouTube, or Twitter to block illegal content in a timely manner or to remove it from their platforms and to report on it on a regular basis (Schulz 2018; Chenou and Radu 2019).

However, the scope of these political interventions has so far remained extremely limited. Paradoxically, these attempts by the state to intervene in the social regulatory sovereignty of platform operators have tended to strengthen the regulatory power of the platforms, namely, by delegating sovereign functions of jurisdiction and enforcement to private sector actors and by providing this shift with political legitimacy. Germany's Network Enforcement Act, for example, has done little to change the fact that companies such as Facebook, Google, or Twitter largely decide for themselves which content they delete and which they do not, yet has, at the same time, strengthened the companies in their role as content moderators and as decisive instances of content evaluation or selection. Further, the enforcement of the right to be forgotten has also been assigned to the platforms themselves, which have thus become more integrated into the legal system and, as private-sector organisations, have been entrusted by government with quasi-sovereign tasks. Chenou and Radu (2019: 74 and 96f.) have accurately described this as the "outsourcing of important governance practices to private intermediaries." The authors have also pointed out the dependence of state regulation on the willingness of platform operators to cooperate:

> In creating new rights, public actors foster strong regulations they may not be able to implement themselves without the collaboration of private actors. More than a transformation of the state, the resulting hybridization of governance also entails a transformation of private actors. In the process, some private actors are given new responsibilities in the governance of technologies and technology-enabled markets. As the case of the "right to be forgotten" showed, Google becomes inserted in the European legal system as a first instance to look at cases of online privacy protection triggered by individual requests.

Overall, the political regulatory approaches, to date, are not suitable for substantially correcting or controlling the regulatory sovereignty of the platform operators. However, the presentation of proposals for a Digital Markets and Services Act by the EU Commission at the end of 2020 (European Commission 2020a, 2020b) and a lawsuit filed by the US Federal Trade Commission against Facebook, which aimed for nothing less than a split-up of the group, show that the question of how the overwhelming power of Internet corporations and their platforms can be limited and more publicly controlled is no longer being considered only in Europe but now also in the United States. In this context, two more far-reaching directions in which considerations about stronger political regulation of Internet corporations should develop are becoming increasingly apparent. These include:

3 The *radical unbundling of the widely networked platforms of the internet corporations* – such as the decoupling of YouTube and other platforms from the Google corporation, or the splitting

up of the ecosystem of Facebook, Instagram, and WhatsApp ((Nadler and Cicilline 2020: 378–382). However, such considerations, which would, admittedly, involve a rather brutish dismantling, should be justified less by a limitation of these corporations' economic market power than by the aim of limiting their extraordinary socio-political structuring and regulatory power.

4 Setting up *public supervisory and regulatory bodies*, for example, at the European and US levels. Controlled by parliament and staffed with recognised and publicly appointed experts, these authorities should be set up as democratically legitimate alternatives to the corporate supervisory bodies (such as Facebook's Oversight Board) and be equipped with far-reaching information, control, and sanctioning powers. They could also be tasked to disclose, control, and impose conditions on algorithmic filtering functions, ranking, and rating principles, as well as community standards, and the search and selection criteria based upon them (Dolata 2018b).

However, even the proposal for public supervisory and regulatory authorities would not, if implemented, lead to a private-state co-regulation of platforms on an equal footing – if only because of the extreme information and knowledge asymmetries of the parties involved. Indeed, political regulators are much less knowledgeable about the extensive socio-technical systems and systemic contexts they are supposed to regulate than those who have developed and now operate these systems. Hence, in this case, too, the responsibility for structuring and regulating economic and social processes on the internet would remain primarily with the platform operators. But at least then their activities could be regularly evaluated, controlled, and sanctioned by a democratically legitimised body.

References

Abbate, J. (1999): *Inventing the internet.* Cambridge, London: MIT Press.

Ahrne, G., Aspers, P. and Brunsson, N. (2015): The Organization of Markets, *Organization Studies* 36(1), 7–27.

Aspers, P. (2011): *Markets.* Cambridge, Malden: Polity Press.

Barbrook, R. and Cameron, A. (1996): The Californian Ideology, *Science as Culture* 6(1), 44–72.

Barefoot, K. et al. (2018): Defining and Measuring the Digital Economy. *Working Paper.* Washington, DC: Bureau of Economic Analysis.

Barlow, J. P. (1996): A Declaration of the Independence of Cyberspace. Electronic Frontier Foundation, 8 February 1996. Available at: www.eff.org/cyberspace-independence [30.04.2021]

Barwise, P. and Watkins, L. (2018): The Evolution of Digital Dominance: How and Why We Got to GAFA, in Moore, M. and Tambini, D. (eds.): *Digital Dominance. The Power of Google, Amazon, Facebook, and Apple.* Oxford: Oxford University Press, 21–49.

Barzilai-Nahon, K. (2008): Toward a Theory of Network Gatekeeping: A Framework for Exploring Information Control, *Journal of the American Society for Information Science and Technology* 59(9), 1493–1512.

Beer, D. (2017): The Social Power of Algorithms, *Information, Communication & Society* 20(1), 1–13.

Bell, E. (2018): The Dependent Press. How Silicon Valley Threatens Independent Journalism, in Moore, M. and Tambini, M. (eds.): *Digital Dominance. The Power of Google, Amazon, Facebook, and Apple.* Oxford: Oxford University Press, 241–261.

Chenou, J.-M. and Radu, R. (2019): The "Right to Be Forgotten": Negotiating Public and Private Ordering in the European Union, *Business & Society* 58(1), 74–102.

Couldry, N. and Hepp, A. (2017): *The Mediated Construction of Reality.* Cambridge, Malden: Polity.

Currah, A. (2007): Managing Creativity: The Tensions between Commodities and Gifts in a Digital Networked Environment, *Economy and Society* 36(3), 467–494.

Davis, G. F. (2016): What Might Replace the Modern Corporation? Uberization and the Web Page Enterprise, *Seattle University Law Review* 39, 501–515.

Dolata, U. (2013): *The Transformative Capacity of New Technologies. A Theory of Sociotechnical Change.* London, New York: Routledge.

Dolata, U. (2018a): Internet Companies: Market Concentration, Competition and Power, in Dolata, U. and Schrape, J.-F. (eds.): *Collectivity and Power on the Internet. A Sociological Perspective*. Cham: Springer, 85–109.

Dolata, U. (2018b): Big Four – Die digitale Allmacht? *Blätter für deutsche und internationale Politik* 63(5), 81–86.

Dolata, U. (2019): Privatization, Curation, Commodification. Commercial Platforms on the Internet, *Österreichische Zeitschrift für Soziologie* 44(Suppl 1), 181–197.

Dolata, U. and Schrape, J.-F. (2014): App-Economy: Demokratisierung des Software-Marktes? *Technikfolgenabschätzung – Theorie und Praxis* 23(2), 76–80.

Dolata, U. and Schrape, J.-F. (2018): *Collectivity and Power on the Internet. A Sociological Perspective*. Cham: Springer.

Dyson, E., Gilder, G., Keyworth, G. and Toffler, A. (1994): *Cyberspace and the American Dream: A Magna Carta for the Knowledge Age*. Release 1.2, 22 August 1994. Available at: www.pff.org/issues-pubs/futureinsights/fi1.2magnacarta.html [16.04.2021]

Elder-Vass, D. (2016): *Profit and Gift in the Digital Economy*. Cambridge: Cambridge University Press.

European Commission (eds.) (2019): *Antitrust: Commission Fines Google €1.49 Billion for Abusive Practices in Online Advertising*. Press Release 20 March 2019. Brussels: European Commission. Available at: https://ec.europa.eu/commission/presscorner/detail/en/IP_19_1770 [16.04.2021]

European Commission (2020a): Proposal for a Regulation of the European Parliament and of the Council on a Single Market for Digital Services (Digital Services Act) and amending Directive 2000/31/EC. *COM(2020)825 Final*. Brussels: European Commission.

European Commission (2020b): Proposal for a Regulation of the European Parliament and of the Council on Contestable and Fair Markets in the Digital Sector (Digital Markets Act). *COM(2020)842 Final*. Brussels: European Commission.

Evans, D. S. and Schmalensee, R. (2005): The Industrial Organization of Markets with Two-Sided Platforms. *NBER Working Paper 11603*. Cambridge, MA: National Bureau of Economic Research.

Evans, D. S. and Schmalensee, R. (2016): *Matchmakers. The New Economics of Multisided Platforms*. Boston: Harvard Business Review Press.

Feick, J. and Werle, R. (2010): Regulation of Cyberspace, in Baldwin, R., Cave, M. and Lodge, M. (eds.): *The Oxford Handbook of Regulation*. Oxford: Oxford University Press, 523–547.

Floyd, C. (1992): Software Development as Reality Construction, in Floyd, C., Züllighoven, H. Budde, R. and Keil-Slawik, R. (eds.): *Software Development and Reality Construction*. Berlin: Springer, 86–100.

Gerlitz, C. and Helmond, A. (2013): The Like Economy: Social buttons and the Data-intensive Web, *New Media & Society* 15(8), 1348–1365.

Gillespie, T. (2010): The Politics of "Platforms," *New Media & Society* 12(3), 347–364.

Gillespie, T. (2014): The Relevance of Algorithms, in Gillespie, T., Boczkowski, P. and Foot, K. (eds.): *Media Technologies. Essays on Communication, Materiality, and Society*. Cambridge: MIT Press, 167–194.

Gillespie, T. (2016): Regulation of and by Platforms, in Burgess, J., Poell, T. and Marwick, A. (eds.): *The SAGE Handbook of Social Media*. Los Angeles: Sage, 254–278.

Gorwa, R. (2019): What is Platform Governance? *Information, Communication & Society* 22(6), 854–871.

Gorwa, R., Binns, R. and Katzenbach, C. (2020): Algorithmic Content Moderation: Technical and Political Challenges in the Automation of Platform Governance, *Big Data & Society* 7(1). https://doi.org/10.1177/2053951719897945

Harris, B. (2020): Preparing the Way forward for Facebook's Oversight Board. Available at: https://about.fb.com/news/2020/01/facebooks-oversight-board/ [15.08.2020]

Haucap, J. and Stühmeier, T. (2016): Competition and Antitrust in Internet Markets, in Bauer, J. M. and Latzer, M. (eds.): *Economics of the Internet*. Cheltenham, Northampton: Edward Elgar, 183–210.

Helmond, A. (2015): The Platformization of the Web: Making Web Data Platform Ready, *Social Media & Society* 1(2), 1–11.

International Monetary Fund (2018): *Measuring the Digital Economy*. Washington, DC: IMF.

Just, N. and Latzer, M. (2017): Governance by Algorithms: Reality Construction by Algorithmic Selection on the Internet, *Media, Culture & Society* 39(2), 238–258.

Katzenbach, C. and Ulbricht, L. (2019): Algorithmic Governance, *Internet Policy Review* 8(4), 1–18.

Kenney, M. and Zysman, J. (2016): The Rise of the Platform Economy, *Issues in Science and Technology* 32(3), 61–69.

Khan, L. M. (2018): Amazon – An Infrastructure Service and its Challenge to Current Antitrust Law, in Moore, M. and Tambini, M. (eds.): *Digital Dominance. The Power of Google, Amazon, Facebook, and Apple*. Oxford: Oxford University Press, 98–129.

Kirchner, S. and Beyer, J. (2016): Die Plattformlogik als digitale Marktordnung. Wie die Digitalisierung Kopplungen von Unternehmen löst und Märkte transformiert, *Zeitschrift für Soziologie* 45(5), 324–339.
Kitchin, R. (2014): *The Data Revolution. Big Data, Open Data, Data Infrastructures & their Consequences*. Los Angeles, London, New Delhi, Singapore, Washington, DC: Sage.
Langley, P. and Leyshon, A. (2016): Platform Capitalism: The Intermediation and Capitalisation of Digital Economic Circulation, *Finance and Society* 2, 1–21.
Lessig, L. (1999): *CODE and Other Laws of Cyberspace*. New York: Basic Books.
Libra Association (2019): *An Introduction to Libra. White Paper*. Libra Association. Available at: https://libra.org/en-US/wp-comntent/uploads/sites/23/2019/06/LibraWhitePaper_en_US.pdf [15.08.2020]
Lobigs, F. and Neuberger, C. (2018): *Meinungsmacht im internet und die Digitalstrategien von Medienunternehmen*. Leipzig: Vistas.
Mai, H. (2019): Libra – A Global Challenger in Payments and for Central Banks? in Deutsche Bank (ed.): *EU-Monitor – Digital Economy and Structural Change*. 22 July 2019.
Mazzucato, M. (2013): *The Entrepreneurial State. Debunking Public vs. Private Sector Myths*. London, New York, New Delhi: Anthem Press.
Misterek, F. (2017): Digitale Souveränität. Technikutopien und Gestaltungsansprüche demokratischer Politik. *MPIfG Discussion Paper 17/11*. Cologne: MPIfG.
Nadler, J. and Cicilline, D. N. (2020): *Investigation of Competition in Digital Markets*. Washington, DC: United States House. Subcommittee on Antitrust, Commercial and Administrative Law of the Committee on the Judiciary.
Nieborg, D. B. and Helmond, A. (2019): The Political Economy of Facebook's Platformization in the Mobile Ecosystem: Facebook Messenger as a Platform Instance, *Media, Culture & Society* 41(1), 196–218.
Nieborg, D. B. and Poell, T. (2018): The Platformization of Cultural Production: Theorizing the Contingent Cultural Commodity, *New Media & Society* 20(11), 4275–4292.
Nielsen, R. K. and Ganter, S. A. (2018): Dealing with Digital Intermediaries: A Case Study of the Relations between Publishers and Platforms, *New Media & Society* 20(4), 1600–1617.
O'Reilly, T. (2005): *What Is Web 2.0. Design Patterns and Business Models for the Next Generation of Software*. O'Reilly Network. 30 September 2005. Available at: http://oreilly.com/pub/a/web2/archive/what-is-web-20.html [15.08.2020]
Parker, G. G., Van Alstyne, M. W. and Choudary, S. P. (2016): *The Platform Revolution*. New York, London: W.W. Norton.
Pasquale, F. (2015): *The Black Box Society. The Secret Algorithms That Control Money and Information*. Cambridge, London: Harvard University Press.
Rahman, K. S. and Thelen, K. (2019): The Rise of the Platform Business Model and the Transformation of Twenty-first Century Capitalism, *Politics & Society* 47(2), 177–204
Rochet, J.-C. and Tirole, J. (2003): Platform Competition in Two-sided Markets, *Journal of the European Economic Association* 1(4), 990–1029.
Schmeling, M. (2019): *What is Libra? Understanding Facebooks Currency*. SAFE Policy Letter, No. 76. Frankfurt a.M.: Goethe Universität Frankfurt.
Schrape, J.-F. (2019): The Promise of Technological Decentralization. A Brief Reconstruction, *Society* 56(1), 31–38.
Schulz, W. (2018): *Regulating Intermediaries to Protect Privacy Online – the Case of the German NetzDG*. HIIG Discussion Paper Series 2018–01. Berlin: HIIG.
Srnicek, N. (2017): *Platform Capitalism*. Cambridge, Malden: Polity Press.
Staab, P. (2019): *Digitaler Kapitalismus. Markt und Herrschaft in der Ökonomie der Unknappheit*. Berlin: Suhrkamp.
Sundararajan, A. (2016): *The Sharing Economy. The End of Employment and the Rise of Crowd-based Capitalism*. Cambridge, London: MIT Press.
Taskinsoy, J. (2019): *Facebook's Project Libra: Will Libra Sputter Out or Spur Central Banks to Introduce Their Own Unique Cryptocurrency Projects?* Sarawak: University Malaysia Sarawak.
Thelen, K. (2018): Regulating Uber: The politics of the platform economy in Europe and the United States, *Perspectives on Politics* 16(4), 938–953.
Van Dijck, J. (2013): *The Culture of Connectivity. A Critical History of Social Media*. Oxford. Oxford University Press.
Van Dijck, J. (2020): Seeing the Forest for the Trees: Visualizing Platformization and its Governance, *New Media & Society*. First Published 8 July 2020. https://doi.org/10.1177/1461444820940293.
Van Dijck, J., Poell, T. and de Waal, M. (2018): *The Platform Society. Public Values in a Connective World*. Oxford: Oxford University Press.

Van Eeten, M. J. G. and Mueller, M. (2012): Where is the Governance in Internet Governance? *New Media & Society* 15(5), 720–736.

Viscusi, W. Kip, Harrington Jr., J. E. and Sappington, D. E. M. (2018): *Economics of Regulation and Antitrust*. 5th ed. Cambridge: MIT Press.

Winner, L. (1980): Do Artifacts Have Politics? *Daedalus* 109(1), 121–136.

Yeung, K. (2018): Algorithmic Regulation: A Critical Interrogation, *Regulation & Governance* 12(4), 505–523.

Zerdick, A. et al. (2001): *Die internet-Ökonomie. Strategien für die digitale Wirtschaft*. Berlin, Heidelberg: Springer.

Zuboff, S. (2019): *The Age of Surveillance Capitalism: The Fight for the Future at the New Frontier of Power*. London: Profile Books.

Zysman, J. and Kenney, M. (2016): *The Next Phase in the Digital Revolution. Platforms, Abundant Computing, Growth and Employment*. ETLA Reports No 61. Helsinki: The Research Institute of the Finnish Economy.

23

NEW MISSION-ORIENTED INNOVATION POLICY IN THE DIGITAL ERA

How policy-based social technologies fuel the development of smart technologies

Marlies Schütz and Rita Strohmaier

Introduction

For several years, the manufacturing sector has undergone an all-embracing restructuring towards the digitisation of production. Yet, what came to be known by terms such as "Industry 4.0" or "Smart Industry" is just a small part of a far greater transformation of the socio-economic system: the digital revolution. From a technology-centred perspective, the digital revolution reflects a wave of disruptive innovations in the field of micro- and nanoelectronics that started off in 1971 with the invention of the Intel 4004 microprocessor, and in recent years has regained momentum due to technical breakthroughs in complementary fields such as artificial intelligence, advanced robotics, monitoring and analytics, and augmented and virtual reality. Interlocked, these technologies reveal the core characteristics of the digital age – networking, cognition, autonomy, virtuality, and the explosion of knowledge (WBGU 2019) – and are re-shaping almost every sphere of society. They enable new concepts in transport and mobility, such as autonomous driving, electric and plug-in hybrid vehicles, and shared mobility models (on-demand mobility); revolutionise the health sector through the development of portable image technologies, telemedicine, 3D-printing of organs and tissue, robotic surgery, or genome editing; have the potential to significantly contribute to the decarbonisation of the economy, as they allow for innovation in the renewable energy sector and for overall energy efficiency improvements (e.g., through smart grids); support dematerialisation in production, as physical products and services can be substituted by digital services; and facilitate new manufacturing methods such as 3D-printing and the full automation of production processes, via cyber-physical systems and smart systems integration. And the list of application areas gets even longer: they transform the way we live, as automated control systems make our buildings, spaces, and cities "smarter", and in the long run can be expected to create leeway for the co-existence of human and artificial agents (see, e.g., TWI2050 2019 for a detailed discussion).

For Schumpeter (2006[1912]), technological change and other types of innovation are the driving force of socioeconomic change, and today, still, innovation studies frequently rely on

DOI: 10.4324/9780429351921-28

this perception. Discussing different perspectives of innovation, Edler and Fagerberg (2017: 4) describe it "as the introduction of new solutions in response to problems, challenges, or opportunities that arise in the social and/or economic environment". According to them, the various existing perceptions of innovation are equally reflected in the interpretation of innovation policy. Consequently, there exist both narrow and more holistic views on innovation policy, as the "policies affecting innovation" (ibid.: 4).

Over the years, several types of innovation policy models have followed each other, differing not least in their design and rationale. Cantner and Vannuccini (2018: 843 ff) consider them as "mutual and not exclusive", distinguishing in chronological order between "mission orientation", "diffusion orientation", "cluster and network orientation", and the "new mission orientation".[1] New mission-oriented innovation policy disposes of a transformative character and aims to tackle today's "grand societal challenges", such as environmental and demographic challenges as well as issues around security, health, and societal well-being (see Schot and Steinmueller 2018 for a comprehensive account of the transformative character of innovation policy).

Against this background, we argue that innovation policy needs to be captured as a dynamic process of learning that is subject to change. Our claim is owed not least to the fact that innovation is fraught with uncertainty, and innovation policy is itself innovative in the most general sense, as it ought to provide solutions to problems. Elaborating on this claim, we address the following research question: *How can we conceptualise the endogenous and dynamic nature of new mission-oriented innovation policy?* Our research aim is to present a conceptual framework that does justice to the aforementioned claim. Therefore, we build on Nelson and Sampat (2001), who introduced the notion of "physical and social technologies", drawing on the "routine" as the basic unit of economic activity. Social technologies initially were described as constituting the aspect of a routine that involves some division of labour and human agency (ibid.: 44). We suggest transferring this to the policy space and contextualising new mission-oriented innovation policy initiatives as "policy-based social technologies", which we define as practice devised by the state, consisting of a set of different measures for organising innovation. To be effective, such policy-based social technologies need to be operated, which in turn involves the coordination of agency of multiple actors from diverse institutional backgrounds. We further argue that such policy-based social technologies co-evolve with "physical technologies", which in their original meaning were identified by Nelson and Sampat (2001: 44) as the aspect of a routine that has some recipe character and requires a specific combination of inputs that are transformed into some output. In our study, we focus on those "physical technologies" that are key drivers of the digital revolution, viz., micro- and nanoelectronic components and systems, as discussed previously.

Referring to earlier discussions of how innovation policy could be mapped as an endogenous process, Flanagan et al. (2011: 4) stress that "if public policy is part of the system then the agency of actors must be acknowledged both in relation to innovation processes and to processes shaping policy problems and solutions". To the best of our knowledge, this has remained an under-investigated issue, especially in light of current new mission-oriented innovation policy. In seeking to fill the existing voids, we adopt a process-oriented view on the topic at hand.

To put analytical rigour on the concept of "policy-based social technologies", we use the three core characteristics of a technology introduced by Carroll (2017): (1) purpose, (2) functions, and

1 This type of innovation policy model appears in the extant literature under different labels, such as "transformative innovation policy" (Schot and Steinmueller 2018) or "Type-2 mission-oriented innovation policy" (see, e.g., Robinson and Mazzucato 2019).

(3) benefit. To describe agency, we follow Capano and Galanti (2018: 27), who define it as "collective patterns of action in which different individuals interact to affect reality". As regards our research design, we proceed in a deductive way. To collect evidence for the proposed conceptual framework of policy-based social technologies, we carry out a case study of a specific innovation policy initiative in the European Union, viz. ECSEL JU – a joint technology initiative (JTI) in the field of electronic components and systems. More specifically, we look at whether ECSEL JU satisfies the three aforementioned core characteristics of a technology. From a methodological viewpoint, our case study relies on comprehensive desk research. We further present some empirical evidence and, to account for the agency of multiple actors involved in this policy initiative, we make use of social network analysis. Due to the organisational form of ECSEL JU as a public–private partnership, which so far has not been frequently implemented in European Union innovation policy, we consider ECSEL JU as an exemplary and very interesting case for illustrating our notion of policy-based social technologies.

The remainder of this chapter is structured as follows. The following section positions our contribution in the extant literature and then defines the conceptual framework that we use for our case study. The successive section includes the case study. After a brief overview of the research design used in the case study, we continue in each subsection with exploring the single key elements of ECSEL JU as a policy-based social technology. The section after this is dedicated to a concluding discussion of our analysis, and our final remarks as well as an outlook on future research needs in this field of study are provided in the final section.

Literature background and conceptual framework

About the "new mission-oriented innovation policy" – a brief account

As previously stated, in the past decades, innovation research has witnessed several paradigmatic shifts in innovation policy models,[2] going hand in hand with changes in the corresponding policy rationale and design. For a long time, innovation policy was legitimised based on the correction of different types of market failure, and then, under the scope of "cluster and network orientation", dealing with systemic failure constituted the main rationale.[3] For the new mission-oriented innovation policy, which is defined by Mazzucato (2018: 804) as "systemic public policies that draw on frontier knowledge to attain specific goals", the principal role and legitimisation is to tackle today's "grand societal challenges", and at its core lies the directionality of innovation. Therefore, policy measures under this framework are tied to some mission – goals of high societal relevance. In this context, Kattel and Mazzucato (2018: 789) stress that "policy missions . . . are by definition about direction – about concrete problems to be solved". Related to this, another defining characteristic of new mission-oriented innovation policy is their focus on the expected outcome. This requires, however, that innovation policy is not considered as a static field of action, where "a one-size-fits-all" approach is used; rather, innovation policy needs to be understood as a dynamic process that is path-dependent and moulded by past experience and action and that involves learning in each of its stages (cf. Schot and Steinmueller 2018 and Kattel and Mazzucato 2018).

2 See e.g., Mazzucato (2016) or Cantner and Vannuccini (2018) for an overview of the various innovation policy models.

3 See e.g., Metcalfe (1995) and more recently Edler and Fagerberg (2017) and Robinson and Mazzucato (2019) for a discussion of the different innovation policy rationales.

The key elements of new mission-oriented innovation policy are summarised by Robinson and Mazzucato (2019: 939) as follows:

- Broad challenges with a complex mix of goals and objectives
- Multiple sources of financing stemming from a variety of innovation system actors
- Decentralised control with a large number of agents involved across many value chains and innovation ecosystems
- The direction of technical change is influenced by a wide range of actors, including government, private firms, and consumer groups
- Emphasis on the development of both radical and incremental innovations in order to permit a large number of firms to participate
- Diffusion of the results is a central goal and is actively encouraged
- The mission is defined in terms of economically feasible technical solutions to particular societal problems
- Complementary policies vital for success and close attention paid to coherence with other goals

New mission-oriented innovation policies perceive further of a process-oriented nature. Setting a mission not only implies the definition of goals and strategies for their implementation – it also requires effective policy coordination at each level of public decision-making: the state, the policy, and the administrative (Kattel and Mazzucato 2018).[4] The way that innovation activities are organised (in terms of not only policy-mixes but also the processes that coordinate the network of actors) eventually determines the policy outcome (see also Ergas 1987).

On the other hand, the new mission-oriented innovation policy can be viewed to some extent as reconcilable with previous innovation policy models: To mention but one example – policy initiatives taken under the umbrella of the new mission-oriented innovation policy bring together multiple actors from both the public and private sector, and this multi-actor nature is also at the heart of the innovation policy model of cluster and network orientation. However, what makes the new mission-oriented innovation policy special is its ground-breaking nature, as it is designed to act as a catalyst for transformative change of the whole socioeconomic system or parts thereof.

Despite a growing body of literature published on topics related to the new mission-oriented innovation policy – primarily empirical studies and case studies[5] – there are blind spots with regard to contributions to this field of research that aim at conceptualising specific aspects of new mission-oriented innovation policy or providing an analytical framework for its study. There are a few exceptions to the rule. After a conceptual and theoretical overview on mission-oriented innovation policies, Kattel and Mazzucato (2018) argue that innovation policy needs to turn its back on the "support and measure approach" and instead follow a "lead-and-learn approach" (ibid: 797). Thus, to be successful, it requires a set of "dynamic capabilities" from the public sector, including leadership and appropriate mechanisms for incentivising private actors to engage in innovation. Based on this article and other work of theirs, Mazzucato et al. (2020) develop a new policy framework ("ROAR") that should help policymakers to effectively engender new mission-oriented innovation policy. Along similar lines, Cantner and Vannuccini (2018) introduce the notion of "Schumpeterian catalytic R&I policy": building on Mazzucato's (2013) argument

4 This is also echoed in the innovation system literature, where the organisation of technology development and diffusion among a diverse set of actors is centre stage (Schot and Steinmueller 2018).

5 Most recently see, for instance, Cappellano and Kurowska-Pysz (2020); Klerkx and Begemann (2020) or Deleidi and Mazzucato (2021).

in favour of an entrepreneurial state and integrating some further key elements of the new mission-oriented innovation policy, the authors present a conceptual framework that accounts for both the directionality and intensity of innovation and sets up a new typology of new mission-oriented innovation policy. To mention a few further examples, Hekkert et al. (2020) advance in a policy brief the idea of "mission-oriented innovation systems", which connects this current innovation policy model with various systems of innovation approaches and offers to both analysts and policymakers a further option for the study of new mission-oriented innovation policies. Moreover, Borrás and Edler (2020) conceptualise the various roles that the state plays in different governance contexts under new mission-oriented innovation policy, with a specific focus on its transformative nature. Most recently, Wanzenböck et al. (2020) have attempted to improve analytical clarity about the plurality and complex nature of societal challenges. In order to deepen the understanding of possible designs of new mission-oriented innovation policy, the authors define three different trajectories that policy may follow to deal with grand societal challenges, distinguishing between problem-led, solution-led, and hybrid pathways.

We add to the existing literature in the following way. In the next section, we present a conceptual framework for capturing the dual aspects of new mission-oriented innovation policy. As a catalyst for transformative change, it aims to facilitate the development and diffusion of new technological breakthroughs powerful enough to tackle current societal challenges. On the other hand, the high uncertainty involved in the emergence of disruptive technologies makes innovation policy itself subject to change. Thus, new mission-oriented innovation policy needs to be understood as an endogenous, dynamic process, not only inducing but also co-evolving with its technological counterpart.

Conceptualising new mission-oriented innovation policy

For deriving our conceptual framework, we draw on literature mainly in evolutionary and innovation economics. The point of departure is the distinction between "physical and social technologies" introduced by Nelson and Sampat (2001). Originally, the basic unit for this distinction was a routine as "a collection of procedures" that is rationalised in the specific context in which it is applied (Nelson and Sampat 2001: 42; see also Nelson 2002: 267). Routines can be dismantled into one part that involves "a recipe that is anonymous regarding any division of labor" (labelled as "physical technology") and another that consists of "a division of labor plus a mode of coordination" (a "social technology"), which is described as the way in which a routine is structured and coordinated (Nelson and Sampat 2001: 44).[6] Though initially developed at the firm level, the concept of social technologies can be extended to other levels of analysis, and as Nelson (2002: 268, emphasis added) remarks, this "is broad enough to encompass both ways of organizing activity within particular organizations . . . and ways of transacting *across* organizational borders". In the following, we will frame the notion in the context of the new mission-oriented innovation policy and define a policy-based social technology as practice devised by the state that includes a set of specific measures to organise innovation and to coordinate activities between multiple actors involved in the development and diffusion of a new physical technology.

The conceptualisation of innovation policy as a social technology offers a straightforward framework of analysis, namely along the core characteristics of technology itself. Recently, Carroll (2017) approached the term from a rather holistic perspective, defining technology as

6 See also Nelson (2002), Nelson (2005), Ch. 7, and Nelson (2008) for a further discussion of "physical and social technologies".

"something that is *organized* (implying creation of order) whose aspects *function* with a *purpose* that can provide some *benefit*" (ibid.: 6). In contrast to former definitions, the author stresses the metaphysical feature: it is "a significant beneficiary of rationally-derived knowledge that is 'used for' a purpose, without itself necessarily being translated into something physical or material that 'does'" (ibid.: 18).

Each of these core features – purpose, function, and benefit – can be contextualised to the concept of new mission-oriented innovation policy. In this regard, policy initiatives that aim at transformative change can be designed (evaluated) by specifying (investigating) their purpose, function, and benefit, that can be linked to each other. Through such feedback effects, these policy-based social technologies become a dynamic and endogenous process. In the following, each of the components and links is briefly outlined. The corresponding framework is illustrated in Figure 23.1.

1 **Purpose.** Targets the overall question of "how to use policy to actively set the direction of change" (Mazzucato 2018: 810). It therefore states *why* these measures are taken. This includes both *directionality* (as the broad vision of *what* to achieve in the future) and the specific *mission* (*how* to achieve it in the presence). The latter can be defined in terms of goals that show the direction of the policy. New mission-oriented innovation policy overall follows two goals: market creation and market shaping (see, e.g., Mazzucato 2016 and Schot and Steinmueller 2018). In order to induce the desired transformative change, missions should, according to Mazzucato (2018: 812–813), be framed in such a way that they (1) are bold and inspirational, with wide societal relevance; (2) have a clear direction – they

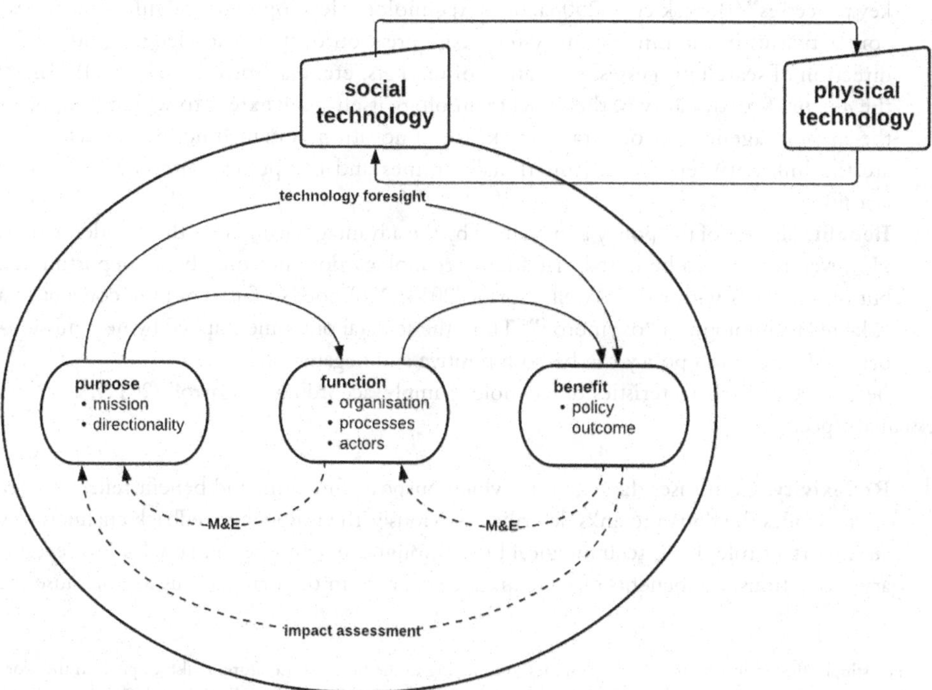

Figure 23.1 Conceptual framework of policy-based social technologies
Source: Authors' illustration

are targeted, measurable, and time-bound; (3) involve ambitious but realistic research and innovative actions; (4) are cross-disciplinary and cross-sectoral, and enforce cross-actor innovation; and (5) offer multiple, bottom-up solutions. Once the overall directionality and mission are defined, the different goals have to be translated into concrete strategies that can be implemented by the various actors in the innovation system through coordinated action (creating a link from *purpose to function*; see Figure 23.1). These strategies are ideally informed by measures of *technology foresight* (*purpose to benefit*).

2 **Function.** Captures all aspects linked to the operation of the social technology: organisation, process, and actors. Transformative change requires multiple actors from different institutional backgrounds (companies, research institutes, universities, third-party organisations (such as NGOs), and other representatives of civil society). Policy coordination needs to work towards creating dynamic links between these actors, while also focusing on the internal organisation of the initiative and nurturing experimentation and institutional learning (Cohen and Levinthal 1990). This form of organising policy has been dubbed "tentative governance" by Kuhlmann and Rip (2014, referenced in Schot and Steinmueller 2018): provisional, revisable, dynamic, and open, and including experimentation, learning, reflexivity, and reversibility. Putting the mission into action is often a "serendipitous" process (Mazzucato 2018: 809). To do this successfully, it needs an agile organisational structure and sound strategic and operational management. In the face of uncertainty, these tasks require that policy actors are equipped with "dynamic capabilities" (see Kattel and Mazzucato 2018), i.e., not only the skills necessary to execute a prearranged plan but also the expertise to overcome unpredictable hurdles and adapt to an ever-changing environment.[7]

Together with the overall mission, these *operational* functions enable a "second level of key processes" (Bergek et al. 2008a: 6) for technology development and diffusion. These core activities in the innovation system, e.g., creation of new knowledge, guiding the direction of search processes, formation of markets, etc. (Jacobsson et al. 2004),[8] shape the *directive* functionality of the social technology itself– the extent to which it supports the mission agenda. In our framework, the innovation system functions directly create the link between the organisational routines and the policy outcome (*function to benefit*).

3 **Benefit.** Success of the policy as measured by the advantages gained for the broader society. However, it needs to be noted that a new technology does not only bring opportunities, but often comes with risks as well. Floridi (2014: 206) goes so far to call the concept of a risk-free technology an "oxymoron". Thus, the societal outcome implied by new mission-oriented innovation policy can be both positive and negative.

These three core characteristics of technology imply, according to Carroll (2017), a further essential aspect:

4 **Reflexivity.** Comprises the extent to which purpose, function, and benefit reflect *on each other*. Besides the dynamic links described previously, the following feedback channels play an important role. First, technological breakthroughs may unintentionally lead to secondary innovations, the benefits of which induce changes in the original mission or cause the

7 Interestingly, the notion of dynamic capabilities, originating in the business literature, is also implicit in the word technology; see the interpretation of the Greek word *techne* as both "skill" and "art" in Carroll (2017).

8 Note that in systemic approaches to innovation, a functional analysis of the innovation process has a long tradition and several definitions – though ambiguous – as well as different categorisations of functions have been raised. See, e.g., Hekkert et al. (2007) or Bergek et al. (2008b).

launch of entirely new missions.[9] In order to prevent negative benefits from the development and use of a technology, continuous *impact assessment* is crucial in order to ensure that outcomes align with the directions and the mission (*benefit to purpose*). Reflexivity furthermore refers to the feedback effects between agency (actors) and system (policy). These effects include, second, the capacity of actors to reflect on the outcome (*benefit to function*), warranting changes in the functioning (organisation) of the technology. Third, this ability for the structure to adapt is in turn an instrumental factor for the self-governance of transformative change (Weber and Rohracher 2012; Schot and Steinmueller 2018) (*function to purpose*). Both of these links are generated by *monitoring and evaluation* (*M&E*) tools. Finally, reflexivity also manifests in the mutual effects between social technology and physical technology that drive their co-evolution, where "co-evolving" means that there is some channel of "causal influence" between them, as Foxon (2011: 2262) stresses.

In summary, departing from the concept of social technologies, we showed that the core features of technology can be mirrored in the key elements of new mission-oriented innovation policy – mission and directionality, organisation and process, and outcome. The element of reflexivity inherent in technology further allows the recognition of new mission-oriented innovation policy as a dynamic and endogenous process. The differentiation between social technologies and physical technologies, last but not least, facilitates the understanding of how policy and institutional change shape technological trajectories and vice versa. It is important to note that policy-based social technologies, similar to meta-technologies (Floridi 2014), operate *on* and not *with* physical technologies; they do not directly interact with their physical counterpart but shape the environment in which it is developed or used.

The conceptual framework laid out so far is general enough to inspect missions of any scope and can imply all types of policy instruments (hard, soft, supply-, and demand-sided). In the next section, we will illustrate the concept on a well-defined policy initiative based on a technology-specific coalition.

ECSEL JU as a policy-based social technology

In the following, we apply the proposed framework by investigating ECSEL JU – the European Union's joint undertaking for electronic components and systems. This policy programme was selected for various reasons:

1 JTIs have been established by the European Commission in the form of public–private partnerships that focus on transnational research collaboration. They have the goal to leverage pioneering future key technologies, boosting their development across the whole innovation cycle and increasing the international competitiveness of selected industries. This form of long-term cooperation between actors from the private and public sector make JTIs in general an interesting case of a policy-based social technology targeting transformative change.
2 The special field of electronic components and systems is not only relevant for the digital transformation but also crucial for driving the sustainability agenda; for instance, the miniaturisation of embedded systems is a prerequisite for the diffusion of renewable energy

9 An example for this need of dynamic mission planning is the concept of drug repurposing, where existing drugs are used for the treatment of new diseases.

sources such as solar and wind technology as well as for the development of alternative energy carriers (e.g., hydrogen).

3 The strong dynamics involved in the policy initiative ECSEL JU in terms of both policy organisation and outcome allow showcasing the (co-)evolution of a policy-based social technology and its physical counterpart.

Regarding research method, the case study is based on the analysis of secondary data, including inter alia policy declarations, strategy papers, mission statements, research and innovation plans, and project assessment reports. To supplement the qualitative study with quantitative data and explore the element of agency in this policy initiative, we make use of CORDIS, which is a freely accessible database from the EU Open Data Portal (2012). The database contains detailed information on the projects funded by ECSEL JU under Horizon 2020 (FP8) as well as on the actors involved.[10] This data is used for analysing research, development, and innovation (R&D&I) collaboration within the scope of ECSEL JU. We find this approach useful, as it allows the detailed presentation of the interaction patterns in this policy initiative.

Before discussing ECSEL JU as a policy-based social technology – i.e., in terms of its purpose, function, and benefit – we will sketch the history, rationale, and structure of this JTI.

A chronology of key events that led to ECSEL JU

Though micro- and nanoelectronic components and systems – as the physical technology – have been present in the European Union policy agenda since the 1970s, we identify three specific key events that mark the trajectory towards the foundation of the complementary policy-based social technology ECSEL JU in 2014.

Key event no. 1: the European Union declaration on key enabling technologies. In 2009, the EC declared micro- and nanoelectronics, advanced materials, photonics, nanotechnology, and biotechnology as key enabling technologies (KETS). KETS dispose a generic character, and micro- and nanoelectronics were assigned to KETS due to their vast application potential, as they are

> essential for all goods and services which need intelligent control in sectors as diverse as automotive and transportation, aeronautics and space. Smart industrial control systems permit more efficient management of electricity generation, storage, transport and consumption through intelligent electrical grids and devices.
>
> *(EC 2009: 4)*

As such, the *KETS declaration* was already mission-oriented, emphasising not only the rate of innovation but also and especially its directionality. It aimed at setting in motion "a process of identifying the KETS that strengthen the EU's industrial and innovation capacity to address the societal challenges ahead" (ibid.: 3). By means of the *KETS declaration*, the EC provided a unified definition of these generic technologies and created a common basis for the organisation of innovation in micro- and nanoelectronics at the European level and across member states.

Key event no. 2: the European strategy for key enabling technologies. To launch the development of KETS in a coordinated way, it was planned within the KETS declaration to initiate a "high level expert group", that is, among other things, responsible for devising a common long-term

10 Through the ECSEL JU webpage which lists all projects and the variable "topics" included in the data file on FP8 projects, we could extract the single ECSEL projects and the corresponding organisations involved (see www.ecsel.eu/projects).

European strategy for KETS (EC 2009: 11). We identify the publication of this *European strategy for KETS* in 2012 as an important act, paving further the way towards the foundation of ECSEL JU. Its realisation should "allow maximum exploitation of the EU's potential in competitive markets" (EC 2012: 2) in these generic technology fields, and its overarching aim was to "create synergies between EU policies and instruments and ensure coordination of EU and national activities" (ibid.: 6).

Key event no. 3: the European strategy for micro- and nanoelectronics. A more specific policy strategy that solely and explicitly addressed the development and deployment of micro- and nanoelectronics was the introduction of the "European strategy for micro- and nanoelectronic components and systems" in May 2013 (EC 2013a). The aim of the *micro- and nanoelectronics strategy* was "for Europe to stay at the forefront in the design and manufacturing of these technologies and to provide benefits across the economy" (ibid.: 3), and its focus was three-fold: (1) to raise funding and attract investment in micro- and nanoelectronics in order to establish a European strategic technology roadmap; (2) to set up an EU-level mechanism that integrates and concentrates the support of innovative activities in micro- and nanoelectronics by member states, the EU, and the private sector; and (3) to implement measures that increase Europe's competitiveness and contribute to a level global playing field with regard to state aid, business development and SMEs, and the existing skills gap in this technology field (ibid.: 2).

Foundation of ECSEL JU as a public–private partnership in May 2014. Having reflected on the three key events that mark the trajectory towards ECSEL JU, its foundation can be directly attributed to action planned under both the European *KETS strategy* and the *micro- and nanoelectronics strategy*. While in the former it was recommended to establish a joint technology initiative as a public–private partnership, this idea is rendered more precisely in the action plan presented in the latter, stating that "the European Commission will propose a Joint Technology Initiative based on Article 187 TFEU that combines resources at project level in support of cross-border industry-academia collaborative R&D&I [for the area of micro- and nanoelectronics components and systems]" (EC 2013a: 10).

The implementation of the proposed actions led to the foundation of ECSEL JU in May 2014 by the EU Council Regulation No 561 2014 (EC 2014). ECSEL JU was planned to run for a 10-year period – from 2014 until 2024 – and it is organised within a top-down hierarchy, consisting of four bodies; these are (in descending order): the Governing Board, the Executive Director, the Public Authorities Board, and the Private Members Board.

We think that the organisational form of ECSEL JU as a public–private partnership especially engenders a long-run commitment from the entrepreneurial state which the development of disruptive key technologies in the field of micro- and nanoelectronics requires. Through engaging both public and private actors, ECSEL JU is deemed to create a new market where the private sector can subsequently enter. For the specific case of micro- and nanoelectronic components and systems, setting up a public–private partnership was legitimised by the EU as it

> represent[s] an innovative way of implementing the Union's research and innovation policy. [It brings] together the frontrunners in terms of research and innovation in the industrial sectors concerned and allow[s] them to focus and align their efforts around strategic research and innovation agendas.
>
> *(EC 2013b: 4)*

Given this brief introduction to ECSEL JU, in the following sections we will verify its key characteristics as a policy-based social technology.

Purpose – mission and directionality

As proposed in the conceptual framework, we presume that the first characteristic of a policy-based social technology – the purpose – can be contextualised to the mission and directionality of a policy initiative. The mission of ECSEL JU is laid out inter alia in the multiannual strategic plan, henceforth MASP (ECSEL JU 2019a:, 9–10), namely:

> to progress and remain at the forefront of state-of-the-art innovation in the development of highly reliable complex systems and their further miniaturisation and integration, while dramatically increasing functionalities and thus enabling solutions for societal needs.

Based on information gathered mainly from the current MASP (ECSEL JU 2019a), we disentangled the mission of ECSEL JU along the different criteria proposed by Mazzucato (2018) and found the following results:

1 **Cross-disciplinary, cross-sectoral, and cross-actor innovation.** In the MASP (ECSEL JU 2019a), five different application areas are considered to be relevant for micro- and nanoelectronic components and systems, including:

   i transport and smart mobility,
   ii health and well-being,
   iii energy,
   iv digital life, and
   v digital industry.

The different application areas and their detailed descriptions included in the MASP (ECSEL JU 2019a) set a broad vision and specify the direction in which the physical technology complementary to ECSEL JU should be advanced. Reflecting on the wide spectrum of application areas, this already warrants the conclusion that innovative activities as planned out in ECSEL JU are both cross-sectoral and require a cross-disciplinary approach. Beyond that, ECSEL JU brings together different types of actors and fosters cross-actor innovation; it is composed of representatives from multinational companies, members of the three industry associations AENEAS, EPoSS, and ARTEMIS, representatives from academia and research organisations, government advisors, and other public and private stakeholders from the European Commission, 25 EU member states,[11] as well as four countries associated with the European Framework Programmes for research and innovation, viz., Norway, Switzerland, Israel, and Turkey. As detailed further in the following, this multi-actor nature manifests in the operationalisation of this policy-based social technology, too.

2 **Targeted, measurable, and time-bound direction.** There is a list of eight broad objectives tied to ECSEL JU (see EC 2013c: 14 for this list), but there is one goal that stands out among the others, namely to achieve "technology leadership" in micro- and nanoelectronic components and systems. This is thus formulated in a measurable way and strives to better position Europe in the global technology race in this field vis-à-vis other

11 Croatia and Cyprus do not participate in ECSEL JU. As of January 1, 2020, the United Kingdom does not participate either.

Table 23.1 Expected societal benefits of innovation in micro- and nanoelectronic components and systems in the single application areas. Based on ECSEL JU (2019a)

Focus on . . .	*Environmental challenges*	*Demographic challenges*	*Infrastructural challenges*	*Safety and security-related challenges*	*Economic challenges*
Transport and smart mobility	x	x	x	x	x
Health and well-being		x	x	x	x
Energy	x		x		x
Digital life	x	x	x	x	
Digital industry	x		x	x	x

global key players, like China, Japan, or the US. Moreover, ECSEL JU is time-bound as its lifespan is limited to 10 years, and action taken under this policy-based social technology is subject to a clearly specified schedule, getting evident in e.g., the annual calls for project proposals as well as the multiannual strategic plan MASP, which includes inter alia a long-term perspective for investment taken under this policy-based social technology.

3 **Bold and inspirational with wide societal relevance.** The overall mission of ECSEL JU conceives of a wide societal relevance and is strongly framed in the need to tackle grand societal challenges and create benefits for society. To figure out this aspect, we assigned the objectives specific to the single application areas to five groups of societal challenges. As can be seen in Table 23.1,[12] most frequently, environmental issues and safety and security issues are addressed, but the focus is also on demographic, infrastructural, and economic challenges. Targets include, for instance, the reduction of CO_2 from mobility and transportation (environmental challenges), the creation of safe and secure spaces (infrastructural challenges), a reliable healthcare provision and the improvement of lifetime quality and well-being (demographic challenges), or a stable supply of energy (infrastructural challenges).

4 **Research and innovation actions that show a balance between ambition and feasibility.** Objectives pursued under ECSEL JU are ambitious – e.g., with regard to achieving technology leadership in micro- and nanoelectronic components and systems. Needless to say, risk is taken and innovative actors are challenged. However, action planned under this public–private partnership is set in an apparently feasible way as the R&D&I-activities necessary for reaching these objectives are precisely defined: without going into detail, it can be noted that these should cover a wide array of the innovation cycle – from technology development up to the deployment of new technology and close-to-market activities. In other words, action set out under ECSEL JU addresses all "technology readiness levels" – from 1 (basic research) up to 9 (system test and launch operation).

5 **Multiple, bottom-up solutions.** ECSEL JU provides different solution paths for achieving its mission. It builds on an integrative approach (see ECSEL JU 2019a: 18 ff), and the R&D&I-activities covered under this programme are focused along two axes. The first axis are five different technology capability domain arenas, including:

i systems and components: architecture, design, and integration,
ii connectivity and interoperability,

12 Note that the information shown in Table 23.1 is based on the MASP (ECSEL JU 2019a: 66–67, 95–96, 113–114, 132–134, 157–159).

iii safety, security, and reliability,
iv computing and storage, and
v ECS process technology, equipment, materials, and manufacturing.

The second axis is the five different application areas, i.e., transport and mobility, digital life, energy, health and well-being, and digital industry. By means of pursuing such an integrative approach, "projects of the ECSEL programme should not limit themselves to covering only one of these key applications or essential technology capabilities; on the contrary, multi/cross-capability projects will be encouraged wherever relevant" (ibid.: 19).

ECSEL JU is thus clearly framed with a mission and direction for change that accounts for its purpose as a policy-based social technology. Sharing a common vision and mission as outlined in the MASP increases the leverage of the members of ECSEL JU on global markets and helps with bundling resources regarding communication and lobbying activities and finding collaboration partners beyond their own intermediate network.

In a nutshell, laying out the strategy and future plans of ECSEL JU, the MASP serves as a tool for the agenda setting process: Linking the *purpose* of ECSEL JU to the *function* it should provide a coordinating mechanism for the activities involved in the implementation of this agenda.

Moreover, the MASP also includes a list of highly prioritised R&D&I-action for the different capability domain arenas and application areas. These are compiled by the industry associations involved in ECSEL JU, based on expert discussion, foresight studies, and the analysis of existing roadmaps. By showing the potential socioeconomic impact of specific technologies (thus drawing the direct link between *purpose* and (expected) *benefit*), this compilation early on nudges the innovation process in the desired direction.

Function – organisation, process and actors

ECSEL JU as a policy-based social technology serves several operational functions that are integrated in its organisational form as a public–private partnership. In the following, we link selected processes (related to R&D&I-funding, financial management and networking activities) directly to a set of key functions in innovation systems, as distinguished by Jacobsson et al. (2004: 6–7).[13]

The main functions of ECSEL JU are the *supply of resources* used for *knowledge creation* in the field of micro- and nanoelectronic components and systems. Innovative activities should span the whole innovation cycle, and the supply of resources is coordinated through launching a set of calls for project proposals that are published on an annual basis. Interested innovative actors may apply for such R&D&I-projects through a competitive procedure. The topics of these calls are given, such that the *direction of search processes* is influenced a priori. According to the MASP (ECSEL JU 2019a), initially a total budget of 5 billion Euros was planned for the duration of the whole work programme, where actors from the public and private sector so far have contributed almost half–half. Sharing financial expenses in this way ensures that the entrepreneurial state de-risks on innovation, and, on the other hand, that private sector actors are also obliged and incentivised to engage in R&D&I-activities. To handle the supply of resources and organise project calls, ECSEL JU is connected to the European Framework Programme for Research and Innovation, and there are two types of actions that relate to ECSEL: so-called innovative

13 This list includes the following five functions: (1) the creation of knowledge, (2) influencing the direction of search processes, (3) the supply of resources, (4) the creation of positive external economies, and (5) the formation of markets.

Table 23.2 Key figures of R&D&I-efforts under ECSEL JU/Horizon 2020 in the single application areas

	Number of projects	*Project costs (in million Euro)*	*Actors*
Transport and smart mobility	21.57	917.527	687.1
Health and well-being	5.98	295.082	161.35
Energy	7.98	248.715	272.18
Digital life	6.9	314.719	193.93
Digital industry	24.57	1,603.103	873.3
Total	*67*	*3,379.148*	*2,188*

Data source: EU Open Data Portal (2012). Authors' calculations

actions (IA) and research and innovative actions (RIA).[14] Funding rates for the different calls of projects coming from the EU are not fixed but depend on the type of action and the type of actor that applies for it. By the same token, the funding rates that national governments contribute are variable across countries (and even regions). On top of that, ECSEL JU is assigned with another function: it should bring together various actors from different fields and in this regard is expected to fulfil the *creation of positive external economies*. Furthermore, the support of networking between innovative actors such as to enable spill-over effects as well as reduce transaction and information costs takes a central role under this policy initiative. ECSEL JU also puts new technology trends on a coherent institutional basis and through this embarks on the issue of uncertainty involved in innovation.

These core functions in innovation systems should equip ECSEL to act as a mechanism of both market-shaping and market-creating, ultimately creating the link between *function* and *benefit*. Operating the policy-based social technology in turn involves policy coordination as well as agency from the side of actors involved.

Shedding light on the operation of ECSEL, we found the following: to serve the functions of supply of resources and knowledge creation, ECSEL JU provides direct support to R&D&I-activities, which is realised through a competitive selection procedure as previously mentioned. As to some basic figures shown in Table 23.2, under Horizon 2020 between 2014 and 2019, 67 projects have been funded under 22 different calls. Altogether, 2,188 actors[15] were engaged within these projects, either as coordinators or as participants. Total project costs amounted to 3.379 billion Euros, and as this high volume of total project costs and EU-funding shows, projects under ECSEL JU are really "the big fish". Regarding application areas, most projects are carried out in the fields of digital industry and transport and smart mobility, while health and well-being and digital life are only niche fields. Digital industry as well as transport and smart mobility also present the highest share of project costs.[16]

14 Apart from this two-fold categorisation into IA and RIA, there is another category, so-called lighthouse initiatives, that address subject areas of common European interest. See Horizon 2020 Online Manual (n.d.) for details.

15 Note that for sake of completeness we also include those actors that have terminated their participation, which is a share of about 5% of all actors. Note also that for our project sample, including projects between 2014 and 2019, actors from the United Kingdom are accounted for. Note also that Bulgaria, Malta, and Estonia are not engaged in any project of our project sample.

16 Forty-three projects were assigned to the different application areas, while we had to classify the remaining 24 projects ourselves. When classifying the projects to the application areas, we consulted the CORDIS database and used the project description as a starting point. Moreover, we – the authors – assigned the projects independently from each other. In a few cases, there were deviations in the results of our self-assignment. We then discussed these single cases and, on this basis, revised the assignment. If unclear, we left out ambiguous

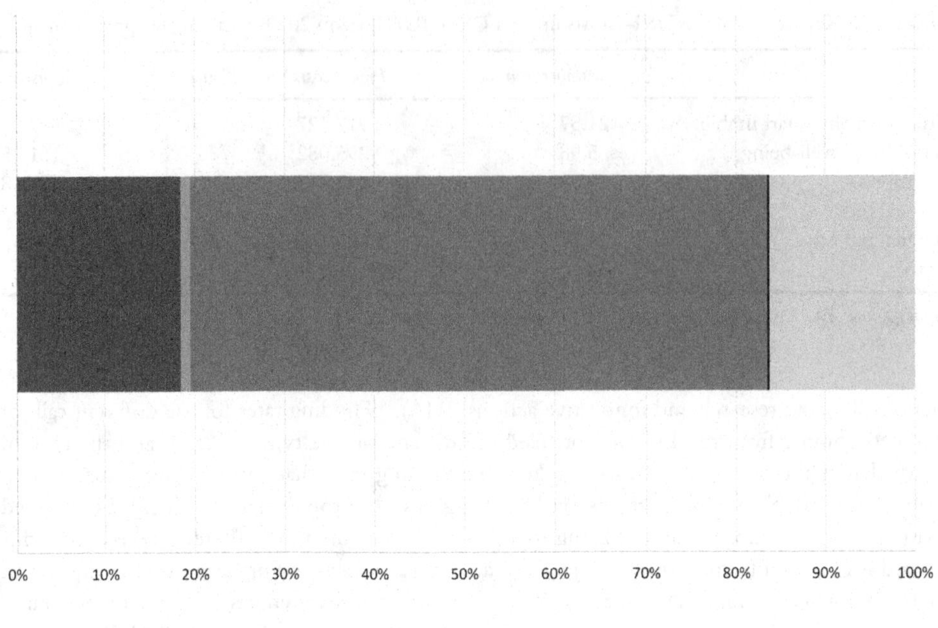

Figure 23.2 Distribution of organisational types in project-based R&D&I-collaboration under ECSEL JU/Horizon 2020. Note the following abbreviations: HES: higher or secondary education establishments; OTH: other; REC: research organisations; PRC: private for-profit entities (excluding higher or secondary education establishments); and PUB: public bodies (excluding research organisations and secondary or higher education establishments). Data source: EU Open Data Portal (2012). Authors' illustration

Irrespective of the application area, each R&D&I-project on average involves more than 30 actors working together and sharing some common goal: to advance knowledge in the respective research topic covered by a specific R&D&I-project. Such R&D&I-projects are thus a textbook example of collective patterns of innovative action. Taking a closer look at Figure 23.2, the distribution of actors involved in R&D&I-projects financed under ECSEL JU reflects the mixed-actor nature of this mission-oriented policy initiative. It shows that the strongest organisational type is private for-profit entities (PRC), followed by higher and secondary education establishments (HES). Notably, actors from the private sector are most frequently represented, indicating that the "entrepreneurial state" sets the organisational framework conditions for innovation and works towards creating the dynamic links between actors under this policy initiative, whereas it is predominantly actors coming from the private sector who in turn engage in R&D&I-activities.

Almost half of the actors (49.8%) were involved in at least two projects, indicating that key processes evolving out of the operation of ECSEL JU are driven by a rather limited set of actors. Such intense interaction of actors across more than one project may facilitate learning and create positive spillovers as well as decrease information and transaction costs. On the other hand,

application areas and decided to better assign projects to fewer (rather than more) application areas. For projects with multiple application areas, we assumed equal weight; thus, if a project has n application areas, then it is counted as $1/n$ in each of them.

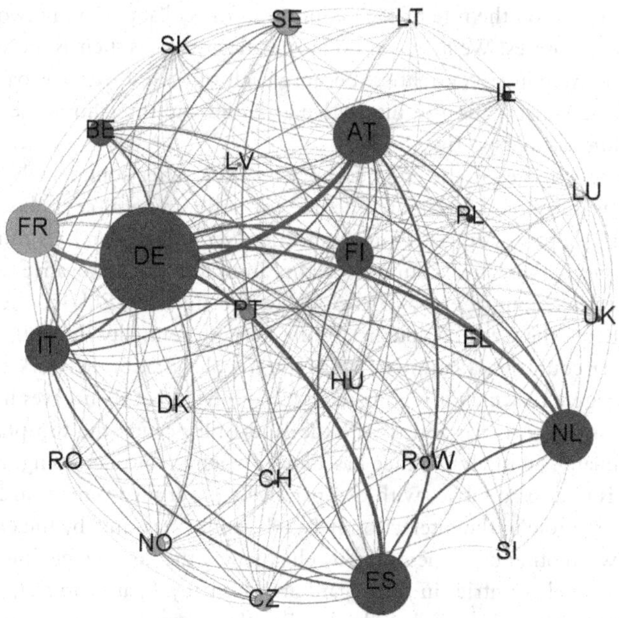

Figure 23.3 Cross-country collaboration network under ECSEL JU (2014–2019). The size of nodes is determined by strength centrality, while the colour represents betweenness centrality; the opacity and width of edges is determined by edge weights. Country abbreviations are according to the ISO code. Note also that we applied a specific version of a force-directed graph layout algorithm developed by Fruchterman and Reingold (1991). Authors' illustration.

this bears the risk that some picking-the-winner structure establishes – instead of "picking the willing", as pursued usually under new mission-oriented innovation policy (Mazzucato 2018).

To finally get an idea of how the collective patterns of action under the operation of ECSEL JU were distributed across participating countries for our project sample between 2014 and 2019, we map the cross-country collaboration network of ECSEL JU and, by means of network analysis, describe its distinct features. Counting for each R&D&I-project all cross-country linkages between actors involved and then aggregating them to the country level,[17] we derive a network with 41,076 (multiple) edges between the 27 countries which account for actors involved in ECSEL projects and one rest group ("RoW"), including project participants the host countries of which are not members of ECSEL JU.[18] Hence, current R&D&I-action under ECSEL JU spans a fairly large network. This multigraph can be represented by a weighted undirected graph, with edges showing the strength of cross-country collaboration between any two countries (see Figure 23.3 for a visualisation). With a network density of 0.89, cross-country cooperation is rather diverse – most of the countries are connected to a multitude of other

17 See, e.g., Amoroso et al. (2018) for a similar approach. For instance, for a project consortium consisting of three actors from different countries, we count three different edges and for one with three actors, where two are from the same country and the third from another one, we count two different edges as we do not account for self-loops and thus ignore collaboration within a country.

18 We refer the reader again to the fact that actors from the United Kingdom are accounted for in the cross-country R&D&I-collaboration network and that Bulgaria, Malta, and Estonia account for neither a participating nor a coordinating actor in our project sample.

countries. To shed light on the role of each country in the collaboration network, we calculate different centrality measures. We focus on node-specific attributes such as the weighted degree, eigenvector, and betweenness centralities, because each of these metrics is based on a specific aspect characterising collaboration in innovative activities and thus shows the importance of a country from a different perspective.

The 27 countries engaged in the cross-country network differ considerably in their intensity of collaboration, as measured by weighted degree (or node strength) centrality. Germany ranks highest, followed by Spain, Austria, France, and the Netherlands, and a strong cross-country collaboration between those five countries' actors indicates that innovative activities tend to cluster. All countries among the top 10 belong to the (former) EU-15, the highest ranked EU-13 member being the Czech Republic. While strength centrality shows the extent of direct linkages with other countries, a country's status in the collaboration network can also be characterised by the degree to which it is connected to countries that themselves have a high score. While Germany, Spain, Austria, and the Netherlands still occupy the top positions (together with Belgium, Finland, and Italy), France has a lower eigenvector than strength centrality score, signifying that it is tied to countries with comparatively lower involvement in ECSEL JU. Furthermore, a country's role in the interaction network can be measured by the extent to which it lies on a path between other countries, thereby channelling the knowledge flows between them. Again, the most central countries in this regard are Germany, Spain, Austria, and the Netherlands, thereby acting as "gatekeepers" or "brokers" in the network when it comes to the transfer of information. However, four other countries – Portugal, Hungary, the UK, and Denmark – have increased their ranking position (relative to the other centrality metrics) and therefore display an important role in bridging knowledge between members in the collaboration network.

Finally, we also calculated an edge-based metric, the so-called edge betweenness centrality, which measures the importance of the linkages themselves. It is defined by the number of weighted shortest paths between any two nodes in the network. Accordingly, an edge with a high edge betweenness score is crucial for the network structure, since removing it would potentially lead to a loss of interaction between different parts of the network (see Girvan and Newman 2002 as well as Yoon et al. 2006, whose algorithm we used for our calculations). The metric highlights again the central position of Germany: almost all its ties to other countries score the highest edge betweenness. The (by far) highest score in this regard represents the connection between Germany and Italy, which is an especially important bridge to Latvia. Another key link is between Germany and Spain, as this edge builds a key connection to Slovenia. Without these (indirect) ties, both Latvia and Slovenia would have a less robust position in the collaboration network. Drawing the link between *function* and *benefit*, we have shed light on the operation of ECSEL JU and the agency this involves. In the next section, we finally explore the third defining characteristic of a policy-based social technology – the (potential) benefit and outcome.

Benefits and outcome

Since ECSEL JU is still running, it is way too early to reflect on the advantages gained for the broader society. Yet, it definitely makes sense to think about the different alternative directions of development that become feasible through operating this policy-based social technology on its physical counterpart. This seems a promising route insofar as we get a first rough idea of the trajectory that direction setting under ECSEL JU implies and the causal influence between the policy-based social technology and physical technology. To trace this latter aspect, we take a closer look at knowledge creation under the scope of R&D&I-projects financed under ECSEL JU: we retrieved from the CORDIS database for the single R&D&I-projects the fields of

Table 23.3 Direction shaping under ECSEL JU/Horizon 2020 in terms of fields of science involved in knowledge creation in the single R&D&I-projects and their respective application areas. Note that fields of science are classified according to EuroSciVoc-taxonomy and the multiple mentions are possible for a single project such that there may be more matrix entries than projects in total. Authors' illustration.

	Agricultural sciences	*Engineering and technology*	*Humanities*	*Medical and health sciences*	*Natural sciences*	*Social sciences*
Transport and smart mobility	2	17	2	0	18	22
Health and well-being	2	7	4	2	7	11
Energy	0	8	2	0	11	11
Digital life	2	8	3	0	9	9
Digital industry	2	11	1	1	24	20

science they are assigned to and linked them in a further step – via project descriptions – to the single application areas. For each project, we counted the different fields of science mentioned, and then connected them to the single application areas.[19] Results are shown in Table 23.3, having as its rows the fields of science and as columns the different application areas. Note further that entries in Table 23.3 reflect the number of mentions.

While calls for projects in the technology capability domain arenas and the application areas identified under ECSEL JU narrow the directionality of actors' R&D&I-activities, it is the actors themselves – their knowledge, skills and experience – who are setting the new pathways for evolution.

What we see at first glance from Table 23.3 is that the matrix is not sparse but has many non-zero entries. Put differently, R&D&I-projects carried out in the different application areas are related to a rather diverse set of fields of science, indicating that micro- and nanoelectronic components and systems are a generic technology field. At the highest hierarchical level (i.e., level 1), surprisingly, it is social sciences, followed by natural sciences and engineering and technology, that are most frequently mentioned within the single application areas. On the contrary, the remaining three level-1 fields of science (agricultural sciences, medical and health sciences, and humanities) are of lesser importance in the process of knowledge creation in this technology field. The dominance of social sciences indicates that the physical technologies being developed are already in a higher stage of the innovation cycle and that collective patterns of action include close-to-market innovative activities.

Altogether, the direction shaping under the single R&D&I-projects is rather broad in its scope: it involves a diverse set of different fields of sciences, and knowledge creation in R&D&I-projects dedicated to the different application areas requires multidisciplinary know-how.

Discussion

So far, we have discussed the main characteristics of ECSEL JU along the core attributes of a policy-based social technology. Figure 23.4 presents a summary account of the case study.

19 Note that for 10 projects fields of science were unavailable, and, different to application areas, it was not possible based on the available information to assign them to the fields of science.

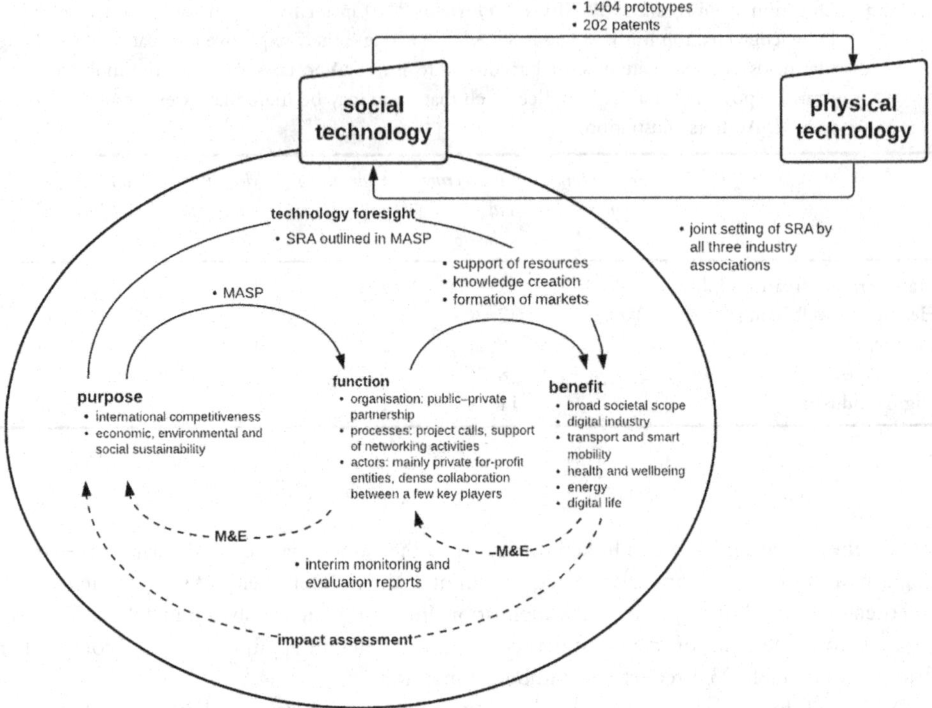

Figure 23.4 Application of the policy-based social technology framework to ECSEL. Note: SRA stands for Strategic Research Agenda. Authors' illustration.

In a nutshell, the *purpose* of ECSEL laid out in its overall mission is to support the industry and research organisations to remain internationally competitive, but it also explicitly reflects on the sustainability agenda. The *function* of ECSEL encompasses not only the various activities and means through which it operates (viz., its organisation as a public–private partnership based on a unique three-party funding [EU, specific member states, and industry]), but also how it organises itself (the funding procedure through project calls) and how it impacts the innovation ecosystem in micro- and nanoelectronic components and systems (illustrated by the collaboration network). As to the *benefits*, while the 67 projects financed under ECSEL have a broad societal scope, application areas such as digital industry or transport and smart mobility feature especially prominently.

These components are connected through dynamic links that channel the feedback effects between one characteristic and another: The MASP operationalises the overall mission of ECSEL JU by defining multiannual targets and strategies (*purpose to function*), whereas its strategic research agenda (SRA) is based on future-oriented technology studies that assess the potential socioeconomic impact of specific innovations to shape the direction corridor of research (*purpose to benefit*). The organisational design of ECSEL regarding both processes and actors influences the extent to which the structural functions – in particular financial support, knowledge creation, and formation of markets – can unfold their effect on the innovation system (*function to benefit*).

While our chapter primarily focused on these aspects of social technology, for the sake of completeness we will briefly remark on the remaining links here as well.

ECSEL managed to generate an enormous innovation output during its time of operation. According to ECSEL JU (2019b),[20] which accounts for 51 projects of our sample of 67, within their scope 202 patents were generated and 1,404 technology prototypes developed. These indicators are only two of a multitude of measures that potentially assess the impact of this policy-based *social technology* on the *physical technology*. Conversely, recent trends in digital technologies also led to changes in the social technology. For example, the merging of three JUs – AENEAS, ARTEMIS and EPoSS – into one ESCEL JU was a direct response to the increased interlocking of the different subtechnologies – micro- and nanoelectronics, smart integrated systems, and cyber-physical systems. To better suit this technological trajectory, the forces of the private sector were bundled in one strategic partnership, showcasing how organisations and institutions change subject to technological development (*physical to social technology*).

Reflexivity is furthermore considered in the various concepts for monitoring the effectiveness, efficiency, research quality, and transparency of the initiative. For monitoring projects and financial management, ECSEL JU uses the ICT tools developed by the European Commission (ECSEL JU 2019b). Interim evaluation reports (see, e.g., ECSEL JU 2017) and other assessment tools have constantly fed back to the strategic outline as well as the organisation of ECSEL, leading inter alia to a stronger focus on the vertical integration chain (*benefit to function*), a better integration of SMEs and start-ups into the joint undertaking, the alignment of national activities, and the harmonisation of participation procedures (*function to purpose*), as well as the development of appropriate metrics for impact assessment (ibid.). The latter is partly reflected in an assessment report that recommends various tools for measuring the impact of ECSEL at the technological, ecosystem, social, economic, and environmental level (Idea Consult 2018). Proposed measures include the evaluation of quantitative data, conducting surveys among and interviews with participants, as well as the analysis of the collaboration network. This is necessary to fully apprehend the wide-reaching effects of this joint undertaking.

To conclude this section, a few observations regarding the reflexivity performance of ECSEL JU – based on our analytical framework – are warranted.

ECSEL JU includes governance aspects, technology foresight (as a form of *up*stream governance, guiding the broad research direction), and impact assessment (as a form of *down*stream governance, informing on the risks and value-added of specific policy projects) (see, e.g., Fisher et al. 2006). This structure itself creates a feedback mechanism that supports the co-evolution of the social and physical technology. When it comes to disruptive innovations, however, it is also important to cover the *mid*stream level by means of an alternative governance form dubbed "innovation assessment" by Hasselbalch (2018): "Before 'technologies' become 'impacts', there is a period of 'innovation', where technologies have taken specific, marketable forms, but are still open to a number of different trajectories of onward development" (ibid.: 4). The salient features of radical breakthroughs and the uncertainty inherent to their development and deployment requires the continuous and systematic evaluation of projects in order to identify technological bottlenecks and steer innovations into the desired societal direction. The institutional architecture of such an assessment ideally involves a broad range of stakeholders that goes beyond the ECSEL JU network, also including civil society. Furthermore, it could be established as an overarching evaluation tool for all the JUs set up by the European Commission to allow a more orchestrated governance of innovation activities. As Hasselbalch (2018) emphasises, such a framework needs to be flexible enough to respond quickly to new technological

20 According to ECSEL JU (2019b), these figures should be interpreted cautiously.

and socioeconomic developments and thus complements – not replaces – more institutionalised evaluation regimes such as impact assessment.

Furthermore, the analysis of the cross-country collaboration network also sheds light on the geopolitical aspect of large-scale policy projects: The central positions of a few Western European countries show the unequal allocation of financial resources and knowledge among its member states. This is owed by default to the varying degree of (financial) engagement of states and industries in this public–private partnership and fostered by the competitive organisation of the innovation process through project calls. While the analysis of the cross-country collaboration network has revealed no particular clusters, there are still countries that run the risk of becoming isolated, as they share fewer ties with the key player, Germany. It is therefore paramount to understand how policy initiatives like ECSEL JU also shape power relations among countries and amplify the status of individual actors in the collaboration network.

Concluding remarks

Re-interpreting the notion of a social technology as the specific aspect of a routine that helps to direct, coordinate, and organise the innovation process among different actors in a limited technology space, we proposed the concept of policy-based social technologies for understanding new mission-oriented innovation policy, targeting transformative change for tackling grand societal challenges.

As a policy evaluation tool, it provides an analytical roster for the study of specific policy initiatives. This was illustrated on the example of ECSEL JU. It is important to note that this chapter only reflects on the first step towards theory-building. Nevertheless, we think that the framework proposed here contributes to the extant literature in the field of innovation research insofar as it shifts the focus from purely regulation-oriented to interaction-oriented policy mixes, the latter having been addressed insufficiently. Aspects of agency and interaction of stakeholders become more and more important, considering the complex nature of today's innovation process. As Mazzucato (2013: 193) stresses in context of the multi-actor nature of innovation processes: "it is now more important than ever to understand the division of 'innovative' labour between the different actors . . . and in particular, the role and the commitment of each actor". The element of reflexivity introduced as a key feature of policy-based social technologies fortifies the notion of new mission-oriented innovation policy as a dynamic and endogenous process.

As a conceptual framework, it is broad enough to cover the different theoretical approaches to effective policy design, such as Mazzucato (2018), Schot and Steinmueller (2018), and Weber and Rohracher (2012). It might be worthwhile in the future to use the notion of policy-based social technologies for a systematic comparison of these and other concepts related to mission-oriented policymaking. As a meta-concept, it can also depict the different failures discussed in the innovation system literature, most notably the four failures attributed by Weber and Rohracher (2012) to hindering transformative change: directionality, coordination, demand articulation, and reflexivity.

An important feature of our framework is the clear differentiation of dynamic processes between technical and non-technical layers of technology. By interpreting innovations as socially constructed phenomena and not distinguishing between the social and physical aspects of technology, we run the risk of encountering "all the pitfalls of technological and social determinisms" (Joerges 1988: 18). In this regard, our concept allows the labelling of social features (institutions, organisations, and interactions) involved in technology development and therefore extends similar studies that focus on social artefacts in technology operation (e.g., Chataway et al. 2010).

In the future, a more systematic delineation of technical and non-technical aspects is necessary. This will also require the conceptualisation of the physical technology. While prototypes clearly represent technical artefacts, technical standards or patents feature both social and physical aspects and could be equally subsumed under social technologies (representing a function and benefit, respectively). Thus, further empirical research is warranted to fine-tune the analytical framework by identifying and defining the set of dimensions for each of the core characteristics. Nevertheless, the case study of ECSEL JU has shown that the concept of policy-based social technologies provides a promising perspective for the systemic study of projects and programmes framed in new mission-oriented innovation policy.

References

Amoroso, S., Coad, A. and Grassano, N. (2018): European R&D Networks: A Snapshot from the 7th EU Framework Programme, *Economics of Innovation and New Technology* 27(5–6), 404–419. https://doi.org/10.1080/10438599.2017.1374037

Bergek, A., Hekkert, M.P. and Jacobsson, S. (2008a): *Functions in Innovation Systems: A Framework for Analysing Energy System Dynamics and Identifying Goals for System-building Activities by Entrepreneurs and Policy Makers* (RIDE/IMIT Working Paper No. 84426–008). Available at: https://imit.se/wp-content/uploads/2016/02/2007_153.pdf [21.04.2021]

Bergek, A., Jacobsson, S., Carlsson, B., Lindmark, S. and Rickne, A. (2008b): Analyzing the Functional Dynamics of Technological Innovation Systems: A Scheme of Analysis. *Research Policy* 37(3), 407–429. https://doi.org/10.1016/j.respol.2007.12.003

Borrás, S. and Edler, J. (2020): The Roles of the State in the Governance of Socio-Technical Systems' Transformation. *Fraunhofer ISI Discussion Papers Innovation Systems and Policy Analysis No. 65*. Karlsruhe.

Cantner, U. and Vannuccini, S. (2018): Elements of a Schumpeterian Catalytic Research and Innovation Policy, *Industrial and Corporate Change* 27(5), 833–850. https://doi.org/10.1093/icc/dty028

Capano, G. and Galanti, M.T. (2018): Policy Dynamics and Types of Agency: From Individual to Collective Patterns of Action, *European Policy Analysis* 4(1), 23–47. https://doi.org/10.1002/epa2.1031

Cappellano, F. and Kurowska-Pysz, J. (2020): The Mission-Oriented Approach for (Cross-Border) Regional Development, *Sustainability* 12(12), 5181. https://doi.org/10.3390/su12125181

Carroll, L. (2017): A Comprehensive Definition of Technology from an Ethological Perspective, *Social Sciences* 6(4), 126. https://doi.org/10.3390/socsci6040126

Chataway, J., Hanlin, R., Mugwagwa, J. and Muraguri, L. (2010): Global Health Social Technologies: Reflections on Evolving Theories and Landscapes, *Research Policy: Policy, Management and Economic Studies of Science, Technology and Innovation* 39(10), 1277–1288.

Cohen, W.M. and Levinthal, D.A. (1990): Absorptive Capacity: A New Perspective on Learning and Innovation, *Administrative Science Quarterly* 35(1), 128. https://doi.org/10.2307/2393553

Deleidi, M. and Mazzucato, M. (2021): Directed Innovation Policies and the Supermultiplier: An Empirical Assessment of Mission-oriented Policies in the US Economy, *Research Policy* 50(2), 104151. https://doi.org/10.1016/j.respol.2020.104151

EC (2009): Preparing for Our Future: Developing a Common Strategy for Key Enabling Technologies in the EU: Communication from the Commission to the European Parliament, the Council, the European Economic and Social Committee and the Committee of the Regions (COM(2009) 512 final). Available at: http://eur-lex.europa.eu/legal-content/EN/ALL/?uri=CELEX:52009DC0512 [21.04.2021]

EC (2012): A European Strategy for Key Enabling Technologies – A Bridge to Growth and Jobs: Communication from the Commission to the European Parliament, the Council, the European Economic and Social Committee and the Committee of the Regions (COM(2012) 341 final). Available at: http://eur-lex.europa.eu/LexUriServ/LexUriServ.do?uri=COM:2012:0341:FIN:EN:PDF [21.04.2021]

EC (2013a): A European Strategy for Micro- and Nanoelectronic Components and Systems: Communication from the Commission to the European Parliament, the Council, the European Economic and Social Committee and the Committee of the Regions (COM(2013) 298 final). Available at: http://eur-lex.europa.eu/legal-content/EN/TXT/PDF/?uri=CELEX:52013DC0298&from=EN [21.04.2021]

EC (2013b): Public–Private Partnerships in Horizon 2020: A Powerful Tool to Deliver on Innovation and Growth in Europe: Communication from the Commission to the European Parliament, the Council, the European Economic and Social Committee and the Committee of the Regions (COM(2013) 494 final). Available at: https://ec.europa.eu/transparency/regdoc/rep/1/2013/EN/1-2013-494-EN-F1-1.Pdf [21.04.2021]

EC (2013c): Proposal for a COUNCIL REGULATION on the ECSEL Joint Undertaking (COM(2013) 501 final). Available at: https://eur-lex.europa.eu/resource.html?uri=cellar:e457e697-eaf1-11e2-a22e-01aa75ed71a1.0002.05/DOC_1&format=PDF [21.04.2021]

EC (2014): COUNCIL REGULATION (EU) No 561/2014 of 6 May 2014 establishing the ECSEL Joint Undertaking (No. 57). Available at: https://eur-lex.europa.eu/legal-content/EN/TXT/PDF/?uri=OJ:L:2014:169:FULL&from=EN [21.04.2021]

ECSEL JU (2017): *Interim Evaluation of the ECSEL Joint Undertaking (2014–2016) Operating under Horizon 2020 – Final Report*. Luxembourg: Publications Office of the European Union.

ECSEL JU (2019a): Multi-Annual Strategic Plan (MASP). Available at: https://ec.europa.eu/*research*/participants/data/ref/h2020/other/legal/jtis/ecsel-multi-stratplan-2020_en.pdf [21.04.2021]

ECSEL JU (2019b): ECSEL Annual Activity Report. Available at: www.ecsel.eu/sites/default/files/2020-06/ECSEL%20Annual%20Report%202019_PUBLISH.pdf [21.04.2021]

Edler, J. and Fagerberg, J. (2017): Innovation Policy: What, Why, and How, *Oxford Review of Economic Policy* 33(1), 2–23.

Ergas, H. (1987): The Importance of Technology Policy, in Dasgupta, P. and Stoneman, P. (eds.): *Economic Policy and Technological Performance*. Cambridge: Cambridge University Press, 51–96.

EU Open Data Portal (2012): CORDIS: EU research projects under Horizon 2020 (2014–2020). Available at: https://data.europa.eu/euodp/en/data/ [21.04.2021]

Fisher, E., Mahajan, R.L. and Mitcham, C. (2006): Midstream Modulation of Technology: Governance From Within, *Bulletin of Science, Technology & Society* 26(6), 485–496. https://doi.org/10.1177/0270467606295402

Flanagan, K., Uyarra, E. and Laranja, M. (2011): Reconceptualising the "Policy Mix" for Innovation. *Research Policy* 40(5), 702–713. https://doi.org/10.1016/j.respol.2011.02.005

Floridi, L. (2014): *The 4th Revolution: How the Infosphere is Reshaping Human Reality*. Oxford: Oxford University Press.

Foxon, T.J. (2011): A Coevolutionary Framework for Analysing a Transition to a Sustainable Low Carbon Economy, *Ecological Economics* 70(12), 2258–2267. https://doi.org/10.1016/j.ecolecon.2011.07.014

Fruchterman, T.M.J. and Reingold, E.M. (1991): Graph Drawing by Force-directed Placement, *Software: Practice and Experience* 21(11), 1129–1164. https://doi.org/10.1002/spe.4380211102

Girvan, M. and Newman, M.E. (2002): Community Structure in Social and Biological Networks, *Proceedings of the National Academy of Sciences of the United States of America* 99(12), 7821–7826. https://doi.org/10.1073/pnas.122653799

Hasselbalch, J.A. (2018): Innovation Assessment: Governing Through Periods of Disruptive Technological Change, *Journal of European Public Policy* 25(12), 1855–1873. https://doi.org/10.1080/13501763.2017.1363805

Hekkert, M.P., Janssen, M.J., Wesseling, J.H. and Negro, S.O. (2020): Mission-oriented Innovation Systems, *Environmental Innovation and Societal Transitions* 34, 76–79. https://doi.org/10.1016/j.eist.2019.11.011

Hekkert, M.P., Suurs, R., Negro, S.O., Kuhlmann, S. and Smits, R. (2007): Functions of Innovation Systems: A New Approach for Analysing Technological Change, *Technological Forecasting and Social Change* 74(4), 413–432. https://doi.org/10.1016/j.techfore.2006.03.002

Horizon 2020 (n.d.): Online Manual. Available at: https://ec.europa.eu/research/participants/docs/h2020-funding-guide/index_en.htm [21.04.2021]

Idea Consult (2018): ECSEL JU Impact Assessment: Strategic Pre-study and Feasibility Study. Prepared for ECSEL JU. Available at: www.ecsel.eu/sites/default/files/users/Anna-zuzanna/ECSEL%20prestudy%20FINAL%20REPORT%202018%2009%2011.pdf [21.04.2021]

Jacobsson, S., Sandén, B. and Bångens, L. (2004): Transforming the Energy System – the Evolution of the German Technological System for Solar Cells, *Technology Analysis & Strategic Management* 16(1), 3–30. https://doi.org/10.1080/0953732032000199061

Joerges, B. (1988): Large Technical Systems: Concepts and Issues, in Mayntz, R. and Hughes, T.P. (eds.): *The Development of Large Technical Systems: Schriften des Max-Planck-Instituts für Gesellschaftsforschung Köln*. Frankfurt am Main: Campus, 9–36.

Kattel, R. and Mazzucato, M. (2018): Mission-oriented Innovation Policy and Dynamic Capabilities in the Public Sector, *Industrial and Corporate Change* 27(5), 787–801. https://doi.org/10.1093/icc/dty032

Klerkx, L. and Begemann, S. (2020): Supporting Food Systems Transformation: The What, Why, Who, Where and How of Mission-oriented Agricultural Innovation Systems, *Agricultural Systems* 184, 102901. https://doi.org/10.1016/j.agsy.2020.102901

Kuhlmann, S. and Rip, A. (2014): The Challenge of Addressing Grand Challenges: A Think Piece on How Innovation Can Be Driven towards the "Grand Challenges" as Defined under the Prospective European Union Framework Programme Horizon 2020.

Mazzucato, M. (2013): *The Entrepreneurial State: Debunking Public vs. Private Sector Myths*. London: Anthem Press.

Mazzucato, M. (2016): From Market Fixing to Market-creating: A New Framework for Innovation Policy, *Industry and Innovation* 23(2), 140–156. https://doi.org/10.1080/13662716.2016.1146124

Mazzucato, M. (2018): Mission-oriented Innovation Policies: Challenges and Opportunities, *Industrial and Corporate Change* 27(5), 803–815. https://doi.org/10.1093/icc/dty034

Mazzucato, M., Kattel, R. and Ryan-Collins, J. (2020): Challenge-Driven Innovation Policy: Towards a New Policy Toolkit, *Journal of Industry, Competition and Trade* 20(2), 421–437. https://doi.org/10.1007/s10842-019-00329-w

Metcalfe, J. S. (1995): Technology Systems and Technology Policy in an Evolutionary Framework, *Cambridge Journal of Economics* 19(1), 25–46.

Nelson, R. R. (2002): Technology, Institutions, and Innovation Systems, *Research Policy: Policy, Management and Economic Studies of Science, Technology and Innovation* 31(2), 265–272.

Nelson, R. R. (2005): *Technology, Institutions, and Economic Growth*. Cambridge, MA: Harvard Univ. Press.

Nelson, R. R. (2008): What Enables Rapid Economic Progress: What are the Needed Institutions? *Research Policy: Policy, Management and Economic Studies of Science, Technology and Innovation* 37(1), 1–11.

Nelson, R. R. and Sampat, B. N. (2001): Making Sense of Institutions as a Factor Shaping Economic Performance, *Journal of Economic Behavior & Organization: JEBO* 44(1), 31–54.

Robinson, D. K. and Mazzucato, M. (2019): The Evolution of Mission-oriented Policies: Exploring Changing Market Creating Policies in the US and European Space Sector, *Research Policy* 48(4), 936–948. https://doi.org/10.1016/j.respol.2018.10.005

Schot, J. and Steinmueller, W. E. (2018): Three Frames for Innovation Policy: R&D, Systems of Innovation and Transformative Change, *Research Policy* 47(9), 1554–1567. https://doi.org/10.1016/j.respol.2018.08.011

Schumpeter, J. A. ([1912] 2006): *Theorie der wirtschaftlichen Entwicklung*. Hrsg. und erg. um eine Einführung von Jochen Röpke/Olaf Stiller (Nachdruck der 1. Auflage von 1912). Berlin: Duncker & Humblot.

TWI2050 (2019): *The Digital Revolution and Sustainable Development: Opportunities and Challenges. Report prepared by the World in 2050 initiative*. https://doi.org/10.22022/TNT/05-2019.15913

Wanzenböck, I., Wesseling, J. H., Frenken, K., Hekkert, Marko, P. and Weber, K. M. (2020): A Framework for Mission-oriented Innovation Policy: Alternative Pathways through the Problem – Solution Space, *Science and Public Policy*, Advance online publication. https://doi.org/10.1093/scipol/scaa027

WBGU (2019): *Towards Our Common Digital Future* (Flagship report). Berlin.

Weber, K. M. and Rohracher, H. (2012): Legitimizing Research, Technology and Innovation Policies for Transformative Change, *Research Policy* 41(6), 1037–1047. https://doi.org/10.1016/j.respol.2011.10.015

Yoon, J., Blumer, A. and Lee, K. (2006): An Algorithm for Modularity Analysis of Directed and Weighted Biological Networks Based on Edge-betweenness Centrality, *Bioinformatics (Oxford, England)* 22(24), 3106–3108. https://doi.org/10.1093/bioinformatics/btl533

24
CRYPTO ASSETS

Tobias Eibinger, Ernst Brudna and Beat Weber

Introduction

The last decade has been characterised by a proliferation of projects that use new technologies to launch various forms of "crypto assets". This development has been inspired by the initial combination of cryptographic methods and economic incentives to construct a decentralised register of data, called "blockchain". The first and most prominent of these crypto assets is the Bitcoin network, which launched in 2008. Bitcoin and a sub-class of blockchain-based projects closely inspired by it have received widespread public attention and are being referred to as "cryptocurrencies".

The term "cryptocurrencies" triggers diverse associations in individuals. It may be seen as a decentralised form of digital money that is severed from the influence of centralised institutions (both public and private). This view is often accompanied by sanguine comments about a society in which the role of the government is reduced to a minimum. By contrast, some might, for the same reason, associate it particularly with illicit behaviour. Indeed, this perspective is what led Nobel Laureate Joseph Stiglitz to argue in an interview with CNBC in 2019[1] that cryptocurrencies should be shut down. Others might simply regard it as yet another speculative asset to increase their wealth. This aspect is understandable, considering that in May 2010 two pizzas were sold for 10,000 Bitcoins (BTC)[2] and that one BTC traded for over 34,000 USD at the beginning of January 2021. A different perspective results from a focus on the underlying technology of cryptocurrencies – the blockchain, the potential of which (and consequently that of cryptocurrencies) can be analysed from several angles. The technology can, for example, be seen from an efficiency lens (i.e. saving transaction costs) or from an institutional lens (i.e. competing with existing institutions in the facilitation of transactions). These two perspectives are likely to be intertwined, as different institutions can exhibit different magnitudes of transaction costs.

In its essence, however, the term cryptocurrencies insinuates some relation to the concept of money (in particular, digital money). Digital money has existed for quite some time now. In fact, the majority of the existing money supply consists of digital money, since (in addition

1 www.cnbc.com/2019/05/02/joseph-stiglitz-we-should-shutdown-the-cryptocurrencies.html
2 https://bitcointalk.org/index.php?topic=137.msg1195#msg1195

DOI: 10.4324/9780429351921-29

to physical notes and coins), the majority of means of payments in official currency consists of funds in bank deposits. But an intriguing question is what makes cryptocurrencies as potential "digital money" so interesting. Part of the answer is related to the concept of transaction costs and the notion of trust. Transaction costs were introduced into the economic doctrine by Ronald Coase in 1937 and later extended by, for example, Oliver Williamson (1985) and Douglass North (1990). To keep things simple, it suffices to note that the coordination of transactions can incur different costs. These include, for example, costs of contract negotiation (and renegotiation), contract writing, contract enforcement, opportunistic behaviour, information asymmetries, price discovery, and unverifiable information. Different institutions may incur these costs to different degrees. Whenever one institution can coordinate transactions at lower costs than another institution, it can be regarded as being more efficient – and more efficient institutions tend to supersede less efficient ones.

Transaction costs in the traditional monetary and financial systems can often be related to dependence on third parties (e.g. central banks, commercial banks, courts) to facilitate transactions: they are needed to safely store money, ensure access to stored money, process and validate transactions, provide security mechanisms in case of unforeseen events, and guarantee the value of money. These third parties have to be trusted to fulfil their legal and contractual obligations correctly and truthfully. Additionally, the legal and economic system have to be trusted to deter or sanction any violations. This seems to be particularly relevant with respect to financial systems. Zingales (2015) raised awareness about the dichotomy between the scientific community's and general public's opinion on the US financial system. The former group is usually characterised by a significantly more positive stance than the latter. The general public commonly associates financial activities with rent-seeking behaviour – and such behaviour increases transaction costs. Cryptocurrencies were arguably born out of a desire to escape the dependence on trusting centralised third parties. This dependence bothered Satoshi Nakamoto, founder of the Bitcoin network, who had the vision to create "a purely peer-to-peer version of electronic cash [that] would allow online payments to be sent directly from one party to another without going through a financial institution" (2008: 1).

While such a decentralised system of record-keeping does indeed eliminate the need to rely on certain centralised third parties to process and validate transactions, this does not imply that this is achieved in a costless manner. In this chapter, we will discuss in what way cryptocurrencies may change transaction costs as well as the need to trust others. Additionally, we argue that associated costs might simply be transformed or borne by other parties than under traditional arrangements. Moreover, even assuming that cryptocurrencies facilitate more transactions at lower costs, the question remains whether a "the more the better" attitude is necessarily a desirable outcome. Be that as it may, since its creation, the Bitcoin network has been at the centre of many debates about the future of our monetary system. This contribution tries to bring some of these debates into perspective.

The rest of this chapter is structured as follows. The next part provides an introduction to the techno-economic characteristics of crypto assets and the blockchain technology. Afterwards, we discuss those crypto assets referred to as cryptocurrencies in the context of money and monetary systems, while the fourth part considers further types of crypto assets designed to provide alternative products and processes in the financial sector; special attention is given to the concepts of tokenisation and decentralised finance (DeFi). The following section discusses the conditions under which individuals choose to transact with each other. It questions the notion that more transactions (due to lower transaction costs) necessarily make all transacting parties better off. Finally, the last section provides concluding remarks.

The techno-economic properties of crypto assets

Before discussing the potential of crypto assets in the context of the monetary and/or financial sector, it is instructive to devote a few pages to the functioning of crypto assets. In particular, this requires an understanding of blockchains – the technology underpinning them. The first blockchain protocol was proposed in the Bitcoin white paper. Nakamoto (2008) noted that previous attempts to create peer-to-peer payment systems that did not rely on centralised institutions failed to solve the double-spending problem. This problem emerges out of inherent properties of digital objects: they can easily be copied. To prevent somebody from copying and subsequently spending the same digital coin twice, transactions have to be monitored and validated. This can be done by traditional third parties. In a decentralised administration system, things get more complicated. Firstly, decentralised consensus on the validity of transactions has to be reached. Secondly, information on processed transactions needs to be distributed in the network and publicly accessible. Finally, transactions must be irreversible. Nakamoto (2008), for the first time, proposed a distributed and decentralised system that meets these requirements and solves the double-spending problem: the Bitcoin blockchain.

Satoshi Nakamoto's innovation utilises a combination of cryptography and economic incentive schemes. A simple decentralised consensus mechanism based on computational effort (proof of work) incentivises and ensures correct validation of transactions by special network participants (miners). Validated transactions are subsequently combined into blocks and added to the system. Every new block is uniquely linked to the previous one, thus creating an ever-growing chain of blocks – the blockchain. Information on the state of the blockchain and information contained within transactions is distributed over myriad storage devices (nodes) of network participants. Cryptography ensures that transactions cannot be reversed. Transactions recorded on the blockchain are thus distributed, decentralised, and immutable. These characteristics are the result of a clever combination of economic incentives and cryptography. The following paragraphs introduce the concepts employed and clarify some of the terminology used in this context.

Decentralised consensus mechanisms are the backbone of the blockchain technology. Their underlying models are based on economics, and since the advent of the Bitcoin network, various concepts have been proposed. The proof-of-work (PoW) based consensus mechanism, proposed by Nakamoto (2008), remains the most prominent up to date, although different concepts are gaining momentum due to inherent shortcomings of this mechanism. But before we elaborate on this topic, we have to understand the concept of the PoW consensus protocol. In the PoW system, "miners" try to solve a mathematical problem that can only be solved by a trial-and-error process through the exertion of computational effort. They are incentivised to exert this effort because the first one to publish a correct solution in the network gets rewarded with a pre-specified amount of BTC, the native cryptocurrency of the Bitcoin network. The protocol publishes a new variant of the problem to be solved in regular time intervals. Once a solution has been found for a new problem variant, recently instructed transactions by Bitcoin owners are validated and get added to the chain. This mechanism creates a competition between miners to validate transactions (to be the first to solve the problem).

To ensure that BTC are not double spent, the network draws on the information contained in all past transactions included in the blockchain. As a result, the newest block is connected to all previous blocks. Such a system is astonishingly robust against malicious attacks. To trick the system (e.g. by altering past transactions to one's benefit), one would have to convince the whole network of a different "truth". The truth in the Bitcoin network (and almost any other blockchain system) is the longest chain of blocks. To change a transaction, a new chain (a fork) would have to be computed that ultimately outpaces the growth rate of the existing chain. This can only be achieved by exerting more computing power than all other miners have together. Thus, manipulating past blocks is extremely difficult, and the task becomes harder as the chain of blocks grows.

This is probably as good a point as any for the introduction of Alice and Bob, two fictional characters that are often used to ease the understanding of concepts related to cryptography (DuPont and Cattapan, n.d.). Imagine Alice wants to send some BTC to Bob, and suppose that both characters are already participating in the Bitcoin network. That means that Alice and Bob possess both a public and a private key. The former identifies the two nodes (although no personal information is provided) and the latter is used to sign transactions. Alice creates a transaction with the amount of BTC she wants to send to Bob together with his public key. She then uses her private key to sign the transaction and her public key to publicly verify that she issued the transaction. Now that the transaction has been proposed to the network, miners start exerting computational effort to validate it. Once this has been achieved, the transaction is added to the blockchain. Now suppose that Alice wanted to trick the system and spend the same BTC again. Since her transaction to Bob has already been validated, she would have to convince the entire network of a different truth. To do this, she would have to validate the transaction containing this alternative truth herself. However, as all transactions are linked to all previous transactions, she would have to redo the PoW on every transaction contained in the currently used blockchain (i.e. the longest chain), which is quite unlikely for a single person to achieve.

However, Nakamoto already acknowledged a potential risk of centralisation in the Bitcoin white paper. He therefore stated that, for the network to function as intended, more than half of the entire network's computing power must be controlled by honest network participants. This potential flaw was later emphasised by Vitalik Buterin, founder of the Ethereum blockchain, the second largest blockchain network. Buterin (2016) pointed out that a consortium of malicious nodes might collude to create a different truth in the blockchain. This consortium might heavily invest in equipment specially developed to perform the required computational tasks most efficiently. Due to economies of scale, such a consortium might gain an advantage over regular nodes and thus be difficult to compete against. Criticisms from a different avenue concern the huge amount of electricity consumed by consensus mechanisms based on computational proof.

Ethereum Wiki (2020) therefore proposed a proof-of-stake (PoS) based consensus mechanism that is more robust against centralisation.[3] The PoS relies on economic stakes rather than computational effort to validate transactions. Nodes can lock some of their native cryptocurrency (put it at stake) in exchange for a right to vote on which transactions are validated and included into the system. The power of a vote is directly related to the economic value of the stake at a 1:1 ratio. This removes a potential centralisation risk due to economies of scale. PoS is additionally strengthened by a combination of economic rewards and penalties. Nodes receive some kind of interest on their capital at stake. They also get a proportion of fees associated with transactions. Penalties can take the form of a loss of the entire capital at stake.[4] Ultimately, PoS is expected to be more robust than PoW due to more robust economic incentive schemes. On top of this, PoS is much more environmentally friendly, as it does not rely on electricity-consuming computational effort.

But the blockchain technology has far more to offer than the possibility of a peer-to-peer payment system (such as Bitcoin). This has inspired the development of the Ethereum blockchain.

3 The proof-of-stake protocol is currently being implemented on the Ethereum blockchain.

4 These are much more severe punishments than in a PoW system, where the source of the miners' power, i.e. computer hardware, cannot be simply taken from them.

The basic idea behind the project was to provide a blockchain system which understood a general-purpose programming language. Put differently, anybody with the required programming skills could create their own blockchain-based application, which would automatically benefit from the specific characteristics of the blockchain technology (i.e. that it is distributed, decentralised, and immutable). A particularly important aspect of the Ethereum blockchain is the extended potential of so-called smart contracts.[5] These are, in essence, scripts that translate obligations of two or more parties with respect to information stored on a blockchain in a transaction as agreed in a contract. These scripts can be encoded in transactions and are executed autonomously. Smart contracts themselves can execute further transactions and even additional smart contracts. Usually, the autonomous execution of smart contracts is tied to the observation of pre-specified events (i.e. information).[6] Since these scripts are implemented on top of the blockchain technology, they also exhibit all of its features: they are immutable, distributed, and decentralised (for example, Buterin 2013). Thus, they have the potential to facilitate trust (e.g. in the fulfilment of future payment obligations within contracts) between two unknown parties who want to transact with each other in a truly decentralised way.

Potential use cases for smart contracts range from automatic execution of financial contracts to the transfer of property rights (we will encounter specific examples later on). They can, for example, be programmed to hold a specified amount of crypto assets in escrow and transfer them to a specified address upon the observation of a specific event, e.g. the transfer of another cryptocurrency or the delivery of a good or service. Theoretically, legal disputes over non-fulfilment of contracts can thus be precluded a priori, as the smart contract – once it is triggered – executes its code inexorably and the transaction cannot be altered ex-post. Smart contracts thus have the potential to economise on transaction costs by making traditional intermediaries (e.g. banks) superfluous and to enable more transactions between individuals by increased trust in contract fulfilment. While smart contracts are neither perfect (complete) contracts in an economic sense (all possible contingencies are accounted for)[7] nor contracts in a legal sense, their ability to automatically enforce contractual obligations referring to information recorded on the blockchain can be considered an alternative to relying on centralised institutions (e.g. courts) as enforcement intermediaries or informal norms. Irreversibility, for example, is beneficial in cases in which one party undertakes asset-specific investments, i.e. investments that have no value outside this relation, and the other party is characterised by rent-seeking behaviour, i.e. by insisting on renegotiating the initial contract ex-post (this point has been persuasively made by Oliver Williamson, for example, 1985). Although it is worth noting that irreversibility of agreements need not always be beneficial, e.g. when outcomes of completed transactions differ from initial expectations held by at least one of the parties involved.

Much of the discussion surrounding the potential of the blockchain and crypto assets is based on a common narrative: trust. The blockchain technology quite generally tries to reduce the need to trust third parties (e.g. banks, courts, etc.) to fulfil their obligations and not to act opportunistically. Additionally, transactions might be executed more efficiently without

5 Even though Ethereum is one of the most prominent platforms for smart contracts, its founder, Vitalik Buterin (2018), seemingly regrets the use of the term "smart contract". His reasoning is that the term relates the concept of self-executing code too much to legal aspects of contracts. He goes on to state that a more technical term, such as "persistent code", might have better defined its purpose.

6 Any real-world event has to be verified by so-called oracles, which can provide off-chain information in a decentralized way. See, for example, Ellis et al. (2017), who discuss the oracle Chainlink.

7 Paul Milgrom and John Roberts (1992: 127) describe a complete contract as one that "would specify precisely what each party is to do in every possible contingency (including those where the contract's terms are violated) so that each party individually finds it optimal to abide by the contract's terms".

additional intermediary steps. This is why many observers see crypto assets as promising and many of their innovations as having the potential to shape our society in the long run. The following sections will discuss two such areas in some detail: the monetary system and finance.

Cryptocurrencies in the context of money and monetary systems

The term "cryptocurrencies" suggests to some degree that the technology might be interpreted as money. After all, Bitcoin contains the term "coin" in its name and was devised as a peer-to-peer electronic cash system, with an elaborate architecture designed to facilitate payments. Its predefined supply limit evokes the limited supply of natural resources like gold, which has a long history of being used as a material to mint coins and backing asset in monetary systems. But whether BTC were intended to challenge traditional money directly is unclear. Moreover, the simple question whether cryptocurrencies in general, and BTC in particular, may be defined as money is not straightforward.

In a capitalist market economy with an elaborate division of labour, most people depend on permanent economic activity in order to earn and spend income. Economic activity under these circumstances requires money to perform three functions (see Krugman et al. 2012, in the place of many). Firstly, it has to function as a unit of account – a stable yardstick to measure and compare prices of goods, services, and contractual payment obligations (wages, debt, taxes). Secondly, it has to be widely accepted as a medium of exchange. Thirdly, money has to be stable enough over time to serve as a basic means to store value in liquid form. Money with these three functions serves as a basic infrastructure for markets to operate. Currently, money takes the form of various national currencies, most of which monopolise monetary functions in their own geographic area, some of them also across borders.

Crypto markets and user behaviour

Now, can cryptocurrencies be described as money according to the three functionalities mentioned? As a first step towards answering this question, we can observe actual user behaviour with regard to cryptocurrencies. Do crypto owners actually treat cryptocurrencies like money, as characterised by the three functions referred to earlier? A key observation about crypto assets like Bitcoin is that their market value is subject to huge fluctuations. The value of one BTC, for example, might change considerably over a single day, making it neither a stable yardstick nor a predictable store of value.[8] Whatever their technical potential as a means of payment, most current crypto users treat cryptocurrencies like other goods and assets priced and traded on markets for stores of value and collector items within existing official currency networks, not as rival currencies: the market value of cryptocurrencies on trading platforms is measured in official currency.

Some merchants may accept them as optional means of payment, but to avoid exchange rate risk, they only rarely set prices of their goods and services in crypto units of account. And except in cases where payment in official currency is no option (e.g. in illicit transactions), spending cryptocurrencies is usually unattractive for their voluntary owners, because it would involve the opportunity cost of missing out on a potential future appreciation of their market value. Additional hurdles may include transaction fees and capacity limits of blockchain payments as well as reduced price transparency, as crypto owners are faced with price in official

8 For a time, series of Bitcoin prices, see: www.coindesk.com/price/bitcoin.

currencies. Ultimately, we can conclude that so far, crypto assets like Bitcoin satisfy the functions of money at best partially and theoretically.

External determinants of cryptocurrencies' competitiveness: network effects on currency markets

To explain the failure of Bitcoin-like cryptocurrencies to rival existing currencies so far, reasons both external and internal to cryptocurrencies can be considered. Central external reasons encompass network effects. Networks characterised by such effects convey various benefits on their users, and these benefits rise with the quantity of membership and the quality of services provided (one may think of languages, telephone networks, social media). That is why the individual choice of joining a network is different to the individual choice concerning competing ordinary consumption goods, which can be chosen and consumed by individuals without regard to what other people are doing. To see, how such effects make big and stable currencies robust to competition in currency markets, note that individual choice among competing currencies is about the decision to join a network with associated qualitative as well as quantitative benefits. The prime quality feature of a currency network is access to the three functions of money outlined earlier.

With respect to quantity of membership, there are benefits for every member of a large network: a large membership in a currency network reduces participants' transaction costs for negotiating and comparing prices, agreeing on a means of payment among transaction partners, exchange rate risks among different means of payment, collecting information to form expectations about the future value of the currency, search costs for transaction partners accepting means of payment in a particular currency, etc. These transaction costs favour a monopoly position of a currency in its own jurisdiction and pose disincentives for members to leave an existing currency network and join a different one. Such networks can be challenged by bigger foreign currency networks of superior quality. But they are particularly robust against completely new potential networks, such as the one offered by several crypto assets.

Money and cryptocurrencies: internal design differences

In addition to the self-supporting features of a currency subject to network effects, the public sector behind official currencies can internally strengthen its own network against competition. Such measures encompass an exclusive acceptance of official currency for tax and administrative payments and legal tender laws normalising official currency as the standard for payments. On top of that, a number of elaborate institutional safeguards have been developed to stabilise the value of an official currency in terms of a predictable evolution of domestic purchasing power of money (as measured by the evolution of prices of goods and services) over time. This necessitates avoiding both excessive inflation and deflation.[9]

In such a system, centralised but controlled intermediaries play a key role. Cryptocurrencies usually try to avoid a dependence on (trusting) such parties. So, it is important to understand what kind of trust contemporary money requires. It is trust in the solidity of a set of interrelated

9 Deflation might be undesirable due to several reasons. In an economy-wide deflation, all or most prices are subject to downward pressure. This can lead to an economy-wide self-feeding downward spiral between falling prices of goods and falling incomes (wages and profits from goods production), hurting economic activity and future planning. At the same time, claims of creditors on debtors remain unchanged, with falling debtor income resulting in an increasing debt burden (see, for example, Krugman 2010).

obligations and responsibilities designed to make any violations of these obligations difficult and costly. In today's major currency areas, a number of legal obligations and economic incentives are in place to promote monetary stability and general acceptance of domestic currencies (Weber 2018).

Legal obligations. While issuing of banknotes and coins in official currency is monopolised by the central bank, issuing deposits in official currency is not. Whatever its form, money is an obligation for its issuer. A central bank is liable for accepting banknotes or deposits issued at nominal value when they are repaid after circulation (to discharge third-party debt obligations acquired in the act of money creation). The public sector is liable for accepting money in domestic currency to discharge tax payment obligations of users. Commercial banks are liable to redeem customer deposits in notes and coins issued by the central bank on demand.

The central bank's objectives as well as their room of manoeuvre is determined by legal mandates. Current mandates in major currency areas focus on stability of prices and macroeconomic performance, granting central banks operational independence to use instruments in a way to achieve their tasks.[10] While central banks accept government debt from their customers in the banking sector in exchange for new money, there are legal prohibitions against direct access of governments to central bank finance in major currency areas. More generally, democratic control procedures are in place (legal constitution and independent courts, parliaments, elections, etc.) to keep government misbehaviour in check. Finally, note that in most jurisdictions, legal tender laws do not outlaw the use of other currencies in private transactions but establish official currency as a standard.

Economic incentives. Before new money can circulate among users in the economy, its creation is conditional on the issuer (central bank) receiving an obligation for future repayment from a debtor at the prevailing interest rate. This generally incentivises responsible economic use of new money. With money creation being demand-determined, there is no way for the central bank to increase the money supply without counterparties willing and able to fulfil the central bank's conditions. Although money is not a legal claim on valuable goods supplied by private producers, its status as being a claim on sizeable and reliable issuers makes it attractive for general acceptance in the economy for private transactions in markets.

The negative fiscal and legitimacy implications of macroeconomic instability associated with high inflation or deflation provide a counterweight to incentives for governments to either promote high inflation as a way to erode the value of government debt and increase tax receipts or to promote deflation by excessive hoarding of tax revenue. Market competition is envisaged to impose discipline on prices, wages, and credit. This also applies to currencies in the global economy: liberalised cross-border movement of capital increases the threat of substitution of domestic currency if the superiority of foreign currencies in terms of membership size and stability becomes large enough for domestic users to incur the transaction costs involved in collective currency switching.

Crypto design compared to official currency

Issuing entities of official currency can adjust the terms and conditions for money creation and backing assets to let supply adjust to changes in demand with the objective to stabilise the value of money. Blockchain-based systems rest on a different set of value propositions, inspired by a desire to avoid the use of centralised intermediaries that can exert an influence on money and its users. To achieve this objective, blockchain protocols largely rely on algorithms, and these may

10 Consolidated Version of the Treaty on the Functioning of the European Union (TFEU) [2016] OJ C202/102 ELI: http://data.europa.eu/eli/treaty/tfeu_2016/art_127/oj

vary substantially. The wave of crypto projects inspired by the appearance of Bitcoin has been characterised by the emergence of new types of crypto assets that are intended to serve a specialised subset of potential use cases and operate with different supply rules. Among these are stable coins and crypto assets intended to serve various non-currency purposes (such as Ethereum).

Bitcoin-like cryptocurrencies depend on reward-based decentralised operations to produce a supply of "coins" that consist of data on blockchain with no external reference (e.g. a backing asset) and no centralised system owner or issuer. In other words, Bitcoin-like cryptocurrencies are no one's liability. The supply of these "coins" is directed by pre-set parameters, which differ in their degree of flexibility across cryptocurrencies. Prices are thus solely left to the interplay between the fraction of the supply of "coins" that is currently offered for sale and the current demand on crypto markets.

In the case of the Bitcoin protocol, the supply is limited to a total of 21 million BTC that can be mined. At the end of 2020, around 18.5 million BTC have been created, and it is estimated that the last subunits of BTC will be created around the year 2140. Once this last unit is created, the supply of Bitcoin cannot be expanded according to the protocol, unless miners agree on changing it. For the blockchain to continue to work after this point, it is expected that Bitcoin users must be prepared to pay considerable fees for transactions to enable continued cost recovery by miners (Hayes 2020). Although future demand for Bitcoin is highly uncertain, supporters hope that continued new demand for Bitcoin on crypto markets after the supply has reached its upper limit will contribute to continued appreciation of its market value in official currency.

A predetermined supply schedule of a good, asset, or currency with no claim or backing asset behind it and no information about (or anchor for) the future evolution of demand almost inevitably results in fluctuating market value in an economic system where market value is determined by the interaction of supply and demand. This is a typical feature of markets for collector items and other assets in limited supply and may incite users to include crypto assets in their portfolio. But it will in all likelihood undermine the asset's competitiveness against stable major currencies.

Stable coins explore mechanisms to improve on the value stability of crypto assets. Some such mechanisms provide services close to an official currency within the crypto sector; some of them aim at entering the global market for mainstream payment services. Some stable coins have decided to recede on the decentralisation principle on which Bitcoin is based in order to profit from improvements requiring centralised responsibility. Most variants of these are vehicles for tying the value of crypto currencies to official currencies, trying to import or free ride on their stability mechanisms instead of offering an alternative currency. Furthermore, some corporate actors and even central banks have started to explore the possible integration of some features of blockchain and crypto assets into potential future payment instruments issued by centralised intermediaries.

The supply management of stable coins is based on more stability-oriented supply rules compared to volatile cryptocurrencies (such as Bitcoin). The mechanisms governing the supply of such stable coins can broadly be categorised as asset-backed and algorithmic. Asset-backed systems either aim at maintaining a peg to official currencies (e.g. USD) or to crypto-assets (e.g. ETH), while algorithmic stable coins rely on smart contract protocols to minimise price volatility (Hileman 2019). One prominent asset-backed stable coin is Tether, which aims to maintain a 1:1 peg to the USD. According to the company issuing the crypto coin, each native USDT is backed by an official USD in reserves.[11] Such a system relies on centralised third parties to maintain the peg and can therefore be described as a centralised stable coin.

11 According to Tether's homepage, reserves "include traditional currency and cash equivalents and, from time to time, may include other assets and receivables from loans made by Tether to third parties, which may include affiliated entities" (Tether 2020).

An example of a decentralised stable coin is the DAI, created and managed by MakerDAO. The stable coin envisages a fixed ratio of 1 DAI to 1 USD.[12] Stability is attempted through a combination of economic incentives and "monetary policy" tools. As soon as the DAI market price deviates from the parity, users are incentivised to buy/sell DAI due to arbitrage possibilities. Additionally, the protocol can be updated to adjust the "DAI Savings Rate", a sort of interest payment on locked DAIs. These fluctuations in supply and demand should steer the DAI towards parity with the USD (MakerDAO n.d.).[13]

Although stable coins that are referenced to official currency do not pose a competitive challenge to their reference currency, they may increase foreign currency access in other currency jurisdictions. Thus, potentially increasing the threat of currency substitution among official currencies, and in some sectors of economic activity where access to official currency is hampered or costly (e.g. unregulated trading of crypto assets against official currency, cross-currency payments).

Crypto assets with additional non-currency functionalities: Ethereum was the starting point for a whole genre of crypto assets. It introduced a blockchain-based platform for projects built on top of it, allowing the launch of crypto assets covering various use cases. Contrary to the Bitcoin network, the Ethereum protocol ties the supply of Ether to the amount of mining activity without determining a predefined total amount of coins. The supply is, for example, determined by block rewards or the difficulty of mining new blocks, where increased difficulty increases the time it takes to validate transactions and thus decreases the amount of newly minted coins (EthHub n.d.a). Such decisions on the Ethereum blockchain more closely resemble traditional policy decision processes. Stakeholders of the network (e.g. developers, miners, users) can propose changes to the protocol, which can then be evaluated and voted upon off-chain. Once all stakeholders agree on an update to the protocol, the changes will be implemented (EthHub n.d.b).

The supply implications depend on the protocol and on the policy decisions the network participants make, offering some degree of potential to adjust supply to demand on crypto markets. Of course, such a system is to some degree susceptible to the influence of stakeholders with a strong influence on other users (e.g. experienced developers or simply well-known individuals). In the case of Ethereum, its more flexible supply governance compared to Bitcoin did not have the effect of stabilising its market value. But is has been used as a platform to both initiate new crypto projects and inspire others. Meanwhile, a subsector of crypto assets has emerged that seeks to retain and refine the decentralisation principle and apply it to areas of the financial sector beyond currencies (we will explore this aspect in the following sections).

Doing away with the trust and costs of intermediation?

Many observers associate blockchain-based crypto assets with the ability to replace the need for trust in transactions. The *Economist*, for example, devoted one of its cover issues in 2015 to the blockchain technology; it was titled "The Trust Machine". It was argued that the blockchain enables two parties to transact with one another without having to trust each other and/or a centralised third party. Scientific papers often refer to the technology as being "trustless" (for example, Davidson et al. 2018). They argue that the blockchain substitutes for trust – trusting another party in a transaction involving only information that can be verified on the

12 It actually maintains a ratio of 1 DAI to the ETH-equivalent of 1 USD.

13 For a time, series of the DAI against the USD, see https://coinmarketcap.com/currencies/multi-collateral-dai/.

blockchain – are no longer a necessity to facilitate efficient transactions. Both views of the blockchain technology – creating trust and substituting for trust – are misleading.

Cryptocurrencies may come without centralised intermediaries, but not without intermediation functions. Some of these functions are decentralised (the register of transactions and holdings is administered by competing miners instead of banks); some of them are not offered within the system (it is technically possible to join the network without asking a responsible bank-like entity for permission or disclosing your identity; no entity is responsible for guaranteeing the value of coins or backing them with assets).

For those intermediary functions that are decentralised in blockchain-based crypto assets, the need to trust in a centralised intermediary is replaced by the need to either trust in the intended functioning of rules, algorithms, and incentives contained in the software, or to trust in one's personal ability and incur the cost of verifying the state of the blockchain with a personal computer. Fees for user transactions apply, depending on the amount of network traffic, and current miner compensation from newly produced crypto units.

Those intermediary functions that are eliminated in the blockchain-based set-up require additional arrangements to be performed outside the cryptocurrency's blockchain. Individual users are faced with the choice of either placing trust in their own ability and personally incurring the cost for the following services, or trusting fee-based intermediaries in the crypto sector to service them: acquiring the knowledge and resources how to access crypto currencies and information about them; finding someone else to exchange and measure cryptocurrencies against something valuable (an official currency, or a good or service); storing and protecting the personal access key to individual crypto ownership against theft or loss, etc.

Additionally, cryptocurrencies' decentralised mechanisms fail to replace the role of intermediaries in contributing to monetary stability – a feature of money that is associated with saving transaction costs for individual money users. In the absence of monetary stability, transaction costs in a currency network rise, making them unattractive: unpredictability of money's value implies the lack of a yardstick for individual and collective planning of the economic future, an undermining of price transparency and comparability as the basis of functioning markets, additional efforts required for acquiring information relevant for economic activity, and the lack of a basic means to store value with minimal risk.

Decentralised cryptocurrencies come without centralised intermediaries, but they eliminate neither the need for intermediation nor the costs and trust required for economic transactions and storing economic value (Campbell-Verduyn 2019). A key point in this context is that the internal consistency of the data produced and stored on the blockchain may be credible but would be meaningless and irrelevant unless referring to, and evaluated by, the outside world according to rules and mechanisms prevailing in this context (e.g. markets where people are prepared to spend official currencies to purchase Bitcoin according to their personal value judgements, contributing to the emergence of a market value for Bitcoin). So, the technology's ability to transform trust requirements is always limited (Frolov 2021).

Crypto use cases beyond currency status

While current cryptocurrencies offer no incentives to use them wherever the use of stable official currency is possible and attractive (for measuring, transferring, and storing liquid predictable value), they have other features which may attract users in special use cases. A major use case for cryptocurrencies has become their use as a speculative asset based on expecting the future appreciation of their market value. Crypto assets can meet user demand on markets for risky assets, complementing a market that includes collector items (antiques, used stamps and records, art, etc.), and other financial and non-financial assets (gold, silver, land, etc.) in limited supply that bear the chance of financial returns or changes in market value.

A further potential use case is the use of cryptocurrencies as means of payment in transactions related to illicit activity (digital markets for drugs and other illegal goods, illegal online gambling, tax evasion, etc.) and for transactions where users have a particularly strong preference for privacy for various reasons (Tzanetakis 2018). Unlike bank accounts, it is technically possible to access and transfer cryptocurrencies without disclosing user identity to intermediaries. In these cases, such users might be willing to incur the additional transaction costs involved in using cryptocurrencies as means of payment.

If such uses became widespread, they could ultimately indeed undermine taxation and regulation of economic transactions as important pillars of the governance of official currencies and economies. To limit the possible contribution of cryptocurrencies to the expansion of dark markets, tax evasion, and money laundering, regulators have agreed in international fora to submit intermediaries serving as access points to crypto assets (e.g. crypto exchange platforms) to Know-Your-Customer (KYC) and Anti-Money-Laundering (AML) rules, thus creating a level playing field with regulated banks in this respect. Tax authorities have submitted a number of crypto uses to taxation and have started to use the services of private firms specialised in identifying users behind crypto accounts to enforce it (Houben and Snyers 2018; OECD 2020a).

While crypto assets seem to be ill suited to compete against traditional currency systems, they might impact other sectors more strongly. In fact, we have already encountered crypto assets that are not targeted towards being used as currency, but rather to facilitate non-currency uses: on top of the Ethereum blockchain, a subsector of crypto assets has emerged that seeks to retain and refine the decentralisation principle and apply it to areas of the financial sector beyond currencies.

Crypto assets as financial instruments

One way to conceptualise crypto assets designed for non-currency functionalities is to consider them as belonging to a more general class of phenomena referred to as "tokens". The private cryptographic keys to identify and transfer a particular sum of a crypto asset on the blockchain might be considered as tokens. These are physical items, tools, or data packages that are built to authorise access to a certain function of the system they belong to: for instance, a casino chip gives access to services within the casino and can be redeemed for a certain sum of cash at the cashier. A banknote represents the value guarantee of the central bank and gives access to goods and services in all those transactions where it is accepted as a means of payment. A supermarket voucher gives access to goods available at the supermarket of a certain quantity or of a certain monetary value.

Both tokens and the phenomena they represent can be recorded in registers, sometimes referred to as ledgers. The blockchain is a ledger or register of tokens where owners of keys can induce changes that find the agreement of all other users based on their trust in the solidity of the operating rules of the blockchain, resulting in transfers among key holders without a single entity able to interfere. Whereas the initial idea of Bitcoin was to offer what was called a "peer-to-peer cash system" with a vague analogy to gold being the only reference to phenomena outside the blockchain, a number of subsequent crypto projects started to explore the idea of whether tokens could be constructed as precise representations of various non-currency uses.

Tokenisation

Various kinds of tokens on blockchains can be distinguished. The most common tokens are native tokens, such as BTC or ETH. Private keys held by owners give them access to the native cryptocurrency, which is stored on the blockchain (i.e. ledger). Buterin (2013), however, already noted in the beginning of the Ethereum white paper that the blockchain technology offers the potential to record more than just the ownership of native cryptocurrencies. Theoretically, information on any physical and non-physical asset can be stored on the blockchain ledger. Consequently, any asset can be tokenised on the blockchain. The concept of tokenisation has already been proposed by Nick Szabo in 1998, who was bothered by the fact that traditional ledgers can be compromised by third parties (e.g. by malicious actors or simply through errors).[14]

The general-purpose blockchain technology provided by Ethereum enables the implementation of any conceivable tokenisation. Ethereum also provides popular standards that include common features for the implementation of new tokens.[15] The ERC-20 standard regulates fungible tokens, i.e. tokens of the same value (Vogelsteller and Buterin 2015). Fungible tokens indicate ownership of an underlying homogenous good with a universal value attached to it, such as gold. They are thus interchangeable. The ERC-721 standard regulates non-fungible tokens, which are characterised by non-interchangeability (Entriken et al. 2018). An early example of such tokens are "CryptoKitties". The idea was that a smart contract autonomously creates unique digital cats every 15 minutes which represent tokens and are stored on the Ethereum ledger.[16]

Other (arguably more useful) scenarios include the tokenisation of artwork, real estate, and securities. For illustration, we draw on two examples given in Voshmgir (2020). To invest in a fraction of a piece of art, a specific token could be issued (e.g. on the ERC-721 standard), where one token is associated with the ownership of a fraction of a specific piece of art. This would enable investment in a painting with considerably lower funds than would otherwise be required to buy an entire piece of art. Similarly, real estate could be tokenised and, for example, finance a credit to buy a house. Tokens (e.g. on the ERC-721 standard) that resemble ownership of a fraction of a house could be created and sold to investors. These investors would then be compensated by fractional rent payments from the owner of the house. This would foster penny-stock-like investments in specific real estate objects with low amounts of capital.

The two examples are built on an interconnection between the blockchain world and the real world. The tokenisation of assets with a real-world representation (such as art and real estate) is usually referred to as asset-tokenisation, to distinguish it from tokens purely related to the blockchain space. The OECD (2020b) issued regulatory concerns about asset-backed tokenisation because it requires trusted centralised authorities that guarantee that the blockchain token has its claimed value in the real world. Whenever this connection is contested (e.g. motivated by opportunism), contracts have to be enforced by legal institutions. In any case, it seems that asset tokens contradict the design-philosophy of a purely decentralised system that wants to minimise dependence on centralised parties.

As a general observation, a properly functioning blockchain system creates, contains, and updates token data according to its internal rules in a credible way. But that does not include

14 Szabo's (1998) concept was, however, never implemented due to technological limitations at the time.

15 The indication of ownership can be either attached to existing native tokens, implemented on a new layer on top of an existing blockchain, or based on smart contracts. The latter method is most common (for example, Schär 2020).

16 www.cryptokitties.co

the creation of consensus around the interpretation of this data, its application and consequences for the world outside the blockchain. As a result, any potential transaction cost savings resulting from blockchain-operated token systems must be considered in the context of transaction costs involved in mechanisms referencing blockchain data to the outside world in terms of valuation and enforcement (Narayanan et al. 2016).

There may be demand for tokenisation of indivisible assets for facilitating their shared use over time among multiple owners (car sharing, sharing a country house etc.). If tokens are tradeable on markets, this contributes to the commodification (expansion of markets) and financialisation (creating claims on marketable assets that are tradeable on secondary markets) of non-financial asset classes. This might lead to the creation of possible positive as well as negative externalities with regard to their accessibility, price, and quality for consumption purposes of these assets among non-owners.

Tokens directly or indirectly related to traditional securities (e.g. debt, equity) are particularly interesting from a financial as well as legal perspective. Two terms commonly used in this context are *security tokens* and *tokenised securities*. The latter simply refers to tokens that represent traditional securities on a new technology. Associated benefits could include increased transparency, lower transaction costs, and lowered access barriers. *Security tokens* focus more directly on the blockchain technology and encompass, for example, any native tokens that pay out dividends. Several use cases will be discussed further later. Put differently, tokenised securities wrap an existing product into a new package, while security tokens represent a new product themselves (Acheson 2019).[17] Tokenised securities are less problematic from a legal point of view because traditional securities are a well understood concept. Security tokens may be more complicated in this regard. Legal institutions have to define which tokens mimic the behaviour of securities and should thus be regulated.

Decentralised finance (DeFi)

Recently, decentralised finance (DeFi) has become the new hype in the blockchain community. The term builds around the concept of providing traditional financial services, empowered by tokenisation (as discussed previously), in a decentralised version. Examples include borrowing, lending, investment, trading, and payment services in a decentralised, peer-to-peer version. All these services are provided by decentralised Apps (dApps) on a blockchain system, each of which may feature distinct tokens. DeFi promises more security, higher efficiency, and more inclusiveness (Sandner 2019). This stands in stark contrast to related traditional systems. The financial sector mainly produces promises to pay (shares, bonds, credits, etc.), and financial services are built around these promises (e.g. creation, pricing, trading, assessment, enforcement). DeFi starts from the basic crypto idea that trust in promises is best avoided (crypto coins are no one's liabilities) and tries to create financial instruments and services from a different angle. Although DeFi thus seems to be a particularly intriguing development in the crypto world, a complete discussion of all DeFi applications is clearly outside the scope of this contribution. The following paragraphs will therefore feature a short discussion of selected concepts that seem most promising to us.

Centralised cryptocurrency exchanges. Cryptocurrencies can either be traded against each other or against official currency. The latter version creates a direct interface between the off-chain and on-chain world and is therefore of particular importance. It is, however, also very controversial.

17 The terms are, however, still used inconsistently in the literature.

Any direct connection to the off-chain world makes the central idea of blockchain – being decentralised and independent from centralised institutions – less feasible. Imagine Bob wants to enter the blockchain world and take advantage of DeFi services. First, he has to convert some official currency into cryptocurrencies. Centralised exchanges (CEX) facilitate exchanges between official and cryptocurrencies as well as between cryptocurrencies. However, CEX are akin to traditional intermediaries and thus have to be trusted not to mismanage funds and to provide security. Private keys are often managed by centralised institutions, thus giving up full control over one's tokens. In any case, native blockchain-like security is, due to the centralised nature of such services, generally not achievable (see for example, Voshmgir 2020). Furthermore, CEX are subject to legal regulation, such as KYC and AML requirements.[18]

Decentralised cryptocurrency exchanges. Decentralised exchanges (DEX) alleviate some of the issues plaguing CEX, at least in the case of exchanges between cryptocurrencies (see, for example, Voshmgir 2020). Such systems avoid the intermediary characteristic of centralised exchanges. This is achieved, for example, through the use of Atomic Swaps. Imagine Bob successfully exchanged official currency for BTC. Now, he wants to exchange BTC for ETH. Conveniently, Bob knows Alice, who wants to trade the equivalent amount of ETH for BTC. A special smart contract can ensure that both parties receive their respective cryptocurrencies after the transaction is completed. DEX would utilise this concept and enhance it with matching algorithms to minimise the coincidence-of-wants problem associated with Atomic Swaps. However, such DEX cannot bridge the off-chain/on-chain gap in a decentralised way. The exchange of official currency for cryptocurrency requires coordination with banks or other third parties who handle cash exchanges. One option for this gap to be bridged in a decentralised version might be the tokenisation of official currencies. That is the business model behind a number of stable coin projects discussed earlier.

Decentralised payment systems. Bitcoin was initially developed with the aim of providing a decentralised, peer-to-peer alternative to traditional payment systems (Nakamoto 2008). However, compared to traditional payment systems, the Bitcoin network seems rather inconvenient. A block is completed about every ten minutes on average, and the system manages 3–4 transactions per second.[19] Transaction fees on the Bitcoin network fluctuate widely.[20] However, dApps, such as the Lightning Network, promise fast transactions with low to zero transaction costs. The application is built on top of the Bitcoin network, on which two users can exchange BTC without the need to validate each transaction. Only when two users have stopped their transactions, is the final amount of BTC each holds added to a block and validated on the original network layer. While this may compromise security to some degree, the added speed and low fees make this dApp an interesting solution for day-to-day transactions of small size (Cointelegraph n.d.).

Decentralised credit systems. Theoretically, blockchain-based decentralised lending and borrowing services offer lower operational costs, more control, and more security compared to traditional systems that rely on banks as intermediaries (Voshmgir 2020). In decentralised credit systems, users can provide (lend) cryptocurrencies to a pool and earn interest on their assets. This pool serves as the basis for borrowers to withdraw cryptocurrencies from. However, to borrow, users have to provide collateral to the pool, as traditional KYC processes are not part of decentralised credit systems. Still, borrowers can potentially profit from lower interest rates compared to traditional systems. Thus, both lenders and borrowers might benefit from arbitrage

18 In the US, the regulation of such exchanges has recently been reinforced by the director of the Financial Crimes Enforcement Network (FinCEN), Kenneth Blanco (Chavez-Dreyfuss 2019).

19 www.blockchain.com/charts/transactions-per-second.

20 Currently at around 4 USD. See www.blockchain.com/charts/fees-usd-per-transaction.

possibilities (Sandner 2019). Furthermore, tokenisation coupled with decentralised credit systems allows for any form of tokenised asset to serve as collateral. This would further bridge the gap between the traditional and decentralised world.

Token sales. The initial sale of tokens has become a popular alternative to equity-crowdfunding and for young companies to accrue capital beyond the usual FFF group (family, friends, and fools). Token sales are usually issued via cryptocurrency exchanges and promise various benefits, some of which inherently stem from the blockchain technology as such (e.g. increased security, no need to trust third parties). Other benefits include efficiency gains over traditional secondary markets that facilitate crowdfunded equity shares trading (e.g. lower costs, lower counterparty-risk) (Schär 2020). Token sales can be divided into several categories. Initial Coin Offerings (ICOs) started in 2013 and many of these were issued as utility tokens, which face lax regulations, as they are not counted as securities. This inevitably invited malicious and opportunistic behaviour, which became most evident in 2017, when 80% of ICOs were identified as scams (Satis Group 2018). This fraudulent behaviour invited harsh criticism and strengthened the public perception of cryptocurrencies as a purely speculative and shady technology. More recently, Security Token Offerings (STOs) gained traction. These tokens are explicitly designed with securities in mind and are, thus, subjected to financial regulation, which helps to strengthen trust in non-opportunistic behaviour from issuers. STOs may nevertheless be prone to the use of cheap signals, e.g. exaggeration or faked information and promises (Ante and Fiedler 2019).

A note on background conditions

Before we conclude our contribution, we want to elaborate on an often overlooked but instrumental concept in the evaluation of how cryptocurrencies might shape society: background conditions, or, put differently, the rules of the game (North 1990). It is sometimes argued that crypto assets (and the blockchain technology in general) reduce certain transaction costs – even though different transaction costs might arise. For now, assume that crypto assets do in fact reduce overall transaction costs. Whenever these costs can be reduced, the amount of mutually beneficial transactions in an economy increases. However, is this always socially beneficial?

According to the economic principle of free contract, this question could be answered in the affirmative. It states that a contract between two adults adequately represents each party's evaluation of the outcome and should thus not be interfered with. There are at least two main reasons for why such a conclusion would be premature. Firstly, even though a transaction might be mutually beneficial for the contracting parties, it might negatively affect other individuals. Secondly, the principle of free contract neglects the potentially unequal distribution of the underlying background conditions.

In essence, background conditions constitute the rules of the game under which contracts (transactions) are signed. For the purpose of our contribution, it suffices to say that these conditions may encompass the prevailing regulatory framework, norms, and standards in an economy as well as the distribution of economic property rights. In particular, background conditions shape feasible alternatives to a contract (i.e. outside options) for participants in an exchange. For an excellent and more sophisticated discussion of background conditions, we refer the interested reader to Sturn (2009).[21]

21 Although Sturn (2009) focused on the labour market, many analogies can be drawn in the context of this contribution. Indeed, Sturn (2020) put background conditions in the context of blockchain technology and network effects in a German text.

Whenever background conditions are unequally distributed, one contracting party is usually in the position to extract a major share of the gains from a transaction. Due to a lack of attractive alternatives, the true preferences of the other (exploited) party may not be revealed. The rules of the game should then ultimately not only be designed to maximise the amount of mutually beneficial transactions, but also to minimise the number of transactions that result out of unequally distributed background conditions. Traditionally, these conditions have been shaped by the public sector to a large degree and provided a fruitful environment for the private sector to unfold its productive potential (see Sturn 2009, 2020).

So far, the most popular crypto projects reveal a preference for the ideal of a pure market economy. They create new asset classes whose creation, accession, and transfer are governed by market mechanisms and principles. Additionally, these rely on both asset and market designs intended to avoid the involvement of entities belonging to or being regulated by the public sector. In this libertarian vision, public sector activity and democratic forms of decision making on economic affairs are perceived as both illegitimate and economically distortive, whatever the initial distribution of background conditions and whatever the results produced by markets (Berg et al. 2019; Golumbia 2016; Scott 2014).[22]

Concluding remarks

Following the structure of this chapter, we will start our concluding remarks with the monetary system. Crypto assets are often regarded and treated as a speculative asset, due to the wide fluctuation of their value vis-à-vis traditional currencies. Therefore, they are not particularly suited as a predictable store of value, which is a crucial concept defining money. Stable coins are a different matter in this regard (as they aim to uphold a stable peg to some commodity, often official currencies) and have to be distinguished from Bitcoin-like crypto assets. In any case, both types of crypto assets as of today cannot be regarded as a medium of exchange nor a unit of account due to the absence of their wider adoption.

This raises the question of what stifles this wider adoption. Transaction costs in the form of costs associated with switching currencies may provide an answer. Established currencies offer huge network effects and currency users are characterised by some degree of inertia resulting from switching costs. These might be quite large. Cryptocurrencies are built on the philosophy to abstract from centralised third parties to facilitate transactions. The blockchain technology that underpins crypto coins provides the techno-economic tool kit to achieve this. Blockchain-based decentralisation of governance poses limits to an asset producing the kind of stability of economic value that would be competitive compared to official currencies. Technically increasing the predictability of a marketable asset's supply does not automatically result in a stable market value. Therefore, some successful stable coins try to join existing official currency networks by attempting to tie the value of their assets to official currency with various techniques.

A key raison d'être for crypto assets is their decentralised architecture. Indeed, the motivation behind the first decentralised peer-to-peer payment system (Bitcoin) came from the mistrust vis-à-vis traditional centralised third parties (e.g. banks). Crypto assets, underpinned by the blockchain technology, should eliminate the need to trust such centralised institutions. The blockchain indeed provides the techno-economic functionalities to facilitate transactions in a decentralised way. Nonetheless, the technology is not entirely "trustless". Trust is simply

22 Nevertheless, there are also efforts to explore the potential of the blockchain technology for projects inspired by a broader range of governance visions (Berryhill et al. 2018; Scott 2016).

shifted to other dimensions; now trust has to be put in the consensus mechanisms as well as the robustness of the specific blockchain architecture employed. But even the decentralised nature of transactions might, in reality, fail to emerge as an indirect consequence of transaction costs.

Cryptocurrency users have to incur search costs for finding someone accepting cryptos as payments, costs of monitoring adequate price conversion (from crypto to official currencies), the cost of protecting and the risk of losing the private key, learning costs, the risk of making an irreversible transaction, etc. For all these risks, centralised intermediary services have developed in the crypto sector that offer to absorb some of these risks if you trust them – for a fee. In any case, the question whether a more decentralised monetary system would even be beneficial to our society remains. Clearly, a decentralised monetary system would bestow certain privileges on certain stakeholders. They might use their privileges opportunistically and try to extract extra profits for themselves.

Having these transaction costs in mind, crypto assets still do offer the potential to impact the financial sector. One aspect to mention in this regard is the concept of the tokenisation of finance or entirely new financial instruments that are native to the blockchain technology. Another buzzword in this context is decentralised finance (DeFi). The concept promises, for example, increased inclusiveness, stronger security, and higher profitability. The first two examples are more closely related to the decentralised nature of the blockchain technology. It is not dependent upon centralised intermediaries, such as banks. Thus, anybody with the required experience can theoretically participate in DeFi. Moreover, there is no need to trust these intermediaries, which strengthens security (in addition to increased cryptographic security). Higher profitability seems to be a more ambivalent topic. DeFi enables the creation of entirely new forms of finance in a sandbox environment. But it is also a fertiliser for speculative and fraudulent behaviour motivated by a lack of accountability. Therefore, regulatory issues seem to continue being a prominent topic surrounding decentralised finance.

Finally, we want to highlight that the assessment of crypto assets also encompasses spheres that tend to be addressed only rarely. Political as well as moral aspects highlight a tension between a decentralised world and equally distributed background conditions, which are usually publicly provided. Private provision (by crypto projects) of these conditions can elicit opportunistic behaviour and ultimately lead to increased inequality. Historically, it has been shown that increased inequality tends to strengthen undemocratic or illiberal responses. Under consideration of these aspects, the role of the government in the public provision of background conditions should not be undermined too strongly. Ultimately, a movement that implicitly threatens the legitimacy of centralised authorities (blockchain, crypto assets) may even necessitate a stronger public sector to defend democratic values.

References

Acheson, N. (2019): Security Tokens vs. Tokenized Securities: It's More Than Semantics. Coindesk. Available at: www.coindesk.com/security-tokens-vs-tokenized-securities-its-more-than-semantics [10.03.2021]

Ante, L. and Fiedler, I. (March 2019): Cheap Signals in Security and Token Offerings (STOs), *Quantitative Finance and Economics* 4(4), 608–639. https://doi.org/10.2139/ssrn.3356303

Berg, C., Davidson, S. and Potts, J. (2019): *Understanding the Blockchain Economy: An Introduction to Institutional Cryptoeconomics*. Cheltenham, UK: Edward Elgar Publishing.

Berryhill, J., Bourgery, T. and Hanson, A. (2018): *Blockchains Unchained: Blockchain Technology and its Use in the Public Sector* (OECD Working Papers on Public Governance No. 28), OECD Publishing. https://doi.org/10.1787/19934351

Buterin, V. (2013): Ethereum Whitepaper. A Next Generation Smart Contract and Decentralized Application Platform. Ethereum. Available at: https://ethereum.org/en/whitepaper/ [05.10.2020]

Buterin, V. (2016): A Proof of Stake Design Philosophy. *Medium*. Available at: https://medium.com/@VitalikButerin/a-proof-of-stake-design-philosophy-506585978d51 [05.10.2020]

Buterin, V. [@VitalikButerin] (2018): To Be Clear, at this Point I Quite Regret Adopting the Term "Smart Contracts". I Should Have Called Them Something More Boring and Technical, Perhaps Something Like "Persistent Scripts". [Tweet; thumbnail link to article], *Twitter*. Available at: https://twitter.com/VitalikButerin/status/1051160932699770882?ref_src=twsrc%5Etfw%7Ctwcamp%5Etweetembed%7Ctwterm%5E1051160932699770882%7Ctwgr%5Eshare_3&%3Bref_url=https%3A%2F%2Fbitcoinist.com%2Fvitalik-buterin-ethereum-regret-smart-contracts%2F [05.10.2020]

Campbell-Verduyn, M. (2019): Blockchains, Trust and Action Nets: Extending The Pathologies of Financial Globalization, *Global Networks* 19(3), 308–328.

Chavez-Dreyfuss, G. (2019): U.S. to Strictly Enforce Anti-Money Laundering Rules in Cryptocurrencies: FinCEN Chief, *Reuters*. Available at: www.reuters.com/article/us-crypto-currencies-fincen/u-s-to-strictly-enforce-anti-money-laundering-rules-in-cryptocurrencies-fincen-chief-idUSKBN1XP1YR [05.10.2020]

Coase, R. H. (1937): The Nature of the Firm, *Econometrica* 4(16), 386–405.

Cointelegraph (n.d.): What Is Lightning Network And How It Works. Available at: https://cointelegraph.com/lightning-network-101/what-is-lightning-network-and-how-it-works [05.10.2020]

Davidson, S., De Filippi, P. and Potts, J. (2018): Blockchains and the Economic Institutions of Capitalism, *Journal of Institutional Economics* 14(4), 639–658. https://doi.org/10.1017/S1744137417000200

DuPont, Q. and Cattapan, A. (n.d.): Alice & Bob: A History of The World's Most Famous Cryptographic Couple. Cryptocouple. Available at: http://cryptocouple.com/#synopsis [05.10.2020]

Ellis, S., Juels, A. and Nazarov, S. (2017): ChainLink: A Decentralized Oracle Network. Available at: https://link.smartcontract.com/whitepaper [05.10.2020]

Entriken, W., Shirley, D., Evans, J. and Sachs, N. (2018): EIP-721: ERC-721 Non-Fungible Token Standard. Ethereum Improvement Proposals. Available at: https://eips.ethereum.org/EIPS/eip-721 [05.10.2020]

Ethereum Wiki (2020): Proof of Stake FAQs. Available at: https://eth.wiki/concepts/proof-of-stake-faqs [05.10.2020]

EthHub (n.d.a): What is Ethereum? Available at: https://ethhub.eth.link/ethereum-basics/monetary-policy/ [05.10.2020]

EthHub (n.d.b): Governance. Available at: https://ethhub.eth.link/ethereum-basics/governance/[05.10.2020]

Frolov, D. (2021): Blockchain and Institutional Complexity: An Extended Institutional Approach, *Journal of Institutional Economics* 17(1), 21–36.

Golumbia, D. (2016): *The Politics of Bitcoin. Software as Right-Wing Extremism*. Minnesota: University of Minnesota Press.

Hayes, A. (2020): What Happens to Bitcoin After All 21 Million are Mined? *Investopedia*. Available at: www.investopedia.com/tech/what-happens-bitcoin-after-21-million-mined/ [05.10.2020]

Hileman, G. (2019): State of Stablecoins (2019). https://doi.org/10.2139/ssrn.3533143

Houben, R. and Snyers, S. (2018): Cryptocurrencies and Blockchain. Legal Context and Implications for Financial Crime, Money Laundering and Tax Evasion, *European Parliament*. https://doi.org/10.2861/263175

Krugman, P. R. (2010): Why is Deflation Bad? *New York Times*. Available at: https://krugman.blogs.nytimes.com/2010/08/02/why-is-deflation-bad/ [05.10.2020]

Krugman, P. R., Obstfeld, M. and Melitz, M. J. (2012): *International Economics. Theory and Policy*. 9th ed. New York: Pearson Education Limited.

MakerDAO (n.d.): The Maker Protocol: MakerDAO's Multi-Collateral Dai (MCD) System. Available at: https://makerdao.com/de/whitepaper/ [05.10.2020]

Milgrom, R. and Roberts, J. (1992): *Economics, Organization and Management*. Englewood Cliffs: Prentice Hall International.

Nakamoto, S. (2008): Bitcoin: A Peer-to-peer electronic Cash System, *Bitcoin.org*. Available at: https://bitcoin.org/bitcoin.pdf [05.10.2020]

Narayanan, A., Bonneau, J., Felten, E., Miller, A. and Goldfeder, S. (2016): *Bitcoin and Cryptocurrency Technologies. A Comprehensive Introduction*. Princeton: Princeton University Press.

North, D. C. (1990): *Institutions, Institutional Change, and Economic Performance*. Cambridge, MA: Cambridge University Press.

OECD (2020a): Taxing Virtual Currencies. OECD Tax Policy Analysis. Available at: www.oecd.org/tax/tax-policy/taxing-virtual-currencies-an-overview-of-tax-treatments-and-emerging-tax-policy-issues.htm [05.10.2020]

OECD (2020b): The Tokenisation of Assets and Potential Implications for Financial Markets. OECD Blockchain Policy Series. Available at: www.oecd.org/finance/The-Tokenisation-of-Assets-and-Potential- Implications-for-Financial-Markets.htm [05.10.2020]

Sandner, P. (2019): Decentralized Finance (DeFi): What Do You Need To Know? *Medium*. Available at: https://medium.com/@philippsandner/decentralized-finance-defi-what-do-you-need-to-know-9cd5e8c2a48 [05.10.2020]

Satis Group (2018): Cryptoasset Market Coverage Initiation: Network Creation. Available at: https://research.bloomberg.com/pub/res/d28giW28tf6G7T_Wr77aU0gDgFQ [05.10.2020]

Schär, F. (2020): Decentralized Finance: On Blockchain- and Smart Contract-based Financial Markets. https://doi.org/10.13140/RG.2.2.18469.65764

Scott, B. (2014): Visions of a Techno-Leviathan: The Politics of the Bitcoin Blockchain, *E-International Relations*. Available at: www.e-ir.info/2014/06/01/visions-of-a-techno-leviathan-the-politics-of-the-bitcoin-blockchain/ [05.10.2020]

Scott, B. (2016): How Can Cryptocurrency and Blockchain Technology Play a Role in Building Social and Solidarity Finance? *UNRISD Working Paper 2016–1*. United Nations Research Institute for Social Development. Available at: www.unrisd.org/80256B3C005BCCF9/(httpPublications)/196AEF663B617144C1257F550057887C?OpenDocument [05.10.2020]

Sturn, R. (2009): Volenti Non Fit Iniuria? Contract Freedom and Labor Market Institutions, *Analyse & Kritik* 31(1), 81–99. https://doi.org/10.1515/auk-2009-0104

Sturn, R. (2020): Überwachungskapitalismus, Überwachungsstaat und Öffentlichkeit: Politische Ökonomie der Digitalisierung, *Normative und institutionelle Grundfragen der Ökonomik* 18, 245–278.

Szabo, N. (1998): Secure Property Titles with Owner Authority. Satoshi Nakamoto Institute. Available at: https://nakamotoinstitute.org/secure-property-titles/ [05.10.2020]

Tether (2020): Digital Money for a Digital Age: Global, Fast, and Secure. Available at: https://tether.to/ [05.10.2020]

The Economist (2015): The Trust Machine. Available at: www.economist.com/leaders/2015/10/31/the-trust-machine [05.10.2020]

Tzanetakis, M. (2018): Social Order of Anonymous Digital Markets. Towards an Economic Sociology of Cryptomarkets, in Potter, G., Fountain, J. and Korf, D. (eds.): *Place, Space and Time in European Drug Use, Markets and Policy*. PABST Science Publishers, 61–80. https://doi.org/10.25365/phaidra.52

Vogelsteller, F. and Buterin, V. (2015): EIP-20: ERC-20 Token Standard. Ethereum Improvement Proposals. Available at: https://eips.ethereum.org/EIPS/eip-20 [05.10.2020]

Voshmgir, S. (2020): *Token Economy: How the Web3 reinvents the Internet*. 2nd ed. Berlin: Shermin Voshmgir, BlockchainHub Berlin.

Weber, B. (2018): *Democratizing Money? Debating Legitimacy in Monetary Reform Proposals*. Cambridge, MA: Cambridge University Press.

Williamson, O. E. (1985): *The Economic Institutions of Capitalism: Firms, Markets, Relational Contracting*. New York: The Free Press.

Zingales, L. (2015): Does Finance Benefit Society? *NBER Working Paper 20894*. National Bureau of Economic Research. https://doi.org/10.3386/w20894

25

BLOCKCHAIN AND THE "SMART-IFICATION" OF GOVERNANCE

The last "building block" in the smart economy

Brendan Markey-Towler

Introduction: blockchain as the last necessary building block in the smart economy

Blockchain has an almost romantic beginning. On 31 October 2008 (Halloween, in point of fact), a mysterious white paper was released to the internet entitled "Bitcoin: A Peer-to-Peer Electronic Cash System" under the authorship of Satoshi Nakamoto (2009). To this day, nobody knows for sure who Satoshi Nakamoto is. Whoever he/she/they is/are, Satoshi had good reason to be jealous of their anonymity: they had just invented a highly secure form of currency tradeable peer-to-peer over a decentralised anonymised network. In other words, they had invented a cryptocurrency that needed neither government nor bank nor any centralised authority to operate and handed possibly the greatest enabling technology since the internet to the cryptoanarchist movement (Ludlow 2001).

In order to build Bitcoin, however, Satoshi Nakamoto needed to invent an underlying technology for a distributed ledger of currency holdings to be held and updated by a decentralised network of computers on the internet through a consensus algorithm. It is this technology that became known as blockchain (Swan 2015; Swan et al. 2019; Mouyagar 2016; Tapscott and Tapscott 2016), so called for the way it enables a decentralised network of computers to come to consensus on a "block" of new records to be added to a "chain" of such blocks held by all nodes on the network. This was revolutionary, for while the first application and, at the time of writing, majority application of this technology was to recording purchasing power holding and transfers – i.e. supporting cryptocurrencies – there is no restriction on the kind of records that can be kept in such a ledger. In particular, blockchain allowed for the incorporation of a "smart contract" that executes algorithmically upon the obtention of certain conditions (Szabo 1994) into the distributed, decentralised ledger of records held and updated by the network. Three particularly notable blockchains emerged to support such contracts – Ethereum (Buterin 2013), EOS (Grigg 2017) and NEO (NEO 2018) – and facilitate the construction of, among other things, Decentralised Autonomous Organisations (DAOs) of interlinked smart contracts.

Our argument in this chapter is that Satoshi Nakamoto, perhaps inadvertently, invented the last essential "building block" in the smart economy – the last technology necessary for a nearly fully decentralised, distributed and automated economic system to emerge. Satoshi did

DOI: 10.4324/9780429351921-30

this by inventing a new institutional technology that allows for decentralised, distributed and automated governance of internet-based platforms. With blockchain Satoshi invented, in other words, a new smart institutional technology for governance. Where robotics and artificial intelligence, the internet of things and platform technologies combine into smart technologies that enable decentralisation, distribution and automation of production and exchange (Brynjolfsson and McAfee 2014; McAfee and Brynjolffson 2017; Agarwal et al. 2019; Sullivan and Zutavern 2017; Parker et al. 2016; Schwab 2016), blockchain decentralises, distributes and automates the governance of production and exchange. With blockchain added to the bundle of smart technologies available to us, therefore, we may decentralise, distribute and automate not only production and exchange systems, but also the governance of those production and exchange systems. We therefore have the smart technologies that enable a nearly complete smart economy that was hitherto impossible.

To establish this argument, and uncover avenues for future research, we first will introduce an evolutionary-institutional perspective on the economy as a complex evolving network of rules for production, exchange and governance activities. We then introduce the argument that pre-existing smart technologies – robotics, artificial intelligence, the internet of things and platforms among them – have the effect of decentralising, distributing and automating this network structure primarily at the levels of production and exchange. We then introduce blockchain as an institutional technology that enables a new kind of governance for internet-based socioeconomic platforms. We discuss how this provides the last "building block" that allows for the near-complete decentralisation, distribution and automation of economic systems through the application of smart technologies by enabling smart governance. We conclude by discussing how a new kind of economics must emerge to understand such systems, one that integrates institutional cryptoeconomics into the core of an evolutionary-institutional perspective on economies as complex evolving networks of rules.

Conceptualising economies as networks of rules: an evolutionary-institutional perspective

If we are to assess the form of an economy enabled by smart technologies, with them integrated into the core of its technological base, we ought to adopt a conceptualisation of economies that is designed for the purpose of considering the impact of technology on their form. This is something neoclassical economics struggles with, being less an exercise in understanding the response of the structure of economic systems to changes in technology as an exercise in understanding their impact on the relationship between macroeconomic aggregates of output, labour and capital (Nelson and Winter 1982). Instead of neoclassical economic perspectives, therefore, we will adopt a fused evolutionary-institutional perspective that is designed to consider the way that technologies enable new forms of economic interaction and organisation. We will adopt, in particular, the micro–meso–macro framework (Dopfer et al. 2004; Dopfer and Potts 2007) introduced by Kurt Dopfer, John Foster and Jason Potts in a seminal 2004 article that effectively fuses the evolutionary economics tradition (Nelson and Winter 1982; Metcalfe 1998) with the institutional economics tradition (Hodgson 1998; Williamson 1998).

The micro–meso–macro framework begins with the economic agent as a cognitively constrained human being who applies comparatively simple but adapted rules to guide their behaviour in economy and society (Dopfer 2004). Where these rules guide interaction with other human beings and artefacts, they create connections in a network that agglomerates across the population to form the economy. As new technologies and organisational and governance rules are introduced, these rules adapt to them and cause behaviour, and thus the connective structure of the economy, to change and evolve. The economy becomes a complex evolving network structure created by rules that evolve and change over time (Potts 2001; Foster 2005). The economy becomes a complex evolving network of rules.

As we try to conceptualise the complex network of rules that the economy comprises at a greater and greater scale, we require a conceptual device that allows us to categorise subpopulations usefully in order to see the macro-scale structure of the economy without loss of generality. This conceptual device is the meso-rule. A meso-rule is a rule that takes a form of sufficient commonality across a subpopulation of the economy such that we can agglomerate the members of that population into a whole – a meso-population – and observe their relationships with other meso-populations treated as wholes. Hence, the micro–meso–macro framework allows us to understand economic structures at the micro-scale of the individual rule-following human being and their immediate network neighbourhood in the economic system, and at the macro-scale of interconnected meso-populations.

For our purposes, we may think of the economic system as categorised into three superordinate sets of meso-rules: meso-rules for production of outputs by the transformation of inputs; meso-rules for exchange of those outputs and inputs between agents; and meso-rules for governance (organisation and dispute resolution) of those activities. We may think in these terms as, when considering smart technologies, we are considering not an individual technology that will disrupt the rule structure and interaction behaviour of a particular meso-population within the macroeconomic structure but rather general purpose technologies which disrupt the rule structure of *all* meso-populations and disrupt the whole of the macroeconomic rule network structure (Lipsey et al. 2005). Smart technologies may have effects across the whole of the production rule network of the economy, the whole of the exchange rule network of the economy and the whole governance rule network of the economy.

Thus, we arrive at a conceptualisation of the economy as a complex evolving network of rules that can be categorised into interconnected meso-populations at a macroeconomic level. This network structure evolves at micro, meso and macro levels as new technologies emerge, and rule structures adapt to harness the expansion of human capabilities they enable. With general purpose technologies such as smart technologies, the network structure evolves across *all* meso subpopulations as rules change in response to the expansion of human capability in production, exchange and governance.

The smart economy as automation and distribution of economic rule networks: robotics, artificial intelligence and the internet of things

For our purposes, we could say that smart technologies arise from the confluence of three general purpose technologies: information processing technology (IPT), information communication technology (ICT) and cybernetics. Information processing technology is enabled by the computer and its capacity for expanding the range of human capability in the transformation of input information into useful output information. Information communication technology is enabled most powerfully by the internet and its capacity for expanding the range of human capability in the transportation of information between computers and the information processing technology manifest within them. Cybernetics allows for each of these technologies to be embedded within physical machines and a loop created whereby that machine, and others it is connected to, produces information inputs for computations that then control the function of that machine. Smart technology allows us to all but fully automate physical processes with machines, distribute those physical processes across a network of machines and effectively decentralise that network.

We are more familiar with manifestations of this confluence of IPT, ICT and cybernetics in smart technologies by other names. These technologies include robotics, artificial intelligence, the internet of things and platforms. Robotics is a direct outgrowth of IPT, ICT and cybernetic technology and is significantly enhanced by the infusion of artificial intelligence within its computational control systems. Robotics expands the range of human production capability to an extraordinary extent to the point of requiring human intervention only for maintenance and oversight (Brynjolfsson and McAfee 2014; Schwab 2016). Artificial intelligence is information processing technology taken to its logical conclusion of automating and exceeding the human capacity for cognitive information processing (von Neumann 1958; Simon 1968), especially once machine learning is overlaid that allows for the processing system to be updated in response to performance. Artificial intelligence drastically expands the range of human capability in information processing, calculation and computing to an extraordinary extent to the extent of only requiring human intervention to program a basic predictive objective and base updating algorithm into a computer (Agarwal et al. 2019; Sullivan and Zutavern 2017). The internet of things allows us to connect artificially intelligent cybernetic control systems to a network of similar systems and distribute highly integrated production systems across unprecedented scales (McAfee and Brynjolffson 2017; Schwab 2016). Indeed, it allows us to connect entire cities or electricity grids into one smart grid of cybernetic control.

All these technologies combine to automate, decentralise and distribute the myriad technologies underlying production and exchange rules in socioeconomic systems, gradually transforming it into a largely (but before blockchain, not yet wholly) smart economy. We observed the rise of smart technologies in production perhaps most notably with the rise of Japanese car manufacturing in the latter 20th century, with artificially intelligent cybernetic control systems dramatically scaling their capabilities and putting them at the core of global manufacturing. Naturally, this trend has spread across the world so that not only manufacturing is increasingly enabled by smart technology (Brynjolfsson and McAfee 2014; McAfee and Brynjolffson 2017), but service delivery as well (Agarwal et al. 2019; Sullivan and Zuvatern 2017). With the rise of e-commerce (most famously, eBay), and their next-generation descendants in the internet platforms such as Amazon, Uber, Spotify, Google and Facebook (especially Facebook "Marketplace"), smart production technologies can now be "plugged" into highly automated, distributed interfaces for human users to interact and exchange the outputs of those technologies (McAfee and Brynjolffson 2017; Parker et al. 2016).

What, then, does an economy, a smart economy, that integrates these smart technologies into its technological base look like? Firstly, a smart economy is one that at least integrates robotics, artificial intelligence, the internet of things and platforms into its technological core, such that that core becomes characterised by advanced interconnected cybernetic control systems embedded within work mechanisation technologies inherited from the first industrial revolutions (Landes 1969). This means the range of human capability in production and exchange can be expanded to cornucopian extents through near complete automation of physical work transforming inputs into outputs and cognitive work transforming base into useful information. It also means, because of the power of the communication technology created by the internet, that that capability can be distributed over integrated production and exchange systems at a scale hitherto unimagined.

A smart economy is therefore a network of production and exchange rules that is nearly fully automated and distributed over a substantially decentralised system. It has production and exchange capabilities in supply chains that require vanishingly little human physical or cognitive work and that can extend over vast geographic surfaces and integrate with other systems at virtually unlimited scale. To the extent that human beings need to interact with these automated, distributed and decentralised systems and with each other, they can do so within internet enabled platforms that effectively lift geographical constraints on said interaction. It is, increasingly, an economy where a human being need hardly leave their home but for leisure and work. They hardly even need interact with economic exchange networks at all – they need only intervene when artificial intelligences incorrectly predict their preferences. The economy becomes a network of automated, distributed and decentralised interconnected machines subject to automated, distributed and decentralised cybernetic control.

Yet, prior to the advent of blockchain, existing smart technologies (including robotics, artificial intelligence, the internet of things and platforms) could not yet affect the *governance* of economic production and exchange. The major platforms, Amazon, Uber, Facebook, Google and so on, are notable for their providing an infrastructure for the coordination of vast, automated, distributed and decentralised systems of production and (particularly) exchange. But they have been noted – especially in the realm of the "culture wars" – for the centralised governance they exert over an otherwise vast, distributed, decentralised network of production and exchange. It is possible to analyse organisational structures and dispute resolution scenarios and even to automate their function once decided using a confluence of IPT, ICT and cybernetics to some extent. But the governance structures themselves which decide the form organisations will take, grant authorisations to perform actions in those structures and resolve disputes remained inherently human and hierarchical. With existing smart technologies, in other words, production and exchange may have become smart and thus automated, distributed and decentralised, but governance did not. Hence, it was necessary for blockchain to be invented to complete the "smart-ification" – the automation, distribution and decentralisation – of this final superordinate category of meso-rules.

Blockchain: an institutional technology enabling governance of internet-based socioeconomic platforms

To understand how blockchain is an institutional technology that enables governance of internet-based socioeconomic platforms, we first need to understand what an institutional technology is, and then some of the technical aspects of blockchain that allow us to see that it is one such technology. We will focus on the most salient aspects of the technology here; a fuller discussion (from an economic perspective) of the technical aspects of the technology and their general importance is provided by Bheemaiah (2015) and Mouyagar (2016). We can then see that blockchain is a new kind of institutional technology characterised by its form as an intermediate between existing institutional technologies, displaying significant contractual and constitutional control of interaction like firms, governments and clubs, yet also distributing and decentralising authority like markets and commons. Because, in this, blockchain automates, distributes and decentralises governance, we can say that it is a smart governance technology, and thus provides the last "building block" necessary for the nearly complete smart economy.

An institutional technology as differentiated from an industrial technology has traditionally been obscured as an object of study, for the former tends to emerge and evolve over centuries and millennia, while the latter emerges and adapts within decades (Allen et al. 2020). Blockchain's unusually rapid evolution is, indeed, what brought the concept of institutional technology to the interest of institutional economists. But, as a technology, institutional technology shares in common with industrial technology that it is, properly defined in the economic sense, a set of rules for the operation of a set of artefacts to expand the range of human capability. However, where industrial technology such as electrical engineering, mechanical engineering, cybernetics and robotics increases the range of human capability in production and exchange, institutional technology increases the range of human capability in the governance (organisation and dispute resolution) of socioeconomic interaction. One may think of institutional technologies as the infrastructure of trust – verifying and validating that certain interactions have or will take place (Novak et al. 2018; Davidson et al. 2018).

Of such institutional technologies, there have traditionally been accounted five. Markets are characterised by the agglomeration of distributed and decentralised mutual exchange contracts between consenting individuals or groups (Hayek 1945; Williamson 1985). Firms are characterised by contracts which establish a command and control hierarchy that obviates the need for extensive and costly negotiation over the performance of certain actions (Coase 1937; Williamson 1985). Governments are characterised by the monopoly granted to them on the use of coercive force by a sufficiently powerful proportion of the population (Downs 1957; Buchanan and Tullock 1962). Commons are characterised by the use of a shared resource subject to emergent social norms emergent from voluntary interaction and association (Ostrom 1990). Clubs are characterised by the use of low-cost goods and services by members of a group who exclude access to non-members through the imposition of membership costs (Buchanan 1965). Blockchain is considered the sixth and is characterised by its manner of establishing governance of interaction on internet-based platforms (Davidson et al. 2018).

Blockchain is, in the pure technological rather than economical sense, a distributed ledger technology (Swan 2015; Swan et al. 2019; Mouyagar 2016; Tapscott and Tapscott 2016). It consists of a set of protocols for the holding and updating of a ledger of records by a distributed and decentralised network of computers within the internet. The data concerning socioeconomic interactions that the parties thereto wish to include in the ledger are broadcast by one of these computers to the network as they occur and are made known to that computer. Periodically, the protocol directs a set of computers, called "consensus nodes" in the network, to collect a set of such records into a block that will be added to the "chain" of such blocks in the ledger (thus, a "block-chain"). If one such block is successfully selected by a consensus algorithm, of which there are many variants but which all ultimately require a majoritarian or super-majoritarian consensus, it is added to the ledger held by the consensus nodes. This is, naturally, automated by computers in the network which support and action the blockchain protocols. If consensus nodes disagree with the block to be incorporated, or the underlying protocols, they may lobby the network as a whole, or engage in what is known as a "fork" whereby they in effect "cryptosecede" from the main network and form a new network with an alternate ledger from that point in time onward.

Why would parties to some interaction wish to include data thereon in a blockchain? The initial reason was that it rapidly becomes an extremely secure and immutable record that cannot be manipulated or tampered with without the malign agent incurring massive costs that would make such activity unprofitable. "Hacking" a blockchain requires a bad actor to not only overpower or game the consensus algorithm, but also to convince the entire network to continue using the modified blockchain rather than "fork" the blockchain at the point the bad actor intervened and erase their gains and influence. By keeping an automated, distributed and decentralised ledger held across nodes in the internet, blockchain creates records that are very difficult to tamper with, and they are additionally, typically, heavily encrypted so that the portions of the distributed ledger pertaining to any given actor can only be read by the holder of their private key. In principle, the data held in a blockchain are therefore vastly more secure, immutable and anonymous than those under the custody of a government or major corporation. It magnifies the effect of the two traditional nonviolent methods for keeping governments or large organisations in check: democracy and secession.

The original data recorded in blockchain was, of course, holdings and transfers of purchasing power, and thus the first blockchains supported cryptocurrencies. But it was quickly realised by Vitalik Buterin particularly that the technology underlying Satoshi Nakamoto's Bitcoin could be used as a basis for implementing Nick Szabo's concept of a "smart contract". With the Ethereum blockchain, and soon after EOS, NEO and a host of others, blockchains thus became ledgers of interconnected smart contracts executing algorithmically upon realisation of certain conditions. Blockchains became the basis for the Decentralised Autonomous Organisation, a set of nodes interconnected into a web of smart contracts for the execution of certain actions under particular conditions.

Blockchain thus facilitates the emergence of protocols that govern the formation, validation and execution of socioeconomic interactions and delimit what can and cannot be included in the record of verified and validated interactions. It facilitates the establishment of authorisations and obligations to perform actions as part of a socioeconomic interaction. Thus, we can say that it is an institutional technology, because it provides for the governance of interactions that take place on internet-based platforms for socioeconomic interaction.

As an institutional technology for the governance of socioeconomic interaction, blockchain is distinguished by the intermediate form it takes, blending aspects of existing institutional technologies. Blockchain reflects the automated, distributed and decentralised nature of the internet for which it provides governance. Like a government or firm or club, it establishes substantial and extensive contractual and constitutional control of socioeconomic interaction through the operation of the verification and validation protocols as well as the formulation and execution of contracts. But, like a market, the web of extensive contractual control emerges through the agglomeration of distributed and decentralised voluntary mutual exchange of free agents. And, like a commons, the constitutional protocols establishing the base norms emerge from the voluntary association of a community organised around a common resource. Extensive obligations and substantial authority are established by blockchain protocols and smart contracts, but these are distributed and decentralised across the network of consensus nodes and socioeconomic agents. The institutional governance of internet platforms for socioeconomic interaction that is enabled by blockchains is, like the internet from which it emerges, highly controlled but in an automated, decentralised and distributed manner.

Thus, we can establish that blockchain is not only an institutional technology that reflects the automated, distributed and decentralised nature of the internet platforms it provides governance for but also a smart technology. Specifically, it is an institutional technology that enables smart governance. The governance structures that are established by blockchain protocols enable, in the first instance, vastly more sophisticated and extensive contracts to be written because they will execute algorithmically. This vastly expands the range of organisational production and exchange interaction that can be automated, and it vastly expands the capacity for the structure of that organisation to be distributed and decentralised because it is automated. But beyond enabling the "smart" organisation – the Decentralised Autonomous Organisation of interlocking smart contracts – the blockchain protocols themselves enable smart governance. Verifying and validating records of the formation and execution of socioeconomic interaction, and thus effectively dispute resolution as well as the overall actioning of institutional governance becomes automated and distributed across a decentralised network to an extent hitherto unimagined. Blockchain enables a form of smart governance that automates, distributes and decentralises the creation and actioning of institutional authority.

The complete smart economy: decentralised, distributed and automated rule networks for production, exchange and governance

Thus far, we have adopted a conceptualisation of the economy as a complex evolving network of rules that can be categorised into interconnected meso-populations, and which evolves as new technologies emerge to harness the expansion of human capabilities they enable. General purpose technologies, of which smart technologies are a kind, cause the network structure of the economy to evolve across *all* meso-subpopulations in the superordinate categories of production, exchange and governance. We have established that prior to blockchain, a smart economy which integrated smart technologies deeply into the core of its technological bases is a network of production and exchange rules that is nearly fully automated and distributed over a substantially decentralised system. It has supply chains that integrate at virtually unlimited scale and require vanishingly little human physical or cognitive work. The economy becomes almost completely a network of automated, distributed and decentralised interconnected machines.

Blockchain provides the final "building block" that is necessary to complete the "smart-ification" of the economy by providing the smart technology for the final superordinate category of governance of the rule networks of complex evolving economies. The governance structures that are enabled by blockchain protocols enable smart contracts that form the basic unit for Decentralised Autonomous Organisations, but they themselves are also automated, distributed and decentralised. Blockchain reflects the structure of the internet whence it emerges in its governance of socioeconomic interactions. It enables a form of smart governance that automates, distributes and decentralises institutional authority.

Thus, if we combine pre-existing smart technologies, including robotics, artificial intelligence, the internet of things and platforms, and introduce blockchains, we have a set of smart technologies that can form the technological base of a nearly completely smart economy. Meso-rules across the superordinate categories of production, exchange and governance become nearly completely automated, highly distributed and substantially decentralised. The human physical and cognitive work that is required to run an economy at global scale is reduced, in principle, to an absolute minimum as human capability is expanded exponentially by automation of production and exchange, but also governance.

What does such an economy "look" like structurally? Present theory in the emerging sub-discipline of institutional cryptoeconomics (led by scholars at the RMIT University Blockchain Innovation Hub) predicts that it will be substantially less hierarchical than existing economic systems as smart technologies infuse into the technology base and find an equilibrium with existing institutional technologies. But it will also be more continually disrupted within the superordinate category of governance meso-rules as the ability for smart governance technologies to be developed using blockchain will be expanded and distributed significantly to the private sector.

Smart technologies infused into meso-rules across production, exchange and consumption are likely to contribute to the emergence of what has been called the V-form organisation (Allen et al. 2019) and de-hierarchicalisation of corporate and regulatory structure more generally (Berg et al. 2018a). Where previously corporate and regulatory structure was M-form and deeply hierarchical with multiple multi-level management departments across major conglomerates and regulatory bodies, smart technologies, and especially smart governance, allow for a V-form organisation between two parties striking one contract registered with one custodian of the verified and validated ledger. The automation of trust by the application of blockchain-enabled smart governance allows layer upon layer of human labour dedicated to verifying and validating to be reallocated (Novak et al. 2018).

Indeed, government, broadly, as an institutional governance structure becomes far less critical in a smart economy, at least in its existing form (Berg et al. 2020). Blockchain-enabled smart governance facilitates the privatisation of large parts of not only institutional governance hitherto actioned by government in the realm of dispute resolution but also its social and distributional functions. Even voting (Allen et al. 2019), identity (Berg et al. 2017) and social welfare (Novak 2018) can be automated, distributed and decentralised by private actors with blockchain-enabled smart governance.

In terms of statics, we may therefore expect an economy that is characterised by a far less hierarchical and human directed, and far more automated, distributed and decentralised rule network structure. But we can expect also novelty in terms of dynamics of this structure. Blockchain smart governance technology expands the range of private entrepreneurial capabilities and competition beyond production and exchange technology to the level of institutional governance (Markey-Towler 2018; Berg and Berg 2017; Berg et al. 2018b; Allen et al. 2020). Institutional entrepreneurship and competition in the past would typically take the form of mass emigration, political revolution or war. The institutional smart governance enabled by blockchain, however, allows for constitutional protocols for institutional governance systems to be drafted by a few individuals with laptops connected to the internet. The accretion of activity onto the platform it supports is entirely voluntary, and not only agents, but also consensus nodes, can "crypto-secede" (MacDonald 2019) from one governance system to another if they become sufficiently dissatisfied. With the scope for institutional discovery and competition thus greatly expanded by blockchain-enabled smart governance technologies, we can expect for the governance structure of economic rule networks to be subject to evolutionary dynamics comparable if not equivalent to those of production and exchange technologies.

The near-complete smart economy that has been made possible by the introduction of blockchain to the bundle of existing smart technologies may lead to an economy that looks quite different to that which we observe today, although traces of its structure can be found. It is complex evolving rule network structure that reflects the internet upon which it is built. Across production, exchange and governance, it is nearly completely automated, distributed and decentralised and subject to substantial evolutionary dynamics across all three categories. It is an economic system that is de-hierarchicalised relative to today's economy, and it is an economy that requires far less human labour to achieve substantially more production, exchange and governance across it. It is an economy where not only the structure of production and exchange meso-rules but also governance meso-rules are subject to disruptive evolutionary pressure through entrepreneurial activity developing and applying smart technologies. The nearly complete smart economy enabled by robotics, artificial intelligence, the internet of things, platforms and blockchains in combination is flatter, faster and fantastical in comparison to the structure which presently exists.

Conclusion: integrating institutional cryptoeconomics into a new kind of economics

We began with a romantic image of a mysterious white paper released on Halloween by an author shrouded in mystery to this day, who handed the cryptoanarchist movement its greatest enabler since the invention of the internet. What we have argued following this image was that, by having to invent blockchain technology, that shadowy author, Satoshi Nakamoto, did more than introduce the world's first true cryptocurrency. We have argued that Satoshi provided the final "building block" technology for the smart economy. Satoshi invented the institutional technology that made smart governance possible and expanded the potential for "smart-ification" beyond the superordinate categories of production and exchange. Satoshi made the nearly complete smart economy, a nearly fully automated, distributed and decentralised economy, possible.

To establish this argument, we first introduced an evolutionary-institutional perspective on the economy as a complex evolving network of rules for production, exchange and governance activities. We then introduced the argument that existing smart technologies – robotics, artificial intelligence, the internet of things and platforms among them – have the effect of decentralising, distributing and automating this network structure primarily at the levels of production and exchange. We then introduced blockchain as an institutional technology that enables a new kind of governance for internet-based socioeconomic platforms. We showed how this provides the last "building block" that allows for near-complete decentralisation, distribution and automation of economic systems through the application of smart technologies by enabling smart governance.

A new kind of economics will be needed to understand the complete smart economy. Existing economic theory, even evolutionary-institutional economic theory, tends to view the economy as either a completely decentralised emergent network of individual voluntary mutual exchanges, or a hierarchical command and control structure. The new economy, even more than the existing one, will be a fuzzy blend of both. It will be characterised by extensive cybernetic control, but that control will be distributed over a decentralised structure in the internet of things, subject to blockchain governance.

Evolutionary-institutional economics is well placed to address the challenge of studying this new smart economy. Indeed, even preliminary formal modelling has begun (Allen et al. 2020), and initial textbooks have been written (Berg et al. 2019) within the subfield of institutional cryptoeconomics. Institutional cryptoeconomics at present is an application of, in particular, Williamsonian institutional economics (Allen et al. 2020) where it focuses on the implications of blockchain for transaction cost structures across the economy, and Ostromian institutional economics (Markey-Towler 2018) where it focuses on the emergence and evolution of different rules for coordinating socioeconomic interaction from blockchain protocols. Much is to be done to integrate the field of institutional cryptoeconomics and its study of smart governance enabled by blockchain with the evolutionary economics of smart technologies and evolutionary economics proper. New models are required that explain the intricacies of the contractual networks that comprise the smart economy, the constitutional protocols that govern it and how these technologies map into the production and exchange rules that actually produce and distribute goods and resources. These new models need to be integrated with our existing models of the evolution of complex economic networks in response to new industrial technologies so that the micro–meso–macro framework may be extended to understand the evolution of meso-rules for *governance* as well as production and exchange. Empirical and policy implications need to be derived and appreciative, history-friendly models must be developed.

The invention of blockchain provided the last building block necessary for the "smart-ification" of the economy. The final piece of the technological puzzle preventing the emergence of a nearly completely smart economy, a nearly fully automated, distributed and decentralised economy, slotted into place when Satoshi released his Bitcoin white paper. To understand the economic world to come, we will need a new kind of economics.

References

Agarwal, A., Gans, J. and Goldfarb, A. (2019): *Prediction Machines*. Cambridge, MA: Harvard Business Review Press.

Allen, D., Berg, A. and Markey-Towler, B. (2019): Blockchain and Supply Chains: V-form Organisations, Value Redistributions, De-commoditisation and Quality Proxies, *Journal of the British Blockchain Association* 2(1), 1–8.

Allen, D., Berg, C. and Lane, A. (2019): *Cryptodemocracy*. Lanham: Rowman and Littlefield.

Allen, D., Berg, C., Markey-Towler, B., Novak, M. and Potts, J. (2020): Blockchain and the Evolution of Institutional Technologies, *Research Policy* 49(1), 103865.

Berg, A. and Berg, C. (2017): Exit, Voice and Forking. Available at: https://ssrn.com/abstract=3081291 [15.02.2021]

Berg, A., Berg, C., Davidson, S. and Potts, J. (2017): The Institutional Economics of Identity. Available at: https://ssrn.com/abstract=3072823 [15.02.2021]

Berg, A., Markey-Towler, B. and Novak, M. (2020): Blockchains = Less Government, More Market, *Journal of Private Enterprise* 35(2), 1–21.

Berg, C., Davidson, S. and Potts, J. (2018a): Capitalism after Satoshi: Dehierarchicalisation, Innovation Policy and the Regulatory State. Available at: https://ssrn.com/abstract=3299734 [15.02.2021]

Berg, C., Davidson, S. and Potts, J. (2018b): Institutional Discovery and Competition in the Evolution of Blockchain Technology. Available at: https://ssrn.com/abstract=3220072 [15.02.2021]

Berg, C., Davidson, S. and Potts, J. (2019): *Understanding the Blockchain Economy*. Cheltenham: Edward Elgar.

Bheemaiah, K. (2015): Why Business Schools Need to Teach About the Blockchain. Available at SSRN: https://ssrn.com/abstract=2596465 [15.02.2021]

Brynjolfsson, E. and McAfee, A. (2014): *The Second Machine Age*. New York: WW Norton & Company.

Buchanan, J. (1965): An Economic Theory of Clubs, *Economica* 32(125), 1–14.

Buchanan, J. and Tullock, G. (1962): *The Calculus of Consent*. Ann Arbor: University of Michigan Press.

Buterin, V. (2013): A Next-Generation Smart Contract and Decentralized Application Platform. Available at: https://github.com/ethereum/wiki/wiki/White-Paper [16.04.2020]

Coase, R. (1937): The Nature of the Firm, *Economica* 4(16), 386–405.

Davidson, S., De Filippi, P. and Potts, J. (2018): Blockchains and the Economic Institutions of Capitalism, *Journal of Institutional Economics* 14(4), 639–658.

Dopfer, K. (2004): The Economic Agent as Rule Maker and Rule User: Homo Sapiens Oeconomicus, *Journal of Evolutionary Economics* 14, 177–195.

Dopfer, K., Foster, J. and Potts, J. (2004): Micro-meso-macro, *Journal of Evolutionary Economics* 14, 263–279.

Dopfer, K. and Potts, J. (2007): *The General Theory of Economic Evolution*. London: Routledge.

Downs, A. (1957): *An Economic Theory of Democracy*. New York: Harper.

Foster, J. (2005): From Simplistic to Complex Systems in Economics, *Cambridge Journal of Economics* 29(6), 873–892.

Grigg, I. (2017): EOS – An Introduction. Available at: www.iang.org/papers/EOS_An_Introduction-BLACK-EDITION.pdf [16.04.2020]

Hayek, F. (1945): The Use of Knowledge in Society, *American Economic Review* 35(4), 519–530.

Hodgson, G. (1998): The Approach of Institutional Economics, *Journal of Economic Literature* 36(1), 166–192.

Landes, D. S. (1969): *The Unbound Prometheus*. Cambridge: Cambridge University Press.

Lipsey, R., Carlaw, K. I. and Bekar, C. T. (2005): *Economic Transformations*. Oxford: Oxford University Press.

Ludlow, P. (2001): *Crypto Anarchy, Cyberstates and Pirate Utopias*. Cambridge, MA: MIT Press.

MacDonald, T. (2019): *The Political Economy of Non-Territorial Exit*. Cheltenham: Edward Elgar.

Markey-Towler, B. (2018): Anarchy, Blockchain and Utopia: A Theory of Political-Socioeconomic Systems Organised Using Blockchain, *Journal of the British Blockchain Association* 1(1), 1–14.

McAfee, A. and Brynjolffson, E. (2017): *Machine, Platform, Crowd*. New York: WW Norton & Company.

Metcalfe, J. S. (1998): *Evolutionary Economics and Creative Destruction*. London: Routledge.

Mouyagar, W. (2016): *The Business Blockchain*. Hoboken: Wiley.

Nakamoto, S. (2009): *Bitcoin: A Peer-to-Peer Electronic Cash System*. Available at: https://bitcoin.org/bitcoin.pdf [16.04.2020]

Nelson, R. and Winter, S. (1982): *An Evolutionary Theory of Economic Change*. Cambridge, MA: Harvard University Press.

NEO (2018): *Neo White Paper*. Available at: https://docs.neo.org/docs/en-us/basic/whitepaper.html [16.04.2020]

Novak, M. (2018): Crypto-Altruism: Some Institutional Economic Considerations. Available at: https://ssrn.com/abstract=3230541 [16.04.2020]

Novak, M., Potts, J. and Davidson, S. (2018): The Cost of Trust: A Pilot Study, *Journal of the British Blockchain Association* 1(2), 1–7.

Ostrom, E. (1990): *Governing the Commons*. Oxford: Oxford University Press.

Parker, G. G., Van Alstyne, M. and Choudary, S. P. (2016): *Platform Revolution*. New York: WW Norton & Company.

Potts, J. (2001): *The New Evolutionary Microeconomics*. Cheltenham: Edward Elgar.

Schwab, K. (2016): *The Fourth Industrial Revolution*. New York: Random House.
Simon, H. A. (1968): *Sciences of the Artificial*. Cambridge, MA: MIT Press.
Sullivan, J. and Zutavern, A. (2017): *The Mathematical Corporation*. La Vergne: Ingram.
Swan, M. (2015): *Blockchain*. Newton: O'Reilly.
Swan, M., Potts, J., Soichiro, T., Frank, W. and Paolo, T. (2019): *Blockchain Economics*. Singapore: World Scientific.
Szabo, N. (1994): Smart Contracts. Available at: www.fon.hum.uva.nl/rob/Courses/InformationInSpeech/CDROM/Literature/LOTwinterschool2006/szabo.best.vwh.net/smart.contracts.html [16.04.2020]
Tapscott, D. and Tapscott, A. (2016): *Blockchain Revolution*. London: Portfolio.
von Neumann, J. (1958): *The Computer and the Brain*. New Haven: Yale University Press.
Williamson, O. (1985): *The Economic Institutions of Capitalism*. New York: Free Press.
Williamson, O. (1998): The Institutions of Governance, *American Economic Review* 88(2), 75–79.

PART 5

Smart technologies and grand societal challenges

26

"BACK TO THE FUTURE"

Smart technologies and the sustainable development goals

J. Carlos Domínguez, Claudia Ortiz Chao and Simone Lucatello

Introduction: from the MDGs to the SDGs

In the year 2000, the UN General Assembly agreed upon a broad international agenda focused on eradicating poverty and tackling some of the most pressing development challenges globally. At the core of this agreement were eight Millennium Development Goals (MDGs). These encompassed concrete targets to be achieved by 2015 (e.g., halving extreme poverty), but more than anything, they were part of an inspirational discourse that attempted to align global efforts for removing some of the main factors hampering the establishment of a "peaceful, prosperous, and just world" (UN General Assembly 2000).

There were remarkable achievements over the following years, including poverty reduction in Asia (particularly China), increasing access to clean water, or HIV/AIDS prevention and treatment worldwide. However, by the time the post-2015 process began in 2010, it was also clear that most efforts would fall short of the original expectations. Progress was concentrated in a few countries, and the international community was still lagging behind in key areas, including environmental sustainability and gender equality (Domínguez and Lucatello 2014: 10–12). Thus, it was necessary to define what would be next: To extend the deadline for achieving the MDGs or to come up with a new agreement, and what kind?

Discussions covered many issues, such as: Should the new agenda be as general as the old one, preserving the inspirational sentiment, providing a common language, and inviting for a broad consensus; or more specific, risking the possibility that many governments did not endorse it and jeopardising the possibilities for a broad consensus? Should the new agenda emphasise goals, despite the limited means that some countries had to achieve them, or vice versa? Should sensitive subjects, such as peace and international security or human rights and governance, be included, or not? Should indicators and targets be defined by each government or by the international community beforehand?

Unlike the first agenda, the SDGs were the result of a multi-layered process involving international agencies, traditional and emerging donors, civil society, and academia, as well as individual countries and country coalitions (e.g., G20, G77, BRICS, or G7+). The process also overlapped and nourished other forums such as Río +20 and climate change negotiations. If the MDGs had been narrowly discussed and agreed upon by heads of states and governments, the SDGs would result from long and extensive deliberations, studies, and massive efforts, including

DOI: 10.4324/9780429351921-32

public consultations in many countries and the involvement of a broad collection of stakeholders and working groups, such as thematic task forces or the UN Secretary-General's High-Level Panel of eminent persons, among others.

Agenda 2030 and the respective Sustainable Development Goals (SDGs) stemmed from these long and strenuous debates. And thus, in comparison to the MDGs, the SDGs are designed with an all-encompassing, maximalist view, trying to tackle development challenges with a more comprehensive approach. It encompasses 17 goals, 169 unique indicators, and 231 targets.

And yet, despite the scope and nuances of the SDGs, there was a noticeable absence during the post-2015 debates: the opportunities and challenges that digitalisation and smart technologies (STs) might create for achieving Agenda 2030. While the early MDGs include a short mention on the need to "make available the benefits of new technologies, especially information and communications" (Target 8.F), the SDGs increased the role of technological adoption but remained mostly within a similar "have vs have not" approach.

The SDGs include the proportion of "youth and young adults with information and communications technology (ICT), skills, and kind of skills" (Indicator 4.4.1), and the proportion of schools with access to internet (Indicator 4.a.1), under the umbrella of Goal 4 (quality education). Other indicators include the number of scholarships available for developing countries in a number of professional areas, including ICT (also as part of Goal 4); "enhancing the use of enabling technology, in particular information and communication technology, to promote the empowerment of women" (Goal 5; Indicator 5.b); quick references to technology upgrading in general (not necessarily digital or smart), particularly in Goals 6, 7, 8, 9, 13, and 14, and a subsection of Goal 17 (global partnership for sustainable development), comprising three short paragraphs, mainly focused on knowledge sharing and technological capacity building, with an emphasis on Least Developed Countries (LDCs).

The lack of a more systematic and integral approach to the impact of digital technologies is striking in the case of both the MDGs and the SDGs, but particularly in the latter, if we consider that the digital revolution had been happening for two decades by the time Agenda 2030 was approved. The World Wide Web was launched in the early 1990s, the euphoria around the dot-coms occurred between 1995 and 2002, and many visionary works suggesting how digital and STs would eventually change our daily lives had already been published in the late 1990s and early 2000s (Castells 1995, 2001a, 2001b, 2002; Mitchell 2000; Negroponte 1995, just to cite a few examples).

The project One Laptop per Child (OLPC) was launched by academics from the Massachusetts Institute of Technology (MIT) in 2005, but it was mainly promoted by private companies and foundations interested in fostering the education and knowledge of children through information. Development institutions and professionals hardly endorsed the initiative, despite the fact that a few heads of state had mentioned the possibility of a digital divide already in the late 1990s (Gore 1998).

The World Bank did not issue a report on digitalisation and development until 2016, and the United Nations Development Programme (UNDP) did not launch a comprehensive digital transformation strategy until 2019. Other instances include the TWI 2050 initiative, where digitalisation constitutes one of six major societal transformations impacting the overarching goal of sustainability (its first report published in 2018) and, more recently, the Technology Executive Committee (TEC) at the United Nations Framework Convention for Climate Change (UNFCCC).

How is it possible that aid and cooperation agencies, international financial institutions, academics, and development consultants remained relatively oblivious to the debate on digitalisation and STs for so long? What are the implications of such omission, where are we now, and where is the debate likely to head in the future?

There are three likely reasons why development agendas have lagged behind the technological debate. First, the Millennium Declaration was drafted with a focus on tackling the needs of the most vulnerable, fighting extreme poverty, and helping developing countries and economies in transition in an increasingly globalised world (UN 2000). The belief that digital technologies (particularly the internet) was a kind of luxury and that basic human development should be guaranteed first (education, health, nutrition, income) likely permeated the philosophy of development professionals in those days. Second, the full potential of STs (the popularisation of mobile apps, AI, big data, machine learning, smart cities, blockchain, among other innovations) had not been properly revealed beyond a small clique of technical experts. And, third, many innovations were (and continue to be) market-driven rather than development-driven (Domínguez 2018).

Looking at these factors from a historical perspective is useful to understand where and why the SDGs stand regarding technology in general and STs in specific. As the international agenda moved from an emphasis on the poorest and least developed societies (MDGs) to include vulnerable populations in developed and industrialised countries, as well as encompassing a broader and more comprehensive set of goals and indicators (SDGs), the role of technology also increased slightly.

Conversely, as technological change has accelerated greatly in the last ten years, its potential to tackle development challenges has also become more evident and development professionals tend to be more enthusiastic regarding these possibilities. In fact, one of the main challenges in the near future lies in avoiding a complete swing from the argument that basic needs should be tackled first, before technology is adopted, towards the argument that technology can be introduced first and solve everything. The development community needs to find a healthy middle point. It needs to go "back to the future" and imagine how STs could shape development in the years to come, but this should not be done blindfolded. In catching up with technological innovations and their impacts on development processes, it is crucial to weigh both threats and opportunities. The following sections offer a few examples.

Can STs help with reducing poverty and eradicating hunger while remaining oblivious to structural inequalities?

The zeal to employ STs as means to achieve the SDGs is reflected in many reports and official documents by international development agencies (FAO 2018; UNDP 2019; World Bank 2016). In one way or another, these reports also adopt different definitions of the "have" vs "have not" problem, as if solving the access to the digital world and the availability of STs constituted the main societal challenge. National policies in many countries have also been designed with such a narrow approach in mind, taking indicators such as the population with access to broadband connections, the number of schools with internet connections, the number of small businesses that use STs, or the number of public services online as key policy indicators. The so-called emerging powers, such as Brazil, India, or Mexico, are a few examples of this trend. No doubt these indicators are necessary to measure one side of the technological equation (access to and availability of infrastructure), but they are not enough.

Thus, from being relatively absent at the dawn of the development agenda, STs are often referred to as a kind of panacea, disregarding the possibility of negative scenarios. These include the possibility that smart devices and apps might not be enough for solving development problems, that the adoption and introduction of STs may actually have side-effects and negative impacts over the long term, or that such strategy might hugely increase inequalities domestically and internationally, leaving the poorest and the most vulnerable behind. Over-optimistic views are mostly held by industry players, but development professionals and academics may also fall into this trap. In fact, in analysing side-effects and negative impacts, these latter often lag behind discussions in the areas of internet and media studies, technology assessment (TA), or sociology of science and technology in general.

Take, for example, the case of the internet. Scholars within sociology and internet studies have researched different aspects of the digital divide in the last 20 years, coming up with new perspectives "on the rise and persistence of digital inequalities" (Ragnedda and Muschert 2018: 2). A bird's-eye view of this research suggests, both theoretically and empirically (Ragnedda and Ruiu 2018; van Deursen and van Dijk 2015), that digital divides can't be pinned down satisfactorily; that there are structural variables, determined first in the offline world (for example, income and education), which greatly limit the potential that online activities have to translate into offline benefits; and, therefore, that digitalisation and STs can actually increase inequality significantly if they are not accompanied by the right policies. A scenario where digitalisation systematically increases inequality inside societies and between countries could inhibit the progress in many indicators that are part of SDGs 1 to 7 (i.e., those related to social outcomes).

The internet is perhaps one of the most widespread and studied technologies in the last two decades, but the same conclusions seem applicable to other STs. For example, as the research published by Vinuesa et al. (2020: 2) suggests, the application of AI has positive prospects as an enabler of different SDGs, such as 1 (no poverty), 4 (quality education), 6 (clean water and sanitation), 7 (affordable clean energy), and 11 (sustainable cities and communities), by supporting the provision of food, education, water, and energy services more efficiently and by contributing to save resources that could – hypothetically – be employed to tackle other social and economic challenges.

And yet, these authors also warn that AI could also trigger deeper inequalities (SDG 10), inhibiting or cancelling out some of these positive achievements in other areas. Some issues include the possibility that AI innovations are often market-driven and respond to market demands, which are located mainly in rich countries or driven by socioeconomically privileged minorities in low and middle-income countries. Moreover, given some features of AI (data analysis, pattern recognition, prediction, interactive communication), these technologies are highly sensitive to specific sociocultural backgrounds. Thus, it is not clear whether legal and ethical dilemmas regarding privacy, surveillance, bias and discrimination, or social control could be exacerbated in LDCs or simply hinder the adoption and usefulness of these technologies.

In the first case (exacerbated dilemmas), the misuse of AI could have an impact on SDG 16 (peace, justice, and strong institutions) given the threat to human rights and the social tensions which could arise from the outcomes mentioned earlier. In the second case (limited adoption), there is a risk that societies with a stronger technological aversion exacerbate their rejection towards development solutions that rely heavily on the use of STs in general. Overall, this could contribute to a persistent gap between those who employ AI to produce socioeconomically meaningful outcomes and those who are not in a position to take advantage of it.

AI, combined with other STs, may increase inequality mainly through its impacts on the labour market. Given the trends towards automation in different economic sectors (mainly services and commerce, but increasingly more in industry), it is likely that a large share of low-skilled jobs will be displaced, whereas some middle-skilled workers might suffer from re-taylorisation and a few high-skilled jobs will be better paid.

These trends will vary across different countries, depending on historical backgrounds, sociocultural attitudes towards technology, and socioeconomic inequalities, as well as the kind of policy interventions that are put in place by their respective governments. For example, it is unclear whether a high-income, industrialised country, like Germany, which has adopted a precautionary principle with regard to the adoption of AI, will enjoy higher technological yields in the long term in comparison to a low middle-income country, like India, which favours a much more aggressive adoption of AI. Despite enjoying a competitive advantage, based on a solid and buoyant software industry, the latter country exhibits persistent inequalities (particularly between urban and rural population) and a gigantic surplus of unskilled labour.

It is likely that the comparative impact on SDGs 8 (decent work and economic growth), 9 (industry, innovation, and infrastructure), and 10 (reduced inequalities) will depend on the mix of government policies, together with the active participation of other stakeholders. For example, Germany has put a lot of emphasis on working together with the labour unions on retraining, skill development, and collective bargaining as ways to balance between the benefits of technological innovation and the need for compensatory policies. Countries like Brazil or Mexico have not paid enough attention to the importance of a strong safety net, and countries like India have betted on the possibility that speedy innovation will bring about long-term efficiency gains and yield enough economic surplus to be transferred to the poorest and most vulnerable.

Research on the comparative impacts of AI between countries with different levels of income per capita, or between countries with similar socioeconomic development but different historical and sociocultural backgrounds, will be very useful in the near future. Many impacts are unlikely to become fully visible before Agenda 2030 reaches its deadline, but many others might already be available to nourish research on the comparative politics and policies of technological adoption, as well as on its impacts and implications for the SDGs.

The research on digital divides and the research on A.I., automation, and job displacement both suggest that there are structural inequalities that need to be tackled in parallel to the dynamics of technological adoption. There are some basic structural requirements that should be met to boost the impact of STs on the SDGs. The availability of good-quality primary and secondary education, as well as vocational training, are two examples. Both dimensions, considered within SDGs 4 (quality education) and 8 (clean water and sanitation), might be enhanced through STs, but these will not be enough on their own.

The paradox is that those individuals and groups with higher levels of education seem to be better positioned to take advantage of STs to educate themselves further; those better trained are the ones that have better jobs and, therefore, those who have more time and flexibility to keep training themselves; and those with higher status and/or income levels are those that are better placed to accumulate even more economic capital thanks to their usage of digital technologies (Calderón Gómez 2020; Van Deursen and Helsper 2015; Zillien and Hargittai 2009).

There is no doubt that STs can work as enablers of development processes and have significant contributions to Agenda 2030. However, STs are likely to increase inequality and have uneven impacts on different economic, social, and political indicators that are covered by the SDGs, if basic aspects of these SDGs themselves are not tackled at the same time.

The impacts that technological change, innovation, and skills have on efficiency, productivity gains, competitiveness, economic growth, market concentration, and/or wage dispersion have been studied extensively by development economists. This includes research on the role of technological change in general (Lall 2001), as well as more specific studies, focused on digitalisation, automation, and other STs (Allen 2017; Ferschli et al. 2020; World Bank 2016). And yet, more research is needed to understand the different causal mechanisms linking development (broadly speaking, beyond market outcomes) and STs adoption.

The impacts on Goal 2 (zero hunger) are illustrative. International agencies, such as the FAO (2018) and World Bank (2016) have stressed the potential impact of STs such as AI, machine learning, data science, IoT, remote sensing, and blockchain on reduced information and transaction costs, improved crop management, and access to financial services. The impact of these technologies is already patent in many developed countries but, apart from a few pilot projects and case studies, developing and LDCs are still lagging behind.

One example in a low-income country is the case of the Community Knowledge Workers initiative, supported by the Grameen Foundation (2018), which aims to provide advisory services and to build "a network of community agents in Uganda who act as intermediaries between smallholders and smartphone app content developers" (FAO 2020: 10). An example in an upper middle-income country is "Smart Campo" in Brazil (and Paraguay), a platform that has been "developed to help farmers in Brazil and Paraguay optimise field management by incorporating weather and climate information into the decision-making process" (Smart Campo 2020).

And yet, there are longstanding structural inequalities in the rural sector of most of the developing world that cannot be underestimated. Economic dualism, for instance, is a persistent feature which can be broadly described as having two agricultural sectors in one: on the one hand, around 20% of production units that are relatively large, very competitive, market-oriented, risk takers, and technology-friendly, in charge of 80% of agricultural production; and on the other hand, 80% of small-holders, not very efficient, risk averse, not very enthusiastic of technological innovations, in charge of 20% of agricultural production, and mostly focused on self-consumption (Ghatak and Ingersent 1984: 4–25).

The numbers and the proportions might have changed slightly since development economists, such as William Arthur Lewis, first talked about economic dualism (Lewis 1954); they might vary from country to country; and there might be new shades in between these two extremes, but there is little doubt that there is a persistent gap between large-scale units that are linked to global agroindustry chains and a vast universe of very poor and vulnerable small-holders. These structural conditions pose a significant challenge for the adoption of STs intended to enable SDG 2 without increasing inequalities in general, simply for the fact that large firms, which are already productive and competitive, are more inclined to adopt innovations in comparison to small-holders, which tend to be poor and vulnerable.

Moreover, regions such as Latin America and especially Africa have relatively young populations. This is double-edged if we consider both the economic and the sociocultural implications. On the one hand, this demographic group might be more willing to adopt and use STs. However, on the other hand, the access to digital technologies, such as the internet, can also feed new models and social expectations that may (apparently) be satisfied when the youth migrate from rural to urban areas. Thus, the challenge lies in promoting STs that are meaningful in multiple ways: that increase economic efficiency but also take into account demographic and sociocultural factors.

There are a few examples of how the poor in both rural and urban areas have benefited from STs. One case is increasing financial inclusion thanks to the widespread use of mobiles and financial apps. However, these benefits should not be taken for granted. There is ample scope for more research on the interplay between the above-mentioned dimensions in specific local contexts and the ways that the impact of STs on SDGs 1 and 2 are maximised.

Can STs help with achieving healthy lives while tackling the challenges of surveillance and social control?

Income inequality is often correlated with geographical location and has an impact on different indicators of well-being and sustainable development, including the ability to enjoy healthy lives (SDG 3). Not surprisingly, the levels of poverty and inequality worldwide are also reflected in the fact that, in 2017, less than half of the global population was covered by essential health services (United Nations 2021).

Eligibility to public or private medical services or access to cultural or social amenities are two areas that enhance human development. However, low income is often correlated with living in segregated areas, a condition that limits citizens' ability to reach goods, places, facilities, and services that are necessary to promote different aspects of well-being. In this respect, STs offer opportunities to bridge physical distance and enhance indicators such as the coverage of essential health services (Indicator 3.8.1) or the number of people covered by health insurance or a public health system per 1,000 population (Indicator 3.8.2), contributing to achieving the target of universal health coverage.

Living in segregated areas means not only that basic health services are far away, but that time invested in reaching them also limits the spare time available for cultural and social activities, decreasing the ability of citizens to enjoy healthy and culturally meaningful lives. Even large urban areas, with thriving and consolidated economies, and with above-average provision of health services and facilities, exhibit unequal access.

A recent paper on mobility highlights that access to jobs, services, and people is key to a city's economic vitality and quality of life. However, the trend seems to be a decline in accessibility in the global South, showing that up to half of urban population experiences restricted access, leading to travel burdens or exclusion from opportunities. The most affected are low- to medium-income groups in suburbs and peripheral settlements, as well as low-income groups in other areas of the city (Venter et al. 2019).

In Mexico City, for example, someone in an upper-income neighbourhood will have 28 times more access to jobs within a 30-minute trip by public transport, than someone living in a lower-class neighbourhood. The same pattern persists for other goods and services, such as medical attention, education, food provision, or culture and entertainment (WRI 2019). The outlook gets grimmer if mere access is also pondered according to quality of provision.

Moreover, structural inequalities can easily worsen in the context of difficult global conjunctures. Due to the COVID-19 pandemic, progress in the provision of health services has been jeopardised. Achievements in the last few years could be reversed or, at best, temporarily disrupted. How can governments tackle these inequalities and deal with these challenges? Could STs offer innovative ways to bring medical services and attention closer to people? The COVID-19 health crisis is itself illustrative of both costs and benefits, threats and opportunities.

A comparison between political measures carried out by governments around the world to face the effects of the pandemic suggests four kinds of measures: public health, digital surveillance, social distance, and repression (Ojeda 2020). According to this analysis, countries like South Korea and Taiwan showed the smartest approach, indeed strongly guided by data and technology, coupled by an early reaction to the crisis. Health interventions included key elements, such as the massive application of tests, the development of new test techniques, and the development of smart apps to retrace the movement of infected people and to make public health services more accessible. These latter were effective thanks to the collaboration between government and civil society to generate data massively.

Countries like China were also very effective in flattening the COVID-19 curve, and STs were also strategic. Medical technology included test development, heat sensors, and industrial technology to adapt innovations to medical needs. The use of advanced technology for digital monitoring and surveillance was remarkable, but it also encompassed a strong state presence and citizen control. Some features included: monitoring of people's smartphones with real-time GPS, accessing personal data massively, using millions of cameras for face-recognition, obligatory reporting of body temperature, and employing apps to warn citizens about proximity to infected patients. Social distance and confinement became mandatory, with enforcement measures that were borderline repressive: the instalment of permanent control posts, the tracking of times and frequency that citizens were going in and out of their homes, as well as restrictions of movement between provinces (Ko 2020; Zhong and Mozur 2020).

Scholars have warned about the dangers of this extreme surveillance developing into a deeper level of 24/7 biometric surveillance or "under-the-skin surveillance" (Harari 2020; Schneider 2020; and Zuboff 2019). They have also warned about the possibility that temporary measures outlast emergencies, as there is always a new emergency hanging out on the horizon (a second wave or a different epidemic or pandemic or other).

Thus, while digitalisation and STs offer enormous opportunities for improvements in the health sector and for achieving indicators within SDG 3, they also pose significant societal threats and challenges that could undermine or inhibit achievements within SDG 16 due to their impact on privacy and human rights (the protection of fundamental freedoms, in accordance with national legislation and international agreements, is part of this SDG). Smart technologies may underpin the development of health treatments in record times and contribute to monitor populations and contain the spread of diseases; to improve medical diagnoses; and/or to deploy massive immunisation efforts more efficiently (Berkley 2017). But the fact that health apps rely heavily on AI and big data analytics means they can easily become threats in "regions where ethical scrutiny, transparency, and democratic control are lacking" (Vinuesa et al. 2020: 3).

The contrast between opportunities and threats suggests that, while STs are helpful in this kind of crisis, their net impacts will depend on specific political contexts and governance models. STs can either be used vertically, feeding into strong surveillance and control models; or more horizontally, through social participation at different levels, together with long-term strategies that do not jeopardise basic citizen rights and liberties. Comparative research on the relation between STs, political governance, sociocultural contexts, and their impact on diverse outcomes in the health sector will prove useful to identify cross-case lessons and design better policies to enhance this area of Agenda 2030.

It should also be noticed that, beyond specific health crises, the long-term usefulness of STs to tackle health challenges depends on other structural factors that were described previously, when discussing SDGs 1 and 2 (no poverty and zero hunger). STs cannot always compensate for the lack of basic medical materials, deficient infrastructure, untrained doctors, or patients' limited education, which is often the case in developing countries and LDCs.

Similar to the case of other SDGs, STs hardly ever offer a magic bullet to solve all deficiencies and all inequalities in the health sector. Without tackling educational gaps or properly increasing the availability of smartphones and other devices, STs are likely to benefit privileged populations and to leave the poorest and most vulnerable behind. Without solving proper infrastructure deficiencies, STs in the health sector might reinforce operational bottlenecks, rather than solving them. Without designing adequate user interfaces, STs are likely to reinforce the exclusion of vulnerable populations that need quality healthcare the most, such as the elderly or citizens with disabilities. In this respect, the research that has been conducted on the provision of different e-government services (Park and Humphry 2019; Ranchordás 2021; Yates et al. 2013) might also be applied to healthcare.

Promoting widespread data and algorithm literacy among the general population is important to maximise the positive impact of STs on SDGs; they are necessary to prevent negative impacts on citizens' privacy and better governance models. In the case of health and medical care, they are particularly important if we consider that the access to massive biometric data constitutes an essential input for the efficiency of STs.

Can STs help to build sustainable cities and communities without falling into the trap of technological fixations?

As we live in an increasingly urban world, the impact of STs on SDGs cannot be fully grasped without understanding the particular dynamics of large and medium cities. Already in 2018, the UN reported that 55% of the world's population lived in urban areas, a proportion expected to increase to 68% by 2050 (UN 2018). Cities also concentrate economic activity, contributing about 60% of global GDP. However, they also account for about 70% of global carbon emissions and over 60% of resource use (UN 2021).

In this respect, the development of new tools and technologies has certainly enabled cities to improve some of their biggest challenges. Efficient transport and mobility, for example, are essential to consolidate more sustainable cities. Both are considered as part of SDG 11 (sustainable cities and communities) and constitute key areas that need to be tackled if inclusive economic growth is to be achieved (SDG 8).

Although one might immediately think of high-speed trains and self-driving vehicles, which can certainly contribute to some extent, integrated transit systems are at the centre of expanding the coverage of public transport networks. This in turn can make public alternatives more convenient and flexible vis-à-vis the extensive use of private cars, contributing to SDG 13 (climate action). If we consider that a citizen of New Delhi or Mexico City might spend around 200 hours a year in traffic, this means that more efficient transport can also contribute to SDG 3 by increasing spare time and reducing unnecessary psychological stress.

The International Transport Forum (Kager and Harms 2017) talks about seamless integration of all modes of transport (bus, bus rapid transit systems, light rail, metro and rail systems, as well as walking and cycling) as the key point and the main challenge to achieve better mobility. This includes proper intermodal infrastructure but also smart integrated ticketing systems and information systems (IS) that inform users about availability, modes, and route efficiency in real time, preferably through personalised interfaces and apps. This is the case of cities such as London, Berlin, or Vienna, where IS also include information on access for users with special requirements such as stations with elevators, ramps, tactile guidance for the visually impaired, etc.

The inclusion of cycling as part of a multimodal integrated scheme can also extend the reach of public transport networks. In this respect, digitalisation has played an important role in increasing bike use thanks to bike-share systems, both dock-based and dock-free, as these are often linked to mobile applications. Shared electric scooters work the same way. Other cities, like San Francisco, have launched programs to improve parking availability and efficiency: the *SFpark Pilot Program* uses wireless sensors to monitor parking-space occupancy in real time, distributing this information among drivers so they can find free spaces faster, reducing the amount of time vehicles circle and, hence, congestion and gas emissions (www.sfmta.com). The city then uses this information to adjust parking prices periodically to encourage more parking in underused areas and control it in areas of high demand.

STs have also helped to increase and improve communication, as well as the involvement of society in a series of urban processes such as the so-called planning support systems (PSS) which have existed since the 1980s. These are digital tools, notably geographical information systems (GIS), that support planning processes of different types and at different stages (Batty 2007; Geertman 2006). Examples of PSS vary widely in complexity, from interactive viewers to digital platforms capable of modelling geospatial alternatives and scenarios based on variables and parameters defined by users, but all of them contribute in different measures to SDG 16 (peace, justice, and strong institutions) by enhancing indicators on democracy and political participation.

Technological advances such as open-source GIS, big data, cloud computing, and the growing capacity of web navigators with a reduction in costs have pushed the development of PSS in recent years. In terms of societal involvement, GIS- and web-based tools have allowed for a faster, broader, more democratic, and more significant access to and use of data. These in turn have also supported the work of different initiatives, including observatories, whether academic, governmental or citizen in origin, and thus also contribute to SDG 16.

Clearly, the proliferation of individual mobile devices and sensors of different kinds has leapfrogged the production of data in cities while also making it extremely detailed, both in time-scale and spatial resolution (Batty 2012). In this respect, big data could potentially be used to improve cities and make them more efficient. The big tech giants are already mining this data for different purposes: marketing, advertising, entertainment. However, this raises the following question: How can we integrate big data and technological innovations that are mainly market-driven with traditional data on planning and design, in ways that improve the management of cities, making them safer, more equitable, and, overall, better and more liveable?

The answer is that STs constitute only one of many ingredients necessary to achieve better city management and administration models. Advances and data will only be useful if used smartly, in the context of interconnected and aligned strategies, with effective coordination, both vertical and horizontal. Linear and fragmentary urban policies have long been shown to fail, even if they are supported by technological innovations. As stated by Batty already in 2012, "The key is no longer technological, as ever it is organisational" (192).

Can STs contribute to climate action and to achieving more environmentally sustainable ways of living without falling into the trap of the Promethean Myth?

Climate change is the most important threat to global welfare of our times. Increasing temperatures due to greenhouse gas emissions (GHG) are stressing ecological and social systems, jeopardising their long-term sustainability like never before. In this respect, in the same year that Agenda 2030 was approved, the international community also signed the Paris Agreement (PA), which commits all signatory countries to keeping global warming to "below 2° C" and, if possible, below 1.5° C above pre-industrial levels (IPCC 2018; United Nations 2015; Goal 13 of the SDGs). Achieving these targets will require fundamental changes to our transportation, agricultural, building, and energy sectors. Can digitalisation and STs offer significant opportunities to enhance both mitigation and adaptation strategies? What are the opportunities and threats, benefits and costs?

Climate science is already relying heavily on massive data gathering and AI, which is applied to improve the monitoring of GHGs and their sources. As stated by Vinuesa et al. (2020: 4), "benefits from A.I. could be derived by the possibility of analysing large-scale interconnected databases to develop joint actions aimed at preserving the environment". AI is also used to downscale climate effects in particular areas of the world as well as to improve weather forecasts or understand different variables such as temperature and humidity, and their interaction with GHGs. At the same time, the number of satellites launched into space over the past decade has increased exponentially, so that AI and satellite imaging can now be used to monitor and analyse the state of forests and oceans (Goals 14 and 15).

Another clear application of digital technologies lies in their potential to improve climate modelling and predictions. For example, meteorology and earth science have largely benefited from machine learning and other STs to improve accuracy and predictability of natural hazards such as hurricanes and tropical storms. This can in turn have positive impacts on SDG 2 because farmers in rural areas could benefit greatly from these applications, reducing risk and improving crop management (as discussed earlier). Some examples include a closer cooperation between United Nations Food and Agriculture Organisation (FAO) and the World Meteorological Association (WMO); the platform SERVIR, developed together by NADA and USAID; or AfricaAdapt in Senegal, all of which aim at offering better meteorological information to help poor farmers face the risks of climate change (FAO 2018: 11).

AI, deep learning, robotics, big data, IoT, and automated decision-making systems, among other STs, are increasingly recognised as important tools for societal transformation (Graglia et al. 2018), and even though the interrelations between climate change action and digital technologies has not yet been fully explored, it is likely that these latter will work not as mere "instruments" for solving sustainability challenges, but rather as fundamental, disruptive, and multi-scalar drivers of change (TWI2050 2018). STs may have substantial positive impacts by supporting the decarbonisation of society in areas such as energy, transportation (see discussion earlier on sustainable cities; Stain 2020), agriculture (see discussion earlier on SDGs 1 and 2), and buildings. It can also help to monitor global environmental indicators such as biodiversity loss (SDGs 14 and 15).

And yet, digitalisation and STs can also be double-edged. Digitalisation narratives frequently emphasise the benefits and enabling potential of digital technologies to solve environmental problems, but they often remain oblivious to their negative impacts. One example is the increasing digital carbon footprint, given the fact that data centres and computing power consume increasing amounts of energy. Another example is the extraction and depletion of valuable resources such as cobalt, palladium, tantalum, silver, gold, indium, copper, lithium, and magnesium, as well as the growing volume of electronic waste which needs to be collected, recycled, or disposed of.

To date, based on the available literature, it remains unclear whether the positive indirect environmental impacts can outweigh the negative direct ones (or the other way round) given the interplay between many economic, political, social, and cultural factors. Further and more detailed analysis of the trade-offs between positive and negative effects will be needed in the near future. For example, according to a pessimistic scenario, GHG emissions may not be reduced overtime but increased instead. Whether this or more moderate scenarios are likely to materialise in the long term, it is clear that STs will not solve climate change and environmental problems on their own.

To believe so would be to fall into the trap of the Promethean Myth (Dryzek 1997); that is, the wrong belief that "technology can be used to overcome any problem facing humanity, including those related to climate change and to the environment in general" (Domínguez and Karaisl 2012: 103). For STs to have lasting impacts on SDGs 13, 14, and 15, it is also necessary to find ways to change broader production and consumption patterns (SDG 7 and 12).

At the same time, the extraction of valuable resources used to manufacture batteries and digital devices is a phenomenon that is already causing serious environmental impacts (such as soil acidification, human toxicity, and groundwater pollution) and human rights violations locally (including the forced displacement of populations and the murdering of environmental activists), in both Africa and Latin America. It is not clear whether STs, together with satellite and drone surveillance, are also contributing to increasing the efficiency of large corporations to screen, to localise, to image, and to target the extraction and commercialisation of natural resources in areas that were previously unexplored and unexploited, but the impacts of these market-oriented dynamics should also be studied with more detail.

These situations seriously risk the progress in other aspects of Agenda 2030, such as SDG 16 (peaceful and inclusive societies for sustainable development) and other indicators and targets of

SDGs 14 (life below water) and 15 (life on land). All these aspects need to be examined, both to understand the large-scale impact of current technological trends and to find ways to gear innovation towards more comprehensive development solutions.

Can STs contribute to achieve SDGs that have positive impacts across the whole Agenda 2030, while minimising negative "side effects"?

One of the most significant challenges (and sources of criticism) regarding the application of Agenda 2030 is that the 17 goals, together with 169 indicators and 231 targets, are closely intermingled and co-determined by each other. Compared to the more general and aspirational MDGs, the SDGs try to be all-encompassing. This offers some advantages, like trying to tackle development challenges with a more comprehensive perspective. But it also poses challenges, such as the difficulty of monitoring the systematic impacts of concrete policies across different areas of Agenda 2030.

For example, as we have argued throughout this work, to talk about STs and their possibilities as enablers or inhibitors of different SDGs constitutes a difficult task, not only because of the variety of STs themselves, but also because positive effects on some indicators might inevitably bring about negative effects on others. STs that have positive uses in the context of some SDGs (example: monitoring citizens' health to contain pandemics; SDG 3) may also have undesirable applications in the context of others (surveillance and social control; SDG 16).

In this respect, we consider that there are at least three SDGs that require special attention: SDG 4 (quality education), SDG 5 (gender equality), and SDG 8 (decent work and economic growth). The main reason for this emphasis is that these three goals are key to building a relation between STs and Agenda 2030 which is mutually enhancing. Paying attention to the positive impacts in these three areas can have significant spill-over effects, whereas not paying attention to the negative effects could cancel out the benefits of potential technological innovations.

Take the example of SDG 4 and SDG 8. Can STs foster education and training to promote inclusion and sustainability? According to the UN, before the COVID-19 crisis, the proportion of children and youth out of primary and secondary school had declined from 26% in 2000 to 17% in 2018. Still, the estimation back then was that 260 million children would be out of school by 2030; that is, 20% of global population in that age group (UN 2021). And yet, already before the pandemic, access and availability of education was not the only problem, as the quality of it was also a concern: around 50% of children and youth lacked basic proficiency standards.

During the COVID-19 pandemic, a majority of schools around the globe closed temporarily, impacting 1.5 billion children (91% of students worldwide). Nearly 700 million were in developing countries. Of course, both rich and poor students are affected by this disruption. However, the consequences in poor places will be far worse. As some studies have shown, the wealth of a country affects exam results as much as the wealth of the pupil's household: students with the same household income score significantly higher if they live in richer countries (Patel and Sandefur 2020).

The economic impact of the pandemic has forced many to abandon studies in favour of work, plus increased the risk of child marriage, sexual assault and violence at home, teenage pregnancy, and exploitation (The Economist 2020). The Save the Children charity suggests that almost 10 million children may never return to school because of the economic impact of the pandemic (Warren 2021).

In this context, governments, international organisations, and many education institutions, such as universities and private schools, have promoted online or remote models of education. In fact, a wide range of (already existing) apps and online platforms have been further developed, released freely, or experienced a boom in users after the world had to suddenly stop the young going to school at the beginning of 2020. The World Economic Forum's Future of Jobs Report (2020) found that between April and June 2020, Coursera, a virtual training platform developed a decade ago by researchers at Stanford University, saw a fourfold increase of people seeking education and training opportunities. The number of employers looking for online training opportunities for their workers increased fivefold, and government online programmes enrolment increased ninefold. These trends show that STs have been essential to advancing in SDG 4 despite the adverse COVID-19 scenario.

Nonetheless, around half of the world's population (more than 4 billion people) does not have access to the internet. Most of the gap is located in LDCs. In Eritrea, Somalia, Guinea-Bissau, the Central African Republic, Niger, or Madagascar, fewer than 5% are online (Roser et al. 2015). The Internet for All Report (WEF 2016) points out four main reasons for the digital divide: 1) infrastructure, a good fast connection or even electricity is not available; 2) affordability, relating to the cost of devices and connectivity, particularly those below the poverty line; 3) skills, awareness, and cultural acceptance, that is, illiteracy and cultural gender issues; and 4) local adoption and use, referring to online content available in only ten languages. The pandemic is widening the pre-existing gap between how much rich and poor children learn.

Predictions refer to a need for reskilling of about half of the working population in the next five years due to the "double disruption" caused by the pandemic and increasing automation. Although the estimation is that 85 million jobs may be displaced, 97 million are likely to emerge in order to adapt to a new division of labour. We are talking about a redefinition of the job market, not about the massive unemployment wave that some fatalists predict. We should be preparing students for the requirements of this new work environment. Tech skills will naturally be welcomed by employers, but they are not the top skills that will be needed. Critical thinking and problem-solving skills are still the most valued by employers together with emergent skills related to self-management, such as active-learning, resilience, stress tolerance, and flexibility (WEF 2020).

It is feasible to think that tech skills could and should be taken as a means rather than an end, modifying how things are taught and how to make students think in different ways. Teaching computer sciences, for example, is sometimes just incorporating the use of computers and other devices, but it could encompass computer thinking, data analysis, interface design, artificial intelligence (algorithms), and cybersecurity, among others. Besides digital abilities, it is necessary to promote competences like problem solving, creative thinking, and collaboration. Digitalisation, especially internet connectivity, has brought some of us closer but has also widened the gap between others. Without tackling these needs, it is unlikely that the positive impact of STs on other SDGs will reach its full potential.

Take the example of SDG 5 (gender equality). Can STs contribute to gender equality without reproducing structural patterns of violence and exclusion? Whereas there has been important progress on this SDG, such as laws being reformed in several countries to support and advance gender inequality, more girls being educated than ever before, and more female representatives in national parliaments, huge challenges remain. Women aged 25 to 34 are 25% more likely than men to live in extreme poverty; in the workplace, women are paid 16% less than men and hold only 1 in 4 managerial positions. While 39% of employed women are working in agriculture, forestry, and fisheries, only 14% of them appear to be landholders. Such inequalities pose significant challenges in the context of climate change and its impacts given the fact that the climate emergency is more likely to affect those groups that lack access to land and resources (UN 2021).

Can STs contribute to tackling the challenges of gender equality? What are the concrete threats and opportunities? There is no doubt that STs can contribute to women's participation in general. For example, social media was key for the organisation of the massive demonstrations of women on 19M (19 March 2020) or the viral spread of women performing and singing "A rapist on your path" all over the world as a protest against gender violence and a sign of solidarity among women of all ages. There have been a number of initiatives of bottom-up organisation, from hashtags to communicate when you board a taxi to local groups of feminist solidarity where women offer and look for services to and for women. In this respect, technology has definitely facilitated processes of economic and political participation that would otherwise be very hard. Digitalisation and STs in general have contributed to raising international attention on the topic.

However, the roots of the problem are much deeper, and technology alone will not solve gender inequalities. The IT industry is itself not free from criticisms. A recent article in a tech industry magazine states that while women make up 47% of the employed adults in the US, they hold only 25% of computing roles as of 2015, despite the fact that science, technology, engineering, and maths (STEM) jobs have grown 79% since 1990 (White 2020). In addition, only 38% of women who majored in computer science are working in the field, compared to 53% of men, and 24% of women with an engineering degree work in engineering compared to 30% of men. IT workplaces are still strongly male-dominated, making it common for women to experience gender discrimination. According to a Pew Research Centre survey in the US (Funk and Parker 2018), this accounts for 50% of women in STEM against only 19% of men.

This has implications for the overall relation between STs and SDGs. Without proper representation of women in the IT industry, it is likely that technological solutions will not reflect their needs and their point of view adequately. STs that are designed from a predominantly male point of view are unlikely to reflect the whole complexity of development problems that are encompassed by Agenda 2030; they will tend to be market-oriented rather than development-oriented. The so-called societal goals (1–6) are just a few examples where women's perspective becomes vital, but this argument applies to all SDGs.

Moreover, the impacts of the COVID-19 pandemic make clear that the relations between STs, digitalisation, gender, and development are much more complex than expected. On the one hand, STs can help to bridge physical distance, improving the provision of basic services and benefiting women who are often the main person responsible for aspects such as family health and care. On the other hand, the increasing importance of distance work and the home office also poses significant challenges, including the blurring of boundaries between public and private life. Palomar Verea (2021) refers to these situations as "unlocalised virtual realities": work, social life, school, shows and entertainment, even museum visits or religious and funerary rituals, have suddenly come to the same house as part of the same private realm thanks to videoconferences and virtual sessions, calls, and meetings.

Women have suffered a higher toll from these unlocalised realities. Since they are usually in charge of domestic work and most care-taking activities within families (children, elderly, sick, etc.), this means that women have to deal simultaneously with their own jobs on top of housework and family duties. Thus, although the possibility of remote work might be good news in the context of the COVID-19 pandemic, the resulting overlap between times and spaces may limit women's productivity and represent risks to their economic security. In Mexico, for example, female employment was equivalent to 45.9% before COVID-19, yet this was reduced to 36.4% from January to April 2020 (UNAM 2020). Future solutions that focus on maximising the benefits of STs while limiting the disproportionate burden that women carry in the context of an extensive home office will require technological innovation, but also more extensive research on gender, law, ethics, and development in general.

And this is not the end of the story. Beyond unequal economic opportunities, lockdown measures, social distancing policies, and 24/7 cohabitation, together with economic and social stress caused by the pandemic, have triggered an increase in gender-based violence. In a few words, many women are being forced to lockdown at home with their abusers. Technology, in particular social media and digital platforms, have played a crucial role in documenting and exposing violence against women. STs have also helped the female population to organise, communicate, and, to a certain extent, protect each other with a number of bottom-up initiatives, before and after confinement.

A remarkable example of the use of digital resources against gender violence is a map of femicides developed by the Mexican geophysicist María Salguero. This is an open access web-based map where Salguero has logged femicides that were registered in Mexico between 2016 and 2020, based on documentation of the National System of Public Security, as well as newspaper notes and reports using geolocation. As Salguero has herself explained, the web map is called "I name you", to remember the fact that victims are not numbers, but real women, with real stories. It presents geolocated data, which can be filtered through different categories such as age range or murderer–victim relationship, together with a description of how it occurred. The map is also collaborative as one can add new data, including description and place of murder (Loaiza 2020; Mendoza 2020). Apart from publicising the problem of femicides, Salguero's work has also helped to understand the causes of gender-based violence.

Conclusions

This chapter has explored the interrelations between STs and SDGs, considering both threats and opportunities, costs and benefits. It has been argued that the international development agenda remained oblivious to the digital revolution for a long time. Neither the MDGs nor the SDGs considered adequately the impact of digital and smart technologies on development outcomes. And yet, almost six years after Agenda 2030 was approved, development professionals, academics, consultants, cooperation agencies, and IFIs are gradually starting to become more enthusiastic about technological innovations. The danger is that this sudden swing, from being relatively oblivious to becoming over-optimistic and "back to the future", is accompanied by the danger of emphasising benefits without closely looking at the negative implications and side-effects of STs.

Given the diversity of STs themselves and the complexity of SDGs, it is difficult to make generalisations. In this respect, although this chapter has sparsely mentioned most of the 17 SDGs, it has mainly focused on a few goals and areas that are illustrative of the complex trade-offs that accompany the application of STs as tools to achieve Agenda 2030. The main lesson is that STs might be significant enablers but are not magic bullets to solve all development problems.

Despite the zeal, it is necessary to keep a cautious approach, balancing between market-oriented and development-oriented devices; taking into account legal and sociocultural conditions locally; and considering the systematic impact of innovations in a broad set of SDGs (not only one or two at a time). Future research should tackle these tensions, contributing to developing STs that are adequate for various local contexts and feeding into the design and implementation of more comprehensive policies. Achieving this might sometimes be possible thanks to local and national resources, but international cooperation and joint technological development might be necessary in other cases.

References

Allen, J. (2017): *Technology and Inequality: Concentrated Wealth in a Digital World*. Palgrave MacMillan.

Batty, M. (2007): *Planning Support Systems: Progress, Predictions, and Speculations on the Shape of Things to Come*. Available at: http://discovery.ucl.ac.uk/15175/ [05.02.2021]

Batty, M. (2012): Smart Cities, Big Data, *Environment and Planning B: Planning and Design* 39, 191–193.

Berkley, S. (2017): Immunization Needs a Technology Boost, *Nature*. Available at: www.nature.com/articles/d41586-017-05923-8 [05.02.2021]

Calderón Gómez, D. (2020): The Third Digital Divide and Bourdieu: Bidirectional Conversion of Economic, Cultural, and Social Capital to (and from) Digital Capital among Young People in Madrid, *New Media and Society* 1(20). https://doi.org/10.1177/1461444820933252

Castells, M. (1995): *La ciudad informacional. Tecnologías de la Información, reestructuración económica y el proceso urbano-regional*. Madrid: Alianza Editorial.

Castells, M. (2001a): *La Era de la Información. Vol. III: Fin de Milenio*. México: Siglo XXI Editores. 2001.

Castells, M. (2001b): *La Galaxia Internet. Reflexiones sobre Internet, empresa y sociedad*. Madrid: Areté.

Castells, M. (2002): *Vol. I: La Sociedad Red. México, Distrito Federal: Siglo XXI Editores. 2002*. México: Siglo XXI Editores.

Domínguez, C. and Karaisl, M. (2012): Climate Change, Infrastructure, and the Promethean Myth, *Voices of Mexico* 95, 101–105.

Domínguez, J. C. (2018): Digitalisation and Exponential Technological Change: Challenges and Opportunities, *Future of Globalisation*, 8 November 2018. Available at: https://blogs.die-gdi.de/2018/11/08/digitalisation-exponential-technological-change/ [15.12.2020]

Domínguez, J. C. and Lucatello, S. (2014): Introducción: Los objetivos de desarrollo del milenio y el debate post-2015. Retos y tensiones en la definición de una nueva agenda, in Domínguez, C. and Lucatello, S. (eds.): *Desarrollo y Cooperación Internacional*. Miradas Críticas y Aportes para la Agenda Post-2015, Instituto Mora, Mexico City.

Dryzek, J. (1997): *The Politics of the Earth*. 2nd ed. Oxford: Oxford University Press.

FAO (2018): Tackling Poverty and Hunger through Digital Innovation, Food and Agriculture Organization of the United Nations. Available at: www.fao.org/policy-support/tools-and-publications/resources-details/es/c/1179279/ [03.02.2021]

Ferschli, B., Rehm, M. and Zilian, S. (2020): Marktmacht, Finanzialisierung, Ungleichheit. Wie die Digitalisierung die deutsche Wirtschaft verändert, Project Report, Friedrich-Ebert-Stiftung. Available at: www.fes.de/index.php?eID=dumpFile&t=f&f=47856&token=3c4e18889ce7d025180e267485cd230e02c0bc09 [30.11.2020]

Funk, C. and Parker, K. (2018): *Women and Men in STEM Often at Odds Over Workplace Equity*. Washington, DC: Pew Research Centre, Social & Demographic Trends.

Geertman, S. (2006): Potentials for Planning Support: A Planning-conceptual Approach, *Environment and Planning B, Planning and Design* 33(6), 863–881.

Ghatak, S. and Ingersent, K. (1984): *Agriculture and Economic Development*. Baltimore: The John Hopkins University Press.

Gore, A. (1998): Remarks by President Al Gore, Digital Divide Event, 28 April 1998. Available at: https://clintonwhitehouse2.archives.gov/WH/EOP/OVP/speeches/edtech.html [06.02.2021]

Graglia, M., Annoni, A., Benczúr, P. and Bertoldi, P. (2018): *Artificial Intelligence: A European Perspective*. Luxembourg: European Commision, Publications Office.

Harari, Y. N. (2020): The World after Coronavirus, *Financial Times*. Available at: www.ft.com/content/19d90308-6858-11ea-a3c9-1fe6fedcca75 [30.11.2020]

IPCC (2018): 1.5 Degrees. UN Special Report on Climate Change. Available at: www.ipcc.ch/sr15/[30.11.2020]

Kager, R. and Harms, L. (2017): *Synergies from Improved Cyclist-Transit Integration: Towards an Integrated Urban Mobility System*. International Transport Forum.

Ko, J. (2020): Cómo ha empleado China la tecnología para luchar contra la COVID-19 y afianzar su control sobre la ciudadanía. *Aministía Internacional Noticias*. Available at: www.amnesty.org/es/latest/news/2020/04/how-china-used-technology-to-combat-covid-19-and-tighten-its-grip-on-citizens/[30.11.2020]

Lall, S. (2001): *Competitiveness, Technology, and Skills*. Cheltenham, UK: Edward Elgar Publishing Limited.

Lewis, W. A. (1954): Economic Development with Unlimited Supplies of Labour, *The Manchester School* 22(9), 139–191.

Loaiza, L. (2020): Mapa de feminicidios muestra patrones complejos de violencia en México. InSight Crime. *Investigación y Análisis de Crimen Organizado*. Available at: https://es.insightcrime.org/noticias/analisis/mapa-de-feminicidios-muestra-patrones-complejos-de-violencia-en-mexico/ [30.11.2020]

Mendoza, V. (2020): Mujeres Poderosas 2020: María Salguero ubica y pone nombre a víctimas de feminicidio, *Forbes México*.

Mitchell, W. (2000): *e-topia*. Cambridge, MA: MIT Press

Negroponte, N. (1995): *Ser digital*. Buenos Aires, Argentina: Atlántida Océano.

Ojeda, A. (2020): *Salud pública, tecnología y participación*. Lab Tecnosocial. Available at: https://lab-tecnosocial.github.io/pol_coronavirus/ [30.11.2020]

Palomar Verea, C. (2021): La academia desde casa. Ciencia, género y cuidados en el contexto del confinamiento por COVID-19, *Debate feminista* 31(61).

Park, S. and Humphry, J. (2019): Exclusion by design: Intersections of Social, Digital, and Data Exclusion, *Information Communication and Society* 22(7), 934–953.

Patel, D. and Sandefur, J. (2020): A Rosetta Stone for Human Capital. *CGD Working Paper*.

Ragnedda, M. and Muschert, G. W. (eds.) (2018): *Theorizing Digital Divides*. Oxford, New York: Routledge.

Ragnedda, M. and Ruiu, M. L. (2018): Social Capital and Three Levels of Digital Divide, in Ragnedda, M. and Muschert, G. W. (eds.): *Theorizing Digital Divides*. Oxford, New York: Routledge, 21–34.

Ranchordás, S. (2021): Connected but Still Excluded? Digital Exclusion beyond Internet Access, in Ienca, M., Pollicino, O., Liguori, L., Stefanini, E. and Andorno, R. (eds.): *The Cambridge Handbook of Life Sciences, Informative Technology and Human Rights*, forthcoming. Available at: https://papers.ssrn.com/sol3/papers.cfm?abstract_id=3675360, [23.10.2020]

Roser, M., Ritchie, H. and Ortiz-Ospina, E. (2015): Internet. *Published online at OurWorldInData.org*. Available at: https://ourworldindata.org/internet [30.11.2020]

Schneider, I. (2020): Democratic Governance of Digital Platforms and Artificial Intelligence? Exploring Governance Models of China, the US, the EU and Mexico, *JeDEM- eJournal of eDemocracy and Open Government* 12(1), 1–24.

Smart Campo (2020): Engineering Solutions for Agriculture (EnsoAg LLC). Available at: http://ensoag.com/smart-campo/ [07.02.2021]

Stain, A. (2020): Artificial Intelligence and Climate Change. Yale Journal on Regulation, Vol. 37, No. 890. *Research Paper No. 20–39*

The Economist (2020): Learn Today, Earn Tomorrow. School Closures in Poor Countries Could be Devastating, *The Economist*. Available at: www.economist.com/international/2020/07/18/school-closures-in-poor-countries-could-be-devastating [30.11.2020]

TWI2050 (2018): Transformations to Achieve the Sustainable Development Goals. Laxenburg, Austria. Available at: www.researchgate.net/publication/331158235_TWI2050_-_The_World_in_2050_2018_Transformations_to_Achieve_the_Sustainable_Development_Goals_Report_prepared_by_The_World_in_2050_initiative_International_Institute_for_Applied_Systems_Analysis_IIASA/link/5c68c9bca6fdcc404eb5c1b5/download [30.11.2020]

TWI2050 (2019): The Digital Revolution and Sustainable Development: Opportunities and Challenges. Laxenburg, Austria. Available at: https://pure.iiasa.ac.at/id/eprint/15913/ [30.11.2020]

UNAM Global (2020): Disminuye la participación laboral remunerada de mujeres en México. *Boletín UNAM-DGCS-833*. México.

United Nations (2015): Paris Agreement. Available at: https://unfccc.int/sites/default/files/english_paris_agreement.pdf [06.05.2021]

United Nations (2018): *Revision of World Urbanization Prospects*. London: UN Department of Economic and Social Affairs.

United Nations (2021): Sustainable Development Goals. United Nations. Available at: www.un.org/sustainabledevelopment/ [06.02.2021]

United Nations Development Program (2019): Future forward. UNDP Digital Strategy. Available at: https://digitalstrategy.undp.org/assets/UNDP-digital-strategy-2019.pdf [02.02.2021]

United Nations General Assembly, United Nations Millennium Declaration, General Assembly resolution 55/2 of 8 September 2000. Available at: www.ohchr.org/EN/ProfessionalInterest/Pages/Millennium.aspx [06.02.2021]

Van Deursen, A. and Helsper, E. J. (2015): The Third-Level Digital Divide: Who Benefits from Being Online, Communication and Information Technologies Annual: Digital Distinctions and Inequalities, *Studies in Media and Communications* 10, 29–53.

Van Deursen, A. and van Dijk, J. (2015): Towards a Multifaceted Model of Internet Access for Understanding Digital Divides: An Empirical Investigation, *The Information Society* 31(5), 379–391.

Venter, C., Mahendra, A. and Hidalgo, D. (2019): From Mobility to Access for All: Expanding Urban Transportation Choices in the Global South. *Working Paper.* Washington, DC: World Resources Institute. Available at: www.citiesforall.org [30.11.2020]

Vinuesa, R., Azizpour, H., Leite, I., Balaam, M., Dignum, V. Domisch, S., Felländer, A., Langhans, S. D., Tegmark, M. and Nerini, F. (2020): The Role of Artificial Intelligence in Achieving the Sustainable Development Goals, *Nature Communications*, January 2020. Available at: www.nature.com/articles/s41467-019-14108-y [06.05.2021]

Warren, H. (2021): *Five Urgent Investments to Get all Children Back to School and Learning.* Save the Children International. Available at: www.savethechildren.net/blog/five-urgent-investments-get-all-children-back-school-and-learning [15.02.2021]

White, S. (2020): Women in Tech Statistics: The Hard Truths of an Uphill Battle. *CIO.* IDG Communications. Available at: www.cio.com/article/3516012/women-in-tech-statistics-the-hard-truths-of-an-uphill-battle.html [30.11.2020]

World Bank (2016): *Digital Dividends. World Development Report 2016.* World Bank Group. Available at: http://documents1.worldbank.org/curated/en/896971468194972881/pdf/102725-PUB-Replacement-PUBLIC.pdf [15.12.2020]

World Economic Forum (2016): *Internet for All. A Framework for Accelerating Internet Access and Adoption.* Geneva: World Economic Forum.

World Economic Forum (2020): *Future of Jobs Report.* Geneva: World Economic Forum.

WRI México (2019): Desde los empleos hasta la educación, la desigualdad en la Ciudad de México es un asunto de accesibilidad. WRI México. Available at: https://wrimexico.org/resources/maps/desde-los-empleos-hasta-la-educaci%C3%B3n-la-desigualdad-en-la-ciudad-de-m%C3%A9xico-es-un [30.11.2020]

Yates, S., Kirby, J. and Lockley, E. (2013): Digital by Default: Reinforcing Exclusion Through Technology, in Foster, L., Brunton, A., Deeming, C. and Haux, T. (eds.): *In Defence of Welfare 2.* Bristol: Policy Press. Available at: www.social-policy.org.uk/wordpress/wp-content/uploads/2015/04/39_yates-et-al.pdf [15.10.2020]

Zhong, R. and Mozur, P. (2020): China recurre a un control social al estilo de Mao para frenar el coronavirus, *The New York Times.* Available at: www.nytimes.com/es/2020/02/17/espanol/mundo/corona virus-vigilancia-china.html [30.11.2020]

Zillien, N. and Hargittai, E. (2009): Digital Distinction: Status-specific Types of Internet Usage, *Social Science Quarterly* 90(2), 274–291.

Zuboff, S. (2019): *The Age of Surveillance Capitalism. The Fight for a Human Future.* New York: Public Affairs.

27

NORTH–SOUTH DIVIDE IN RESEARCH AND INNOVATION AND THE CHALLENGES OF GLOBAL TECHNOLOGY ASSESSMENT

The case of smart technologies in agriculture

Andreas Stamm

Introduction

The following chapter analyses smart technologies in agriculture from the point of view of technology assessment (TA). The guiding question is what responsible action related to disruptive innovations in agriculture might look like and how responsible and irresponsible action can be assessed systematically. Three observations and assumptions guide the analyses. First, TA has been conceptualised in the global North in times where a critical approach to new technologies was mainstream thinking, and TA was mainly seen as an early warning system for risks and unintended side effects of new technologies. Under the conditions of eroding planetary boundaries, the focus might need to be shifted towards a more balanced assessment of opportunities and risks, considering innovation not so much as a driver of economic growth but rather a way of finding new ways to address global challenges. Second, TA has been implemented mainly on the national level; this is no longer adequate in a globalised and networked world, where technological developments in one part of the world may have impacts in any other. Third, from an ethical point of view, industrialised countries (including new science and technology hubs, such as China and India) have an obligation to support the development of technologies which may help developing countries in shaping their development under the conditions of environmental limits to conventional economic growth. Low and middle income countries are especially affected by global environmental changes but do not have full-fledged innovation systems and have fewer resources available to develop solutions on their own. International science, technology and innovation (STI) partnerships between the global North and the South should be given preference to traditional modes of technology transfer.

These issues are discussed with regard to two recent disruptive agricultural innovations: digital agriculture as an umbrella term for ICT-enabled innovations in agriculture and genome editing (CRISPR-Cas9) as a new breeding technology. In the first case, we will show that the outcome of technological innovation might have positive impacts but may also threaten job

 DOI: 10.4324/9780429351921-33

opportunities in commercial agriculture in the global South. In the second case, it is shown how legal rulings in the global North, based on "ad hoc" rather than explicit TA, may delay and negatively shape the development of technical solutions with potentials for the global South.

The structure of the chapter is as follows: we describe how TA has developed in a specific historical (1960s and 1970s) and regional (Germany, Europe) context and how implicit TA has had an important impact in the development of a specific clean coal technology (carbon capture and storage, CCS). To place the topic in the global context, we describe the remaining and increasing rather than decreasing North–South divide in research and development (R&D), which implies that developing countries still need international cooperation in STI to address societal challenges. Subsequently, the challenges that global agriculture is facing for the coming decades are briefly described, as are the two mentioned possible responses from the technological frontier to address these challenges. It is, in addition, shown that an EU court ruling at first sight addressing a technical issue is having far-reaching social and economic consequences in the developing world and, thus, far from the hubs where innovation processes are mainly developed. The concluding section of the chapter sketches some general lines of how TA might need to adapt to the imminent global challenges (climate change, food security) and develop from a basically national to a global TA.

Framing of technology assessment in its national and historic contexts

Some ad-hoc technology assessment is done, most of the time, when mankind develops a new product or process, raising expectations of improved benefits and weighing these with the expenses of making it happen and risks associated with the innovation. However, only in the 1960s was the term technology assessment (TA) coined and the underlying concept developed in the USA and Europe. In its practical implementation, TA was first seen as a means of mitigating information asymmetries between the legislative and executive branches of the US government. While the latter could directly rely on a range of agencies to access up-to-date information about technologies, this was not the case for legislation. TA was thus framed as a form of science-based policy advice to parliament (Grunwald 2010: 66). In Europe, the need to create dedicated offices for policy advice through TA has been discussed since the early 1970s, while concrete implementation took effect only starting in the 1980s, e.g. with the Rathenau-Institute in the Netherlands, the *Büro für Technikfolgenabschätzung* (TAB) in Germany and the *Scientific and Technological Options Assessments* (STOA) at the European parliament.

Since its beginning, TA was mainly considered as an "early warning" system that was central to the identity of TA, as it was seen as a means to identify potential hazards and minimise their effects (Hahn and Ladikas 2019: 2). It is important to note that since the 1960s and 1970s, TA in its infant stages has been embedded in an increasingly technologically critical discourse. Grunwald (2010: 66) sees the collapse of progress optimism as a constitutive element of the emerging technology assessment. Since the early 1970s, the incompatibility of continuous growth processes with planetary boundaries became increasingly obvious. In most of the relevant works, existential threats were mainly seen as arising from high rates of population and economic growth and the overburdening of local, regional and global ecosystems (Meadows 1972; Barney 1980). Technological advancements were in this context held responsible for being drivers of unsustainable growth processes.

Among the few authors criticising technological developments directly was Herbert Marcuse, with his concept of "technological rationality", first coined as early as 1941 (Marcuse 1941) and taken up again 1964 in his most influential work (One Dimensional Man). Marcuse criticises mainly the one-dimensionality and linearity of thought on both individual and collective levels in societies under a technological rationality. While Marcuse strongly influenced the new left movements in Western societies, a less-known thinker with probably a broader impact in analysing possible consequences of new technologies was Hans Jonas, with his magistral work "Imperative of Responsibility", first published 1979 in German as "Das Prinzip Verantwortung". In Jonas' work, technological development today has potentially far-reaching, sometimes global effects (Jonas 1984). Responsible action related to technology, in his approach, would always have to start from a worst-case assumption. If the use of a particular technology poses the threat of catastrophe – however uncertain the threat and positive the other possible outcomes – Jonas recommends that we give precedence to "the bad over the good prognosis" (Coyne 2018: 232). This implies that without full and undisputable knowledge about the direct and indirect effects of applying a new and potentially risky technology, implementation should not be allowed, much less promoted.

The book had a strong first impact in the German debate on the responsible approach to technological development and, more generally, on the responsibility of present generations for the integrity of life in the future (Schütze 1994). While this "heuristics of fear", as Jonas coins his basic recommendation, has never been adopted by science-based TA, it has influenced pre-scientific thinking and writing (mass media) about disruptive technologies, especially in Germany, but also beyond. Hoffmann (2019) traces back the precautionary principle in the European Union, as laid down in Article 191 of the European Treaty, to a large extent to the work of Hans Jonas. His ontological-ecological imperative, "act so that the effects of your action are compatible with the permanence of genuine human life" (Jonas 1984: 11), influenced the definition of sustainable development as coined by the Brundtland commission in 1987: "sustainable development is development that meets the needs of the present without compromising the ability of future generations to meet their own needs".[1]

In real life, an implicit "heuristic of fear", translated into wide-spread fears of possible catastrophic events related to the large-scale application of new technologies, has strongly influenced the anti-nuclear energy movement in Europe and led to phasing-out a potentially important technology development at a very early stage in Germany. Carbon capture and storage (CCS) is considered by the international climate change research community as an indispensable element of promising strategies to keep global warming below 1.5 degrees or at least 2 degrees above pre-industrial levels (IPCC 2005, 2019: 15). In 2009, the European Union issued a Directive with the goal of incentivising the member states to upscale research and demonstration (R&D) in CCS technologies. Germany approved a CCS law in the same year and revised versions in 2011 and 2012. The law provided for a veto of the German states to rule out CCS R&D and implementation on its territories. Without delays, all states with relevant capacities for carbon dioxide storage enacted this veto, prohibiting CCS-related activities within their jurisdiction. With the withdrawal of a large-scale CCS demonstration project by the Swedish energy company Vattenfall, CCS technology development in Germany was "clinically dead" in 2012. It was mainly the opposition of the regional population in areas with storage potential that impeded CCS R&D in Germany. This fear was stirred up by local NGOs, who lobbied against CCS as a high-risk technology and campaigned with an implicit equalisation of CO_2 underground storage with final disposal of nuclear waste (Herrenbrück 2015: 13). Phasing out CCS R&D in Germany before it had even entered the demonstration phase may challenge the achievement of national climate commitments; it makes the country, traditionally strong in energy research

1 https://sustainabledevelopment.un.org/content/documents/5987our-common-future.pdf

and process engineering, drop out as a possible R&D partner for countries with weaker innovation systems but high dependency on fossil fuel energy generation, such as South Africa or Indonesia.

During the past 20 or so years, debates on technology assessment have mostly focused on conceptual and methodological aspects, such as the needs to combine in an effective way expert knowledge and stakeholder involvement, and the question of whether the outcomes of TA should consist of giving single recommendations on the best technologies to foster or a policy to implement or, rather, of communicating multiple priorities and preferences of different groups within society effectively and transparently to decision-makers (Ely et al. 2011).

This chapter proposes to take one step back and suggests that technology assessment

- should partially refocus and strengthen its role in analysing potentials of technological developments for multi-dimensional development and
- should pursue more ambitious goals and develop formats to influence technology-related policy making and regulations on multilateral and global levels instead of addressing mainly the national level.

The North–South divide in research on global challenges

Among the multiple global divides, the one related to capacities in science, technology and innovation (STI) is frequently overlooked, even in development research. Without disregarding the role of indigenous or tacit knowledge and informal modes of knowledge generation and transfer, key to the solution of global challenges is the systematic generation of new knowledge and its application in innovation systems, most often conceptualised as national or sectoral innovation systems, or local innovation and production systems (Cassiolato and Martins (2020). That there is no global innovation system is not a gap in the terminology but rather reflects the fact that systematic knowledge generation still happens to an overwhelming degree on the national level, with some transnational regional innovation systems at least at an embryonic stage, e.g. in the form of the Horizon Europe program of the European Union. Multilateral cooperation in science, technology and innovation is still more the exception than the rule, even when global challenges are concerned (Figueroa and Stamm 2012; Stamm et al. 2012).

In all indicators referring to the input side of national innovation systems (NIS), the gap between the global South and the North is evident. Gross economic spending on research and development (GERD) as a percentage of GDP is the indicator most often used for comparing innovation capacities of countries. Here, many countries of the EU strive to reach 3% or achieve this value (Germany, Denmark), while the global top performers invest 5% of GDP in R&D (Israel, South Korea). Few countries in the global South approach or reach the 1% level (South Africa, Brazil), whereas many range between 0.5% and 1% (e.g. Costa Rica, Kenya) and others remain below this level, among them even some OECD countries, e.g. Mexico (0.3%) and Colombia (0.2%). In many developing countries, the value is not measurable in an adequate way.

In terms of absolute GERD figures, a clear picture is formed. The UNESCO Science Report "Towards 2030" shows that the global divide in R&D spending between high-income countries (HICs) and the other income groups is shifting slightly, with upper-middle income countries (UMICs) gaining larger shares at the expense of HICs. This shift is to a large extent related to the fact that GERD has increased very significantly in China and that this country still falls into the category of UMICs. The global position of LICs (low-income countries) and LMICs has largely remained stable and on a low level over the years (see Table 27.1).

Table 27.1 Strengths of different country groups in global R&D efforts

	Share in global population		*Share in global GDP*		*Share in global GERD*	
	2007	*2013*	*2007*	*2013*	*2007*	*2013*
High income countries	18.9	18.3	57.7	51.0	79.7	69.3
Upper middle-income countries	34.8	34.1	27.6	32.1	19.9	25.8
Lower middle-income countries	35.1	35.7	13.2	13.7	4.3	4.6
Low-income countries	11.2	11.9	1.4	1.7	0.2	0.3

Source: UNESCO 2015: 24–27

Table 27.2 Scientific and technical journal articles 2018 by World Bank country group

Country group	*Absolute number of publications*	*Publications per 1 million people*
Low income countries (LICs)	5,308	8
Middle income countries (MICs)	1,106,517	192
MIC w/o China	580,254	133
China	528,263	377
High income countries (HICs)	1,450,446	1,177

Source: World Development Indicators online April 2021

For the NIS performance indicator "researchers in R&D per million people", the average was 4,116 in HICs and 737 in MICs (2015, last available data), while for LICs, these data are not available (World Development Indicators, online October 2020).

It does not come as a surprise that the gap is more than evident on at the output side of NIS as well. Patent indicators only reveal relevant numbers in a few countries that do not belong to the global North. More relevant for research on global challenges are the number of scientific and technical papers published in journals. The relevant data are shown in Table 27.2. Here it is clear that it is important to separate China from the statistical group of MICs, as 48% of the total number of publications from the MICs group are from China.

One could argue that the global divide in input and output factors of the national innovation systems does not directly reflect differences in capacities for science-based responses to societal challenges. However, there is evidence that even in fields highly relevant for the global South and global challenges, most science is carried out and its agenda defined in the North. Blicharska et al. (2017) analyse the institutional affiliation of a large number of scientific publications explicitly dealing with climate change issues and find that during the period 2000–2014, more than 85% of the author affiliations of relevant scientific papers published (93,584 publications) were from OECD countries, while less than 10% were from other high-income economies or any Southern country income category (only 1.1% in the case of low-income economies). They suggest that the North–South divide in environmental research deprives the scientific community of considerable intellectual capital, influences research priorities and "most likely confines approaches to narrow paradigms from a few cultural settings and perspectives". Similarly, only 10% of funding for health research is spent in the South, where 90% of the world's burden of disease resides (Blicharska et al. 2017: 22).

The North–South divide in innovation performance is very pronounced, and there are no indications that it might get narrower over time if we leave the rapid rise of China as global technology hub out of the equation. Even if we cannot yet assess the long-term effects of the 2020–2021 COVID-19 pandemic, it is clear that many of the relatively good performers of the global South in R&D and innovation are heavily affected, such as Chile, Colombia, Peru or South Africa. It cannot be expected that in the near future they will have the financial resources for a broad-based strategy of catching-up in STI. This implies that many developing countries will need strong support from the global North in identifying and implementing innovations to tackle global challenges and, here, assuring food security for a growing population under conditions of climate change.

Innovation to address food security under conditions of climate change – responding to a global challenge

Challenges of agricultural development for the next decades

Agriculture is a core sector to address several of the Sustainable Development Goals (SDGs) to which the world community committed in the Agenda 2030, namely SDG2 (Zero Hunger), SDG3 (Good Health), SDG13 (Climate Action) and SDG15 (Life on Earth). More than any other sector, agriculture is under extreme performance stress due to several interlinked factors: population growth, the exhaustion of land reserves and climate change. The United Nations expect global population to grow from 7.8 billion today to 9.7 billion in 2050, an increase of around 24% in 30 years. Due to competing uses of agricultural output (food, feed and biofuel), FAO (2017a: 46) estimates that global agriculture in 2050 will need to produce almost 50% more than it did in 2012. In sub-Saharan Africa and South Asia, agricultural output would need to more than double by 2050 to meet increased demand, while in the rest of the world the projected increase would be about one-third above current levels (FAO 2017a: 46). This happens under difficult conditions. Expansion of agricultural land has slowed down and nearly come to a halt; between 2000 and 2016, it increased by barely 1%.[2] In areas where the agricultural frontier advances, it comes virtually always at the expense of natural habitats and biodiversity, e.g. in the Amazonas region or in many mountainous areas of the global South. This implies that food production has to increase substantially in the coming decades on basically the same extension of land as is available today.

In addition, climate change is increasing the abiotic stress (higher temperature, more frequent droughts, salinity) for food crops. Reports by the Intergovernmental Panel on Climate Change (IPCC) and other studies have shown that climate change may result in significant yield losses which are likely to become more severe in the future (Challinor et al. 2014; Aggarwal et al. 2019). In the short term, these studies project a loss of up to 25%, depending on the crop and region. By 2080, these losses are estimated to be as high as 50%. Climate change impacts on crop yields are projected to be geographically differentiated, with developing countries in the tropics being more affected (and hence more vulnerable) than developed countries in temperate latitudes. South Asia and Africa are supposed to witness yield losses of up to 25% by the 2020s compared to Europe and North America, where there could even be some yield gains (Knox et al. 2016; Aggarwal et al. 2019). Food security will thus increasingly be a challenge for developing countries, where the main percentage of population growth will happen, where climate

2 Own calculations, based on World Development Indicator data, accessed 25 November 2020.

change impacts are stronger and where financial resources for acquiring food on global markets are limited. In addition, their resilience to global warming is lower due to the lack of technical and economic resources to establish, for instance, buffers against water scarcities.

Food security is especially an issue for sub-Saharan Africa, where land productivity and accessibility of external inputs are low, a significant proportion of land is degraded and a rapidly growing population needs to be supplied with food under conditions of increasingly noticeable climate change (WBGU 2021: 128–134). Already since the mid-1970s, Africa has been a net food importer (Rakotoarisoa et al. 2011). It is well understood that assuring food security under such conditions requires a well-coordinated package of actions, from reducing post-harvest losses, renaturation of terrestrial ecosystems, a sustainable intensification and diversification of food systems and transition to more plant-based dietary habits (WBGU 2021). Also, expansion of irrigated land areas can play an important role.

However, a core challenge will be to increase the amount of food that can be harvested on available agricultural land and avoid yield fluctuations due to abiotic stress for crop plants (Zafar et al. 2019). In the past decades, yield increases were significant, mainly based on higher usage of external inputs in combination with improved plant species. The annual growth rate of nitrogen fertilizers applied is almost 1% globally and up to 3.5% in developing countries (Aggarwal et al. 2019: 45). There are, however, at least two reasons as to why relying exclusively on higher external inputs for yield increases is not a feasible option in the context of an overall commitment to maintaining "a safe operating space for humanity" (Rockström et al. 2009): First, fertilizer and pesticide production and application are important sources of GHG emissions, both CO_2 and the more powerful driver of global warming, NO_2. In 2015, synthetic fertilizers alone emitted 0.6 $GtCO_2e$ along their value chain. Second, an often-overlooked planetary boundary, already in the "red zone" of clearly "beyond a safe operation space", are biochemical flows:

> Modern agriculture is a major cause of environmental pollution, including large-scale nitrogen- and phosphorus-induced environmental change. At the planetary scale, the additional amounts of nitrogen and phosphorus activated by humans are now so large that they significantly perturb the global cycles of these two important elements.
>
> *(Rockström et al. 2009: 474)*

This implies that from an earth system perspective, further intensification of external input usage in agriculture cannot be a recommended option. Thus, the question of how to produce more food with the same amount or reduced usage of land and external inputs will be high on the agenda during the upcoming decades. At the technological frontier, two important innovations seem to have the potential to offer new opportunities for assuring food security under the difficult challenges described: digital agriculture and new breeding technologies, specifically CRISPR-Cas9 and other techniques of genome editing of organisms.

Digital agriculture, precision agriculture and robotics

Digital agriculture is the umbrella term for a series of applications of information and communication technologies (ICTs) in the agricultural sector (FAO 2017b). The body of literature on various aspects is swiftly increasing, reflecting the rapid development of the system's elements (e.g. sensors, drones and robots), their merging into more complex systems, their spread across different applications and their respective ethical implications (Van der Burg et al. 2019): Depending on what kind of data is used and processed and for which concrete application, other concepts are used, such as smart farming, precision agriculture or precision farming, decision agriculture or agriculture 4.0 (Klerkx et al. 2019). Regardless of the exact term used, digitalisation implies that management tasks on-farm and off-farm are increasingly based on the use of different sorts of data (on location, weather, phytosanitary status, etc.). This goes hand in hand with using sensors, unmanned aerial vehicles (UAM, drones) and satellites to monitor animals, soil, water, plants and humans. UAMs can also be used for pollination and spraying of herbicides, pesticides and fertilizers (Frankelius et al. 2019). More recently, they are being tested as an instrument to spray insecticides to combat locust swarms heavily affecting food crops in East Africa (Armstrong 2020). Agricultural robotics is an additional application of digital agriculture, with potentially far-reaching socio-economic impacts for developing countries (Jha et al. 2019; Sparrow and Howard 2021).

The data obtained by several devices is used to interpret the past and predict the future, to make more timely and accurate decisions through constant monitoring or specific big data science enquiries (Klerkx et al. 2019). In 2015, the terms "fourth agricultural revolution" or "agriculture 4.0" were proposed. These notions referred to the impact of sensors, satellites, digital technology and robotics, not least in terms of paving the way for precision farming (Frankelius et al. 2019: 682). Digital agriculture may lead to more stable or increased yields, with lower usage of external inputs:

> Big data, the Internet of things, drones, and artificial intelligence may catalyse precision farming, requiring fewer agrochemical inputs for existing agricultural processes.
>
> *(UNCTAD 2017: 22)*

From the perspective of technology assessment, however, it is important to stress that digital agriculture does not automatically cause agriculture to contribute to the above-mentioned SDGs. Potential trade-offs may be exemplified in the field of agricultural robotics. Current development in this area is concerned with the substitution of the human workforce by field robots or mechanised systems that can handle the tasks more accurately and uniformly at a lower cost and higher efficiency (Shamshiri et al. 2018: 3). On the one hand, automatized work in commercial agriculture might have some positive impacts on the ecological dimension of sustainability, as e.g. fertilizers and pesticides might be applied more precisely, lowering overall quantities and avoiding leakages into the environment. On the other hand, harvesting and weeding are two agricultural activities providing a high number of jobs for low- to semi-skilled laborers in developing countries. As most of the R&D projects related to smart farming are conducted in industrialised countries, it can be assumed that they address the specific factor endowments in the global North, e.g. relatively high wages and sometimes absolute scarcity of agricultural laborers. Automation of weeding and harvesting may, thus, be high on the R&D agenda, mainly following the purpose of substituting manual work by robotics. A paper by Shamshiri et al. (2018), for instance, describes the first fully automatised harvesting platform for sweet pepper, developed in the context of an EU Horizon 2020 project. It can be expected that these innovations will not be in reach of smallholder farmers in developing countries. They may, however, be of interest for large-scale agricultural business in developing countries, once the efficacy, efficiency and robustness of the systems have reached acceptable levels.

Potential ethical issues related to digital agriculture are manifold and to date only partially framed. Van der Burg et al. (2019) mention the following themes, each with several and partially interlinked issues: 1) data ownership, accessibility, sharing and control, 2) distribution of power and 3) impacts on human life and society. Considering the big potential of digitalisation for a sustainable intensification of agriculture, and taking into consideration the North–South divide in innovation performance, TA should see its role, at this early stage, in identifying potential threats of technological developments for disadvantaged stakeholder groups, e.g. the risk of losing low- to semi-skilled job opportunities in developing countries via AI and robotics. It should at the same time contribute to identifying potential applications of digital agriculture in the context of the SDGs, which might not be addressed through innovation processes driven by the commercial interests of companies in the North. For instance, enabling organised smallholder farmers to use drones and remote sensing to implement precision farming might not be a business-case for big ICT or agricultural engineering companies, but might very well be in the global public interest and could be supported by development cooperation.

Gene-editing in plant breeding

Few decisions of the Court of Justice of the European Union (CJEU) have triggered as strong a response from academia and media as in the case of the court's ruling of July 25, 2018,[3] concerning the regulation of so-called new plant breeding technologies (NBT). In its essence, the court ruling classifies organisms created by genome editing techniques (such as CRISPR-Cas9) as genetically modified organisms (GMOs) within the meaning of the EU GMO Directive (Schleissing et al. 2019: 9f). GMOs are the product of a direct manipulation of an organism's genome, via the transfer of genes from either species with which a traditional breeding would be possible (*cisgenesis*) or from species with which traditional breeding would be impossible (*transgenesis*).

This at first sight merely technical decision may have far-reaching consequences for the further development of CRISPR-Cas9 and other genome editing technologies, both in terms of speed and with regard to the potential drivers of the innovation process, limiting especially participation of economically less powerful actors. It implies nothing else than that all future innovation cycles in the field of gene editing in the European Union have to follow the strict regulations for GMO crops. In order to prevent any possible dangers related to the development and commercialisation of GMO products, EU directives stipulate risk assessment, approvals, labelling and traceability requirements. This process is lengthy (4–6 years) and costly (7–15 million euros) (Van Belle et al. 2019: 34). EU directives foresee three phases with rather strict rules until each single GM crop (case-by-case approach) may be commercialised (step-by-step approach, Dederer 2019: 79):

1 work in a closed system (laboratory, growing chamber, greenhouse),
2 deliberate release into the environment (field trials, first small-scale, then large-scale),
3 placing on the market (i.e. marketing and use e.g. as seed, food or feed).

The step-by-step approach means that GMOs have to be handled in closed systems before open field trials can be applied and approved. From the TA perspective, one of the guiding concepts behind the strict regulation of steps 2 (field trials) and 3 (placing on the market) of GMOs is the potential irreversibility of releasing genetically modified organisms, as laid down in point 4 of the Preamble of Directive 2001/18/EC of the European Parliament and of the Council:[4]

> Living organisms, whether released into the environment in large or small amounts for experimental purposes or as commercial products, may reproduce in the environment

3 https://curia.europa.eu/jcms/upload/docs/application/pdf/2018-07/cp180111en.pdf
4 https://eur-lex.europa.eu/resource.html?uri=cellar:303dd4fa-07a8-4d20-86a8-0baaf0518d22.0004.02/DOC_1&format=PDF

> and cross national frontiers thereby affecting other Member States. The effects of such releases on the environment may be irreversible.

The decision by the CJEU responded to the request of French farming associations that gene edited plants should not be considered equal to those produced by existing breeding technologies. Since around the 1960s, traditional crossbreeding has partially been substituted by technologies which trigger mutations at a much faster pace than what happens in nature. Traditional crossbreeding aims to improve the characteristics of organisms (yield, nutritional value) by deliberately crossing related species. The technique is limited to the naturally existing genetic variations. "Traditional" mutagenesis breeding increases genetic diversity using ionising radiation (X-rays, gamma rays) and chemical mutagens such as Ethyl methanesulfonate (EMS). It induces quasi-natural mutations but does so at a much higher rate than nature (Van Belle et al. 2019: 27). Thus, while traditional mutagenesis breeding could also be considered as a form of genetic modification, it is considered conventional breeding and is regarded as "safe" as it has been in use for decades, with no evidence of harm. Mutants created via traditional mutation breeding are therefore exempt from the strict approval procedures of GMO regulations.

CRISPR-Cas9 is a molecular biological method for cutting and subsequently modifying DNA. In this way, individual genes or DNA building blocks can be rewritten or "edited" (Kawall et al. 2020). Such procedures are collectively known as genome editing (or gene editing), and increasingly as "gene scissors" or "gene surgery". Gene editing techniques do not involve permanent transfer of genes between species. Rather, they offer the possibility of inducing genetic modifications with high precision and efficiency, using the cell's own repair system. Unlike "traditional" mutation breeding, genome editing relies on targeted mutations rather than accidentally occurring desired mutations among a relatively high number of off-target mutations. For instance, in rice, only 1 out of 16 to 2,800 EMS mutations is located in the desired gene (Van Belle et al. 2019: 28).

The fact that genome editing does not include transfer from one species to another, alien one could imply that it is closer to crossbreeding or traditional mutagenesis breeding than to GMO. It can also be seen as a more sustainable and effective technology compared to both crossbreeding and traditional mutagenesis: As the desired changes in the organisms can be achieved in a directed way and not randomly, it allows saving of energy and other inputs for the (much larger) part of induced mutations which do not have the desired characteristics and will thus not be further used in breeding and instead discarded.

The CJEU decision to classify genome editing as GMO has been followed by a series of statements from the science community, criticising the ruling for technical and ethical reasons and calling for amendments to the European genetic engineering law and for the consideration of scientific dynamics in the application of the precautionary principles. Important voices in this direction came from the Leopoldina, the Union of the German Academies of Sciences and Humanities together with the German Research Foundation DFG and the Union of German Science Academies (Leopoldina et al. 2019), followed by the European Academies' Science Advisory Council (EASAC 2020) and a number of scientists e.g. from plant breeding (Van Belle et al. 2019), European and international Law (Voigt 2019) and even ethics and theology (Hoffmann 2019).

From the technical perspective, Leopoldina criticises that the CJEU ruling does not take into consideration "the type of genetic modification present in the genome edited organism and whether this modification could have occurred accidentally or through traditional breeding methods" and whether the origin of the genetic modification can be identified and traced to a particular breeding method (Leopoldina et al. 2019: 46). The last point, as technical it may appear, has potentially far-reaching consequences for international trade and may affect developing countries and their trade opportunities with the EU.

Van Belle et al. (2019: 36) describe a certain "Catch 22"-type situation regarding the approval of gene edited products under the current regulation:

> the approval procedure of genome edited plants for cultivation will be too lengthy and expensive until such time as there is absolute certainty about their safety. At the same time, because no genome edited plants will be cultivated, they cannot be conventionally used in numerous applications, nor can a long safety record can be built.

From an ethical point of view, a series of arguments are put forward. For Leopoldina (2019: 48), the strict regulation, which covers all genome edited plants indiscriminately, substantially restricts the freedom of research in the EU without substantial justification (Leopoldina et al. 2019); this sits against the backdrop that in many countries outside the EU, genome edited organisms are exempt from the strict GMO regulations. Argentina is seen as pioneering tailored regulation. In 2015, the country issued a specific regulation for products of so-called new breeding technologies. Developers are asked to present their project to the biosafety commission in order to decide, case by case, whether the product to be developed should be treated as GMO (Lema 2019).

Possible impacts of the CJEU ruling on developing countries

Compared to traditional mutagenesis working with radiation or chemicals inducing random mutations, gene editing can be much more directed, thus saving time and resources in the breeding process. The time argument is especially important in times of rapidly progressing climate change. Leopoldina et al. (2019: 55) list a series of examples of gene edited crops which might be of great importance for climate change adaptation in the global South, due to improved fungus, bacterial or drought resistance and salt tolerance, e.g. in maize, tomato, rice, wheat and banana. The lead time for bringing adapted varieties to application will be much higher with the complicated procedures under GMO rules. In addition, the time consuming and costly approval process will lead to a further monopolisation in the already concentrated plant breeding and seeds markets. Normatively, this is contrary to the objective of equal access to technological opportunities and a "level playing field". Bartkowski et al. (2018: 170) see the potential of genome editing to diversify the market for biotechnology, especially compared to traditional genetic engineering with high upfront costs in equipment, know-how and lead times to bring innovations to practice. These advantages might be partially lost by high regulatory barriers imposed on genome editing.

Innovation research indicates that with a higher number of researchers, research groups and entrepreneurs involved in developing new technological solutions, the quantity and variety of innovations will be higher. This is especially the case when a technology is still "in flux" and many different paths are still open, which describes well the situation of CRISPR-Cas9 and related technologies. With a more limited number of actors able to invest financial resources and time, the risk exists that further developments are streamlined according to the interests of big business. Solutions involving specific food crops in the South might not be developed if large companies cannot see a sufficiently interesting business case in it.

For developing countries, the CJEU might have serious impacts. Considering the abovementioned required yield increases to assure food security of their growing population, they may have to opt for an increasingly high external input strategy. For most developing countries, this might imply higher expenditures for imports of intermediate goods, fertilisers and pesticides.

In addition, increased chemical inputs will raise the carbon footprint of agricultural production and lead to other negative environmental impacts, e.g. on the nitrogen and phosphorus cycle. Genome-edited plants may not be a sufficiently strong tool to prevent a high-input path for agriculture in developing countries, but they might offer options for innovative solutions, such as combining accelerated breeding via CRISPR-Cas9 with principles of organic agriculture and increased biodiversity on the field. Some fungi, for instance, are an unresolved problem in organic agriculture, and breeding fungus-resistant varieties by traditional crossbreeding may take several decades.[5]

Zaidi et al. (2019: 1390f) see a great potential for genome editing to contribute to a "world without hunger, if properly managed". It is unlikely, however, that many developing countries severely affected by lack of food security will be able to develop their own gene edited organisms for their agriculture, with the exception of countries like Brazil, Argentina or India, which have relatively developed agricultural innovation systems and already have 25 years of experience in growing GM crops (Lema 2019: 147). The entry barrier may not be so much the hardware in laboratories and greenhouses, as the requirements of gene editing make it a relatively inexpensive field of research. However, for most LDC and many MICs, it is evident that they need international partners for technology transfer and joint research and development. While the EU regulation acts as a brake for relevant research, developing countries might see the need to approach partners, e.g. in world areas, where gene editing is explicitly exempt from GMO regulation (USA) or where regulation is still in flux (China).[6]

Even if countries might acquire the capabilities to develop their own gene edited crops, they might hesitate to do so, mainly due to possible trade problems. The ruling of the CJEU requires "that the import of genome edited plants, . . . be subject without exemption to the authorization requirements of European genetic engineering law" (Leopoldina et al. 2019: 68). One difficulty arises from the fact that genome editing processes leave a final product, which is not easily rendered to tracing the technical process of its genesis. Thus, easy border inspections will not allow for determining gene edited products on their way to be sold in Europe. Developing countries with traditional trade ties to Europe could face the dilemma that they might have to either refrain from applying gene editing techniques for their domestic market, or try to go for a lengthy and insecure (case by case) approval of products which shall be sold on the EU market.

This implies that a ruling meant to regulate an emerging technology for the EU has far-reaching and potentially significant negative impact for developing countries. In an increasingly globalised world, technology assessment, whether explicit or implicit as in the case of gene editing and the verdict of the CJEU, will have to widen its perspectives and include the potential benefits and risks for stakeholders who might reside far away from the innovation hubs.

Considerations for a future-oriented TA under precautionary principle criteria

Not in all cases, policies and regulations which shape technological development and innovations are based on sound science- and evidence-based TA. As in the case of CCS and CRISPR-Cas9, decisions are sometimes taken based on an ad-hoc understanding of risks and opportunities related to technologies. In these cases, perception of high risks has led to breaking the innovation cycle at an early state (CCS in Germany) and heavy regulation, which may delay

5 https://taz.de/Oekoforscher-ueber-neue-Gentech-Methode/!5290509/

6 Canada follows a precautionary approach (risk assessment applies to novel products), yet to a lesser extent than the EU (Van Belle et al. 2019).

innovation and reduce the variety of solutions which may be identified (genome editing in the EU). On the other hand, not all product and process innovations have been subject to any kind of significant TA before they were rolled out on the markets. For instance, Harremoës et al. eds. (2002) list a series of innovations of past decades, e.g. asbestos in buildings and antifoulants for ships, which had negative impacts on public health and the environment but where science was not able to extend early warnings, or where warnings were not taken up by policy makers at a sufficiently early stage. The complexity of the final disposal of radioactive waste could be mentioned as an additional technology field where TA did not provide sufficient evidence in a convincing way and at an early stage of the usage of nuclear energy.

Thus, there seems to be room for improvement in the practice of TA. Rather significant efforts have been made in advancing in methodological aspects (from classical to participatory approaches) and increasing the policy relevance and impact (Hahn and Ladikas 2019: 2–4). Considering the several "blind spots" in technological developments and the very fast innovation cycles in many fields, there might also be the need to discuss issues related to agenda and priority setting in TA. How (and by whom) can assessment exercises be put on the TA agenda and how (and by whom) can priorities be set, in the face of limited resources available for TA?

Complexity of TA has increased both internally (including additional stakeholders in the TA process) and externally. Technological developments are increasingly interdisciplinary processes (e.g. Fiedeler 2008 for the case of nanotechnology). Innovation cycles are becoming increasingly shorter and lead to diverse products and services with varying social and environmental impacts; as the case of ICT demonstrates, the array of both possibilities (e.g. e-education, e-government, digital agriculture) and risks (from protection of rights of personality to fast affluence of e-waste) have increased and diversified in very short times.

From the analyses in this chapter, there seem to be more challenges for TA under the conditions of eroding planetary boundaries and the need to find answers to global challenges. The early history of TA saw it mainly as an early warning system stressing the potential risks and non-intended side-effects of a technology, which led to caricaturing the term as "Technology Arrestment" (Grunwald 2010: 287). Considering the urgency of addressing climate change and other global challenges by new technological solutions,[7] there might today be a need to better balance assessments of risks and opportunities. Von Schomberg (2019) sees the need to refocus the assessment of innovation processes away from an exclusive focus on the risks of new technologies and to the question of directing or redirecting research and innovation towards societally desirable ends. We discussed that technological advancements in digital agriculture might have negative or positive outcomes on the SDGs, and TA could take over a stronger role in identifying routes for technological developments desirable from a global public goods perspective. Institutionally, Von Schomberg (2019: 14) proposes the separation of the two processes, which seems a possible but not only option:

> This implies that we not only need to have professional bodies for risk assessment but also professional bodies that should look into the type of outcomes we want to obtain from research and innovation processes, and the establishment of governance mechanisms that should give some direction to – or steer- the innovation process.

In detail, this also calls for discussing some elements of the precautionary principle as one ethical point of reference for TA at least in Europe. The PP, introduced to EU rules in the Maastricht

7 We assume that in most cases, technological solution will only be one element in a "package of measures" to address a specific global challenge.

Treaty in 1992, entails three main procedures, the first being "the fullest possible scientific evaluation, identifying at each stage and as far as possible the degree of scientific uncertainty". This definition of a responsible approach to technological development is fully in line with Hans Jonas' concept of "heuristics of fear", which we presented earlier.

There might be an important trade-off between the goal of achieving the "fullest possible scientific evaluation" and the need to apply new technologies early enough, e.g. to avoid tipping points of the earth system to pass. This raises a series of difficult ethical questions. While a massive roll-out of a not sufficiently understood technology can clearly be considered irresponsible, it might be also irresponsible to unduly delay the R&D process. Until outcomes of R&D become an innovation (knowledge applied in societal practices), it may take several years. If R&D is delayed until "the fullest possible scientific evaluation" of potential risks of a technology has been achieved, the emerging time lag might be costly in terms of addressing global challenges.

The obvious solution would be to frame TA differently at the various steps of the innovation cycle. This can be exemplified taking the case of genome editing. Even if in Europe the societal consensus may be to apply genome editing in agriculture extremely cautiously, the ethical obligation may still be to contribute to advance directed innovations in this field, e.g. innovations which will benefit the interests of smallholders and other poor strata of the population in developing countries. Public money could be invested in segments of agriculture which would otherwise be neglected, because they do not provide an interesting business case for the technological powerhouses in the private sector. Technically, this might work in the way that relevant R&D efforts are concentrated in multilateral formats, such as the CGIAR research centres, with sufficient support from, in our case, Europe. This does not mean to "outsource" technological risks to low-income countries, but rather to assure that there are locations where technological advancements can be implemented. In the case of clean coal technologies, something similar could be considered with regard to up-scaled multilateral funding of the Global CCS Institute, based in Australia.

Adequate governance of multilateral research could make sure that ethical guidelines are followed and risks are minimised, e.g. when genome edited plants are passed from the lab to open-field trials. Efforts would also have to be made to avoid anti-ethical usage of new technologies, like, in our case, genome editing to manipulate human germlines for non-essential medical purposes. International codes-of-conduct would have to be developed to define ethical bottom lines.

Countries and regions (EU) participating in multilateral research and development would still be in conditions to apply a formerly not accepted technology at home, once an adequately long safety record has been established. This might help to overcome the described "Catch 22" situation in the admission of genome edited organisms. Separating R&D from innovation (application of a new technology, whether commercially or in the public interest) in TA could help in the urgently needed task of developing new technologies without unnecessarily delaying the innovation process. Even relatively easy to develop innovations, e.g. with CRISPR-Cas9, need time to mature and be both effective and safe to apply. Develop now – regulate later could mean that technologies are developed which at a later stage will either be applied or not, depending on the outcome of TA exercises. This might imply that some pre-investments in R&D might be lost, but this seems to be a less serious problem compared to time lags in developing innovations which could help to address global challenges.

From the perspective of innovations to address global challenges, an additional aspect of the precautionary principle is the "participation of all interested parties in the study of precautionary measures, once the results of the scientific evaluation and/or the risk evaluation are available" (EC 2017). One could criticise the separation of providing scientific evidence from stakeholder participation as two successive and not interlinked processes. What seems more relevant and poses open questions in our context is how the term "all interested parties" can be reasonably operationalised when the task is evaluation of technologies with a global outreach and which might have a much more significant effect in regions far away from where the capacities for technological R&D are located, as in the case of genome editing.

We want to emphasise again the very significant North–South divide in inputs to and outputs from innovation systems. We explained, taking the example of agriculture and food security, that it is exactly the population in developing countries, weak in research and innovation, which is likely to suffer from the consequences of climate change, mainly caused by action and inaction of the industrialised countries. There is thus a clear ethical obligation to mobilise existing research strengths in identifying new options for tackling climate change impacts and other global challenges anywhere on the globe.

As a consequence, it is important and urgent to achieve a global technology assessment. It is a somehow surprising fact that only in 2019 was a comprehensive book published with the purpose of laying the foundation for a global TA (Hahn and Ladikas 2019), including comparative analyses of science and technology governance systems and national values in traditional (Europe) and non-traditional (China, India) global technology hubs. How developing countries with less-developed innovation capacities might be included in relevant efforts would be the next step in the analyses and discourse.

Institutionally, in 2019 a global TA network of not-for-profit organisations was established to further advance the idea and developing frames for global TA.[8] The three core elements of its missions are:

- develop a global framework and code of conduct for the assessment of impacts of (new) technologies,
- facilitate global cooperation for the assessment of emerging technologies in order to maximise their benefits and minimise the risks,
- support adequate anticipatory governance of new technologies that may have significant impacts on the attainment of the UN Sustainable Development Goals.

Among the (at the end of 2020) 29 members of the global TA network, most are based in the traditional and new technological hubs, mainly Europe, the USA, Australia, China, South Korea and India. Some middle-income countries from Latin America (Chile, Mexico) have joined the initiative, but low-income countries and Africa are still white areas on the global TA map. This might change with UNCTAD having recently taken up the global TA topic. UNCTAD's Division on Technology and Logistics and the Commission on Science and Technology for Development (CSTD), in 2020, launched a project to anchor the TA concept in stakeholders related to the energy and agricultural sectors in Africa. This could be an important step to systematically involve a larger number of developing countries in processes relevant for global TA.

References

Aggarwal, P., Vyasa, S., Thorntona, P., Campbell, B. and Kropff, M. (2019): Importance of Considering Technology Growth in Impact Assessments of Climate Change on Agriculture, *Global Food Security* 23, 41–48.

Armstrong, L. (2020): Huge Locust Swarms are Threatening Food Security, But Drones Could Help Stop Them, *The Conversation*. Available at: https://theconversation.com/huge-locust-swarms-are-threatening-food-security-but-drones-could-help-stop-them-140625 [02.04.2021]

8 https://globalta.technology-assessment.info/

Barney, G. (1980): *Global 2000 – Report to the President.* New York et al.: Pergamon Press.
Bartkowski, B., Theesfeld, I., Pirscher, F. and Timaeus, J. (2018): Snipping around for Food: Economic, Ethic and Political Implications of CRISPR-Cas Genome Editing, *Geoforum* 96, 172–180.
Blicharska, M. et al. (2017): Steps to Overcome the North – South Divide in Research Relevant to Climate Change Policy And practice, *Nature Climate Change* 7, 21–27. https://doi.org/10.1038/NCLIMATE3163
Cassiolato, J. E. and Martins, H. M. (2020): The Framework of "Local Productive and Innovation Systems" and its Influence on STI Policy in Brazil, *Economics of Innovation and New Technology* 29(7), 784–798.
Challinor, A. J., Watson, J., Lobell, D. B., Howden, S. M., Smith, D. R. and Chhetri, N. R. (2014): A Meta-analysis of Crop Yield under Climate Change and Adaptation, *Nature Climate Chance* 4(4), 287–291.
Coyne, L. (2018): Responsibility in Practice: Hans Jonas as Environmental Political Theorist, *Ethics, Policy & Environment* 21(2), 229–245.
Dederer, H.-G. (2019): Options for the Regulation of Genome Edited Plants – Framing the Issue, in Dürnberger, C., Pfeilmeier, S. and Schjeissing, S. (eds.): *Genome Editing in Agriculture*. Baden-Baden: Nomos, 77–122
EASAC (European Academies' Science Advisory Council) (2020): *The Regulation of Genome-edited Plants in the European Union.* Halle, Brussels: EASAC.
EC, European Commission (2017): The Precautionary Principle: Decision-Making under Uncertainty. *Science for Environment Policy, Future Brief 18.* Luxembourg.
Ely, A., van Zwanenberg, P. and Stirling, A. (2011): New Models of Technology Assessment for Development. *STEPS Working Paper 45*, Brighton: STEPS Centre.
Fiedeler, U. (2008): Technology Assessment of Nanotechnology: Problems and Methods in Assessing Emerging Technologies, in Fisher, E., Selin, C. and Wetmore, J. M. (eds.): *Presenting Futures. The Yearbook of Nanotechnology in Society*. Vol. 1. Dordrecht: Springer, 241–263.
Figueroa, A. and Stamm, A. (2012): Effective International Science, Technology and Innovation Collaboration: From Lessons Learned to Policy Change, in OECD (ed.): *Meeting Global Challenges through Better Governance: International Co-Operation in Science, Technology and Innovation.* Paris: OECD Publishing, 207–231.
Food and Agricultural Organisation of the United Nations, FAO (2017a): *The Future of Food and Agriculture – Trends and Challenges.* Rome: FAO.
Food and Agricultural Organisation of the United Nations, FAO (2017b): *Information and Communication Technology (ICT) in Agriculture – A Report to the G20 Agricultural Deputies.* Rome: FAO.
Frankelius, P., Norrman, C. and Johansen, K. (2019): Agricultural Innovation and the Role of Institutions: Lessons from the Game of Drones, *Journal of Agricultural and Environmental Ethics* 32, 681–707.
Grunwald, A. (2010): *Technikfolgenabschätzung – Eine Einführung.* Berlin: edition sigma.
Hahn, J. and Ladikas, M. (eds.) (2019): *Constructing a Global Technology Assessment. Insights from Australia, China, Europe, Germany, India and Russia. Karlsruhe Institute of Technology.* Karlsruhe: KIT.
Harremoës, P. et al. (eds.) (2002): *Late Lessons from Early Warnings: The Precautionary Principle 1896–2000.* Copenhagen: European Environment Agency.
Herrenbrück, R. (2015): CCS in Deutschland, *IPW Selected Student Paper*, 52. Aachen: Rheinisch-Westfälische Technische Hochschule Aachen, Philosophische Fakultät, Institut für Politische Wissenschaft.
Hoffmann, A. F. (2019): The Precautionary Principle: Origin, Ethical Implications, and Theological Outlook, in Dürnberger, C., Pfeilmeier, S. and Schjeissing, S. (eds.): *Genome Editing in Agriculture.* Baden-Baden: Nomos, 251–258
IPCC (Intergovernmental Panel on Climate Change) (2005): *Carbon Capture and Storage. IPCC Special Report.* Cambridge, UK et al.: Cambridge University Press.
IPCC (Intergovernmental Panel on Climate Change) (2019): *Global Warming of 1.5°C. An IPCC Special Report on the Impacts of Global Warming of 1.5°C above Pre-industrial Levels and Related Global Greenhouse Gas Emission Pathways, in the Context of Strengthening the Global Response to the Threat of Climate Change, Sustainable Development, and Efforts to Eradicate Poverty.* Geneva: Intergovernmental Panel on Climate Change.
Jha, K., Doshi, A., Patel, P. and Shah, M. (2019): A Comprehensive Review on Automation in Agriculture Using Artificial Intelligence, *Artificial Intelligence in Agriculture* 2, 1–12.
Jonas, H. (1984): *The Imperative of Responsibility. In Search of an Ethics for the Technological Age.* Chicago: University of Chicago Press.

Kawall, K., Janet, C. and Then, C. (2020): Broadening the GMO Risk Assessment in the EU for Genome Editing Technologies in Agriculture, *Environmental Science Europe* 32, 106. https://doi.org/10.1186/s12302-020-00361-2

Klerkx, L., Jakkub, E. and Labarthe, P. (2019): A Review of Social Science on Digital Agriculture, Smart Farming and Agriculture 4.0: New Contributions and a Future Research Agenda, *NJAS – Wageningen Journal of Life Sciences* 90–91, [100315]. https://doi.org/10.1016/j.njas.2019.100315

Knox, J. et al. (2016): Meta-analysis of Climate Impacts and Uncertainty on Crop Yields in Europe, *Environmental Research Letters* 11, 113004.

Lema, M. A. (2019): Regulatory Aspects of Gene Editing in Argentina, *Transgenic Research* 28, 147–150.

Leopoldina (Nationale Akademie der Wissenschaften), DFG (Deutsche Forschungsgemeinschaft), Union der deutschen Akademien der Wissenschaften (2019): *Towards a Scientifically Justified, Differentiated Regulation of Genome Edited Plants in the EU. Halle, Bonn, Mainz*. Köthen (Anhalt): druckhaus.

Marcuse, H. (1941): Social Implications on Technology, *Zeitschrift für Sozialforschung* 9(3), 414–439.

Meadows, D. L. (1972): *The Limits to Growth. A Report for the Club of Rome's Project on the Predicament of Mankind*. New York: Universe Books.

Rakotoarisoa, M., Iafrate, M. and Paschali, M. (2011): *Why has Africa Become a Net Food Importer? Explaining Africa Agricultural and food trade deficits*. Rome: FAO.

Rockström, J. et al. (2009): Planetary Boundaries: Exploring the Safe Operating Space for Humanity, *Ecology and Society* 14(2), 32.

Schleissing, S., Pfeilmeier, S. and Dürnberger, C. (2019): Genome Editing in Agriculture: Between Precaution and Responsibility. An Introduction: Between Precaution and Responsibility, in Dürnberger, C., Pfeilmeier, S. and Schleissing, S. (eds.): *Genome Editing in Agriculture, Between Precaution and Responsibility*. Baden-Baden: Nomos, 9–22

Schütze, C. (1994): The Political and Intellectual Influence of Hans Jonas, *Hastings Center Report* 25(7), 40–44

Shamshiri, R. R., Weltzien, C., Hameed, I. A., Yule, I. J., Grift, T. E. and Balasundram, S. K. et al. (2018): Research and Development in Agricultural Robotics: A Perspective of Digital Farming, *International Journal of Agricultural and Biological Engineering* 11(4), 1–14.

Sparrow, R. and Howard, M. (2021): Robots in Agriculture: Prospects, Impacts, Ethics, and Policy, *Precision Agriculture* 22, 818–833.

Stamm, A., Figueroa, A. and Scordato, L. (2012): Addressing Global Challenges Through Collaboration in Science, Technology and Innovation, in OECD (ed.): *Meeting Global Challenges through Better Governance: International Co-Operation in Science, Technology and Innovation*. OECD Publishing, 25–42.

UNCTAD (United Nations Conference on Trade and Development) (2017): *The Role of Science, Technology and Innovation in Ensuring Food Security by 2030*. Geneva: UNCTAD.

UNESCO (United Nations Educational, Scientific and Cultural Organisation) (2015): *UNESCO Science Report: Towards 2030*. Paris: UNESCO.

Van Belle, J., Schaart, J. and van Loo, R. (2019): The Logic of Exempting CRISPR/Cas9-Plants from Strict Approval under European Union GM Regulations: A Case Study on Genome, in Dürnberger, C., Pfeilmeier, S. and Schjeissing, S. (eds.): *Genome Editing in Agriculture*. Baden-Baden: Nomos, 25–45

Van der Burg, S., Bogaardt, M.-J. and Wolfert, M.-J. (2019): Ethics of Smart Farming: Current Questions and Directions for Responsible Innovation towards the Future, *NJAS – Wageningen Journal of Life Sciences* 90–91. https://doi.org/10.1016/j.njas.2019.01.001

Voigt, B. (2019): Genome Editing: Why go beyond Precaution? The (Non-) Place for "Irrational" Public Opinion in EU Risks Regulation, in Dürnberger, C., Pfeilmeier, S. and Schjeissing, S. (eds.): *Genome Editing in Agriculture*. Baden-Baden: Nomos, 123–134

Von Schomberg, R. (2019): Why Responsible Innovation? in von Schombert, R. and Hankins, J. (eds.): *International Handbook on Responsible Innovation – A Global Resource*. Cheltenham, Northampton: Edward Elgar Publishing, 12–32.

WBGU (German Advisory Council on Global Change) (2021): *Rethinking Land in the Anthropocene: from Separation to Integration*. Berlin: WBGU.

Zafar, S. A. et al. (2019): Engineering Abiotic Stress Tolerance via CRISPR-Cas-mediated Genome Engineering, *Journal for Experimental Botany* 71(2), 470–479.

Zaidi, S., Vanderschuren, H. Qaim, M., Mahfouz, M. M., Kohli, A., Mansoor, S. and Tester, M. (2019): New Plant Breeding Technologies for Food Security, *Science* 363(6434), 1390–1391.

28

SMART TECHNOLOGIES, ENERGY DEMAND AND VULNERABLE GROUPS; THE SCOPE FOR 'JUST' METERING?

Dan van der Horst

Introduction

Across the world, analogue electricity meters are being replaced by digital meters which offer extended functionality. These digital meters are sometimes referred to as 'advanced metering infrastructure' (AMI) because they come as a package with customised communication hardware, data analytics software, data storage and management systems, regulations etc. In this chapter, they are referred to by their more widespread and colloquial name: smart meters.

Many governments are supporting energy smart meter deployment, recognising the need for both smart meters as key components of smart grids that can accommodate more distributed and intermittent (renewable) energy generation, and standardised meters to ensure competition and inter-operability in privatised energy markets (Darby 2012; Pitì et al. 2017). In short, it is a technology that needs to fit within a heavily regulated market where the government is seeking to balance a diverse set of objectives also known as the Energy Trilemma (e.g. Gunningham 2013): affordable, reliable and clean energy provision. The focus of this chapter is more specific still; can smart meters help vulnerable groups of energy users?

Smart meters are not just a technical tweak. They collect very granular data on our domestic energy use, making it possible to monitor our behaviour in the privacy of our homes in real time and in unprecedented detail. The ownership and governance of this data is thus of huge importance to public policy in a democratic society. Unlike the old analogue meters, smart meters are not a single type of product with standard capabilities. Smart meters can be designed or configured to accommodate a range of features, including variable frequency of measurement, different storage, manipulation and data transmission options. Moreover, they are often designed for two-way transmission, allowing such features as software updates and remote switch-off, but as a consequence also open themselves up to potential abuse, e.g. through hacking (McHenry 2013; Depuru et al. 2011; Anas et al. 2012). And, finally, smart meters have the potential to be 'disruptive technologies' (Danneels 2004), enabling new business models and new value propositions, potentially at the expense of incumbents. Smart meters can be linked to sensor technologies, e.g. for monitoring indoor air quality, securing the property against break-ins or monitoring people within the house (e.g. dependents you care about, visitors you don't fully trust). Finally, in the context of the low carbon transition, smart meters open the door to

DOI: 10.4324/9780429351921-34

time of use (ToU) tariffs. This will open a market for companies to programme and teleswitch many of our (smart) domestic devices, thus shifting electricity demand away from peak times and better utilising periods of abundant renewable electricity generation (e.g. McKenna 2012; Di Cosmo et al. 2014; Darby 2018).

Smart energy metering has been introduced in many countries. The US committed federal investments to smart metering as part of the recovery plan after the 2008 financial crash (Bugden and Stedman 2019). At the same time, European directive 2009/72/CE stipulated that all EU member states had to undertake a cost–benefit analysis on the large-scale deployment of smart meters, and if the finding was positive, then 80% of consumers should have been equipped by 2020. Most EU countries, including all larger states, reported positive cost–benefit results – except for Germany, which is now taking a more selective approach (JRC 2021). Italy pioneered the (near) universal roll-out of electricity smart meters, installing over 30 million meters in the early 2000s. This success story has helped save ENEL, the largest Italian electricity utility, a reported $750m annually, not only from labour savings (cutting back on their employment of door-to-door meter readers) and efficiency gains in running their power plants but also by addressing (alleged) 'rampant electricity theft and other forms of fraud' (Scott 2009; see also Pitì et al. 2017); suggesting that the latter was a key driver for this case of early adoption. Sweden and Finland achieved a rapid universal roll-out a few years later (for a consumer perspective, see Vassileva and Campillo 2016). With digital meters having a lifespan of about 15 years, these early adopters are already looking to roll out second generation smart meters (Pitì et al. 2017). In other countries, roll-out was much slower or affected by public protests (e.g. in California; Hess 2014) and even resistance from intermediaries (e.g. French local authorities; Chamaret et al. 2020). The UK roll-out, aiming for no fewer than 53 million smart meters, has been plagued by technical problems and concerns about the unfair distribution of benefits and impacts (see Sovacool et al. 2017; Jenkins et al. 2018) and is yet to be completed.

As the deployment of smart technologies advances, it becomes more important to ask how these technologies affect society: who stands to benefit and who might be negatively affected? Most technologies are not inherently good or bad. The societal value of smart meters depends on how they are used, which is linked to who is using them and for what purpose. Given the highly regulated nature of energy distributed through national grids, i.e. nation-wide electricity and gas networks, the question of 'who benefits' is largely in the hands of state policies, and thus exposed to political processes. Different countries have taken different approaches to smart metering. It is beyond the ambitions of a single book chapter to provide a full international snapshot of the status of smart metering and offer a detailed mapping of all the impacts of smart metering on society. Pitì et al. (2017) identify no fewer than 26 'real time services' that second generation smart meters could potentially provide. But technical potential and real-life experiences are inherently different, due to not only implementation gaps (e.g. actual functionality being constrained by costs, design flaws, legal restrictions, influence of powerful stakeholders) but also perception gaps between the techno-optimistic views of experts in labs, and the more complex lived experiences of people in their own homes. Implementation gaps are often the subject of critical policy reviews by academics (e.g. Sovacool et al. 2017; Jenkins et al. 2018 about the UK smart metering programme) or formal government-commissioned ex-post evaluations. The perception gap has been the focus of more theoretically anchored social science research, seeking to characterise for example the socio-technical imaginaries inherent in government support for particular energy technologies (e.g. Jasanoff and Kim 2013; Ballo 2015), or to unpack the visions of 'home' inherent in the designs and words of technical experts developing 'smart home' technologies (Strengers and Nicholls 2017; Gram-Hanssen and Darby 2018). Strengers (2014) evocatively introduces us to 'resource man', the ideal smart energy consumer whose daily life is driven by the desire to efficiently micro-manage his resource use and who will maximise all the potential uses of smart technologies offered to him. She then compares this man with the messy daily realities of domestic life as most of us know it, with heads of households having to juggle many different priorities, negotiate social relations and function under time pressure within a crowded material environment that is a polar opposite to the tidy diagrams representing smart homes in technical papers.

Resource man could be perceived as a two-dimensional cartoon figure, but it is more useful to think of him as an archetype. The problem is that the smart energy system is designed with him in mind, while he represents only a very small section of society. This illustrates not only the potential difference in attitudes that different sections of the population may have towards smart energy technologies, but also that there are big question marks about the potential benefits of the smart energy revolution for most residents and households. This is also reflected in the recurring disconnect between public expectations of smart metering and the technical reality of what smart meters are designed to do (e.g. Krishnamurthi et al. 2012; Vassileva and Campillo 2016). The technological trend is neither driven by public demand nor co-designed with (different) end users. This underlines the importance of asking the question 'who benefits'. Table 28.1 attempts to summarise how the key benefits from the introduction of smart meters are likely to be distributed between the private sector, the population as a whole and vulnerable consumers in particular. As this table illustrates, the benefits to consumers are often contingent on the behaviour of the suppliers. Also, there are clear differences in the position of vulnerable groups vis-à-vis the position of more affluent groups; the latter stand to gain more from the smart metering 'revolution' because they can afford smart appliances that interface with the

Table 28.1 Key claims about the benefits of smart metering, and the author's interpretation of how these are likely to be distributed

Benefits or risks (from the literature)	*Private sector benefits?*	*Population benefits?*	*Vulnerable consumers benefit?*
Reduces cost of consumer data collection [a]	Yes: (after installation) distributors could cut workforce	(only if cost savings are redistributed)	(only if cost savings are redistributed)
Reduces energy theft [b]	Yes: (traditional) producers and distributors	(only if cost savings are redistributed)	(only if cost savings are redistributed)
Yields more detailed consumer data [c]	Yes: new market opportunities (esp. for consumer data analytics beyond energy)	Maybe in theory (will be more effectively targeted to buy new goods and services)	Maybe in theory (could reduce indiscriminate targeting, but more likely to be ignored or preyed upon)
Allows more distributed (low carbon) generation [d]	Yes: small-scale tech providers and installers	Yes: more citizens can become prosumers	No: lack of money to invest, also likely increase of electricity costs
Helps to reduce (wasteful) consumption [e]	No (unless they are mandated to help)	Yes, in theory (explored in this chapter)	Yes, in theory (explored in this chapter)

Sources: van Aubel and Poll 2019 [a, b, c, e]; Zheng et al. 2013 [a, c, d, e]; Bugden and Stedman 2019 [c, d, e]

smart meter to make the most of those 'real-time services'. More vulnerable groups are more likely to be affected by the costs of introducing smart metering, especially if this is recouped through utility bills, as is the case in the UK.

In addition to potential benefits, we should also consider potential risks associated with the introduction of new technologies. As mentioned earlier, hacking and abuse of personal data are key risks associated with smart metering. More vulnerable groups may use older appliances that are less safe or may lack the skills or technical literacy to protect themselves. Wealthier sections of the population are more likely to buy and own multiple and novel smart appliances, which may bring new forms of risk exposure but they are also far more resilient to coping with situations arising; they can afford to take more financial risk, to purchase back-up equipment or to pay for good technical advice or insurance. In other words, it can be argued that we should not so much concern ourselves with the level of risk in terms of likelihood of something bad happening with the technology, but rather worry about vulnerabilities and coping mechanisms, i.e. the severity of impacts on people's lives as a result of such events. This brings us to the topic of vulnerability, which is a key focus for this chapter: (how) can smart technologies help the vulnerable? But before we explore this further, I first want to 'zoom out' to see the bigger picture and ask how this particular advancement in metering sits within the wider landscape of sustainable resource management.

Allocating resource flows: the role of metering

Before discussing how energy meters may be able to serve the vulnerable, it would be useful to place not just smart metering, but the whole notion of resource metering within a wider socio-economic and socio-technical context.

The term 'goods and services' is usually a reference to tangible and intangible commodities, respectively. But there is also a hybrid category of goods that flow, like water and gas, and more modern and even less-tangible categories like electricity and data. They may be held in solid-state and tangible storage devices, but due to the ways and quantities in which we produce and utilise them, we want them to flow to the user. So, how can society allocate goods that flow? I propose that we think of the basic models as a two-by-two matrix: unlimited versus limited use and metered use versus unmetered use (see Table 28.2).

Limited, unmetered use is the oldest and most low-tech way to allocate such resources. Allocation rules, often understood as social contract, are typically based on equal access principles,

Table 28.2 Four allocation models of goods that flow, i.e. delivered to users through network technologies

	Unmetered use	*Metered use*
Limited use	Rule-based access, usually equal access (oldest, low-tech solution, rarely relevant for power distribution beyond a single prosumer)	Found in supply limited mini-grids and in prepayment meters (i.e. credit limited) – typically for poorer households. Could be used to help (e.g. through social tariffs) but raises practical dilemmas about fair limits and redistribution
Unlimited use	Free for all: good for everyone in the short-term, but risk of creating shortages in the long term, affecting vulnerable groups first/worst	Market/profit driven model: good for suppliers, fine for the wealthy, but problematic for vulnerable groups.

using physical limits or time limits. For example, village scale canal irrigation systems were managed by routing the water to each farmer's field for a set period of time (Fernald et al. 2012). We can even find a version of this model in electricity provision, namely in (supply-constrained) mini-grids where grid stability is threatened during periods of peak demand. This is where energy demand over time needs to be managed by allocating maximum momentary user rights. For example, for the community-owned renewable energy mini-grid on the Scottish island of Eigg, residents agreed on a 5-Kilowatt cap on domestic use. Residents have a locked fuse box in the home and if it is triggered because the 5KW cap is exceeded, they have to call the islander with the key come to their home to flip the switch back on, and pay a small fine (see Chmiel and Bhattacharyya 2015; Gardiner 2017). Other examples can be found in locations where mini-grids depend on diesel generators, e.g. in Lebanon where the national grid has been damaged and apartment blocks are served by entrepreneurs who have set up diesel generators in the street below (Abi-Ghanem 2018).

In the case of unlimited and unmetered use, everybody can expect to pay a standard fee. This is for example the case for households without a water meter, or for people on an unlimited mobile phone contract for data or calls. This model works in particular circumstances, e.g. when metering is expensive/difficult, the unit cost of the service is low or when companies have other ways to reduce the costs of consumption or raise revenues (e.g. encouraging water-saving devices, slowing down the internet service at moments of peak demand, selling more expensive handsets). This model is typically associated with a contract and regular fixed fees.

Limited, metered allocation may be appropriate in situations where supply is limited or where other policies are in place to limit consumption. Metering can then be a technological improvement of equal access allocation (see limited, unmetered) or a commercial method to sell limited amounts to customers who want to control their expenditure or who may be perceived as at risk of failing to pay their bills. Especially in the latter case it is associated with pre-payment and no contract. The higher the speed and convenience of buying more 'blocks' of the resource, the less 'limited' and the more commercial this model becomes. Although prepaid meters are far from convenient, qualitative studies have shown that many people in fuel poverty appreciate prepaid meters, because it helps them to better control and manage their energy use without falling into payment arrears (Darby 2010; Middlemiss and Gillard 2015). Unfortunately and unfairly, the unit price of electricity sold through pre-payment meters is often higher (the opposite of social tariffs) and their automatic disconnect can leave poor and vulnerable households deprived of essential services (Snell et al. 2015).

For suppliers, prepaid electricity and gas meters are more expensive to purchase but cheaper to use with customers who might be perceived as less likely to pay their bills. Prepaid meters automatically disconnect the consumer if they do not pay, thus saving the supplier the costs of sending reminders, debt collectors etc. But even if the customer does pay the bills, prepaid meters are cheaper to run for suppliers, as they do not have to send their meter readers around.

Prepaid meters are sometimes presented as beneficial because they prevent electricity 'theft' through unpaid bills which suppliers would off-set by charging more to paying customers. Research attention to electricity theft by households is mostly found in the global South (e.g. Dike et al. 2015; Yakubu et al. 2018), whereas stories from the global North seem to focus more on illegal, energy intensive indoor cannabis plantations which would be exposed by metering (Depuru et al. 2011; Paoli et al. 2015). The argument that addressing theft (through smart metering) is automatically in the public interest could be interrogated more critically. Theft is not so easily quantifiable or distinguishable from other losses. Secondly, what if some 'thieves' are poor households who would otherwise not be able to access such crucial services? Thirdly and perhaps most critically, the public benefits of reducing electricity theft depend entirely on the willingness of the company to set aside their increase in revenue and redistribute it to their customers in the form of lower bills.

Finally, there is the unlimited, metered model, allowing consumers to consume as much as they can, and the companies to maximise their sales. It is a 'on tap, on tab' model (Buchanan et al. 2018), providing an uninterrupted service ('on tap', 24/7) that is paid for through direct debit from the consumer's bank account ('on tab'). Customers are only informed retrospectively of their consumption and the cumulative costs incurred. Generations of Western electricity consumers have become accustomed to a transactional arrangement that is radically different from almost anything else they buy. The widespread model of payment through (automated) direct debit for the total consumption over a previous period of weeks or months removes customer control and encourages higher levels of consumption. Smart meters with 'real time' digital consumer feedback have the technical potential to reduce this disconnect and thus serve as a tool to assist vulnerable consumers. But they still depend on the ability and commitment of consumers to engage proactively with this information, and subsequently (feel empowered to) act to reduce energy use or avoid falling into debt.

As we can see from Table 28.2, limited use has the tendency to raise equity questions about fair distribution and thus draws attention to the needs of the vulnerable. Metering as such tends to be less beneficial to vulnerable groups. In the hands of progressive authorities, it can be a tool for redistributing resources and for accurate allocation of social tariffs, but unlimited, metered use (i.e. the current market situation) is potentially the worst of all for vulnerable groups.

But metering is a basic element in the current socio-technical configuration of the grid and is an essential component of the neoliberal vision of privatised utilities. Even without state policy drivers to move towards a smart grid, metering is likely to move from analogue to digital. All meters wear out and need to be replaced at some stage. And since digital and remote metering technologies present cost-saving opportunities to suppliers, it makes sense for the private sector to develop new smart meters that will serve their own commercial interests. This is why state interventions are needed: to ensure that the design, roll-out and use of this technology will also provide public benefits, including consumer protection and support for vulnerable groups.

Vulnerable groups

Given long-standing government commitments with respect to social inclusion and poverty alleviation, one of the biggest challenges is how to ensure that the benefits of a state supported smart meter roll-out are realised by all sections of society (Citizens Advice Bureau 2016). This requires an understanding of who is vulnerable and how, before considering how smart meter design and implementation can be helpful in alleviating this vulnerability.

Academic literature on vulnerability is vast, and only a very small section examines vulnerability in the context of energy services (e.g. Middlemiss and Gillard 2015). For this reason, it is useful to anchor some of the discussion in policy documents, rather than academic papers. The Office for Gas and Electricity Markets (Ofgem; the UK regulator) defines 'vulnerable energy consumers' as people who are less able to represent or protect their interest in the market or are more likely than others to suffer detriment (Ofgem 2013). Broadly speaking, when looking for ways in which smart technologies can help to reduce the vulnerability of energy users, the focus is to avoid detrimental effects and help vulnerable people to make the most of the opportunities that smart metering and smart technologies can bring. In this context, it is important to recognise that these technologies also have significant *indirect* potential to support the vulnerable by 'helping the helpers'; carers, support groups, families, state and third sector institutions (e.g. Royston et al. 2014; Baker et al. 2018).

Vulnerability is a concept that is multi-dimensional and scalable; different types of vulnerability and thresholds of vulnerability may be defined for the purpose of policy making (see Delor and Hubert 2000). In the context of electricity and gas consumption in the home (i.e. assuming that people have a home and that their electricity and gas supplies are functional and uninterrupted), vulnerable customers may struggle with one or more of the following six categories (Thomson et al. 2017): access to market (choice, competition); affordability (including debt); flexibility (household's capacity to engage with different technologies or suppliers); energy efficiency (status of the house and key appliances); practices (including experienced control over energy use, such as energy rationing and self-disconnection); and needs (linked to personal health, other vulnerabilities, thermal comfort). Perhaps the simplest model of vulnerability, and a prominent one in IPCC climate adaptation reports (Cardona et al. 2012), consists of exposure (living in a hard-to-heat home, or facing unaffordable energy bills), susceptibility (to the negative social, physical and mental health impacts of being unable to afford sufficient energy services) and adaptive capacity (the ability to affect change and thus reduce exposure or susceptibility). As this chapter sets out to examine how smart metering could help the vulnerable, the focus shifts from who is vulnerable to how certain categories of vulnerable people may benefit (or may fail to benefit), through smart meter deployment, from improved energy consumption feedback. This brings the issue of (digital) energy literacy into the picture, not as a category of vulnerability, but as a factor that might help or constrain people's adaptive capacity. This is recognised in the UK smart metering installation code of practice (SMICoP 2021), which mentions that customers needing extra

Table 28.3 Examples of potentially vulnerable groups and their likely exposure to several forms of vulnerability (affordability, health) or barriers to coping (limited literacy and adaptability)

Group characteristic	*Financial affordability*	*Adaptability*	*Energy literacy*	*Other concerns*
Elderly	Fixed pension; home too large to heat	Less likely/able to embrace change	Lower adoption of ICT	Expected deterioration of physical health over time
Poor health & limited mobility	Dependent on benefits; other financial needs	Home bound; help dependent		Poor health + cold = worse health
Mental health problems	Limited income opportunities	Risk of anxieties; social isolation	Risk of misinterpretation of information provided	Risk of self-harm/ deterioration of physical health
Learning difficulties	Limited income opportunities	Limited ability to plan	Limited literacy	Risk of under-reporting of health problems
Immigrants with limited English language skills	Limited income opportunities	Risk of social isolation	Limited literacy in English	Cultural adaptation & mental health; comfort levels in a colder climate
People in poor quality private rental homes	Limited housing options	Can't improve their home		

support to make the most of their smart meter may be older, of poor health, or with a disability or severe financial insecurity. Vulnerability experienced by such groups can also vary in nature and severity over time, posing extra challenges to those seeking to provide assistance (Harrison and Gray 2010). Table 28.3 provides some examples of how particular groups may be analysed in terms of exposure to categories of vulnerability, and enablers like adaptiveness and energy literacy.

Before examining the potential benefits that a smart meter rollout could bring to addressing vulnerability, it is useful to revisit the current problems faced by vulnerable groups. Despite the existing provision of various forms of generic advice and support, vulnerable people currently face a range of different problems, from difficulties in financial planning for fluctuating, cumulative or retrospective energy bills to identifying and implementing those energy saving measures that are not only significant from a cost perspective, but also personally achievable and socially acceptable. This can result in financial difficulties and anxieties, as well as perverse strategies like self-disconnecting, which can have detrimental health impacts (O'Sullivan et al. 2013). Despite important efforts by help providers (fuel poverty NGOs, local councils, social land lords), many vulnerable people remain 'hard to reach' and are therefore not receiving support that is responsive to their particular needs and takes account of their particular dimensions of vulnerability (e.g. Stewart and Habgood 2008; Mould and Baker 2017). On the one hand, vulnerable people often struggle to access relevant information or find relevant support, and on the other hand, support organisations lack accurate, timely and joined-up data to identify vulnerable households, so they can reach out to them (Royston et al. 2014).

(How) can smart metering benefit vulnerable people?

In-home displays (IHDs) have been issued alongside smart meters in several national roll-outs, which is beneficial to people without easy access to tablets, computers or smart phones. In the UK, these mandated IHDs rarely exceed their required minimum functionality. Using both energy units and financial costs, they show a household's current and historic energy use, allowing some comparison with previous time periods. As many social science studies have shown, energy is typically consumed unconsciously, in the process of carrying out regular and habitual chores, social and leisure activities in the home, such as cleaning, washing, cooking and entertaining. IHDs are therefore considered an important tool in making energy 'visible' by helping people identify the impact that different activities are having on their energy consumption and, therefore, energy bills (e.g. Boomsma et al. 2016; Darby 2006; Hargreaves et al. 2010).

This has the *potential* to support vulnerable consumers on low budgets to save energy and money on their bills. Furthermore, for those who are particularly vulnerable to cold, a better understanding of the 'real time' cost of heat could help them plan for and manage these costs, and allay concerns about a future 'bill shock' (Grubb 2015). Nevertheless, smart metering tends to yield rather modest energy savings for consumers (Darby 2006; Delmas et al. 2013; Buchanan et al. 2015), for example through the adoption of more energy efficient light bulbs (Smart Energy GB 2017).

When considering vulnerable groups, as Table 28.3 illustrates, it is important to recognise that a certain level of digital literacy is essential to access, analyse and make sense of the kind of information provided by IHDs. There is scope for innovation and tailoring to support vulnerable people, for example, through foreign language translations, pedagogical software and voice-controlled and audio-visual apps, which can be of significant assistance to people who struggle to make sense of written English text or graphics on the screen. Further innovations are required in the area of feedback and control, co-designed with and especially tailored for different vulnerable user groups. This would be a systemic change from the current supply of cheap IHDs with limited capabilities (see Anderson and White 2009; Wood et al. 2019).

Smart meters enable suppliers to communicate with meters – and take meter readings – remotely. In countries where the meter is located indoors, households with analogue meters have to either provide meter readings themselves or allow the meter (wo)man to come into their home to read their meter. Those with restricted mobility or impaired vision may find it physically challenging to take meter readings, while others may feel uncomfortable having a stranger enter their home to take a reading.

Automated meter reading also takes away the need for estimated bills, as up-to-date energy use data will always be available to suppliers, enabling consumers to be billed for their actual – rather than estimated – use. Inaccurate estimated bills can be a particular cause of stress and financial hardship for consumers on limited budgets, as overpayments can cause cash flow problems, and underpayment can lead to debt build-up (Owen and Ward 2008). It also allows suppliers to identify vulnerable customers who under-heat their homes. This could be done simply by comparing energy consumption with outside temperature, but this indication that the home is cold needs to be corroborated by evidence that the home is actually occupied. For households that have both gas and electricity smart meters, ongoing electricity use will indicate occupancy, which then can be compared with gas consumption which indicates heating.

Smart meter energy data also provides the basis for the development of pre-payment meters that are more convenient to use (e.g. top-up payments by phone; tailored IHD with more functionality) and the potential design of new and better targeted social tariffs. Abovementioned concerns about pre-payment meters remain, but an alternative and more equitable scenario could be envisaged where all houses get pre-payment meters that can (also) be topped up by smart phone. That would have the potential to stimulate more affluent consumers to engage more actively with their energy needs and consumptive habits, as well as with the sources of energy. For example, information about abundance of wind and solar on the grid could motivate environmentally conscious consumers to shift their consumption patterns to low carbon electricity production periods, even if that does not bring financial benefits. This kind of flexibility by relatively motivated and affluent groups of the population may free up other groups, e.g. vulnerable groups who cannot afford such temporal flexibility, from the need to shift their consumption over time. While many people say that they care about the cost of energy and the damage to the environment, this is often insufficiently reflected in behavioural change, like switching to a cheaper provider or cutting back on wasteful forms of energy consumption. This particular 'value-action gap' (Barr 2006) is not only due to the abovementioned invisibility of energy consumption in social practices and opaque billing systems, but also because of the traditional market dominance of selling goods (the appliances) rather than (energy) services. There are proven digital technologies and the business models to run energy hungry domestic appliances like fridges, freezers, dishwashers and washing machines in ways that reduce and shift energy demand without significantly impacting on our wellbeing, but government interventions are needed to speed up their adoption. And while better public engagement and communication is required to explain that some impacts may be felt in terms of (perceived or measurable) reduction in consumer autonomy, it would be useful to bring more attention to the many disservices people are tolerating from 'normal' domestic appliances, such as slow, poor or patchy functionality, negative side-effects and automated nudging signals (for more critical thoughts on the suitability of appliances, see Dini 2020; Tu et al. 2018).

From smart meters to smart support for the vulnerable

The influence and consequences of the rollout of energy smart meters should not only be seen in isolation; it represents just the first wave in live, digital feedback on everyday life. In addition to the growth in smart meters which monitor consumption and cost (electricity and gas but also water or heat), we can expect to see a further proliferation of environmental monitors (e.g. temperature, humidity, CO_2 concentration) and health monitors (e.g. mobility, heart rate) in the home. The combined data feedback from these devices opens the door to a potential revolution in service provision by the private, state or third sector. Domestic technologies can double up, e.g. motion sensors can ensure that lights are only on when needed, as well as monitor for break-ins. Interesting energy-related developments extend beyond the home, such as cheaper car insurance offered in conjunction with a dashboard sensor which provides evidence of careful driving (e.g. Jarvis et al. 2019). And in the field of education, coaching and mental support, there are many ICT innovations addressing different forms of literacy as well as providing mental health support (Barak and Grohol 2011). We can see the emergence of three interrelated trends that have potential relevance for energy vulnerability, namely assisted living, digital health and social or group support.

Assisted living

Assisted living enables people to live on their own with a little help, regular or optional, available on the side (for a history, see Brown-Wilson 2007; for the role of ICT, see Rashidi and Mihailidis 2012). It is cheaper than a care home and can provide frail elderly people or people with disabilities a more homey environment and a better quality of life. There is a trend towards building bespoke new homes, physically and electronically designed for assisted living, but this supply is unlikely to be sufficient to meet demand in an aging society. Automation and monitoring can create assisted living conditions in ordinary homes, allowing people to stay and feel comfortable in their familiar surroundings (IMECHE 2018). While these developments are very positive in terms of managing a number of vulnerabilities, it is useful to point out that these adjustments can be costly in their own right and may further increase domestic electricity consumption. Where such an improvement in quality of life is indeed more energy intensive in a (relatively affluent) global North setting, there is a need to look for ways to off-set it, notably by adding more renewable energy generation to the (local?) grid. Assisted living with extra solar panels may still be cheaper than living in a care home.

Digital health

In terms of government expenditure, national healthcare systems in the global North easily dwarf the national energy systems. Providing universal and affordable healthcare is a universal challenge, made more difficult by demographic shifts in society (more elderly people, relatively fewer people of working age, fewer people per household) and effective but expensive technological advances in healthcare provision. Digital healthcare has the potential to cut the costs of effective health provision through, among others, innovative prevention programmes, automation of health monitoring, more effective early warnings and more timely interventions (see Fell et al. 2017). With regard to prevention, there is a key role to play for engaging apps, used by peer support groups or by progressive employers to encourage colleagues and peers, or even pay employees to participate in sport and exercise (Almendrala 2014; Fessler 2017).

Physical vulnerability means that exposure to cold is very likely to have detrimental physical health impacts (NICE 2016), and the same goes for heat waves or for overheated homes due to a combination of high insulation and poor ventilation. The UK has a high number of excess winter deaths, estimated at 200 deaths per day over the winter period.[1] Also per capita, these figures are bad; the UK excess winter death rate stands at 18%, compared to 10% for Finland (Age UK 2012), which is a much colder country with a longer winter season. These comparative statistics suggest more underlying health conditions, a weaker healthcare system and a lack of affordable warmth for many elderly people in the UK. Most people physically affected by the cold do not die, but many do end up in hospital and their treatment is estimated to add £1.36 billion/year to the national healthcare costs (ibid.). Prevention is better than a cure and cheaper, too; the daily cost of a hospital bed (>£200; NICE 2016) far exceeds the cost of heating an average UK home for a month (<£100; Sykes 2019). Monitoring people at risk through sensors in the house or wearable technology (e.g. Klein 2017) and providing state support for keeping them warm or providing early intervention treatment could be very cost effective for the NHS (Carrington 2014). Cheap and fast interventions could include making direct payments for heating services.

Social networks and peer-support

Most innovations in assisted living and digital health are likely to be driven internationally by the private sector but will also require the state to play a key role in creating a supportive legislative environment as well as providing safeguards for the public good. In the key policy domain of public health, the state is likely to be the main customer, commissioning digital health services that will provide better value for taxpayers' money. However, we should not overlook the role of the third sector and civil society.

Social networks are essential for the dissemination of domestic energy 'know-how' (Burchell et al. 2016) and may mitigate the effects of low literacy by providing opportunities for peer-to-peer learning. Moreover, the electronic sharing of personal data can help friends and relatives to overcome geographical distance and keep an eye out for each other. This is where the concept of the 'quantified self' (Swan 2013) is expanded into a surveillance of the 'quantified other' (Olson and Tilley 2014). Notwithstanding the benign intentions, it is clear that this example of technological progress can open complicated and dynamic ethical discussions about consent and agency of the person being monitored.

Falling within the wider realm of citizen science, it is clear that data collection and data sharing by lay people is key to enhanced data literacy, societal engagement and behavioural change for citizens in the 21st century. Most people are willing to share personal information under particular circumstances, for particular purposes, with particular, trusted partners (Tolmie and Crabtree 2018; Watts 2019). This can be a spontaneous, bottom-up process, but given the transaction costs and technical expertise required, there is clearly an important facilitating role to play for a competent and trusted 'third party' such as new data cooperatives or existing fuel poverty NGOs (Blasimme et al. 2018; Pentland et al. 2019). The rules and practices of data protection need to evolve so that people are able to share personal data more easily within social and support networks, while still being actively protected against potential abuse.

1 Figure from Age UK (2012), relating only to England and Wales, representing an average over the period 2000–2010. Excess winter death in England and Wales has fallen since, by some 10–15%, and this is strongly associated with warmer winters (ONS 2019).

Conclusions

Helping the vulnerable is a societal challenge, but it is clear that innovations in ICT have a considerable potential to assist. This will not happen by itself; government has a key role to play in encouraging and facilitating innovations that benefit the weaker sections of society, while protecting citizens against the risks of disruption of services, data misuse or an unfair distribution of costs. Drawing on the UK context for this chapter, the policy of energy smart metering with mandated feedback to all households (the in-home displays) and clear rules on data protection (consumers own their own data) is a step in the right direction. By itself, a government-led smart metering programme has the potential to create new opportunities to identify or support vulnerable people. Perhaps more importantly in the longer term, it is a key early step towards not only a cleaner smart electricity grid, but also progressive governance of the ICT revolution in the domestic and personal sphere, encouraging innovations that are beneficial for the public purse and sensitive to the needs of different vulnerable groups. This encouragement should target both the private sector and civil society to design, test and improve digital services and, in doing so, generate data on the nature and fate of domestic vulnerability in the 21st century: providing new and up-to-date evidence for policy making. This challenge may share similarities with 'innovation at the bottom of the pyramid' in developing countries (see Hilmi 2012; Sarkar and Pansera 2017).

The ways in which digital technologies can help deliver greater resource efficiency, on the one hand, and measure human health and wellbeing, on the other, opens the door not just for more accurate and appropriate definitions, assessments and provisions of 'energy services'. It also has the potential to open up wider conceptual debates about rights to knowledge, personal freedoms (which may include the right *not* to know, or the right not to be monitored), social duties connected with greater knowledge of self and other and the ethics of surveillance and control.

So far, this chapter has struck an optimistic tone on how benign and progressive institutions could use ICT to better help the vulnerable. The obvious counter-story is that these positive interventions may fail to materialise, allowing unscrupulous companies and actors to defraud or abuse all but the most resilient of consumers. Ubiquitous and pervasive metering and monitoring technologies in the internet (of things) age may open the door to other emergent and uncomfortable developments in algorithmic governance (Danaher et al. 2017; Katzenbach and Ulbricht 2019). The emergence of cheap digital mass surveillance technologies to quantify the other is providing companies with new profit-making opportunities and providing both liberal and authoritarian governments with powerful new tools to monitor and intervene in people's lives. Without smart meters and smart grids, the low carbon energy transition will be even harder to achieve, but for moral and practical reasons alike, this needs to be a just transition, narrowing current inequalities and avoiding new injustices. Harnessing the technical potential of smart monitoring and metering specifically for socially beneficial and fair outcomes may well become one of the biggest and most persistent governance challenges of the 21st century.

Acknowledgements

The author would like to thank the editors for the feedback on earlier drafts and acknowledge financial support from the UK Engineering & Physical Sciences Research Council (EPSRC) through the TEDDINET project (ref. EP/L013681) and the Smarter Households project (ref. EP/K002635).

References

Abi Ghanem, D. (2018): Energy, the City and Everyday Life: Living with Power Outages in Post-war Lebanon, *Energy Research & Social Science* 36, 36–43.

Almendrala, A. (2014): Does Getting Paid to Exercise Work? This App Says Yes, *Huffington Post*, 9 May. Available at: www.huffingtonpost.co.uk/entry/paid-to-exercise_n_5283708 [02.04.2021]

Age, U. K. (2012): The Cost of Cold. Available at: www.ageuk.org.uk/Documents/EN-GB/Campaigns/The_cost_of_cold_2012.pdf?dtrk=true [02.04.2021]

Anas, M., Javaid, N., Mahmood, A., Raza, S. M., Qasim, U. and Khan, Z. A. (2012): Minimizing Electricity Theft Using Smart Meters in AMI, *2012 Seventh International Conference on P2P, Parallel, Grid, Cloud and Internet Computing*. IEEE, November, 176–182.

Anderson, W. and White, V. (2009): Exploring Consumer Preferences for Home Energy Display Functionality, *Report to the Energy Saving Trust* 123, 1–49.

Baker, K. J., Mould, R. and Restrick, S. (2018): Rethink Fuel Poverty as a Complex Problem, *Nature Energy* 3(8), 610–612.

Ballo, I. F. (2015): Imagining Energy Futures: Sociotechnical Imaginaries of the Future Smart Grid in Norway, *Energy Research & Social Science* 9, 9–20.

Barak and Grohol (2011): Current and Future Trends in Internet-Supported Mental Health Interventions, *Journal of Technology in Human Services* 29, 155–196.

Barr, S. (2006): Environmental Action in the Home: Investigating the 'Value-Action' Gap, *Geography* 91(1), 43–54.

Blasimme, A., Vayena, E. and Hafen, E. (2018): Democratizing Health Research Through Data Cooperatives, *Philosophy & Technology* 31(3), 473–479.

Boomsma, C., Goodhew, J., Goodhew, S. and Pahl, S. (2016): Improving the Visibility of Energy Use in Home Heating in England: Thermal Images and the Role of Visual Tailoring, *Energy Research & Social Science* 14, 111–121.

Brown-Wilson, K. (2007): Historical Evolution of Assisted Living in the United States, 1979 to the Present, *The Gerontologist* 47(suppl_1), 8–22.

Buchanan, K., Russo, R. and Anderson, B. (2015): The Question of Energy Reduction: The Problem(s) with Feedback, *Energy Policy* 77, 89–96.

Buchanan, K., Staddon, S. and van der Horst, D. (2018): Feedback in Energy Demand Reduction, *Building Research Information* 46(3), 1–7.

Bugden, D. and Stedman, R. (2019): A Synthetic View of Acceptance and Engagement with Smart Meters in the United States, *Energy Research & Social Science* 47, 137–145.

Burchell, K., Rettie, R. and Roberts, T. C. (2016): Householder Engagement with Energy Consumption Feedback: The Role of Community Action and Communications, *Energy Policy* 88, 178–186.

Cardona, O. D., Van Aalst, M. K., Birkmann, J., Fordham, M., Mc Gregor, G., Rosa, P., Pulwarty, R. S., Schipper, E. L. F., Sinh, B. T., Décamps, H. and Keim, M. (2012): Determinants of Risk: Exposure and Vulnerability, in Field, C. B. et al. (eds.): *Managing the Risks of Extreme Events and Disasters to Advance Climate Change Adaptation: Special Report of the Intergovernmental Panel on Climate Change*. Cambridge, UK, New York: Cambridge University Press, 65–108.

Carrington, D. (2014): 'Boiler on Prescription' Scheme Transforms Lives and Saves NHS Money, *Guardian*, 9 December. Available at: www.theguardian.com/environment/2014/dec/09/boiler-on-prescription-scheme-transforms-lives-saves-nhs-money [02.04.2021]

Chamaret, C., Steyer, V. and Mayer, J. C. (2020): 'Hands Off My Meter!' when Municipalities Resist Smart Meters: Linking Arguments and Degrees of Resistance, *Energy Policy* 144, 111556.

Chmiel, Z. and Bhattacharyya, S. C. (2015): Analysis of Off-grid Electricity System at Isle of Eigg (Scotland): Lessons for Developing Countries, *Renewable Energy* 81, 578–588.

Citizens Advice Bureau (2016): Early Consumer Experiences with Smart Meters. Available at: www.citizensadvice.org.uk/Global/CitizensAdvice/Energy/Energy%20Consultation%20responses/Accent%20full%20report%20-%20Early%20consumer%20experiences%20of%20smart%20meters.pdf [02.04.2021]

Danaher, J., Hogan, M. J., Noone, C., Kennedy, R., Behan, A., De Paor, A., Felzmann, H., Haklay, M., Khoo, S. M., Morison, J. and Murphy, M. H. (2017): Algorithmic Governance: Developing a Research Agenda through the Power of Collective Intelligence, *Big Data & Society* 4(2), 2053951717726554.

Danneels, E. (2004): Disruptive Technology Reconsidered: A Critique and Research Agenda, *Journal of Product Innovation Management* 21(4), 246–258.

Darby, S. (2006): The Effectiveness of Feedback on Energy Consumption, *A Review for DEFRA of the Literature on Metering, Billing and direct Displays* 486(2006), 26.

Darby, S. (2010): Smart Metering: What Potential for Householder Engagement? *Building Research & Information* 38(5), 442–457.

Darby, S. J. (2012): Metering: EU Policy and Implications for Fuel Poor Households, *Energy Policy* 49, 98–106.

Darby, S. J. (2018): Smart Electric Storage Heating and Potential for Residential Demand Response, *Energy Efficiency* 11(1), 67–77.

Delmas, M. A., Fischlein, M. and Asensio, O. I. (2013): Information Strategies and Energy Conservation Behavior: A Meta-analysis of Experimental Studies from 1975 to 2012, *Energy Policy* 61, 729–739.

Delor, F. and Hubert, M. (2000): Revisiting the Concept of 'Vulnerability', *Social Science & Medicine* 50(11), 1557–1570.

Depuru, S. S. S. R., Wang, L. and Devabhaktuni, V. (2011): Electricity Theft: Overview, Issues, Prevention and a smart Meter Based Approach to Control Theft, *Energy Policy* 39(2), 1007–1015.

Di Cosmo, V., Lyons, S. and Nolan, A. (2014): Estimating the Impact of Time-of-use Pricing on Irish Electricity Demand, *The Energy Journal* 35(2), 117–136.

Dike, D. O., Obiora, U. A., Nwokorie, E. C. and Dike, B. C. (2015): Minimizing Household Electricity Theft in Nigeria Using GSM Based Prepaid Meter, *American Journal of Engineering Research (AJER) e-ISSN, 23200847*, 2320–0936.

Dini, R. (2020): 'The Lamentation of a Vacuum Cleaner': Appliance Disappointments in John Cheever and Richard Yates, *Textual Practice*, 1–36.

Fell, M., Kennard, H., Huebner, G., Nicolson, M., Elam, S. and Shipworth, D. (2017): *Energising Health: A Review of the Health and Care Applications of Smart Meter Data.* London: SMART Energy GB.

Fernald, A., Tidwell, V., Rivera, J., Rodríguez, S., Guldan, S., Steele, C., Ochoa, C., Hurd, B., Ortiz, M., Boykin, K. and Cibils, A. (2012): Modeling Sustainability of Water, Environment, Livelihood, and Culture in Traditional Irrigation Communities and their Linked Watersheds, *Sustainability* 4(11), 2998–3022.

Fessler, L. (2017): What if Your Employer Paid You to Exercise and Sleep? Inside One Startup's Wellness Experiment, *Quartz Media*, June 29. Available at: https://qz.com/1012333/some-companies-are-paying-their-employees-to-workout-and-sleep-creepy-or-incredible/ [02.04.2021]

Gardiner, K. (2017): The Small Scottish Isle Leading the World in Electricity, *BBC Future*, 30 March. Available at: www.bbc.com/future/article/20170329-the-extraordinary-electricity-of-the-scottish-island-of-eigg [02.04.2021]

Gram-Hanssen, K. and Darby, S. J. (2018): 'Home is Where the Smart is'? Evaluating Smart Home Research and Approaches against the Concept of Home, *Energy Research & Social Science* 37, 94–101.

Grubb, M. D. (2015): Consumer Inattention and Bill-shock Regulation, *The Review of Economic Studies* 82(1), 219–257.

Gunningham, N. (2013): Managing the Energy Trilemma: The Case of Indonesia, *Energy Policy* 54, 184–193.

Hargreaves, T., Nye, M. and Burgess, J. (2010): Making Energy Visible: A Qualitative Field Study of How Householders Interact with Feedback from Smart Energy Monitors, *Energy Policy* 38(10), 6111–6119.

Harrison, P. and Gray, C. (2010): The Ethical and Policy Implications of Profiling 'Vulnerable Customers, *International Journal of Consumer Studies* 34(4), 437–442.

Hess, D. J. (2014): Smart Meters and Public Acceptance: Comparative Analysis and Governance Implications, *Health, Risk & Society* 16(3), 243–258.

Hilmi, M. F. (2012): Grassroots Innovation from the Bottom of the Pyramid, *Current Opinion in Creativity, Innovation and Entrepreneurship* 1(2). https://doi.org/10.11565/cuocient.v1i2.5

IMECHE (2018): Healthy Homes: Accommodating an Aging Population. Institution of Mechanical Engineers, UK. Available at: www.imeche.org/docs/default-source/1-oscar/reports-policy-statements-and-documents/imeche-healthy-homes-report.pdf?sfvrsn=2 [02.04.2021]

Jarvis, B., Pearlman, R. F., Walsh, S. M., Schantz, D. A., Gertz, S. and Hale-Pletka, A. M., Spireon Inc (2019): Insurance Rate Optimization through Driver Behavior Monitoring. *U.S. Patent 10,169,822.*

Jasanoff, S. and Kim, S. H. (2013): Sociotechnical Imaginaries and National Energy Policies, *Science as Culture* 22(2), 189–196.

Jenkins, K. E., Sovacool, B. and Hielscher, S. (2018): The United Kingdom Smart Meter Rollout through an Energy Justice Lens, in Jenkins, K. E. and Hopkins, D. (eds.): *Transitions in Energy Efficiency and Demand.* London: Routledge, 94–109.

JRC (2021): Smart Metering Deployment in the European Union. Joint Research Centre (JRC). Available at: https://ses.jrc.ec.europa.eu/smart-metering-deployment-european-union [02.04.2021]

Katzenbach, C. and Ulbricht, L. (2019): Algorithmic Governance, *Internet Policy Review* 8(4), 1–18.

Klein, A. (2017): Smart Watches Know You are Getting a Cold, Days Before You Feel Ill, *New Scientist*, 12 January. Available at: www.newscientist.com/article/2117854-smartwatches-know-youre-getting-a-cold-days-before-you-feel-ill/ [02.04.2021]

Krishnamurthi, T., Schwartz, D., Davis, A., Fischhoff, B., de Bruin, W. B., Lave, L. and Wang, J. (2012): Preparing for Smart Grid Technologies: A Behavioral Decision Research Approach to Understanding Consumer Expectations about Smart Meters, *Energy Policy* 41, 790–797.

McHenry, M. P. (2013): Technical and Governance Considerations for Advanced Metering Infrastructure/Smart Meters: Technology, Security, Uncertainty, Costs, Benefits, and Risks, *Energy Policy* 59, 834–842.

McKenna, E., Richardson, I. and Thomson, M. (2012): Smart Meter Data: Balancing Consumer Privacy Concerns with Legitimate Applications, *Energy Policy* 41, 807–814.

Middlemiss, L. and Gillard, R. (2015): Fuel Poverty from the Bottom-up: Characterising Household Energy Vulnerability through the Lived Experience of the Fuel Poor, *Energy Research & Social Science* 6, 146–154.

Mould, R. and Baker, K. J. (2017): Documenting Fuel Poverty from the Householders' Perspective, *Energy Research & Social Science* 31, 21–31.

NICE (2016): Preventing Excess Winter Deaths and Illness Associated with Cold Homes. National Institute for Health and Care Excellence, QS117. Available at: www.nice.org.uk/guidance/qs117/chapter/introduction [02.04.2021]

Ofgem (2013): Consumer Vulnerability Strategy. Available at: www.ofgem.gov.uk/about-us/how-we-work/working-consumers/protecting-and-empowering-consumers-vulnerable-situations/consumer-vulnerability-strategy [02.04.2021]

Olson, P. and Tilley, A. (2014): The Quantified Other: Nest and Fitbit Chase a Lucrative Side Business, *Forbes. com* 5.

ONS (2019): Excess Winter Mortality in England and Wales: 2018 to 2019 (provisional) and 2017 to 2018 (final). Office for National Statistics. Available at: www.ons.gov.uk/peoplepopulationandcommunity/birthsdeathsandmarriages/deaths/bulletins/excesswintermortalityinenglandandwales/2018to2019provisionaland2017to2018final [02.04.2021]

O'Sullivan, K. C., Howden-Chapman, P. L., Fougere, G. M., Hales, S. and Stanley, J. (2013): Empowered? Examining Self-disconnection in a Postal Survey of Electricity Prepayment Meter Consumers in New Zealand, *Energy Policy* 52, 277–287.

Owen and Ward (2008): Consumer Implications of Smart Meters. Sustainability First, London. Available at: www.sustainabilityfirst.org.uk/images/publications/other/The%20Consumer%20Implications%20of%20Smart%20Meters%20-%20Report%20for%20NCC%20by%20Sustainability%20First.pdf [02.04.2021]

Paoli, L., Decorte, T. and Kersten, L. (2015): Assessing the Harms of Cannabis Cultivation in Belgium, *International Journal of Drug Policy* 26(3), 277–289.

Pentland, A., Hardjono, T., Penn, J., Colclough, C., Ducharme, B. and Mandel, L. (2019): Data Cooperatives: Digital Empowerment of Citizens and Workers. MIT Connection Science. Available at: https://connection.mit.edu/sites/default/files/publication-pdfs/Data-Cooperatives-final_0.pdf [02.04.2021]

Pitì, A., Verticale, G., Rottondi, C., Capone, A. and Lo Schiavo, L. (2017): The Role of Smart Meters in Enabling Real-time Energy Services for Households: The Italian Case, *Energies* 10(2), 199.

Rashidi, P. and Mihailidis, A. (2012): A Survey on Ambient-assisted Living Tools for Older Adults, *IEEE Journal of Biomedical and Health Informatics* 17(3), 579–590.

Royston, S., Royston, S. and Guertler, P. (2014): *Reaching Fuel Poor Families*. London: Association for the Conservation of Energy and The Children's Society.

Sarkar, S. and Pansera, M. (2017): Sustainability-driven Innovation at the Bottom: Insights from Grassroots Ecopreneurs, *Technological Forecasting and Social Change* 114, 327–338.

Scott, V. M. (2009): How Italy Beat the World to a Smarter Grid, *Der Spiegel*, 17 November. Available at: www.spiegel.de/international/business/energy-efficiency-how-italy-beat-the-world-to-a-smarter-grid-a-661744.html [02.04.2021]

Smart Energy GB (2017): Smart Meters and Energy Usage: A Survey of Energy Behaviour before and after Upgrading to a Smart Meter. Available at: www.smartenergygb.org/en/~/media/SmartEnergy/

essential-documents/press-resources/Documents/Smart-meters-and-energy-usage-October-2017.ashx [02.04.2021]

SMICoP (2021): Smart Metering Installation Code of Practice. Available at: www.smicop.co.uk/code-of-practice-pdf/ [15.04.2021]

Snell, C., Bevan, M. and Thomson, H. (2015): Justice, Fuel Poverty and Disabled People in England, *Energy Research & Social Science* 10, 123–132.

Sovacool, B. K., Kivimaa, P., Hielscher, S. and Jenkins, K. (2017): Vulnerability and Resistance in the United Kingdom's Smart Meter Transition, *Energy Policy* 109, 767–781.

Stewart, J. and Habgood, V. (2008): Benefits of a Health Impact Assessment in Relation to Fuel Poverty: Assessing Luton's Affordable Warmth Strategy and the Need for a National Mandatory Strategy, *The Journal of the Royal Society for the Promotion of Health* 128(3), 123–129.

Strengers, Y. (2014): Smart Energy in Everyday Life: Are You Designing for Resource Man? *Interactions* 21(4), 24–31.

Strengers, Y. and Nicholls, L. (2017): Convenience and Energy Consumption in the Smart Home of the Future: Industry Visions from Australia and beyond, *Energy Research & Social Science* 32, 86–93.

Swan, M. (2013): The Quantified Self: Fundamental Disruption in Big Data Science and Biological Discovery, *Big Data* 1(2), 85–99.

Sykes, D. (2019): What is the Average Energy Bill in the UK? Available at: https://octopus.energy/blog/what-is-the-average-energy-bill-in-the-uk/ [02.04.2021]

Thomson, H., Bouzarovski, S. and Snell, C. (2017): Rethinking the Measurement of Energy Poverty in Europe: A Critical Analysis of Indicators and Data, *Indoor and Built Environment* 26(7), 879–901.

Tolmie, P. and Crabtree, A. (2018): The Practical Politics of Sharing Personal Data, *Personal and Ubiquitous Computing* 22(2), 293–315.

Tu, J. C., Nagai, Y. and Shih, M. C. (2018): Establishing Design Strategies and an Assessment Tool of Home Appliances to Promote Sustainable Behavior for the New Poor, *Sustainability* 10(5), 1507.

Van Aubel, P. and Poll, E. (2019): Smart Metering in the Netherlands: What, How, and Why, *International Journal of Electrical Power & Energy Systems* 109, 719–725.

Vassileva, I. and Campillo, J. (2016): Consumers' Perspective on Full-scale Adoption of Smart Meters: A Case Study in Västerås, Sweden, *Resources* 5(1), 3.

Watts, G. (2019): Unicorns and Cowboys in Digital Health: The Importance of Public Perception, *Lancet Digital Health* 1, e319–e374.

Wood, G., Day, R., Creamer, E., Van Der Horst, D., Hussain, A., Liu, S., Shukla, A., Iweka, O., Gaterell, M., Petridis, P. and Adams, N. (2019): Sensors, Sense-making and Sensitivities: UK Household Experiences with a Feedback Display on Energy Consumption and Indoor Environmental Conditions, *Energy Research & Social Science* 55, 93–105.

Yakubu, O., Babu, N. and Adjei, O. (2018): Electricity Theft: Analysis of the Underlying Contributory Factors in Ghana, *Energy Policy* 123, 611–618.

Zheng, J., Gao, D. W. and Lin, L. (2013): Smart Meters in Smart Grid: An Overview, *2013 IEEE Green Technologies Conference (GreenTech)*, 57–64.

29

SMART HEALTH

Thomas Czypionka and Susanne Drexler

Introduction

Health and long-term care consist mainly of services that are delivered in person to people in need. The sector is therefore not an obvious candidate for digitalisation, and indeed, it took quite a while for digital technologies to be adopted in everyday processes. However, the pace of transformation is very slow, and a lot of institutional obstacles seem to impede it further. This is deplorable, as without doubt, digital technologies have a lot to offer to patients, service providers and even informal carers. A very recent example is the difference in Covid-19 vaccine roll-out between EU countries that resembles a kind of natural experiment. In 2020, all EU-countries received the same amount of vaccine doses due to joint negotiations with pharmaceutical companies. Notwithstanding, countries like Denmark with a well-established digital platform, were much faster bringing the vaccine to people than other European countries because prioritisation could be performed on existing databases, and invitation to vaccination as well as appointment scheduling were digitalised. The same can be observed without the common starting point in Israel or the UK.

In this chapter, we will give an overview of how digital technologies change the shape of healthcare. As a starting point, we will first discuss the impact of digital technologies on the relationships between stakeholders that might help explain why implementation is not always smooth. We argue that digitalisation in healthcare has many benefits and the potential to increase effectiveness, efficiency and equity in healthcare delivery when implemented in a way that acknowledges its impacts on healthcare as a social system and includes professionals and patients in the implementation. However, applications in some areas are arguably more advanced than in others. In order to illustrate this, we will give examples from various fields that are representative of the current state of play. On the policy side, we will also investigate why it is difficult for digital technologies to take hold in healthcare systems.

Digital transformation of the healthcare system

Transformation processes in healthcare can impact the relationship between patients, service providers, insurers (or the organisation fulfilling the insurance function like a national health service) and the system as a whole. While some applications are targeted and therefore often

DOI: 10.4324/9780429351921-35

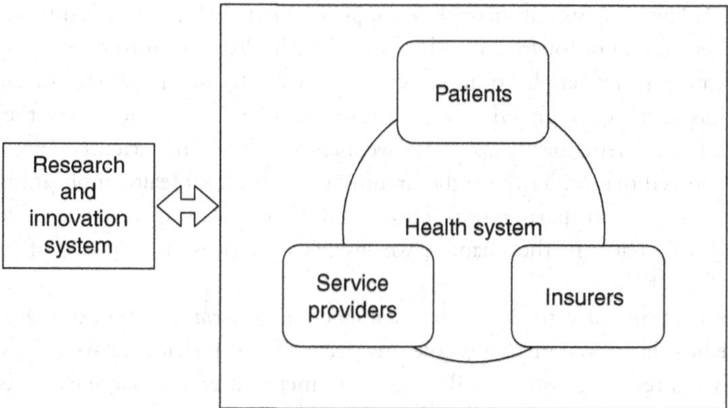

Figure 29.1 Stakeholders in healthcare

Source: Authors' display

marketed to only one of these stakeholders, they will always influence the relationship with other stakeholders as well. Take an app for monitoring diabetes, for example. Even if it is used without connecting to service providers, it *empowers* the patients and gives them more control over their disease, thus impacting the relationship with their doctors. Another example is the use of big data consisting of health services linked with outcomes data. Such an analysis has the potential to reveal quality problems, which is important for insurers, regulators and also patients, but might be opposed by professionals.

It is helpful to use a diagram showing the main stakeholders and their relationships to better categorise digital solutions and also to deliberate the impact of transformation (Figure 29.1).

Overview of applications at different stakeholder levels

On the level of *service providers*, digital solutions include hospital information systems or medical practice management systems used to organise everyday processes and data. Such systems can be connected to an electronic health record or an electronic portal like *Sundhed.dk* (see later) on the level of the health system, but also to insurers to process payments or provide a portal to patients for them to exchange documents like clinical reports or to book an appointment. Service providers may also want to analyse their own data to improve processes and receive quality certificates, and even to market such achievements. They may also want to use telemedicine to improve the continuity of care. We will discuss this field in greater detail later.

Insurers might want to give their insurees the opportunity to manage their insurance contract online, submit invoices for reimbursement, find providers (sometimes called *navigation information*) and access information on the quality of providers or on other health-related topics. Such systems become increasingly smart by connecting different types of information in a context-sensitive way, such that a patient searching for information on their respective disease is also pointed to potential healthcare providers in their vicinity that offer specialised services for it, to apps for disease management for their smartphone and to chat forums in order to share experiences with other patients. There are also projects to make access even easier using chatbots like the primary care app *GP at hand* used by the National Health Service (NHS) in England, which we discuss further later. Both insurers and the health system as a whole also want to analyse data to improve their processes and generate evidence for their decisions.

At the *patient* level, the use of smartphone apps is widespread, as there is little regulation and firms often offer their apps for free in order to retrieve health data or to offer an upgraded paid version. Apart from simple applications tracking physical activity, mood or calories, several apps are now used as part of integrated care programmes. They sometimes carry the label *serious health* in order to set them apart from more wellness and free time-oriented apps. A review of the literature showed that under the right circumstances, such mHealth applications can potentially improve information sharing, self-efficacy and the relationship with the provider (Qudah and Luetsch 2019). Later in the chapter, we discuss *mySugr* as an example of the increasing applicability of this field.

A close connection exists to the *research and innovation system* in a country. Decision makers governing the healthcare system as a whole, insurers and also providers have their data analysed by universities and research institutes. Besides econometric methods, large data sets are increasingly investigated using artificial intelligence. An example of large data sets creating benefit for health systems is the OpenSAFELY platform, which uses the electronic health records of 40% of all patients in England. In a recent study (Williamson et al. 2020), data of over 17 million patients were linked to Covid-related deaths. It was thus possible to quantify the hazard ratios of serious Covid-19 for risk factors like diabetes, hypertension and also socioeconomic factors. Another example of an innovation in this field is the national Danish e-health portal *Sundhed.dk*, which we describe in greater detail in what follows.

Numerous other examples could be mentioned at all levels, and many of them span more than one, as is the case in telemedicine.

Telemedicine and related concepts

One of the largest fields connecting several stakeholders is e-health, which is an umbrella term for the utilisation of the internet and related technologies to deliver health services (Eysenbach 2001; Triberti et al. 2019). As visible in Figure 29.2, telehealth is a subcategory of e-health covering the use of telecommunication tools for exchanging health-related services and information. A form of telehealth is mHealth, which is the provision of medical care and public health

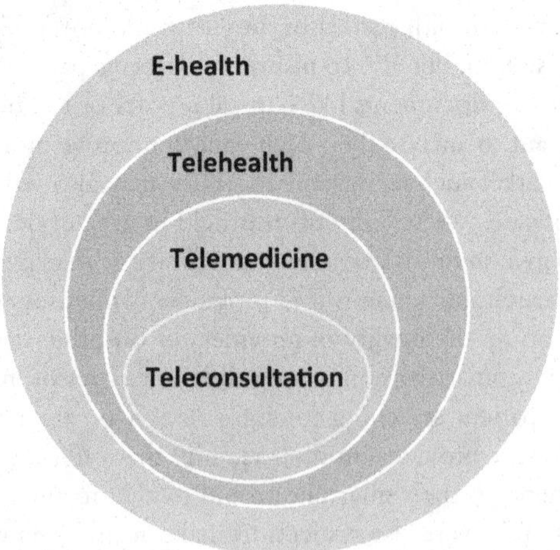

Figure 29.2 Telemedicine as a subset

practice through mobile devices such as mobile phones, patient monitoring devices and other wireless tools (WHO Global Observatory for eHealth & World Health Organization 2011). Telemedicine is a subcategory of telehealth describing the remote provision of clinical services through telecommunication tools, i.e. excluding the provision of pure information or education. A remote provider contact is called a teleconsultation. Covid-19 has practically boosted this area, as it can help avoid physical contact, while current, widely available technologies such as smart phones allow for a stable audio-visual communication between providers and patients (Greenhalgh et al. 2018; Rockwell and Gilroy 2020).

Telemedicine is nowadays widely used in care for chronic conditions, which constitute the majority of today's disease burden in high- and middle-income countries. In contrast to acute care, which is mainly provided by a single provider for a limited period of time (e.g. a hospital admission to treat erysipelas), caring for patients with chronic illness requires consultations with numerous providers at different occasions. It has been shown that continuity, coordination and timeliness of care are paramount to outcome quality in such cases (Czypionka et al. 2020; Kringos et al. 2010). Therefore, care for the chronically ill is ideally organised along predefined patient pathways. Such care is called *integrated care*, and use of digital tools is practically a sine qua non, regardless of whether integrated care takes the form of a *disease management programme*, a *population health management programme* or a *public health management programme*. Disease management programmes are coordinated healthcare interventions following an explicit structure and integrating several carers in chronic care, often using digital solutions as a support. Examples for disease management support services are the in-ear-sensor *EPItech*, the diabetes app *mySugr* and the remote monitoring and consultation programme *TeleCare Nord*, which are later discussed in more detail. Population health management programmes are set up to increase health outcomes for a specific part of the population (Swarthout and Bishop 2017), while public health management programmes are interventions rolled out to improve the health experience of the whole population (Hunter 2001). The electronic health platform *Patients Know Best* (see later) is an example of how population health management programmes make use of smart technology to coordinate their efforts. Within such programmes, providers exchange diagnostic and therapeutic data, monitor clinical parameters or consult with the patient remotely (Baltaxe et al. 2019). Examples for the effective utilisation of e-health in a public health management programme is the Danish platform *Sundhed.dk* and the primary care app *GP at hand*.

Very well-known examples of telemedicine include the Mercy Virtual Care Center (mercyvirtual.net) in St. Louis, USA. Founded in 2015, it once labelled itself "hospital without patients", providing telemonitoring services and teleconsultation to patients and also expert advice to physicians who work in remote areas. Some countries, like Denmark, aspire to expand telehealth solutions throughout their healthcare system (Healthcare Denmark 2018). Their successful programme *TeleCare Nord* for COPD patients (discussed later in more detail) is being rolled out for other diseases like heart failure. Hospitals are supplied with the means to provide *hospital-at-home* solutions. The rationale behind hospital-at-home is the fact that hospital stays are expensive and at the same time tend to isolate the patient from their social support system. Providing the necessary technical equipment for their care, the patients' parameters are monitored remotely when receiving e.g. chemotherapy or in cases of pregnancy complications. Necessary adaptions can be made by home care personnel, GPs or sometimes even by informal carers or patients themselves. Emergency situations can be predicted as well, allowing for hospital admission possibly even before critical events occur.

Telemedicine is also used in the field of rehabilitation, which aims to restore functional ability after an acute event or maintaining functions in a chronic illness. A recent study

(Amorim et al. 2020) gives a good example of how new technologies can be harnessed for the benefit of patients. It reviews virtual reality-based serious games as a means to improve functional ability in stroke patients. In small -scale studies, this strategy has proved effective for improving upper limb functions.

Technology in healthcare becomes smarter

Besides telecommunication and standard information technology, increased computing power made new, "smarter" technologies available for use in healthcare. One major field is artificial intelligence, which is increasingly used in medical areas such as radiology, pathology or dermatology, where visual data can be analysed by software that uses pattern recognition e.g. with deep neural networks (Topol 2019). Concerns have been raised, however, about the black box nature of strong AI (Castelvecchi 2016). Under established clinical standards, diagnosis and therapy decisions need to be documented so that they can be reproduced in a transparent way. What is more, medical data are sensitive, underlying special data privacy requirements that are naturally anathema to black box type systems.

However, apart from the black box issue, further problems arise for the wider application of artificial intelligence in clinical medicine. For example, while the format and structure of relevant documents may be standardised in one location, it is probably not across several sites (e.g. hospitals), which limits the transmission of information generated by AI. Another challenge arises whenever *natural language processing* is used. Abbreviations can be very context sensitive or have different meanings in different specialities or even regions. E.g. "IM" can mean "internal medicine" but also "intramuscular", "intramucosal" etc. Even with words and phrases there is a kind of jargon specific to different specialities in medicine or even local variants that are used in written reports and patient records, making them surprisingly difficult for machine learning to comprehend. For example, in the case of an admission to a hospital due to a heart attack, clinicians might use the words "myocardial infarction", be more general using "acute coronary syndrome" or be more specific clinically using "NSTEMI" (non-ST-elevation myocardial infarction) or "STEMI" (the opposite) or refer to anatomical structures using "posterior myocardial infarction" (PMI). Adding to this complication, terms may mean different things depending on their position in the document, e.g. as a diagnosis as opposed to a complication that needs to be avoided. A recent study discusses such challenges in the context of gastroenterology (Nehme and Feldman 2020).

In neurology, artificial intelligence can help in multiple ways. With stroke, it can help perform risk stratification after minor forms (Chan et al. 2019) or help monitor atrial fibrillation together with mobile ECG in order to avoid embolisms (Li et al. 2019). Artificial intelligence can also support the analysis of electrical activities of the brain and help predict and avoid falls due to epilepsy, as in the *EPItect* project discussed later. In an ongoing project, a deep learning system is trained to diagnose Parkinson's disease and essential tremor from hand movements and drawings (Varghese et al. 2019).

Artificial intelligence combined with *robotics* enable systems that understand the objects around them and take actions which maximise the likelihood of success to also endorse physical change in the world (Murphy 2019; Riek 2017). In healthcare, intelligent robotics can be used inside, on and outside the body of the patient, support doctors, formal and informal caregivers as well as provide decisionmakers with relevant information (Riek 2017). Robots are already used to perform micro-surgery like unclogging blood vessels and as smart prothesis that are sometimes even more reactive than the original body parts. (Marius 2019). Furthermore, robots can act as personal assistants for the elderly population by performing routine checks such as taking blood pressure and sugar levels, but also by conversing with them. Those robots sometimes come with the ability to analyse the personality and sentiment of the patients, which is especially helpful for people with mental conditions (Marius 2019). Service robots are used in hospitals to support healthcare staff by delivering items, restocking and cleaning.

Apart from improving robotics, artificial intelligence can also help analyse genomic sequences and their relation to disease, as has been shown e.g. in oncology (Shimizu and Nakayama 2020). Analyses of genomic material have become comparably cheap in recent years due to the rise of *next generation sequencing* techniques. Due to the sheer volume of data that this creates, strong AI systems have become indispensable in their analysis. While these are very helpful in research, they are increasingly used to support decisions in clinical medicine as well, e.g. to recommend therapies given certain mutations in cancers. Artificial intelligence is further used to make *gene editing* faster, cheaper and more accurate (Marr 2018). In the past years, several advances in the development of algorithms that predict repairs made to altered DNA have been reported, which in the future could improve research models to study genetic illnesses (Yeager 2019). Those models can subsequently be used for the development of *personalised medicine*, which is the creation of interventions that are tailored to a specific person or a genetically similar population. To date, this form of treatment is extremely costly, but automatic gene data analysis can significantly lower expenses that are necessary for the development of personalised medicine, such as predicting an individual's chance to develop an illness or their response to a certain treatment (Marr 2018).

Another innovation that makes personalised medicine more realisable are *3D printing technologies*, as it enables the creation of drug delivery services that are explicitly tailored to the patient's need (Shaqour et al. 2020). The application of 3D printing in healthcare is mostly still in the research stage, but is apart from drug delivery services already used to produce surgical tools, guides and implants, external prosthetics or orthotics and tools for preoperative planning or simulation as well as medical education (Diment et al. 2017; Lau and Sun 2018; Tack et al. 2016).

The application of *virtual reality* in medical education to train and assess surgical skills has proven to be efficient (Pfandler et al. 2017) and has recently been extended to train healthcare staff in non-technical skills such as teamwork and situation awareness (Bracq et al. 2019). Currently, the ability of virtual reality to improve diagnosis and treatment of mental health conditions, to complement the management of Parkinson's disease and to support rehabilitation of spinal cord injury is being explored (de Araújo et al. 2019; Rus-Calafell et al. 2018; Triegaardt et al. 2020).

Drug development is another field where artificial intelligence is increasingly used in a number of ways (Shaqour et al. 2020). Designing a drug needs an understanding of complex three-dimensional molecular structures and their change in shape under certain biochemical and physiological conditions. The strength of effect, toxicity and interactions with other substances can be simulated with the help of artificial intelligence. *Computational drug repurposing or repositioning* is an emerging field in which AI is used to increase the efficiency of data analysis to enhance the discovery of new applications for existing drugs (Park 2019). This method is for example used to accelerate the detection of treatments for Covid-19 (Zhou et al. 2020) and has been successful in identifying a drug for people with advanced pancreatic cancer that is currently being assessed in clinical trials (Fleming 2018). Existing drugs can be screened for possible application in new contexts (drug repurposing). Even the production process can benefit from predicting the yield of different synthesising procedures. These are just a few examples of this rapidly evolving area.

Examples

In order to illustrate the multiple ways in which digital and smart technologies are used in healthcare, we will present some examples of telemedicine for the aforementioned areas.

EPItect

EPItect is an in-ear-sensor that provides nursing support for people with epilepsy through sensory seizure detection. The device was jointly developed by cosinuss GmbH, Fraunhofer ISST and multiple experts for epilepsy. It was mainly funded by the German Federal Ministry of Education and Research (*Epitect – Sensorische Anfallsdetektion* 2016).

The device aims to solve three major problems faced regularly by people who suffer from epilepsy. First, few seizures are recognised early enough, meaning that carers cannot take safety measures in time (Baldinger n.d.). By measuring heart rate and blood pressure fluctuations, the sensor can anticipate seizures and automatically contact alarm services in case of critical attacks (Baldinger, n.d.). Consequences of injuries and premature mortality can thus be reduced (*Epitect – Sensorische Anfallsdetektion* 2016). Second, the documentation of epileptic seizures requires a lot of resources when done manually and is in many cases incomplete, incorrect or illegible, which complicates the prediction of seizures (Houta 2018). *EPItect* solves this issue by continuously and automatically collecting the patient's health data (*Epitect – Sensorische Anfallsdetektion* 2016). The data is then both anonymised and used for further research and sent to the patient's mobile app via Bluetooth, where personal information can be accessed anytime (*Epitect – Sensorische Anfallsdetektion* 2016). The availability of more precise data can be used as the basis for individual therapy decisions and scientific-pharmaceutical research. Third, the involved actors are currently not connected, which hampers efficient exchange of information that is needed to optimise care (*Epitect – Sensorische Anfallsdetektion* 2016). To increase the accessibility and the exchange of information, a web portal for the carers was developed, on which treatment and therapy plans of the patients can be monitored at any time. When used properly, *EPItect* has the ability to increase patient autonomy as well as the quality of life and a sense of safety for both carers and patients (*Epitect – Sensorische Anfallsdetektion* 2016).

In practice, the in-ear-sensor needs to be improved further according to users, because the battery power is too low, the appearance of the in-ear-sensor is perceived as too visible and the device impairs hearing. Additionally, users reported discomfort connected to wearing the sensor, like increased stigmatisation, a feeling of surveillance and constant reminders of one's own illness (Houta 2018).

The accuracy of seizure detection is currently 40%, but may increase if more data is gathered (Baldinger, n.d.). However, users criticised that even if the seizure is correctly diagnosed, the notification is given only shortly before. Therefore, despite the limitations and problems caused by the sensor, seizures still persist and sometimes help comes too late (Houta 2018).

MySugr

MySugr is a certified medical device that supports users in the dosage of insulin and facilitates the documentation of all diabetes data. The app was developed by four entrepreneurs in 2012 and acquired by Roche in 2017 (About Us | MySugr Global n.d.; Roche Acquires mySugr 2017). Since then, patients can order a starter-kit when signing up, which in addition to access to the app and online coaching includes unlimited test strips, a blood glucose meter and a lancing device made by Roche. At the moment, there are more than 2 million registered users (mySugr – Diabetes Tracker Log n.d.).

The device aims to simplify the lives of diabetes patients by quickly gathering their therapy data on their smartphone. This is achieved by connecting the blood glucose meter with the smartphone via Bluetooth, where patients can access and analyse their own values to identify patterns more easily. In addition, the app is equipped with a diabetes diary and an insulin calculator and gives access to diabetes advisors that provide assistance between visits to the doctor (Diabetes App, Blood Sugar and Carbs Tracker | mySugr Global n.d.).

MySugr can be purchased individually from the medical devices made by Roche, since various blood glucose meters from different manufacturers can be seamlessly integrated (Kukla 2018). Also, patients are provided with a demand-oriented supply of test strips, whereby both the calculation of the test strip consumption and its dispatch directly to the patients is carried out automatically (Diabetes Bundle All-in-one Package | mySugr Global n.d.).

Research has shown that apps can in general support changes in lifestyle and glucose monitoring for people with type 1 and type 2 diabetes (Kebede and Pischke 2019). A study evaluating *mySugr* found that blood sugars significantly decreased for high-risk patients and that low blood glucose indices were reduced (Debong et al. 2019). In addition, a reduction of estimated average blood sugar levels over the past three months (eHbA1c[1]) and improved glucose control was observed ("Science & Research | mySygr Global" n.d.).

Patients seem to be satisfied with the product, as the app received a user experience rating of 4.7 out of 5 stars based on 8,000 evaluations (Diabetes App, Blood Sugar and Carbs Tracker | MySugr Global n.d.). Kebede and Pischke (2019) evaluated multiple diabetes apps and found that *mySugr* was the most popular one.

This stands in contrast to a usability evaluation of commercially available diabetes apps of Fu et al. (2020). The researchers tested individual app functions and found that *mySugr* has the most shortcomings compared to the apps *OnTrack Diabetes*, *MyNetDiary* and *Glucose Buddy*. *MySugr* had the greatest number of violations of heuristic principles, especially in the area "Help and documentation", followed by "Error prevention" and "Aesthetic and minimalist design". Examples for those detected deficiencies are that the help function was unavailable or not easy to use or that the screen was too busy (Fu et al. 2020). The authors also criticised that *mySugr* lacked a blood glucose analysis report sorted by meal, which is necessary to support patient autonomy when modifying meals. However, all diabetes apps were rated poorly by the authors and in their opinion needed improvement in user friendliness.

GP at hand

GP at hand is an app which supports the provision of primary care. Patients can write requests to a chatbot that delivers context-sensitive information, and if the problem is more serious, a practice nurse is connected via video conference ("Product" n.d.). The app was developed by Babylon Health and integrated into the British National Health Service (NHS) in 2017, meaning that patients covered by the national health insurance can use it for free. The service has expanded rapidly over the past two years and has currently more than 90,000 users ("Why choose us" n.d.).

1 HbA1c is a parameter for monitoring the effect of elevated blood glucose and its levels are a good proxy for therapeutic success as well as for prognosis of organ damage. With continuous measurement of blood glucose, this parameter can be estimated (eHbA1c), reducing the need for additional blood sampling.

The key innovation of the service is the development of an AI system that can read, understand and learn from anonymised and consented medical datasets, patient health records and consultation notes ("AI" n.d.). This information can be used by healthcare staff to make faster decisions based on more accurate information about triage, causes of symptoms and future predictions of health. As a result, the process is speeded up and capacities are freed to see more patients. The AI system also supports patients to address symptoms and get faster information about diseases ("AI" n.d.).

The app was created to give more power to the patient by increasing the accessibility and affordability of healthcare ("About" n.d.-a). Compared to traditional GP services, consultations through *GP at hand* are available around the clock every day of the year, which lowers the barriers to attending appointments. Online appointments are available within hours or days ("How we work" n.d.), and if patients need services like vaccinations that definitely cannot be done over the phone, they can book an appointment through the app (Rachael 2018).

In case a doctor prescribes a medicine after an online consultation, the prescription is sent via the NHS online service to a pharmacy that is near the patient, where it can be collected (Noble 2018). Consultations are recorded, meaning that patients can replay the video, which is especially helpful for patients who have difficulties with hearing or remembering information ("How we work" n.d.).

The app currently has a satisfaction score of 96% based on 40,300 ratings ("Babylon" n.d.). However, the proportion of patients that deregister is 28%, where half of them returned to their original office (Burki 2019). Although *GP at hand* simplifies consultations, experts are concerned that this way of delivering service is not sustainable in the long run, as it leads to "cherry-picking" patients, where younger and healthier patients use *GP at hand* and older and more deprived patients are left to traditional GP facilities (Crouch 2017). Crouch (2017) points out that this development will increase the pressure on GPs, who already suffer from staff shortage. In addition, an evaluation of the service has shown that although the patient group using *GP at hand* is generally healthier compared to the rest of the population, the usage is higher (Iacobucci 2019). This could be caused by supplier-induced demand, meaning that the easy access to appointments generates unsustainable health-seeking behaviours. Another explanation is that people who could not keep a doctor's appointment before have now the possibility to receive consultations (Burki 2019).

Sundhed.dk

Sundhed.dk is the national Danish e-health portal that provides access to personal health data of hospitals, general practitioners and communities as well as general information about health-related topics. It is part of the public health sector and its strategies, and was therefore developed and funded by different governmental institutions (Jensen and Thorseng 2017). The portal is used by 1.7 million citizens each month and is the most used application among healthcare providers in Denmark (Petersen 2018).

The aims of the portal are to support general practitioners in Denmark in their function as "gatekeepers" of the healthcare system, to increase the self-determined involvement of patients, to promote cross-sectoral and integrated care and to provide coherent care across the national, regional and local level. Health service providers can use the portal to access their patients' health data. Citizens can access their medical records and medication but are also provided with quality-assured information about health topics and an overview of the Danish healthcare system. Patients can book doctor's appointments via the platform. Furthermore, the data that is collected is also used to inform decisions about future investments, strategies and solutions for healthcare services (Jensen and Thorseng 2017).

It is not the purpose of *Sundhed.dk* to create data, but rather to present already existing data extracted from data sources such as hospital information systems, practice information systems, prescription databases and lab systems. It also integrates other healthcare services, such as booking of doctor's appointments. These services can be located and run somewhere else (Jensen and Thorseng 2017). The platform can be accessed by the national electronic identifier scheme (NemID), which is a common secure login that is also used to access online banking or tax files ("Introduktion til NemID – NemID" n.d.).

The portal empowers patients by providing them with transparency, supports healthcare staff, facilitates communication between patients and healthcare providers and supplies decision makers with relevant information. According to the CEO of the platform, Danish society is well aware of problems regarding data security, but considers those benefits to be higher than the risks of damage through a possible hacker attack (Petersen 2018). Danes have in general lower data protection concerns and are less concerned about contact with the industry. Since it is normal to make health data available, it is not possible to opt-out of the e-health portal (Petersen 2018).

Results of a study evaluating differences in electronic health literacy between users and non-users of the platform showed that there are no differences regarding age, sex, educational level and self-rated health (Holt et al. 2019), indicating that the platform has a high user-friendliness, which is important considering that 36% of the users are above 60 years old and that this amount will rise in future. Other upcoming obstacles are keeping the platform up to date, as the rapid development of health technologies constantly creates further features that need to be implemented (Jensen and Thorseng 2017).

Patients Know Best

Patients Know Best is a patient-controlled electronic health record developed by a social enterprise ("About" n.d.-b) and funded by the British National Health Service (NHS) ("Commissioners and Payers" n.d.). The platform has newly expanded to other European providers and is also available in Germany and the Netherlands. Currently, it is used by 8 million people, which makes it the most widely used patient-controlled electronic health record in Europe ("Carepoint" 2020).

The platform combines personal health data such as test results or entries in a health diary with information provided by the healthcare staff, like appointment letters and multidisciplinary treatment plans. Patients always have access to the data, but can decide if and how much they want to share with health service providers or family and carers ("About" n.d.-b). It is especially helpful for people suffering from multiple conditions who often need treatment in several facilities, but can be used by any patient.

The aim of *Patients Know Best* is to provide patients with a tool to become more active in managing their health and wellbeing, to enhance patient-centred care and to increase safety and efficiency among healthcare providers ("About" n.d.-b). If a patient seeks care at a new facility, the staff will have insight into the patient's medical history and can therefore build a better understanding of what type of care is needed. This is especially helpful during emergencies ("The Personal Health Record Founded for Patients" n.d.).

When the impact of the platform on patients, clinics and financial resources was evaluated, benefits that arise by using *Patients Know Best* could be found in all of those areas (Gamet et al. 2016). Clinical outcomes were improved and reductions of waiting-times and non-attendances of doctor's appointments could decrease costs substantially. Compared to similar services, *Patients Know Best* has the highest participation rate, which after the authors can be explained by the fact that it was not built around a particular group of patients and that it is useful for both healthcare professionals and patients (Gamet et al. 2016).

The results of a different study analysing *Patients Know Best* found that, contradicting common assumptions, only a minority of patients who were offered to use the service showed interest in taking more responsibility for and control over their health management. Furthermore, the willingness to use the platform depended on the patient's coping style (Schneider et al. 2016). The authors criticised that it is not enough to "activate" patients and suggested the service be improved by including this and other patient attributes.

TeleCare Nord

TeleCare Nord is the common telemedical healthcare service in Northern Denmark. It currently offers monitoring for people who suffer from heart failure or chronic obstructive pulmonary disease (COPD). The service was developed by 11 municipalities in North Jutland, general practitioners and Aalborg University. It was funded by different private and public Danish funds and the EU Social Fund (Udsen 2015).

The system is an integral part of the health service in North Jutland, where citizens with COPD can be referred to telemedical home monitoring via a general practitioner or the hospital ("Telemedicin til borgere med KOL" 2020). The aim of the program is to increase the safety of patients by simplifying monitoring of their condition. It is expected that timely interventions decrease hospital admissions, which saves costs in the long run ("Til patienter og pårørende – KOL" 2020).

When patients are referred to the program, they receive a TeleKit which includes a tablet and measurement equipment used for measurements of oxygen saturation, heart rate and blood pressure ("Til patienter og pårørende – KOL" 2020). The data is sent from the measuring devices to the tablet via Bluetooth. There are two apps installed on the tablet. One of them, OpenTele, contains questionnaires about patients' health status, displays measurements from the devices and gives access to a message function which enables communication with the health professional (Udsen 2015). The other app is called OpenTele Info and contains a digital user guide, videos about how to use the equipment and training videos tailored for COPD patients.

The measurements are also sent to the regional care centre or the hospital, where the healthcare staff monitors the parameters. If these change, medical personnel will assess with the patient if the treatment needs to be adjusted or if additional interventions are necessary (Udsen 2015).

A study assessing *TeleCare Nord* showed that 88% of the participants found the TeleKit very easy to use and that citizens were in general very satisfied with applying telemedicine (Udsen 2015). The patients reported several advantages of using the system, such as a better control of the disease, an increased feeling of security, higher awareness of their own symptoms, more frequent responses to exacerbations and a higher level of freedom. In addition, the blood pressure of patients using *TeleCare Nord* dropped significantly compared to the control group (Udsen 2015).

Since the results of the project were positive, the local governments and Danish regions have decided to extend the service to COPD patients in other parts of the country ("Forskning viser positive resultater for TeleCare Nord Hjertesvigt" 2020). Furthermore, the system is currently tested on patients with heart failure, but the permanent integration is still debated ("Nordjyske patienter med hjertesvigt får fortsat tilbud om telemedicin" 2019). An assessment of *TeleCare Nord* as an add-on to the standard of care for patients with heart failure showed that it was highly cost-effective, as there were significant reductions in expenditures for hospitalisation and primary care contacts, as well as a decrease in total costs (Vestergaard et al. 2020). Additionally, patients who were referred to the program reported that they feel safer when they are monitored between consultations and after leaving hospital, which increases their mental health and thus their quality of life ("Nordjyske patienter med hjertesvigt får fortsat tilbud om telemedicin" 2019).

Socioeconomic implications

We can see that digital technologies have changed healthcare delivery in numerous ways so far. Efficiency and effectiveness of healthcare delivery may be improved. But what does the digital transformation mean for equity in healthcare?

Indeed, the literature shows that e-health may increase access to care, reduce healthcare costs and simplify self-management (Wynn et al. 2020), which has the potential to improve health outcomes for underserved populations, the elderly and citizens of low- and middle-income countries (LMICs). In high-income countries, people who live in remote areas, have low income, low education or immigrant status or are of high age often lack adequate access to healthcare and are therefore referred to as underserved population (Chesser et al. 2016). E-health could increase their access to services, as individual costs may decrease, while the amount of services offered may rises (Wynn et al. 2020). However, disadvantaged communities usually have lower levels of e-health literacy (Chesser et al. 2016), meaning that they lack digital expertise to seek and apply health knowledge via electronic sources to address or solve a health problem (Norman and Skinner 2006). Therefore, additional interventions such as provision of devices and support in using them are necessary until the benefits of e-health reach the underserved population, which in turn has led to concerns about increasing inequity due to differences in digital literacy (Azzopardi-Muscat and Sørensen 2019).

The elderly population uses the healthcare system most frequently and could therefore benefit immensely from e-health interventions that lower the barriers to access (Vancea and Solé-Casals 2015). Additionally, societal costs can decrease significantly if access to electronic self-care programmes is increased. However, the elderly often struggle with using complex devices compared to younger generations (Ziefle and Bay 2005) and are significantly less likely to use the internet compared to the average population (Niehaves and Plattfaut 2014). Until now, many health apps have failed to accomplish their goal of supporting self-care for the elderly because of their complicated design (Vancea and Solé-Casals 2015), which can be partially explained by the age gap between the users and the developers (Hawthorn 2007). To increase user-friendliness and therewith the efficiency of e-health applications for the elderly population, researchers suggest including the older generation during the development of e-health interventions (Hawthorn 2007; Vancea and Solé-Casals 2015).

E-health has the potential to increase the provision of healthcare in LMICs, where the distribution of providers is often poor and staff shortages are common (Schweitzer and Synowiec 2010). An essential driver of this development is the increasing supply of mobile devices to remote areas and LMICs (Mechael 2009; Schweitzer and Synowiec 2010). Positive outcomes caused by the use of e-health in LMICs include higher adherence to drug regimes, increased efficiency of healthcare workers and remote delivery of healthcare services and information as well as the facilitation of data-based budgeting practices (Elder et al. 2013; Schweitzer and Synowiec 2010). Negative consequences could be an increased workload for healthcare staff and delays in service delivery due to bottlenecks in training (Elder et al. 2013). Furthermore, there is still limited evidence about the cost-effectiveness of e-health, which is necessary to inform future investment decisions (Fritz et al. 2015; Lau 2017; Schweitzer and Synowiec 2010). This uncertainty hampers the introduction of e-health in LMICs, where resources are significantly more restricted compared to high-income countries. It is likely that, as with many technologies in healthcare, the potential recipients need to be stratified in order to identify groups for which the benefits outweigh the costs.

Reasons for slow pace of transformation

While we hope to have illustrated the benefits of digital and smart technologies in healthcare, it is surprising that the actual adoption in healthcare systems is comparably slow. Indeed, numerous obstacles can be identified on different levels. We will discuss them one by one, but they also interact with each other, contributing to the phenomenon.

Health systems are complex social systems

In healthcare, numerous stakeholders interact with each other. They have developed rules of interaction over time that are adopted by individuals by following them over and over in their everyday dealings. In the social sciences, this process is called *habitus formation* and the summative result are *institutions*, i.e. the sets of rules that structure social interactions. Such institutions stabilise a social system but also change slowly. Implementation of new processes, as is the case in smart technologies, tends to disrupt these institutions, creating a reaction and requiring legitimation (Hinings et al. 2018). While the phenomenon of *institutions* and *disruption* is present in other sectors as well, the degree of complexity seems to be higher in healthcare (Martínez-García and Hernández-Lemus 2013).

Large parts of health systems are not market-oriented

Many European countries organise their healthcare systems with a strong reliance on public financing or even public provision. Even when provision is private, it is tightly regulated due to quality and equity concerns, but also prone to rent-seeking behaviour by professional organisations (Tuohy and Glied 2011). Thus, many of its moving parts are not exposed to market forces, and therefore the need for process innovation in general is low. This is the reason why so many countries have experimented with more market-orientation, most notably the Netherlands with their managed-competition approach.

While the absence of competition can reduce the adoption of new technologies, countries may also opt explicitly to foster digital solutions on a central level, like Denmark or the UK. It bears mentioning that there seems to be a cultural divide concerning the topic of digitalisation that we will explore further later. Compared to these cultural aspects, financial issues seem to play a minor role; while Estonia is strong in digitalisation, Germany and Austria are lagging behind.

Rigidity in public health systems

Another reason complementary to low competition is the rigidity of regulation in healthcare, which is an obstacle to digitisation (Meskó et al. 2017). In political economy, there is a constellation often called an *iron triangle*, i.e. a rather rigid relationship between legislation, payers and providers (Buse et al. 2012) that hinders innovation. As in public healthcare systems, there are normally few real market prices but rather fee schedules that are negotiated between payers and provider representatives, often subject to tight regulations; introducing digital solutions that go beyond e.g. a practice management system will need to be reflected in such a fee schedule. The question then arises whether this solution will reduce or increase costs for the provider and thus requires an increase in fees, and to what extent. Risk-averse payers, even more so without competition for insurees, will naturally be hesitant given that revenue will not increase, in line with legislators and providers seeing little gain in change. This problem is exacerbated when cost-effectiveness studies yield ambiguous results, which is not uncommon given the complexity of digital solutions and their impact on health systems.

Fragmentation in healthcare

Many healthcare systems suffer from fragmentation between sectors, service providers and insurers. This problem has become more prominent with the shift from acute disease to chronic disease, the latter requiring several providers and payers working together along the patient pathway, while organisational setups often vary between and within providers and payers. While digitalisation can help mitigate the problem, this requires several stakeholders to agree on standards, interfaces or common platforms. In his seminal work, Leutz formulated five laws of integrated care, one of which reads "your integration is my fragmentation" (Leutz 1999), which can easily be applied to digitalisation as well. If a consortium of providers like in the Dutch chronic care programmes decides on one joint system, other providers that use a different system may be excluded. Agreeing on standards while keeping up competition and innovation is therefore an essential task for health policy.

Data privacy and transparency

Much of the data in healthcare falls under the special category of personal data according to the General Data Protection Regulation (GDPR), i.e. is very sensitive in nature. This alone explains why establishing digital processing of such data requires time and the implementation of strong safeguards. While the technical aspect seems to be more or less under control, there is also the aspect of trust on the individual and societal level. The black-box nature of some AI applications does not help make things easier in this respect.

Different countries seem to have different approaches to this. While some like England or the Scandinavian countries seem to embrace digital solutions in healthcare more easily, Central and Eastern Europe seem to be more cautious. It is likely that societies put different weights on disadvantages like possible data breaches, on the one hand, and benefits like improved continuity of care, on the other. Policy makers therefore need to acknowledge people's need for data privacy and explain carefully the advantages and safeguards repeatedly. It seems advisable to actively and openly discuss the possibilities and risks of data breach or misuse of data.

Notwithstanding, digital solutions unequivocally increase the transparency of healthcare delivery. With a good electronic health record, it is theoretically possible to assess the quality of different ways of healthcare delivery from diagnosis to therapy, but also of individual healthcare providers. In a "blame and shame" culture, such providers may be opposed to increased transparency through digitalisation.

But even when there is a culture of continuous quality improvement and learning from mistakes, digitalisation means a paradigm shift for healthcare providers. Patients that are "in control" of their disease through digital platforms, mHealth applications and self-help chat-groups are a completely different counterpart to health professionals. Thus, digital solutions can also redefine the whole physician–patient relationship, and health professionals may need support in adapting to new roles.

Digital health literacy

Marked differences between countries also exist in terms of health literacy and digital literacy, often jointly referred to as *e-health literacy* or *digital health literacy* (Quaglio et al. 2017). Without sufficient digital health literacy, patients may not feel comfortable using digital solutions or having their health data digitally processed. Without increasing digital health literacy, digital transformation of healthcare may fail to yield expected results and give rise to a decrease rather than an increase in health equity (Azzopardi-Muscat and Sørensen 2019).

Conclusion

The digital transformation of healthcare has an impact on nearly all relationships between stakeholders and levels of care, but the speed of implementation in healthcare falls short of the speed in other sectors. We explored several possible explanations for this phenomenon, and in order to facilitate further change, decision makers may want to give these a closer look.

According to the literature, smart health in its various forms has the potential to increase effectiveness, efficiency and also equity in healthcare delivery – given the implementation is accompanied by measures ensuring the acceptance by professionals and patients alike. It is a sine qua non in efforts to integrate chronic care but affects many other areas as well. We can see that the maturity of solutions varies by field. While there exist powerful information systems and telemedicine applications, there are still e.g. challenges for artificial intelligence to grasp the complex contexts in medicine.

The Covid-19 pandemic made it very obvious that the pace of digital transformation varies widely between countries. Various reasons can lead to slow adoption of digital solutions in healthcare. It seems vital to understand the institutional fabric of a country's system in order to facilitate digital transformation. Without paying heed to the specific layout, and possibly changing some regulation, digitalisation can face considerable resistance by stakeholders and is affected by all factors that also hamper innovation. Successful implementation therefore hinges on the careful inclusion of people affected, i.e. patients and professionals alike, but also e.g. administrative personnel. Ethical and data protection issues seem to be valued differently in different countries, contributing to variations in uptake speed, so broad and upfront public discussions are needed. It seems clear that – as with all other industries – digital transformation leads to disruption of traditional roles and relationships. This needs to be acknowledged and it is an important task for health policy to manage such issues. Lastly, for people being able to make use of digital solutions and avoid inequitable access to healthcare, digital health literacy needs to be increased across the population. Thus, numerous control knobs need to be turned in order to pave the way for a successful digital transformation of healthcare.

References

About. (n.d.-a): Babylon Health. Available at: www.babylonhealth.com/about [07.12.2020]

About. (n.d.-b): Patients Know Best. Available at: https://patientsknowbest.com/about/ [07.12.2020]

About us | mySugr Global. (n.d.): Available at: www.mysugr.com/en/about-us/ [04.12.2020]

AI. (n.d.): Babylon Health. Available at: www.babylonhealth.com/ai [07.12.2020]

Amorim, P., Santos, B. S., Dias, P., Silva, S. and Martins, H. (2020): Serious Games for Stroke Telerehabilitation of Upper Limb – A Review for Future Research, *International Journal of Telerehabilitation* 12(2), 65–76. https://doi.org/10.5195/ijt.2020.6326

Azzopardi-Muscat, N. and Sørensen, K. (2019): Towards an Equitable Digital Public Health Era: Promoting Equity through a Health Literacy Perspective, *European Journal of Public Health* 29(Supplement_3), 13–17. https://doi.org/10.1093/eurpub/ckz166

Babylon: 24/7 Appointments. (n.d.): App Store. Available at: https://apps.apple.com/gb/app/babylon-24-7-appointments/id858558101 [07.12.2020]

Baldinger, M. (n.d.): Study: Early Detection of Epileptic Seizures (EPItect). *Cosinuss.* Available at: www.cosinuss.com/en/portfolio-items/epileptic-seizure-detection/ [07.12.2020]

Baltaxe, E., Czypionka, T., Kraus, M., Reiss, M., Askildsen, J. E., Grenković, R., Lindén, T. S., Pitter, J. G., Molken, M. R., Solans, O., Stokes, J., Struckmann, V., Roca, J. and Cano, I. (2019): Digital Health Transformation of Integrated Care in Europe: An Overarching Content Analysis of 17 Integrated Care Programmes, *Journal of Medical Internet Research* 21(9). https://doi.org/10.2196/14956

Bracq, M.-S., Michinov, E. and Jannin, P. (2019): Virtual Reality Simulation in Nontechnical Skills Training for Healthcare Professionals: A Systematic Review, *Simulation in Healthcare: Journal of the Society for Simulation in Healthcare* 14(3), 188–194. https://doi.org/10.1097/SIH.0000000000000347

Burki, T. (2019): GP at Hand: A Digital Revolution for Health Care Provision? *The Lancet* 394(10197), 457–460. https://doi.org/10.1016/S0140-6736(19)31802-1

Buse, K., Mays, N. and Walt, G. (2012): *Making Health Policy.* 2nd ed. Maidenhead: McGraw-Hill Education.

Castelvecchi, D. (2016): Can We Open the Black Box of AI? *Nature News* 538(7623), 20. https://doi.org/10.1038/538020a

Chan, K. L., Leng, X., Zhang, W., Dong, W., Qiu, Q., Yang, J., Soo, Y., Wong, K. S., Leung, T. W. and Liu, J. (2019): Early Identification of High-Risk TIA or Minor Stroke Using Artificial Neural Network, *Frontiers in Neurology* 10. https://doi.org/10.3389/fneur.2019.00171

Chesser, A., Burke, A., Reyes, J. and Rohrberg, T. (2016): Navigating the Digital Divide: A Systematic Review of eHealth Literacy in Underserved Populations in the United States. *Informatics for Health and Social Care* 41(1), 1–19. https://doi.org/10.3109/17538157.2014.948171

Commissioners and Payers. (n.d.): Patients Know Best. Available at: https://patientsknowbest.com/commissioners/ [09.12.2020]

Crouch, H. (2017): *GP at Hand App Which Promises to Cut Waiting Times is Launched.* Digital Health. Available at: www.digitalhealth.net/2017/11/a-24-hour-video-chat-app-aiming-to-slash-waiting-times-launches-across-london/ [09.12.2020]

Czypionka, T., Kraus, M., Reiss, M., Baltaxe, E., Roca, J., Ruths, S., Stokes, J., Struckmann, V., Haček, R. T., Zemplényi, A., Hoedemakers, M. and Rutten-van Mölken, M. (2020): The Patient at the Centre: Evidence from 17 European Integrated Care Programmes for Persons with Complex Needs, *BMC Health Services Research* 20. https://doi.org/10.1186/s12913-020-05917-9

de Araújo, A. V. L., Neiva, J. F. de O., Monteiro, C. B. de M. and Magalhães, F. H. (2019): Efficacy of Virtual Reality Rehabilitation after Spinal Cord Injury: A Systematic Review, *BioMed Research International 2019*, 7106951. https://doi.org/10.1155/2019/7106951

Debong, F., Mayer, H. and Kober, J. (2019): Real-World Assessments of mySugr Mobile Health App, *Diabetes Technology & Therapeutics* 21(S2), S2–35. https://doi.org/10.1089/dia.2019.0019

Diabetes App, Blood Sugar and Carbs Tracker | mySugr Global. (n.d.): Available at: www.mysugr.com/en/diabetes-app/ [04.12.2020]

Diabetes Bundle All-in-one Package | mySugr Global. (n.d.): Available at: www.mysugr.com/en/diabetes-bundle/ [04.12.2020]

Diment, L. E., Thompson, M. S. and Bergmann, J. H. M. (2017): Clinical Efficacy and Effectiveness of 3D Printing: A Systematic Review, *BMJ Open* 7(12), e016891. https://doi.org/10.1136/bmjopen-2017-016891

Eerste 1.000 patiënten van Reumazorg Zuid-West Nederland aangesloten op patientsknowbest.nl | Carepoint. (2020): Available at: www.carepoint.nl/klantenervaringen/eerste-1-000-patienten-van-reumazorg-zuid-west-nederland-aangesloten-op-patientsknowbest-nl/ [04.12.2020]

Elder, L., Emdon, H., Fuchs, R. and Petrazzini, B. (2013): *Connecting ICTs to Development: The IDRC Experience.* New York: Anthem Press.

Epitect – Sensorische Anfallsdetektion. (2016): Fraunhofer Institut für Software- und Systemtechnik ISST. Available at: www.isst.fraunhofer.de/content/dam/isst-neu/documents/Projekte/Gesundheitswesen/Epitect/Farunhofer-ISST_PB_EPITECT-DE_Web.pdf [04.12.2020]

Eysenbach, G. (2001): What is e-health? *Journal of Medical Internet Research* 3(2), e20. https://doi.org/10.2196/jmir.3.2.e20

Fleming, N. (2018): How Artificial Intelligence is Changing Drug Discovery, *Nature* 557(7707), S55–S57. https://doi.org/10.1038/d41586-018-05267-x

Forskning viser positive resultater for TeleCare Nord Hjertesvigt (2020): Region Nordjylland. Available at: https://rn.dk/service/nyhedsbase-rn/2020/02/forskning-viser-positive-resultater-for-telecare-nord-hjertesvigt [04.12.2020]

Fritz, F., Kebede, M. and Tilahun, B. (2015): The Need for Cost-benefit Analyses of eHealth in Low and Middle-income Countries, *Studies in Health Technology and Informatics* 216. https://doi.org/10.3233/978-1-61499-564-7-981

Fu, H., Rizvi, R., Wyman, J. and Adam, T. (2020): Usability Evaluation of Four Top-Rated Commercially Available Diabetes Apps for Adults With Type 2 Diabetes, *Computers, Informatics, Nursing: CIN* 38(6), 274–280. https://doi.org/10.1097/CIN.0000000000000596

Gamet, K., Al-Ubaydli, M., Humphreys, L. and Boerner, D. (2016): *The Effectiveness and Impact of Patient-controlled Records: A Report on Patients Know Best.* Research Paper.

Greenhalgh, T., Shaw, S., Wherton, J., Vijayaraghavan, S., Morris, J., Bhattacharya, S., Hanson, P., Campbell-Richards, D., Ramoutar, S., Collard, A. and Hodkinson, I. (2018): Real-World Implementation of Video Outpatient Consultations at Macro, Meso, and Micro Levels: Mixed-Method Study, *Journal of Medical Internet Research* 20(4), e150. https://doi.org/10.2196/jmir.9897

Hawthorn, D. (2007): Interface Design and Engagement with Older People, *Behaviour & Information Technology* 26(4), 333–341. https://doi.org/10.1080/01449290601176930

Healthcare Denmark. (2018): *Denmark – A telehealth nation.* Available at: www.healthcaredenmark.dk/media/r2rptq5a/telehealth-v1.pdf [04.12.2020]

Hinings, B., Gegenhuber, T. and Greenwood, R. (2018): Digital Innovation and Transformation: An Institutional Perspective, *Information and Organization* 28(1), 52–61. https://doi.org/10.1016/j.infoandorg.2018.02.004

Holt, K., Karnoe, A., Overgaard, D., Nielsen, S., Kayser, L., Røder, M. and From, G. (2019): Differences in the Level of Electronic Health Literacy Between Users and Nonusers of Digital Health Services: An Exploratory Survey of a Group of Medical Outpatients, *Interactive Journal of Medical Research* 8(2), e8423. https://doi.org/10.2196/ijmr.8423

Houta, S. (2018): Sensorbasierte Anfallsdetektion und Vernetzungsinfrastruktur in pflegerischen und medizinischen Prozessen bei Epilepsie. Vienna Healthcare Lectures 2018. WU Executive Academy. Vienna. 19 September 2018. Available at: https://www.sozialversicherung.at/cdscontent/load?contentid=10008.713285&version=1540461587 [11.10.2021]

How we work. (n.d.): GP at Hand. Available at: www.gpathand.nhs.uk/how-we-work [07.12.2020]

Hunter, D. J. (2001): *Public Health Management.* WHO. Available at: www.who.int/chp/knowledge/publications/PH_management7.pdf?ua=1 [07.12.2020]

Iacobucci, G. (2019): GP at Hand: Patients are Less Sick than others but Use Services More, Evaluation Finds, *BMJ* l2333. https://doi.org/10.1136/bmj.l2333

Introduktion til NemID – NemID. (n.d.): Introduction to NemID. Available at: www.nemid.nu/dk-en/about_nemid/introduktion_til_nemid/index.html [07.12.2020]

Jensen, T. B. and Thorseng, A. A. (2017): Building National Healthcare Infrastructure: The Case of the Danish e-Health Portal, in Aanestad, M., Grisot, M., Hanseth, O. and Vassilakopoulou, P. (eds.): *Information Infrastructures within European Health Care: Working with the Installed Base.* Cham: Springer, Chapter 13. Available at: www.ncbi.nlm.nih.gov/books/NBK543679/; https://doi.org/10.1007/978-3-319-51020-0_13 [07.12.2020]

Kebede, M. and Pischke, C. (2019): Popular Diabetes Apps and the Impact of Diabetes App Use on Self-Care Behaviour: A Survey Among the Digital Community of Persons With Diabetes on Social Media, *Frontiers in Endocrinology.* https://doi.org/10.3389/fendo.2019.00135

Kringos, D. S., Boerma, W. G., Hutchinson, A., van der Zee, J. and Groenewegen, P. P. (2010): The Breadth of Primary Care: A Systematic Literature Review of its Core Dimensions, *BMC Health Services Research* 10, 65. https://doi.org/10.1186/1472-6963-10-65

Kukla, M. (2018): Technische Umsetzung und Hürden bei der Implementierung von digitalen Services im österr. Gesundheitswesen. Vienna Healthcare Lectures 2018. WU Executive Academy. Vienna. 19 September 2018. Available at: www.sozialversicherung.at/cdscontent/load?contentid=10008.713287&version=1540461589 [04.12.2020]

Lau, F. (2017): Chapter 5. eHealth Economic Evaluation Framework, in Lau, F. and Kuziemsky, C. (eds.): *Handbook of eHealth Evaluation: An Evidence-based Approach* [Internet]. Victoria, BC: University of Victoria. Available at: www.ncbi.nlm.nih.gov/books/NBK481593/ [15.01.2021]

Lau, I. and Sun, Z. (2018): Three-dimensional Printing in Congenital Heart Disease: A Systematic Review, *Journal of Medical Radiation Sciences* 65(3), 226–236. https://doi.org/10.1002/jmrs.268

Leutz, W. (1999): Five Laws for Integrating Medical and Social Services: Lessons from the United States and the United Kingdom, *The Milbank Quarterly* 77(1), 77–110. https://doi.org/10.1111/1468-0009.00125

Li, K. H. C., White, F. A., Tipoe, T., Liu, T., Wong, M. C., Jesuthasan, A., Baranchuk, A., Tse, G. and Yan, B. P. (2019): The Current State of Mobile Phone Apps for Monitoring Heart Rate, Heart Rate Variability, and Atrial Fibrillation: Narrative Review, *JMIR MHealth and UHealth* 7(2), e11606. https://doi.org/10.2196/11606

Marius, E. (2019): 6 Ways AI and Robotics Are Improving Healthcare, *Robotics Business Review*. Available at: www.roboticsbusinessreview.com/health-medical/6-ways-ai-and-robotics-are-improving-health care/ [07.12.2020]

Marr, B. (2018): The Wonderful Ways Artificial Intelligence Is Transforming Genomics and Gene Editing, *Forbes*. Available at: www.forbes.com/sites/bernardmarr/2018/11/16/the-amazing-ways-artificial-intelligence-is-transforming-genomics-and-gene-editing/ [07.12.2020]

Martínez-García, M. and Hernández-Lemus, E. (2013): Health Systems as Complex Systems 2013, *American Journal of Operations Research* 3(1A), 113–126. https://doi.org/10.4236/ajor.2013.31A011

Mechael, P. N. (2009): The Case for mHealth in Developing Countries, *Innovations: Technology, Governance, Globalization* 4(1), 103–118. https://doi.org/10.1162/itgg.2009.4.1.103

Meskó, B., Drobni, Z., Bényei, É., Gergely, B. and Győrffy, Z. (2017): Digital Health is a Cultural Transformation of Traditional Healthcare, *MHealth* 3. https://doi.org/10.21037/mhealth.2017.08.07

Murphy, R. R. (2019): *Introduction to AI Robotics*. 2nd ed. Cambridge, MA, London: MIT Press.

mySugr – Diabetes Tracker Log. (n.d.): App Store. Available at: https://apps.apple.com/us/app/mysugr-diabetes-tracker-log/id516509211 [04.12.2020]

Nehme, F. and Feldman, K. (2020): Evolving Role and Future Directions of Natural Language Processing in Gastroenterology, *Digestive Diseases and Sciences* 66, 29–40. https://doi.org/10.1007/s10620-020-06156-y

Niehaves, B. and Plattfaut, R. (2014): Internet Adoption by the Elderly: Employing IS Technology Acceptance Theories for Understanding the Age-related Digital Divide, *European Journal of Information Systems* (6), 708–726. https://doi.org/10.1057/ejis.2013.19

Noble, M. (2018): How Do I Collect My Prescription? Available at: https://vimeo.com/301628049/301225868 [07.12.2020]

Nordjyske patienter med hjertesvigt får fortsat tilbud om telemedicin (2019): Region Nordjylland. Available at: https://rn.dk/service/nyhedsbase-rn/2019/09/nordjyske-patienter-med-hjertesvigt-faar-fort sat-tilbud-om-telemedicin [07.12.2020]

Norman, C. D. and Skinner, H. A. (2006): eHealth Literacy: Essential Skills for Consumer Health in a Networked World, *Journal of Medical Internet Research* 8(2). https://doi.org/10.2196/jmir.8.2.e9

Park, K. (2019): A Review of Computational Drug Repurposing, *Translational and Clinical Pharmacology* 27(2), 59–63. https://doi.org/10.12793/tcp.2019.27.2.59

Petersen, M. (2018): The National Danish e Health Portal. Vienna Healthcare Lectures 2018. WU Executive Academy. Vienna. 19 September 2018. Available at: https://www.sozialversicherung.at/cdscon tent/load?contentid=10008.713289&version=1575391493 [11.10.2021]

Pfandler, M., Lazarovici, M., Stefan, P., Wucherer, P. and Weigl, M. (2017): Virtual Reality-Based Simulators for Spine Surgery: A Systematic Review, *The Spine Journal: Official Journal of the North American Spine Society* 17(9), 1352–1363. https://doi.org/10.1016/j.spinee.2017.05.016

Product (n.d.): Babylon Health. Available at: www.babylonhealth.com/product [09.12.2020]

Quaglio, G., Sørensen, K., Rübig, P., Bertinato, L., Brand, H., Karapiperis, T., Dinca, I., Peetso, T., Kadenbach, K. and Dario, C. (2017): Accelerating the Health Literacy Agenda in Europe, *Health Promotion International* 32(6), 1074–1080. https://doi.org/10.1093/heapro/daw028

Qudah, B. and Luetsch, K. (2019): The Influence of Mobile Health Applications on Patient – Healthcare Provider Relationships: A Systematic, Narrative Review, *Patient Education and Counseling* 102(6), 1080–1089. https://doi.org/10.1016/j.pec.2019.01.021

Rachael, J. (2018): How Does It Work for Things that Definitely Can't Be Done over the Phone – Like Vaccinations or Smear Tests? Available at: https://vimeo.com/301628049/301628049 [07.12.2020]

Riek, L. D. (2017): Healthcare Robotics, *Communications of the ACM* 60(11), 68–78. https://doi.org/10.1145/3127874

Roche Acquires mySugr to Form a Leading Open Platform for Digital Diabetes Management. (2017): Available at: www.roche.com/media/releases/med-cor-2017-06-30.htm [07.12.2020]

Rockwell, K. L. and Gilroy, A. S. (2020): Incorporating Telemedicine as Part of COVID-19 Outbreak Response Systems, *The American Journal of Managed Care* 26(4), 147–148. https://doi.org/10.37765/ajmc.2020.42784

Rus-Calafell, M., Garety, P., Sason, E., Craig, T. J. K. and Valmaggia, L. R. (2018): Virtual Reality in the Assessment and Treatment of Psychosis: A Systematic Review of its Utility, Acceptability and Effectiveness, *Psychological Medicine* 48(3), 362–391. https://doi.org/10.1017/S0033291717001945

Schneider, H., Hill, S. and Blandford, A. (2016): Patients Know Best: Qualitative Study on How Families Use Patient-Controlled Personal Health Records, *Journal of Medical Internet Research* 18(2), e43. https://doi.org/10.2196/jmir.4652

Schweitzer, J. and Synowiec, C. (2010): The Economics of eHealth. Results for Development Institute. Available at: www.r4d.org/wp-content/uploads/Economics-of-eHealth.final_.3Nov1010_0.pdf [07.12.2020]

Science & Research | mySygr Global. (n.d.): Available at: www.mysugr.com/en/science-and-research/ [04.12.2020]

Shaqour, B., Samaro, A., Verleije, B., Beyers, K., Vervaet, C. and Cos, P. (2020): Production of Drug Delivery Systems Using Fused Filament Fabrication: A Systematic Review, *Pharmaceutics* 12(6), 517. https://doi.org/10.3390/pharmaceutics12060517

Shimizu, H. and Nakayama, K. (2020): Artificial Intelligence in Oncology, *Cancer Science* 111(5), 1452–1460. https://doi.org/10.1111/cas.14377

Swarthout, M. and Bishop, M. A. (2017): Population Health Management: Review of Concepts and Definitions, *American Journal of Health-System Pharmacy* 74(18), 1405–1411. https://doi.org/10.2146/ajhp170025

Tack, P., Victor, J., Gemmel, P. and Annemans, L. (2016): 3D-printing Techniques in a Medical Setting: A Systematic Literature Review, *BioMedical Engineering OnLine* 15(1), 115. https://doi.org/10.1186/s12938-016-0236-4

Telemedicin til borgere med KOL. (2020): Region Nordjylland. Available at: https://rn.dk/sundhed/til-sundhedsfaglige-og-samarbejdspartnere/telecare-nord/telemedicin-kol [07.12.2020]

The Personal Health Record Founded for Patients. (n.d.): Patients Know Best. Available at: https://patientsknowbest.com/patients-and-carers/ [07.12.2020]

Til patienter og pårørende – KOL. (2020): Region Nordjylland. Available at: https://rn.dk/sundhed/til-sundhedsfaglige-og-samarbejdspartnere/telecare-nord/telemedicin-kol/patienter-og-paaroerende [07.12.2020]

Topol, E. J. (2019): High-performance Medicine: The Convergence of Human and Artificial Intelligence, *Nature Medicine* 25(1), 44–56. https://doi.org/10.1038/s41591-018-0300-7

Triberti, S., Savioni, L., Sebri, V. and Pravettoni, G. (2019): eHealth for Improving Quality of Life in Breast Cancer Patients: A Systematic Review, *Cancer Treatment Reviews* 74, 1–14. https://doi.org/10.1016/j.ctrv.2019.01.003

Triegaardt, J., Han, T. S., Sada, C., Sharma, S. and Sharma, P. (2020): The Role of Virtual Reality on Outcomes in Rehabilitation of Parkinson's Disease: Meta-analysis and Systematic Review in 1031 Participants, *Neurological Sciences: Official Journal of the Italian Neurological Society and of the Italian Society of Clinical Neurophysiology* 41(3), 529–536. https://doi.org/10.1007/s10072-019-04144-3

Tuohy, C. H. and Glied, S. (2011): The Political Economy of Health Care, in Glied, S. and Smith, P. C. (eds.): *The Oxford Handbook of Health Economics*. https://doi.org/10.1093/oxfordhb/9780199238828.013.0004

Udsen, F. W. (2015): *Forskningsresultater i TeleCare Nord. I TeleCare Nord – Afslutningsrapport.* Aalborg: Aalborg University.

Vancea, M. and Solé-Casals, J. (2015): Population Aging in the European Information Societies: Towards a Comprehensive Research Agenda in eHealth Innovations for Elderly, *Aging and Disease* 7(4), 526–539. https://doi.org/10.14336/AD.2015.1214

Varghese, J., Niewöhner, S., Soto-Rey, I., Schipmann-Miletić, S., Warneke, N., Warnecke, T. and Dugas, M. (2019): A Smart Device System to Identify New Phenotypical Characteristics in Movement Disorders, *Frontiers in Neurology* 10. https://doi.org/10.3389/fneur.2019.00048

Vestergaard, A. S., Hansen, L., Sørensen, S. S., Jensen, M. B. and Ehlers, L. H. (2020): Is Telehealthcare for Heart Failure Patients Cost-effective? An Economic Evaluation Alongside the Danish TeleCare North Heart Failure Trial, *BMJ Open* 10(1), e031670. https://doi.org/10.1136/bmjopen-2019-031670

WHO Global Observatory for eHealth and World Health Organization. (2011): *MHealth: New Horizons for Health Through Mobile Technologies*. World Health Organization. Available at: www.who.int/goe/publications/goe_mhealth_web.pdf [07.12.2020]

Why choose us. (n.d.): GP at Hand. Available at: www.gpathand.nhs.uk/why-choose-us [07.12.2020]

Williamson, E. J. et al. (2020): Factors Associated with COVID-19-related Death Using OpenSAFELY, *Nature* 584, 430–436. https://doi.org/10.1038/s41586-020-2521-4

Wynn, R., Gabarron, E., Johnsen, J.-A. K. and Traver, V. (2020): Special Issue on E-Health Services, *International Journal of Environmental Research and Public Health* 17(8). https://doi.org/10.3390/ijerph17082885

Yeager, A. (2019): Could AI Make Gene Editing More Accurate? *The Scientist Magazine*®. Available at: www.the-scientist.com/the-literature/could-ai-make-gene-editing-more-accurate-65781 [07.12.2020]

Zhou, Y., Wang, F., Tang, J., Nussinov, R. and Cheng, F. (2020): Artificial Intelligence in COVID-19 Drug Repurposing, *The Lancet Digital Health* 2(12), e667–e676. https://doi.org/10.1016/S2589-7500(20)30192-8

Ziefle, M. and Bay, S. (2005): How Older Adults Meet Complexity: Aging Effects on the Usability of Different Mobile Phones, *Behaviour & Information Technology* 24(5), 375–389. https://doi.org/10.1080/0144929042000320009

30
CYBERSECURITY AND ETHICS
An uncommon yet indispensable combination of issues

Karsten Weber

If there is anything we can now regard as solidly established, it is that we don't know how to build secure systems of any real complexity.

(Odlyzko 2019: 4)

Introduction: from dangerous computers to endangered computer systems

Today, one can read it in almost every newspaper nearly every day: computers or any other devices that are programmable and can be connected to the Internet are potential targets for a hacker attack. In a rather harmless case, such an attack might allow access to the camera and microphone of a laptop or smartphone, so that images and sound recordings can be made without the user's knowledge. However, the potential damage an attack on computers might cause can be increased to virtually any degree. Examples of attacks might be illegal access to financial data, encryption of important data and extortion for a ransom, theft and misuse of personal data or trade secrets, purposeful sabotage of industrial plants, damage to or destruction of computer systems, shutdown, malfunction or destruction of critical infrastructures such as energy or water supplies. One even finds reports on the hacking of live supporting implantable devices like pacemakers or insulin pumps (e.g. Baranchuk et al. 2018: 1285f.; Coventry and Branley 2018: 48f.; Mohan 2014: 372; Ransford et al. 2014: 158ff.; Woods 2017). The list of threats and assaults likely could be extended almost indefinitely. Some attacks are carried out because the people running them want to show that they are able do it – this was probably the original starting point for many hackers. Many committing these assaults have a criminal background; the attacks, then, represent the virtual version of a bank robbery or extortion and are usually referred to as cybercrime and, at times, cyberespionage (c.f. Connolly and Wall 2019; Dunn Cavelty 2014; Nadir and Bakhshi 2018). Sometimes, attacks on computer systems must be understood as terrorist actions; the term for this is cyberterrorism (see, for instance, the papers in Chen et al. 2014; Jarvis and Macdonald 2015; Weimann 2005). If such attacks are carried out by state authorities or by groups that are in close relationships to state actors, we would have to speak of state terrorism or even of a virtual variant of war – often called cyberwar or cyberwarfare (e.g. Liff 2012; Robinson et al. 2015).

DOI: 10.4324/9780429351921-36

In public perception, however, because there was hardly any information about the actual capabilities of computers readily available to the public in the early years of computer development, computers were not always seen as targets of attacks, but as the starting point of a threat. In the 1960s and 1970s, when computers were increasingly visible to the public but laypeople were poor at assessing their capabilities, stories were circulated, especially in the science fiction genre, in which giant "electronic brains" subjugated humans and sought world domination – the movie *Colossus: The Forbin Project* from 1970 can be seen as paradigmatic here (Weber 2018a). More recent films, such as the *Terminator* franchise with its Skynet computer unleashing a global thermonuclear war against humanity, tie in with such threat scenarios.

The discussion about the security of computer systems underwent a major change with the 1983 film *War Games*. In his book *Dark Territory* (2016), which provides an excellent account of the history of cyber threats, journalist Fred Kaplan writes that this film was instrumental in raising awareness among political and military leaders in the US about the vulnerability of military computer systems. The public, or at least parts of it, particularly in Germany and the US, probably first became aware of the hacking of (military) computer systems through Clifford Stoll's book, *The Cuckoo's Egg* (1989).

As far as computers are concerned, the 1980s are characterised by the speedy dissemination of home computers and the first personal computers. This is associated with the first large-scale attacks on these devices by computer viruses, which also attracted public attention. In fact, computer viruses appeared several years earlier (for the history of computer viruses, see Szor 2005). With the rapid spread of the Internet in the 1980s and 1990s, these malicious programs (or malware) were no longer spread through the exchange of data storage media such as floppy discs, but increasingly occurred because many computers were networked with each other locally and globally. The recognition that computers can therefore be targets for attacks probably became generally accepted by this time at the latest. Today, it is most likely that almost everybody living in an industrialised country has heard about some sort of cyberattack, since these are ubiquitous. As Fred Kaplan (2016: 5) puts it:

> Once the workings of almost everything in life were controlled by or through computers – the guidance systems of smart bombs, the centrifuges in an uranium-enrichment lab, the control valves of a dam, the financial transactions of banks, even the internal mechanics of cars, thermostats, burglary alarms, toasters – hacking into a network gave a spy or cyber warrior the power to control those centrifuges, dams, and transactions: to switch their settings, slow them down, speed them up, or disable, even destroy them.

Cybersecurity begins and ethics (more or less) follows suit

However, the discussion about the security of computer systems started much earlier. Although at the beginning of the corresponding debates the terms "computer security" or "information security" were used rather than "cybersecurity", the basic issues and perspectives were shaped very early. With regard to cybersecurity, probably two texts published by Willis H. Ware (1967a and 1967b) mark the beginning of the debates. Both papers are titled "Security and privacy" – the first text was written for the Rand Corporation (which primarily advises the US Air Force), and the second text was presented at an ACM conference. The ACM paper includes Figure 30.1, which presents possible attack vectors and threats:

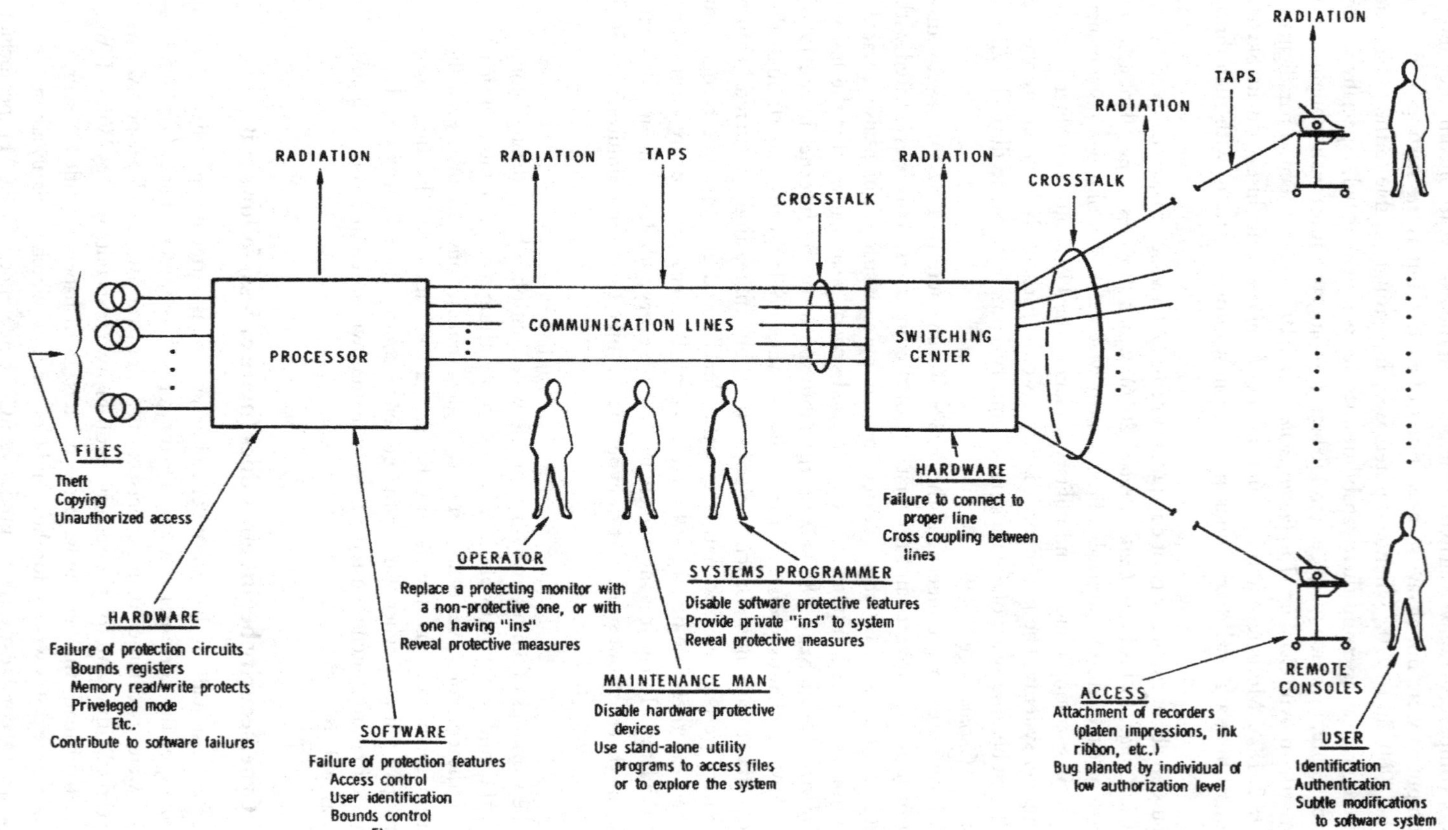

Figure 30.1 Typical configuration of resource-sharing computer system

Source: Republished with permission of Association for Computing Machinery, from Ware 1967b; permission conveyed through Copyright Clearance Center, Inc.

It is also worth taking a closer look at the text even today, as Ware lists key challenges, as early as 1967, which arise in connection with the protection of computer systems against unauthorised access and which are still relevant nowadays (Ware 1967b: 281f.):

> To summarize, there are human vulnerabilities throughout; individual acts can accidentally or deliberately jeopardize the protection of information in a system. Hardware vulnerabilities are shared among the computer, the communications system, and the consoles. There are software vulnerabilities; and vulnerabilities in the system's organization, e.g., access control, user identification and authentication. How serious anyone of these might be depends on the sensitivity of the information being handled, the class of users, the operating environment, and certainly on the skill with which the network has been designed. In the most restrictive case, the network might have to be protected against all the types of invasions which have been suggested plus many readily conceivable.

Certainly, computer technology has changed significantly since Willis Ware wrote that paragraph, but the comments he makes in this very early paper regarding attack vectors, threats and possible motives for attack as well as the human factor still hold true. Actually, the definition of cybersecurity, provided by the International Telecommunications Union (2008: 3), somehow echoes the list of vulnerabilities Willis Ware mentions by enumerating the "collection of tools, policies, security concepts, security safeguards, guidelines, risk management approaches, actions, training, best practices, assurance and technologies that can be used to protect the cyber environment, organisation and user's assets" as means to provide cybersecurity.

Almost as soon as the debates concerning cybersecurity gained momentum, moral, legal and political questions were raised that have shaped the discussion up to the present day regarding the extent to which the widespread use of computer systems will have an impact on individuals, groups and entire societies. In this regard, one of the most important contributions was certainly Alan F. Westin's book *Privacy and Freedom* (1967), which established the connection between the protection of personal data and the preservation of individual freedom. Since then, scholarly work explicitly linking computer security or cybersecurity to ethics has continued to appear (e.g. Campbell 1988; Cooper 1995; Leiwo and Heikkuri 1998). Quite prominently, Diffie and Landau (1998) pointed out political aspects of computer system security in the late 1990s; Dittrich et al. (2011) called for the formation of a (scientific) community that should deal with computer security or cybersecurity and ethics. For some years now, more and more papers and books have been published that deal with such issues (see, for instance, the papers in Christen et al. 2019; see also Christen et al. 2017; Domingo-Ferrer and Blanco-Justicia 2020; Loi et al. 2019; Manjikian 2018; Pattison 2020).

The lack of cybersecurity and its negative impact on societies

Since there are many different types of threats to computer systems, whether attacks of cybercrime, cyberterrorism or cyberwar, that can have very different ramifications, these consequences can be measured with different scales (cf. Gandhi et al. 2011). For example, if SCADA systems of a power plant are attacked (cf. Chhaya et al. 2020; Mazzolin and Samueli 2020), this may cause malfunctions that could trigger the collapse of the power supply; this in turn may result in economic losses, but also in ecological damage, injury to people or, in the worst case, loss of life. Attacks on computers belonging to a country's administration or government, in turn, can result in political instability, increase distrust of citizens in state institutions or limit a government's ability to act (for psychological and political effects of cyberattacks see, for instance, Gross et al. 2016; Herzog 2011; Iasiello 2013).

Although these are only two examples, it is probably already becoming apparent that the possible consequences that might arise from attacks on computer systems affect very different dimensions of individual, societal, corporate or political life. This makes it difficult to compare the amount of damage on a quantitative scale in these different dimensions. For this reason, the various reports that shed light on the consequences of computer attacks generally refer to economic damage in the sense of the costs incurred because of such attacks. While this facilitates comparisons, for example, between different sectors of the economy (cf. Tripathi and Mukhopadhyay 2020), between countries or even between the defender and adversary of an cyberattack (e.g. Derbyshire et al. 2021), it also hides the fact that a lack of cybersecurity not only causes monetary costs, but can also result in far-reaching damage that is difficult to measure.

The costs of cyberattacks worldwide have now grown to orders of magnitude that pose significant challenges to even high-performing economies. Moreover, these costs seem to know only one direction, increasing year by year. Cashell et al. (2004) report:

> Several computer security consulting firms produce estimates of total worldwide losses attributable to virus and worm attacks and to hostile digital acts in general. The 2003 loss estimates by these firms range from $13 billion (worms and viruses only) to $226 billion (for all forms of overt attacks).

However, the authors of this rather early report also emphasise that "the reliability of these estimates is often challenged; the underlying methodology is basically anecdotal". Nearly a decade and a half later, the Center for Strategic and International Studies (CSIS 2018: 4) writes: "Our current estimate is that cybercrime may now cost the world almost $600 billion, or 0.8% of global GDP". Other metrics from CSIS used to measure the scope of cyberattacks are as impressive as they are worrisome:

> One major internet service provider (ISP) reports that it sees 80 billion malicious scans a day, the result of automated efforts by cybercriminals to identify vulnerable targets. Many researchers track the quantity of new malware released, with estimates ranging from 300,000 to a million viruses and other malicious software products created every day.

Two years later, the cybersecurity company McAfee (2020: 3) writes that "we estimated the monetary loss from cybercrime at approximately $945 billion. Added to this was global spending on cybersecurity, which was expected to exceed $145 billion in 2020. Today, this is $1 trillion dollar drag on the global economy". Now, the problem with such claims lies in not only the ambiguity of the methodology and thus the evidence on which these claims rest, but also in the fact that different reports seem to contradict each other, or at least operate with very different figures. For example, Klon Kitchen (2019), Director of the Heritage Foundation's Center for Technology Policy, claims that "by 2021, cybercriminals are projected to cost the global economy more than $6 trillion annually, up from $3 trillion in 2015". The figures he mentions are much higher than McAfee's – but they, as well as probably all other statistics with regard to cybersecurity and threat analysis, are difficult or impossible to verify, at least for most parts of the audience (cf. Brito and Watkins 2011; Quigley et al. 2015; Stevens 2020).

From both a business and an economic perspective, not only direct costs (e.g. costs of repair, fraud-related losses) but also indirect costs (e.g. costs of preventive measures) and implicit costs (including lower productivity gains due to lower trust in digital transactions) are attributable to cybersecurity breaches (Bauer and van Eeten 2009). Given the above-mentioned figures, it is hardly surprising when the Center for Strategic and International Studies writes (CSIS 2018: 12): "Over the last 20 years, we have seen cybercrime become professionalised and sophisticated. Cybercrime is a business with flourishing markets offering a range of tools and services for the criminally inclined". If one knows where to look, one can hire cybercriminals to do the job, or one might find all the tools needed on the web to carry out cyberattacks, without needing any expert knowledge in this business; one simply buys "cybercrime-as-a-service".

One can indeed draw radical conclusions from this rather convoluted situation, as Andrew Odlyzko (2019) does when he writes that

> there is a rising tide of security breaches. There is an even faster rising tide of hysteria over the ostensible reason for these breaches, namely the deficient state of our information infrastructure. Yet the world is doing remarkably well overall, and has not suffered any of the oft-threatened giant digital catastrophes. This continuing general progress of society suggests that cyber security is not very important.

For the chapter at hand, however, another statement by Odlyzko (2019: 4) seems much more important, as he suggests that in addition to the aim of ensuring cybersecurity, there are other objectives that are at least as important:

> There is another factor that is not discussed here, namely that even if we could build truly secure systems, we probably could not live with them, as they would not accommodate the human desires for flexibility and ability to bend the rules.

What follows will therefore attempt to demonstrate that it makes little sense to set the aim of ensuring cybersecurity as absolute, while ignoring that there are, for instance, other economic and/or social goals whose importance is at least as significant as cybersecurity. Moreover, it will be argued that making cybersecurity absolute would pose a threat to numerous moral values as well as to many design requirements for technology.

The challenge of cybersecurity: balancing competing and conflicting aims and values

This text will not serve as an introduction to the many aspects of cybersecurity, or as an introduction to (applied) ethics. Actually, this is not necessary, because there are many excellent textbooks for both purposes. Those who want to familiarise themselves with the subject area of cybersecurity might use, for example, Bruce Schneier's *Secrets and Lies* (2000). Those who want to get acquainted with the theories of ethics can use, for instance, William K. Frankena's book *Ethics* (1963), as it offers a very easy-to-understand approach. Yet, any introductory book would also be suitable; the same can be said with regard to applied ethics (cf. Morscher et al. 1998). In the end, however, dealing with the details of both cybersecurity and (applied) ethics will not be required to understand the following remarks; the coming sections place emphasis on showing that the provision of cybersecurity is accompanied by numerous conflicts of aims and values. These values are not necessarily only moral values but can also include, for example, economic values as well. The conflicting aims can be technical design requirements, organisational aspects or usability issues, to name but a few.

At the end of this chapter, an attempt will be made to at least give some hints on how such conflicts could be resolved. However, it must remain an attempt, because if conflicts of aims and values are to be resolved, then one cannot avoid delving deeper into the fundamentals of both cybersecurity and ethics. As far as ethics is concerned, one then can no longer do without more precise statements about what one considers morally good or evil or which ethical theory one presupposes. The same applies with regard to cybersecurity, because without more precise knowledge of possible threats, attack vectors, risks and damage levels, as well as the respective technology, it is not possible to make any robust statements about a functioning balance between – in the broadest sense of the word – technical requirements and moral claims. Finally, if cybersecurity is to be established while maintaining ethical standards, one lesson that has to be learned is that this can only be achieved through interdisciplinary cooperation among, for instance, software engineers, legal experts, computer scientists, economic experts and – indeed – ethicists.

Moral aims and values in ICT

An exhaustive enumeration of all relevant moral values in a society is, for many reasons, rather impossible, not least because it would first need clarification as to what is regarded as relevant. If one tries to develop a standard here, one will very quickly find that different moral values are considered relevant in different ways in different societies. Emphasis on social justice and equality, for example, is more pronounced in countries of continental Europe like France, Germany or the Scandinavian countries than in Great Britain or the US. In all Western societies, individual autonomy is highly appreciated as a moral value, but this value is moderated in different ways by the importance of social ties to, for instance, religious associations, the family and certain social groups as well as other institutions, organisations or associations. In short, it is difficult to draw up a generally accepted list of moral values across societies and cultures (cf. Barry 2001; Fleischacker 1999). Moreover, it has to be emphasised that, so far, we have only been talking about Western societies, but it must be understood that the value systems that actually exist there are not dominant in other parts of the world. We must therefore assume that there is a great pluralism of (moral) values worldwide.

If one wants to be a little more modest and just explore which moral values are relevant to cybersecurity, then one might examine the relevant scholarly debates to see which moral values are explicitly mentioned there (e.g. Kruger et al. 2011; Sawaya et al. 2017). However, it must be clear that this is an inescapably narrow perspective, as it does not take into account the various political or ideological attitudes towards the relationship between cybersecurity and ethics, which shape the ways cybersecurity is dealt with. This narrow focus, however, is unavoidable if one wishes to achieve some understanding of the topic at all.

If one analyses the existing scholarly literature on information and communication technology (ICT) in general and on cybersecurity in particular as comprehensively as possible with regard to moral values (cf. Christen et al. 2017; Yaghmaei et al. 2017: 9–17), one will find, among others, privacy and trust, freedom and (informed) consent, fairness and equality (which can be translated to social justice), as well as dignity and solidarity (see also Weber and Kleine 2020: 145f.). Although this list is by no means complete, it is already clear that a wide range of moral values is involved in the development and use of ICT and the provision of cybersecurity. In particular, when considering ICT in healthcare, one can expand these values to include the principles of Beauchamp and Childress (2019): autonomy, beneficence, non-maleficence and justice. While these principles originate in biomedical ethics, they can be applied in other professional domains as well.

This means that the above-mentioned values and principles should guide the professional behaviour of, for instance, physicians, computer scientists, or engineers – especially when they are concerned with ensuring cybersecurity. In fact, it can be argued that the adherence to these values and principles must be the focus of respective professional behaviour, because they constitute the core of at least some professions. Accepting this last statement, the implication is that there may be circumstances in which the core moral values of a profession may compete or even conflict with technical or other requirements.

Technical aims and values in ICT

If we try to aggregate the numerous technical requirements for ICT, the list might look as follows: efficiency and quality of services, privacy of information and confidentiality of communication, usability of services, safety, integrity, availability (Yaghmaei et al. 2017). Some of these aims or requirements for ICT relate to cybersecurity, while others are more general in nature.

Regardless of whether these technical objectives compete or conflict with moral values and principles, even a cursory examination of the aforementioned requirements reveals that they already cannot always be realised together. For example, the establishment of a very strict security architecture to protect the confidentiality of data and communication often has the consequence that the usability of corresponding systems suffers (cf. Garfinkel and Lipford 2014). Thus, from the user's point of view, the efficiency and quality of the service may also decline (e.g. Al Abdulwahid et al. 2015; Dhamija and Dusseault 2008; Kainda et al. 2010). Another example would be the strong encryption of communication to provide confidentiality, which in the case of mobile devices, and particularly IoT devices, may compete with available energy and thus ultimately with availability of services and systems. In fact, a considerable amount of research and development is being carried out in this regard (cf. Elhoseny et al. 2016; Gupta et al. 2020; Pranit Jeba Samuel et al. 2020), so it is to be expected that, adapted to the respective requirements, appropriate solutions might be found. Nevertheless, this example in particular illustrates how different, in the narrower sense of the word, technical requirements might compete with each other.

Other factors, aims and values in ICT

Without doubt, moral values and technical requirements do not exhaust the factors effective in shaping ICT and thus influence the question of what level of cybersecurity can be achieved. Even a brief look at the scholarly literature shows that economic considerations in particular play a crucial role. Cybersecurity is expensive and, like all preventive measures, its benefits are difficult to quantify – if cybersecurity is successful or cyberattacks are not successful or even do not occur at all, then damages or costs prevented can at best be credibly estimated, but not quantified unequivocally. To put it somewhat bluntly: successful cybersecurity is no good advertisement for more cybersecurity. In any case, however, there must be an economic payoff to investing in cybersecurity (e.g. Anderson and Fuloria 2010; Dynes et al. 2007; Ekelund and Iskoujina 2019; Moore 2010; Wirth 2017). Nevertheless, even if the utility of cybersecurity measures can be demonstrated, it is still true that the resulting costs have to be paid. Often, attempts are made to pass these costs on to the end users, which can only succeed if they are willing to pay and to bear these costs (cf. Blythe et al. 2020; Johnson et al. 2020; Rowe and Wood 2013).

Political aspects undoubtedly also play a role in the design of computer systems and thus in the question of what measures are taken to strengthen cybersecurity (cf. Christensen and Liebetrau 2019; Dunn Cavelty and Egloff 2019; Harknett and Stever 2011; Liebetrau and Christensen 2021). For example, strong encryption methods have often been associated with export restrictions (e.g. Buchanan 2016; Diffie and Landau 2007; Herr and Rosenzweig 2016; Manpearl 2017; Sundt 2010; Swire and Ahmad 2012), or governments demand that encryption methods contain backdoors that allow law enforcement or intelligence agencies, for example, to break the encryption (cf. Ahmad 2009). In these cases, the respective actors pursue interests at the expense of the interests of other stakeholders. The problem here, however, is that such measures reduce cybersecurity in general, because actors with criminal intentions, for instance, can also use backdoors. Export restrictions, in turn, lower the level of protection that can be achieved for all stakeholders, which makes attacks easier and can lead to mistrust among stakeholders.

Cybersecurity as a multi-dimensional challenge

If one now tries to draw some tentative conclusions from what has been mentioned up to now, it is well worth listening to those who professionally work on the analysis of cyberattacks and the establishment of cybersecurity (Wirth 2017: 52):

> The world is facing a growing prevalence of increasingly sophisticated, targeted, and malicious cyberattacks. This new reality forces us to continually evolve our understandings of our cyberenvironment and to re-evaluate and update our security posture in ways that minimize the growing cyberrisks to ourselves, our infrastructure, and our businesses. In doing so, we must recognize that any comprehensive cybersecurity strategy includes more than just technical elements. It must include aspects of leadership, societal, and corporate culture and encompass larger economic and even sociopolitical elements (e.g. national security).

With Quigley et al. (2015) as well as Stevens (2020), there are sound arguments to at least critically examine the first part of the quoted statement, but the conclusion that establishing cybersecurity is not a solely technological challenge, or that technology does not even come first, appears more than plausible based on what has been said so far. Cybersecurity has to be understood as a multi-dimensional task. Without claiming to be exhaustive, one can at least identify moral values, technical requirements and other factors that influence the design of ICT and thus have a tremendous impact on the conditions under which cybersecurity can be provided. The assignment of existing values, aims, requirements and factors to the above-mentioned categories is of course arbitrary. For example, it would be conceivable and presumably reasonable to place aspects such as usability, acceptance or willingness to pay in an additional category that could be called "human factors". Similarly, political and social conditions could be subsumed into a category of their own. Ultimately, however, such categorisations do not really matter, because the decisive consideration is that the chance or impracticality of the provision of cybersecurity depends on numerous values, aims and requirements that are interrelated but also in competition or even conflict with each other. If one accepts this finding, the question that remains to be answered is how it is then possible to strike a balance and combine these values, aims and requirements in such a way that all stakeholders can agree to the compromise that eventually is found – if there is actually any such thing.

Methods to balance competing and conflicting aims and values

The question of what should be morally required or forbidden when technology is developed usually does not find easy answers if one is willing to broaden one's view and not only take one's own interests into account. However, even if one tries to consider other stakeholders and their interests, judgement about the morally appropriate design of technology depends on numerous other aspects. Which conception of human beings is presupposed? Which ethical theory is being considered, which normative assumptions are being made with regard to the relationship between different generations, which and whose normative claims are being prioritised, how should norm conflicts or norm competition (whereby these norms are not limited to moral norms) be resolved? Which understanding of the profession is present? All these and probably many more (normative) considerations already affect the ethical evaluation of technology at the theoretical level. Yet, if one wants to give an answer not only on an abstract or theoretical level, but for the actual use of technology in a (more or less) clearly defined environment, further influencing factors are added. Ethical considerations are "contaminated" by personal involvement of stakeholders and their interests and (mostly implicit and often unconscious) subjective attitudes as well as external constraints, which may make appear unfeasible what is normatively desirable, unsuitable for practice or inappropriate from a professional point of view.

Reijers et al. (2018a) describe a plethora of methods that could be used to first identify the competing and/or conflicting claims of the various stakeholders involved in the provision of cybersecurity, and then possibly find a solution in the form of balancing the different claims. Many of the methods mentioned by Reijers et al. are based on a central intuition drawn from discourse ethics: what is morally right or wrong cannot be answered by recourse to universal norms and values, some concept of utility or the idea of virtue, but must be negotiated among the stakeholders. Only in this way, so the argument goes, can the various interests and perspectives be adequately taken into account (e.g. Reijers et al. 2018b; Schuijff and Dijkstra 2020; Thorstensen 2019).

The idea of negotiating a balance or compromise among different stakeholders, their claims and other relevant factors also has the advantage that cultural preconditions and other social aspects are automatically taken into account, as they will be included in the arguments of the stakeholders involved in the negotiation. However, this is also a decisive weakness (similar arguments are also put forward against the moral principles of Beauchamp and Childress, e.g. Clouser and Gert 1990; Hine 2011; Sorell 2011) because the result of such negotiations is contingent and is not based on universally valid and accepted norms and values. However, good arguments can be given that in practice it is still better to find a compromise between the different (normative) stakeholder claims that can be accepted by all, rather than a solution based on a universal moral theory, but which only a few stakeholders would accept.

A well-tested tool, for example, is MEESTAR (*M*odel for the *E*thical *E*valuation of *S*ocio-*T*echnical *A*rrangements; see Weber 2018b). The basic idea of this tool is that those stakeholders who might be affected by technology (MEESTAR was developed for the ethical evaluation of ambient assisted living technology) carry out their own ethical evaluation of the technology in question in a well-defined process. The results of this evaluation are then fed into the development process. At best, this would take place several times in an iterative process during the whole R&D process. As already mentioned, the ethical evaluation and development of solutions for moral problems caused by the respective technology ultimately represent some kind of negotiation process – MEESTAR thus represents a method based on discourse ethics. In the original version of MEESTAR, the evaluation of technology is carried out with regard to seven so-called moral dimensions. Six out of the seven moral dimensions can be mapped onto Beauchamp and Childress' (2019) four principles (Table 30.1).

Yet, the seventh moral dimension, "self-perception of stakeholders", finds no correlation in Beauchamp and Childress' principlism. The idea behind this dimension is that the use of

Table 30.1 Conceptual relation of Beauchamp and Childress' principlism and MEESTAR

Principlism	*MEESTAR*
Autonomy	Autonomy and privacy
Beneficence	Care and social participation/inclusion
Non-maleficence	Safety/Security
Justice	Justice

technology alters the self-perception or self-image of stakeholders and is intended to encourage stakeholders to think about possible changes caused by technology, for example, with regard to their own profession or their own self-image and to evaluate these changes from a moral point of view. Most important, it is possible to modify MEESTAR in a way that dimensions can be added or removed, depending on the technology or application to be evaluated; this, of course, also applies to technology for the provision of cybersecurity.

It is crucial to note that MEESTAR, like presumably all tools based on discourse ethics, cannot provide a permanently valid answer to the question of how technology can be designed to meet all (normative) claims of all stakeholders; rather, this can only be stated for a concrete system or for an actual situation or application. This limitation can hardly be avoided without recourse to universally accepted moral theories, norms and values.

Conclusion

Even though the term "cybersecurity" was not used at the beginning of computer development, the problems associated with this term already existed at a very early stage and became increasingly important in parallel with the rapid spread of computers, for example for economic, military or organisational purposes. At the same time, it also became clear that establishing cybersecurity is not only a technical challenge with technical solutions, but also that legal, economic or organisational aspects play at least as important a role. Finally, it turned out to be apparent, also very early in this development, that the provision of cybersecurity raises ethical questions, since cybersecurity can affect moral values such as autonomy, freedom or privacy – to name just a few. If one wants to achieve that the establishment of cybersecurity is accepted on an individual, organisational and societal level, then it is essential to find a balance between the different claims of the many stakeholders involved in the provision of cybersecurity. The aim of this chapter was to present this complex situation and to give at least some hints on how this balance could be achieved.

Acknowledgments

This text could not have been written without the work done as part of the project "Constructing an Alliance for Value-driven Cybersecurity" (CANVAS, https://canvas-project.eu/). I am deeply indebted to my colleagues with whom I had the privilege of working on this project and whom I would like to thank for their excellent collaboration. CANVAS has received funding from the European Union's Horizon 2020 research and innovation program under grant agreement No. 700540; additionally, it was supported (in part) by the Swiss State Secretariat for Education, Research and Innovation (SERI) under contract number 16.0052-1.

References

Ahmad, N. (2009): Restrictions on Cryptography in India – A Case Study of Encryption and Privacy, *Computer Law & Security Review* 25(2), 173–180. https://doi.org/10.1016/j.clsr.2009.02.001

Al Abdulwahid, A., Clarke, N., Stengel, I., Furnell, S. and Reich, C. (2015): Security, Privacy and Usability – A Survey of Users' Perceptions and Attitudes, in Fischer-Hübner, S., Lambrinoudakis, C. and López, J. (eds.): *Trust, Privacy and Security in Digital Business*. Vol. 9264. Heidelberg: Springer International Publishing, 153–168. https://doi.org/10.1007/978-3-319-22906-5_12

Anderson, R. and Fuloria, S. (2010): Security Economics and Critical National Infrastructure, in Moore, T. Pym, D. and Ioannidis, C. (eds.): *Economics of Information Security and Privacy*. Heidelberg: Springer, 55–66. https://doi.org/10.1007/978-1-4419-6967-5_4

Baranchuk, A., Refaat, M. M., Patton, K. K., Chung, M. K., Krishnan, K., Kutyifa, V., Upadhyay, G., Fisher, J. D. and Lakkireddy, D. R. (2018): Cybersecurity for Cardiac Implantable Electronic Devices, *Journal of the American College of Cardiology* 71(11), 1284–1288. https://doi.org/10.1016/j.jacc.2018.01.023

Barry, B. (2001): *Culture and Equality: An Egalitarian Critique of Multiculturalism*. Cambridge, MA: Harvard University Press.

Bauer, J. M. and van Eeten, M. J. G. (2009): Cybersecurity: Stakeholder Incentives, Externalities, and Policy Options, *Telecommunications Policy* 33(10–11), 706–719. https://doi.org/10.1016/j.telpol.2009.09.001

Beauchamp, T. L. and Childress, J. F. (2019): *Principles of Biomedical Ethics*. 8th ed. Oxford: Oxford University Press.

Blythe, J. M., Johnson, S. D. and Manning, M. (2020): What is Security Worth to Consumers? Investigating Willingness to Pay for Secure Internet of Things Devices, *Crime Science* 9(1), 1–9. https://doi.org/10.1186/s40163-019-0110-3

Brito, J. and Watkins, T. (2011): Loving the Cyber Bomb? The Dangers of Threat Inflation in Cybersecurity Policy. Mercatus Center, Georg Mason University, Working Paper No. 11–24. Available at: www.mercatus.org/system/files/Loving-Cyber-Bomb-Brito-Watkins.pdf [27.01.2021]

Buchanan, B. (2016): Cryptography and Sovereignty, *Survival* 58(5), 95–122. https://doi.org/10.1080/00396338.2016.1231534

Campbell, M. (1988): Ethics and Computer Security: Cause and Effect, *Proceedings of the 1988 ACM Sixteenth Annual Conference on Computer Science – CSC'88*, 384–390. https://doi.org/10.1145/322609.322781

Cashell, B., Jackson, W. D., Jickling, M. and Webel, B. (2004): The Economic Impact of Cyber-attacks. CRS Report for Congress. Retrieved January 27, 2021. Available at: https://archive.nyu.edu/bitstream/2451/14999/2/Infosec_ISR_Congress.pdf [30.03.2021]

Chen, T. M., Jarvis, L. and Macdonald, S. (eds.) (2014): *Cyberterrorism*. New York: Springer. https://doi.org/10.1007/978-1-4939-0962-9

Chhaya, L., Sharma, P., Kumar, A. and Bhagwatikar, G. (2020): Cybersecurity for Smart Grid: Threats, Solutions and Standardization, in A. K. Bhoi, K. S. Sherpa, A. Kalam, and G.-S. Chae (eds.): *Advances in Greener Energy Technologies*. Singapore: Springer, 17–29. https://doi.org/10.1007/978-981-15-4246-6_2

Christen, M., Gordijn, B. and Loi, M. (eds.). (2019): *Ethics of Cybersecurity*. Cham: Springer International Publishing.

Christen, M., Gordijn, B., Weber, K., van de Poel, I. and Yaghmaei, E. (2017): A Review of Value-Conflicts in Cybersecurity, *The ORBIT Journal* 1(1). https://doi.org/10.29297/orbit.v1i1.28

Christensen, K. K. and Liebetrau, T. (2019): A New Role for "the Public"? Exploring Cyber Security Controversies in the Case of WannaCry, *Intelligence and National Security* 34(3), 395–408. https://doi.org/10.1080/02684527.2019.1553704

Clouser, K. D. and Gert, B. (1990): A Critique of Principlism, *Journal of Medicine and Philosophy* 15(2), 219–236. https://doi.org/10.1093/jmp/15.2.219

Connolly, L. Y. and Wall, D. S. (2019): The Rise of Crypto-ransomware in a Changing Cybercrime Landscape: Taxonomising Countermeasures, *Computers & Security* 87, 101568. https://doi.org/10.1016/j.cose.2019.101568

Cooper, H. A. (1995): Computer Security, Ethics, and Law, *Journal of Information Ethics* 4(1), 41–53, 95.

Coventry, L. and Branley, D. (2018): Cybersecurity in Healthcare: A Narrative Review of Trends, Threats and Ways Forward, *Maturitas* 113, 48–52. https://doi.org/10.1016/j.maturitas.2018.04.008

CSIS – Center for Strategic and International Studies (2018): Economic Impact of Cybercrime – No Slowing Down. Available at: www.mcafee.com/enterprise/en-us/assets/reports/restricted/rp-economic-impact-cybercrime.pdf [01.04.2021].

Derbyshire, R., Green, B. and Hutchison, D. (2021): "Talking A different Language": Anticipating Adversary Attack Cost for Cyber Risk Assessment, *Computers & Security* 103, 102163. https://doi.org/10.1016/j.cose.2020.102163

Dhamija, R. and Dusseault, L. (2008): The Seven Flaws of Identity Management: Usability and Security Challenges, *IEEE Security & Privacy Magazine* 6(2), 24–29. https://doi.org/10.1109/MSP.2008.49

Diffie, W. and Landau, S. (1998): *Privacy on the Line: The Politics of Wiretapping and Encryption*. Cambridge, MA: MIT Press.

Diffie, W. and Landau, S. (2007): The Export of Cryptography in the 20th and the 21st Centuries, in de Leeuw, K. and Bergstra, J. A. (eds.): *The History of Information Security: A Comprehensive Handbook*. Amsterdam: Elsevier, 725–736. https://doi.org/10.1016/B978-044451608-4/50027-4

Dittrich, D., Bailey, M. and Dietrich, S. (2011): Building an Active Computer Security Ethics Community, *IEEE Security & Privacy Magazine* 9(4), 32–40. https://doi.org/10.1109/MSP.2010.199

Domingo-Ferrer, J. and Blanco-Justicia, A. (2020): Ethical Value-centric Cybersecurity: A Methodology Based on a Value Graph, *Science and Engineering Ethics* 26(3), 1267–1285. https://doi.org/10.1007/s11948-019-00138-8

Dunn Cavelty, M. (2014): Breaking the Cyber-security Dilemma: Aligning Security Needs and Removing Vulnerabilities, *Science and Engineering Ethics* 20(3), 701–715. https://doi.org/10.1007/s11948-014-9551-y

Dunn Cavelty, M. and Egloff, F. J. (2019): The Politics of Cybersecurity: Balancing Different Roles of the State, *St Antony's International Review* 15(1), 37–57.

Dynes, S., Goetz, E. and Freeman, M. (2007): Cyber Security: Are Economic Incentives Adequate? in Goetz, E. and Shenoi, S. (eds.): *Critical infrastructure protection*. Vol. 253. Hanover: Springer, 15–27. https://doi.org/10.1007/978-0-387-75462-8_2

Ekelund, S. and Iskoujina, Z. (2019): Cybersecurity Economics – Balancing Operational Security Spending, *Information Technology & People* 32(5), 1318–1342. https://doi.org/10.1108/ITP-05-2018-0252

Elhoseny, M., Yuan, X., El-Minir, H. K. and Riad, A. M. (2016): An Energy Efficient Encryption Method for Secure Dynamic WSN: An Energy Efficient Encryption Method for Secure Dynamic WSN, *Security and Communication Networks* 9, 2024–2031. https://doi.org/10.1002/sec.1459

Fleischacker, S. (1999): From Cultural Diversity to Universal Ethics: Three Models, *Cultural Dynamics* 11(1), 105–128. https://doi.org/10.1177/092137409901100107

Frankena, W. K. (1963): *Ethics*. Englewood Cliffs, NJ: Prentice-Hall.

Gandhi, R., Sharma, A., Mahoney, W., Sousan, W., Zhu, Q. and Laplante, P. (2011): Dimensions of Cyber-attacks: Cultural, Social, Economic, and Political, *IEEE Technology and Society Magazine* 30(1), 28–38. https://doi.org/10.1109/MTS.2011.940293

Garfinkel, S. and Lipford, H. R. (2014): Usable Security: History, Themes, and Challenges, *Synthesis Lectures on Information Security, Privacy, and Trust* 5(2), 1–124. https://doi.org/10.2200/S00594ED1V01Y201408SPT011

Gross, M. L., Canetti, D. and Vashdi, D. R. (2016): The Psychological Effects of Cyber Terrorism, *Bulletin of the Atomic Scientists* 72(5), 284–291. https://doi.org/10.1080/00963402.2016.1216502

Gupta, N., Jati, A. and Chattopadhyay, A. (2020): MemEnc: A Lightweight, Low-power and Transparent Memory Encryption Engine for IoT, *IEEE Internet of Things Journal*, 1–1. https://doi.org/10.1109/JIOT.2020.3040846

Harknett, R. J. and Stever, J. A. (2011): The New Policy World of Cybersecurity, *Public Administration Review* 71(3), 455–460. https://doi.org/10.1111/j.1540-6210.2011.02366.x

Herr, T. and Rosenzweig, P. (2016): Cyber Weapons and Export Control: Incorporating Dual Use with the PrEP Model, *Journal of National Security Law & Policy* 8, 301–319. https://doi.org/10.2139/ssrn.2501789

Herzog, S. (2011): Revisiting the Estonian Cyber Attacks: Digital Threats and Multinational Responses, *Journal of Strategic Security* 4(2), 49–60. https://doi.org/10.5038/1944-0472.4.2.3

Hine, K. (2011): What is the Outcome of Applying Principlism? *Theoretical Medicine and Bioethics* 32(6), 375–388. https://doi.org/10.1007/s11017-011-9185-x

Iasiello, E. (2013): Cyber attack: A Dull Tool to Shape Foreign Policy, *2013 5th International Conference on Cyber Conflict (CYCON 2013)*, 1–18.

International Telecommunications Union (2008): ITU-TX.1205: Series X: Data Networks, Open System Communications and Security: Telecommunication Security: Overview of Cybersecurity. Available at: www.itu.int/rec/T-REC-X.1205-200804-I [27.01.2021]

Jarvis, L. and Macdonald, S. (2015): What is Cyberterrorism? Findings from a Survey of Researchers, *Terrorism and Political Violence* 27(4), 657–678. https://doi.org/10.1080/09546553.2013.847827

Johnson, S. D., Blythe, J. M., Manning, M. and Wong, G. T. W. (2020): The Impact of IoT Security Labelling on Consumer Product Choice and Willingness to Pay, *PLOS ONE* 15(1), e0227800. https://doi.org/10.1371/journal.pone.0227800

Kainda, R., Flechais, I. and Roscoe, A. W. (2010): Security and Usability: Analysis and Evaluation, *2010 International Conference on Availability, Reliability and Security*, 275–282. https://doi.org/10.1109/ARES.2010.77

Kaplan, F. M. (2016): *Dark Territory: The Secret History of Cyber War*. New York: Simon & Schuster.

Kitchen, K. (2019): A Major Threat to Our Economy – Three Cyber Trends the U.S. Must Address to Protect Itself. Available at: www.heritage.org/cybersecurity/commentary/major-threat-our-economy-three-cyber-trends-the-us-must-address-protect [27.01.2021]

Kruger, H. A., Drevin, L., Flowerday, S. and Steyn, T. (2011): An assessment of the Role of Cultural Factors in Information Security Awareness, *2011 Information Security for South Africa*, 1–7. https://doi.org/10.1109/ISSA.2011.6027505

Leiwo, J. and Heikkuri, S. (1998): An Analysis of Ethics as Foundation of Information Security in Distributed Systems, *Proceedings of the Thirty-First Hawaii International Conference on System Sciences* 6, 213–222. https://doi.org/10.1109/HICSS.1998.654776

Liebetrau, T. and Christensen, K. K. (2021): The Ontological Politics of Cyber Security: Emerging Agencies, Actors, Sites, and Spaces, *European Journal of International Security* 6(1), 25–43. https://doi.org/10.1017/eis.2020.10

Liff, A. P. (2012): Cyberwar: A New "Absolute Weapon"? The Proliferation of Cyberwarfare Capabilities and Interstate War, *Journal of Strategic Studies* 35(3), 401–428. https://doi.org/10.1080/01402390.2012.663252

Loi, M., Christen, M., Kleine, N. and Weber, K. (2019): Cybersecurity in Health – Disentangling Value Tensions, *Journal of Information, Communication and Ethics in Society* 17(2), 229–245. https://doi.org/10.1108/JICES-12-2018-0095

Manjikian, M. (2018): *Cybersecurity Ethics: An Introduction*. New York: Routledge.

Manpearl, E. (2017): Preventing Going Dark: A Sober Analysis and Reasonable Solution to Preserve Security in the Encryption Debate, *University of Florida Journal of Law & Public Policy* 28, 65–99.

Mazzolin, R. and Samueli, A. M. (2020): A Survey of Contemporary Cyber Security Vulnerabilities and Potential Approaches to Automated Defence, *2020 IEEE International Systems Conference (SysCon)*, 1–7. https://doi.org/10.1109/SysCon47679.2020.9275828

McAfee (2020): The Hidden Costs of Cybercrime. Available at: www.mcafee.com/enterprise/en-us/assets/reports/rp-hidden-costs-of-cybercrime.pdf [27.01.2021]

Mohan, A. (2014): Cyber Security for Personal Medical Devices Internet of Things, *2014 IEEE International Conference on Distributed Computing in Sensor Systems*, 372–374. https://doi.org/10.1109/DCOSS.2014.49

Moore, T. (2010): The Economics of Cybersecurity: Principles and Policy Options, *International Journal of Critical Infrastructure Protection* 3(3–4), 103–117. https://doi.org/10.1016/j.ijcip.2010.10.002

Morscher, E., Neumaier, O. and Simons, P. (eds.) (1998): *Applied Ethics in a Troubled World*. New York: Springer. https://doi.org/10.1007/978-94-011-5186-3

Nadir, I. and Bakhshi, T. (2018): Contemporary Cybercrime: A Taxonomy of Ransomware Threats & Mitigation Techniques, *2018 International Conference on Computing, Mathematics and Engineering Technologies (ICoMET)*, 1–7. https://doi.org/10.1109/ICOMET.2018.8346329

Odlyzko, A. (2019): Cybersecurity is Not Very Important, *Ubiquity*, 1–23. https://doi.org/10.1145/3333611

Pattison, J. (2020): From Defence to Offence: The Ethics of Private Cybersecurity, *European Journal of International Security* 5(2), 233–254. https://doi.org/10.1017/eis.2020.6

Pranit Jeba Samuel, C., Dharani, K. G. and Bhavani, S. (2020): Power Algorithm to Improve the IoT Device for Lightweight Cryptography Applications, *Materials Today: Proceedings*, S2214785320389392. https://doi.org/10.1016/j.matpr.2020.11.326

Quigley, K., Burns, C. and Stallard, K. (2015): "Cyber gurus": A Rhetorical Analysis of the Language of Cybersecurity Specialists and the Implications for Security Policy and Critical Infrastructure Protection, *Government Information Quarterly* 32(2), 108–117. https://doi.org/10.1016/j.giq.2015.02.001

Ransford, B., Clark, S. S., Kune, D. F., Fu, K. and Burleson, W. P. (2014): Design Challenges for Secure Implantable Medical Devices, in Burleson, W. and Carrara, S. (eds.): *Security and privacy for implantable medical devices*. New York: Springer, 157–173. https://doi.org/10.1007/978-1-4614-1674-6_7

Reijers, W., Koidl, K., Lewis, D., Pandit, H. J. and Gordijn, B. (2018b): Discussing Ethical Impacts in Research and Innovation: The Ethics Canvas, in Kreps, D., Ess, C., Leenen, L. and Kimppa, K. (eds.): *This Changes*

Everything – ICT and Climate Change: What Can We Do? Bd. 537. Cham: Springer International Publishing, 299–313. https://doi.org/10.1007/978-3-319-99605-9_23

Reijers, W., Wright, D., Brey, P., Weber, K., Rodrigues, R., O'Sullivan, D. and Gordijn, B. (2018a): Methods for Practising Ethics in Research and Innovation: A Literature Review, Critical Analysis and Recommendations, *Science and Engineering Ethics* 24(5), 1437–1481. https://doi.org/10.1007/s11948-017-9961-8

Robinson, M., Jones, K. and Janicke, H. (2015): Cyber Warfare: Issues and Challenges, *Computers & Security* 49, 70–94. https://doi.org/10.1016/j.cose.2014.11.007

Rowe, B. and Wood, D. (2013): Are home Internet Users Willing to Pay ISPs for Improvements in Cyber Security? in Schneier, B. (ed.): *Economics of Information Security and Privacy III.* New York: Springer, 193–212. https://doi.org/10.1007/978-1-4614-1981-5

Sawaya, Y., Sharif, M., Christin, N., Kubota, A., Nakarai, A. and Yamada, A. (2017): Self-confidence Trumps Knowledge: A Cross-cultural Study of Security Behavior, *Proceedings of the 2017 CHI Conference on Human Factors in Computing Systems*, 2202–2214. https://doi.org/10.1145/3025453.3025926

Schneier, B. (2000): *Secrets and Lies: Digital Security in a Networked World.* New York: Wiley.

Schuijff, M. and Dijkstra, A. M. (2020): Practices of Responsible Research and Innovation: A Review, *Science and Engineering Ethics* 26(2), 533–574. https://doi.org/10.1007/s11948-019-00167-3

Sorell, T. (2011): The Limits of Principlism and Recourse to Theory: The Example of Telecare, *Ethical Theory and Moral Practice* 14(4), 369–382. https://doi.org/10.1007/s10677-011-9292-9

Stevens, C. (2020): Assembling Cybersecurity: The Politics and Materiality of Technical Malware Reports and the Case of Stuxnet, *Contemporary Security Policy* 41(1), 129–152. https://doi.org/10.1080/13523260.2019.1675258

Stoll, C. (1989): *The Cuckoo's Egg: Tracking a Spy through the Maze of Computer Espionage.* New York: Doubleday.

Sundt, C. (2010): Cryptography in the Real World, *Information Security Technical Report* 15(1), 2–7. https://doi.org/10.1016/j.istr.2010.10.002

Swire, P. P. and Ahmad, K. (2012): Encryption and Globalization, *The Columbia Science & Technology Law Review* 13, 416–481. https://doi.org/10.2139/ssrn.1960602

Szor, P. (2005): *The Art of Computer Virus Research and Defense.* Hagerstown, MD: Addison-Wesley.

Thorstensen, E. (2019): Stakeholders' Views on Responsible Assessments of Assistive Technologies through an Ethical HTA Matrix, *Societies* 9(3), 51. https://doi.org/10.3390/soc9030051

Tripathi, M. and Mukhopadhyay, A. (2020): Financial Loss Due to a Data Privacy Breach: An Empirical Analysis, *Journal of Organizational Computing and Electronic Commerce* 30(4), 381–400. https://doi.org/10.1080/10919392.2020.1818521

Ware, W. H. (1967a): *Security and Privacy in Computer Systems.* RAND Corporation.

Ware, W. H. (1967b): Security and Privacy in Computer Systems. Proceedings of the April 18–20, 1967, *Spring Joint Computer Conference – AFIPS'67 (Spring)*, 279–282. https://doi.org/10.1145/1465482.1465523

Weber, K. (2018a): Computers as Omnipotent Instruments of Power, *The ORBIT Journal* 2(1), 1–19. https://doi.org/10.29297/orbit.v2i1.97

Weber, K. (2018b): Extended Model for Ethical Evaluation, in Karafillidis, A. and Weidner, R. (eds.): *Developing Support Technologies.* Vol. 23. Cham: Springer International Publishing, 257–263. https://doi.org/10.1007/978-3-030-01836-8_25

Weber, K. and Kleine, N. (2020): Cybersecurity in Health Care, in Christen, M., Gordijn, B. and Loi, M. (eds.): *The Ethics of Cybersecurity.* Vol. 21. Cham: Springer International Publishing, 139–156. https://doi.org/10.1007/978-3-030-29053-5_7

Weimann, G. (2005): Cyberterrorism: The Sum of All Fears? *Studies in Conflict & Terrorism* 28(2), 129–149. https://doi.org/10.1080/10576100590905110

Westin, A. (1967): *Privacy and Freedom.* New York: Atheneum.

Wirth, A. (2017): The Economics of Cybersecurity, *Biomedical Instrumentation & Technology* 51(s6), 52–59. https://doi.org/10.2345/0899-8205-51.s6.52

Woods, M. (2017): Cardiac Defibrillators Need to Have a Bulletproof Vest: The National Security Risk Posed by the Lack of Cybersecurity in Implantable Medical Devices, *Nova Law Review* 41(3), 419–.447

Yaghmaei, E., van de Poel, I., Christen, M., Gordijn, B., Kleine, N., Loi, M., Morgan, G. and Weber, K. (2017): Canvas White Paper 1 – Cybersecurity and Ethics. Available at: https://doi.org/10.2139/ssrn.3091909 [27.01.2021]

PART 6

Smart technologies

Case studies

31

A DIGITAL SOCIETY FOR AN AGEING POPULATION

The Japanese experience

Yuko Harayama and René Carraz

Introduction

According to a UN (2020) report, the number of persons worldwide aged 65 or above is projected to double by 2050 to reach 1.5 billion people. Japan currently has the highest old-age dependency ratio (OADR) in the world, with 51 persons aged 65 years or over per 100 persons aged 20 to 64 years. UN (2020) projections indicate that in 2050, Japan will remain the country with the highest OADR (81 persons). Demographers have long written about the declining birth rates, on the one hand, and increasing longevity of the Japanese population, on the other, yet it could be argued that the discourse about Japan ageing started in the 1980s, amplifying when the term chōkōrei shakai (super-ageing society) began to appear in the late 1990s (Coulmas 2007). The topic of a super-ageing society is still prevalent today in policy discussions (Vogt 2008), academic debates (Coulmas et al. 2008; Tamiya et al. 2011; Heinrich and Galan 2018) and in the news media. Indeed, with Japan having the highest proportion of older adults in the world, the consequences of an ageing society are not limited to the health care and social security systems, but impact all levels of society.

It could be argued that the well-functioning and sustainability of Japan's social system as a whole would be challenged by the economic costs and social implications of a rapidly ageing population. There is a need for comprehensive measures to tackle these issues, and the Ministry of Health, Labour and Wellbeing (MHLW), the primary responsible Ministry in relation to the ageing population, affirmed its commitment to move into the direction of community-based integrated care for the elderly in the report "The Japan Vision, Health Care 2035", launched in 2015.

These people-centred, comprehensive and inclusive approaches to care for an elderly population clearly constitute an important pillar of any remedial measures. Alongside and complementing these health and care policies, what we observe today is the emergence of challenge-driven science, technology and innovation (STI) policy backed by digital technologies and data-infrastructure, with a particular focus on the issues of an ageing population. In this regard, Japan is an interesting benchmark for at least three dimensions. Firstly, it has spent heavily on STI over the last two decades, investing a high proportion of its GDP in research and development (R&D) (3.26% in 2018[1]). Secondly, since 2016 the Japanese government in

1 Data accessed at https://stats.oecd.org/.

 DOI: 10.4324/9780429351921-38

tandem with the Keidanren (Japanese Business Federation) has been promoting the concept of a "Society 5.0" as a response to the rapid evolution of information and communications technology (ICT), proposing a coordinated, forward-looking strategy that could ensure Japan's leadership in an era of digital transformation (Carraz and Harayama 2018; Holroyd 2020). Just as "Industry 4.0" was a tentative response to the digital transformation in the manufacturing sector backed by Germany (Kagermann et al. 2013; Schwab 2017), Society 5.0 emerged from the need to master the challenges of digitisation and connectivity across a wide range of platforms at all levels of society, that is, to achieve the digital transformation of society itself. Thirdly, these investments and policy initiatives are mobilised to address ageing population issues, and some policy measures are already implemented in the field. For instance, the use of robots for healthcare to alleviate the shortage of care workers for seniors is already implemented and culturally accepted (Ishiguro 2017), even though the deployment of new digital technologies is sometimes blocked due to a lack of clarity on the regulatory framework for the collection and use of personal data and privacy issues.

The fact is that the Japanese government has decided to explore the potential of digital transformation and is opting for a whole-of-government approach to address the societal challenges of an ageing population.[2] On this point, the Tsukuba Communiqué,[3] launched at the end of the G7 Science and Technology Ministers' Meeting, which took place in 2016 under the G7 Japanese Presidency, is insightful. It states:

> We recognized the importance of helping promote a society with active ageing, where elderly citizens continue to be fully engaged within their societies in ways befitting their capacity and interests. We also recognized the role of Science, Technology and Innovation in contributing to this thorough well-designed health system for elderly care, including prevention, timely diagnosis, treatment, assistance and care of age-related health issues, and the social and physical infrastructure that enhanced inclusion.
>
> *(Cabinet Office 2016)*

In our view, this statement translates perfectly the standpoint of the Japanese government vis-à-vis its ageing population. On the one hand, it is striving to promote a society where active ageing is favoured through people-centred approaches. On the other hand, it is aiming to harness the contributions of science, technology and innovation empowered by the digital transformation.

More concretely, in the elderly care and nursing sectors, the use of ICT and robotics is expected to contribute to enhancing quality of life and ensuring mobility, as well as to mitigate the on-site burden of healthcare and caregiving. The potential is there, and several R&D programs have been launched by the Japanese key funding agencies, such as Japan Science and Technology Agency (JST), New Energy and Industrial Technology Development Organization (NEDO) and Japan Agency for Medical Research and Development (AMED). These programs provide the opportunity to explore new ideas to address the societal challenges of ageing and the means to test on-the-ground prototypes or models derived from these ideas.

This brief overview of how Japan is addressing the challenge of an ageing population shows the complexity of taking multi-dimensional and multi-stakeholder issues, and there is a need for

2 Annual reports on the Aging Society: https://www8.cao.go.jp/kourei/english/annualreport/index-wh.html

3 All Ministerial documents in this chapter were accessed on 21 April 2020; additionally, we have tried to use an English version when available. G7 Science and Technology Ministers' Meeting. 2016 May 15–17. Tsukuba, Ibaraki (Cabinet Office 2016).

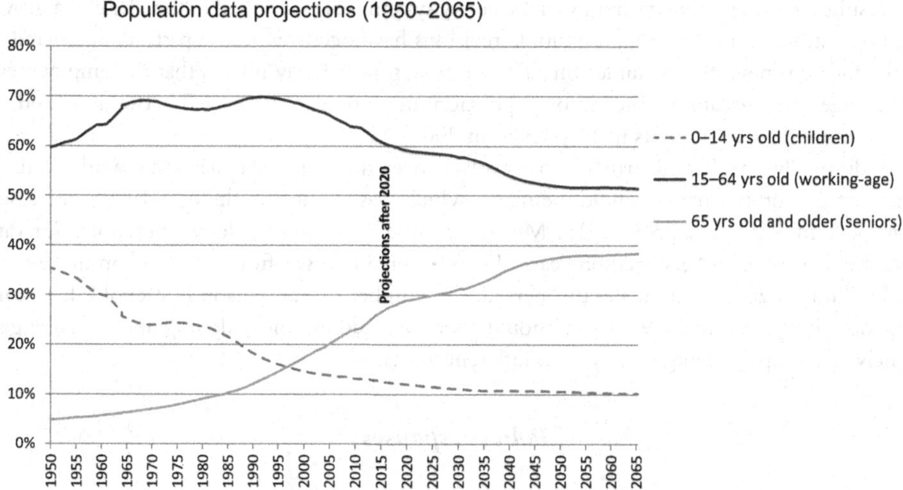

Figure 31.1 Ageing trends in Japan (1950–2065)
Source: OECD Stat and Statistical Handbook of Japan

a policy response beyond the well-established health and welfare policy. The Japanese initiative has a value of social experimentation, which could be studied by other countries, as the country is investing massively in STI and is at the forefront of the ageing trend, and this chapter will be developed from this perspective.

After this introduction, in the following section, we describe the challenges of an ageing population in the Japanese context and the government's policy responses. Afterwards, we introduce an STI policy perspective, and in the successive section, we revisit ageing population issues from this perspective and highlight the main STI policy programs implemented by JST, NEDO and AMED, the three principal policy-oriented funding agencies. We end with our conclusion, which summarises key lessons learned from the Japanese experience.

The challenge of ageing populations

Ageing society

Japan, with demographic pressures characterised by a declining population under the age of 14 and the rapid growth of those over 65 (Figure 31.1), is entering into a phase never experienced before by any other country around the world. Data[4] show that the life expectancy at birth continues to increase (81.25 for men and 87.32 for women in 2018), while the total fertility rate remains low (1.42 in 2018); people tend to marry later (31.1 years for men and 29.4 years for women in 2018) and the first birth tends to come later as well (average age 30.7 years in 2018).

Following these trends, there is no indication that these demographic pressures will decrease. Also, despite this prospect of longer life, the gap between healthy life expectancy at birth and life expectancy at birth remains at around 8 years for men and 13 years for women (Cabinet Office 2019).

4 Statistics gathered from multiples MHLW documents.

Despite the fact that the current social security system was established in the phase of a growing population, and even though multiple revisions have occurred in the past, its sustainability in the future is nevertheless under threat. A positive sign we may note is that the employment rate by age increases among the senior population; in particular, those aged between 65 and 69 gained more than 10% points in 10 years from 2008.

Additionally, the family structure has evolved over time, in particular after World War II. The average number of household members, which was about 5 in the mid-1950s, decreased from 3.41 in 1970 to 2.33 in 2015. Meanwhile, the number of elderly households (private households with members aged 65 years old and over) increased from 12.7 million in 1995 to 21.7 million in 2015, and during this period, the number of one-person households has more than doubled (Bureau 2019). Japan should therefore address these demographic challenges, namely of a "super-ageing society", on different fronts.

Policy responses

Guideline of measures for ageing society

In this vein, the Japanese government promulgated the Basic Law on Measures for the Ageing Society in 1995, and according to this law, the Conference on the Measures for the Ageing Society, chaired by the Prime Minister, was established. This Basic Law also stipulated the formulation of the Guideline of Measures for Ageing Society, which constitutes the basis for national policies related to ageing society; the first version was adopted in 1996 (Cabinet Office 2018). Since 1997, the white paper[5] Annual Report on the Ageing Society has been published on the topic on an annual basis to accompany this policy framework.

The Guideline was revised in 2001, 2012 and 2018 in order to reflect the changing socio-economic environment, as well as to expand the fields of R&D related to the ageing society. While the changing socio-economic environment has been considered one of the principal pillars in the Guideline from the beginning, the priority areas have evolved considerably over time.

Indeed, starting with the initial focus on gerontology, assistive devices and universal design products in the 1996 Guideline, the 2001 revised Guideline shed a particular light on dementia, cancers and lifestyle-related diseases as priority research areas. The 2012 revision, however, was initiated with a view to alignment with the five-year Strategy for Medical Innovation (2012), on one hand, and putting greater emphasis on research centred on Intelligent Transport Systems (ITS) to facilitate the mobility of the elderly, on the other.

This alignment with other policy areas became even more apparent with the latest revision of the Guideline (2018), which includes not only R&D components but also employment and income, health and welfare, learning and social participation and living environment, demonstrating that the entire Guideline has been adapted to the concept of Society 5.0[6] as the government's response to the rapid evolution of ICT and digitisation of the economy. By providing policy directions with a view to creating an "age-free society", consolidating local community and promoting innovation – in other words advocating for an inclusive and elderly centred policy, backed by the digital transformation – this revision illustrates the intention of the Japanese government to consider the "ageing society" as a privileged field for the implementation of Society 5.0.

5 The white papers can be accessed here (in Japanese): https://www8.cao.go.jp/kourei/whitepaper/index-w.html.

6 This concept will be discussed in detail in the next section.

Ministry of Health, Labour and Wellbeing (MHLW) policies

The Ministry of Health, Labour and Wellbeing (MHLW), one of the ministries on the front line of the ageing transition, has been implementing national health care reforms in order to respond to the present and future needs of a Japanese health care system, with a particular focus on the ageing populations. An important document to guide these reforms has been "The Japan Vision, Health Care 2035", launched in 2015. It has marked a fundamental paradigm shift from "system inputs to patient value" (MHLW 2017: 1), from the quantity of services provided to patients to quality, from cure to care and from specialisation of services to integrated approaches across medical and social service sectors, while putting more emphasis on a community-based integrated care for the elderly. The guide also advocates for mobilising innovation in health care and for a more efficient use of information. More concretely, it proposes to develop a health care database to support telemedicine applications and to build and utilise a health care network that links data using unique identifiers, signalling a first move toward data-driven health care (MHLW 2015).

Another important topic addressed by the MHLW is that of dementia. Even though the Japanese government's action on dementia goes back to the mid-1980s with the creation of centres for the elderly with dementia, the most comprehensive approach was initiated by the Ministry in 2012 with the launch of the Five-Year Plan for the Promotion of Dementia Measures (2013–2017). This so-called Orange Plan advocated establishing four main objectives: a standardised dementia care pathway; early detection of dementia; the establishment of local structures for the provision of care services; and the creation of training schemes for medical and care staff.

As a follow-up to the G8 Dementia Summit (2013) initiated by the UK government, Japan organised a Legacy Event on prevention and care in 2014. This event led Japan to move to a more comprehensive approach on dementia, and thus the Orange Plan has been adjusted accordingly to become the "New Orange Plan" (2015). The idea behind this updated version of the Plan was to focus actions on people-centred and community-based approaches, aiming at promoting the "development of dementia-friendly communities and to improve the living environments of people with dementia by enabling them to continue living in familiar spaces and environments as long as possible".[7] Many policy measures have since been initiated, such as "Alzheimer Cafés" – events to gather people with dementia and their families, and to converse with the community, their supporters and specialists.

Alongside the New Orange Plan, MHLW is taking policy measures in line with the Framework for Promoting Dementia Care, a cross-ministerial initiative launched in 2019 by the Ministerial Council on the Promotion of Policies for Dementia Care chaired by the Chief Cabinet Secretary. The framework aims at promoting further collaboration across ministries and agencies around the guiding principles of "inclusiveness", that is, that persons with dementia may live with dignity in our society and "risk reduction" i.e. delaying or slowing the offset of dementia and not necessarily targeting its eradication. The framework recommends policy actions such as the promotion of research on dementia, or expanding the Project for Supporting Community Preventive Long-Term Care Activities. This last initiative, called "Kayoinoba" in Japanese, is worth noting in that it is a bottom-up initiative. It focuses on creating a space in a given local area designed and managed by its residents with the aim of bringing together local citizens and people suffering from dementia, contrasting with still prevalent traditional top-down decision-making procedures.

7 Information about the New Orange Plan can be found here: http://japanhpn.org/en/1-2/.

Policy considerations

As we have mentioned in this section, ageing is a longstanding trend in Japan, with widespread implications for the whole society, including increases in health and social budgets and an almost flat nominal economic growth prospect. In the last 20 years, all government expenses have been decreasing proportionally except for national debt services and social security expenses, the latter going from 23.6% in 1997 to 34.6% of the total in 2017.[8] That is to say, almost all other items are in competition with social security expenses. From the STI policy perspective, enhancing innovation capacity has been considered a key to addressing the societal challenges of an ageing population; in the longer-term, in fact, it is expected to contribute to the reduction of social security expenses. Yet, the Japanese government still has to proceed to a certain arbitrage between R&D and social security. The question that remains unanswered is whether the proposal of Society 5.0 will contribute to a move beyond this arbitrage and reconcile these two policy priorities: innovation-led economic growth and financing of an ever-increasing social security budget led by the ageing trend.

The Japanese government has enacted a series of laws and regulations to deal with the ageing issue and its ballooning costs. Nevertheless, situations where alignment to the Cabinet's guidance would not be the priority may happen, given that each ministry has its own policy goals, which could partially conflict with this alignment. In the past, Johnson (1982), for instance, showed in his seminal book about the Ministry of International Trade and Industry (MITI) how tensions between competing ministries could unfold and hamper the government's responses. Reflecting this thinking in terms of policy actions, the Cabinet Office is expected to play the role of coordinator through its aforementioned Conference on the Measures for the Ageing Society chaired by the Prime Minister, and distribute policy measures among ministries, in particular MHLW; Ministry of Education, Culture, Sports, Science and Technology (MEXT); and Ministry of Economy, Trade and Industry (METI). Thus, the structure is in place to put into practice the whole-of-government approach, with a strong leadership and coordination role by the Cabinet Office.

Moreover, the paradigm shift from "system inputs to patient value" mentioned earlier has a significant implication on the practice of policy making. Indeed, the traditional top-down policy making process has a limited capacity to fully capture the information on the ground. Therefore, the government has recognised the need for a more decentralised way of decision-making, with a view to soliciting community-based actions, and to engaging stakeholders with grounded experiences – in short, to supplement a bottom-up channel in the loop of the policy making process.

In our view, another important dimension to addressing ageing issues is the nexus with science, technology and innovation (STI) policy, and there too a decentralised process is necessary. In terms of the expected contribution of STI, in particular the deployment of digital technologies and data analytics, there are extended fields to be explored, as noted in the Fifth Science and Technology Basic Plan (2016–2020);[9] these include the development of new drugs, medical devices and medical technology; the establishment of medical ICT infrastructures for use in examination, treatment and medication; and the greater utilisation of data in the fields of medical and long-term care in order to improve their quality. This exploration cannot be conducted alone at the ministerial level with centralised decision-making processes. The collective

8 Statistics gathered from multiples Cabinet Office documents.

9 A five-year comprehensive plan for the promotion of science and technology: https://www8.cao.go.jp/cstp/english/basic/5thbasicplan.pdf.

challenge of ageing has to be addressed collaboratively by universities, private companies and government agencies to meet the best interests of society at large. The overarching strategy to respond to the pressing ageing challenges is to be found in the concept of a Society 5.0, which aims to improve the innovation environment in Japan; we will define it more precisely in the next section.

Mobilising science, technology and innovation (STI)

STI policy framework[10]

In the 1990s, Japan's economy was struggling and the government was seeking additional ways to recover momentum, besides its monetary and fiscal policies. One such action was to invest in science and technology, expecting that this investment would contribute to a boost in economic growth. In 1995, the Science and Technology Basic Law was promulgated as a legal foundation of this policy engagement, and the Council for Science Technology Policy (CSTP), chaired by the Prime Minister, was founded to administer this policy within the Cabinet Office. Also, based on the Basic Law, the five-year so-called Science and Technology Basic Plan has in place since 1996.

In the 2000s, the mainstreaming of innovation became apparent in the science and technology policy arena, as illustrated by the launch of OECD's Innovation Strategy (2010). Japan was no exception; in 2014, CSTP was renamed "Council for Science, Technology & Innovation" (CSTI), and henceforth became responsible for formulating science, technology and innovation policy and ensuring its sound implementation. As a result, we have observed the convergence of growth strategy and science, technology and innovation (STI) strategy.

In fact, the main focus of the Basic Plans has evolved over time, starting with the consolidation of science and technology infrastructures (First Basic Plan – Fiscal Year (FY) 1996–FY 2000[11]), passing through a more technology-focused policy formulation (Second and Third Basic Plans) and moving to a challenge-driven approach (Fourth Basic Plan). Notably, the Fifth Basic Plan (FY 201–FY 2020) marked a leapfrog over the previous ones.

Observing that the world is increasingly interconnected beyond traditional national and institutional borders, that this connectivity is evolving at an accelerated rate never experienced before and backed by the digital transformation empowered with artificial intelligence (AI) and internet of things (IoT), CSTI recognised the limitations of being able to predict the technological trends even just five years ahead. Therefore, CSTI decided, instead of identifying key technological areas or key challenges as before, to place emphasis on the "preparedness" for this unpredictable and unforeseeable near future with the preparation of the Fifth Basic Plan during 2015. Therefore, the capacity to design future industry and society has been considered as instrumental, and the prospect of investing in people and providing the space to experiment their ideas emerged.

Furthermore, 2015 was the year the Sustainable Development Goals (SDGs) were adopted at the United Nations. Recognising that innovation will play a central role in addressing these global challenges, CSTI's stance was to ensure that the Fifth Basic Plan would be compatible by design with SDGs, with the consequence of expanding the scope of innovation policy into the sphere of sustainability and wellbeing.

10 For a historical account of Japan's STI policy, refer to Carraz and Harayama (2018).

11 In Japan the government's fiscal year (FY) runs from April 1 to March 31.

What we observe here is the shift from the traditional technology-driven to a more society-centred innovation policy, and from planning to an experience-based approach in its implementation. With this in mind, the Fifth Basic Plan was composed of four main pillars:

1 Preparing the next: future industry and society;
2 Addressing socio-economic and global challenges;
3 Investing in "fundamentals": people and excellence;
4 Better-functioning STI systems.

These four pillars are formulated around Society 5.0 and offer new guiding principles for innovation, which could be considered as a concept proposal from Japan, with the expectation that this concept would inspire the global community.

Society 5.0

In 2015, the CSTI took into consideration how the coming digital transformation could become a dominant driver for societal transformation. Given that many innovations driven by technology have been a factor for social change in the past, the main question under debate by the executive members of the CSTI at the early phase of preparation of the Fifth Basic Plan was, why should we formulate STI policy from a different perspective this time?[12]

What we observe today is arguably just a tip of the digital transformation. CSTI executive members' opinions converged on scenarios around the potential magnitude in scale and scope of its impacts on not only the economy and society, but also, and more particularly, human beings. This human-centred tropism largely outdoes the sphere of what can be decided alone by STI professionals, such as scientists, engineers, entrepreneurs, technocrats or politicians.[13] Therefore, CSTI considered that humans must remain central actors, active subjects to the contingencies of technological systems, and that the engagement of all stakeholders, including citizens, was critical in designing our future society backed by STI. This idea is widely debated in the literature; recent changes in how we do science, for instance, are in part driven by powerful collaborative tools enabled by the internet (Nielsen 2020); innovation is more democratised (Von Hippel 2005) and helped by collective intelligence processes (Tovey 2008).

Also, within the context of unstable and ever-changing socio-economic and political conditions, which characterise the first half of the last decade, it became imperative to identify key values on which future society would be founded, and in this vein, the SDGs advocating for the value of sustainability and inclusiveness have been considered as a reference. Consequently, "human-centred" and "value-based" dimensions became the guiding principles of this new breed of STI policy.

So, why "5.0"? Conceptually, the CSTI took a very long look at human history in the exercise of foreseeing our future, starting with a "hunting and gathering" society (1.0), where people were living in a perceived symbiosis with nature in order to ensure the continuity of our species. Given that curiosity and exploration are integral parts of human nature, humans were

12 Even though it is out of the scope of this chapter to delve into these ontological questions, a worthy read on the topic is certainly the seminal book written by Mokyr (2002) on the historical origins of the knowledge economy.

13 This statement reflects the convergence of views among the executive members of the CSTI during the preparation of the Fifth Basic Plan, and it underlines two CSTI documents: "Toward the Fifth S&T Basic Plan" (October 2014) and "Guiding Framework for the Formulation of the Fifth S&T Basic Plan" (April 2015). All translations are by the authors, unless otherwise specified.

not satisfied being entirely dependent on nature, and a "farming" society (2.0) emerged, leading to a more structured form of society. Then, starting with the invention of the steam engine, fossil fuel-powered machines opened the door to mass production and tech-driven manufacturing – the so called "Industrial" society (3.0). Today we are experiencing digital transformation – the "Information" society (4.0). Increasingly, values are created by deploying and combining information, and in our everyday life, at work, in public space, as well as at home and in private, we are already strong users and consumers of information, as illustrated by the intensive use of the smartphone as a tool to access ubiquitously digital content. Our dependency on connectivity seems to be irreversible, transforming the way we interact. "Society 5.0" is expected to be the next stage: a human-centred society, still unknown, yet to be built by all stakeholders based on these "human-centred" and "value-based" guiding principles.

Behind the eye-catching titles and programme initiatives – such as Third Industrial Revolution (Rifkin 2011), Fourth Industrial Revolution (Schwab 2017), Industry 4.0 (Kagermann et al. 2013) or the FIWARE open source platform supported by the European Union and Japan's Society 5.0 – lies a fundamental shift in how economies may be structured in the future as industries, academia and governments create, store and integrate various data streams into daily production processes to provide goods and services. For the last 30 years, Japan's government has been actively promoting its national innovation strategies. Society 5.0 represents its latest effort to reboot its policy perspective by putting societal needs at the centre of scientific and technological transformations. Yet, since its inception, Society 5.0 has had to be nurtured, tested and developed in order to become an operational concept, in particular to serve as the basis for formulating the Fifth Basic Plan, knowing that the latter expects to meet the four following overarching goals: the advancement of science, technology and innovation (STI); the achievement of economic growth and wellbeing; addressing societal challenges; and contributing to global prosperity.

Data and digital technologies at the heart

The Basic Plan also indicated the approach to orient policy actions toward Society 5.0. More concretely, it proposed 11 focus areas – among them, Intelligent Transport System, Energy Value Chain, Monozukuri (manufacturing) System, Regional Inclusive Care System, Disaster Reduction System, Hospitality System – where advanced digital technologies and data infrastructure would be developed and tested. This would take shape along the lines of the system of systems perspective and the integration of a wide variety of data sources.

The 11 focus areas are those which have been investigated previously by the Cross-Ministerial Strategic Innovation Promotion Program (SIP), a program initiated by the CSTI in 2013, ahead of the Fifth Basic Plan, with the aim of confronting the most important societal challenges facing Japan, as well as contributing to economic growth through collaboration among ministries and agencies and between the private sector and public research institutions. By doing so, CSTI expected to capture and explore the early results of this program and to advance one step further in the direction of systems backed by the digital infrastructure. A common data and digital technology platform was conceived in the Basic Plan, which consists of a set of working frameworks, such as the standardisation of data formats and interfaces, cybersecurity technology and open data systems to serve all these focus areas and to facilitate data exchanges across areas. Among them, three areas, specifically, Intelligent Transport System, Energy Value Chain and Monozukuri System, have been identified as a test bed to realise data integration.

Beside these data infrastructures, the Basic Plan recommended investing in key technologies, such as cybersecurity, internet of things (IoT) system architecture technology, big data analytics, artificial intelligence (AI), device technology, network technology and edge computing, on the one hand, and to consider ethical, legal and social implications (ELSI) before any implementation of these digital technologies and the use of data sets, and to revisit the regulatory framework to secure the protection of personal information, on the other hand.

Challenges of ageing populations revisited

MHLW's policy alignment[14]

Plan for reforming healthcare and welfare service

At the MHLW level, a "headquarter meeting for social security and work style reform for the year 2040" was established in 2018 to identify the direction of the Japanese social system reform in general and healthcare reform in particular, and the Plan for Reforming Healthcare and Welfare Service was reported in 2019, advocating for the "effective use of robotics, artificial intelligence (AI) and ICT, and Data Health Reform (DHR)". For this purpose, smart assisting devices and systems are expected to be developed and tested in collaboration with the Ministry of Economy, Trade and Industry (METI) and the Ministry of Education, Culture, Sport, Science and Technology (MEXT), with the support of the Japan Agency for Medical Research and Development (AMED). In addition, pilot programs are recommended in order to attract the interest of those who are in the elderly care business in the use of robotics and sensing devices. With regard to the DHR, efforts will be made by MHLW to better equip data infrastructure, including Personal Health Records (PHR), to ensure the interoperability of different medical and health-related data sets and to promote the use of national databases.

This policy orientation fully reflects the prospect of Society 5.0 and indicates the alignment of MHLW to the latter. However, MHLW did not wait until 2019 to move in this direction. Within the Ministry, the ground has been prepared by the Advisory Board on the Use of AI in Health and Medical Care (formed in 2017), which was consolidated as a Consortium for Accelerating Development of AI in Health and Medical Care, on one hand, and by Data Health Reform Promotion Headquarters (also formed in 2017), on the other.

Use of artificial intelligence (AI)

In its final report, the Advisory Board on the Use of AI in Health and Medical Care identified six fields where AI would be introduced relatively early and recommended investigation of these fields to develop and make use of AI technologies. "Nursing and dementia care" was one of them, while the Advisory Board insisted that the direction R&D takes should be guided by the needs of those working in the nursing and dementia care services, instead of seeking for application once the technology has been developed. Based on this recommendation, the Consortium for Accelerating Development of AI in Health and Medical Care took over to determine more precisely those issues to be addressed with a view to accelerating the development of AI technologies and facilitating the adoption of these technologies by healthcare service providers. The Consortium was also charged with identifying potential obstacles for implementation, and proposed a roadmap listing actions to be taken to remedy them.

14 Information in Japanese was retrieved from the MHLW internal documents and website.

In the case of nursing and dementia care, the Consortium proceeded, in its early phase, to a hearing of the director of the Center of Assistive Robotics and Rehabilitation for Longevity and Good Health to have a grasp of the current situation and challenges of the use of AI and robotics. The latter reported the preliminary results of the research studies under investigation at the Center on the use of devices such as robotic canes, humanoid robots, conversational robots, urination sensor systems; in the context of nursing and dementia care services, the Center expressed the need to create a data platform in order to collect and manage data generated by these devices alongside other vital and relevant environmental data, which would enable them to extract value from these devices with AI, network technologies and the internet of things (IoT).

After a series of hearings and a review of regulations related to the health and medical data, and taking into account observations from in the health and medical care fields, the Consortium attempted to identify potential roadblocks to the development, implementation and adoption of AI. Their report, published in June 2020, specified a list of issues to be addressed, which we summarise as follows.

In the development phase of AI, the availability of data for learning and training AI is essential. Given that in the field of health and medical care, most of these data rely on personal information, which may contain medical records, genetic information and other vital data, the collection and use of data should respect not only the Act on the Protection of Personal Information, but ethical guidelines as well, with data security and data management guidelines in place. However, the Consortium expressed a particular concern about whether patients were appropriately informed on the use of their data and how their usage could potentially benefit public interest in terms of better health and medical services, particularly when the development takes place in partnership with the private sector. They thus suggested revisiting ethical guidelines, in particular informed consent regulations, and collecting and analysing cases and making them accessible to the public.

As regards transfer and security management related to health and medical data, there are already three Guidelines from three related ministries in place, namely Guidelines for the Security Management of Medical Information Systems (published in 2005, revised in 2017) by MHLW; Security Management Guidelines for Cloud Service Providers Handling Medical Information (2018) by the Ministry of Internal Affairs and Communication (MIC); and Security Management Guidelines for Information Processing Providers Dealing with Medical Information (2012) by METI. Meanwhile, the Consortium recommended a better coordination of these Guidelines and expressed the need to continuously revisit them in order to adjust them to the advancement of technologies as well as the standardisation of electronic health and medical records.

On the use of cloud systems for data processing and storage, the Consortium noted that the need to ensure transparency, to improve speed of data transfer and to design data management systems according to the characteristics of data are commonly recognised among stakeholders. However, the Consortium observed that consensus on the requirements has not yet been achieved; therefore, it suggested work on prototype developments as a first step.

Data Health Reform

Capturing the concept of Society 5.0, and recognising that health-related data will be playing a critical role in exploring the potential of digital transformation in the field of health, medical care and the nursing care system, the MHLW in 2017 established the cross-sectoral Data Health Reform Promotion Headquarters within the Ministry, with two main objectives: firstly to extend healthy life expectancy and secondly to improve the efficiency in medical and nursing care service provision.

In 2018, eight categories of service[15] as well as an institutional reform of the health insurance system were identified as areas to be investigated to move forward with digitalisation. This was with a view to offering accurate options to patients and creating a one-stop health data management platform, deploying evidence-based care and supporting the fight against cancer. After this kick-off, the strategy to promote Data Health Reform was reshaped in 2019, based on the observation that digital transformation is occurring at an accelerated rate in all policy areas, and that health, and ageing in particular, was becoming a privileged field for the implementation of Society 5.0, as noted in the previous section. Henceforth, the following four goals of the Data Health Reform have been clearly defined and the services mentioned above are expected to be initiated to support to realise these goals:

1 To advance genomic medicine and the use of AI with a view to realising personalised medicine and making health and medical services more efficient.
2 To implement operational personal health record (PHR) system, which will allow all citizens to have access to check their health data from their own devices.
3 To promote the use of data in the medical and nursing communities, aiming at enhancing the quality of services.
4 To promote the use of national databases, in particular Japan Diagnosis Procedure Combination Database (DPC) and National Database of Health Insurance Claims and Specific Health Checkups of Japan (NDB), by ensuring interoperability; adjusting the institutional and regulatory framework; improving the quality of data; and integrating databases under construction, such as CHASE (Care, Health Status and Events: data related to nursing care called).

To move forward and to ensure that efforts are made collectively and in a coordinated way, the Plan for intensifying health data reform was formulated by the MHLW in July 2020 (MHLW 2021). It consists of three focus areas for action. The first goal is to expand further the framework for health and medical data sharing at the national level. The second goal is to put in place a reformed regulatory framework to implement an online system for electronic prescriptions. Finally, there is a need to standardise medical and health data and the related legal framework, in order to allow access and use by individual citizens and patients of their own health and medical records. There is a precise timeline of two years for implementing these goals. It is worth noting that these actions intend to make better use of "My Number System",[16] which consists of assigning a single 12-digit identifier to every resident in Japan. It was introduced in 2016, but it is still at the early phase of deployment.

The realisation of this ambitious plan is not only technically challenging in terms of data management and data security but also relies on the capacity of MHLW to mobilise key stakeholders – in both the public and private sectors – to collaborate with local governments and to attract the interest and engagement of ordinary citizens.

15 Such as "sharing of health records", "health scoring", "health data analytics", "cancer genomics" and "artificial intelligence".

16 "My number" is used in administrative procedures in the areas of social security, taxation and disaster response.

Funding agencies' response[17]

Research and development on dementia by AMED

The Japan Agency for Medical Research and Development (AMED) was established in 2015 under the joint effort of MEXT, MHLW and METI, with the aim of supporting R&D seamlessly in the medical field from basic research to clinical trials with the prospect of fostering practical application and medical innovation. It had an overall budget for FY 2020 of around ¥127 billion (€1.1 billion[18]) and allocated a budget of ¥9 billion (€78.2 million) for basic and applied research in the field of geriatrics and dementia and psychiatric and neurological disorders.

The number of dementia and Alzheimer sufferers increased greatly over the last quarter century in the general population of the Japanese elderly (Ohara et al. 2017), and its worldwide prevalence is expected to triple by 2050 (Prince et al. 2016). Therefore, the development of effective measures is an important public health concern in Japan. Yet currently, no simple and minimally invasive method for making an objective diagnosis of dementia is available, and thus making a differential diagnosis of dementia is difficult. In these circumstances, AMED is promoting intensive integrated R&D in the field of medicine, from basic research to clinical trials, focusing on the five aspects of dementia: clarification of actual condition, prevention, diagnosis, treatment and care.

For instance, the "dementia initiative" that spans from 2016 to 2020 aimed to integrate state-of-the art basic neuroscience and clinical research to elucidate the mechanisms by which Alzheimer's disease originates and develops new antibody therapies (AMED 2019). On a more applied level, the dementia technology development group is focusing on the development of new technologies for brain science such as new neural network manipulation technologies, physiological measurement technologies and new behaviour analysis methods.

In June 2019, a Framework for Promoting Dementia Care has been adopted by the Ministerial Council on the Promotion of Policies for Dementia Care at the Prime Minister's Office. In order to support the independence of people with dementia and reduce the burden on caregivers, the framework recommended AMED to investigate the develop devices that utilise Japan's cutting-edge robot technology, sensors and ICT technology, with a strong emphasis on drawing upon participatory design principles, and with users of the technology being seen as active partners in the project. Indeed, the development of practical equipment suitable for the needs of nursing care sites requires the demands of the nursing care facilities to be reflected from the early stages of development through design, testing and prototyping, as a user's intention to adopt a particular technology is often determined by its perceived ease of use and perceived usefulness (Davis 1989). To take a first move into the direction of this recommendation, a call for proposal for grants on "Research and Development on AI and IoT Systems for Dementia" was launched in January 2020.

Realisation of smart society through AI technology application by NEDO

The New Energy and Industrial Technology Development Organization (NEDO) is a funding agency under the jurisdiction of METI that aims to promote industrial technology. By funding R&D projects to explore advanced new technology as well as practical application and promoting standardisation of technology and services, NEDO acts as an "innovation accelerator".

17 Documents and information were retrieved from the funding agencies' websites; the documents were in Japanese.
18 1 euro = 115 Yen | 21 April 2020

NEDO's overall budget for FY 2020 is around ¥159 billion (€1.38 billion[19]), while it has allocated a budget of¥45 billion (€39.13 millions) for R&D in the field of industrial technology.

To align to the concept of Society 5.0, NEDO has developed a "Strategy for Realization of Smart Society through AI Technology Application" in FY 2017. Three domains – (1) productivity; (2) health and medical and nursing care; (3) mobility in space – have been identified as the primary focus areas, and NEDO proposes to establish a five-year R&D program based on public and private partnership. The program, Realization of Smart Society through AI Technology Application, was launched in 2018, and in the domain of health, medical and nursing care, "AI-powered ambient sensor"; "daily life function of the elderly"; "life phenomenon modelling"; "health promoting behaviours"; and "IoT and AI-assisted health and nursing care service system" are the areas to be explored initially.

NEDO's report "Realization of Smart Society through AI Technology Application – R&D Projects" (2020) gives us an overview of ongoing projects, with their key technological challenges and expected outcomes. Among them, four projects are related to nursing care:

1 Designing and testing data-driven nursing care assisted by robotic devices and promotion of its implementation.
2 Prototyping a program of physical exercises leading to a self-sustaining motivation.
3 Designing and testing in the field of an effective care service provision system using data from wearable sensors and assistive robots and capturing the mental and psychological state of the person in care.
4 Assisting and supporting caregivers with the use of digital tools and data analytics.

What is common to all of these projects is that they not only explore data resources and data infrastructure, but also focus on the motivation of elderly people and working in close collaboration with nursing care service providers. It is worth noting the presence of large private companies from outside the care business, such as Panasonic, Takenaka and Seiko Instruments.

Improving quality of life by JST

The Japan Science and Technology Agency (JST), under the jurisdiction of MEXT, whose aim is to promote basic research and development activities linked to innovation and commercialisation activities, has been actively engaged to address the challenges of an ageing society by creating new programs through two channels: Creation of Science, Technology and Systems that Enriches the Ageing Society and Redesigning Communities for Aged Society, which was proposed and is supervised by the Research Institute of Science and Technology for Society (RISTEX)" – JST's internal body, established in 2001 – in order to promote R&D to address Science, Technology and Society (STS) issues. The first program was initiated in 2010 for ten years, the main objective being to help the elderly stay active professionally and in their everyday lives, through mobilising the power of STI with a particular eye on information and communication technology (ICT) and information-robotics technology (IRT).

More concretely, this program focused on the sensory, brain and physical functions of the elderly and proposed to work on the circulation of information connecting these functions. Four projects have been selected with a budget of approximately ¥70 million (€610,000) per year for each project:

19 1 euro = 115 Yen | 21 April 2020.

1 Development of robot support systems for the elderly with memory and cognitive decline.
2 Intelligent car-driving system to support the autonomy of the elderly.
3 R&D to create an ICT platform operating as a depository of experiences and knowledge of the elderly population.
4 R&D to create the technology for assisting the elderly's social participation and establishment of an assessment system.

The program was completed in FY 2019, and the ex-post evaluation is planned during the course of FY 2020. However, the mid-term evaluation report, which was published in 2017,[20] gives us a taste of what has been the challenge of this program. The activity on basic research contributing to a better understanding of the phenomena and development of component technologies in assistive devices have been positively assessed, although the report noted that to move ahead from this stage to the preparation of prototypes and their implementation, the needs on the ground – such as feedback from the elderly, regulatory issues and social acceptance – should be carefully examined.

The second program, Redesigning Communities for Aged Society (2010–2015), reflects the guiding principles of RISTEX, which place value on interdisciplinarity, the development of methodologies to implement a multi-stakeholder approach and "interactive practice in in the field of humanities and social sciences, as well as the natural sciences". It therefore aims to realise "community-based co-design/co-production of science & technology for ageing community development". This multi-stakeholder and community-based program contrasts with the more technology-driven approach of the program Creation of Science, Technology and Systems that Enriches the Ageing Society.

More concretely, a call for proposals was carried out with the aim of developing a methodological framework to address societal issues of ageing while at the same time encouraging the prototyping of solutions. A total of 15 projects were selected. Given that these projects place more emphasis on community-based actions, the focus on the development of digital tools is less apparent compared to the first program. However, the project "Ageing in place with ICT" constitutes an interesting case, since it bridges these two approaches. In fact, its strategy consists of creating a monitoring network system called the "HOW ARE YOU network". This was developed by the Iwate Prefectural University and supported by an Iwate prefecture's organisation; it links elderly citizens with monitoring centres managed by local social welfare councils and social welfare workers by using a telephone system equipped with an "I want to talk" button operational at any time, and by applying existing digital technologies and digital infrastructure.

The program was completed by March 2016. In order to capitalise on the outcomes of this program, a general incorporated association, "Co-Creation Center for Active Ageing", was established in April 2017, offering a training program for community-based, problem-solving skill development and a support program for the creation of living labs.

Conclusion

In this chapter, we described in detail the policy initiatives led by the government in Japan. As such, we are placing ourselves in the footsteps of Johnson's concept of a "developmental state" – an interventionist state that combines private ownership with state guidance – to characterise the role played by the Japanese state in the unexpected post-war high economic growth

20 The interim evaluation covers the first three projects, the last one having been completed in FY 2013.

(Johnson 1982, 1999; Boyer and Yamada 2000). We face a political system where the state bureaucracy is given a sufficient timeframe and scope to take initiative and operate effectively, while at the same time there are multiple and continuous interactions between the bureaucracy, universities and companies through various informal and formal forums, associations and platforms for exchanging ideas and propositions. For instance, Keidanren (Japan Business Federation), the largest representative of the business community in Japan, has been in constant discussion with the CSTI in developing the concept of a Society 5.0, with the aim of promoting an innovation ecosystem that contributes to solving social problems while creating business opportunities (Holroyd 2020). In that respect, the Data Health Reforms engaged by the government to allow the collection and utilisation of anonymously processed medical information have the potential to advance medical research, innovative drug discovery and new business creation. Already, 260 million results of the specific health check-ups for the population aged 40 to 75 have accumulated in the National Database of Health Insurance Claims and Specific Health Checkups and could be used to study lifestyle diseases (Kumar 2019). Japanese data sets, with their specific demographic trend, could be valuable information and increasingly in high demand as other countries follow the ageing trend.

As already mentioned, three ministries have been at the forefront of the policy momentum, reforming the legislative structure to adapt the ageing society, the Ministry of Health, Labour and Welfare (MHLW); the Ministry of Economy, Trade and Industry (METI); and the Ministry of Education, Culture, Sports, Science and Technology (MEXT). The social and healthcare dimensions have been combined with scientific and economic considerations from the mid-1990s, with a guiding vision from the Cabinet Office. In our view, the Japanese policy experience we have described in this chapter can be summarised from three perspectives: policy convergence; paradigm shift in health system; and digital transformation.

Firstly, we have seen in the previous sections that the framework to respond to an ageing society and to reform the STI policies was greatly overhauled in the 1990s, with the enactment of the Basic Law on Measures for the Ageing Society, on one hand, and Science and Technology Basic Law, on the other. Facing major challenges in the 1990s, the Japanese government had to adjust its policy responses. What is interesting to note here is that, 25 years later, although initially these two Basic Laws were founded independently, we now observe the convergence of these two policy frameworks, the driving force being the recognition of the central role played by STI policy, one of the principal components of Society 5.0, as a horizontal policy tool across ministries.

Secondly, as we mentioned, "The Japan Vision, Health Care 2035" drafted by the MHLW suggests a paradigm shift for the health system, redirecting its focus from "system inputs to patient value", from the quantity of services provided to patients to quality, from cure to care and from specialisation of services to integrated approaches across medical and social service sectors. Here again, the concept of Society 5.0 plays the role of a unifier by putting people at the heart of societal transformation.

Thirdly, it could be argued that Japan is in a strong position for mobilising innovation in health care and for putting in place measures toward data-driven health care. Indeed, based on the successful experiences in the country's post-WWII technology-led industrial development, Japan sees the potential of mobilising and deploying technological capacity, in particular digital technologies in the health care sector. The Guideline of Measures for Ageing Society revised in 2018, as well as the Digital Health Reform revisited in 2019, reflect this approach, again under the one roof of Society 5.0. It is worth noting that Keidanren published in April 2019 its policy recommendations on the social security system, titled "For Securing a Social Security System With a View to 2040". It encourages the promotion of the "digital transformation" by investing in technologies such as robotics, sensing devices, ICT and AI and mobilising the "creativity of people in its diversity" to address this challenge, with the goal of making Society 5.0 a reality (Keidanren 2019).

What we observe here is that the concept of Society 5.0 frames these three perspectives, leading and in some cases supporting changes in policy making with regard to this multi-dimensional challenge of an ageing population. The policy implementation is underway, and it is too early to assess whether these measures are inducing societal change, as advocated by Society 5.0. Even though we reported in the last section of this chapter on the emergence of community-based actions led by the different funding agencies that accompany the top-down types of policy measures, we still need to see whether the concept of Society 5.0 will gather enough momentum to translate from policy elaboration to societal implementation.

In that respect, the implementation of Intelligent Transport Systems (ITS) will be interesting to follow; mobility services constitute a fundamental element of social life and therefore impact the quality of life of the elderly and in particular those living in rural or suburban areas. Indeed, the Guideline of Measures for Ageing Society 2018, as we already mentioned in this chapter, recommends advancing R&D in ITS, which includes Traffic Signal Prediction Systems (TSPS); Driving Safety Support Systems (DSSS); and Electronic Toll Collection System (ETC) 2.0; it also recommends R&D in advanced automated driving systems, with a particular eye to unmanned automated driving transport services for the elderly. It is worth noting that all these systems are heavily reliant on the use of data, data analytics and AI technologies. A successful implementation will need to combine legal reforms (Imai 2019), upgrading digital and physical infrastructure (Mahmassani 2016), access and sharing of data and investments in STI. Countries are developing ITS with the goal of attaining greater mobility for the elderly and handicapped persons, improving safety and maintaining the competitiveness of their transport industry (West 2016). Japan is in a leading position in developing the technology as it has a powerful industrial base, government and legislative support and a strong basis for experimentation with its ageing population. If the technology is successfully developed and socially accepted, Japan could use its comparative advantage to export it.

To conclude this chapter, we raise the following issues to be addressed further in view of better exploring the Japanese experiences:

1 We have observed that Japan is testing new approaches in the field, in particular using digital technologies and data analytics to address ageing issues and cumulating experiences. However, even in the case where expected results are obtained, the problems of scaling up, gaining social acceptance and ensuring the sustainability of the new schemes remain; these need to be addressed for further implementation.
2 The existence of an overarching policy framework such as Society 5.0 was necessary to ensure a policy coherence when confronted with multi-dimensional and multi-stakeholder issues, particularly in our case of an ageing population. The remaining question is whether this overarching policy framework facilitates experience-sharing and cross-learning among stakeholders.
3 In Japan, community-based actions are advocated alongside top-down policy making, such as Guidelines, Strategies, and Plans, expecting that both approaches complement each other. However, given the diversity and multiplicity of communities, there is no one-size-fits-all solution. The articulation should be designed carefully to take advantage of this potential complementarity, and this piece is still missing in this complex policy system in place in Japan.

Finally, we would like to note that by addressing the societal challenges of an ageing population, we are all invited to reconsider the sense of "community" to leave a better future for the generations to come.

References

AMED (2019): Project for Psychiatric and Neurological Disorders. Japan Agency for Medical Research and Development. Tokyo, Japan. Available at: www.amed.go.jp/content/000060744.pdf [01.04.2020]

Boyer, R. and Yamada, T. (eds.) (2000): *Japanese Capitalism in Crisis: A Regulationist Interpretation*. London: Routledge.

Bureau (2019): The Statistical Handbook of Japan 2019. Statistics Bureau of Japan, Tokyo. Available at: www.stat.go.jp/english/data/handbook/ [01.04.2020]

Cabinet Office (2016): G7 Science and Technology Ministers' Meeting. May 15–17, 2016. Tsukuba, Ibaraki. Cabinet Office, Tokyo, Japan. Available at: https://www8.cao.go.jp/cstp/english/others/20160517communique.pdf [01.04.2020]

Cabinet Office (2018): The Guideline of Measures for Ageing Society. Cabinet office, Tokyo, Japan. Available at: https://www8.cao.go.jp/kourei/measure/taikou/pdf/p_honbun_h29e.pdf [01.04.2020]

Cabinet Office (2019): Annual Report on the Ageing Society. Cabinet Office, Tokyo, Japan. Available at: https://www8.cao.go.jp/kourei/english/annualreport/2019/pdf/2019.pdf [01.04.2020]

Carraz, R. and Harayama, Y. (2018): Japan's Innovation Systems at the Crossroads: Society 5.0, *Panorama: Insights into Asian and European Affairs*, 33–45.

Coulmas, F. (2007): *Population Decline and Ageing in Japan-the Social Consequences*. London, New York: Routledge.

Coulmas, F., Conrad, H., Schad-Seifert, A. and Vogt, G. (eds.) (2008): *The Demographic Challenge: A Handbook about Japan*. Leiden, Boston: Brill.

Davis, F. (1989): Perceived Usefulness, Perceived Ease of Use, and User Acceptance of Information Technology, *MIS Quarterly*, 319–340.

Goto, A. (2000): Japan's National Innovation System: Current Status and Problems, *Oxford Review of Economic Policy* 16(2), 103–113.

Heinrich, P. and Galan, C. (eds.) (2018): *Being Young in Super-ageing Japan: Formative Events and Cultural Reactions*. London, New York: Routledge.

Holroyd, C. (2020): Technological Innovation and Building a "Super Smart" Society: Japan's Vision of Society 5.0, *Journal of Asian Public Policy*, 1–14.

Imai, T. (2019): Legal Regulation of Autonomous Driving Technology: Current Conditions and Issues in Japan, *IATSS research* 43(4), 263–267.

Ishiguro, N. (2017): Care Robots in Japanese Elderly Care. Cultural Values in Focus, in Christensen, K. and Pilling, D. (eds.): *The Routledge Handbook of Social Care Work around the World*. London, New York: Routledge, 256–269.

Johnson, C. (1982): *MITI and the Japanese Miracle: The Growth of Industrial Policy, 1925–1975*. Stanford: Stanford University Press.

Johnson, C. (1999): The developmental state: Odyssey of a Concept, *The developmental state* 12, 32–60.

Kagermann, H., Wahlster, W. and Helbig, J. (2013): *Recommendations for Implementing the Strategic Initiative Industrie 4.0 – Final Report of the Industrie 4.0 Working Group, Communication Promoters Group of the Industry-Science Research*. Frankfurt a.M.: acatech.

Keidanren (2019): For securing Social Security System with a View to 2040. Keidanren, Tokyo, Japan. Available at: www.keidanren.or.jp/policy/2019/033.html [01.04.2020]

Kumar, S. (2019): Capitalising on Japan's Digital Healthcare Economy during this Era of Ageing Societies. *White Paper*. Healthcare Information and Management Systems Society, Inc. (HIMSS)

Mahmassani, H. (2016): 50th Anniversary Invited Article – Autonomous Vehicles and Connected Vehicle Systems: Flow and Operations Considerations, *Transport. Science* 50(4), 1140–1162.

MHLW (2015): The Japan Vision: Health Care 2035. Ministry of Health, Labor and Welfare. Available at: www.mhlw.go.jp/seisakunitsuite/bunya/hokabunya/shakaihoshou/hokeniryou2035/assets/file/healthcare2035_proposal_150703_summary_en.pdf [01.04.2020]

MHLW (2017): OECD Policy Forum on the Future of Health, Speech by the Minister of Health, Labour and Welfare, Yasuhisa Shiozaki. Ministry of Health, Labor and Welfare. Available at: www.mhlw.go.jp/file/06-Seisakujouhou-10500000-Daijinkanboukokusaika/0000137631_1.pdf [01.04.2020]

MHLW (2021): Data-based Health Management Initiative Roadmap. Available at: https://www.mhlw.go.jp/english/policy/health-medical/data-based-health/dl/211124-01.pdf [28.01.2022]

Mokyr, J. (2002): *The Gifts of Athena: Historical Origins of the Knowledge Economy*. Princeton: Princeton University Press.

Nielsen, M. (2020): *Reinventing Discovery: The New Era of Networked Science (Vol. 70)*. Princeton: Princeton University Press.

OECD (2010): *The OECD Innovation Strategy: Getting a Head Start on Tomorrow*. Paris: OECD.

Ohara, T., Hata, J., Yoshida, D., Mukai, N., Nagata, M., Iwaki, T. and Ninomiya, T. (2017): Trends in Dementia Prevalence, Incidence, and Survival Rate in a Japanese Community, *Neurology* 88(20), 1925–1932.

Prince, M., Comas-Herrera, A., Knapp, M., Guerchet, M. and Karagiannidou, M. (2016): World Alzheimer Report 2016: Improving Healthcare for People Living with Dementia: Coverage, Quality and Costs Now and in the Future. LSE Research Online. Available at: http://eprints.lse.ac.uk/67858/ [15.01.2021]

Rifkin, J. (2011): *The Third Industrial Revolution. How Lateral Power is Transforming Energy, the Economy and the World*. New York: Palgrave MacMillan.

Schwab, K. (2017): *The Fourth Industrial Revolution*. New York: Penguin.

Tamiya, N., Noguchi, H., Nishi, A., Reich, M., Ikegami, N., Hashimoto, H. and Campbell, J. (2011): Population Ageing and Wellbeing: Lessons from Japan's Long-term Care Insurance Policy, *The Lancet* 378(9797), 1183–1192.

Tovey, M. (ed.) (2008): *Collective Intelligence: Creating a Prosperous World at Peace*. Oakton, VA: Earth Intelligence Network.

UN (2020): World Population Ageing 2019 (ST/ESA/SER.A/444). United Nations Department of Economic and Social Affairs, Population Division.

Vogt, G. (2008): Talking Politics: Demographic Variables and Policy Measures in Japan, in Kohlbacher., K. and Herstatt, C. (eds.): *The Silver Market Phenomenon*. Berlin, Heidelberg: Springer, 17–29

Von Hippel, E. (2005): *Democratizing Innovation*. Cambridge, MA: MIT Press.

West, D. (2016): *Moving Forward: Self-driving Vehicles in China, Europe, Japan, Korea, and the United States. Center for Technology Innovation at Brookings*. Report. Washington, DC: Brookings Institution.

32

DIGITALISATION AND DEVELOPMENT IN INDIA

An overview

Syed Mohib Ali Ahmed

Introduction

Digitalisation is expected to have a profound influence on our economy and society by transforming the nature of economic activity and everyday life. The effects of digitalisation are visible across different areas, such as governance, social relations, economy and communications, among other spheres, to the point where we now live in a 'knowledge society'. The digitalisation of the economy holds tremendous potential in transforming the economic structure by improving productivity and creating value through new products and markets. The present chapter gives an overview of the ongoing process of digitalisation in India and evaluates its potential in achieving economic development.

The information and communications technology (ICT) revolution was set off by innovations in microprocessor technology in the 1970s that led to the widespread diffusion of computers in households and businesses. Digitalisation as the process of converting information into bits represents a continuation of that trend but is also a radical breakthrough because of the exponential growth in computing power and the availability of huge amounts of data, with application to a wide variety of activities. Thus, the emerging digital revolution is in its synergetic phase, where interrelated innovations in technologies, markets and business models are creating significant feedback effects among each other. A wide variety of interrelated technological breakthroughs, such as advanced robotics, artificial intelligence (AI), 3D printing, blockchain technology, data analytics etc. have digitalisation at their core and feed off each other. Thus, advances in AI and machine learning are contributing to the growth of data generation and vice versa. The surge in technical innovations is also facilitated by the gradual changes in business models and practices. Most multinational corporations today coordinate their operations digitally and have a global production network that is flexible and supported by just-in-time supply chain management, unlike the mass production systems characteristic of the Fordist era. Such changes in business practices are both a cause and consequence of the widespread diffusion of ICT technology. Similarly, the recent rise of 'platform capitalism', where customers, businesses and suppliers interact over a digital platform, has enabled rapid digitalisation of markets. These intermediating platforms are characterised by indirect network effects, where more end-users create positive externalities for new users to join on the other side of the market. Again, these indirect network effects are based on the availability of huge data and the ability of the platform

DOI: 10.4324/9780429351921-39

provider to collect and monetise the data, thus incentivising new users to join the platform. More significantly, with the advent of Internet of Things (IoT), digitalisation is expected to revolutionise manufacturing and transform production, heralding 'connected industry' and 'smart enterprises' (Schwab 2016). The possibility of communication between machines through the internet in an interconnected manufacturing network blurs the distinction between virtual and physical systems and allows for physical processes like production to be monitored digitally and become self-correcting. It is due to the specific nature of the technology – such as zero marginal costs, low costs of entry due to minimal physical investment, large network effects, increasing returns to scale and strong synergies with other innovations – that digitalisation is occurring at a rapid pace. The rapid pace and the (potential) pervasiveness of digitalisation in its interrelationship with other technologies, and the scope for further improvements, thus lends itself to be characterised as a general-purpose technology (Brynjolfsson and McAfee 2014; Schwab 2016). The ongoing Covid-19 pandemic promises to further accelerate digital adoption, with most businesses managing their operations digitally and core services such as education and healthcare now being offered digitally.

India, as an emerging market economy, has a large user base of nearly half a billion people connected to the internet and has significant potential for 'catching-up' in the process of digitalisation. The Indian economy has grown steadily in the last decade at an average of 7% compound annual growth rate (CAGR), one of the biggest exporters of ICT services. An ever-increasing number of goods and services are going digital to create value and harness the potentialities of digitalisation in India. The rapid rise of technology start-ups and digital platforms have become predominant in services, especially e-commerce, communications, healthcare, service-sharing and electronic payments among other activities. Despite this rapid growth, there are still 800 million people in India who are not connected to the internet, and as of 2011–12, there were nearly 270 million people below the poverty line with 80% of the workforce employed in the low productivity, unorganised sector (Dutta 2018). In order to bridge the digital and real divide, the state launched the 'Digital India' initiative in 2016, which seeks to hasten the transition towards digital transformation by providing the necessary infrastructure and incentives for digital adoption. In doing so, digitalisation is expected to address the twin challenges of promoting economic growth and social inclusion by increasing productivity and strengthening the welfare system.

The aims of this chapter are twofold. First, it describes the ongoing process of digitalisation with its accompanying technical innovations, their applications and impact on the economy. Second, the article critically evaluates the potential of digital technologies for economic development and the role of the state in this regard. More explicitly, the article queries the reasoning underlying state policy in its push for digitalisation as a driver of economic growth by *potentially* reducing transaction costs, rationalising government expenditure and promoting financial inclusion. Not only is such a rationale tenuous, the digitalisation of services portends to continue the regime of jobless growth. It should be noted here that there are important caveats to the perspective adopted in this chapter. The focus on development as structural transformation does not fully capture the broader changes brought about by digitalisation through the improvements in quality, availability of new products, services, experiences and creation of new value. Moreover, the lack of appropriate data on digitalisation at the industry level coupled with the informality of the Indian economy restricts the analysis of diffusion by producers, and instead warrants a descriptive account of adoption by consumers.

The chapter is structured as follows. First, the evolving digital landscape in India as a continuation of the ICT revolution and a constituent part of its national innovation system is described. It also describes the efforts taken by the state to bridge the digital divide through its umbrella program of Digital India. Afterwards, the impact of ICT technology on the economy in terms of its effect on output, employment and productivity is elucidated. With regard to productivity, the section focuses on specific examples of digitalisation in manufacturing and services to underscore the lack of diffusion. The following section focuses on the rise of digital payments that have become ubiquitous in recent years and are transforming the nature of governance and welfare delivery. The expansion of digital payments is expected to promote financial inclusion and foster economic growth, a claim that is queried. Finally, the role of the state in directing technical change towards economic development is critically evaluated and the need to build complementary assets and absorptive capacities in the economy so as to allow digital technologies to achieve their full potential is emphasised.

Digital India: a national innovation system

The digital landscape in India presents a mixed picture, with few sectors experiencing accelerated digitalisation and other sectors completely unaffected. The spread of digital technologies is accelerated by the intervention of the state, which seeks to transform India into a digital knowledge economy. Such a vision is consistent with the objective of India's national innovation system (NIS), 'linking contributions of science, research and innovation system with the inclusive economic growth agenda' (STIP 2013: 4). To this end, the state has initiated many programs, most notably Digital India, an umbrella program that covers various ministries, policies, schemes and stakeholders to promote digitalisation. The state has also actively fostered the development of financial-technology (fintech) and digital payments in the country apart from embarking on a project of e-governance. The following section presents the variegated landscape of digitalisation in India as a constituent part of its national innovation system by first giving an account of the prior ICT revolution and its outcomes, and then the recent efforts towards greater digitalisation that build on the existing infrastructure.

The ICT revolution

India's NIS is constituted by many institutions, elements and relationships which interact in the production, use and diffusion of knowledge. The NIS as it exists is influenced by India's experience with development planning and has historically been characterised by the need for self-reliance and a strong role for the state in technological upgradation. The various technological revolutions witnessed in the country have thus involved the visible hand of the state. ICT received considerable support by the state in the 1980s and achieved tremendous dynamism after the liberalisation reforms in 1991, which deregulated markets and allowed greater foreign investment. Particularly, it was the telecommunications sector that was prioritised by the state as it set up the Centre for Development of Telematics and the Telecommunications Commission in the late 1980s to install the necessary infrastructure for socio-economic development. Information technology (IT) also emerged in the late 1980s with big multinational corporations such as Microsoft, Motorola and Hewlett Packard setting store in India and utilising the available pool of engineers from national technical institutes. To promote IT sector growth, the state set up the Software Technology Park of India as an autonomous body which provides land for technology parks, incubation centres, tax exemptions, marketing support, statutory and consultancy services. As a result, the ICT sector has emerged as one of the most dynamic sectors in the economy, with IT contributing significantly to exports and GDP.

Telecommunications have become an important aspect of the ICT revolution and have made deep inroads in India, with almost 1,181 million mobile phone connections and a tele-density of about 90% in 2019, up from 18% in 2007. The policy of liberalising the telecom sector and auctioning spectrum drastically lowered tariffs, to about 2$ a month, to the point where India has one of the most affordable telecom services[1] (World Bank 2016: 218). On the other hand, internet connections in the country are still low compared to tele-density and the total number of connections (broadband and narrowband) are 636 million, of which wireless mobile subscribers are 615 million and fixed internet connections only 21 million (TRAI 2020).

A significant element of the ICT revolution is the global competitive advantage in the export of ICT services, especially software, afforded by the availability of cheap, skilled software professionals and the growing global demand by businesses for software services. A relatively strong system of higher technical education enabled the software boom by providing the necessary skills. Software exports have grown steadily to a value of 73$ billion and contributed 45% of all services exports in 2016–17 (Economic Survey 2017–2018: 157). Apart from production in the ICT *sector*, the service sector too has been transformed by the *use* of ICT technology. These services include business process outsourcing and management (BPO, BPM) in finance, banking, insurance, product and content development, maintenance of information bases, information services, medical transcription and a host of other activities. The spread of ICT in India signals systemic change where the gains made by IT-enabled services (ITes) were facilitated by the establishment of communications infrastructure, affordable connectivity, technical education that produced skilled IT workers and greater FDI flows in telecom. The socio-economic consequences of the ICT revolution are far-reaching and have transformed the urban landscape in India, creating new demand for technical education and driving urban consumption growth through higher incomes.

Digital India

By building on the gains made through ICT, the state has embarked on a strategy of further digitalisation as a means for economic and social transformation. In this direction, the state under the charge of Ministry of Electronics and Information Technology (MeitY) has initiated important programs such as the National E-Governance plan (NeGP) in 2006 and the Digital India program in 2015, which aim to provide digital infrastructure and e-governance to the last mile and 'transform India into a digitally empowered society and knowledge economy'. The roll-out of *Aadhaar*, a unique biometric identity number, to nearly 1.2 billion people has transformed e-governance and the welfare system. As part of the NeGP, the government undertook core infrastructural projects such as the State Wide Area Network (SWAN), which establishes a secure network for state-level government administrations; State Data Centre (SDC), which provides a repository for data storage; and Common Service Centres (CSCs), which serve as multipurpose service delivery centres for citizens across the country.[2] Another ambitious scheme is the National Optical Fibre Network (NOFN), which aims to connect 250,000 Gram Panchayats or local villages across the country with an outlay of 200 billion rupees or 2.9$ billion.

Apart from investment in physical infrastructure, there is also an effort by the government to develop institutional capacities and build linkages with the private sector and research institutions. The majority of the R&D expenditure in India, about 42%, is done by public enterprises to develop capacities and innovate (EAC-R&D 2019). For instance, Bharat Heavy Electronics

1 The Department of Industrial Policy and Promotion (DIPP) reports that the telecom sector received substantial foreign equity inflow, to the tune of 6.1$ billion in 2017–2018, eager to capitalize on the growing market.

2 The outlay for SWAN is 3.3 billion rupees (48$ million) over five years, 17 billion rupees (250$ million) for SDC, and for CSC, 57 billion rupees (0.8$ billion) over four years, including private funding (Dutta 2018).

Limited has established a Centre for Excellence for Intelligent Machines and Robotics that researches the areas of computer integrated manufacturing, radio frequency identification technology and 3D printing, among other things. In addition, the government is promoting research in recent technologies by funding incubation centres at universities and research institutions. Thus, the various organs of the state, like the National Institution for Transforming India (NITI), which piloted a National Strategy for Artificial Intelligence; the National Productivity Council, which established a centre for excellence on IT for Industry 4.0; and several public universities, which have set up incubation centres, are involved in developing an ecosystem for establishing smart technologies in India.[3] These are crucial components in building a robust national innovation system that tries to develop technical and organisational capacities through consultations with various stakeholders and establishes strong user–producer linkages (Lundvall 1988).

The private sector, too, has shown tremendous dynamism by creating a vibrant start-up ecosystem, facilitated by the state. India has become the second largest start-up ecosystem by utilising its ICT capacities with nearly 20,000 start-ups, mostly in the areas of cloud computing, data analytics, machine learning, AI and fintech. Corporate-run incubators and accelerators try to catalyse new start-ups by creating products with a strong technology focus that can serve consumers. For instance, the Confederation of Indian Industry (CII) in partnership with private firms like Tata Communications and Dell Technologies have set up the Centre for Digital Transformation, which incubates companies to adopt digital practices and offers consultancy, training and advisory services. The growth of start-ups is also greatly facilitated by the rise of the e-commerce market in India, which is estimated to be worth 33$ billion and is growing rapidly at 19.1% per annum in 2016–17 (Economic Survey 2017–2018: 160). This is on the back of increased internet connectivity as a growing number of goods and services are delivered online.[4]

Digital divide

The Digital India program was initiated to digitally empower India and bridge the digital divide in the country. The extent of digital divide becomes clear when it is recognised that there are nearly 800 million people or 81% of the population in India who are offline (World Bank 2016: 8). While the total number of internet connections (broadband and narrowband) is high at 636 million and is mostly connected through mobile phones, the demand for internet, i.e. individuals who have access and use it, is quite low, with the percentage of individuals using the internet only 34% (ITU 2019). The picture becomes even starker when it is seen that the total number of fixed broadband subscribers in India is only 21 million or an average of 1.4 people per 100 in the population (ITU 2019).

The digital divide is quite deep between rural and urban India. While the average tele-density across the country is high at 90%, the disparity between rural and urban tele-density is substantial, with 57% in the former and 160% in the latter. The government estimates that there are still nearly 55,600 villages in India that have no mobile coverage yet.[5] Similar to the total 636 million internet connections, urban areas show a high concentration of internet

3 National Strategy for AI (2018), https://niti.gov.in/writereaddata/files/document_publication/NationalStrategy-for-AI-Discussion-Paper.pdf, Annual Productivity Report (2018), www.npcindia.gov.in/NPC/Files/Publication/Annual-Productivity-Report-2017-18.pdf.

4 E-commerce has existed since the Information Technology Act in 2000 provided legal recognition to electronic transaction and e-governance. However, it has grown significantly in the last few years.

5 www.digitalindia.gov.in/content/universal-access-mobile-connectivity.

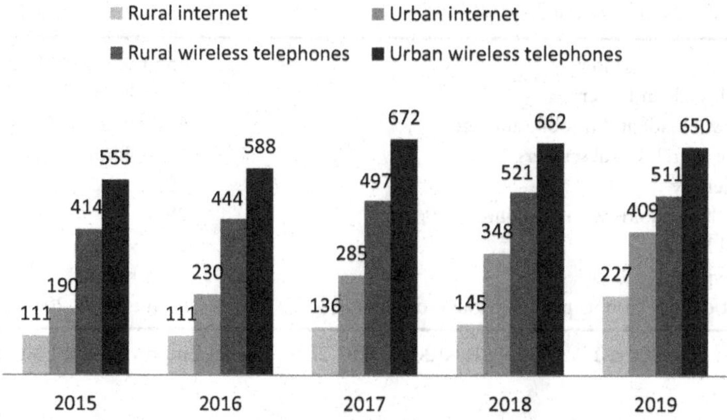

Figure 32.1 Urban and rural, mobile and internet subscriptions in millions

Source: TRAI

connections with 409 million or 64% of the total. Figure 32.1 shows the disparity between urban and rural areas with respect to wireless telephones and internet subscriptions. Digital literacy, defined as the ability to receive, process and send information via digital means, is quite low, with the Digital Empowerment Foundation estimating that 90% of Indians lack basic digital literacy.[6] To address this problem, the government launched the National Digital Literacy Mission in 2016 with an aim of providing digital literacy to at least 60 million rural households or 40% of all households by 2020.

As regards the digital infrastructure, the NOFN scheme has managed to provide working commercial broadband to only 2.5% of the 250,000 intended villages.[7] Similarly, SWAN, which provides a minimum bandwidth of 2 mbps at the lowest administrative level and a maximum of 600 mbps at the state headquarter level, is using only about 60% of the available bandwidth. This clearly points to a demand issue where the uptake and usage of internet is low due to various reasons, such as lack of computers, adequate servers and electricity.[8] Thus, despite the recent push towards digitalisation, most of the economy remains unaffected by the ICT revolution. India ranks 91 out of 139 countries on the World Economic Forum's (WEF) Networked Readiness Index with significant gaps in its digital infrastructure and low individual usage. Table 32.1 summarises the status of digitalisation in India.

India's NIS, particularly its ICT capabilities, are rooted in the institutional and production structure of the economy. The institutional structure is characterised by the predominance of the state in catalysing innovation. The ICT sector received a major boost after liberalisation reforms in 1991 which made newer technologies and foreign capital available and imparted dynamism to the service sector. As a result, the last two decades have seen the creation of an ICT technological base from scratch, and the present thrust towards digitalisation promises to build on this momentum. Despite the initial rapid growth, India's ICT capabilities still fall short

6 www.financialexpress.com/education-2/a-look-at-indias-deep-digital-literacy-divide-and-why-it-needs-to-be-bridged/1323822/.

7 As opposed to test connections, working connections are functional and are billed indicating that cables have been installed but are not yet functional https://thewire.in/government/narendra-modi-government-digital-india-village-broadband-connections.

8 https://scroll.in/article/808622/for-800-million-citizens-modis-digital-india-highway-is-a-bridge-to-nowhere.

Table 32.1 Status of digitalisation

Individuals using internet	455 million (ITU 2019)
Fixed broadband internet	21 million
Wireless broadband (mobile internet)	615 million (TRAI 2020)
Mobile (wireless) subscribers	1.18 billion
Tele-density	90%
Urban households with computers	20%
Digital literacy	10%
ICT exports	111$ billion
Robot density (robots per ten thousand workers)	3 units (IFR 2017)

Sources: ITU 2019; TRAI 2020; IFR 2017; ESC 2018; Digital Empowerment Foundation

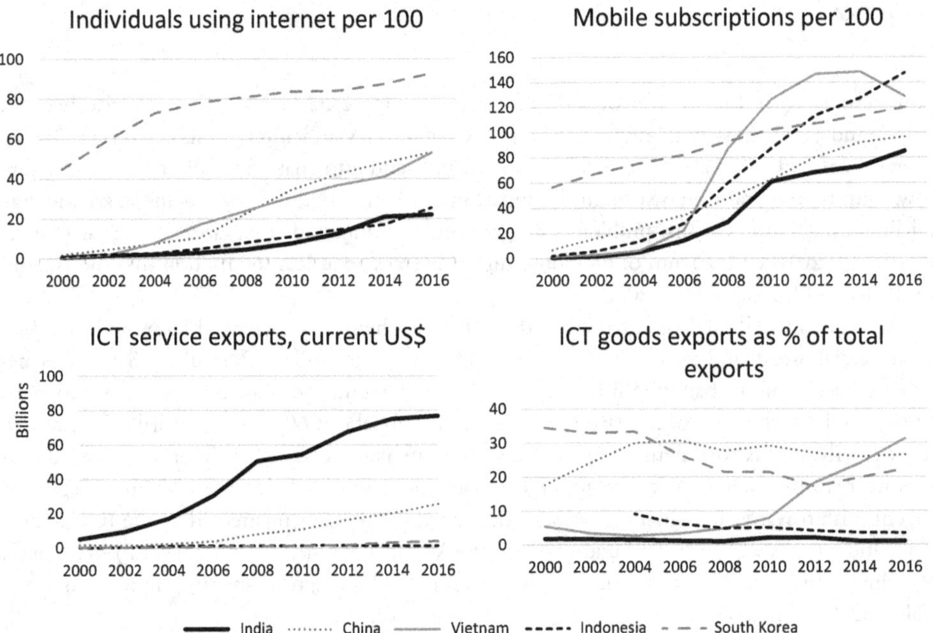

Figure 32.2 Select ICT indicators for India and other Asian economies

Source: World Bank

compared to some of the other Asian economies. Figure 32.2 shows a comparison with four Asian economies across four select indicators. As is evident, the only indicator on which India has performed consistently well over the last two decades is the export of ICT services. Internet usage by the population remains low and the abysmally low share of ICT goods exports, i.e. hardware, telecom etc., reflects the lopsided nature of technological development being biased heavily towards services. The uneven diffusion of ICT and its varied impact on the different sectors of the economy will be considered in the next section.

Digitalisation and the economy

To understand the impact of digitalisation on the economy, it is instructive to distinguish between the direct and indirect effects of ICT.[9] The *direct* effect refers to the contribution of the ICT *sector* to the economy in terms of output and employment. The production and growth of ICT goods and services resulting from product innovation and diversification contribute to the economy directly. The *indirect* effect refers to the *productivity effect* of ICT on the non-ICT sectors of the economy, whereby the production process is transformed by the *use* of ICT technology and leads to increases in productivity and growth via economies of scale, reducing transaction costs and increasing firm efficiency. Both of these effects, direct and indirect, are necessary for the realisation of potential benefits of digitalisation and economic transformation through the widespread diffusion of technology across sectors.

ICT, output and employment

In 2016–17, the total size of the ICT sector was 180$ billion, and it contributed to nearly 8% of the GDP. The ICT sector includes IT services such as software and BPO, as well as manufacturing like computer hardware, electronics and telecom equipment. Within the ICT sector, the largest share is of IT services i.e. software and BPO/ITes (nearly 75%) which contributed 13 5$ billion in 2016. This is followed by electronics and telecom equipment in manufacturing whose share within the ICT sector is 7% or 12.6$ billion each, respectively. The sectoral composition of the ICT sector is shown in Figure 32.3. The IT industry (software and ITes) is a major driver of economic growth, and the industry has grown tremendously in the last 15 years from a low base of 12$ billion in 2001 to 146$ billion in 2017, as shown in Table 32.2. Software services alone contribute to 5% of the GDP at 113$ billion. Almost two-thirds of its revenue is from exports, which benefits from a variety of reasons such as cost competitiveness and provision of end-to-end services. The IT industry employs nearly 4 million people directly and is estimated to provide another 8 million indirect jobs in IT-related occupations (NASSCOM 2017).

Despite the rapid growth in IT, the percentage share of employment in ICT-related occupations is less than 1% in India (World Bank 2016: 107). The growth rate of the IT industry has slowed down in the last decade (6.33% CAGR), showing signs of weakening global demand, automation and re-shoring of ICT and knowledge intensive activities. Automation is most likely to affect BPO services by substituting low-skilled manual jobs with automated bots. More worryingly, the analogue component of the IT industry, i.e. computer hardware, has not kept in line with the growth of software service exports and has been stagnant since 2014–15 at 3$ billion (ESC 2018).

In assessing the direct contribution of the IT sector to the economy, two trends stand out. The first is the export orientation of the sector and the second is the dominance of software exports. Both of these trends entail weak backward linkages with the rest of the economy where much of the gains achieved as part of the ICT revolution are channelised in exports without harnessing them for economic development (Joseph and Abraham 2005: 147; Chandrasekhar 2005). The next sub-section deals with the indirect effects or the use of ICT technology across the economy and its contribution to productivity.

9 It is assumed here that digital technologies are embedded in the ICT infrastructure and represent a continuation of innovations in ICT.

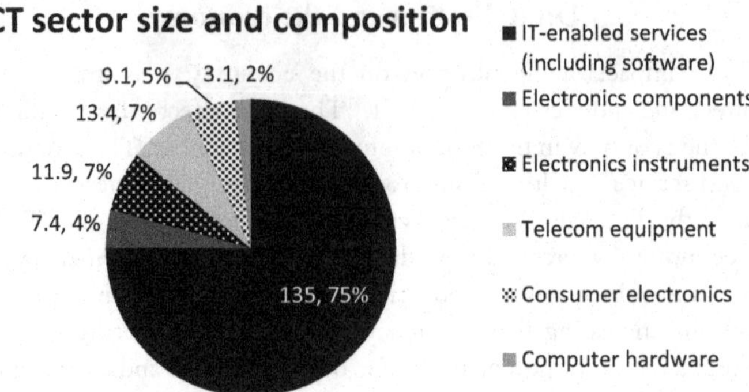

Figure 32.3 Size and share of the ICT sector in billion$ at current prices (2016–17)

Source: Electronics and Computer Software Export Promotion Council (ESC 2018)

ICT and economic productivity

There are several problems in analysing the impact of ICT and digital technologies on economy-wide productivity. The primary problem is that the Indian economy is not an industrialised mass production economy but is characterised by petty commodity production and informal economic arrangements. Consequently, any analysis of technical innovation and its diffusion and adoption across the different sectors and industries of the economy is severely limited. Given the structural heterogeneity of the economy and the lack of appropriate data on ICT adoption at the industry level, it is prudent to focus on specific cases of application of smart technologies drawing from reliable surveys and reports in light of previous studies that have tried to estimate the impact of ICT.

At the macro level, there is little evidence to suggest economy-wide diffusion of ICT or digital technology that is necessary for it to qualify as a general-purpose technology.[10] The tangible gains that were achieved as part of the ICT revolution were channelised in the export of ITes and not towards enhancing productivity across all sectors. In their study on the impact of ICT investment, Erumban and Das (2016) argue, using a supply-side growth accounting framework, that the indirect productivity effects i.e. productivity gains or spill-overs from the use of ICT, have been negligible over the years. They argue that while ICT investment is increasingly contributing to economic growth, it is largely limited to the services sector, especially ICT using services (ITes). ICT using services outperform both ICT producing sectors (electronics, computer hardware and telecom) and ICT using manufactures in their respective contributions to total factor productivity growth[11] (Erumban and Das 2016: 16). Their findings are consistent with the fact that the services, particularly IT-enabled services, have been the engine of growth in the last two decades, unlike the manufacturing sector, which has stagnated.

10 There are numerous micro studies estimating, for instance, the impact of mobile phones and internet on productivity gains to micro-entrepreneurs. The advantage of such connectivity is seen to lie in access to information, employment opportunities, credit and marketing facilities, among other things. See Jensen (2007) for a seminal study on the impact of mobile phones in a fish market in Kerala.

11 Using a similar framework, Jorgenson and Vu (2016) find that non-ICT capital accumulation has contributed more towards economic growth in India than either ICT capital or total factor productivity.

Digitalisation in manufacturing

The extent of digitalisation or use of ICT technology in the manufacturing sector is extremely limited despite some growth in automation and industrial robots from a low base which has raised the total stock of industrial robots to 16,000 in the country. However, the economy-wide robot density is still very low at 3 units per 10,000 workers, reflecting the very low levels of diffusion (IFR 2017). The state has undertaken several initiatives to promote Industry 4.0 and to make India a manufacturing hub of smart enterprises (see section 'Digital India'). One of the industries where automation is expected to have a definitive impact is the automobile industry, which has an above-average robot density of 54. The use of collaborative robots (co-bots) is expected to generate significant economic value in the automobile industry by increasing revenues through efficiency gains and reducing costs and to help the 74$ billion industry increase in size to 300$ billion by 2026 and potentially employ 60 million workers (Grant Thornton 2017).

An example of industry 4.0 in the automobile industry is the German electronics and auto parts manufacturer *Bosch*, which has adopted smart manufacturing across its 15 plants in India. As part of this 'connected industry' approach where human resources, machines and materials are interconnected in a seamless manner through IoT, there is a real time transfer of know-how and data to shorten the throughput time and improve productivity. Thus, each single unit is connected to other units in the line, managed by a monitoring system which allows online benchmarking and quicker reactions to deviations in the production process and thus continuous improvements in quality.[12] Similar such examples of smart enterprises have been reported across manufacturing but are too few and far between to suggest any knowledge spill-overs to domestic firms and any significant increase in productivity.[13]

The employment effects of digitalisation and automation in the future remain uncertain. From a purely technological point of view, the World Bank expects automation and digitalisation to affect nearly two-thirds of all jobs in developing countries, especially the middle-skilled routine manual jobs such as clerks, operators etc. (World Bank 2016: 23). However, the low levels of robot density per worker and low wages in India counteract the tendency of firms to substitute workers with machines, making the net effect uncertain in the long run (Brynjolfsson and McAfee 2014). As it stands, the automobile sector, especially manufacturers of auto-parts, was expected to shed about a million jobs even before the pandemic struck due to low aggregate demand, with Bosch expected to cut a few thousand jobs as sales slump.[14] Thus, there is no presumption against the fact that technical change can lead to persistent technological unemployment in the long run and any increases in labour productivity will have to be accompanied by the necessary growth in aggregate demand to ensure high rates of employment (Cesaratto et al. 2003).

Digitalisation of services

Perhaps the most evident and tangible impact of digitalisation is in the services sector, which has seen the convergence of different innovations driven by big data, platform-based services, networked products and AI. Many services across diverse activities such as retail, education,

12 www.business-standard.com/article/companies/bosch-rolls-out-industry-4-0-approach-to-manufacture-two-wheeler-parts-across-india-116080400810_1.html

13 https://economictimes.indiatimes.com/tech/internet/automation-is-sweeping-across-indias-manufacturing-space/articleshow/69262579.cms?from=mdr

14 www.reuters.com/article/us-india-autos-jobs/indias-auto-parts-makers-warn-of-1-million-job-cuts-if-slowdown-continues-idUSKCN1UK132, www.bloombergquint.com/business/bosch-india-to-cut-jobs-as-worst-auto-slowdown-in-a-decade-hurts

transport, finance etc. are now offered through a digital platform. By improving the efficiency of service delivery and providing solutions to social problems, digitalisation of services promises to create efficiency gains and imbue further dynamism in the service sector.

The shift towards the digitalisation of services in India is primarily driven by consumers with the growing use of internet, smartphones and decreasing internet tariffs while businesses have shown slow and uneven adoption (McKinsey 2019). Several mobile phone-based applications (apps) serve as online marketplaces or aggregators for home-grown start-ups like OYO, which is an accommodation aggregator, or OLA, a ridesharing service. Most of these digital service companies try to differentiate their product according to the local market and build an ecosystem of interrelated products to capture economic value in new segments (Francis 2018). The growing digital consumer base has enabled the rapid growth of e-commerce, which grew to 33$ billion in 2017 and is expected to grow further as a growing network of retailers are going digital to sell their products and minimise the costs incurred on marketing. With regard to digital adoption by businesses, progress is uneven and skewed. A survey of CEOs and managers of large enterprises (above 70$ million) found that most firms in India are likely to adopt digital practices such as big data analytics, app-based services and cloud computing as part of their operations (WEF 2018). This is more so the case for IT firms for whom data storage, analysis and sharing via cloud is an integral part of their business model (NASSCOM 2018). It must be noted here that regardless of the industry, most big businesses tend to adopt digital practices such as digital marketing, online customer service or inventory management systems purely due to the size of the organisation. On the other hand, most small businesses are beginning to accept digital payments as part of their day-to-day operations following demonetisation (see section 'Digital Payments'), which removed high denominated currency from circulation (McKinsey 2019: 6)

Social services such as healthcare and education are also now being offered digitally, holding the promise to overcome the chronic lack of infrastructure in developing economies. In this direction, the government has piloted several initiatives on its own and in partnership with private agencies to deliver basic health and educational services. In healthcare, the government has partnered with NGOs to improve health outcomes using digital tools such as video-calling, health ATMs or mobile apps that help in diagnosis, monitoring and delivery. Similarly, there are many flourishing education technology companies that produce education content and deliver it digitally via smart phones. Various government services are also now offered digitally, ranging from women's safety to hospital management.[15] In fact, it is the government that has provided the necessary impetus for digitalisation of services through core infrastructural measures such as the national digital identity card *Aadhaar*, e-governance initiatives and creating the digital payments infrastructure which has enabled the rapid rise of Fintech industry. The ongoing Covid-19 pandemic will only act as a positive external shock in digital adoption, although its permanent effects in the digitalisation of services remain to be seen.

To sum up, there is little evidence for the diffusion of digital technology in manufacturing leading to an overall increase in productivity. However, services, especially ITes and finance, have seen rapid digitalisation and show potential for future value creation. The growth of ITes is symptomatic of the broad trajectory of Indian economic development since liberalisation characterised by jobless growth, where services increasingly contribute to a large share of value-added but have been unable to absorb the surplus labour from the low productivity sectors in the economy, retarding the process of structural change.[16] The service sector contributed to

15 For a full range of government services provided digitally, see www.digitalindia.gov.in/services.

16 There are two distinct aspects of structural change. The first is the reallocation of surplus labour from agriculture to industry. Second, it also entails movement from the unorganized to the organized high productivity sector.

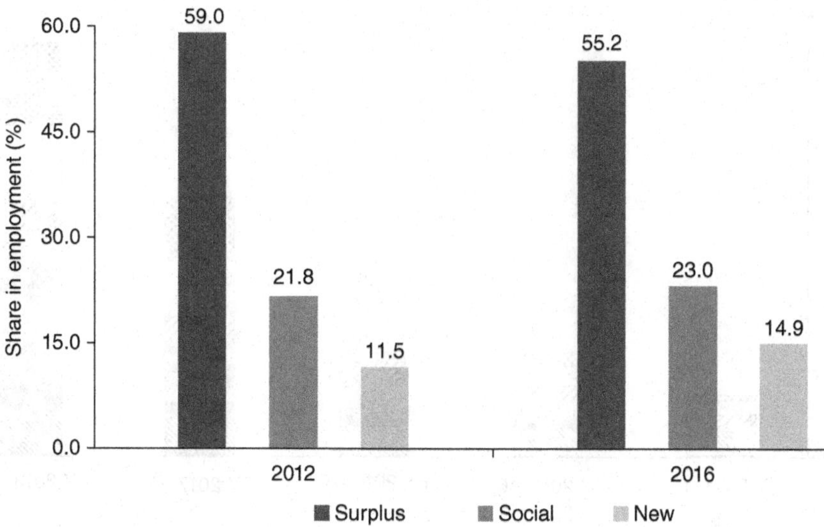

Figure 32.4 Share of service sector employment

Source: Basole (2018). State of Working India report

55% of the value-added and employed only 30% of the workforce in 2017. Figure 32.4 shows that even within the services sector, employment expansion has been high in 'surplus' industries characterised by self-employment and own-enterprises compared to 'new services' such as IT, finance and retail, which have high productivity (and low employment elasticity) and are most likely to use ICT technology as part of their core operations.[17] The next section will deal with the rise of Fintech as an example of digitalisation of services that has radically transformed the nature of businesses and governance.

Digital payments, financial inclusion and welfare

Digitalisation has enabled the remarkable rise of digital payments and the transformation of businesses and governance in the last few years. Digital payments have become the backbone of a new welfare regime that uses digital IDs and mobile phones for direct cash transfers to bank accounts. By actively pushing for financial inclusion via digital payments, the state expects the potential benefits to be huge due to the likely reduction in leakages and increased savings channelised through the financial system (Economic Survey 2016–2017: 62). The digitalisation of important economic activities like transactions, governance and welfare is acting as a catalyst for further change.

Digital payments

Digital payments are those that use digital or electronic means to send or receive money. They are a part of broader innovations in finance which use emerging technologies like digital

17 Surplus occupations in the services sector refer to amorphous activities such as petty retail, hotel and transport, whereas social industries are education, health and public administration.

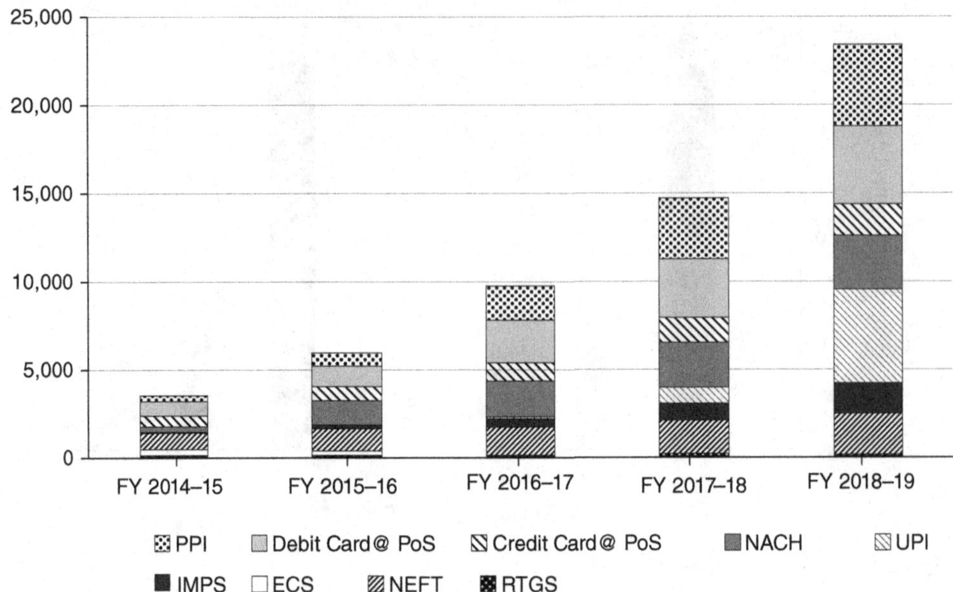

Figure 32.5 Digital transactions in volume in millions

Source: RBI 2019

platforms, blockchain and data analytics to deliver financial services. The widespread use of digital payments is expected to reduce the high transaction costs of cash payments and promote financial inclusion. The explicit and implicit costs associated with cash, such as the cost of printing currency, maintaining currency chests or ATMs and the implicit costs, such as foregone tax revenues due to illegal cash transactions, can be minimised by the use of digital payments.[18]

Digital payments are still at a nascent stage in India since it is predominantly a cash-based economy. The total number of digital payment users in India is estimated to be about 100 million, and the average number of digital transactions undertaken by a citizen in 2019 is still very low at 22, rising from two transactions a year in 2014 (RBI 2019: 3–24). Digital transactions include high value payment systems such as RTGS (real time gross settlement), NEFT (national electronic fund transfer), retail payment systems such as IMPS (immediate payment service), mobile payments through UPI (unified payment interface), digital wallets or PPI (prepaid payment instruments) such as Google Pay and card payments at PoS (point of sale). In terms of volume, there has been a significant growth of digital payments since 2016, especially in UPI, PPIs and debit card payments at PoS, indicating a rise in consumer transactions through mobile phones and digital wallets. Figure 32.5 shows the growth of the distinct types of digital transactions in volume.

Digital payments in India received a big push through the policy shock of demonetisation in November 2016, which demonetised high currency bills of 500 and 1000 rupees. The

18 The cost to consumers in terms of average transit time and access costs is higher in India than the global average. The 11 million residents in New Delhi collectively spend 6 million hours per month in getting cash from ATMs (Mazzotta et al. 2014). The adoption of digital payments for retail transactions is expected to save 0.7% to 0.9% of the GDP by 2025 (McKinsey 2019: 59)

policy announced by the state with the objective of tackling corruption and black money had the unintended consequence of promoting digital payments as there was practically no cash in circulation. The ex-post rationale for the decision was that 'demonetisation has the potential to generate long-term benefits in terms of reduced corruption, greater digitalisation of the economy, increased flow of financial savings, and greater formalisation of the economy, all of which would eventually lead to higher GDP growth' (Economic Survey 2016–2017: 2). Following demonetisation, various measures were introduced by the government to incentivise digital payments. These included the waiving of transaction charges on digital payments made to government agencies, service tax waiver for high value digital payments and the setting up of the united payments interface (UPI), which has greatly facilitated interoperability.[19] However, the widespread use of digital payments is hampered by the lack of digital infrastructure such as smartphones and PoS machines which are inadequate in India (BIS 2016). The problem is accentuated by the lack of demand for digital payments, as low acceptance is the key barrier to digital adoption since customers and merchants are locked in a cash ecosystem (Ligon et al. 2019).

Financial inclusion

An important reason cited for the shift to digital payments is its ability to engender financial inclusion by helping individuals overcome physical infrastructural barriers and providing them access to a range of personalised financial services such as financial management, incentives for savings and innovation, credit facilities and risk management (World Bank 2016: 96). With the widespread reach of mobile phones and internet, financial inclusion of all adults in developing economies becomes a real possibility. Due to the presumed long-term benefits of financial inclusion on economic growth, digital finance and fin-tech have now become a major part of the development agenda in emerging economies and are supposed to help countries achieve sustainable development goals (Beck and Frame 2018).[20]

The state is thus increasingly relying on solutions provided by fin-tech and digitalisation to tackle some of the long-standing problems of the economy through digital payments and financial inclusion. For instance, digital payments are expected to revitalise SMEs (small and medium enterprises) by bringing them under the ambit of the financial system and facilitating easier access to credit through innovative solutions such as peer-to-peer lending and invoice trading on a digital platform (GoI 2019: 32). Similarly, the state hopes that by tapping into the small savings of a vast number of individuals, it can spur economic growth and the formalisation of economic activity. In this direction, the government launched the *Pradhan Mantri Jan Dhan Yojana* (PMJDY) or the 'people's wealth' scheme in 2015 and opened nearly 300 million bank accounts in rural areas.

Despite the active push, the status of financial inclusion in India remains poor since a majority of the population relies on informal lending and are outside the financial system. The World Bank's Global Findex (2018) report finds that at least 47% of the total adult population is financially excluded in terms of not having a basic transaction account, and the percentage of adults who have made or received a digital payment in the last year is even less, at 29 %. The use of mobile money accounts i.e. those which do not require linkage with a bank account, is

19 http://cashlessindia.gov.in/promoting_digital_payments_people.html

20 The celebrated example of M-Pesa in Kenya is widely cited as the success of mobile money in developing economies. See Suri and Jack (2016), who estimate that the use of mobile money led to long-run increase in consumption levels and lifted about 2% of Kenyan households out of poverty.

abysmal at only 2%.[21] The primary reason for individuals not having a bank account is due to insufficient available funds, and it is unclear how digital payments will help solve this problem. It is evident, however, that digital payments have aided the phenomenal growth of the fin-tech industry, which is expected to grow to 130$ billion by 2025 by making forays into agriculture, e-commerce and micro-loans (McKinsey 2019: 59). More significantly, the expansion of digital payments and bank accounts have laid the foundations for a new system of social security and welfare.

Welfare and e-governance

Digitalisation has become the backbone of the new welfare regime where social security benefits and citizen entitlements are now being delivered digitally. The new welfare regime is characterised by a shift from transfers in kind to cash transfers. Under the *Jan Dhan, Aadhaar, Mobile* (JAM) initiative, the government is linking the bank accounts opened under its financial inclusion drive with the unique digital IDs and mobile phones of the beneficiaries to provide direct benefit transfers (DBT). Such an initiative allows the government to target beneficiaries and ensure that the benefits go to their bank account in a secure and transparent manner. By doing so, the government can plug leakages and rationalise expenditure by identifying duplicate and fake beneficiaries. In fiscal year 2018–19, nearly 369 central government schemes were delivered through DBT, including compensation for the workfare program, Mahatma Gandhi national rural employment guarantee scheme (MNREGS) and other financial assistance schemes. The government estimates that it saved 1,70,000 crores (25$ billion) or about 1.5% of the GDP by plugging leakages.[22] While several studies have shown that the use of biometric IDs has reduced leakages, the most insidious problem with *Aadhaar*-enabled DBT is the exclusion of genuine beneficiaries and increase in transaction costs due to design problems (Khera 2017; Muralidharan et al. 2020). Thus, the use of digital IDs and electronic payments by themselves have no benefits, and impact depends on the specific ways in which it is used.

Apart from social security benefits, a wide range of citizen entitlements and public services are also now accessible digitally as part of the e-governance program. The initiative aims to simplify and ease the interaction between the state and citizens and to deliver services in a seamless and effective manner. A wide range of public services across different government departments and jurisdictions are digitally made available to the public through front-end delivery points like the CSC. The services offered range from birth certificates, industrial licenses, land records etc. The DigiLocker initiative of the government offers free and secure cloud-based platform for the storage, verification and sharing of such official documents. Thus, digitalisation is going a long distance in terms of strengthening state capacity by building a secure payments infrastructure, digitising government records and radically transforming service delivery for citizens.

The role of the state

The trajectory of economic development in India suggests that there is very little diffusion of ICT technology across the broad sectors of the economy and instead an expansion of the

21 The use of digital payments is skewed along the lines of gender, employment, region, education and income, where adults in the richest 60% income category are more likely to have made or received digital payments (Global Findex 2018).

22 https://dbtbharat.gov.in/

ICT sector in terms of output, especially IT enabled services and 'new services' such as fintech (Chandrasekhar 2005; Joseph and Abraham 2005; Erumban and Das 2016; Francis 2019b). While important aspects of the economy are undergoing rapid changes through the digitalisation of services, the genuine potential of digitalisation as a technological transformation remains unrealised due to the low diffusion of technology and negligible productivity effects, especially in manufacturing. In contrast, the services sector, especially ICT exports and 'new services', has flourished due to state support. This coincides with the period of accelerated growth witnessed between 2005 and 2013, driven by services concentrated in pockets of urban India. However, the services (and manufacturing) sector has been unable to absorb surplus labour, reflecting a failure of structural change, and has furthered the regime of 'jobless growth'. By actively pursuing an ICT export-led growth strategy, the state has neglected the potential of the domestic market in building linkages, promoting innovation and using ICT-induced productivity for economic development (Joseph 2002).

Furthermore, the development strategy of the state has instead implicitly relied on *potential* gains to be had by reducing transaction costs and focusing on the digitalisation of services. There is an increasing reliance by the state on mobile apps and digital payments to solve long-standing economic problems ranging from poor education and healthcare, to the agrarian distress and industrial sickness in the economy. It is taken for granted that digitalisation can improve efficiency by reducing costs and create market access and gains for the poor by overcoming the physical barriers of infrastructure. Reliance on digital technology coupled with financial inclusion promoted by private sector actors signifies a shift to a global, finance-led development strategy where providing financial services to the 'unbanked' is a new high risk/return frontier for financial capital (Gabor and Brooks 2017). A prime example of a policy based on such a strategy was demonetisation that was premised on cashlessness and the 'short term costs and long-term benefits' accruing from transitioning to digital payments via financial inclusion (Economic Survey 2016–2017: 19).

In actuality, demonetisation led to an aggregate demand and supply shock, especially in the cash-intensive informal sector, which led to a *permanent* fall in both actual and potential output. The Indian economy has not returned to the pre-demonetisation trend rate of growth, and recent performance suggests that the fall in GDP growth is due to a lack of demand engendered by demonetisation. The slump is particularly evident in the technology-intensive sectors that are most likely to see digitalisation and automation, such as the automobile and banking industries. Thus, an uncritical dependence on digital technology as a driver of economic growth by reducing costs and rationalising expenditure can be inimical, especially in a period of chronic under-utilisation of capacity. Moreover, the hope that the use of digital technologies by the state and private actors to provision basic services in education and healthcare in an affordable and accessible manner has not seen any long-term observable outcomes at the scale that is required, and is not a substitute for physical investment in social infrastructure.

As an enabler of new technologies, the state has a crucial role in determining the direction of technology towards economic development and ensuring technological 'catching-up' since there are significant discontinuities in the process of development (Perez and Soet 1988; Francis 2018). The objective of state policy must be geared towards the rapid diffusion of digital technology and increasing the *use* of ICT, keeping in mind the long-term goals of development by increasing absorptive capacities and innovation capabilities (Karo and Kattel 2011). In this regard, the two key areas where the state can intervene are, first, creating complementarities in the production and use of digital technology; and, second, the protection and regulation of data, which is an essential input for most digital technologies (Francis 2019b).

Building complementarities in the systems of production and innovation is crucial as it leads to feedback effects and increases in productivity. The growing embeddedness of sensors and software in the production of goods following innovations in IoT requires the state to promote the domestic computer hardware and electronics industry to build synergies with the software capabilities India already possesses (Francis 2019b). This will correct India's ICT growth trajectory, which has been heavily skewed towards software to the neglect of hardware and give a fillip to the ailing manufacturing sector (Chandrasekhar 2005). The emphasis on the analogue foundations of digitalisation is critical because it contributes to the building of downstream linkages and capabilities essential for diffusion (World Bank 2016). Not only do complementary assets help in generating and appropriating economic value from innovation, but they also help with moving up the global value chain and preventing erosion of the manufacturing base (Teece 1986; Francis 2019a). Thus, the possibility of India 'leapfrogging' in productivity rests on the absorptive capacities developed for ICT use in the economy (Steinmueller 2001).

The second area where the state must intervene is the protection and regulation of data. The development of innovations in AI, IoT and blockchain is contingent on the availability of big data, and the state has to regulate its use, especially the privacy and security of consumers whose data is increasingly monetised. Regulation of data is important since data ownership is a primary barrier to the entry of new players in the digital market. The ownership of data tends to entrench the position of the dominant firm since the multifold benefits from network effects feed the greater amount of data generated from a growing number of users (Francis 2018). More importantly, regulation of data is paramount for individual privacy and national security.

There are a host of other measures that the state can take to ensure that digitalisation contributes to the goals of economic development. While continuing with its policy of deepening digital infrastructure, it also has to ensure a more digitally skilled workforce. The sole emphasis on technology will be redundant if workers lack the appropriate skills and firms do not restructure their organisations appropriately (Brynjolfsson and McAfee 2014). In this regard, the continuous training, skilling and retraining of India's large labour force through vocational, digital literacy programs acquires importance (Srija 2018). Regulating competition and creating a level playing field in these new services and digital markets is also important given the nature of digital technologies that tend to create winner-takes-all outcomes and market concentration.

Conclusion

A preliminary survey of the ongoing process of digitalisation, along with the interrelated processes of automation, data generation and monetisation, emergence of e-commerce and platform capitalism, the drive towards financial inclusion and new modes of service delivery reveal a complex picture of the digital landscape in India. Most of these advances have built on the previous (and ongoing) ICT revolution and have interacted with the structure of the economy in complex and roundabout ways. Thus, while digital technologies have opened up tremendous opportunities in terms of value creation, the digital divide threatens to exacerbate the unbalanced structure of the economy, furthering the ongoing trend of jobless growth. While the government has actively pushed for digitalisation and tried to bridge the digital divide, it is imperative to guide the direction of technical change in a manner that is productivity enhancing and employment generating instead of reducing costs. Crucially, the success of widespread diffusion depends on creating complementary goods and infrastructure that can leverage the skills and potential of a digitally empowered workforce. Without the necessary policy action, the possibility of unbalanced and unequal development looms large.

Acronyms

AI:	artificial intelligence
BPM:	business process management
BPO:	business process outsourcing
CAGR:	compound annual growth rate
CSC:	Common service centre
DBT:	direct benefit transfer
ESC:	Electronics and Computer Software Export Promotion Council
Fintech:	financial technology
ICT:	information and communication technology
IFR:	International Federation for Robotics
IoT:	Internet of Things
IT:	information technology
ITes:	information technology enabled services
ITU:	International Telecommunication Union
JAM:	Jan Dhan, Aadhaar and Mobile
MeitY:	Ministry of Electronics and Information Technology
MNREGS:	Mahatma Gandhi national rural employment guarantee scheme
NASSCOM:	National Association for Software and Service Companies
NeGP:	National e-Governance plan
NIS:	national innovation system
NOFN:	National Optical Fibre Network
PoS:	point of sale
PPI:	prepaid payment instrument
RBI:	Reserve Bank of India
SDC:	State data centre
SME:	small and medium enterprises
STIP:	science, technology and innovation policy
SWAN:	State Wide Area Network
TRAI:	Telecom Regulatory Authority of India
UNCDF:	United Nations Capital Development Fund
UPI:	unified payments interface
WEF:	World Economic Forum

References

Basole, A. (2018): *State of Working India Report 2018*. Azim Premji University.

Beck, T. and Frame, W. S. (2018): Technological Change, Financial Innovation, and Economic Development, in Beck, Thorsten, Levine, Ross (eds.): *Handbook of Finance and Development*. Cheltenham, UK, Northampton, MA: Edward Elgar Publishing, 369–390.

BIS (2016): Bank of International Settlements. Available at: https://stats.bis.org/statx/srs/table/t5?c=IN&p=2018 [15.02.2021]

Brynjolfsson, E. and McAfee, A. (2014): *The Second Machine Age: Work, Progress, and Prosperity in a Time of Brilliant Technologies*. New York, London: WW Norton & Company.

Cesaratto, S, Serrano, F. and Stirati, A. (2003): Technical Change, Effective Demand and Employment, *Review of Political Economy* 15(1), 33–52.

Chandrasekhar, C., P. (2005): The Diffusion of Information Technology and Implications for Development: A Perspective Based on the Indian Experience, in Saith, A. and Vijayabaskar, M. (eds.): *ICTs and Indian Economic Development: Economy, Work, Regulation*. New Delhi, India: SAGE Publications.

Demirguc-Kunt, A., Klapper, L., Singer, D., Ansar, S. and Hess, J. [Global Findex] (2018): Global Findex Database 2017: Measuring Financial Inclusion and the Fintech Revolution. Washington, DC: World Bank. Available at: https://openknowledge.worldbank.org/handle/10986/29510 [15.01.2021]

Dutta, D. (2018): Development under Digital Divide in India, in Dutta, D. (ed.): *Development under Dualism and Digital Divide in Twenty-First India*. Singapore: Springer, 155–197.

EAC-R&D (2019): *R&D Expenditure Ecosystem: Current Status and Way Forward, Economic Advisory Council*. Available at: http://psa.gov.in/sites/default/files/pdf/RD-book-for-WEB.pdf [15.02.2021]

Economic Survey (2016–2017): Ministry of Finance, Government of India.

Economic Survey (2017–2018): Ministry of Finance, Government of India.

Erumban, A. and Das, D. K. (2016): Information and Communication Technology and Economic Growth in India, *Telecommunications Policy* 40(5), 412–431.

ESC (2018): Electronics and Computer Software Export Promotion Council. Available at: www.escindia.in/resource-center/ [15.02.2021]

Francis, S. (2018): *Evolution of Technology in the Digital Arena: Theories, Firm-level Strategies and State Policies*. CWS/WP/200/47, Centre for WTO Studies, Indian Institute of Foreign Trade.

Francis, S. (2019a): Playing Catch-up in the Digital Economy, *Business Today*. Available at: www.businesstoday.in/opinion/columns/playing-catch-up-in-the-digital-economy/story/309284.html [15.02.2021]

Francis, S. (2019b): Digitalization and the Role of the State, *The Hindu BusinessLine*. Available at: www.thehindubusinessline.com/opinion/digitalisation-and-the-role-of-the-state/article28101969.ece [15.02.2021]

Gabor, D. and Brooks, S. (2017): The Digital Revolution in Financial Inclusion: International Development in the Fintech Era, *New Political Economy* 22(4), 423–436.

Government of India (GoI) (2019): Report of the Steering Committee on Fintech Related Issues. Available at: https://dea.gov.in/sites/default/files/Report%20of%20the%20Steering%20Committee%20on%20Fintech_1.pdf [15.02.2021]

Grant, T. (2017): India's Readiness for Industry 4.0. Available at: www.grantthornton.in/globalassets/1.-member-firms/india/assets/pdfs/indias_readiness_for_industry_4_a_focus_on_automotive_sector.pdf [15.02.2021]

IFR (2017): International Federation of Robotics. *Industrial Robots.*

ITU (2019): *International Telecommunications Union*. Yearbook of Statistics: Telecommunications/ICTindicators-2009–2018. Available at: www.itu.int/en/ITU-D/Statistics/Pages/stat/default.aspx [15.02.2021]

Jensen, R. (2007): The Digital Provide: Information (Technology), Market Performance, and Welfare in the South Indian Fisheries Sector, *The Quarterly Journal of Economics* 122(3), 879–924.

Jorgenson, D. W. and Vu, K. M. (2016): The ICT Revolution, World Economic Growth, and Policy Issues, *Telecommunications Policy* 40(5), 383–397.

Joseph, K. J. (2002): Growth of ICT and ICT for Development: Realities of the Myths of the Indian Experience (No. 2002/78). *WIDER Discussion Paper.*

Joseph, K. J. and Abraham, V. (2005): Moving Up or Lagging Behind? An Index of Technological Competence in India s ICT Sector, in Saith, A. and Vijayabaskar, M. (eds.): *ICTs and Indian Economic Development: Economy, Work, Regulation*. New Delhi, India: SAGE Publications, 131–153.

Kaka, N., Madgavkar, A., Kshirsagar, A., Gupta, R., Manyika, J., Bahl, K. and Gupta, S. [McKinsey] (2019): *Digital India: Technology to Transform a Connected Nation*. McKinsey Global Institute, Mumbai, India. March.

Karo, E. and Kattel, R. (2011): Should 'Open Innovation' Change Innovation Policy Thinking in Catching-up Economies? Considerations for Policy Analyses, *Innovation: The European Journal of Social Science Research* 24(1–2), 173–198.

Khera, R. (2017): Impact of Aadhaar in Welfare Programmes, *Economic & Political Weekly* 52(50), 61–70.

Ligon, E., Malick, B., Sheth, K. and Trachtman, C. (2019): What Explains Low Adoption of Digital Payment Technologies? Evidence from Small-scale Merchants in Jaipur, India, *PloS One* 14(7), e0219450. https://doi.org/10.1371/journal.pone.0219450

Lundvall, B. A. (1988): Innovation as an Interactive Process: From User-producer Interaction to the National System of Innovation, in Dosi, G. et al. (eds.): *Technical Change and Economic Theory*. London, New York: Pinter publishers, 349–369.

Mazzotta, B. et al. (2014): *Cost of Cash in India*. The Fletcher School, Tufts University. Available at: https://sites.tufts.edu/ibgc/files/2019/01/COC-India-lowres.pdf [15.02.2021]

Muralidharan, K., Niehaus, P. and Sukhtankar, S. (2020): Identity Verification Standards in Welfare Programs: Experimental Evidence from India (No. w26744). National Bureau of Economic Research.

NASSCOM (2017): The IT-BPM Industry In India 2017: Strategic Review. Available at: www.nasscom.in/knowledge-center/publications/it-bpm-industry-india-2017- strategic-review [15.02.2021]

NASSCOM (2018): CEO Survey: 2019–20 Industry Performance: 2018–19 and What Lies Ahead. Available at: https://nasscom.in/sites/default/files/Industry-Performance2018-19-and-what-lies-ahead_0.pdf [15.02.2021]

Perez, C. and Soet, L. (1988): Catching Up in Technology: Entry Barriers and Windows of Opportunity, in Dosi, G. et al. (eds.): *Technical Change and Economic Theory*. London: Francis Pinter, 458–479.

RBI (2019): Reserve Bank of India. Report of the Committee on Deepening of Digital Payments. Available at: https://rbidocs.rbi.org.in/rdocs//PublicationReport/Pdfs/CDDP03062019634B0EEF3F7144C3B65360B280E420AC.PDF [15.02.2021]

Schwab, K. (2016): *The Fourth Industrial Revolution*. New York: Crown Business Books.

Srija, A. (2018): The Fourth Industrial Revolution: Realizing India's Demographic Dividend. *Working Paper*, Centre for Sustainable Employment, Azim Premji University, 2018.

Steinmueller, W. E. (2001): ICTs and the Possibilities for Leapfrogging by Developing Countries, *International Labour Review* 140(2), 193–210.

STIP (2013): Science, Technology and Innovation Policy, Ministry of Science and Technology. Available at: http://dst.gov.in/sites/default/files/STI%20Policy%202013-English.pdf [15.02.2021]

Suri, T. and Jack, W. (2016): The Long-run Poverty and Gender Impacts of Mobile Money, *Science* 354(6317), 1288–1292.

Teece, D. J. (1986): *Profiting from Technological Innovation: Implications for Integration, Collaboration, Licensing and Public Policy, Research Policy* 15(6), 285–305.

TRAI (2020): Telecom Regulatory Authority of India, Yearly Performance Indicators of Telecom Sector, Various years. Available at: https://main.trai.gov.in/release-publication/reports/performance-indicators-reports [15.02.2021]

World Bank (2016): *World Development Report 2016: Digital dividends*. Washington, DC: World Bank. https://doi.org/10.1596/978-1-4648-0671-1

World Economic Forum (WEF) (2018): *The Future of Jobs Report 2018*. Geneva: WEF. Century.

33

INDUSTRY 4.0 IN CHINA

Han Li and Wei Zhang

Introduction

Germany's launch of an Industry 4.0 initiative triggered a fashion imitated by countries around the globe. In 2013, the German government introduced the concept of Industry 4.0 in its High-Tech Strategy 2020 and identified it as one of ten major future projects to support a new generation of research and development in industrial technology. German scholars and businessmen consider the concept of Industry 4.0 to be the fourth industrial revolution, dominated by intelligent manufacturing and revolutionary means of production; people can make great use of the combination of computerised information technology and advanced technologies, such as cyberspace and cloud computing, to transform the whole manufacturing industry into an intelligent production system.

Thereafter, many countries – especially those developed countries that intend to regain manufacturing capacity and promote employment, as well as developing countries with ambitions of upgrading their industry structure – have expected to find paths to achieve their goals with the contributions of Industry 4.0 construction. Industry 4.0 has thus gradually become a popular national strategy around the world.

China publicly followed this path of industry upgrading in 2015 with the announcement of the Made in China 2025 strategy. Made in China 2025 is a development strategy that aims to enhance the competitiveness of domestic manufacturing and to foster economic growth driven by advanced manufacturing industries and technical innovations.

For the last 70 years, since the founding of the People's Republic of China in 1949, China has developed from a large, traditional agricultural economy into an important industrialised one. So far, China possesses one of the most integrated industrial systems in the world and lives up to its reputation as the "world factory" by its total output, providing nearly all kinds of manufactured goods, and its significant role in the global supply chain. At present, all 39 divisions of modern industries are located in China, including 191 subgroups of manufacturing industries and 525 even lower classes of manufacturing industries, according to the International Standard Industrial Classification of All Economic Activities (ISIC) published by the United Nations. This integrated industry system enables China to supply global consumers with almost every kind of goods.

In 1949, agriculture contributed the most to China's GDP. The secondary industry accounted only for 15.5%, of which heavy industry only took a share of 4.5%. In 2018, however, the

DOI: 10.4324/9780429351921-40

value-added of China's manufacturing sector was 26.5 trillion Yuan, which is about US $4 trillion (calculated at the exchange rate of that year), and it accounted for almost 30% of China's GDP (Li 2019).

Moreover, according to the databank of the United Nations Industrial Development Organisation, there are 22 major manufacturing industries in China, whose annual value-added rank the highest in the world. In addition, the production of textile, clothing, leather and metals products all account for even more than 30% of the world's total value-added. Currently, China ranks as the top producer of hundreds of manufacturing products, such as steel, copper, cement, chemical fertilisers, chemical fibres, shipbuilding, automobiles, computers, laptops, printers, television sets, air conditioners, washing machines and so on.

In 2018, China's total value-added of its manufacturing industry accounted for 28% of the whole global manufacturing industry; this is the highest of all countries, with the United States accounting for 17% and Japan for 9% during the same period. In 2018, the export volume of China's manufactured goods amounted to 235.2 billion Yuan, that is, nearly 17.5% of the world export aggregate. In the same year, the export of manufactured goods in China accounted for 85.4% of its total exports of goods and services (China Economic Information Service 2019).

However, there are also evident weaknesses in China's industry system. On the whole, China's manufacturing sector is still in the transitional process from "Industry 2.0" to "Industry 3.0", characterised by electrification and automation respectively, not to mention the higher level of Industry 4.0. There are great differences in not only the automation level among different firms, industries and regions but also great variation among firms as to the role of digitisation and ICT (information and communications technology) used in production processes. China has made remarkable progress in promoting the new type of industrialisation and coordination between industrialisation and ICT, and some domestic industries and firms have increased their investment in information, intelligent technology and human capital. However, while China has a few leading firms in some high-tech industries, the majority of Chinese manufacturing firms are generally latecomers and still in the early stages of applying advanced technologies.

Misalignments of industrial structure

China's manufacturing industries produce the largest quantity of some products worldwide, but its production in high-end chips, electronics, consumer electronics, industrial software and other areas does not meet the basic needs of domestic consumption. The development of high-end equipment manufacturing industry and producer services lags far behind as well. The internationalisation level of Chinese firms is low, as is their management efficiency, which has been amplified in the globalisation age. Moreover, resource- and labour-intensive industries are disproportionally large, while the proportion of technology-intensive industries and production-oriented service industries is relatively undersized for sustainable and coordinated growth for China's manufacturing industrial structure. Moreover, the level of industrial agglomeration and cluster development of domestic firms is insufficient. Redundant investment and overcapacity of production in some industries have made those firms vulnerable to outside challenges.

Poor product quality

Currently, the product quality problem is prominent in China, and it not only leads to concerns of domestic consumers but also keeps foreign consumers away. According to product quality random inspection conducted by China's State Administration for Market Regulation in 2019, the average passing rate for quality sampling inspection of 97 categories of consumer goods is slightly less than 90% (China State Administration for Market Regulation 2019). The manufacturing industry in China bears a direct quality loss exceeding 200 billion Yuan each year, comparing the indirect loss exceeding 1 trillion Yuan; this loss is hitting downstream industries and contributing to loss of market share and pollution control due to inferior product quality (Zhang 2015). Export products made in China also lack a reputation for high quality in the international market, compared to products from Germany or Japan (Cagé and Rouzet 2015). Foreign demand is determined by expected quality, which is driven by the dynamics of consumer learning through shopping experiences and the country of origin's reputation for quality. Therefore, a range of quality Chinese firms are permanently kept out of the market by information friction, which causes net welfare loss to Chinese firms and foreign consumers.

Lack of innovation capability

There is still a big gap between the innovation capability of Chinese manufacturing firms and that of firms in advanced countries. In 2018, China's overall R&D spending amounted to 2 trillion Yuan, in which manufacturing R&D expenditure was more than 1 trillion Yuan; the annual growth rate was stable at 10%, based on the published data of the China National Bureau of Statistics. However, the ratio between R&D expenditure and GDP was 2.19% in 2018. Furthermore, the R&D intensity of manufacturing firms (the ratio of R&D investment to their main business revenue) was only slightly over 1.1%, which is far behind the world's advanced level. For instance, the United States and Japan generally maintain their R&D intensity at around 3%: the indicator for the United States was 2.81% in 2017, and 3.21% for Japan in 2018 (China National Bureau of Statistics 2019). Meanwhile, even in technology-intensive industries such as pharmaceuticals, computers, machinery equipment and electrical equipment, R&D investment of China is lower than that of the United States and Japan. China needs to overcome the bottleneck caused by shortage of core technology that results from backwardness of research and development.

Low revenue share in global value chains

The data show that more than 80% of China's electronic chips depend on imports, which were worth $305.5.1 billion in 2019, and the value is even higher than that of imported crude oil (China General Administration of Customs 2020). China is still entrapped in the "manufacturing–processing–assembly" link with low technical contents and only contributes low value-added to the international division of labour, falling short of competitiveness in sessions with high value-added, such as R&D, design, project contracting, marketing and after-sales service. A much more famous case study reveals that for the wholesale price for an iPhone of the Apple company, which is $178.96, Japan, Germany and South Korea take 34%, 17% and 13% share of the value, respectively, while China has only 3.6%, which is about $6.5 in value, or one thirtieth of the aggregate price (Xing and Detert 2010).

Harsh competitive pressure

China's manufacturing firms have participated extensively in international market competition and have the advantage of providing massive amounts of goods and services to international consumers. Nevertheless, because of the inferiority of the technical capacity and product quality, Chinese firms and products are already at a disadvantage in the world market, especially with regard to the position in global value chains and income share acquired by Chinese firms. Moreover, in recent years, developed countries such as the United States, Germany and the United Kingdom have proposed a "re-industrialisation" strategy focusing on revitalising their domestic manufacturing industry, and their strategies rely on the application of information technology and digital manufacturing technology to seize a new commanding point of manufacturing industry. Therefore, not only do Chinese firms need to adapt to competitive pressure in the global market, but they also need to catch up with the new rival frontier in this emergent tide of internalising manufacturing capacity.

China has already become a manufacturing power evaluated according to multiple dimensions, such as its production and exports and its role in global value chains, but it still has a relatively poor reputation in manufacturing for its relatively low product quality and shortage of technology contents. Additionally, although the overall scale of China's manufacturing industry is huge, there is significant excess production capacity and redundant investment, and thus the constraints of resources, the use of energy and the environment become the main obstacles to the future development of China's manufacturing industry.

China's manufacturing industry has entered a new stage facing two diverging roads, with one way pointing towards a low-level trap and the other pointing towards a bright future. The journey in the second direction needs to be guided by the Industry 4.0 program. This plan will enhance the core competitiveness of enterprises, improve product quality and intelligent manufacturing, because Industry 4.0 will realise the optimisation and automation of the production process. Therefore, the implementation of Industry 4.0 in China is associated with high expectations.

Policy evolution of China's Industry 4.0 programme

Industrial policy in China

China's Industry 4.0 programme started with an industry policy plan known as Made in China 2025, which together with previous 5-year plans has taken an important and active role in China's industrialisation and economic development process for the last 40 years. However, it is always a controversial topic in academic research and policy discussions. It is necessary to explore the role of policy factors in the development of Industry 4.0 in China.

A brief account of the history and features of China's industrial policy will be helpful to understand why China implemented an Industry 4.0 strategy. China studied the experiences made in Japan and South Korea, especially at their take-off stages, in order to design and implement industrial policy. Academic research has suggested that appropriate industrial policy helped these two countries to achieve very rapid economic growth in the 1950s and 1960s (Sakoh 1984). Rodrik (2010) observed that "the real question about industrial policy is not whether it should be practiced, but how".

Since the reform and opening up in 1978, industrial policies have assumed an indispensable role in China's economic development, especially in manufacturing. Behind the evolution of China's industrial policy were two driving forces: one is adjustment of the relationship between government and market in the process of market-oriented reform, which has an important influence on the orientation of industrial policy and the choice of policy tools. T second is a change in the main problems faced by the industrial development and the transformation of industrial structure in the rapid development of China's economy.

Policy portfolio

The "Made in China 2025" strategy is a forerunner of Industry 4.0 policies in China, which puts forward the strategic goal of realising manufacturing power through accomplishing "three steps" by the time of the centenary of the founding of "New China" in 2049. To achieve those strategic goals – of a manufacturing power China has aspired to for several decades – the "Made in China 2025" strategy has defined nine strategic tasks and priorities to be fulfilled and many of them are also the core contents of Industry 4.0 program.

Except for the "Made in China 2025" program, related policies have come out continually. For example, one message of resolution, from a report delivered at the 19th National Congress of the Communist Party of China, is to drive deep integration of the internet, big data, artificial intelligence (AI) and the real side of the national economy. Moreover, five-year plans during those years and yearly work reports from the central government, released on the annual meeting of the National People's Congress, set the targets and basic structure of practical policies to promote Industry 4.0. Those policies have constantly emphasised the major tasks related to Industry 4.0, such as AI, industrial internet platform, big data, digital economy. Table 33.1 lists some important policies and detailed measures to promote the development of Industry 4.0 in recent years.

Logic of policies related to Industry 4.0

Focus on various aspects of Industry 4.0

Table 33.1 reveals that the various policy measures to develop Industry 4.0 involved different Industry 4.0 content, such as AI, industrial internet, intelligent hardware industry, industrial robot and so on. Moreover, these policies often provide time frames and enact specific requirements for their effectiveness to be realised. In addition, these policy programs usually provide specific measures such as subsidies, tax deductions and exemptions.

The role of those policies

The policy adopted usually tries to incentivise the redistribution of resources among industries or to accelerate the development of Industry 4.0 in certain business activities of private firms in various industries. In other words, it is the kind of policy that promotes the production, investment, research and development, modernisation and restructuring of one industry that would impose restrictions on similar activities in other industries, or the policy instruments for Industry 4.0 would cost the investment in or production of other business activities. Those policies bear the gene of selective industrial policy because of the potential extensive intervention in microeconomic operation, with selection bias, distortion of factors and misallocation of resources as well.

The ambiguous effectiveness of those policies

The effects of those policies are awaiting an evaluation, and the results might be highly controversial. With those selective policies, the limitations of China's industrial policy system gradually become visible. Although China is promoting the transition of policy from a selective industrial policy mode to competition policy in recent years, this cannot be accomplished in a single step. Therefore, in the case of entering the new stage of economic development, it is urgent

Table 33.1 Timeline and contents of major policies related to Industry 4.0 in China

Time	*Policy*	*Main content*
July 2015	Guiding Opinions of the State Council on Vigorously Advancing the "Internet Plus" Action	Foster the development of AI in emerging industries; promote the key areas of intelligent product innovation; improve the level of intelligent terminal product etc. The main goal is to speed up the breakthrough of the core technology of AI and promote the application of AI in the fields of smart home, intelligent terminal, intelligent automobile, robotics and so on.
August 2016	Program of Standardization and Quality Improvement in Equipment Manufacturing	By 2020, the standard system in key areas such as intelligent manufacturing and green manufacturing will be fundamentally improved, and the quality and safety standards in those areas will be in line with international standards.
April 2016	Robot Industry Development Plan (2016–2020)	By 2020, the annual output of industrial robots with six axes and more will reach more than 50,000 units. The annual sales revenue of service robots exceeds 30 billion Yuan; the main technical requirements of domestic industrial robots to reach the level of similar products abroad; and to make breakthroughs on key parts such as the precision reducer, servo motor and driver for robots.
September 2016	Special Action on Innovation and Development of Smart Hardware Industry (2016–2018)	To create basic resources and innovation platform for AI. AI industry system is basically established; the overall technology and industry development and international synchronisation, application and system-level technology are partially in a leading position in the world.
November 2016	Guideline on Emerging Strategic Sectors during the 13th Five-Year Plan Period (2016–2020)	Cultivating AI industry ecology, promoting AI technology to all-round integration and penetration of various industries. Specific tasks include: speed up the construction of the AI support system; promote the application of AI technology in various fields, encourage all industries to strengthen the integration with AI and gradually achieve intelligent upgrading.
May 2018	Industrial Internet Development Action Plan	By the end of 2020, the industrial internet infrastructure and industrial system will be preliminarily built, and five or so of the country's top-level marking and resolution nodes will be completed, with a registration capacity of more than 2 billion.
March 2019	Guidance on Promoting Deep Integration of AI and the Real Economy	Grasp the development characteristics of the new generation of AI, combined with the characteristics of different industries and different regions, explore the path and methods of the application and transformation of innovation and construct the intelligent economic form of data-driven, human–computer cooperation, cross-border integration.

for China to give full play to the role of market mechanism, to stimulate innovation, explore the novel direction of future industry and promote the continuous improvement of economic efficiency. China needs to adopt effective industrial policies or competition policies to enhance industrial competitiveness and accelerate the adjustment of industrial structure.

The current status of Industry 4.0 in China

In this section, we use indicators to reflect the status and trend of China's Industry 4.0 program up until now. Based on the definition and connotation of Industry 4.0, which contains the characteristics of making manufacturing industry more digitalised, internet-based and intelligent, and because of the availability of data, the indicators we have chosen here include the coverage of broadband network, the application of industrial robots, the growth of AI industry and the development of big data industry. If Industry 4.0 is conceived as an organic system metaphorically, then AI is the brain, industrial robots are extended limbs, broadband network is much like the bonds that connect every corner of the system and big data is the blood that flows through the meridians.

Broadband network penetration

Broadband is a key infrastructure of the Industry 4.0 program. Both China's "Made in China 2025" and Europe's national Industry 4.0 initiatives have higher demands for broadband networks, because it is essential for intelligent production and smart factories. The construction of Industry 4.0 is based on a continuous stream and large volume of industrial big data, which requires a more powerful capacity of network bandwidth. The information transmission of each link of the industrial value chain or the data transmission of each section within the firm cannot be separated from the broadband network. Additionally, real-time information transmission of numerous industrial equipment, terminals, supply chains and workers in a large network puts forward higher requirements for the stability and the coverage of the network. The closer links between producers and consumers in Industry 4.0, such as the promotion of personalised customisation, the construction of platform economy and digital ecosystem, make the penetration of broadband a prerequisite as well. Due to constraints of data availability, we here use overall broadband penetration as a proxy variable to represent the network infrastructure development of Industry 4.0 in China.

In the past, China's broadband infrastructure was relatively weak compared to that of other countries. This is reflected in many aspects, such as the fact that most industrial parks have a network barely meets the standards of routine office works, and the result is that the bandwidth, coverage or stability of its internet are inadequate to support modern smart production. However, with the progress of economic development, especially the construction of Industry 4.0, the usage of fixed broadband bandwidth has grown very rapidly. The data in Figure 33.1 reveal that both the total number of network users and the size of fixed broadband users show a steady upward trend of growth, and by 2018 they reached a size of nearly 830 million and more than 400 million people, respectively. In 2019, the ratio of fixed broadband penetration, which is the quantity of fixed broadband users divided by the population of China, had attained 31.1%, which is even higher than the average level of OECD countries according to the White Paper on the Development of China's Broadband (CAICT 2019).

Broadband networks are accelerating into 5G, recently featuring ultra-high speeds. The low-time delay features of 5G can be used in real-time instruction control, real-time distribution of capital goods, parts assembly and other management and production scenarios. Therefore,

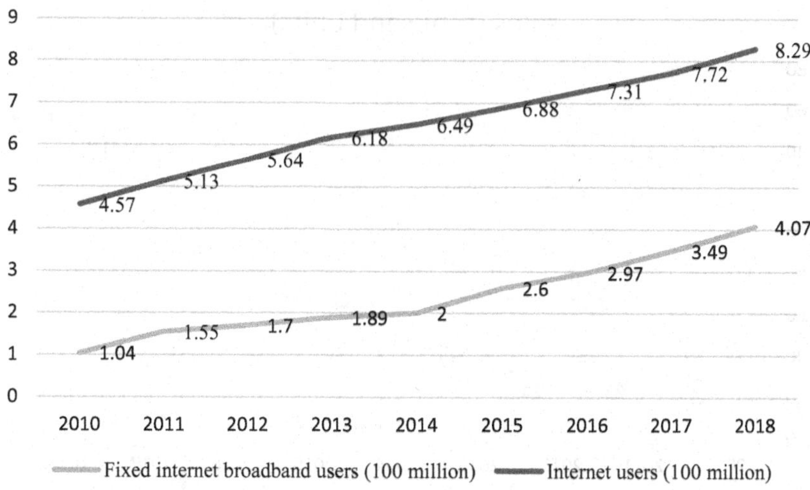

Figure 33.1 Internet penetration in China

Source: Ministry of Industry and Information Technology (2018)

ever-widening industrial application will become an important direction for 5G network construction. Furthermore, the wide connection features of 5G can be used to control the whole life cycle of modern factory's now ubiquitous digital devices.

Broadband application has expanded from consumption to production, opening a new era of interconnectedness of all business activities. The industrial internet has begun to be applied in the automobile, machinery, aviation and other industries and has extended from the peripheral links of production such as security and logistics to the core and internal links of product design, tooling manufacturing and production control, as well as quality inspection. The rapid popularisation of broadband networks has greatly promoted the integration of the network information technology within the economic and social fields, accelerating the quality revolution of China's industry, which stands for not only the enhancement of product quality, but also the upgrading of China's industries.

Usage of industrial robots

The widespread application of industrial robots in various subfields of manufacturing industry has become a remarkable fact, reflecting the growth of Industry 4.0, and it is of great significance for the transformation and upgrading of the manufacturing industry in China, as well as the rapid development of other industries. In China, the growing demand for industrial robots is reinforced by ever-increasing labour costs driven by the elapsing demographic dividend, which means the share of the working-age population (15 to 64) is no longer larger than the non-working age share of the population. Therefore, the application and dissemination of industrial robots to improve production efficiency, ensure product quality and reliability and even enhance the optimisation of corporate organisation and labour relations is the rational choice by entrepreneurs. The growth of industrial robots in China is reflected by not only the huge quantity increase of its usage in production, but also the expansion of application scenarios to provide services.

The statistics in Figure 33.2 show that in China the number of industrial robots sold reached 156,400 units in 2018. China has been the largest industrial robot application market in the

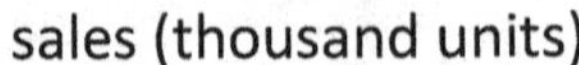

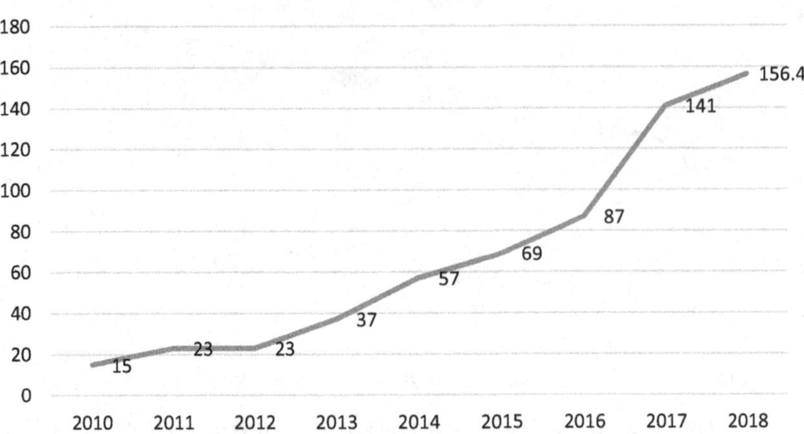

Figure 33.2 Growth of China's industrial robots' application

Source: International Federation of Robotics (2019) and Lu and Liu (2019)

world for six consecutive years. In 2017, the sale of industrial robots in China accounted for 35.6% of the global sales figure. In 2019, the size of China's robot market is expected to reach $8.68 billion, and the average growth rate during the course of 2014–2019 was 20.9%.

Provinces that need a lot of labour have introduced a series of fiscal subsidies and tax abatement to support the application of industrial robots in local firms. For example, Guangdong Province has arranged special funds for the development of industrial robots in its "Special Fund for the Development of Industry and Informationization" and has given the firms the amount of subsidy according to a certain proportion of the purchasing price they paid for industrial robots. Other regions like Shanghai, Hubei and Heilongjiang have also arranged policy support and subsidy for industrial robots' application. Even cities such as Shenyang, Dongguan and Qingdao have established funds to foster industrial robot application.

Market size of AI

AI is a new general-purpose technology, which can be widely applied in the economic and social environment. AI has penetrated different links of production, and it has quietly changed the mode of production process and operations of economic organisation. The application of industrial robots leads to the substitution for human physical strength, while AI technology mainly leads to the substitution of human mental power. Whether for entrepreneurs, middle-level managers or front-line workers, the application of AI technology has liberated labour from cumbersome procedural work, improved productivity and then increased overall TFP (total factor productivity) of the macroeconomy, which ultimately leads to high-quality development of the macroeconomy featuring technology innovation and knowledge-driven progress. High quality is mainly reflected by the enhancement of efficiency, in other words, the growth in labour productivity or TFP.

The data in Figure 33.3 shows that in recent years the AI industry has experienced a strong trend upwards with high growth rates in sales. Its market size has risen from about 11 billion Yuan in 2015 to 57 billion Yuan in 2019, and thus has grown almost five-fold in five years.

The average annual growth rate during this period was about 55%, which is much higher than that of the global average rate of growth in sales, which is about 36% per year in the same period. The total number of AI firms in mainland China (excluding Hong Kong, Macao and Taiwan) is 1,040, which ranks only second to the United States. In addition, the number of AI patents filed by different countries collected by the World Intellectual Property Organization shows that in 2018 (WIPO 2019), China, the United States and Japan already accounted for three-quarters in all patents applied related to AI technology, and China alone took a share of 40%. Moreover, the first half of 2019 has witnessed a total investment of over 47.8 billion Yuan in China's AI sector (Deloitte Report 2019), which projects an even faster growth of AI industry for China in the near future. With the ever-maturing environment for the development of AI industry, the overall market size of China's AI industry will reach 200 billion Yuan. At the same time, AI has a significant spill-over effect, which will promote the sustainable development of other related technologies, accelerating the transformation and upgrading of traditional industries.

Expansion of big data industry

The development of the big data industry itself is driven by data explosion in industrial production and organisational change, and the promising function of big data to improve efficiency. Additionally, big data communicates with all corners of the Industry 4.0 system and virtual connected related facilitates. Although the big data industry contains hardware facilities, its software dimension, which refers to basic services, technology development and integration applications of different levels, data collection and management, as well as the overall solution of data application, is crucial to the development of Industry 4.0.

Industrial big data has great and far-reaching impacts on the mode of production, operation management and ecological complex of manufacturing industry. The role of industrial big data in the manufacturing industry is mainly reflected in intelligent design and production, networked collaborative manufacturing, intelligent service, personalised customisation and so on. Under the impetus of big data, advanced manufacturing industry will achieve more rapid development, and the level of digitisation and intelligence of traditional industries is expected to be further improved as well.

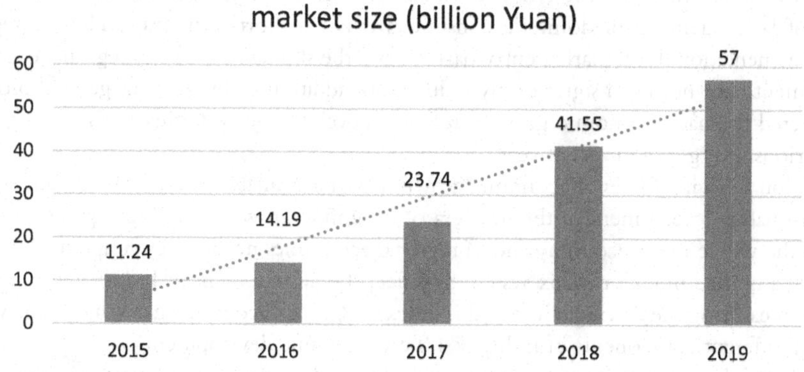

Figure 33.3 Growth of AI industry

Source: CAICT (2020a)

Since 2015, accelerating the growth of big data industry has become a national development strategy in China. Since then, provincial and municipal authorities have established industrial parks targeted at the development of big data. Meanwhile, along with development strategies of the new generation of information technology, smart city, digital China and so on, digital transformation in the social and economic spheres has been enhanced, the support of big data to industries strengthened and the scope of big data application accelerated.

The industrial application of big data is not limited to the manufacturing industry, and it also includes vast fields such as medical treatment, marketing and public safety etc. As far as Industry 4.0 is concerned, the generation and application of big data run through the whole product life cycle from design, research and development, manufacturing and marketing to after-sales service. By 2019, the sales volume of the big data industry in China amounted to 538.6 billion Yuan, continuing the sustained growth momentum since 2014. In five years, the overall size of the industry tripled, with an average annual growth rate of more than 20%. It is expected that the market size will reach 807.06 billion Yuan in 2021 (China Center for Information Industry Development 2019).

Impacts of Industry 4.0 on China' socioeconomic system

China is promoting the development of Industry 4.0 to solve shortcomings of China's industry in the face of competitive international pressure. Those major problems are mainly reflected in the transformation of industry structure, the improvement of labour productivity, technology innovation and the adjustment of labour structure. Industry 4.0 is ultimately aiming at promoting the high-quality development of the entire economy. In the following, we will discuss the effects of Industry 4.0 on different economic and social variables, and ultimately its impact on economic growth.

A key driver of China's economic development

China's economic growth has entered a new era and faces several new challenges. In recent years, the returns to domestic capital investment have been falling; the shift from the demographic dividend to the human capital dividend takes time, and investment in technological innovation has soared due to a lack of breakthrough innovations. International financial and trade uncertainty has increased and trade conservatism has arisen. Demand-pull policies implemented by China are still not fully launched and are waiting to become effective, because the release of latent demand of domestic consumers is not a short-term issue. Hidden regional market segmentation, local market entry barriers and the substantive gaps in regional economic development have not been significantly reduced. In addition to the current general problems encountered by macroeconomic growth, regional development in China is also facing much more serious divergence in recent years.

Traditional driving forces of economic growth have been unable to meet the urgent demand for high-quality development in the new era of economic growth. The high-quality development of the whole macroeconomy and of regional economies needs a new impetus. Instead of designing a unified index or index system to reflect the development of Industry 4.0 in China and then measuring the driving effect of this index on China's economic growth, we analyse the effects of different components of Industry 4.0 on economic development.

As one of the important components of Industry 4.0, network economy is also a vital link for the development of Industry 4.0 in China. The network economy has now become one of the important new driving factors of Chinese economic growth. The Institute of

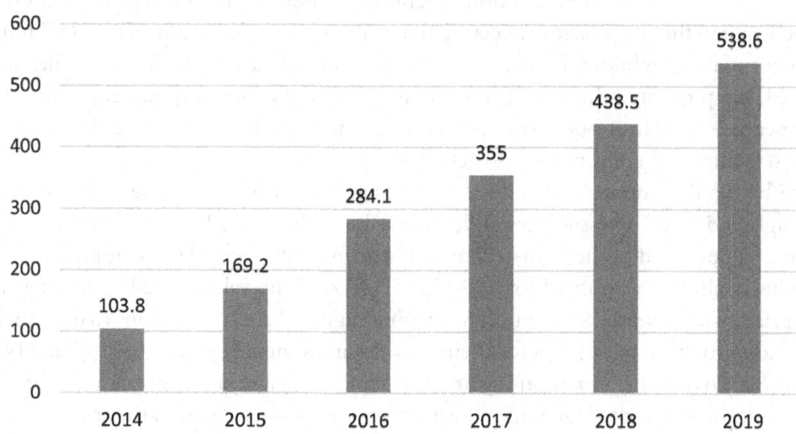

Figure 33.4 Development of big data industry

Data Sources: China Center for Information Industry Development (2019)

Statistical Science (2019) designed a new economic development momentum index system, which reflects the development of new industries, new commercial activities and new business models in China, and the composite index includes five groups of indicators with equal weight regarding innovation activities, network economy, human capital, industry upgrading and structural transition.

According to the report of the Institute of Statistical Science (2019), in 2019 China's network economy index has grown six-fold compared to the data of 2014. At the same time, the contribution rate of the network economy index alone to the growth of new factors of economic development composite index has reached 80.5%, which greatly exceeds the contribution of other indexes of economic transformation, structure upgrading and technology innovation.

Moreover, China's digital economy, which accounted for 36.2% of GDP in 2019, has become a new engine of growth and continues to unleash its dynamism, according to the White Paper on the Development of China's Digital Economy in 2020 (CAICT 2020a). The report defined the digital economy as the sum of the value-added of the electronic information manufacturing industry, basic telecommunications industry, network industry and software service industry, plus the growth of output and enhancement of efficiency brought by the application of digital technology in other traditional sectors of the national economy. The value-added of China's total digital economy exceeded 35 trillion Yuan in 2019, according to the data (CAICT 2020b), and its contribution rate to GDP growth reached about 68%.

Effects on upgrading of industry structure

One of the targets of Industry 4.0 is to achieve a new version of industrialisation which will naturally lead to further upgrading and high-quality development within the manufacturing industry, on the one hand, and to the improvement of the share of high-quality industries with respect to the overall manufacturing industry, on the other hand.

The optimisation of the within and between structure of the manufacturing industry will eventually change the landscape of the macroeconomy.

Since 2012, the upgrading of consumption structure represented by the decrease of Engel's coefficient, advances in internet and other technology, and also the emergence of new economy, have become the main factors affecting the evolution of industrial structure in China. The development of new technologies related to Industry 4.0, including the industrial internet, big data and AI, has promoted the rise of new industries and new business models, which affected not only people's daily lives but also all aspects of production. It is promoting the upgrading of industrial structure and consumption structure in all aspects.

Driven by the development of Industry 4.0, China's industrial structure continues to move towards high-end level. The statistics bulletin of China's National Bureau of Statistics pointed out that the value-added of the high-tech manufacturing industry and the equipment manufacturing industry already accounted for 14.4% and 32.5% of the value-added aggregates in 2019 for enterprises with a yearly main business income higher than 200 million Yuan, which is an average yearly growth rate of 6.9% for their main business income since 1995 (China National Bureau of Statistics 2019). In recent years, the rapid integration of the new generations of information technology into different industries has vigorously promoted the transformation of manufacturing services, resulting in the gradual growth of producer services, and the continued explosive growth of service-oriented manufacturing. The service industry has become a hot spot of innovation and entrepreneurship in the era of Industry 4.0.

Impacts on employment structure

Industry 4.0 is an inevitable choice for a sustainable and high-quality development of China's manufacturing industry after the disappearance of the demographic dividend. The main effect of Industry 4.0 on employment is the substitution of labour with different skill intensities. The type of labour required for Industry 4.0 is quite different from that of the traditional industries as the former is more inclined to employ high-skilled labour. As a result, not only the aggregate numbers but also the ratio of unskilled to skilled labour may be subject to changes. Therefore, demand for low-skilled labour may decrease, and it will be gradually replaced. The

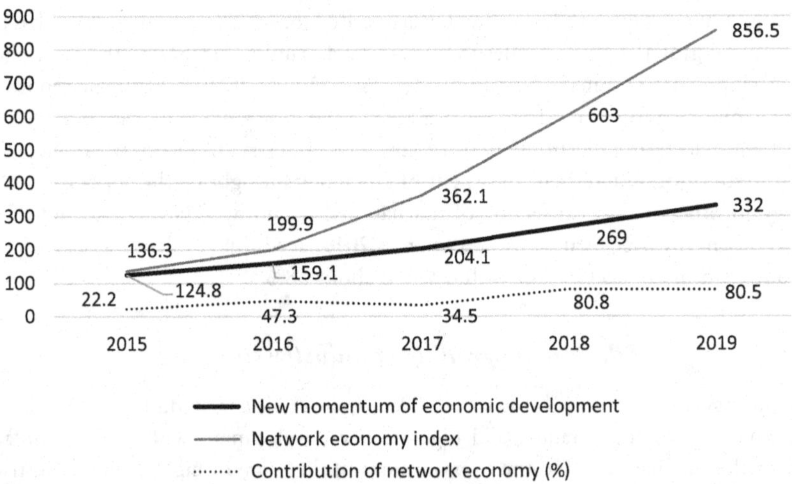

Figure 33.5 Network economy and new economic development momentum

Source: Institute of Statistical Science, China National Bureau of Statistics (2019)

implementation of Industry 4.0, in time and space, will have significant impacts on the employment figures and employment structure in China.

The data from the Economy and Information Department of Zhejiang Province in China (2017) shows that, in Zhejiang Province, with the program of "substitution of labour by robots", the aggregate number of industrial employment was reduced by nearly 2 million in the period 2013–2015. As regards another province, the data from Statistics Department of Guangdong Province (2018) reveals that in 2015, the growth of new robots is 18,000 units, and in 2016 and 2017, the number amounted to more than 20,000 units. Further, in 2018, the growth of new robots rose to 32,100 units. According to the substitution ratio of one unit of robot for 5 to 10 workers, therefore, a considerable number of workers will be replaced.

In addition, the new method of intelligent and automated production makes the job no longer limited by a location such as the factory but instead becomes virtual and geographically flexible, and available to be carried out remotely. Employees will have a high degree of autonomy to adjust and switch work and life status at any time. Achieving a new mode of production also means a need for a different education system. Higher education with narrowly defined majors will not provide enough qualified talent for the new industry, but education that emphasises multidisciplinary cooperation will bear fruitful results.

Increasing productivity

The implementation of Industry 4.0 will enhance labour productivity and TFP significantly. According to the German Electronic and Electrical Industry Association, Industry 4.0 can increase the production efficiency of this industry by 30% (Ding and Li 2014). Industry 4.0 will revolutionise the design, manufacturing, operation and service processes of the global production system, providing multiple benefits such as greater flexibility, swift production, enhanced productivity, quality improvement, etc. Increasing connectivity between components, machines and personnel is expected to bring forth a 30% and 25% increase in the speed and efficiency of production systems, respectively.

According to a sample survey of 1,815 enterprises in ten cities across China, 73% of the enterprises have a strong willingness to implement intelligent manufacturing. Moreover, according to another survey of 308 firms focusing on smart manufacturing, the average productivity of those 308 firms in 2015–2017 increased by 34%, the average energy utilisation rate enhanced by 17.2%, the average operating cost reduced by 22 %, the average product development cycle increased by 32.4% and the average defective product rate dropped by 29.4% (Research Group on "A New Generation of Artificial Intelligence Under The Leadership Of Intelligent Manufacturing Research" 2018).

Promoting overall economic growth

Taking AI as an example, this represents an opportunity not only to revolutionise the manufacturing industry, but also to accelerate economic transformation and economic growth. AI can influence macroeconomic growth in two ways. The first is to utilise its permeability, substitution, synergy and creativity, which are the four major technology–economic features of the new generation of information technology; and to promote the growth of various sectors of the national economy by increasing the efficiency of input–output and accelerating the creation of knowledge. Finally, it will benefit macroeconomic growth. The second is that the industrial ecological system boosted by AI technology is infiltrating and affecting the various fields of the national economy, and its own scale will also grow along with it, boosting macroeconomic growth as well. Following the first path, the growth in input–output efficiency and knowledge creation will lead to an increase in the overall factor productivity of the macroeconomy and support the high-quality growth that China is seeking. With the second path, the growth of AI and other new generations of information technology industries lead to the expansion of the high-tech sector, which is fully in line with the direction of optimising and upgrading China's industrial structure. Therefore, the burgeoning growth of Industry 4.0 will stimulate high-quality development in China.

Future development

From the above survey, we are aware that China's Industry 4.0 has made remarkable achievements across different dimensions, and the overall trend of future development is promising. At the same time, the development of Industry 4.0 has promoted the upgrading of China's industrial structure, improvement of the employment structure and productivity through various innovative functions. Furthermore, China's Industry 4.0 program has a positive effect on its economic development. However, there are a series of problems which need to be tackled.

Problems

Persistence of quantity shortage

Although various fields of Industry 4.0 in China have been improved quantitatively after several years of rapid development, there is still a big gap from the frontier to fulfil the demand for further economic development in China. For example, the penetration rate of industrial robots in China's manufacturing industry remains lower than in advanced countries, and there is much to be done in this new automation age. The data from the World Robotics 2019 International Federation of Robotics revealed that the usage density of industrial robots in China was 140/10,000 in 2018, which stands for 140 units of robots per 10,000 workers in the manufacturing industry. The density of industrial robots in advanced countries, such as South Korea, Japan and Germany, is well above 300/10,000. Especially in South Korea, where the robot density in the manufacturing industry is 774 units per 10,000 workers. Therefore, the penetration rate of industrial robots in China's manufacturing industry still has much room for improvement. Needless to say, there are also a large number of people and economic units in China not connected to the internet, let alone to a broadband network. The development of China's AI industry is also still at a low level compared to developed countries.

Low technology capacity of domestic producers

In a few key areas of Industry 4.0, and for some important infrastructure equipment in manufacturing industries, domestic firms in China have up until now been unable to provide suitable products. Even if they could provide them, the technical level of those products is much lower than the advanced international standards. Take industrial robots as an example: the biggest four producers of industrial robots accounted for 60% of the market share in China. However, all of them are foreign companies, while China's domestic brands account for less than 10% of the market share, and their products are mainly concentrated on the middle and low end of the value chain of the robot industry (China Robot Industry Alliance 2018).

Even in an optimistic environment with booming growth prospects, sales of domestic brand industrial robots rose only 20% in 2017, which is well below the average growth rate of the whole industry (China Robot Industry Alliance 2018). While the foreign capital, joint venture brand industrial robots gradually reduce their prices, most domestic robot enterprises earn profits mainly by providing services of application rather than the sales of final goods of robots. There is another example. At present, most of the AI innovations in China focus on the application of technology, and there is still a big gap from the world's leading level. Moreover, the key components and parts of China's smart manufacturing system rely mainly on imports, such as high-performance servo motors in the field of industrial robots, high-precision reducers, functional components in the field of computer-numeric-control machines and lasers of the core components of 3D printers. On the whole, China's smart manufacturing industry emphasises the production and manufacturing services, but they neglect the design and management components, which leads to obstacles to the further development and application of Industry 4.0.

Negative effects of designed industrial policies

A number of Industry 4.0 policies have been designed and implemented in China, but they are essentially formulated according to the paradigm of selective industrial policy. Those industrial policies are easily influenced by administrative intervention, which is not quite adaptive to the market economy law of Industry 4.0. The room for unnecessary intervention of those industrial policies is vast, at both the industrial and the firm level. All in all, this kind of policy model will even produce adverse incentives to firms' market behaviour and innovation. Therefore, as regards the policy system of Industry 4.0, China should strengthen the cooperation of government, industry and university, maintaining and strengthening firms' basic role in the market economy.

Policy recommendations

In order to overcome important obstacles encountered in the development of Industry 4.0, and to realise its stable and rapid economic growth in the future, China needs to make further efforts in the following respects.

Replacement of selective industrial policies with competition policy

In the present stage of China's economic development, the relationship and weights between industrial policy and competition policy should be adjusted and redefined, to avoid the accumulated resource misallocation caused by the former. The main target is to establish the priority and fundamental role of competition policy, and the key is to install a coordination mechanism between competition policy and industrial policy, with the industrial policy not distorting market competition. The Industry 4.0 policy system of China should pay more attention to functional rather than selective investment programs to avoid the misallocation of production factors and utilise functional industrial policy to stimulate an ideal market mechanism. Even in the fields of natural monopoly, externality, public goods and so on, the application of functional industrial policy should be strengthened, while the scope for selective industrial policy should be continuously and greatly reduced.

Support of small and medium-sized enterprises

Compared with state-owned enterprises and large private enterprises, China's small and medium-sized enterprises (SMEs) lack financial and material resources to take an active role in the course of Industry 4.0 development. At the same time, they can easily be bypassed by various policies. These factors lead to the unfavourable position of SMEs in Industry 4.0 construction. However, SMEs already occupy a very important position in China's national economy. The national income and employment created, as well as exports conducted, by those enterprises are crucial to economic growth in China. Therefore, the inclusion of them in the Industry 4.0 road map amounts to a large part of the success or failure of Industry 4.0.

In an attempt to give full recognition to the role of small and medium-sized manufacturing enterprises, policy support, a certain amount of financial funds and technical assistance are needed. For those SMEs, on the one hand, they should make full use of the policy support to improve the foundation of smart production; on the other hand, they should actively learn the advanced production mode and adopt organisational forms designed by the German Industry 4.0 strategy, actively cultivating their own potential advantages of personalised production and forming their own core competitiveness.

Providing high-quality human capital

Simple and routine mechanical work will progressively be replaced by precision and intellectual work with higher value-added production. At the same time, it sets higher requirements for workers' knowledge and skill ability. Industry 4.0 not only means a substantial replacement of low-skilled labour for high-skilled labour, but its future development also needs more high-quality human capital. Otherwise, it will not be able to establish the software and hardware basis of the network economy, not to mention the application of industrial robots, the promotion of research and development of AI and their industrial applications, and the promotion of big data industry. Consequently, more attention should be paid to the personnel training and actively to promote vocational education, improve and update training system in time, provide retraining for present personnel and build up human resources with advantages in knowledge and skills needed by Industry 4.0.

Acknowledgements

Thanks to the support of National Social Science Fund (15BJL022) and the Fundamental Research Funds for the Central Universities (2018QN058).

References

Cagé, J. and Rouzet, D. (2015): Improving "National Brands": Reputation for Quality and Export Promotion Strategies, *Journal of International Economics* 95(2), 274–290.

The China Academy of Information and Communications Technology (CAICT) (2019): White Paper on the Development of China's Broadband. Available at: www.caict.ac.cn/kxyj/qwfb/bps/201910/t20191031_268469.htm [10.03.2021]

The China Academy of Information and Communications Technology (CAICT) (2020a): White Paper on the Development of China's Digital Economy. Available at: www.caict.ac.cn/english/research/whitepapers/202007/t20200706_285683.html [10.03.2021]

The China Academy of Information and Communications Technology (CAICT) (2020b): World AI Industry Development Blue Book. Available at: www.caict.ac.cn/english/research/whitepapers/202007/t20200706_285683.html [10.03.2021]

China Center for Information Industry Development (2019): White Paper on the Development of China's Big Data Industry. Available at: www.ccidgroup.com/gzdt/12892.htm [10.03.2021]

China Economic Information Service (2019): China Manufacturing Quality Development Report. Available at: www.xinhuanet.com/tech/2019-12/28/c_1125399121.htm [10.03.2021]

China General Administration of Customs (2020): Value and Quantity Table of National Key Commodities Imports. Available at: www.customs.gov.cn/customs/302249/302274/302275/2833746/index.html [10.03.2021]

China National Bureau of Statistics (2019): Statistical Bulletin on the 2019 National Economic and Social Development of People's Republic of China. Available at: www.stats.gov.cn/tjsj/zxfb/202002/t20200228_1728913.html [10.03.2021]

China Robot Industry Alliance (2018): Report of 2018 Industrial Robot Market. Available at: http://cria.mei.net.cn/news.asp?vid=3834 [10.03.2021]

China State Administration for Market Regulation (2019): Tasks and Initiatives of Industrial Product Quality and Safety Supervision. Available at: www.samr.gov.cn/xw/xwfbt/201912/t20191217_309277.html [10.03.2021]

Deloitte Report (2019): Global Artificial Intelligence Industry Whitepaper. Available at: https://www2.deloitte.com/cn/zh/pages/technology-media-and-telecommunications/articles/global-ai-development-white-paper.html [10.03.2021]

Ding, C. and Li, J. (2014): Germany "Industry 4.0": Content, Motivation and Prospects, *Deutschland Studien* 29(4), 49–66 (in Chinese).

The Economy and Information Department of Zhejiang Province (2017): How "Machine Replace Labour" Affects Future Job Markets. Available at: www.xinhuanet.com/mrdx/2017-05/02/c_136249855.htm [10.03.2021]

Institute of Statistical Science, China National Bureau of Statistics (2019): China's Economic Development New Kinetic Energy Index Increased 23.4% Over the Previous Year in 2019. Available at: www.stats.gov.cn/tjsj/zxfb/202007/t20200713_1775420.html [10.03.2021]

International Federation of Robotics (2019): World Robotics 2019 Edition. Available at: https://ifr.org/free-downloads/ [10.03.2021]

Li, X. (2019): The Course and Valuable Experience of China's Manufacturing Industry Development, *Economic Daily*. Available at: http://paper.ce.cn/jjrb/html/2019-10/30/content_404672.htm

Lu, Y. and Liu, C. (2019): The Trends of China's Robots Industry in 2019, *The Internet Economy* 1, 44–49.

Ministry of Industry and Information Technology (2018): *National Telecommunications Statistics Bulletin*. Available at: https://www.miit.gov.cn/gxsj/tjfx/txy/art/2020/art_f9b061284a1646498f135584d8f78757.html [10.08.2021]

Research Group on "A New Generation of Artificial Intelligence under the Leadership of Intelligent Manufacturing Research" (2018): A Study on the Development Strategy of Intelligent Manufacturing in China, *China Engineering Science* 20(4), 1–8.

Rodrik, D. (2010): The Return of Industrial Policy, *Project Syndicate*, 12 April. Available at: www.project-syndicate.org/commentary/the-return-of-industrial-policy [10.04.2021]

Sakoh, K. (1984): Japanese Economic Success: Industrial Policy or Free Market? *Cato Journal* 4(2), 521–548.

The Statistics Department of Guangdong Province (2018): Statistical Bulletin on the 2018 National Economic and Social Development of Guangdong Province. Available at: http://stats.gd.gov.cn/tjgb/content/post_2207563.html [10.03.2021]

The World Intellectual Property Organization (WIPO) (2019): WIPO Technology Trends 2019: Artificial Intelligence. Available at: www.wipo.int/publications/en/details.jsp?id=4396 [10.03.2021]

Xing, Y. and Detert, N. (2010): How the iPhone Widens the US Trade Deficit with PRC. *ADB Institute Working Paper 257*.

Zhang, Q. (2015): Speech at the National Conference on Product Quality Improvement. Available at: www.samr.gov.cn/zljds/fwfh/ldjh/201510/t20151028_299998.html [10.03.2021]

INDEX

Printed in the United States
by Baker & Taylor Publisher Services